Volume 39

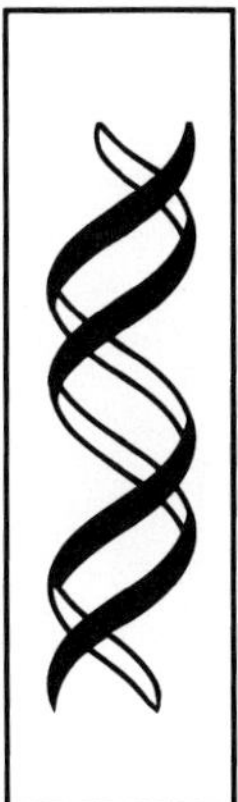

Advances in Genetics

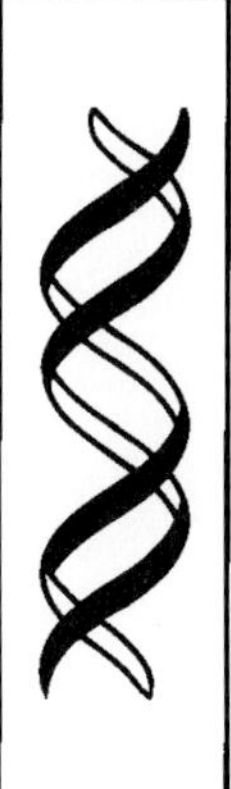

Volume 39

Advances in Genetics

Edited by

Jeffrey C. Hall
Department of Biology
Brandeis University
Waltham, Massachusetts

Jay C. Dunlap
Department of Biochemistry
Dartmouth Medical School
Hanover, New Hampshire

Theodore Friedmann
Department of Pediatrics
Center for Molecular Genetics
School of Medicine
University of California, San Diego
La Jolla, California

Francesco Giannelli
Division of Medical and Molecular Genetics
United Medical and Dental Schools of Guy's and St. Thomas' Hospitals
London Bridge
London, United Kingdom

Academic Press
San Diego London Boston New York Sydney Tokyo Toronto

This book is printed on acid-free paper. 

Academic Press
a division of Harcourt Brace & Company
525 B Street, Suite 1900, San Diego, California 92101-4495, USA
http://www.apnet.com

Academic Press
24-28 Oval Road, London NW1 7DX, UK
http://www.hbuk.co.uk/ap/

International Standard Book Number: 0-12-017639-4

PRINTED IN THE UNITED STATES OF AMERICA
99 00 01 02 03 04 EB 9 8 7 6 5 4 3 2 1

Contents

1 Genetic Organization of Polytene Chromosomes

I. F. Zhimulev
Institute of Cytology and Genetics
Siberian Division of Russian Academy of Sciences
Novosibirsk 630090, Russia

I. GENERAL REMARKS

A. Author's preface

During the past decades, polytene chromosomes have become indispensable in studying many questions relating to cytogenetics, cytology, and molecular developmental biology. This is because of their persistence throughout interphase and their polyteny, which produces their giant size. This allows visualization of the enfoldment of many processes and their interrelationships. The crucial questions pertain to the organization of genetic material in the morphological structures of the interphase chromosome—the chromomeres and interchromomeric regions—and the structural changes occurring during the activation of the chromomeres, the puffs.

The possibility to visually observe the processes occurring during puff formation, their sequence, their topography, and the capacity of the cell to respond by changing the genetic activity of particular loci in response to the action of various environmental and intercellular agents have all made the polytene chromosome essential to understanding in gene action in development.

Polytene chromosomes have also provided answers to questions concerning the second function of the chromosomes, namely reproduction and transmission of exact copies of hereditary material through successive generations.

All these questions and answers are dealt with here, in the last of a three-volume series concerned with polytene chromosomes. The first two vol-

Advances in Genetics, Vol. 39

0065-2660/99 $90.00

umes "Morphology and Structure of Polytene Chromosomes" and "Polytene Chromosomes, Heterochromatin, and Position Effect Variegation" have been published previously (Zhimulev, 1996, 1997).

B. Introduction

The interphase polytene chromosome is morphologically differentiated into tightly compacted (the chromomeric) and decompacted (the interchomomeric) regions, as the chromatid from telomere to telomere has the appearance of a string with threaded beads, the chromomeres. A great deal of discussion has, for decades, revolved around the question of the genetic organization of the chromomeres. The hypothesis that the chromomere of the pachytene chromosome is the gene itself was first suggested by Belling (1928) and has since evolved into distinct questions.

The first question concerns the stability of the chromomere as an element of the structure of the chromosome. It has now become quite obvious that a particular pattern of chromomere differentiation corresponds to each stage of the cell cycle; that is, the genetic organization of the chromomeres is different in chromosomes of different types (Lima-de-Faria, 1975). In chromomeres of a particular type—bands of the interphase polytene chromosomes, for example—the question of their genetic organization has been disputed since the mid-1930s. The dispute (review: Beermann, 1972) can be reduced to the main issue of whether bands are specialized structures where the genes reside (the "lair of the gene," as Painter, 1934, aptly put it). With the discovery of giant puffs large enough to provide the formation of many bands and interbands, interpretation of the genetic organization of the chromomeres became much more complicated. Furthermore, other genes may be located in the introns of these genes (see, for example, Furia *et al.*, 1991). Discussions concerning the genetic organization of interchromomeric regions started long ago. Hypotheses about this organization have been put forward for the past 60 years with scant verifiable facts. This is accounted for by the minute size of the interbands, making them inaccessible to study by methods of microscopy. Interesting results can be obtained by methods of molecular biology, including deciphering the primary structure of interband DNA (e.g., Demakov *et al.*, 1991).

During the activation of transcription of genes located within chromomeres, the constituent material of a chromomere loosens and a large swelling, the puff, is formed. Study of the sequence of the appearance of puffs in the course of development and other experiments have led to the concept that hormones are involved in regulating puff activity: the first wave of puffs arises under the effect of the hormone ecdysterone; expression of genes on the early puffs is required for the induction of the later ones. Thus, the sequence of activation of the hormone-induced puffs conforms to a cascade pattern (Clever,

1964; Ashburner *et al.*, 1974). Data obtained by molecular and genetic studies, providing evidence for the molecular basis of a cascade regulation of puff activity, have been rapidly accumulating (see, for example, Georgel *et al.*, 1991).

A review of the data concerning the molecular and genetic organization of many puffs and the mechanism of their activation, obtained in recent years, allows us to reconsider the general problems of the regulation of gene activity in eukaryotes. The phenomenon of puffing itself remains an enigma, primarily in terms of its fundamental significance. For example, puffing waves are induced by hormones in the salivary gland chromosomes of larvae and prepupae; that is, in organs completely degenerating already after some hours.

In experiments with artificial activation of puffs by various agents, Ritossa has established that certain puffs are inducible under the effects of high temperature (heat shock) and the inhibition of oxidative phosphorylation (Ritossa, 1962–1964). Subsequently, when it became feasible to isolate protein products of the heat-shock-activated genes with the use of electrophoresis, it showed that the cell responds to stress conditions by synthesizing heat shock-proteins as a general biological mechanism, providing the organism with resistance to unfavorable conditions.

The nucleoli, Balbiani rings, and DNA puffs are specific types of active regions in the polytene chromosomes. Their morphology differs to a large extent from that of the usual puffs. The nucleolar organizer contains a gene block, organized in a complex manner and encoding ribosomal RNA (review: Bashkirov, 1989). Balbiani rings are composed of numerous repetitive elements, and they encode RNA molecules of a giant size (Daneholt, 1972). Amplification of DNA takes place in the DNA puffs (Lara *et al.*, 1991).

From these considerations it follows that the body of evidence concerning the genetic organization of chromomeres in active and inactive states warrants analysis. This book is concerned with this analysis.

II. GENERAL ASPECTS OF THE CHROMOMERIC ORGANIZATION OF THE CHROMOSOMES

A. Definition of the chromomere

At the turn of 19th century investigators (E. G. Balbiani, W. Pfitzner, and W. Flemming; see Wilson, 1936) made the observation that the chromosome fiber in early mitotic prophase contains several small intensely staining bodies, or chromomeres, differing in size and shape. In meiotic prophase, the pattern of the chromomeres in homologous chromosomes is completely symmetrical (Figure 1). According to Wilson, each chromatic grain of a fiber has a counterpart

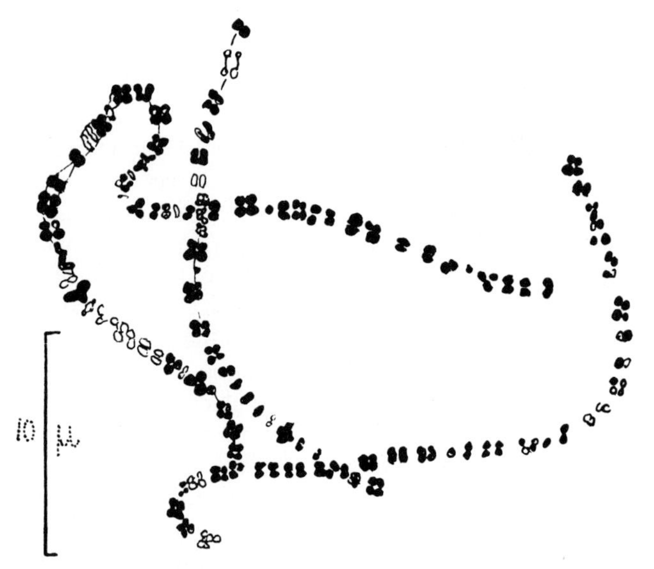

Figure 1. A fragment of a nucleus of *Lilium pardalinum* with pachytene chromosomes bearing chromomeres (from Belling, 1928).

in another fiber, and each and every feature of a fiber is perfectly copied in its counterpart (Wilson, 1936). ". . . Differing in size, shape, DNA content and disposition in the chromosome from one another, the chromomeres have a distinctive individuality and impart a constant and definite pattern fixed hereditarily to the chromosome fibers and its separate regions" (Prokofyeva-Belgovskaya and Bogdanov, 1963, p. 36).

It is only at mitotic and meiotic prophase (at leptotene and pachytene) that all the eukaryotic chromosomes possess a chromomeric organization; the chromomeres are quite rarely seen at the other stages of the cell cycle and only in certain organisms (Ris, 1957; Lima-de-Faria, 1975). The idea that the chromomeres (the granules on chromosome fibers) also exist in interphase nuclei in which the individual chromosomes are not recognized was developed by Flemming as early as 1882.

H. Bauer (1935) suggested that the bands of polytene chromosomes result from lateral fusion of the chromomeres in chromatids arranged in parallel rows. The term "banding pattern" (Querscheibenmuster) has been coined by Heitz (1934).

Gall (1954) described chromomeres in the "lampbrush"-type chromosomes (Figure 2), as observed in the primary oocytes (at prophase; presumably,

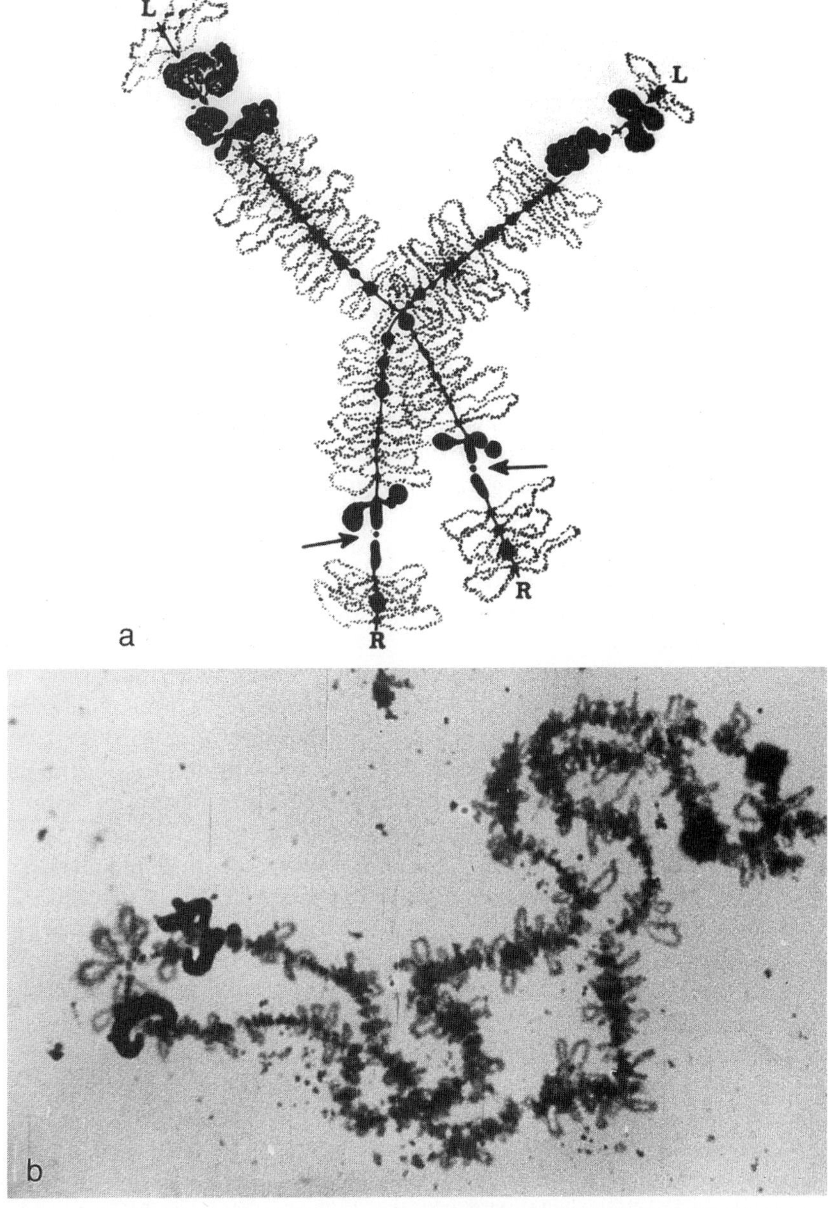

Figure 2. Chromomeres with extending loops in a bivalent of chromosome XII (a) of the triton *Triturus cristatus karelinii* (after: Callan, 1963) and (b) chromosome II of *Gallus gallus domesticus*. (L and R) the left and right arms of the chromosome, respectively. Arrows point to the position of the centromere (redrawn from M.S. Chechik and E.R. Gaginskaya, unpublished).

at diplotene) of many vertebrates and some invertebrates (Callan, 1963). Loops projecting laterally from the chromomeres have the appearance of a brush used for cleaning an oil lamp. In amphibia, the chromomere pattern exhibited by chromosomes of this type is generally the same from preparation to preparation, although it is not always identical (review: Callan, 1986).

The total number of chromomeres for the haploid chromosome set varies from 4 to 10 thousand in three species of salamander (Vlad and Macgregor, 1975).

Chromomeres have been studied at meiotic prophase in insects (Ashton and Schultz, 1971). Lampbrush-type chromosomes have also been described in cells of the primary spermatocytes of *Chironomus pallidivittatus*, spread on an air/water interface. These have loops 33 μm long, on average; they arise from axial fibers 6 nm in diameter and at a distance of 0.07–0.15 μm one from another. There are no chromomeres in the usual sense ("beads-on-a-string") in this chromosome type, although it cannot be ruled out that the chromomere, the band of the polytene chromosomes, originates from the loops (Keyl, 1975).

Therefore, the concept that chromosome organization includes chromomeres as fundamental constituents was established. Thomas (1973) called it the most significant contribution to classical cytogenetics. However, more recent data indicate that the concept is less than universal (review: Lima-de-Faria, 1975).

The chromomeres of different types have two features in common: (1) all are segments of compacted DNA and (2) they show the same number and configuration at a given stage of the cell cycle (Lima-de-Faria, 1975).

Upon closer examination of the features distinguishing the chromomeres of different types, their number proved to be larger.

1. While the chromomeres of meiotic chromosomes occur in germ line cells, those of polytene chromosomes, for example, occur in somatic cells. In terms of development, the salivary glands are in a final stage of differentiation, whereas the meiotic cells give rise to new cell generations (Lima-de-Faria, 1975).
2. Chromomere number varies a great deal from one tissue to another (Figure 3). Thus, in *Agapanthus umbellatus* there are 1600 chromomeres in pachytene chromosomes and 239 at prophase II of meiosis. In *Ornithogalum virens*, the total is 274 at pachytene and 67 at prophase II of pollen mitosis (Lima-de-Faria, 1975).

   The differences in the number of compacted segments of chromatin have been explained by different degrees of compaction of the chromomeres at different stages of the cell cycle. If the chromomeres making up bands of maximally decompacted polytene chromosomes are the shortest fragments of genomic DNA, then the chromomeres seen at prophase of

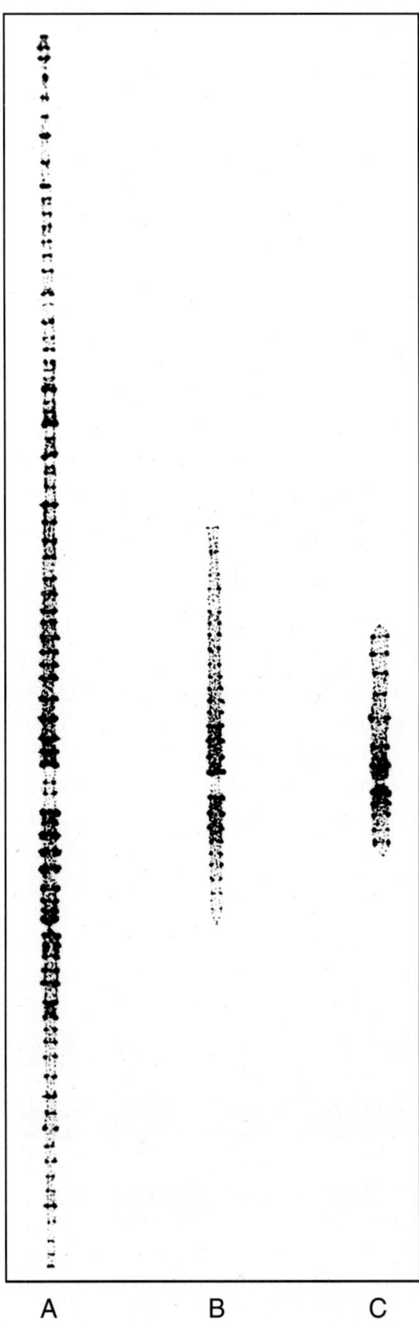

Figure 3. Chromomeric pattern of one of the chromosomes of *Agapanthus umbellatus* in pachytene of (A) meiosis, (B) midprophase II and (C) midlate prophase of mitosis (redrawn from Lima-de-Faria, 1954, 1975).

meiosis or mitosis are regions of extensively packed material and, as some authors believed, of more tightly compacted chromatin. Ris (1960) thought that the chromomeres are turns of a "new helix" arising approximately at this stage.

At later stages of cell division, the chromomeres become still greater in size, the spacing between them shortens and, ultimately, all the material unites to form a single mitotic chromosome (Figure 4).

3. The genetic content of the chromomeres changes in proportion to their number. Belling (1928), from comparisons of chromomeres and genes of *Liliaceae*, concluded that the pachytene chromomere is the gene itself. This conclusion was based on similar behavioral and organizational features of the genes and chromomeres: (1) linear order in the chromosomes; (2) division at pachytene; (3) substantial differences in size from one another; and (4) their pairing only with homologous regions (Belling, 1928).

Notwithstanding the attempts made to find a correspondence between the morphological structure of the chromosome and its information units (the genes), the hypothesis was soon abandoned. This was followed by the realization that chromosomes must have many more genes than originally considered. Lima-de-Faria (1975) expressed the prevailing opinion by stating that the total chromomere number in pachytene chromosomes is of the order of 500. If the one gene–one chromomere hypothesis is correct, humans would have only 500 structural genes, a ridiculously low figure. Given early estimates of the number

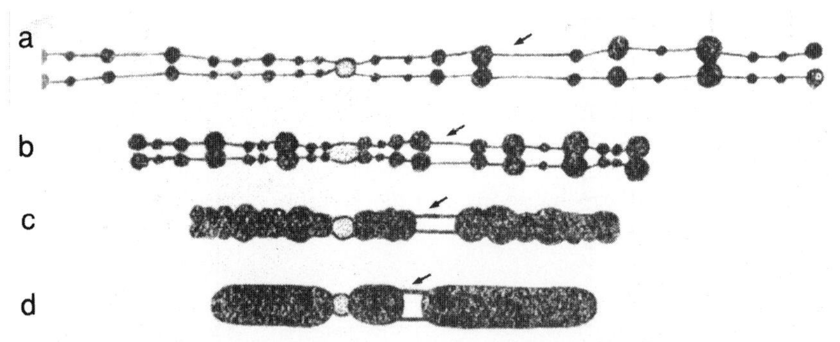

Figure 4. Changes in chromomeric pattern in the nucleolus-forming chromosome of *Allium cepa* in prophase–metaphase of mitosis (a) Early prophase; (b) contraction of the prophase chromosomes, the fibrils between the chromomeres shorten; (c) further reduction in the size of the interchromomere regions and fusion of chromomeres; (d) the metaphase chromosome. The position of the nucleolar organizer is indicated by arrows (from Ellenhorn, 1937).

of structural genes in humans were in the 10 to 100 thousand range (Stern, 1973; Lima-de- Faria, 1975, p. 272).

Callan (1978, p. 382) considered the genetic organization of the chromomeres in lampbrush chromosome type in amphibia and observed that there exist at least several loops projecting from the chromomeres that each contain more than one transcription unit. Furthermore, he noted a fairly large number of chromomeres carrying several pairs of lateral loops. The following is taken from his introduction: "This leads me to the conclusion that the chromomeres of 'lampbrush' chromosomes, far from being themselves functional units, are passively generated by the transcriptional activity of interspersed functional units."

In conclusion, it may be stated that chromomeres are chromosome fragments capable of local compaction. The length of a DNA fragment included in a chromomere is different at different stages of the cell cycle. Therefore, there is every reason to believe that the chromomere is not a constant structural unit of genomic organization and that each stage of development has its own set of chromomeres. Consequently, following Lima-de-Faria (1975), all the known chromomeres fall into four groups: (1) those associated with leptotene; (2) those associated with pachytene; (3) those associated with the "lampbrush" chromosomes; and (4) those associated with the interphase polytene chromosomes. The functional organization and genetic composition of the chromomeres are different in each chromosome group.

The subject of the functional organization of the chromomeres of polytene chromosomes will be deferred until later sections.

B. Development of concepts of the chromomeric organization of polytene chromosomes

From the time Kostoff suggested "that transverse dark bands are packets in which Mendelian genes are located" (Kostoff, 1930, p. 323) discussion has ensued over the genetic organization of bands and interbands in polytene chromosomes. The question as to whether genes are located in heavily staining material (the bands) or whether in the other structures of the chromosomes was raised by Painter in 1934. Although incapable of discriminating between the chromomere or the part of the chromosome between the chromomeres, Koltzoff (1934, p. 313) thought rather that the latter structure corresponds to the gene. He wrote "we can only suggest that the chromatin granules and 'bands' are merely links connecting the genes and sites where crossingover takes place." The hypothesis that several or all of the genes are located in interbands also received support (Demerec and Hoover, 1936; Marshak, 1936).

After Mackensen (1934, 1935) and Muller and Prokofyeva (1935) mapped the *white*, *vermilion*, and *scute* genes to short intervals of the chromosomes—including the single bands, or even portions of them, using translocations, inversions, and short deletions—it became feasible to consider bands as bearers of the genes. In this regard Demerec and Hoover (1936) believed that some genes are located in bands while others reside in interbands. Metz (1937) considered the possibility that a part of a gene may be located in a band with the rest of it being in an interband.

Different opinions were expressed concerning the possible number of genes in the bands. Demerec and Hoover (1936) believed that the band possibly marks the locus of a single gene. In Muller and Prokofyeva's view, at least the regions of "heterochromatic nodes" (bands) in the salivary gland chromosomes contain genes (they left open the question of the composition of unstainable internode regions); and at least some of the "nodes" contain clusters rather than individual genes (Muller and Prokofyeva, 1935, p. 19–20). Bridges's (1935, p. 63) inference as to the relation of the genes to the structures visible in the salivary gland chromosomes was that each of the faint cross bands corresponds to one locus; while some "capsule-forming" bands correspond to three loci (one in each "capsule" wall and one in the spaces between the walls). Thus, the foundation was already laid in the mid-1930s, with discussions centering on the structures (bands or interbands) and number (single or cluster) of genes in the structures.

Subsequently, disputes mainly focused on the bands, presumably because the phenomenon of puffing was regarded, in thought-provoking papers, as a morphological manifestation of gene action (Beermann, 1952a,b, 1956, 1962, 1965, 1967, 1972a; Mechelke, 1953; Breuer and Pavan, 1955). It became evident that puffs are formed when band material becomes decompacted (Beermann, 1952a, 1962; Pelling, 1966). Furthermore, using genetic methods, a few genes were localized to specific chromosome regions, the number of genes in the regions being approximately similar to the number of bands (White, 1954). Ultimately, models appeared concerning chromosome organization, according to which bands were conceived of as functional units composed of structural genes and other sequences controlling gene action. It was thought at first that each band contains one such unit. This point of view was supported by a number of arguments. "The chromomeres or bands of the giant interphase chromosomes of Diptera represent operational units—units of transcription" (Beermann, 1965, p. 378). "Chromomeres, the physical equivalents of Mendelian genes, are thought to be giant operational units combining the functions of transcription with those of messenger RNA stabilization and packaging and perhaps with other modulating functions in the subsequent translation processes" (Beermann, 1967, p. 198). According to Pelling (1966), the chromosomal unit of replication and synthesis was the concept of the chromomere. And according to Lefevere, "there is good reason to believe that genes and bands are related one-to-one" (Lefevre, 1971, p. 509). "Functionally, chromosomes correspond to single genes . . . no facts are known which would be at variance with the conclusion that each band contains a single gene in the informational sense of the word" (Thomas, 1971, p. 249, 250). "We suggest a genetic structure named the 'service unit' for which the band is the cytological counterpart. The simplest unit is composed of the main gene which is the structural gene and all sequences needed for its function. These sequences include initiation and termination sites for replication and transcription, recognition of sequences needed for regulation of the translational process, . . . sequences involved in chromosome condensation, asynapsis and recombination. On the average there are 20 sequences . . ." (Lifschyts, 1971, p. 45; 1973). "We propose that the chromomere is a single complementation unit consisting of a relatively short length of structural DNA together with a variable length of regulatory DNA" (Judd *et al.*, 1972, p. 154). "A number of investigators agree that no more than one genetic function is associated with a single band . . ." (Lefevre and Green, 1972, p. 404). "It is a common belief that generally a single genetic locus corresponds to each band of the polytene chromosomes, i.e., to a single chromomere . . . in further discussion we shall hold the view that the chromomere is a discernal manifestation of the genetic locus" (Mglinets, 1973, p. 192). Bands may be viewed "not only as units of inactivation . . . rather as functional chromosomes capable of being in a state of activity or inactivity" (Kiknadze, 1971, p. 727; 1972, p. 15).

In the mid-1960s, a revision of previous estimates of the amount of DNA present per average band in the haploid chromosome yielded a value of 30–40 kb for *Drosophila* (Rudkin, 1965) and approximately 100 kb in *Chironomus tentans* and *Rhynchosciara* (Daneholt and Edstrom, 1967; M. Cordeiro, 1972; Lara *et al.*, 1991). According to the concepts of the time, this amount of DNA was far more than what would be required for coding a single gene. This was presumably why there was a new interest in ideas of the polygenic nature of the chromomere in the beginning of the 1970s: "Strict ordering of puffs in bands allows to look upon the chromomere . . . as the basis of the general functional organization of the chromosome. This is perhaps because the chromomere (the band) unites complexes of genes that are associated in their action, whereas interbands play the role of the 'landmark region' " (Kiknadze, 1971b, p. 178).

Britten and Davidson (1973) put forward two hypotheses based on the recognition of the polygenic structure of the band: (1) the chromomeres could consist of many structural genes transcribed coordinately and polycistronally; (2) the polycistron transcript of the complementation group is translated to produce regulatory molecules which affect the activity of many structural genes elsewhere in the genome regions.

Olenov (1974, p. 407) stated that the hypothesis suggesting that a single gene may be located in a chromomere "brings us to the old idea that there are as many genes in the genome as bands, or chromomeres, as visible in the microscope....Judging by the amount of DNA there are about 3000 genes in *Escherichia coli*. Thus, the number of genes in *Drosophila* and bacteria is the same on a one gene–one chromomere assumption." This was difficult to believe, given that *Drosophila* is, of course, a complex metazoan.

Without reference to the chromomeric organization of the chromosomes, Korochkin (1976, 1979) suggested that the genes regulating the same stage of development are grouped in the genome and united into blocks.

Another model postulates that the gene may be mapped within a rosette (Figure 5), produced during the spreading of a chromosome on borate buffer by Miller's technique. The introns are in the loops, and the exons are at the bases of the loops. The chromomeres can contain several such rosettes (Reznik *et al.*, 1991a,b) (see Patrushev, 1997, for further details).

An interesting explanation about why DNA is in apparent excess in the chromomere was offered by Akifjev and Edstrom. Akifjev (1974, p. 54) thought of the chromomere as of "an evolutionary melting pot in which genetic novelties are constructed from excess of DNA." In his view, the structural gene, with all its regulatory sequences and a tremendous amount of silent (even nonsense) DNA, lies in the chromomere (Figure 6). Edstrom (1975, p. 164) stated the belief that "most of the eukaryotic DNA performs an exclusively meiotic function. Each gene is associated with the meiotic DNA (Figure 7) in a morphological unit, a chromomere." The gene is transcriptionally active in somatic tissues and the meiotic DNA during meiotic prophase.

All the above models and concepts have a common feature. They assign no functions to interbands, while bands are represented by functional units specially organized to provide the activity of a gene or a gene group.

In the following models a band or an interband is regarded as a entity: "a cytogenetic unit (a chromomere plus an inter-chromomere)" (Ashburner, 1972e; Sokoloff, 1977). The model of Crick assumed that a single genetic complementation group is usually contained either in an interband plus a band (Figure 8) or in an interband plus part of the bands on either side (Crick, 1971, 1972) (see Akifjev and Makarov, 1972, for criticism of this model). Paul positioned the structural genes and controlling elements quite differently in the "cytogenetic unit" (Paul, 1972, p. 446). According to his model (Figure 9) "the interband regions are essentially polymerase-binding sites, and transription moves into the band regions from initiation sites which are likely to be situated near the band–interband junctions." However, Daneholt expressed the view that "on the basis of present knowledge, it cannot be decided whether or not transcription starts and/or terminates in interchromomere regions" (Daneholt, 1973, p. 159).

Sorsa postulated that genetic units are located along the border between the interchromomere and chromomere (Figure 10). His model implies

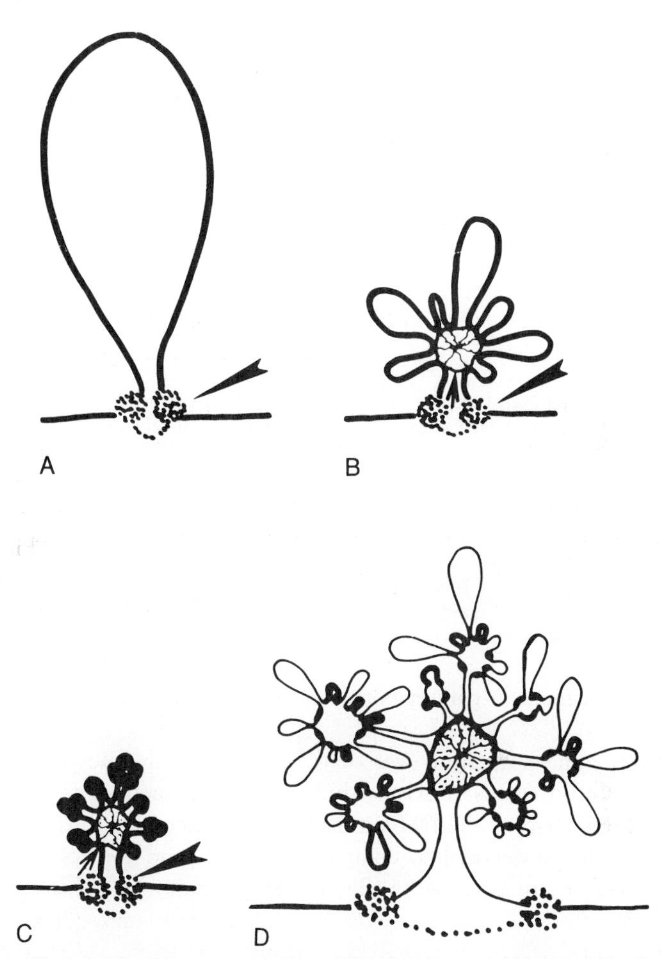

Figure 5. A scheme of the genetic organization of the chromomere. (A) A completely unravelled DNA molecule (60–90 kb) composing the chromomere loop in the region of the attachment to the chromosome scaffold (long arrows); (B) the loops of chromomeric DNA (5 kb) form a rosette composed of decompacted material; (C) the loops are packaged in nucleosomes and nucleomeres in an inactive chromomere; the short arrow points to the core of the chromomere; (D) a scheme for the structure of the chromomere containing several genes (from Reznik *et al.*, 1991a,b).

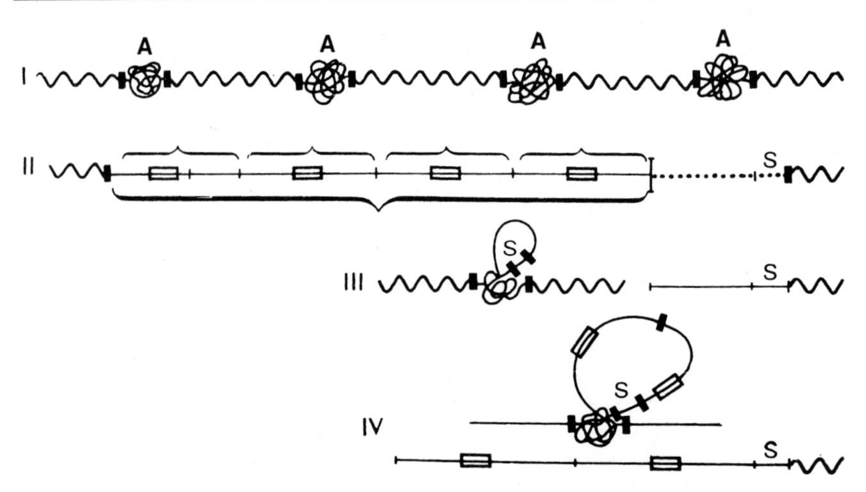

Figure 6. The structural organization of the chromomere. (I) A region of the chromosome composed of (A) individual chromomeres connected by interchromomeric DNA. Dashes delineate the arbitrary boundaries of the chromomeres. (II) The unravelled structure of one of the chromomeres composed of a transcriptional unit containing the structural gene (S) and the zone of "silent" DNA (in brackets). The silent zone consists of several microreplicons. (III) Transcription in somatic cells. The unit containing S is transcribed, the silent DNA is not involved in transcription. (IV) Transcription in gametogenesis. The transcriptons is of larger size than that in somatic cells, a considerable part of DNA is also involved in transcription in cells of the germ line (from Akifjev, 1974).

that the gene occupies a part of both structural elements (Sorsa, 1975, 1976, 1984).

Beermann's suggestion was that, "as a rule, only one protein coding sequence, or cistron, seems to exist within each chromomere–interchromomere complex. The majority of the DNA in each chromomere and thus in the entire genome seems to be devoted to complex regulatory and control functions in addition to merely structural ones" (Beermann, 1972a, p. 30; 1972f).

Chromomere (evolutionary unit)

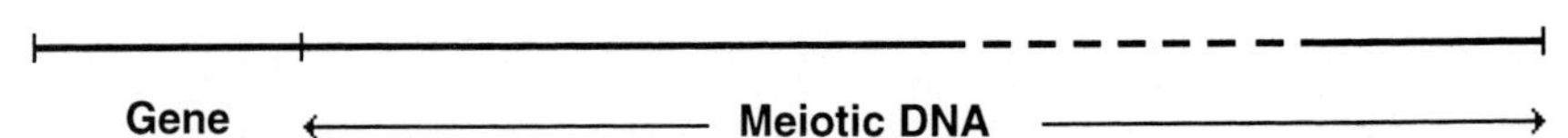

Figure 7. A scheme of the distribution of function in the chromomere. Each chromomere contains a gene and meiotic DNA. The gene is transcribed in somatic tissues, meiotic DNA is transcribed during meiotic prophase (reprinted from *J. Theor. Biol.*, **52**, Edstrom, Eukaryotic evolution based on information in chromosomes on allele frequencies, 163–174, 1975, by permission of the publisher Academic Press Limited, London.).

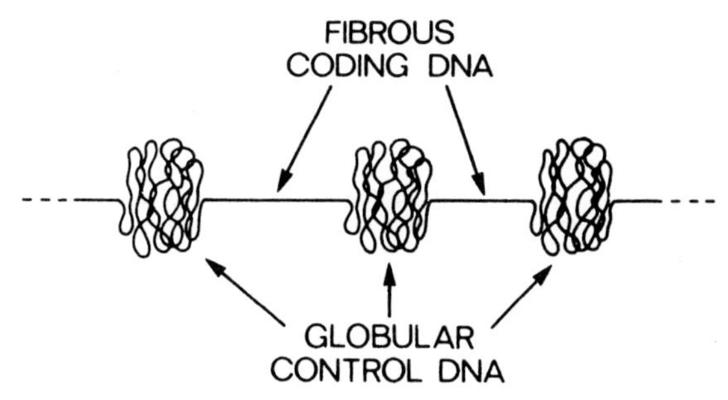

Figure 8. A scheme of DNA structure in the chromatid of the polytene chromosome. It is postulated that the majority, if not all, the genes encoding proteins are located in the interband regions (fibrous coding DNA). The regulatory sites are located in DNA of the bands (globular control DNA). The genetic complementation group corresponds to the interband and neighboring band or to the interband with parts of the neighboring bands [reprinted with permission from *Nature* (Crick, 1971). Copyright (1971) Macmillan Magazines Limited].

In terms of molecular organization, the chromomere–interchromomere complex is organized in a rather complex way. In the model of Bonner and Wu (1973) it contains 30–35 different unique sequences, each about 750 bp long. Such blocks are separated from one another by middle-repetitive sequences, 100–150 bp in length. According to Thomas' model (Figure 11), the greater part of the chromomere is also filled with repetitive sequences.

Models accepting only bands, or bands plus interbands, as functional units share another feature. They all consider that differential activity of the genome manifests itself only as conversion of the band into a puff. Because

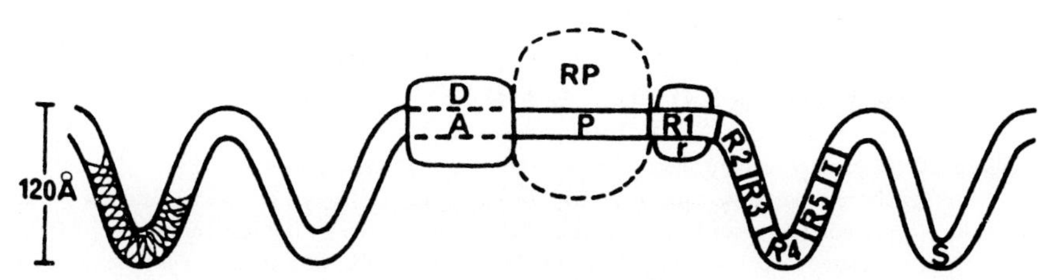

Figure 9. The structure of transcribed DNA in chromatin. Chromatid is organized as a fibril 12 nm in diameter. In the interband there are the "address loci" (A), and the promoter (P), the regulatory (R1-R5), and initiator (1) sites occupy the band–interband boundary. The molecule destabilizing packaging (D) binds to the address locus to cause local relaxation of supercompaction, permiting the molecule of RNA polymerase (RP) to attach to the promoter. If repressors (r) do not bind the regulatory sites, transcription of the structural (S) gene is initiated [reprinted with permission from *Nature* (Crick, 1971). Copyright (1971) Macmillan Magazines Limited].

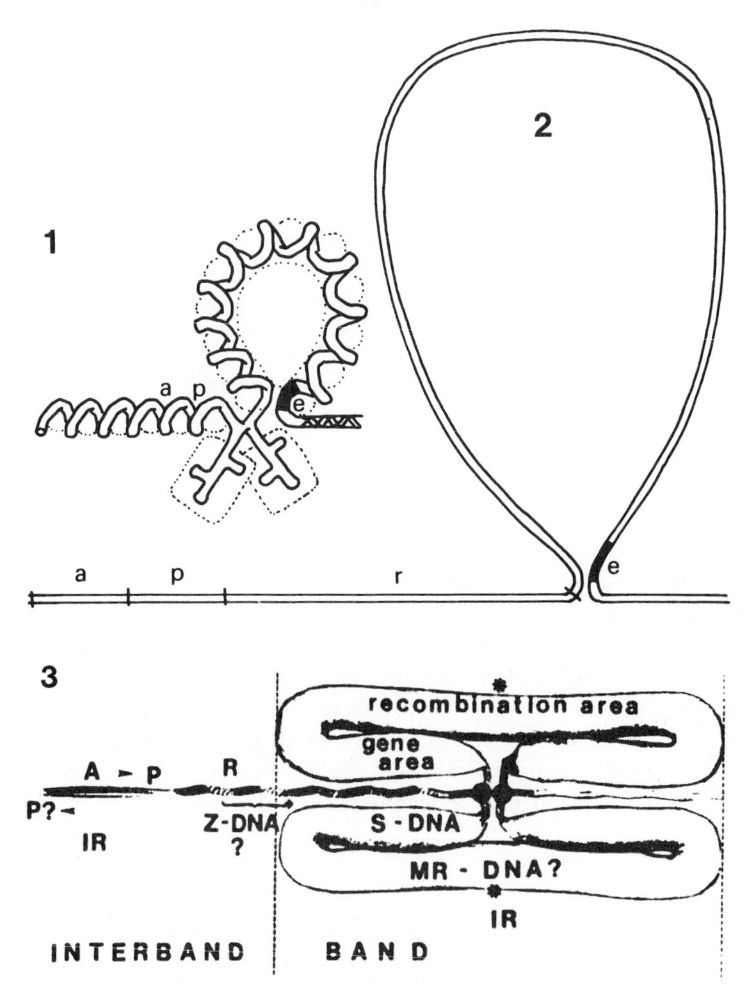

Figure 10. A scheme of the disposition of the genes in morphological units of the polytene chromosomes in inactive (1 and 3) and active (2) states. (1 and 2) Address (a) and promoter (p) loci located at the edge of the interband; the regulatory regions (r) are depicted as palindromes. The gene itself is located in one of the loops making up the chromomere; (e) the termination site of transcription (from Sorsa, 1975). (3) Promoter (P), address (A), regulatory (R) sequences, and Z-DNA are located in the interband; the genes, repetitive sequences (MR-DNA), and recombinational regions are located in the band. The palindrome sequences (IR) are, presumably, located in the address locus of the interband and in the region of replication initiation of the band (asterisks) (from Sorsa, 1984).

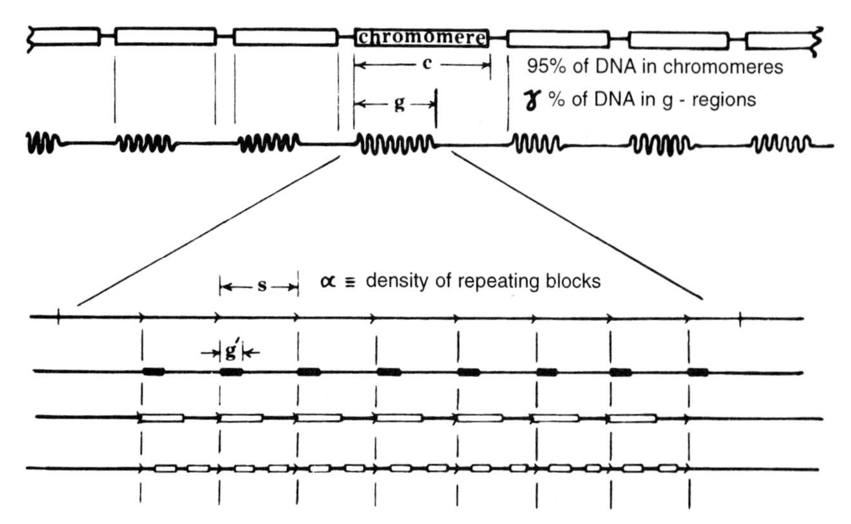

Figure 11. A model of the distribution of DNA sequences in chromomere. A great part of chromatid DNA (>95%) is located in the chromomere (c). A part of this DNA composes a fragment of the chromomere (g) enriched with repeats, such as simple tandem (s) or alternating repeats (g') (from Thomas, 1973).

there are a dozen or so of distinct puffs by the end of the third larval instar in *Drosophila,* and also because those appearing later are needed to involve new genes in metamorphosis (Ashburner, 1972a, 1975), it is difficult to believe that the few puffs are enough to provide the cell with all of the processes to be carried out.

Kosswig and Shengun (1947a, p. 237) were the first to express the view that banding patterns depict the disposition of inactivated genes corresponding to a particular state of the cell. They wrote: "According to the new view which we here propose, the discs seem to represent merely regions of activity (or inactivity) along the chromosome, the pattern of which is conditioned by the function of the cell in which it is housed." Later on, S. Fujita (1965, p. 743) suggested that ". . . the bands and heterochromatic regions of salivary giant chromosomes may be composed of genes which do not function in salivary cells and that the pattern of the distribution of inactivated DNA in the chromosome may be a genetic basis of major differentiation . . . RNA synthesis of the unravelled portions including interbands, puffs, nucleolar organizer regions and fainter bands, can be reversibly regulated by some mechanism . . . ." Akifjev (1974, p. 51) held the opinion that the "most common basic mechanism, which converts 'extra' into 'silent' DNA, acts to incessantly restrict transcription in the somatic cells through packaging of DNA into a compact chromomere structure."

In Gersh's view (1975, p. 425), "the bands also contain inactive genes, but some of these become active at various stages of development. On the other hand, active genes are in the extended regions . . . interbands represent genes in a steady state of activity, the genes of puffs become intensely active over a more limited period of time."

Speiser (1974) suggested the existence of housekeeping genes, which must be active in all cells and continually engaged in transcripton. He thought that these genes are most likely located in the interbands (the "miniature puffs") and the genes needed for differentiation are in the chromomeres.

Zhimulev and Belyaeva (1975a) developed a model based on the idea of the dynamic nature of chromomere pattern: bands contain genes and DNA fragments not engaged in transcription in a given tissue at a given time; the material of these genes is in a compacted state. The interbands are the transcriptionally active chromosome regions responsible for basal metabolism (the sites of the housekeeping genes). The constancy of such processes in the cells of different organs would correlate with invariance of banding pattern in the chromosomes. It follows that the banding patterns of the chromosomes correspond to a certain functional capacity of the cell and that variations in banding patterns, depending on cell metabolic specificity may be brought about by changes in activity and compaction degree of interbands (Figure 12). When transcription is completely inactivated, the material of interbands becomes closely packed, and the bands are brought close together to unite into blocks of inactive chromatin, virtually into new bands. In this model, nonfunctional units are contained in bands, they rather passively originate as DNA fragments that are transcriptionally inactive, and, consequently, compacted between two active interbands. The genetic content of these bands can be highly variable (Zhimulev and Belyaeva, 1975a,c, 1985a,b; Zhimulev *et al.*, 1981a,f,g, 1983).

The ideas of Rayle and Green (1968) provide another variation of these themes. On the one hand, they agree that the especially thick bands, the "doublets," are composed of a set of recombinationally separable subunits with identical or similar informational content. On the other hand, they also suggested that several bands might together constitute a single functional unit. Yet another thematic variation is Ashburner's (1976, p. 87) hypothesis that there may "exist silent bands performing no function, bands with more than one function, and he also envisioned situations where one function spreads over more than one band." Several reviews summarize the information leading to these and other hypotheses about chromomeres (White, 1954, 1973; Welshons, 1965; Kiknadze, 1970; Comings and Okada, 1974; Lefevre, 1974; Sorsa, 1974, 1988; Swift, 1974; Thomas, 1974; Korochkina, 1977; Judd, 1979; Hennig, 1980;

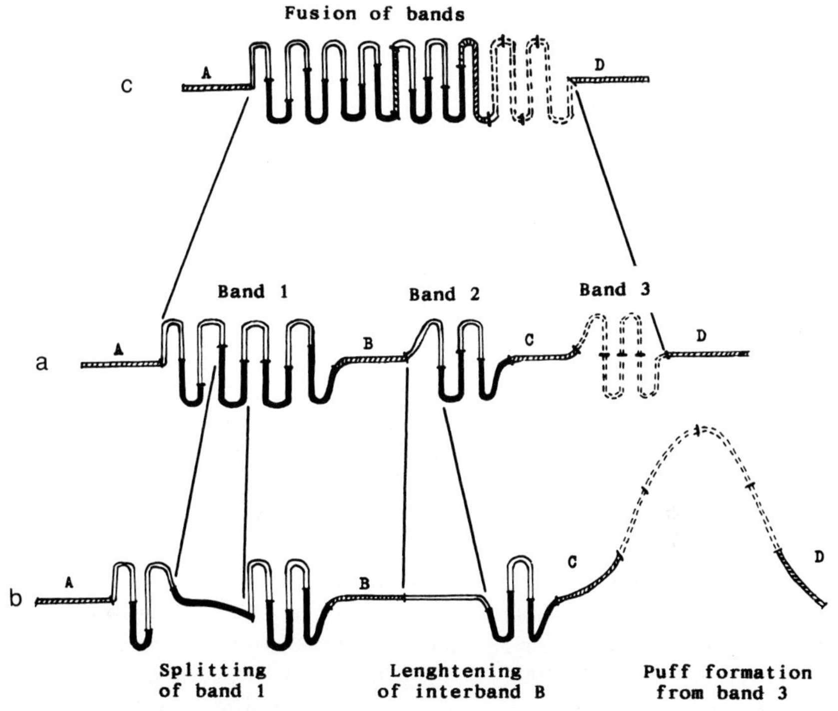

Figure 12. A model of the dynamic organization of the chromomeres of polytene chromosome. (a) The usually observed banding pattern; (A, B, C, and D) the continuously transcribed sequences–interbands. Bands 1, 2, and 3 represent a series of tissue- and stage-specific genes. The sequences are independently activated (in bands 1 and 2) or under common control (band 3). (b) Change in banding pattern produced by activation of some of the sequences; splitting of band 1, lengthening of the B interband, and formation of the puff from band 3. (c) Inactivation of transcription in the interbands leads to fusion of bands and interbands into new chromatin block (reprinted from Zhimulev and Belyaeva, 1975a, copyright © Springer-Verlag; and Zhimulev *et al.*, 1981a. Informational content of polytene chromosome bands and puffs. *CRC Crit. Rev. Biochem.* **11**, 303–340. Copyright CRC Press, Boca Raton, Florida).

Lewin, 1980; Richards, 1985, 1986; Hill and Rudkin, 1987; Zhimulev and Belyaeva, 1991).

It is appropriate to summarize several points:

1. The problem of the informational content of the interchromomeres and chromomeres of polytene chromosomes has been treated theoretically for decades. The host hypotheses, some of them mutually exclusive, reveal that the problem is complex and incompletely resolved.

2. The questions raised are important in the context of current genetic thinking: (a) Are genes located in the interchromomeres or chromomeres? (b) Is the chromomere, or its derivative the puff, a functional unit? (c) How many

genes are contained in the chromomeres? (d) What function does "extra" DNA perform in the chromomere? (e) If there are many genes in the chromomere, to what extent are they functionally related? (f) Are there structures in the chromosomes, besides the puffs, synthesizing RNA?

3. The majority of the models are based on an implicit or explicit assumption that banding patterns are stable, largely ignoring facts to the contrary. In this respect "genetic interpretation is further complicated, if accepting differences in banding patterns in different tissues" (Ashburner, 1976, p. 84).

C. Constancy and variability of banding pattern in polytene chromosomes

Constancy of the chromomeric organization of polytene chromosomes is by definition the accurate reproduction of band (chromomeres) sequences differing in morphology, size, and distance from one another in each particular chromosome region; in cells of different tissues or at different stages of development.

Changes in banding pattern of the polytene chromosomes result from: (1) weakening of the synaptic attraction between the chromatids composing the chromosome; while the chromomere set of each chromatid remains identical, the banded structure disappears; a "pomponlike" chromosome arises instead; chromosomes of this type have been described in detail in the previous reviews of this series (Zhimulev, 1992, 1996). (2) Alterations in chromomere pattern itself are due to differential compaction or decompaction of the chromosome regions, in the context of different models of dynamic organization of the chromosomes (Lima-de-Faria, 1975; Zhimulev and Belyaeva, 1975a; Callan, 1978); thus, the number and morphology of bands in a particular region of the chromosome can be different when the chromosomes from different organs and at different stages of development are compared. This section presents a detailed analysis of these changes in chromosome organization.

The constancy and variability in banding pattern have been mostly discussed 40–50 years ago (Berger, 1940; Kosswig and Shengun, 1947a; Pennypacker, 1950; Slizynski, 1950; Pavan and Breuer, 1952; Sengun, 1954). From their morphological study of the chromosomes from different tissues, Kosswig and Shengun arrived at the conclusion that general morphology of the chromosomes and their banding pattern (Figure 13) in different tissues have nothing in common (Kosswig and Shengun, 1947a-e; Sengun and Kosswig, 1947; Sengun, 1947–1954). Thay elaborated the "Istambul hypothesis," which involves the coils of the chromosome spiral separated from one another by intercalary material of different nature, the intercalations being different in each tissue and occurring in different regions. Hence, the banding pattern in the chromosomes of different organs is conditioned by the functional features of the cell in which the gene is housed (Kosswig and Shengun, 1947a).

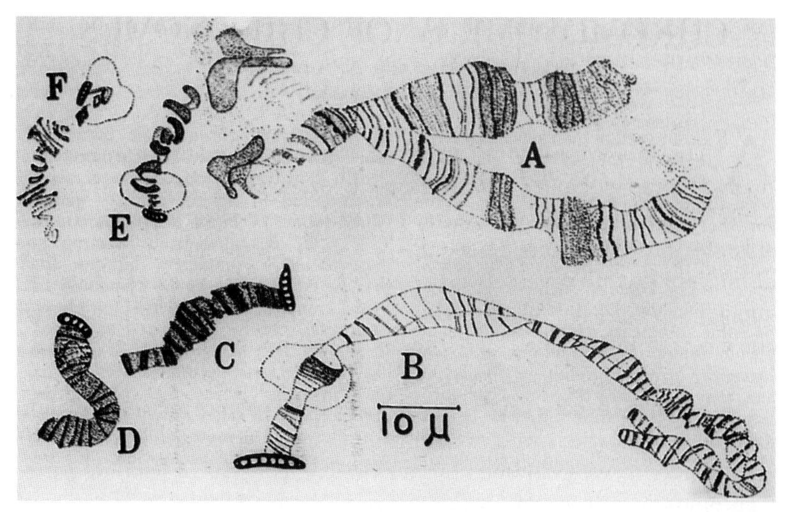

Figure 13. Morphology and banding patterns in the fourth chromosome from the cells of different organs of *Chironomus* sp. (A) Salivary glands; (B) midgut; (C and D) Malpighian tubules; (E and F) rectum (from Kosswig, C., and Shengun, A. (1947). Intraindividual variability of chromosome IV of *Chironomus*. *J. Hered.* **38**, 235–239, by permission of Oxford University Press).

As Beermann and Clever (1964) observed later, this contradicted the rule of linear gene order in the chromosomes and became the subject of debate. Thorough analysis of the puffing of the chromosomes of four organs of *Chironomus tentans* led Beermann (1950, 1952a, 1962, 1972) to the conclusion that a given banding pattern is, to a large extent, the same and "all the visible differences are due to technical difficulties, such as seeming or real fusion of neighboring bands" (Beermann, 1965, p. 376). He distinguished four types of differences seemingly inexplicable in terms of differential puff activity: (1) "Presence or absence"; the band is readily observable in one tissue, not in others. (2) "A singlet-doublet"; two bands are seen in one tissue and only one band in another. (3) "Different distance"; the distance between distinct homologous bands in the chromosomes of different tissues are different (lengthening-shortening of the interbands). (4) "Tissue-specific" increase in staining intensity of the "interbands" joining two or more close thick bands.

Given the sharp changes in morphology and banding pattern, examples of which are shown in Figure 13, as explanation by Beermnann was provided: during growth of polytene chromosomes, along with increase in polyteny level, such chromosomes are subject to morphological changes. At first, the chromosomes are represented by ribbons, coiled into a spiral; next, large blocks of concavities (meanders) are formed in the body of the chromosome, with the result that the chromosomes assume the shape depicted in Figure 13e. With attainment of highest degree of polyteny, the chromatids making up the chromosomes dispose themselves in a cable, with the chromosome appearing as a cylinder (Beermann, 1952a, 1962). This model was considered in greater detail earlier (Zhimulev, 1992, 1996).

The literature dealing with constancy and variability in banding pattern of dipteran polytene chromosomes is extensive. Comparative studies of banding pattern of the chromosomes of various types take into consideration mainly the prominent and medium-sized puffs most easily recognized in the light microscope. However, about a half of bands of the chromosomes are small (Zhimulev, 1992, 1996). They can be accurately identified only when viewed in the electron microscope, and such analysis is not commonly performed. The following conclusions should be treated with this stipulation.

From most comparative studies of chromosomes in different normally functioning organs, it is inferred that their banding pattern is essentially identical. The results were obtained by pairwise comparisons of banding patterns of the chromosomes from the cells of many organs (Table 1). The numerous species and diverse organs, whose banding patterns have been compared, support the conclusion that the most prominent bands are stable. Variations in banding pattern were found in association with high constancy of banding pattern in the cells of normally functioning organs because, in Beermann's view (1956), distinctness and detectability of bands may be different in different tissues. The number of bands was estimated as 380 for the third chromosome of the salivary gland and rectal cells of C. *tentans*, 440 for Malpighian tubule cells, and 480 for those of the anterior part of the midgut (Beermann, 1952a, 1956). In *Melanagromyza obtusa*, 864 bands were identified in salivary gland, 832 in midgut, and 779 in Malpighian tubule chromosomes (Singh, 1991).

There is greater variability in the number of bands of a given region of the homologous chromosomes of the same organ. Thus, the band number varies from 7 to 26, being mainly 9 to 18 in sections 1–4 in any of the 229 examined chromosomes of the salivary gland cells of *Drosophila repleta*. However, 31 bands can be counted on combined maps. The "ideal" reconstructed map of the chromosomes from tracheal cells has 17 bands; individual chromosomes have bands varying in numbers from 6 to 13 (Sengun, 1954). Electronmicrographs (Semeshin *et al.*, 1979b) showed that the detectability of bands in a small region of the X chromosomes of the salivary glands of *Drosophila* correlates with band size: only 4 (the largest) of 20 bands were identified with a 100% frequency. The frequency with which the thinnest bands occurred were in the 15–50% range. Other problems relating to the identification of bands and building of cytological maps have been considered in detail (Zhimulev, 1992, 1996).

Besides variations in banding pattern due to difficulties in their identification, there is ample evidence indicating that their differences correspond to

Table 1. Detection of Similarity in Banding Pattern of Polytene Chromosomes from Cells of Various Organs in Representatives of the Diptera Order

Species	Compared organs	Reference
Acricotopus lucidus	Salivary glands and Malpighian tubules	Staiber and Behnke (1985)
Ceratitis capitata	Thoracic and orbital trichogen cells and hindgut cells	Bedo (1986)
Chironomus tentans	Salivary glands, rectum, midgut and Malpighian tubules	Beermann (1950, 1952a)
Drosophila auraria	Salivary glands, fat body, and midgut	Mavragani-Tsipidou *et al.* (1990b)
Drosophila gibberosa	Salivary gland and midgut	Roberts (1988)
Drosophila hydei	Salivary gland, stomach, and Malpighian tubules	Berendes (1965a,b, 1966a)
Drosophila melanogaster	Salivary glands, stomach, and mid- and hindgut	Zhimulev and Belyaeva (1977)
	Salivary glands and ring gland of the *DTS3* mutant	Holden and Ashburner (1978)
	Salivary glands and fat body	Richards (1980a)
	Ring gland, the middle part of the midgut, hindgut and salivary glands	Hochstrasser, 1987
	Larval salivary glands and pseudonurse cells in *otu* and *fs(2)B* mutants	Heino (1989, 1994); Mal'ceva *et al.* (1995)
Drosophila repleta	Salivary glands and tracheal cells	Sengun (1954)
Melanagromyza obtusa	Salivary glands, midgut and Malpighian tubules	Gupta and Singh (1983); Singh (1988a, 1991)
Rhynchosciara angelae	Salivary glands, Malpighian tubules and midgut	Pavan and Breuer (1952); Guevara and Basile (1973); Pavan *et al.* (1975)
Sarcophaga bullata	Trichogen, tormogen, and footpad cells	Whitten (1969)
Sciara ocellaris	Salivary glands and midgut	Berger (1940)
Simulium ornatipes	Salivary glands of larvae and Malpighian tubules of adults	Bedo (1976)

those Beermann (1952a) has described, that is, fusion–fission of neighboring bands and lengthening of interbands. For example, in *C. tentans*, region 19A of the first chromosome, which starts with one to two thick bands in salivary glands, is represented by five bands in the chromosomes of midgut cells. In region 18A, there are six thin bands in the salivary glands and eight in Malpighian tubules. Small variations are detectable in the righthand part of section

6A of the third chromosome in Malpighian tubules, rectum, and midgut. Regions 2AB, 5B, 9A, 13A, and 13B provide other examples of variations in the pattern of puffing (Beermann, 1952a).

In *Rhynchosciara angelae*, differences in band number in section 13 in chromosome A were revealed by comparing the chromosomes of salivary glands and Malpighian tubules (Pavan and Breuer, 1952, Figure 1).

Polytene chromosomes of *Drosophila* have become the focus of intense study. The problems taken up were variations in banding pattern in different strains and at different developmental stages in salivary glands; furthermore, gastric, hind-, and midgut chromosomes were analyzed in animals homozygous for *giant* mutations. Differences in banding patterns of polytene chromosomes from salivary glands and different tissues were found between strains and at different stages of development for *giant* mutants. A single dense band is consistently located in the regions 31A1-2-B1 of Batumi-L larvae, while two bands are separated by an interband in the majority of chromosomes of wild-type Oregon-R larvae (Figure 14). All the 30 studied larvae were heterozygous for band number in this region in the progeny of the first generation from crosses between males of Batumi-L and females of Oregon-R strains (see Figure 14d).

Other cases are known in which banding patterns change in the third chromosomal regions 61C3-4 and 80A1-2. In the majority of third-instar larvae of different strains, salivary gland cells vary somewhat in their banding pattern for region 79E1-4 (Figure 15): a single band and two almost completely fused bands, hardly discernible in the electron microscope, are encountered. There are three bands and three interbands in *giant* larvae at puff stage 7 (PS7). In Oregon-R prepupae ("zero hour"), the bands 79E1-2 and 79E3-4 are at a considerable distance apart, and a small puff is formed from the 79E1-2 interval.

Band morphology continually varies in certain regions; however, no specificity in variation was detected; thus, band morphology varied from a single dense band to two separated by an interband in the 73A1-4 region in all strains and at all stages of development. No consistent variation was found for the 9B1-2 band, which splits in some larvae heterozygous for *Df(1)v*^{L11}*/FM6* (deletion/multiple inversions) (Zhimulev and Belyaeva, 1977; Zhimulev 1982).

In three regions, the material composing a band in salivary gland chromosomes consists of two, even three, bands in the gastric chromosomes 64C1-5, 93F1-5, and 35D1-4. Cases are known in which the length of interbands in the regions 59F, 61F (Figure 16), and 79E (Figure 15d) changes. In the gastric chromosomes, the interband in the 61F region is so much lengthened that the interband is indistinguishable from the puff (Zhimulev and Belyaeva, 1977). There are obviously fewer dense bands in section 98 (right arm of chromosome 3) in the ring gland than in the homologous region in the salivary gland chromosomes of *Drosophila* (Holden and Ashburner, 1978, Figure 2).

A characteristic feature of region 84AB of the salivary gland chromosomes is the presence of three large bands, while the number of dense bands in

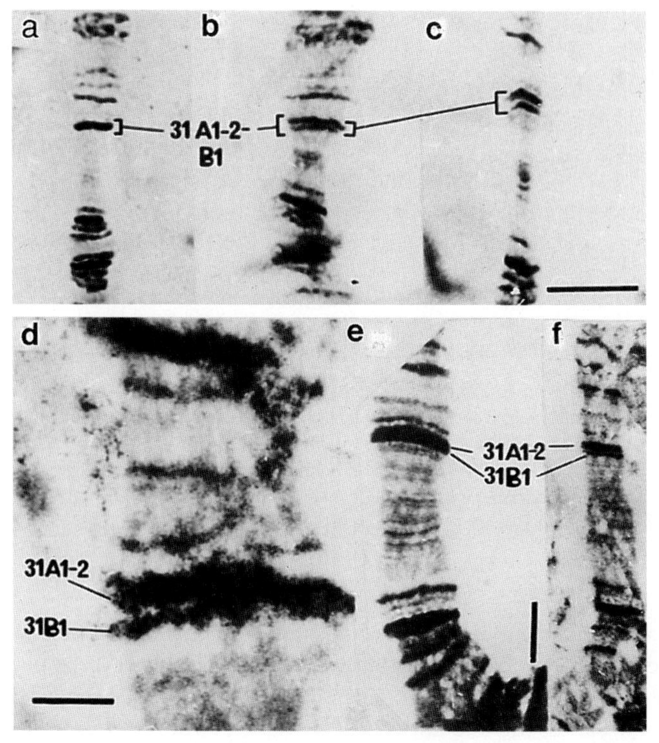

Figure 14. Splitting–fusion of the 31A1-2-B1 bands in the chromosomes of salivary glands of larvae and zero-hour prepupae of *D. melanogaster* of Batumi-L (a and e), Oregon-R (b, c, and f) strains, and Oregon X Batumi (d) hybrids. (a–c) Light microscopy; (d–f) different magnification of electron microscopy. Bar = 10 μm (a–c), 1 μm (d), and 5 μm (e and f) (from Zhimulev and Belyaeva, 1977; Zhimulev, 1982).

the fat-body chromosomes is four (occasionally five). Region 93E-94A is a major landmark of the right arm of chromosome 3, with four heavily staining bands. Seven large bands are prominent in fat-body chromosomes in this region (Richards, 1980a). The author does not document the differences in interband length. He notes that tissue differences in spacing between bands exist and that they require explanation.

In *D. hydei*, a comparison of banding pattern in salivary gland and Malpighian tubule chrosomomes revealed differences in band thickness and interband length (Berendes, 1966).

In the midge *Simulium ornatipes*, differences in band number were detected in certain regions (Bedo, 1976, Figs. 1–3).

Variations in polytene chromosome banding patterns were observed in

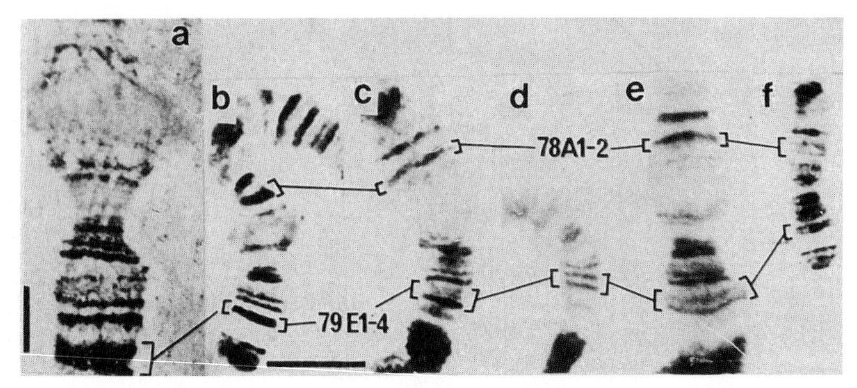

Figure 15. Splitting-fusion of bands in the 78A1-2 and 79E1-4 regions of chromosome 3L of *D. melanogaster* (a, b, c and e) salivary gland chromosomes of larvae at the end of the third instar of Batumi-L (a and b), *giant* (c), Oregon-R strains (e); (d) chromosome of gastric cells of *giant* larvae; (f) salivary gland chromosome of 6-hr prepupa of Batumi-L strain; (a) electron microscopy; (b–f) light microscopy. Bar = 5 μm (a); 10 μm (b–f) (from Zhimulev and Belyaeva, 1977; Zhimulev, 1982).

Drosophila larvae during development. In the early third-instar larvae and prepupae, the 70A1-5 region consists of two dense bands. However, a newly arisen interband in larvae of different strains at PS5-PS8 separates the 70A4-5 band into two new bands, the fine-dotted 70A3 and 70A4-5, which becomes thinner (Zhimulev and Belyaeva, 1977).

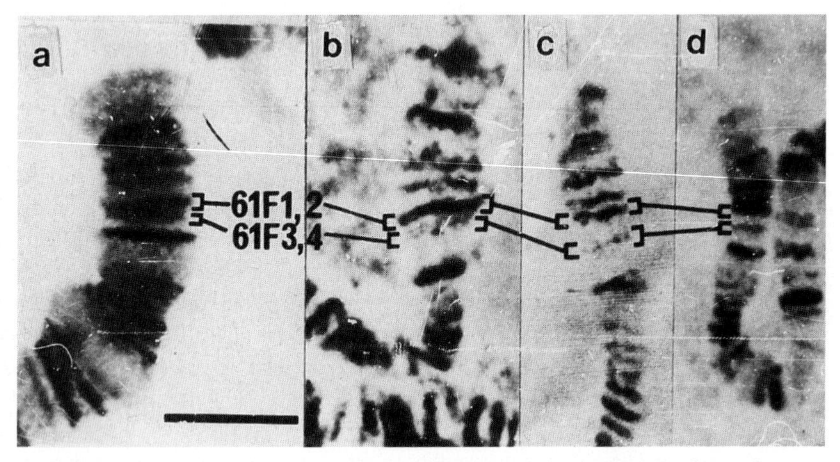

Figure 16. Variation in banding pattern in the 61F region of chromosome 3L of *D. melanogaster*. Salivary gland (a and b) and gastric (c and d) cells. Bar = 10 μm (from Zhimulev and Belyaeva, 1977).

The densely condensed material, composing the single 100B1-2 band at the studied stages of larval development, splits into two bands in 4- to 6-hr prepupae. [^{3}H]Uridine is incorporated into the newly arisen interband and into the two new bands as well (Zhimulev and Belyaeva, 1977; Zhimulev, 1982).

The 1E1-4 region is represented by two dense bands: 1E1-2 and 1E3-4. The bands are so close together that they are barely distinguishable when viewed in the electron microscope (Figure 17). In 4- to 6-hr prepupae the decondensed distal part of the 1E3-4 considerably increases the distance between the bands. This decondensation is associated with heavier labeling of the lengthened interbands by [^{3}H]uridine; when label incorporation is heavier the band 1E3 is more densely labeled (Figure 17f–17h). Lengthening of the 1E1-2/

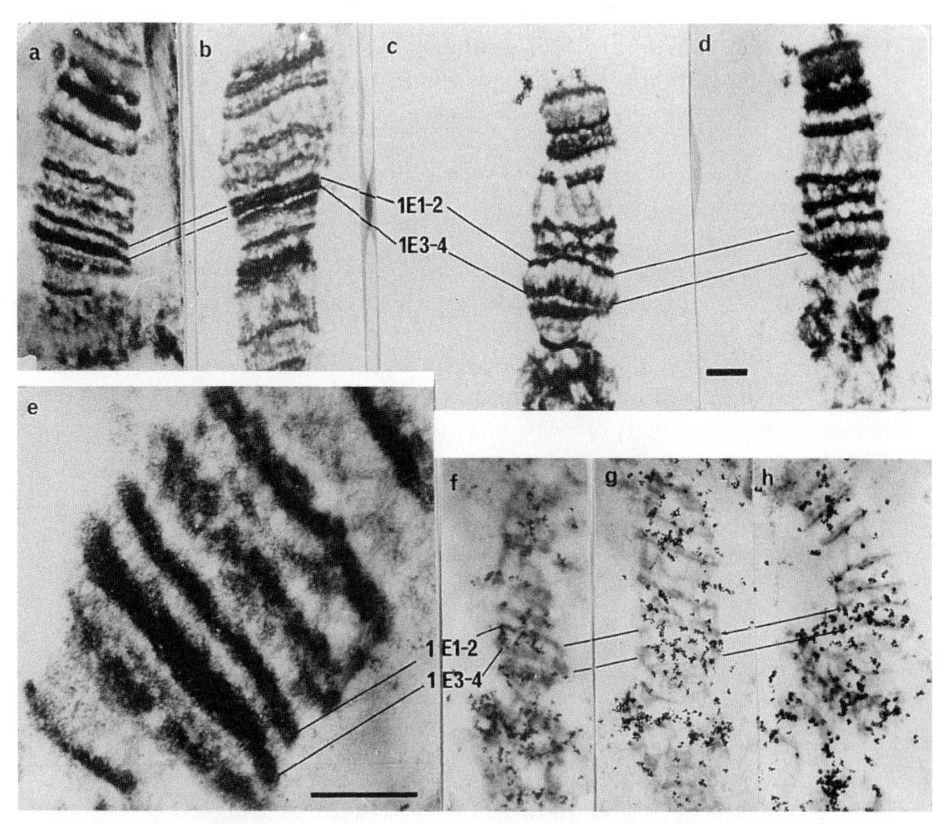

Figure 17. Lengthening of an interband (formation of a micropuff from a part of a band) in the 1E1-4 region of salivary gland chromosomes in *D. melanogaster* during development. (a–e) Electron microscopic data; (f–h) incorporation of [^{3}H]uridine (electron microscopic autoradiography). Larvae (a) and prepupae at the ages of 0 hr (b) and 6 hr (c, d, and f–h) of Batumi-L strain. Bar = 1 μm (from Zhimulev *et al.*, 1983, with kind permission from Kluwer Academic Publishers).

1E3-4 interband was observed in Batumi-L, *yellow*, and *swi*^{t467} mutant strains at this stage (Zhimulev, 1982).

Electronmicroscopic studies demonstrated that, when the ecdysterone puff 2B3-5 is formed at the end of the third larval instar, the 2B3-4 band separates into two parts and the puff originates from the proximal part (Belyaeva *et al.*, 1987).

On Bridges' map, two bands are depicted in the regions 78A1 and 78A2, which are so close in larvae at the end of the third-instar and zero-hour prepupae that they appear as one band in the light microscope (see Figure 15f). In younger larvae (96–110 hr) and in older prepupae (4–6 hr) the two bands are separated by a distinctive long interband. Later still (10- to 12-hr prepupae), both bands come closer again or fuse with each other to form a single band.

Variations consistent with the "presence–absence" or "singlet–doublet" pattern may result from band splitting followed by the formation of a puff from the halves (Figure 18). Splitting of the 2C1-2 band during puff formation was observed in electronmicroscopic studies with *D. melanogaster* (Semeshin *et al.*, 1985a).

Variations in band number in some chromosome regions were disclosed when closely related species or strains within a species were compared (Figure 19).

Comparison of banding patterns in three tissues of *Melanagromyza obtuza* revealed that the banding pattern of the chromomeres is more or less the same, because homologies are detectable for the majority of chromomeres. There were, however, differences for certain regions, which are reduced to five types: tissue differences in the thickness of bands, their compaction, differences in number and stainability of bands, and in distances between the bands (Singh, 1991).

To illustrate, 11D appears as two bands, 14A as one band, 15D as three bands, in chromosome A of *D. subobscura*; band numbers are 1, 2, and 2 in the counterpart regions in *D. guanche*. Similar differences were found between the puffing patterns of the other chromosomes (Molto *et al.*, 1987a).

P-element transposition was shown to cause variation in banding patterns (Semeshin *et al.*, 1989a,b). In *D. melanogaster*, the *vermilion* (*v*) gene was mapped to the left part of band 10A1-2 (Zhimulev *et al.*, 1981c; Kozlova *et al.*, 1994). In a *v* strain with the mutation induced by P-element insertion, 10A1-2 is separated into two unequal parts, with an interband originating between them (Figure 20a,b): the new "left" band is 0.09 ± 0.01 μm thick and the "right" is 0.44 ± 0.03 μm thick; the thickness of the "interband" is 0.09 ± 0.01 μm. The 412 mobile element does not change the structure of the 10A1-2 band (Figure 20c). Two cases are known in which an increase in spacing of bands is also associated with mobile element insertion. *l(1)BP7*^{S3} strain contains P-element inserts into the region 8C1-2. The length of the region increased from

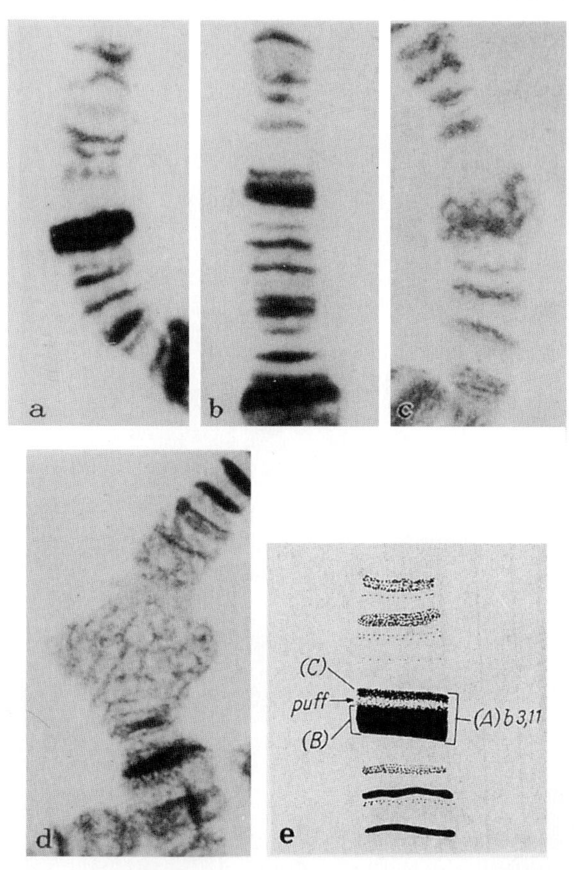

Figure 18. Splitting of the b3,11 band in the third chromosome of *Chironomus thummi thummi* during puff formation. (a and b) Staining by Feulgen's technique; (c and d) orcein; (e) drawing (reprinted with permission from Keyl, 1965).

0.34 ± 0.02 μm in the control strain to 0.43 ± 0.02 μm in the one containing the insert. The *l(1)BP7*^{S3} mutation itself has resulted from insertion of a P element at the *l(1)BP7*$^{+}$ locus, within the region 10A6-7 (Zhimulev *et al.*, 1981c; Pokholkova *et al.*, 1991). The average distance between 10A4-5 and 10A8-9 in the strain carrying the mutation was 0.69 ± 0.06 μm; in the control the average distance was 0.56 ± 0.04 μm. Therefore, the interband resulting from splitting is 0.09 μm in length, and the two regions become 0.09–0.13 μm longer upon insertion (Semeshin *et al.*, 1989a,b, 1990).

Because the P-element promoter is weak and can be transcribed in all cells through development (Laski *et al.*, 1986; O'Kane and Gehring, 1987), it may be thought that both separation of the band 10A1-2 and lengthening of

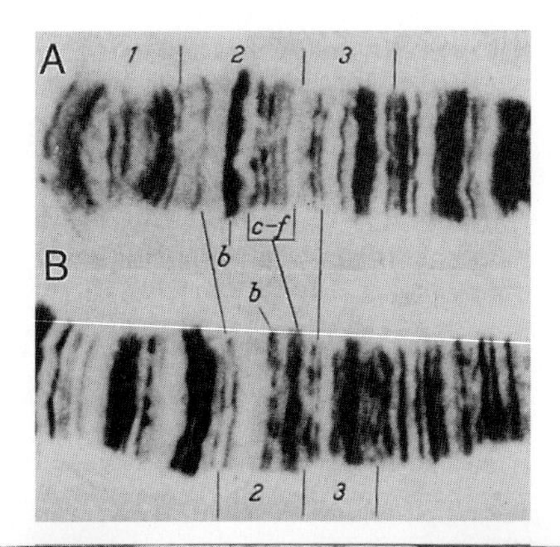

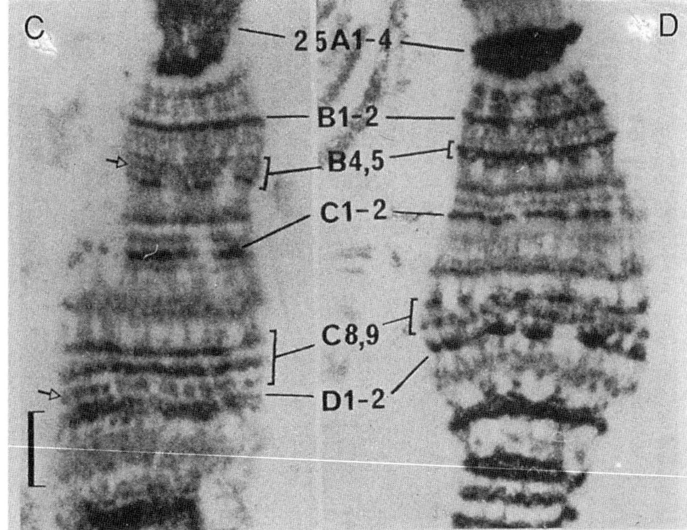

Figure 19. Variation in band number in homologous regions (c–f in A) of the chromosome of related species (A) *Chironomus melanescens* and (B) *C. obtusidens* in the 25B4-5 and 25C8-9 regions in (C) Batumi-L and (D) Oregon-R strains of *D. melanogaster* (arrows). (A and B) reprinted by permission from Keyl (1961); (C and D) reprinted by permission from Semeshin *et al.* (1985a). Bar = 1 μm in C and D.

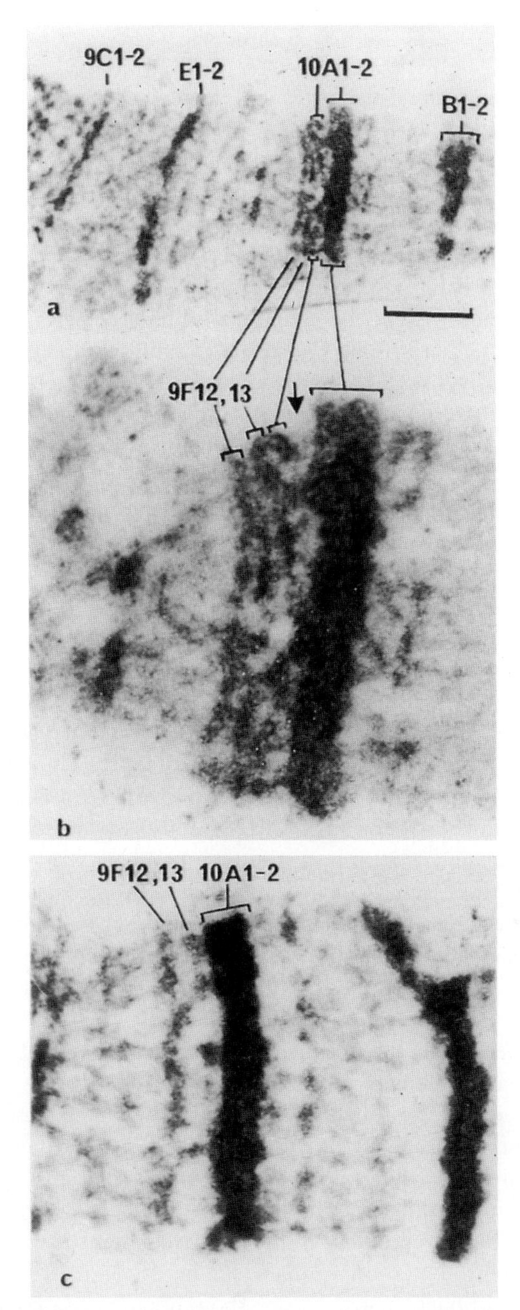

Figure 20. Splitting of the 10A1-2 band and formation of a new interband as a result of insertion of a mobile P element into the *vermilion* gene located in the left part of the band. (a-c) See text for details. Bar = 1.5 μm (reprinted with permission from Semeshin *et al.*, 1989 a,b; 1990).

the 8C and 10A4-9 regions results from P-element DNA and associated decompaction of chromatin.

There are data indicating considerable differences in banding patterns of the chromosomes from cells of normally functioning organs. A comparison of polytene chromosome banding patterns from trichogen cells (Bedo, 1986, 1987) and salivary gland cells (Zacharopoulou, 1987, 1990) failed to reveal homology between banding patterns despite their stability within each examined tissue. The authors believe that local puffing of certain bands in these tissues cannot cause this striking intertissue variability in polytene chromosome banding patterns (Bedo and Zacharopoulou, 1988). It is not clear to what extent the banding patterns in the polytene chromosomes from the ovarian nurse cell differ from those in other cell types. A comparison of the banding pattern from larval salivary glands and ovarian nurse cell of adult females of the mosquito *Anopheles superpictus* disclosed no appreciable differences in distinctive bands in most chromosome sections; consequently, the lack of homology between the majority of bands made their mapping quite tentative (Coluzzi *et al.*, 1970).

In *Calliphora erythrocephala*, chromosome banding patterns of adult ovarian nurse cells and pupal bristle-forming cells differed greatly so that no homology could be established (Ribbert, 1967, 1975, 1979). Ribbert found that the ovarian nurse cells and bristle-forming cells contain polytene chromosomes showing banding patterns that are difficult to correlate. There were differences in band number (802 in the chromosomes from the ovarian nurse cells and 1360 in those from bristle-forming cells) and also in the total lengths of the chromosome (1108 and 1537 μm, respectively) (Ribbert, 1970).

However, the results of a study of the chromosomes from the nurse cells of another mosquito *Anopheles stephensi* could not be interpreted in such a straightforward manner. From a comparison of the banding pattern of the left arm of chromosome 3 and the X from larval salivary glands and ovarian nurse cells, it was concluded that these chromosomes have the same band–interband organization. There were 12 cases in which the appearance of bands or groups of bands between the salivary gland chromosomes and the nurse cell chromosomes were different. The differences fell into two categories: an increase in interband distance and band splitting (Redfern, 1981a). They were attributed to preparative artifacts (differential chromosome stretching). The conclusion does not appear justified, as may be judged by the accompanying photographs representations: the chromosomes from the two different tissues were stretched to the same extent. Later on, from a similar study in this species, it was concluded that, although many regions show homology, a complete band-to-band correspondence was lacking (Mahmood and Sakai, 1985).

In *D. melanogaster*, although the ovarian nurse cells are polyploid, the homologous chromatids are conjoined, and the usual polytene chromosomes are not formed (see Zhimulev, 1992, 1996, for more detail). Certain mutations causing tumors of the ovaries, *otu (ovarian tumor)*, or female sterility, *fs(2)B*, generate pseudonurse cells whose polytene chromosomes are quite suitable for cytological analysis (Dabbs and King, 1980; King *et al.*, 1981; King and Riley, 1982; Sinha *et al.*, 1987; Mal'ceva *et al.*, 1995). There is, as a rule, a two- to threefold increase in spacing between bands in which chromosomal asynapsis occurs, and alterations in banding patterns of chromosomes of this type are quite apparent.

When larvae were grown at 17°C or at lower temperatures, their polytene chromosomes from ovarian pseudonurse cells and salivary gland cells showed remarkably similar banding patterns (H. Gyurkovics, 1986, personal communication; Heino 1989; Mal'ceva and Zhimulev, 1993; Mal'ceva *et al.*, 1995). A comparison of the banding patterns of polytene chromosome *3* from ovarian pseudonurse cells of homozygous *otu* females revealed high homology between the two tissues. The differences in the patterns were only due to changes in interband distance (lengthening of interbands?) in the regions 63C, 65E, 69ABC, 73A, 84A, and 90DE (Heino, 1989).

The phenomenon of seasonal changes in the length of polytene chromosomes, described for some species of *Chiromomus*, is another good example of considerable change in the banding patterns of polytene chromosomes. Before the cold season, the chrosomomes become much shorter, the number of puffs is minimal, and many of the easily recognized neighboring bands fuse to form blocks of heterochromatin. Individual bands become 1.5 to 2 times smaller than in summer. The chromosomes start to lengthen in January–February, and there are 3 to 4 times more bands in March than in September. The chromosomes lengthen because many pale spacings appear between the bands; that is, the "September" blocks of heterohromatin split into separate bands and interbands. The same sort of changes recurred for several years (Ilyinskaya and Maksimova, 1976, 1978; Ilyinskaya, 1994; see Zhimulev, 1992, 1996, for more details and references).

Changes in banding pattern are of a functional nature, as has been demonstrated by special experiments. Thus, when larvae were collected, transported in a thermostat at 2–4°C, and then transferred to a refrigirator, their polytene chromosomes retained features of the "winter" banding patterns for at least a week. However, these features were rapidly lost after maintaining the specimens for 30–60 min at room temperature (Ilyinskaya, 1980). The data indicate that the seasonal changes in banding patterns were not due to the disappearance of bands, to the *de novo* appearance of single bands, or to the development of the usual puffs at the sites of single bands, which might have decreased puff count. The changes were rather due to an increase in the degree of chromosome coiling and fusion of a group of bands into completely and heavily staining blocks in September and band splitting in winter (Ilyinskaya, 1980). The changes were also not due to stretching of the chromosomes to different lengths during their preparation. The chromosomes isolated from the salivary glands of the September larvae did not show more bands when stretched but simply got torn to pieces. In contrast, the chromosomes of winter larvae had already lengthened in unsquash nuclei (Ilyinskaya, 1980).

A study of [^{3}H]uridine incorporation into the salivary gland cells of "wintering" larvae demonstrated that RNA syntheses proceeds and that chromosome cells label uniformly, even at a temperature of 1–3°C. At a higher temperature, the chromosomes shorten, and the label becomes nonuniformly distributed (Ilyinskaya *et al.*, 1991). To obtain quantitative estimates of changes in transcriptional activity, immunofluorescence localization of DNA/RNA hybrids in polytene chromosomes was used. The method is advantageous because, after chromosomal DNA denaturation and reannealing, hybrid molecules are formed between chromosomal DNA and endogenous RNA molecules in their transcriptional sites.

When larvae were grown at low temperature, virtually all the regions of their polytene chromosomes with decompacted heterochromatin fluoresced. The number of fluorescent bands reflecting the number of transcriptionally active sites with high and low compaction was counted in the third chromosome of *Chironomus plumosus*. The number of fluorescent bands varied from 30 to 50 in the shortened chromosomes to 40 to 60 in the lengthened chromosomes (Vlassova *et al.*, 1991).

Morphological and structural changes, including wide variations in banding patterns, are associated with impairment of cell or organ physiology. The state of organs and cells in larvae whose development is impaired by mutation or infection and those subjected to prolonged culture in adult abdomen is pathological. The pathological condition of the dipteran cells has been previously described in detail (for review see Zhimulev, 1992, 1996).

When larval (Hadorn *et al.*, 1963; Berendes and Holt, 1965; Ashburner and Garcia-Bellido, 1973; Zhimulev and Belyaeva, 1976) or embryonic (Staub, 1969) salivary glands were implanted into the abdomens of adult females and cultured up to 50 days, their polytene chromosomes showed changes in both morphology and banding pattern: they loosened, stained paler, contained newly formed meanders, and occasionally decreased much in length. Shortening of the chromosomes can be associated with loosening of chromosome texture and distinctive banding pattern. For example, polytene chromosomes showing a well-defined banding pattern are seen after 13 days of incubation of 7-hr salivary gland anlagen. However, these are two to three times shorter than the normal chromosomes. The number of identifiable bands in such chromosomes is much smaller than in the usual chromosomes (Staub, 1969, Figure 13b). Chromosomes start to contract in the cells of the transplanted salivary glands 19 days after incubation *in vivo,* and chromosomes two to three times shorter than normal are identified 24 days after it (Zhimulev and Belyaeva, 1976). A process approximating neighboring bands underlies shortening. The regions 100AF and 71C-75C of the fourth chromosome are a case in point. The bands occasionally come so close together that a new band is formed. Thus, fusion of bands into blocks may be detected in the region 76E-77A, where, according to Bridges, there are 13 bands and also in the region 62B-D composed of 23 bands (Figure 21a, Figure 23b). Twenty-nine bands of the region 61A-F4 unite to form a wide loose block of chromatin (see Figure 23b).

The mutant *l(3)tl* of *D. melanogaster* serves well to illustrate alterations in banding pattern of polytene chromosomes. Larvae homozygous for the mutation *l(3)tl* fail to pupate, and their larval period is extended to 30 days. By this time, a variety of alterations in morphology occur in the chromosome organization of polytene chromosomes; they become shorter and thicker (Rodman and Kopac, 1964; Kobel and Breugel, 1967; Breugel and Bos, 1969). At this age, when normal larvae initiate pupariation (about 120 hr after egg deposition), the chromosomes are of normal length in *l(3)tl* homozygotes.

Decrease in chromosome length is noticeable already at the age of 144 hr, and it is very conspicuous in 408-hr larvae (Figure 21d–21f). The chromosomes shorten because many of the adjacent bands are brought close together and unite into blocks of chromatin giving rise to new bands (Zhimulev *et al.*, 1974, 1976). For example, there are 19 bands in the region 31F-32C in the left arm of chromosome *2* on Bridges' map; their number is 14 in the region 32E-33A5. These groups of bands are separated by the puff 32D. A good landmark can be puff 31A at the left and puffs 33B and 33E at the right. In as early as 144-hr larvae of the *l(3)tl* mutant, the regions between the dense bands shorten and, as a consequence, the bands are brought close together; they are juxtapositioned in 168-hr larvae.

The transformation of the regions 61A-F4 and 62B-D in the left arm of chromosome 3 into blocks of chromatin is shown in Figure 22c. Only five small regions form puffs in the 61A-63F interval: 61F5-8, 62A, 62C, 62F, and 63BC. The other regions condense. By 408 hr, a block of chromatin is formed from the 61A-F4 region containing 29 bands. By this time, puff 62C becomes inactivated and all 32 bands of the region fuse (Figure 23c). It is noteworthy that this new pattern of the region 61A-63F was observed in 42 of the 50 studied 3L chromosomes; that is, distinct regions presumably condense conforming to a definite pattern and order.

The 76E-77A region provides other example of band fusion. In 144-hr *l(3)tl* larvae, region 76E-77A, usually consisting of 11 bands, is transformed into a block of chromatin. The block becomes so narrow that it bears resemblance to a band that can be found 24 hr later.

The region of chromosome 3R extending from the small 97D puff to the "weak spot" in 98C (Figure 22) contains 45 bands, which come closer together in 264-hr larvae. The transformation of the 93EF region, which is

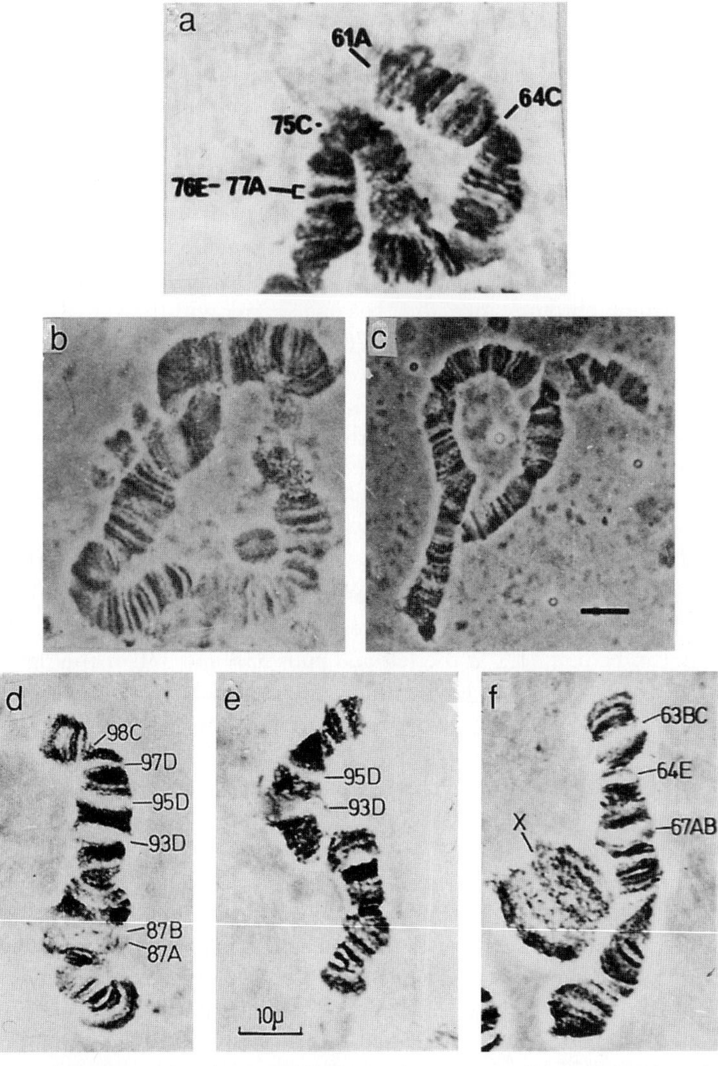

Figure 21. Shortening of the salivary gland polytene chromosomes of *D. melanogaster* incubated for 32 days in (a) adult abdomens, (b and c) in mutants of the *l(1)BP5* locus aged 14 days and (d–f) in larvae homozygous for *l(3)tl* mutation aged 408 hr. (a) from Zhimulev and Belyaeva, 1976; (b and c) from Bgatov *et al.*, 1986; (d–f) reprinted by permission from Zhimulev *et al.*, 1976. Numbers and letters designate the chromosome regions (a) and the heat shock puffs (d–f). Bar = 10 μm.

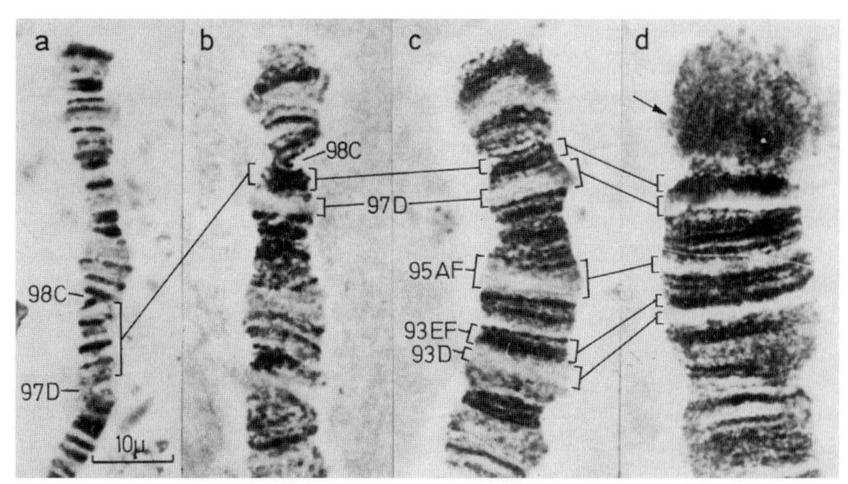

Figure 22. Banding pattern in the distal segment of chromosome 3R in (a) normal salivary gland cells and (b–d) in the *l(3)tl* mutant of *D. melanogaster*. (a) Batumi-L strain; (b–d) *l(3)tl* larvae aged 264 hr (b) and 408 hr (c–d). Arrow points to the pompon-like part of chromosome 3R (reprinted by permission from Zhimulev *et al.*, 1976).

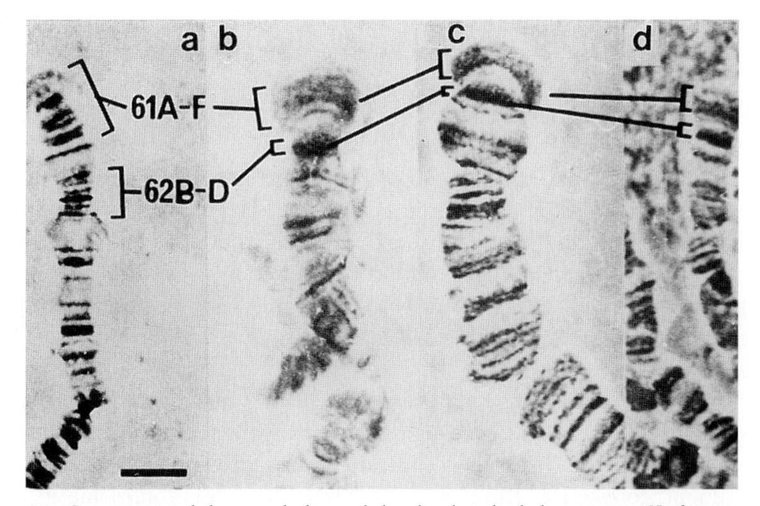

Figure 23. Comparison of the morphology of the distal end of chromosome 3L from normally functioning salivary gland cells of (a) *D. melanogaster* and (b) cells incubated in adult abdomens for 40 days in the (c) *l(3)tl* mutants and in heterozygotes for mutations at the (d) *l(1)BP5* locus (from Zhimulev, 1982).

normally composed of 25 bands, into a single chromatic block, is shown in the figure.

The puffing patterns were conspicuously changed not only in the cells of the salivary glands of 408-hr larvae, but also in those of salivary gland duct, esophagus, fat body, Malpighian tubules, stomach, gastric caeca, and hindgut. Contracted and thickened chromosomes of this sort were already seen in the cells of the salivary gland duct of 144-hr larvae.

Thus, although quite different with respect to chromosome morphology, the 264- 408-hr larvae all show an at least twofold chromosome shortening and a change in banding pattern manifests as fusion of adjacent bands into a block of chromatin (Zhimulev *et al.*, 1974, 1976).

With increasing shortening there is a progressive decrease in transcriptional activity. The intensity of [^{3}H]uridine incorporation into the chromosomes of the *l(3)tl* mutant is quite different in different cells of the same organ. Nuclei with diffuse chromosomes label heavier than those with strongly contracted chromosomes showing distinct banding patterns. The regions of the fused bands label at a much lower intensity, remaining at the level of the wild type in regions with unaltered morphology. The relative rate of label incorporation expressed as the ratio of silver grain number over the chromosome regions with fused bands to that over the regions with constant morphology is tenfold lower for the regions 31F-32C and 32D-33A, where new bands originated from a large number of initially normal bands; and it is 1.5- 3.0-fold lower for regions 61A-F4 and 62B-D (Zhimulev *et al.*, 1976).

Immunofluorescent localization of DNA/RNA hybrids demonstrated that transcriptional activity is considerably decreased in the chromosomes of the *l(3)tl* mutant. While the number of bands differing in fluorescence intensity is 100–110 in 2R chromosomes of normal length, it is about 40–60 in the shortened chromosomal arm (Umbetova, 1991).

The *l(1)BP5*^{E54} and *l(1)BP5*^{G76} mutations are fully complementary at 30°C, partially at 25°C, and not at all (i.e., lethal) at 18°C. At 18°C, heterozygous larvae survive to the end of the third instar, but they fail to pupariate, although living for about 2 weeks longer. Their salivary gland chromosomes become two to three times shorter (Figs, 21b and 21c), because neighboring bands fuse into blocks of chromatin as do regions 61A-F and 62B-D, for example (Figure 23d).

These observations support the following conclusions. The sets of bands in normally functioning cells of different organs and at different stages of development and in individuals of different strains are essentially identical. It is quite difficult to estimate the number of bands in each chromosome to which this conclusion can be applied. Only prominent and medium-sized bands are accessible to analysis. These are in the minority. Their number does not exceed 500–1000, at least in *Drosophila*. The rest of the bands are very slender;

they weakly absorb stain, appearing grey in the phase-contrast microscope. These small bands are seen in nowhere near all the preparations. Consequently, distinction between a constantly present band and one resulting from fusion with its neighbor is hampered. Whatever the real situation, constancy of the banding patterns of even large bands is not the golden rule, as many believe, and their variation can occur even in the most readily analyzed bands. These are manifest as splitting-fusion of bands (a single band in one situation, two to three in another) or as elongation–shortening of interbands (a long interband in one situation, a short in another). Beermann (1952a) thought that these two types of differences are most widespread. The variations were attributed to preparative artifacts (Redfern, 1981a), inappropriate fixatives, different resistance to stretching of chromosomes of different polyteny. However, Zhimulev and Belyaeva (1977) made preparations under standard conditions using the same fixatives, and they chose chromosomes of the same polyteny to compare banding patterns in different strains. Morphological differences were most apparent in the same nucleus in the region 31A of hybrids between Oregon-R and Batumi-L strains (Figs. 14a–14d). These are arguments against the view that variations are artifactual. Richards (1980a) proposed that incomplete local polytenization of interbands (Laird, 1980) may be a cause of tissue differences in interband shortening. However, the foremost lengthening of the short interband in X-chromosomal region 1E at later stages of development is an argument against this possibiblity (Fig. 17).

Incorporation of [^{3}H]uridine into the splitting bands and interband lengthening support the functional explanation of the variation. The most evident explanations are the start of transcription and sharp increase in its intensity in distinct DNA regions composing a band. When DNA sequences lying at the edge of the band are activated, this material can decondense, seemingly lengthening the preexisting neighboring interband.

When transcription is activated only in the sequences located in the middle part of the band, splitting of the band occurs. It is thus quite apparent that the bands showing morphological variations are informationally complex in the sense that they are composed of different sequences, each able to become activated independently of its neighbors. The smallest number of such sequences is three in splitting bands (the remaining decondensed edges of the band and the active "interband" between them) and the two formed when the interband lengthens (the decondensed and condensed parts of the previous band).

The dense band in the region 31A-B of Batumi-L is a specific case of informational complexity. The two large 31A1-2 and 31B1-2 bands fuse into one in most larvae of Batumi-L stocks. There are two possible ways in which this can occur: (1) bands can fuse to form a new "synthetic" band because of interband deficiency (Keppy and Welshons, 1980); (2) the material of three structures can form a new band during inactivation followed by condensation of the 31A1-2/31B1-2 interband.

Thus, a definite banding pattern identical with respect to its major feature in most normally functioning cells corresponds to the normal state of the cells. The small functional differences between the cells are, presumably, provided by functioning of DNA sequences located in few specific puffs (Berendes, 1965a,b, 1966a). The suggestion that "the visible pattern of the chromosome is induced by differentiated cytoplasm of the same cell" (Kosswig and Shengun, 1947a) remains plausible. Indeed, a definite functional state of the cell and polytene chromosome banding pattern is interrelated. When cell activity is of similar type, the sets of active and inactive genes and similar and so is their general banding pattern.

The chromosomes respond similarly to drastic external or internal changes. They sharply contract or shorten. In all the cases, chromosome shortening is due to the condensation of the previously transcriptionally active material of the grey bands, interbands and puffs and also due to band fusion into blocks of chromatin or new bands separated from other identical blocks by transciptionally active regions. It is of interest that during shortening the same blocks fuse first. To illustrate, the end of chromosome 3L assumes the same appearance after long-term culturing the glands of the *l(3)tl* mutant and *G76/E54* heterozygotes for 1(1)BP5 mutations (Figure 21), and so do the regions 76E-77A, 32-33, among others.

Thus, from this survey the following inferences may be drawn: the chromomeric pattern in the cells of normally functioning organs is a relatively stable. The structure of the chromomeres is flexible and the general banding pattern is highly variable when intra and/or extracellular conditions change.

III. TRANSCRIPTIONAL ACTIVITY OF POLYTENE CHROMOSOMES

A. Localization of transcriptionally active regions in the chromosomes

By the 1950s it was accepted that the synthesis of RNA is indicative of gene activity. For this reason, research interests have been focused on identification of chromosome regions where RNA content is increased. There is much information on mapping of regions where RNA is synthesized using histochemical, autoradiographic methods, as well as electron and immunofluorescence microscopy.

1. Histochemical staining specific to RNA

By the 1960s, preferential RNase-sensitive accumulation of RNA-specific dyes (toluidine blue at pH 5.0, azure B at pH 4.1, and pyronine, after staining with methyl green) was localized mainly in nucleoli, Balbiani rings, and puffs (Beermann, 1952a, 1956, 1963a, 1964; Gross, 1957; Pelling, 1959, 1964, 1972a; Schurin, 1959; Swift, 1959, 1965; Kiknadze and Filatova, 1960, 1963; Beermann and Clever, 1964; Gabrusewicz-Garcia and Kleinfeld, 1966; Stevens and Swift, 1966; Clever, 1967; Kiknadze, 1971b, 1972; Kress, 1972, 1981, 1993; Sass, 1980a; Maximova and Ilyinskaya, 1983). Studies of RNA distribution in the chromosomes by UV spectrophotometry led to similar conclusions (Vogt-Kohne and Carlson, 1963).

RNA was detected not only in these easily recognized morphological structures; according to Pelling's (1964) data, the diameter of the chromosome increased (by puffing) only in the minority of the 272 staining regions in the polytene chromosomes of *C. tentans*. The remaining RNA-containing regions were represented by diffuse bands (Pelling, 1964; Kiknadze, 1972) and by light regions reminiscent of interbands. Absence of staining of the very small interbands (Kiknadze, 1971b; Gersh, 1975) is not evidence that RNA is absent; likewise, failure to stain by Feulgen's procedure cannot be taken to mean that DNA is absent. Both failures are due to the high dispersal and low concentration of the material, and there are detection problems with the light microscope.

The polytene chromosomes of *D. virilis* stained for pyronine at the end of the third larval instar in 348–360 regions and only a minority were mapped to puffs. The chromocenter did not stain for RNA (Kress, 1972, 1993).

An indirect method of immunoelectron microscopy was developed based on ultra-rapid freezing of salivary glands in freon 13 at a temperature of −175°C, staining of sections with gallocyanin-chromalum, and treatment with RNase. RNA granules were visualized in all the decompacted regions of the chromosomes, the nucleoli, puffs, interbands, and spacings between the bands (Gersh and Gersh, 1973a,b).

2. Incorporation of [^{3}H]uridine

One can conclude that RNase-sensitive labeling occurs over all the chromosomes, notably over the puffs, Balbiani rings, and nucleolus after *in vivo* and *in vitro* incubation of the cells of salivary glands or other organs with radioactive precursors of RNA synthesis followed by examination of autoradiographs. This conclusion has received ample support (Pelc and Howard, 1956; Gross, 1957; Ficq, 1959; Pelling, 1959, 1962, 1963, 1964, 1969, 1970, 1971, 1972b; Rudkin, 1959; Rudkin and Woods, 1959; Sirlin, 1960; Sirlin and Schor, 1962; Sirlin *et al.*, 1962; Swift, 1962; Beermann, 1963a, 1964, 1966, 1967, 1971; Fujita and Takamoto, 1963; Plaut, 1963; Ritossa, 1964a,b; Arnold, 1965; Berendes, 1965a, 1972; Fujita, 1965; Kiknadze, 1965a, 1966, 1971b; Robert and Kroeger, 1965; Pavan, 1965; Ritossa *et al.*, 1965; Clever and Romball, 1966; Gabrusewycz-Garcia and Kleinfeld, 1966; Gaudecker, 1967; Goodman *et al.*, 1967, 1987a; Desai and Tencer, 1968; Simoes and Cestari, 1969; Basile *et al.*, 1970; Korge,

1970b; Burdette and Carver, 1971; Edstrom *et al.*, 1971; Egyhazi *et al.*, 1972; Zhimulev and Lychev, 1972a; Holmquist, 1972; Panitz *et al.*, 1972a; Poels *et al.*, 1972; Alonso and Grootegoed, 1973; Diez, 1973; Walter, 1973; Berendes *et al.*, 1974; Zhimulev and Belyaeva, 1974, 1975b; Abdelhay and Miranda, 1975; Alonso and Berendes, 1975; Heitz and Plaut, 1975; Belyaeva and Zhimulev, 1976; Pages and Alonso, 1976; Wolf and Sokoloff, 1976; Bonner and Pardue, 1977a,b; Diez *et al.*, 1977; Hameister, 1977; Mukherjee and Chatterjee, 1977; Pardue *et al.*, 1977; Sokoloff and Zacharias, 1977; Bonner *et al.*, 1978; Kress, 1979; Lakhotia, 1979; Semeshin *et al.*, 1979; Eastman *et al.*, 1980; Sass, 1980a,b, 1981; Lakhotia *et al.*, 1981; Miranda *et al.*, 1981; Kiknadze and Valeyeva, 1983; Maximova and Ilyinskaya, 1983; Silva, 1984; Ghosh and Mukherjee, 1986; Visa *et al.*, 1988; Diez *et al.*, 1990).

There is also evidence that radioactive uridine is incorporated into more than puffs, Balbiani rings, and the nucleolus (Fujita and Takamoto, 1963). Thus, of the total number of loci incorporating [^{3}H]uridine into *D. hydei* chromosomes, only 35–50% showed swelling of these chromosome regions into the puffed structure (Berendes, 1965a). In *C. tentans* the label was diffusely distributed all over the chromosomes, not confined to puffs and Balbiani rings, after the salivary glands were incubated for 30 min with the radioactive precursor (Clever and Romball, 1966).

In *D. melanogaster*, in typical puffs not more than 50% of the total number of grains are detected in the chromosomes of zero-hour prepupae (Zhimulev and Belyaeva, 1974, 1975b). When taking into account that the functioning puffs reach their maximum number at this particular stage (see Section VI,C,1), the label percentage for unpuffed regions would be higher at the other stages.

The following should be borne in mind in quantitative studies of the incorporation of [^{3}H]uridine.

1. Puffing pattern changes rapidly and quite expectedly by the end of the larval stage of development (see Section VI,C,1). For this reason, if larvae, differing even slightly in age, are included in a sample, incorporation intensity of the precursor can vary widely. It is advised to choose individuals at the same developmental stage marked by morphological criteria; zero-hour prepupae of *Drosophila* are most appropriate for this purpose. Their anterior spiracles evert within a few seconds, variations in individual age, as well as in the patterns of banding and [^{3}H]uridine incorporation, are minimal (Poluektova, 1969; Burdette and Carver, 1971).
2. Lactoacetic orcein is conventionally used to make squash preparations, because it permits the obtainment of well-spread chromosomes showing distinctive banding patterns. However, lactic acid removes RNA from the chromosomes and, consequently, better preparations can be made using acetic acid alone (Holmquist, 1971).

3. Labeling time must be shorter than the total time required for synthesis, packaging, and transport of the newly synthesized RNA. This makes it certain that label is not redistributed in the nucleus. The chromosomes label even after the glands were incubated for 1.0–2.5 min with the precursor (Plaut, 1963b; Belyaeva and Zhimulev, 1976).
4. There is considerable variation (up to three orders of magnitude) in the intensity of [^{3}H]uridine incorporation into the chromosomes of salivary glands, midgut, and Malpighian tubules in various *C. tentans* larvae. With these variations, labeling intensity of certain organs is independent of the others. Striking differences were found between different cells within an organ. Such a variegated pattern of [^{3}H]uridine incorporation is characteristic of certain regions of the gut, while Malpighian tubules show a definite label gradient. Insofar as these variations are observed on the background of continuous puffing, it was concluded that puffing is the necessary, yet not the sufficient, condition for RNA synthesis to occur (Pelling, 1962–1964). However, in Beermann's (1966) opinion, the reason for the differences may be that exogenous [^{3}H]uridine is not converted into nucleoside triphosphates, because a large pool of the latter has already accumulated in the cell. Therefore, label is not incorporated; concominantly with this, transcription proceeds.

   Wide variations in the level of [^{3}H]uridine incorporation into different cells of a salivary gland have been observed in *D. melanogaster* (Zhimulev and Lychev, 1972a). Circadian rhythms in incorporation of the precursor were described. Cells of the proximal part of the gland label heavier than those of the distal. Maxima of [^{3}H]uridine incorporation occurred at 12–18 hr of the 24-hr period. The distal cells showed a bimodal rhythmicity with maxima at 6–9 and 21 hr (Probeck and Rensing, 1974; Nagel and Rensing, 1974). There may be a 24-hr rhythmicity in the permeability of the nuclear membranes to exogenous [^{3}H]uridine, with a resulting change in the labeling intensity of cells, whose RNA remains unaltered due to the presence of the endogenous precursor.
5. The geometry of a chromosome region (the state of a puff or a band), and the associated differences in radioactivity self-absorption, do not affect detectability of [^{3}H]uridine incorporation (Holmquist, 1972; Zhimulev and Belyaeva, 1974).

Inferences were made from the results obtained after incorporation of [^{3}H]uridine for 5 min into the salivary glands of zero-hour prepupae of *D. melanogaster*, and autoradiography, followed by localization of each grain on photomaps of all the chromosomes:

1. The chromosomes label quite discontinuously; the large compact bands and some other bands do not show any label (Figure 24). By way of contrast,

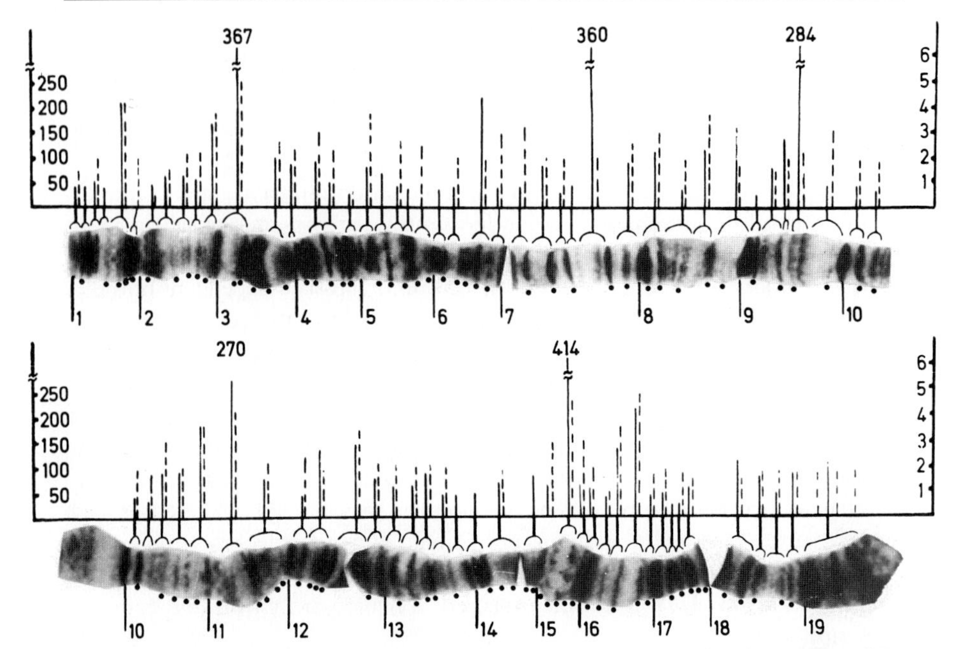

Figure 24. Intensity of [³H]uridine incorporation into distinct regions of X-chromosome of *D. melanogaster* at the stage of zero-hour prepupa. Abscissa, the chromosome region according to Bridges; left ordinate, grain number (summed for 20 chromosomes); right ordinate, puff size, scores. Continuous line is grain number; hatched line is scores (reprinted by permission from Zhimulev and Belyaeva, 1974, 1975b).

the puffs are heavily labeled, most other regions weakly. An alternation of regions with different label intensities produces the discontinuous banding pattern (Figure 24). This makes possible the detection of chromosome regions differing in the intensity of label incorporation after obtainment of an overall estimate of grain number over many chromosome regions. More or less labeled sites are occasionally distinguishable even within single puffs (Zhimulev and Belyaeva, 1974, 1975b; Belyaeva and Zhimulev, 1976).

2. The lack of correspondence between the number of labeled sites and the number of morphologically detectable active regions is of interest. Even when the small puffs of classes 1 and 2 (see Section VI,B,3) in the regions showing no noticeable increase in chromosome diameter are taken into account, the total number of RNA-synthesizing sites turns out to be much greater. The regions with small puffs and those showing no puffing label weakly. Exceptional in this regard are 6DE and 9E1-2 in the X chromosome; 68BC and 77BC in 3L; 42D in 2R; and 82C and 82E in 3R. There

are large loose bands (9D1-2, 68BC) in some cases and blocks of thin dot bands in other ones. The levels of label incorporation into these regions are higher than into many large puffs (Zhimulev and Belyaeva, 1974, 1975b).

Loose bands might be regarded as loci at the early or late stages of puff development (and, consequently, decompaction); then their intense incorporation of [^{3}H]uridine would not be occasion for surprise.

Using electronmicroscopic autoradiography (Kerkis *et al.*, 1975, 1977) with higher resolution, as demonstrated by model experiments (Salpeter *et al.*, 1969), it was shown that the dense compact bands are transcriptionally inactive, while [^{3}H]uridine is very intensely incorporated into the loose bands (Figure 25) (Semeshin *et al.*, 1979). Analysis of the data obtained with the stretched polytene chromosomes (Anaiev and Barsky, 1978) led Semeshin *et al.* (1979) to the same conclusion.

There is uncertainty regarding the alternating pattern of blocks of thin dot bands and interbands. It is difficult to decide whether label is incorporated into a thin band, the spacing between the band, or both when examining preparations by the light microscope. Analysis of electron microscopic preparations (Figure 25) provides evidence for label incorporation of both interbands and loosened bands (Semeshin *et al.*, 1979). Label was not detected in interbands of the stretched chromosomes (Ananiev and Barsky, 1978; Lakhotia, 1979) presumably because exposure did not reach a level that allows it to be observed (see for discussion: Semeshin *et al.*, 1979).

3. The great majority of unpuffed regions label much weaker than puffs. Inasmuch as these constitute most of the labeled regions, their total RNA transcriptional activity makes up about half of that of the chromosomes. The large puffs identified at the zero-hour prepupal stage (see Section VI,C), when their activity is maximal, contain 40% of silver grains in the X chromosome (when discounting label incorporation into the nucleolus). The percentage of silver grains is 34% for *2L*; 40% for *2R*; 64% for *3L*; and 54% for regions within *3R*. About 20% of total label is incorporated into the large puffs, such as 63E1-5 and 71CE in the left arm of chromosome *3* (Zhimulev and Belyaeva, 1975b).

   Nuclear RNA of the salivary glands nuclei hybridized to many sites of polytene chromosomes and more intensely to the puffs. This was taken to mean that the puffs are more engaged in transcription than the other loci, although they are not the only transcriptionally active in nuclear RNA (Bonner and Pardue, 1977b).

4. There is an obvious positive correlation between the intensity of label incorporation and the degree of puff development. In the puffs of classes III–V—whose diameters were 1.2, 1.5, and 1.9-fold larger than those of the neighboring bands—the differences in grain number roughly corresponded to those between puff volumes. This pattern became more prominent

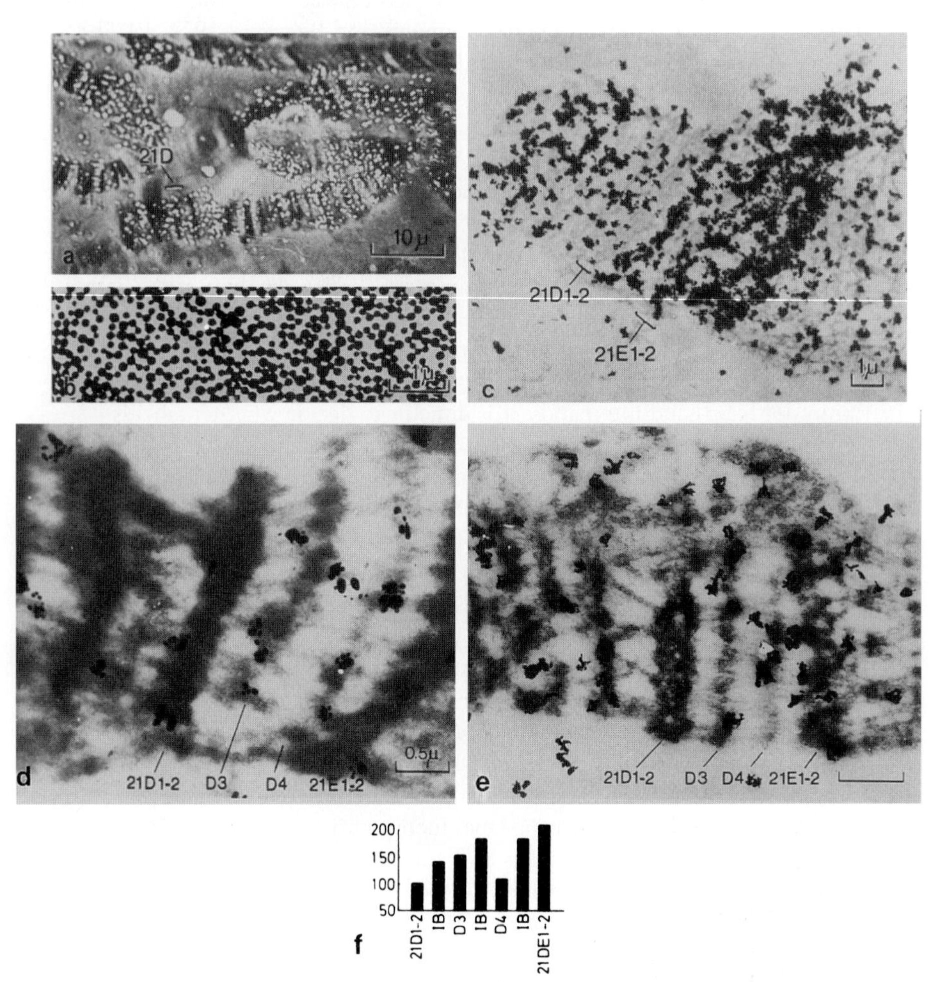

Figure 25. Incorporation of [^{3}H]uridine into the 21D1-4 region of chromosome 2L of *D. melanogaster* at the level of the (a) light and (b–f) electron microscope. (a) Preliminary light microscope autoradiographs of polytene chromosomes embedded in araldite (phase contrast); (b) a monolayer of undeveloped silver crystals in photoemulsion; (c) an EM autoradiography of an intensely labeled chromosome; (d and e) the same, intensity is weaker; (f) total incorporation of [^{3}H]uridine into the 21D1-2-21E1-2 region. Abscissa, regions of the chromosome (IB–interband); ordinate, total number of silver grains in the regions of 70 sections of 30 chromosomes (reprinted by permission from Semeshin *et al.*, 1979).

when the relation between puff size and labeling incorporation intensity was graphically represented: each consecutive class of puffs differed 2-fold, on the average, from the preceding class. There was a great variation in the intensity of labeling. The 86C, 86D, 85D18-27, and 87B puffs of class II labeled 10 times weaker than the 95B puff of the same class; 67F and 63BC of class III labeled very differently; 76A and 76CD of class IV were five times less active than 61AB and 70E of the same size.

There are several reasons for these discrepancies (Zhimulev and Belyaeva, 1974, 1975b). One is that different amounts of DNA are transcribed in a puff. Indeed, estimates of puff size are based on the thickness of a region; in such a case, the size and number of puff-forming bands are disregarded. Other important factors affecting the amount of incorporated [^{3}H]uridine include the density of RNA polymerase II along the DNA axis and the length of RNA molecules synthesized in a puff (Pelling, 1971, 1972a,b). Another factor may be the nonuniform distribution of AT/GC base pairs along the chromosomes. In fact, the regions of antibody binding on 5-methyl-cytosine and [^{3}H]uridine incorporation into certain DNA and RNA puffs and bands and interbands are not coincident in polytene chromosomes of *Sciara coprophila* (Wei *et al.*, 1981).

3. Antibodies against DNA–RNA hybrids

Indirect immunofluorescence offers advantages in studies of the localizaion of the RNA synthesizing regions in the chromosomes. DNA–RNA hybrid molecules containing single-stranded DNA are formed after denaturation–renaturation of chromosomal DNA, while the other strand of the double helix is represented by cell endogenous RNA located at its transcription site. Subsequent localization of the hybrid molecule allows mapping of the transcriptionally active chromosome regions.

The DNA–RNA hybrid molecules are recognized by a specific antiserum raised in rabbits against synthetic hybrids poly(A) · poly(dT). The detection procedure of the hybrid includes exposure of renaturated preparations of polytene chromosomes to goat antiserum against DNA–RNA hybrids followed by incubation with fluorescein isothiocyanate-labeled rabbit antigoat immunoglobulins (Stollar, 1970, 1975; Rudkin and Stollar, 1977a,b; Stuart and Porter, 1978; Stuart *et al.*, 1981; Alcover *et al.*, 1981, 1982; Busen *et al.*, 1982; Kitagawa and Stollar, 1982; Hiraoka *et al.*, 1983; Alonso *et al.*, 1984; Roap *et al.*, 1984; Arbona *et al.*, 1991, Vlassova *et al.*, 1991; Zhimulev *et al.*, 1996; Stocker *et al.*, 1997; Valeyeva *et al.*, 1997). The antiserum acts as a specific reagent detecting RNA–DNA hybrid molecules. Bright chromosomal fluorescence disappears if any of the following steps is omitted: DNA denaturation, exposure to the first or secondary antibody, and if the chromosomes are treated with pancreatic RNase before denaturation (Rudkin and Stollar, 1977b; Alcover *et al.*, 1982; Busen *et al.*, 1982; Vlassova *et al.*, 1984).

Using the indirect fluorescence method, many hybridization sites were revealed in polytene chromosomes (Figure 26): about 200 fluorescent bands in the chromosomes of *C. pallidivittatus* (Diez and Barettino, 1984) and about 350 in zero-hour prepupae of *D. melanogaster* (Vlassova *et al.*, 1984, 1985).

What may be the informational significance of a single fluorescent band? Before answering that question, we consider the morphological structures relevant to the information. Alcover *et al.* (1982) obtained patterns showing that fluorescence is most frequently located in an interband opposition to a band and possibly partly covering a small part of it. However, detailed analysis of the results obtained with a combination of phase-contrast, fluorescent, and autoradiographic images of the same chromosomal preparation demonstrated that a fluorescent spot can coincide with the localization of a puff, a loose region, or a whole region containing a set of thin loose bands and interbands (the small puffs; e.g., the 100DE region in Figure 26). Such regions constitute the majority in the polytene chromosomes of *Drosophila* (Vlassova *et al.*, 1985; Umbetova *et al.*, 1994). Hence, a fluorescence spot is not a functional unit at the morphological level, because it consists of a set of bands and interbands. There is no correlation between the number of bands and interbands in cells. Thus, 5200 poly(A)$^+$RNA sequences were identified in the cytoplasm of cultured cells of *Drosophila* (Izquierdo and Bishop, 1979); that is, at a number one order of magnitude higher than that of fluorescent spots.

Nevertheless, the data on the localization of fluorescent bands are of considerable interest. In *D. melanogaster* all the regions of polytene chromosomes decompacted with varying degrees were revealed by indirect immunofluorescence. They were the prominent, medium-sized, and small puffs showing bright fluorescence and quite intense incorporation of [^{3}H]uridine. Taken together, they constituted half of the bright fluorescent bands. The other half represented the small puff weakly incorporating [^{3}H]uridine and sites showing no swelling and incorporating little or no [^{3}H]uridine; these were mostly loosened bands, which must be probably referred to the small puffs that formerly passed unnoticed. The compact black bands from which Bridges's subdivisions start hardly ever occurred in the bright fluorescent regions (Vlassova *et al.*, 1984, 1985; Umbetova *et al.*, 1994).

Although Alcover *et al.* (1982) believe that fluorescence is mainly detected in the interbands, this does not appear plausible because the "true" interbands are actually visible only in the electron microscope (see Section IV,A). Correct judgments concerning the presence of antibodies against DNA–RNA hybrids can be based on analysis of those interbands, which are located between the large compact nonfluorescent bands and do not contain minibands according to the electron microscopic observations. The large interband be-

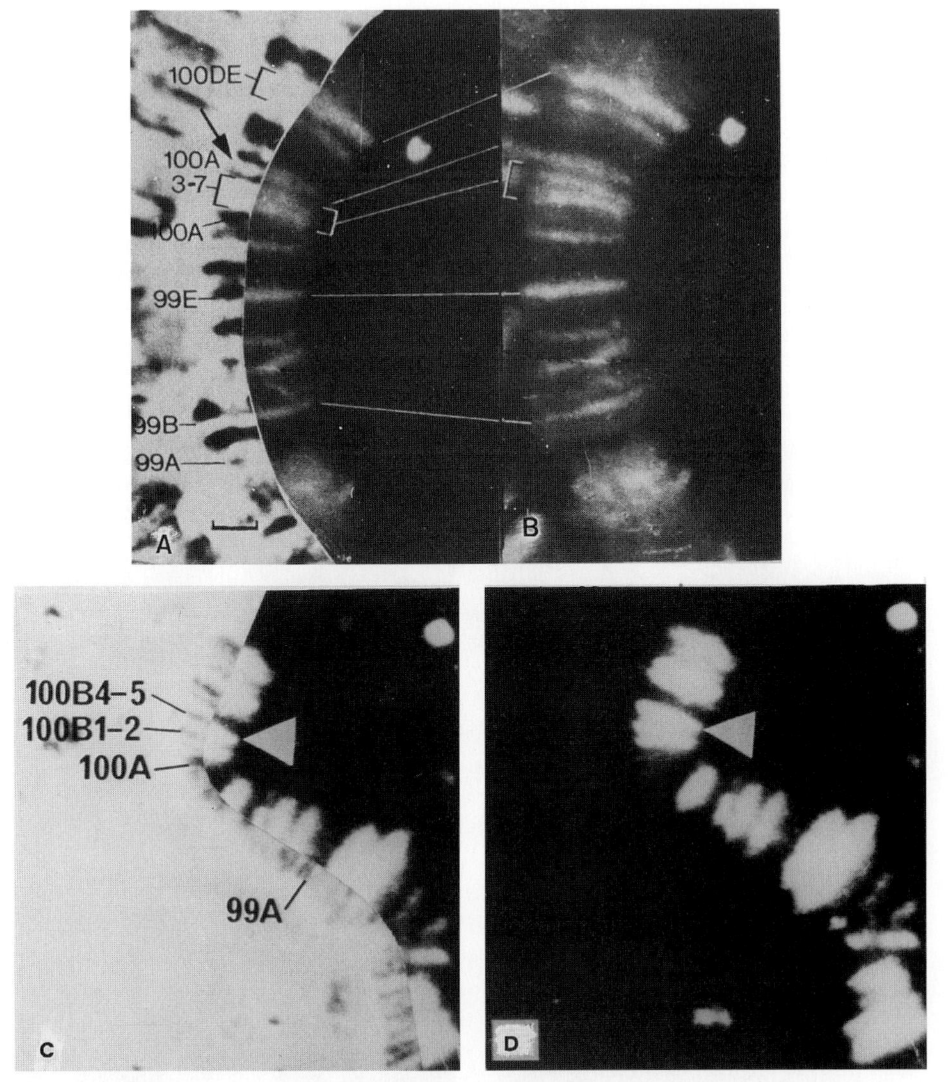

Figure 26. Detection of DNA/RNA hybrids in the 99–100 regions of chromosome 3R of *D. melanogaster* with the use of fluorescent antibodies. (A, C, E, and G) combined photographs; left, phase contrast; right, fluorescence in the same chromosome; (B, D, F, and H) fluorescence in this chromosome. Arrowheads point to the fluorescent interbands in the 100B3/100B4-5 region (from Vlassova *et al.*, 1984, 1985a,b; Umbetova, 1991; and reprinted by permission from Zhimulev *et al.*, 1996).

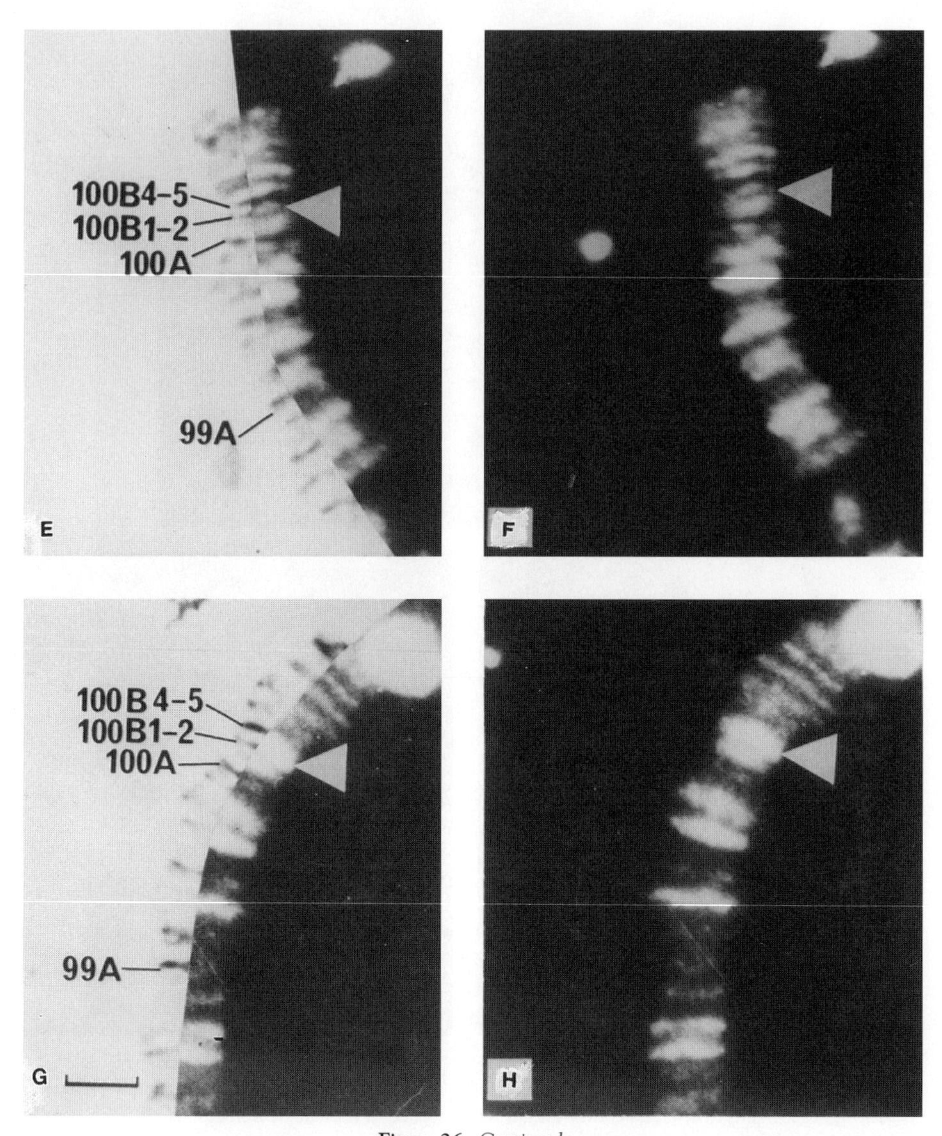

Figure 26. *Continued*

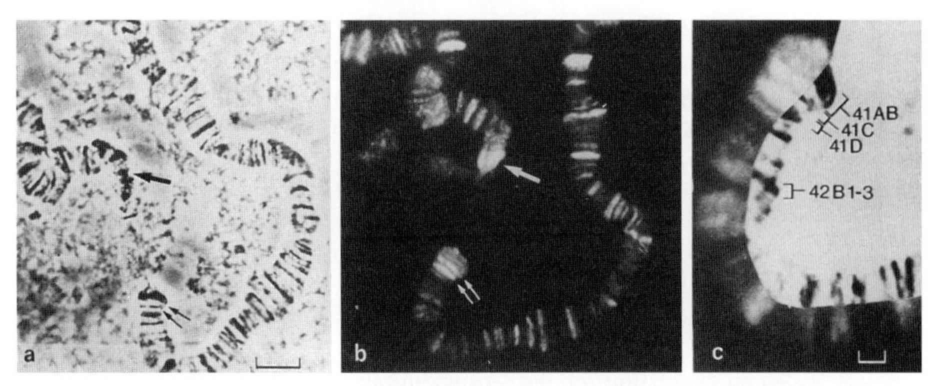

Figure 27. Immunofluorescence of the 40 and 41AB pericentromeric regions. (a) phase contrast; (b) fluorescent microscope; (c) combined photograph. The large arrow points to the 40 heterochromatic region, small arrows point to the unusually bright small 41C and 41D puffs (reprinted by permission from Vlassova *et al.*, 1984, 1985a,b).

tween the two dense 100B3 and 100B4-5 bands fulfills the above requirements (Figure 26). The binding of DNA–RNA hybrids in the 100B3/100B4–5 interbands (Zhimulev *et al.*, 1996), in conjunction with the results obtained with the incorporation of [^{3}H]uridine (Semeshin *et al.*, 1979) and with the immunoelectron microscopic localization of RNA polymerase on polytene chomosomes (Sass and Bautz, 1982b), provide convincing evidence for the transcriptional activity of these regions. Intense hybridization was detected in the regions of centromeric heterochromatin (Figure 27). A direct correlation between the degree of chromatin packing and fluorescence intensity has not been consistently observed.

Variations in the patterns of fluorescence intensity and induced puffing were generally correlated (Alcover *et al.*, 1981, 1982; Busen *et al.*, 1982; Alonso *et al.*, 1984; Diez and Barettino, 1984; Lezzi *et al.*, 1989). However, only low level of fluorescence was seen in a number of very large puffs in *D. hydei*, *Trichosia pubescens*, and *D. melanogaster*, even though they intensely incorporated [^{3}H]uridine (Alcover *et al.*, 1982; Busen *et al.*, 1982; Vlassova *et al.*, 1985).

In the *l(1)npr1* mutant of the *D. melanogaster Broad-Complex* gene, RNA was not detected in the 68C puff which is site of location of *Sgs-3*, *-7*, and *-8* genes (Crowley *et al.*, 1984). In homozygotes for the *l(1)t435* allele, antibodies against DNA–RNA hybrids were not found in this puff site, although it incorporated [^{3}H]urudine (G. H. Umbetova, I. E. Vlassova, E. S. Belyaeva, and I. F. Zhimulev, 1989, unpublished observations).

In the *l(3)tl* mutant, whose characteristic features are shortened chromosomes, fluorescence was absent from some of the heat-shock puffs (see Section VI,H,1). The fluorescence was absent from some of the morphologically similar small puffs contrasting with the weak in the others. No fluorescence was observed in particular chromosome regions of the small puffs, although it was detected in the other regions of the same puffs (Vlassova *et al.*, 1985).

The correlation between intensity of [^{3}H]uridine incorporation and fluorescence did not always hold. As a rule, the largest puffs that incorporated high quantities of [^{3}H]uridine precursors fluoresced brightly; however, many of the small puffs weakly incorporating the precursors showed bright fluorescence (Vlassova *et al.*, 1985).

When explaining the discrepancy in puff size and fluorescence intensity, it should be borne in mind that the latter is dependent on the amount of RNA present at a given time point in a chromosome region. The amount is determined by the balance established between the rate of RNA synthesis, processing, and transport from an active region. Any increase in the rate of one of the processes leads to a change in the amount of puff RNA accessible to hybridization to the DNA strand. As a consequence, the correlation between puff size and fluorscence intensity will not always hold.

The activities of large puffs change, as a rule, in a stage-specific manner during the course of development. The majority of small puffs are active throughout whole development (Belyaeva *et al.*, 1974; Zhimulev, 1974; Belyaeva, 1982). Analysis of fluorescent bands in *Drosophila* chromosome *2R* during puffing stages 1–16 (taking about 12 hr) revealed from 100–110 fluorescent bands in the chromosome at each stage, and 90 of these were localized in regions of the small puffs. Despite this, fluorescence intensity can widely vary in these regions at different developmental stages. However, the bands cease fluorescing altogether in a few exceptional cases (Umbetova, 1991; Umbetova *et al.*, 1994). Thus, there is reason to conclude that the regions of the small puffs constantly function during development.

In the *ecs*^{lt435} and *l(3)tl* mutants, the ecdysterone-induced puffs do not develop, and fluorescence is observed only in the regions of the small puffs (Umbetova, 1991).

4. Proteins involved in transcription

To identify the transcriptionally active regions of polytene chromosomes, it is important to map proteins involved in transcription, processing, and packaging of the transcribed RNA. However, some caution is advised before interpreting the results, because proteins can be removed during the squashing procedure. With this in mind, formalin at different concentrations is used to keep the proteins attached at the site where they must be in their native state. This makes feasible accurate localization of a region where protein components involved in transcription, not necessarily in the process itself, because proteins can be also located in the inactive chromosome regions; these perform the role of a pool.

RNA polymerase I has been mapped to the nucleolus by the indirect immunofluorescence technique (Jamrich *et al.*, 1977b, 1978). With this technique, RNA polymerase II has also been detected in all the decompacted chromosome regions; for example, in Balbiani rings, puffs, loose bands, and interbands. The enzyme has not been localized in the dense compacted bands and the chromocenter (Greenleaf and Bautz, 1975; Plagens *et al.*, 1976; Jamrich *et al.*, 1977a,b, 1878; Greenleaf *et al.*, 1978; Elgin *et al.*, 1978; Kramer *et al.*, 1980; Sass, 1982; Sass and Bautz, 1982a,b; Kabisch and Bautz, 1983; Korge, 1987; Visa *et al.*, 1988).

The results of autoradiographic analysis and of the localization of RNA polymerase II in polytene chromosomes are in agreement. The coincidence of indirect immunofluorescence with the incorporation of [^{3}H]uridine was most conspicuous in the stretched salivary gland chromosomes (Figure 28). Electron microscopy using antibodies directed against RNA polymerase II demonstrated its presence in most, if not all interbands in C. *tentans* chromosomes (Figure 29). It was concluded that the interbands of polytene chromosomes are the binding sites for the enzyme and the start points for transcriptional activity (Sass and Bautz, 1982b).

RNA polymerase III was localized in about 30 regions of polytene chromosomes to which the transport RNA genes and several small nuclear RNAs were assigned either cytogenetically or by *in situ* hybridization (Kontermann *et al.*, 1989).

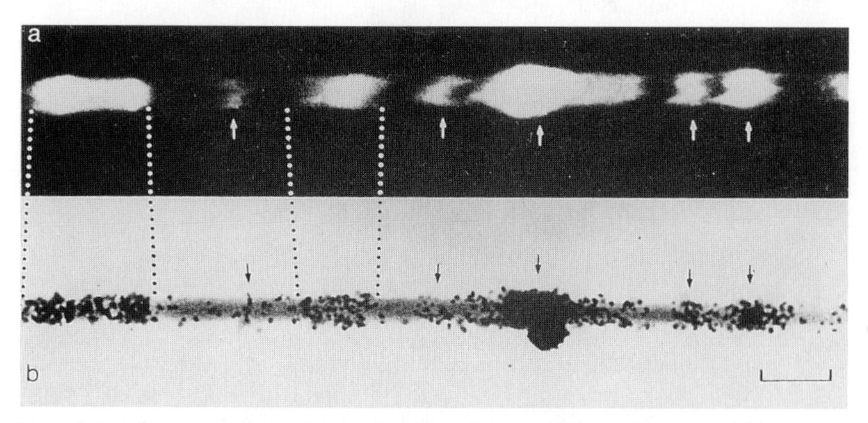

Figure 28. A fragment of one of the stretched chromosomes of *Chironomus tentans*. (a) localization of antibodies against RNA polymerase II; (b) incorporation of [^{3}H]uridine into the same fragment. Arrows and dotted lines designate homologous regions (from Sass, 1982, © Cell Press). Bar = 10 μm.

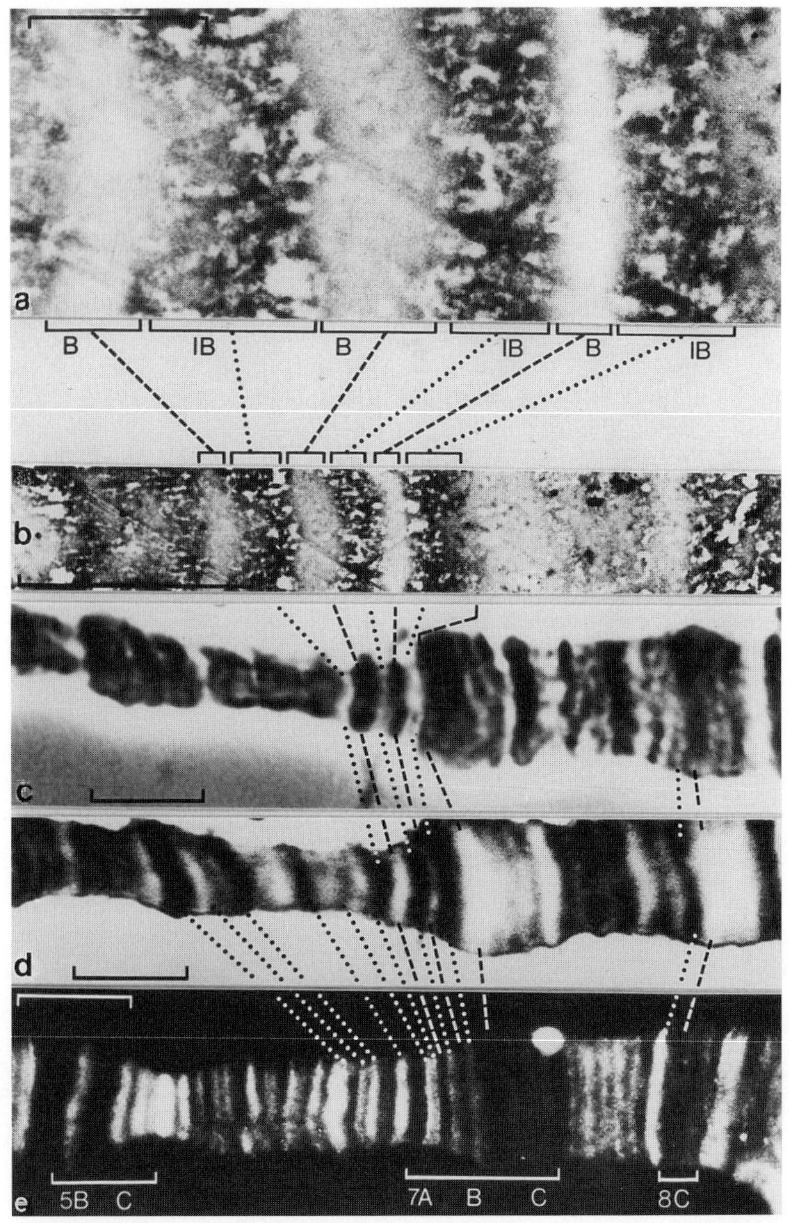

Figure 29. A fragment of a stretched third chromosome of *Chironomus tentans*. (a and b) larger (a) and smaller (b) magnification under the electron microscope; "staining" with antibodies against RNA polymerase II labeled with peroxidase; (c) phase contrast; (d) "staining" with immunoperoxidase, light microscope; (e) immunofluorescence. The bands are connected by hatched lines, the interbands by dotted lines. IB, interband; B, bands. Bar = 1 μm (a), 7 μm (b-e) (reprinted by permission from Sass and Bautz, 1982a).

In *Drosophila,* RNA polymerase II was detected in the 87A and the 87C heat-shock puffs, as well as in the other chromosome regions after exposure of larval salivary glands to 37°C for 30 min. However, the enzyme was localized only in the heat-shock puffs after 45 min of heat shock (Jamrich *et al.*, 1977a,b, 1978). In *C. tentans* after 20–30 min of heat-shock treatment, RNA polymerase II remained only in the heat-shock puffs and Balbiani rings, although they were much reduced in size (Sass, 1982). Thus, heat shock has become a good tool for localizing transcriptionally active RNA polymerases I and III. In *D. melanogaster*, the following puff sites labeled with [^{3}H]uridine after prolonged heat treatment: 2E1-2, 5C3-16, 10B3-16, 10EF, 18D3-13, 22E1-2, 26A3-9, 24D3-11, 37EF, 38F1-6, 43E3-18, 44DE, 45E1-2, 47B1-8, 47D1-4, 48E3-12, 49E1-15; 50C3-7, 50D1-7, 51D3-12, 56E1-6, 61C1-4, 63B, 64E3-13, 67AB, 70B1-7, 73D1-7, 78E1-2, 84E1-6, 85B6-9, 87A7-10, 87B10-15, 88E3-5, 93D1-10, and 95D1-11 (Belyaeva and Zhimulev, 1976).

A comparison of the distribution patterns of topoisomerase I and RNA polymerase II using antibodies demonstrated a generally similar association with the transcriptionally active regions of the genome (Fleischmann *et al.*, 1984; Elgin *et al.*, 1988).

A protein with a molecular weight of 52 kDa was identified in the heat shock puffs. It colocalizes with RNA polymerase II at low heat-shock temperatures, but appears at 3′ flanking regions of the genes at full heat shock (Champlin et al., 1991; Udvardy and Schedl, 1993; Champlin and Lis, 1994).

A boundary element-associated factor of 32 kDa (BEAF-32) is bound to *scs*′ (special chromatin structures located on 3′ end of the genes). Immunostaining localizes BEAF to hundreds of interbands and many puff boundaries on polytene chromosomes (Zhao *et al.*, 1995).

Another chromatin associated protein, CHD1 (chromo-ATPase/helicase-DNA-binding domain), is located in sites of extended chromatin, namely interbands and puffs (Stokes *et al.*, 1996).

Localization of antibodies against proteins binding to synthesized mRNA and, presumably, included in RNP particles (see below), generally agrees with their localization in puffs and interbands. These proteins rarely occur in chromocenters and not at all in densely packed bands (Saumweber *et al.*, 1980; Christensen *et al.*, 1981; Hugle *et al.*, 1982; Kabisch *et al.*, 1982; Dangli and Bautz, 1983; Kabisch and Bautz, 1983; Risau *et al.*, 1983; Schuldt *et al.*, 1989; Amabis *et al.*, 1990; Matunis *et al.*, 1993; Bauren and Wieslander, 1994).

Many antigens were isolated from chromatin, and antibodies against them were detected in the puffs. There was reason to assign the proteins to the active region; for example, the Bj6 and Bx42 proteins were mapped to puffs and they bound to 17- to 19-kb DNA surrounding the *Sgs-4* gene. Puff 3C formed when *Sgs-4* became activated in *D. melanogaster* (Saumweber *et al.*, 1990).

There is a variety of small nuclear ribonucleoproteins, or snRNP's, which are essential for the assembly of a multicomponent complex, termed the spliceosome. The spliceosome is involved in processing events resulting in the production of the mature RNA transcripts and especially in splicing of mRNA precursors (pre-mRNA). The spliceosome contains several types of snRNAs rich in uridine. Thirteen, at least, distinct snRNA species have been, so far, detected. Their number is most frequently five (U_1, U_2, U_4, U_5, and U_6); their size varies from 106–189 nucleotides. The U_3, U_8, and U_{13} are localized in the nucleolus where they may be engaged in the splicing procedure of preribosomal RNA. The U_1 and U_2 snRNPs additionally contain seven different proteins. The spliceosome is a round body, 8 nm in diameter, with two protrusions 4–7 nm long and 3–4 nm thick (Birnstiel, 1988; Steitz, 1988; Luhrmann *et al.*, 1990; Pironcheva, 1990).

It might be expected that antibodies against the proteins of spliceosomes would be primarily localized in the puffs of polytene chromosomes where the intron-containing genes are actively transcribed. However, in *D. melanogaster*, these antibodies were detected in the heat-shock 87A and 87C puffs, where the *hsp70* genes devoid of introns are located (Martin *et al.*, 1987).

Antibodies against the chironomid U_1 and U_2 snRNPs were localized in the major sites of transcription Balbiani rings and puffs (Sass and Pederson, 1984; Vazquez-Nin *et al.*, 1990). It is noteworthy that indirect fluorescence with specific snRNP antibodies at Balbiani rings of *C. tentans* chromosomes correlated with rates of [^{3}H]uridine labeling. Changes in transcriptional activity in Balbiani rings and puffs occurred in parallel with those in intensity of U_1 and U_2 RNP antibody immunofluorescence (Sass and Pederson, 1984).

5. Ribonucleoprotein products of transcription

Observations on the fine structure of the Balbiani ring in *Chironumus* have shown the presence of specific granules (Beermann and Bahr, 1954). Treatment of the chromosome preparations with the enzymes DNase and RNase demonstrated that they contain ribonucleic acid and protein (Stevens, 1964; Swift *et al.*, 1964; Stevens and Swift, 1966; Bernhard, 1969; Perov, 1971; Perov and Chentsov, 1971; Vazquez-Nin and Bernhard, 1971).

The ribonuclein granules are often located in groups: the stalked granules attached to chromatin fibrils make up the scaffold of the Balbiani rings (Figure 30). The granules are either freely dispersed in the nucleoplasm or pass through the pore of the nuclear envelope (Beermann, 1963a; Beermann and Clever, 1964; Stevens, 1964; Stevens and Swift, 1966; Olins *et al.*, 1982; Skoglund and Daneholt, 1986; Bjorkroth *et al.*, 1988; Mehlin *et al.*, 1988; Ericsson *et al.*, 1989, 1990; Kiseleva *et al.*, 1996). Many authors regard RNA granules as the main mode of packing and transport of messenger RNA from the cell nucleus to the cytoplasm (Beermann, 1962; Vazquez-Nin and Bernhard, 1971; Daneholt, 1975, 1992; Skoglund *et al.*, 1986; Matunis *et al.*, 1993; Bauren and Wieslander, 1994; Masich *et al.*, 1997).

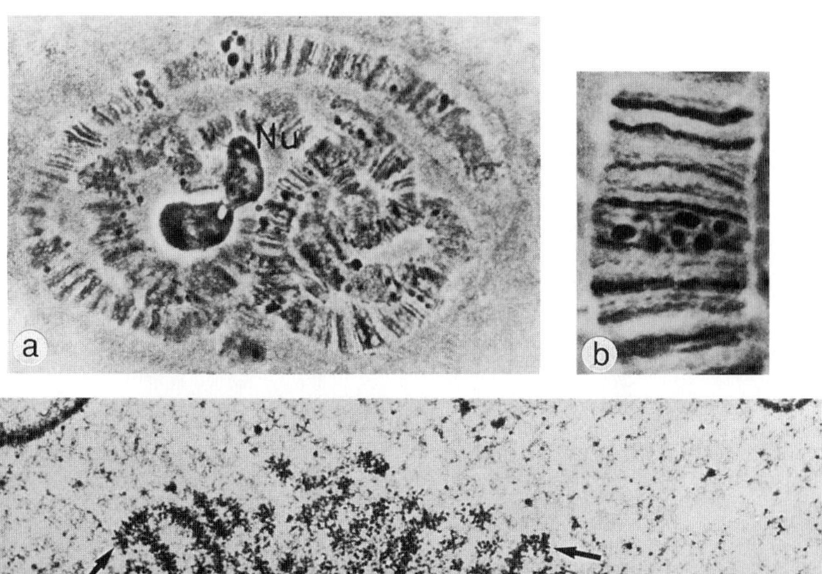

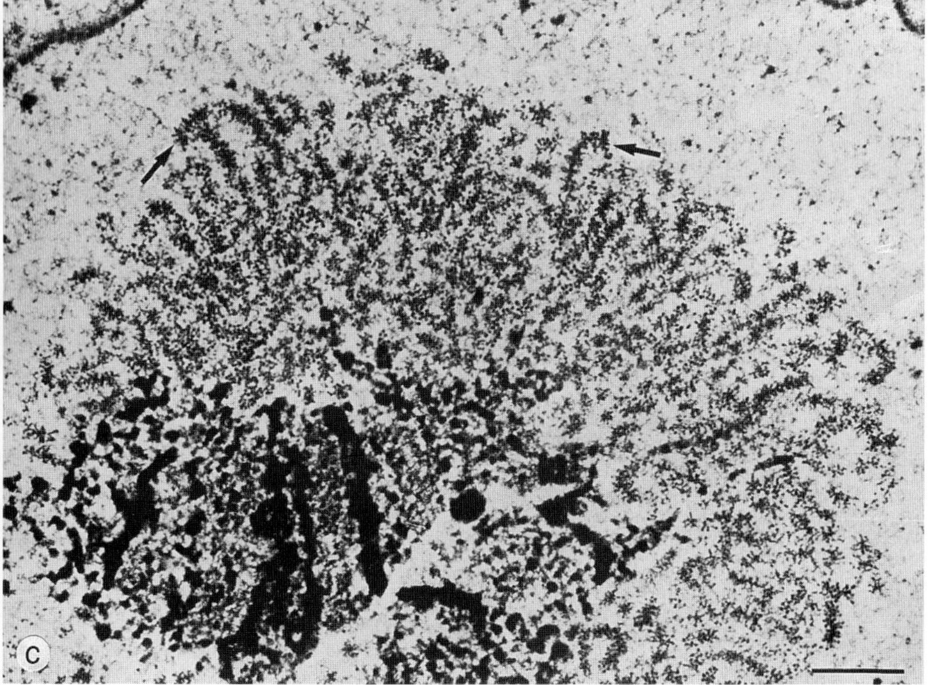

Figure 30. Morphological types of ribonucleoprotein granules occurring in the transcriptionally active regions of polytene chromosomes. (a and b) RNP "drops" in *Chironomus pallidivittatus;* (c) granules in Balbiani rings of *C. tentans;* (d and e) one of DNP fibrils from Balbiani ring; (f and g) giant granules in the 2–48C puff of *D. hydei;* (h) granules in interband; (i and j) granules in centromeric heterochromatin in *D. melanogaster*. Arrows point to the large transcribed fragments of (c) Balbiani rings,(h) supergiant granules in the interband and (i) centromeric heterochromatin. Nu, nucleolus. Bar = 1 μm (c and f), 300 nm (g), 0.5 μm (h). (a and b) reprinted by permission from Stockert and Diez (1979); (c–e) from Anderson *et al.* (1980) and Daneholt *et al.* (1982); (f) reprinted by permission from Berendes (1972a); (g) from Derksen *et al.* (1973), reproduced from *The Journal of Cell Biology*, 1973, **59,** 661–668 by copyright permission of The Rockefeller University Press; (h) from Skaer (1977, *J. Cell Sci.* and Company of Biologists Ltd.; 1978); (i and j) from Lakhotia and Jacob (1974).

d

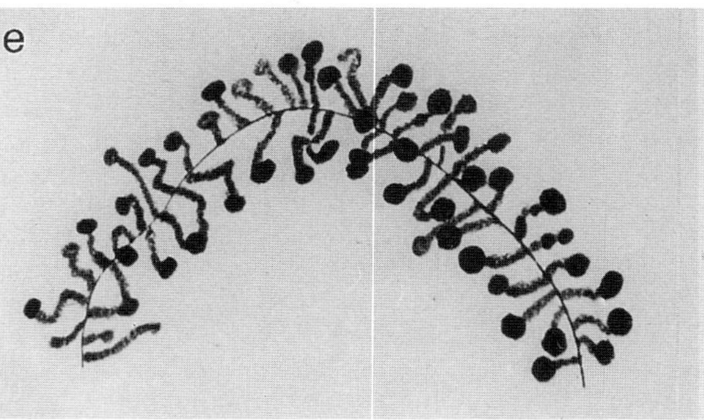

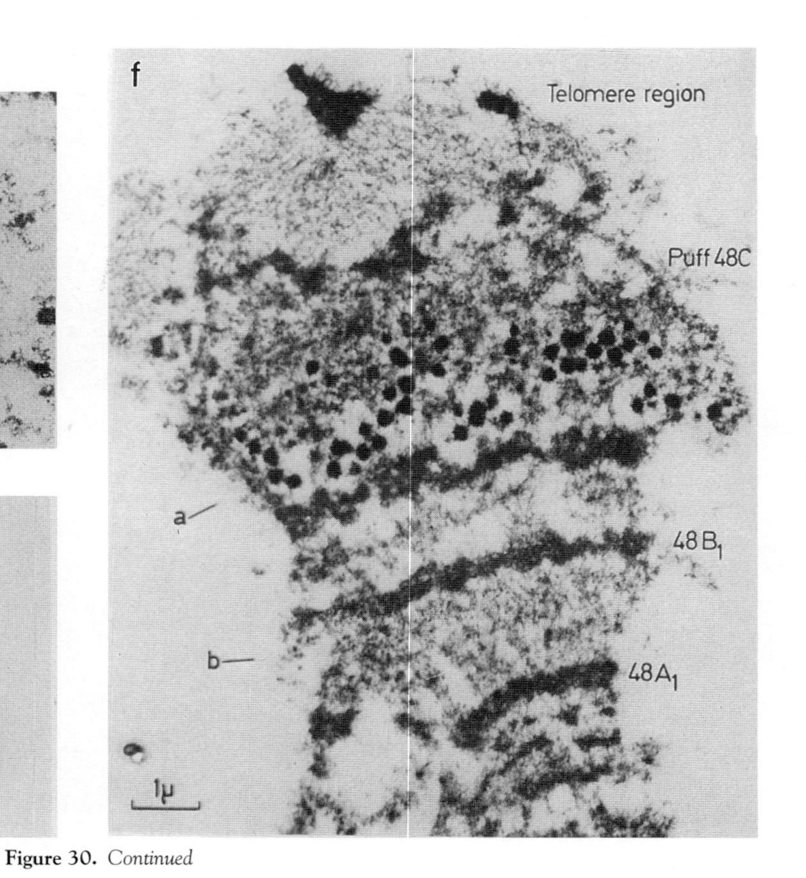

Figure 30. *Continued*

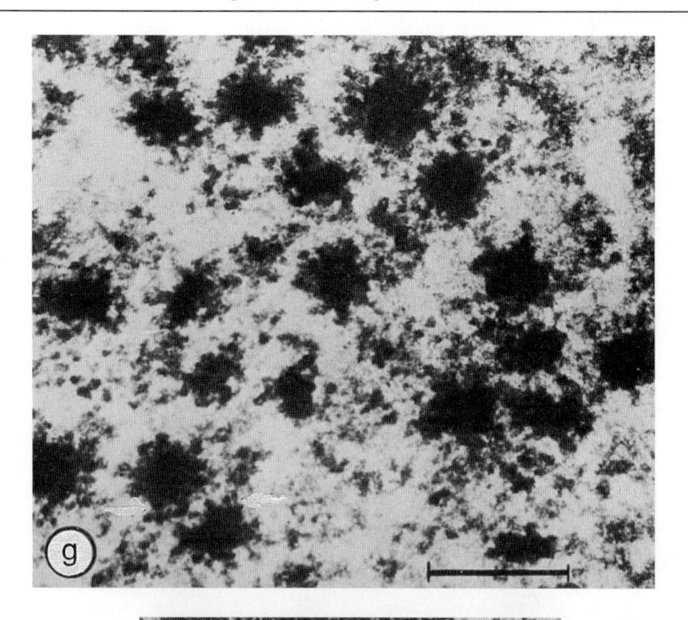

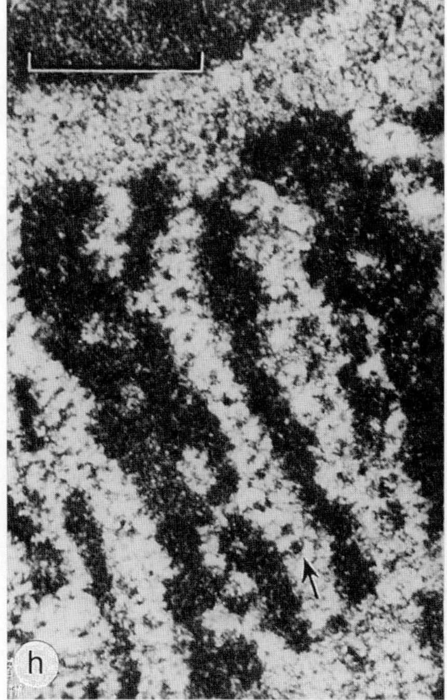

Figure 30. *Continued*

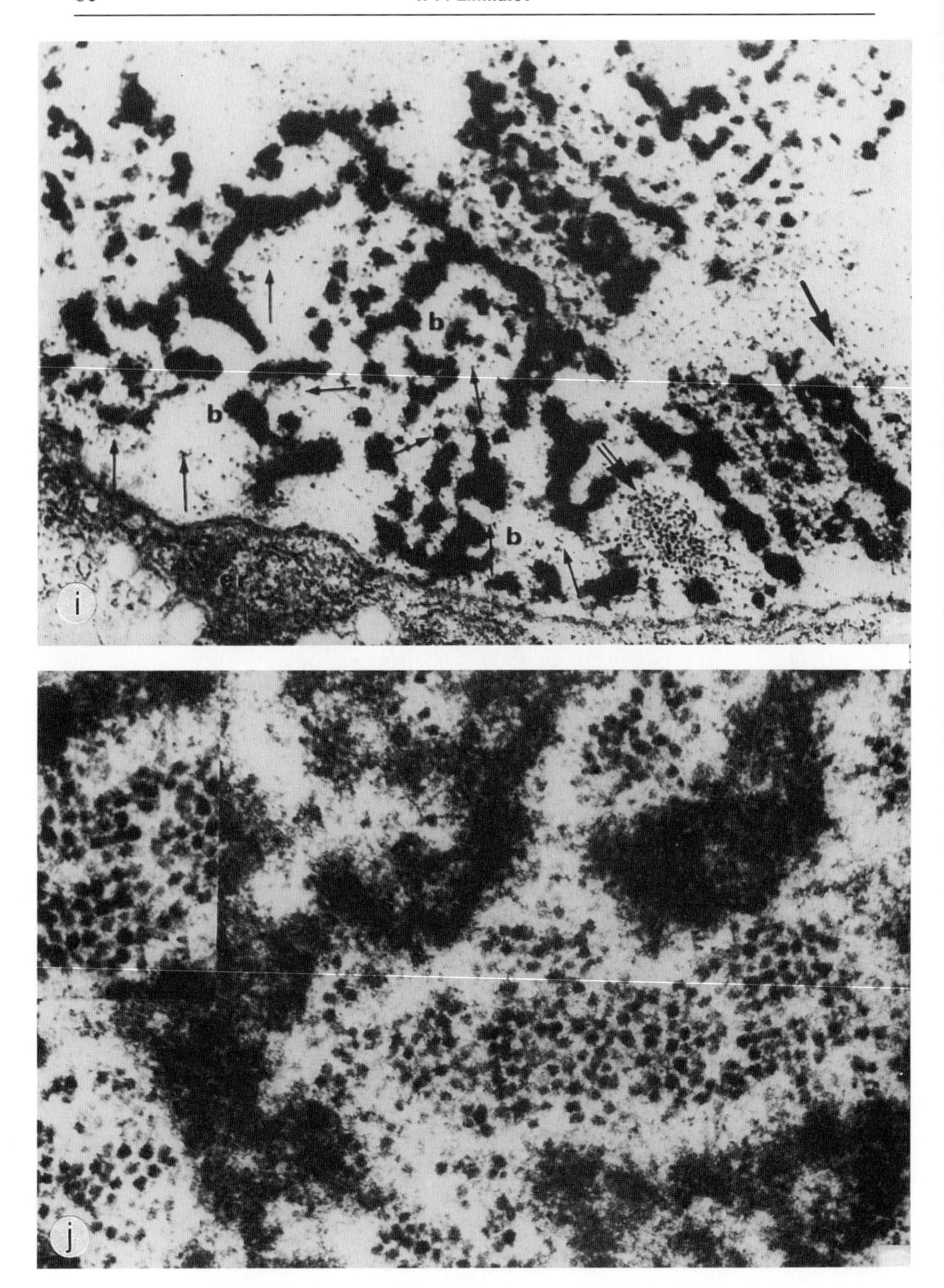

Figure 30. *Continued*

The Balbiani ring granules have been described as a thick ribbon bent into a ring-like structure. Four major domains are demarcated along the ribbon, the first and the fourth domains representing its ends (Figure 31). The ribbon is 5–10 nm in diameter (Wurtz *et al.*, 1990a,b; Lonnroth *et al.*, 1992). Upon passage through the nuclear pore, the ribonucleoprotein particle specifically orientates at the entrance, straightens out and reenters the pore at the 5′ end of messenger mRNA (Mehlin *et al.*, 1991, 1992).

RNP granules were found to be present not only in Balbiani rings; they were also identified in all the chromosome structures decompacted at least to some extent: puffs, interbands, spacing between bands, and loose areas of centromeric heterochromatin (Table 2). It has been estimated that RNP granules occur in about 30% of the interbands, thereby indicating that more than 1500 sites are transcriptionally active at any one time (Skaer, 1977).

Giant RNP granules are formed in the "specific heat-shock puffs" in *Drosophila;* 93D in *D. melanogaster,* 20CD in *D. virilis* and related species, 2-48BC in *D. hydei*, and also in single puffs in Chironomidae (see Figures 30f and 30h; Table 2).

In *D. hydei*, granules, 100–300 nm in diameter, appear in the 2-48B puff 5–8 min after heat shock, and there are uniformly distributed in puff volume 30 min after the exposure (Berendes, 1973; Derksen and Willart, 1976). Granules of this kind also form after treatment of salivary glands with vitamin B6 (Derksen, 1975a,b).

Cytochemical analysis indicated that the core region of the giant granule consists of protein without detectable RNA (Derksen *et al.*, 1973; Derksen and Willart, 1976). According to data obtained in *D. melanogaster*, giant granules contain a specific protein identified only in the 93D puff. Its molecular

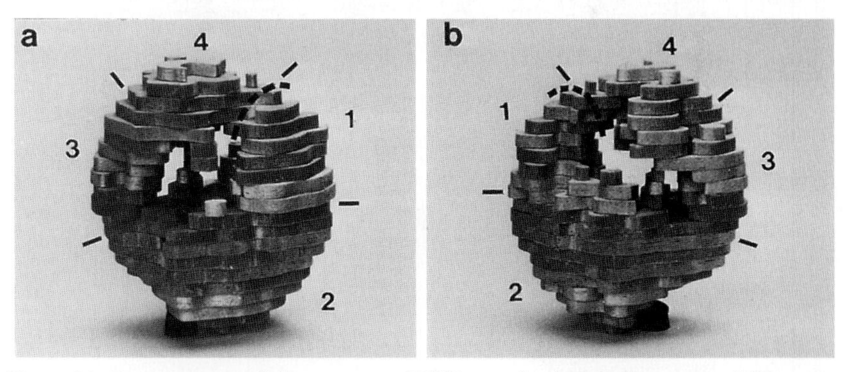

Figure 31. Reconstruction of the structure of RNP particles from Balbiani ring of *Chironomus tentans* with the use of electron microscopic tomography with resolution of 8.5 nm. (a) Frontal view; (b) posterior view. Numbers 1–4 mark the various domains within RNP particles (from Lonnroth *et al.*, 1987, with kind permission from Kluwer Academic Publishers).

Table 2. Location and Characteristics of Ribonucleoprotein Products of Transcription in Polytene Chromosomes of Diptera[a]

Species	Size (nm)	Additional information	Reference
		Balbiani rings	
Chironomus	25	—	Holderegger (1973)
C. *agilis*	31–34	In three Balbiani rings	Kerkis *et al.* (1989); Istomina and Kiknadze (1989)
C. *balatonicus*	38.4–39.2	In four Balbiani rings	Istomina and Kiknadze (1989)
C. *militaris*	30–50	—	Kalnins *et al.* (1964b)
C. *muratensis*	39.2–40.6	In three Balbiani rings	Istomina and Kiknadze (1989)
C. *plumosus*	30–50	—	Kalnins *et al.* (1964b)
	40–80	—	Perov and Chentsov (1971)
	37.2–39.1		Istomina and Kiknadze (1989)
C. *tentans*	30	—	Beermann and Bahr (1954); Beermann (1956, 1964)
	40–50	—	Stevens and Swift, (1966); Sass (1995)
	40–50	Appear 6 hr after hyperthermic (39°C) shock	Yamamoto, 1970
	50	—	Daneholt (1975); Daneholt *et al.* (1982); Skoglund *et al.* (1983, 1986); Lonnroth *et al.* (1987)
	42.4	—	Olins *et al.* (1980)
	30–50	—	Sass (1981)
C. *tentans*	13–20	In Balbiani ring 3	Sass (1981)
C. *thummi*	40–50	—	Stevens (1964); Stevens and Swift (1966)
	45	—	Vasquez-Nin and Bernhard (1971)
	50	—	Perov *et al.* (1975b)
	50–60	—	Perov *et al.* (1976)
	45–50	—	Diez *et al.* (1990)

	40.2	In Balbiani ring 1 of the main lobe of the salivary gland	Istomina *et al.* (1983)
	41.4	In Balbiani ring 2 of the main lobe of the salivary gland	Istomina *et al.* (1983)
	29.7	In Balbiani ring of the special lobe of the salivary gland	Istomina *et al.* (1983)
Glyptotendipes barbipes	—	—	Kalnins *et al.* (1964b)
G. lobiferus	—	—	Kalnins *et al.* (1964b)
		RNA puffs	
Bradysia mycorum	40	—	Jacob and Sirlin (1963)
Chironomus	25	—	Vasquez-Nin and Bernhard (1971)
C. plumosus	30–60	In the chromosomes of the "myocyte" cells	Ilyinskaya and Nilova (1989)
C. tentans	25–35	The ecdysterone-induced IV-2B puff and the IV-5C heat-shock puff	Sass (1981)
	25–100	In IV-5C heat-shock puff	Sass (1995)
	50 and 75	In I-20A heat-shock puff	Sass (1995)
	23–38	Nonheat-shock-induced puff I-17	Sass (1995)
	Small particles	In nonheat-shock induced puff III-11	Sass (1995)
	25 and 75	Heat shock-induced puff III-12B	Sass (1995)
C. thummi	20–50	25 puffs studied	Perov *et al.* (1976)
	20–30 (infrequently 40–50)	The 2A3e puff	Kiknadze *et al.* (1977)
Drosophila hydei	30–40	The heat-shock puffs	Vossen *et al.* (1977)
D. melanogaster	30–40	—	Berendes (1970, 1973)
	20; 35; 40–45	—	Skaer (1977)
	30–45	The 2B3-5 puff	Mott *et al.* (1980)
	20–40	The 68C puff	Mott *et al.* (1980)

(continues)

Table 2. *Continued*

Species	Size (nm)	Additional information	Reference
	20–35	The 74EF, 75B puffs	
	30–45	The 85F puff	
D. virilis	10	—	Swift (1963)
Hybosciara fragilis	50	—	DaCunha *et al.* (1973)
Sciara coprophila	12–23	—	Gabrusewycz-Garcia (1972)
	31	—	Gabrusewycz-Garcia and Garcia (1974)
		"Special" puffs	
Ceratitis capitata	100–300	In one puff only	Semeshin *et al.* (1995)
Chironomus thummi	200–250	In single puffs	Perov (1971)
	200–250	In the 4-E2de puff in prepupae	Perov *et al.* (1975a, 1976)
Drosophila bipectinata	—	Unidentified region	Thomopoulos and Kastritsis (1979)
D. eugracilis		The same	Thomopoulos and Kastritsis (1979)
D. hydei	100–300	In the 2-48BC puff induced by heat shock or arsenite	Berendes (1973); Derksen *et al.* (1973); Berendes *et al.* (1974); Vossen *et al.* (1977); Grond and Derksen (1983)
	28–32	Elementary RNA particles, included in large ones	Berendes (1975)
	250–340	During induction of the 2-48BC puff by vitamin B6 in the presence of actinomycin D	Derksen (1975a,b)
		Detected in the cells of gastric caeca, imaginal disk cells, embryonic cells	Derksen (1975b)
	100–200	The 4-81c heat shock puff	Alanen (1985)
D. lucipennis	—	Unidentified regions	Thomopoulos and Kastritsis (1979)
D. malercotliana	—	The same	Thomopoulos and Kastritsis (1979)
D. melanogaster	—	In the 93D heat shock puff	Dangli *et al.* (1983)

D. montium, D. rajasekari, D. simulans, D. suzuki, D. takahashi	—	Unidentified regions	Thomopoulos and Kastritsis (1979)
D. virilis	40–250	In some puffs	Swift (1959, 1962, 1963, 1965)
	90–300	In the puff on the end of the second chromosome	Schurin (1959); Derksen (1975b)
	200–500	In the 20CD head shock puff	Gubenko *et al.* (1991a,b)
D. yakuba	—	In unidentified region	Thomopoulos and Kastritsis (1979)
		DNA puffs	
Sciara coprophila	23.8	—	Gabrusewycz-Garcia and Garcia (1974)
		Interbands	
Chironomus plumosus	40–80	—	Perov and Chentsov (1971)
C. thummi	20–60	In large interbands, not defected in interbands of 0.05 μm size	Perov *et al.* (1975b)
D. hydei	30	—	Alonso and Berendes (1975)
D. melanogaster	—	—	Skaer (1977, 1978)
	30–45	—	Mott *et al.* (1980); Mott and Hill (1986)
		Bands	
C. thummi	20–60	In band cavities	Perov *et al.* (1975a,b)
D. melanogaster	20–25, 40–45	In band cavities	Lakhotia (1979)
		Perichromocentric region	
D. melanogaster	40–45	In the puff close to heterochromatin	Skaer (1977)
	50–60	—	Mott *et al.* (1980)
	50–60	In the 20C region	Mott and Hill (1986)
		Chromocenter	
D. hydei	25–30	In β-heterochromatin	Lakhotia (1974)

(continues)

Table 2. *Continued*

Species	Size (nm)	Additional information	Reference
D. melanogaster	25	In β-heterochromatin	Lakhotia and Jacob (1974)
	45–50	In chromocenter lacunae	Skaer (1977)
	35	—	Mott *et al.* (1980)
	—	Numerous granules in the chromocenter	V. Sorsa (1979)
	Unidentified Regions of the Chromosomes		
D. virilis (after incubation of salivary glands in RNase solution)	40–50	Granule groups in various chromosome regions	Smith (1969)
Ceratitis capitata	25–30	In the heterochromatinized X chromosome of trichogen orbital cells	Semeshin *et al.* (1995)
	30	In the heterochromatinized X chromosome of salivary glands	Semeshin *et al.* (1995)
	30–50	In the heterochromatinized Y chromosome of trichogen orbital cells	Semeshin *et al.* (1995)

Note. Blanks denote only mention of the presence of granules.
[a]Cells of salivary glands, if not stipulated.

weight is 38 kDa (Dangli *et al.*, 1983; Dangli and Bautz, 1983). Determination of amino acid sequences of protein P11 revealed that its structure consists of two specialized structures in the protein molecule: two RNA binding domains at the aminoterminal half and a glycine-rich region at the carboxy-terminal half (Hovemann *et al.*, 1991). The small RNP granules are located on the surface of the central sphere (Figure 30g).

In some regions of the salivary gland polytene chromosomes of *Chironomus*, accumulations of small bodies well stained with silver were revealed after treatment with silver salts (Faussek, 1913). Later on, staining with hematoxylin revealed small nucleoluslike bodies in the "interband" regions or at the edges of the chromosomes (King and Beams, 1934); droplets of ribonucleoprotein material (additional nucleoli, "Nebennukleolen") were occasionally detected in these puffs (Bauer, 1936). The droplets (Figures 30a and 30b) were detected in the transcriptionally engaged regions both during normal development and after cold or heat shock. The sizes of these droplets varied from 0.1 to 5 μm (Beermann, 1952a, 1956, 1959; Kalnins *et al.*, 1964; Pelling, 1964; Yamamoto, 1970; Perov and Chentsov, 1971; Colman and Stockert, 1975; Stockert and Diez, 1979; Sass, 1981; Semeshin *et al.*, 1983; Stockert, 1985; Bultmann, 1986b; Gubenko *et al.*, 1991a,b). Droplets of this kind appeared in several puffs of *Chironimus thummi* late prepupae ortly before they transformed into pupae (Perov *et al.*, 1976). The number of micronucleoli can be large: a chromosome segment of *Chironimus plumosus* 100 μm in length can accomodate 12–17 regions containing micronucleoli (Perov and Chentsov, 1971). The droplet material is Feulgen-negative, but stains positively for RNA and proteins (Beermann, 1959; Pelling, 1964; Stockert and Diez, 1979).

The data considered provide evidence indicating that the gene products (RNP particles) occur in all the chromosome regions decompacted to different degrees. Localization of the RNA-synthesizing regions in the polytene chromosomes, judging on localization of RNP particles, cannot be completely accurate inasmuch as "the granules can migrate into the spacing between the bands of the polytene chromosomes after they have been synthesized in the other regions, the puffs, for example" (Perov and Chentsov, 1971, pp. 1454–1455). However, the giant granules of "the specific heat shock puffs" (see review (Gubenko, 1989)) occur in the genome in the only puff in each species; this observation was taken as an indication that giant granules are not redistributed in the nucleus.

B. Effect of inhibitors of transcription on the functional organization of polytene chromosomes

A number of compounds used *in vitro* or *in vivo* systems produce a deficiency in total or specific RNAs in cells. These inhibitors differ in the molecular mechanisms of their action. RNase digests synthesized RNA in the simplest case;

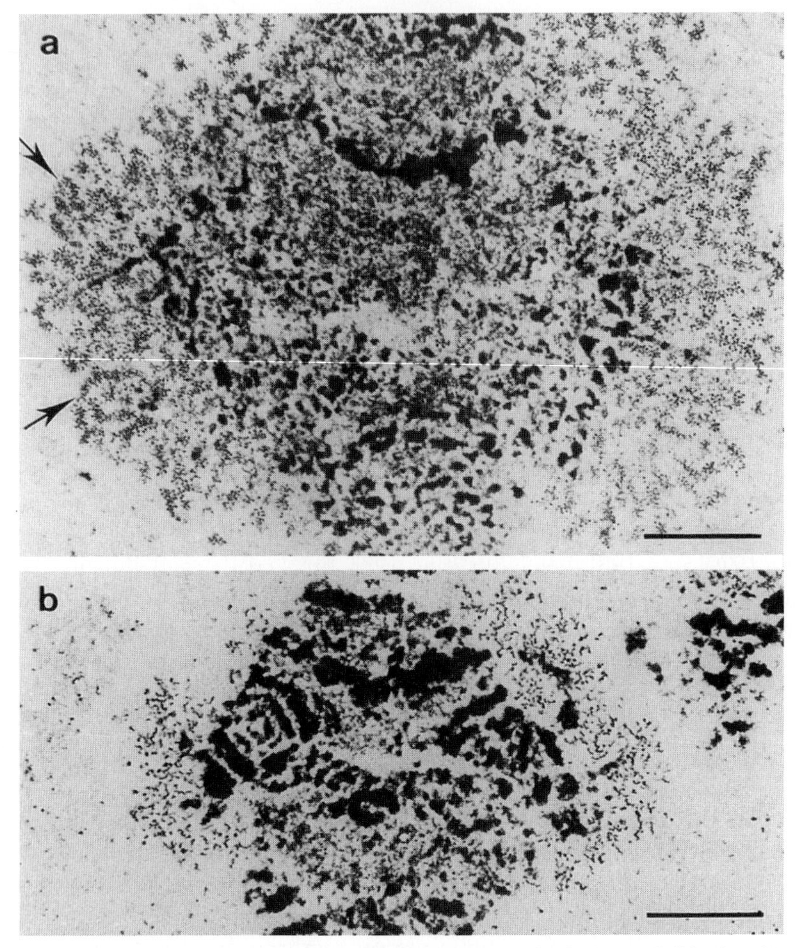

Figure 32. Structure of Balbiani rings of *Chironomus tentans* (a) before and (b) after 60 min treatment with DRB. Bar = 2 μm. Arrows point to the transcriptional units (loops) (reprinted by permission from Andersson *et al.*, 1982).

transcription is blocked at some step in other instances. Tritiated actinomycin D binds to all the chromosome regions, and it completely inhibits RNA synthesis (Camargo and Plaut, 1967; Ebstein, 1967; Desai and Tencer, 1968; Valeyeva *et al.*, 1984; Stepanova *et al.*, 1985). Other inhibitors interact with different RNA polymerases (Table 3) and, accordingly, the effects exerted by the inhibitors on chromosome morphology are different.

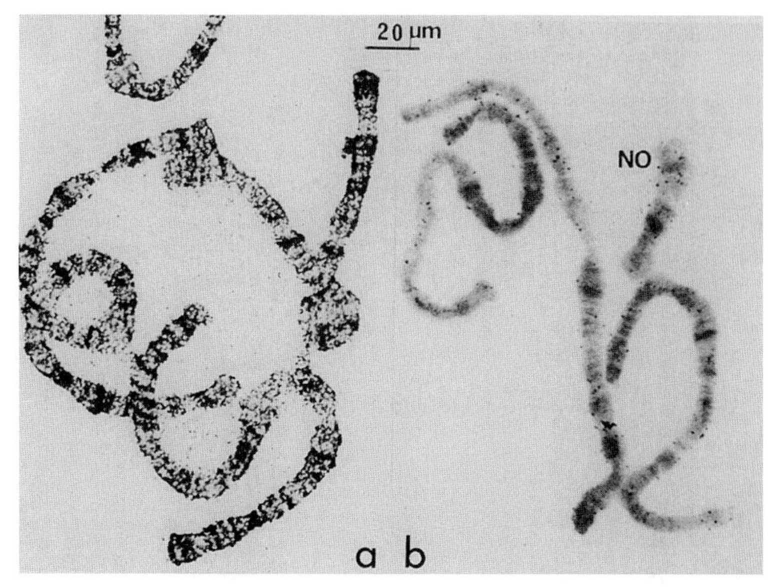

Figure 33. Incorporation of [^{3}H]uridine into the sister glands of the same larva of *Chironomus thummi*. (a) Control; (b) treatment with cordycepin (100 mkg/ml). NO, nucleolar organizer (from Gopalan, H. N. B. (1973). Cordycepin inhibits induction of puffs by ions in *Chironomus* salivary gland chromosomes. *Experientia* **29,** 724–726).

A series of experiments demonstrated that exposure of salivary glands to actinomycin D causes regression of puffs, Balbiani rings, and nucleoli (Table 3). There is some disagreement over the question whether puffs can be induced in larvae cultured in the presence of an antibiotic. Clever (1967) found that puffs are not induced when salivary glands from larvae are incubated in the presence of ecdysterone and actinomycin D. However, subsequent studies showed that there was a slight induction of puffs in heat-shocked *D. hydei* larvae exposed to ecdysterone when RNA synthesis was inhibited. The size that these puffs attained was 15–50% of normal (Berendes, 1968a). There was only a 10% induction of the 74EF puff by ecdysterone in the presence of actinomycin D in *D. melanogaster* (Ashburner, 1972c). The small swelling that appeared in the puffed region might have been produced by accumulation of acidic protein as a consequence of induction (Section VI,I for details).

Puffs and large Balbiani rings did not regress; rather, they "froze" when exposed to high concentrations (20 mg/ml) of antibiotic. When bound to chromosomal DNA, actinomycin presumably affects chromatid structure in such a way that condensation becomes impossible when transcription is blocked (Beermann, 1964, 1966). Intercalation itself is without effect on the possible

Table 3. Effect of Inhibitors of RNA Synthesis on the Transcriptional Activity and Morphology of Polytene Chromosomes

Inhibitor	Mechanism of inhibition	Effect of inhibitor on transcription and morphology of chromosomes	Reference
Adenosine	Initiation is presumably inhibited (Egyhazi and Holst, 1981)	During treatment puffs and [^{3}H]uridine incorporation rapidly inhibited. No inhibition of RNA synthesis in the nucleolus	Beermann (1967)
		[^{3}H]uridine incorporation into heterogenous RNA and 4-5S RNA decreased by 70–90% and 45–60%, respectively	Egyhazi and Holst (1981)
Actinomycin D	Intercalation between G-C pairs and arrest of movement of RNA polymerase molecules	[^{3}H]uridine incorporation inhibited	Ritossa and Pulitzer (1963); Beermann (1964, 1966); Laufer *et al.*, (1964); Clever and Romball (1966)
		The nucleolus and Balbiani rings regress (1–10 mg/ml, 24 hr treatment)	Laufer (1963); Jacob and Sirlin, (1964); Laufer *et al.* (1964); Clever and Romball (1966); Kiknadze (1966); Clever (1967)
		Concentration 0.1–2 μg/ml inhibits puffs, concentration 20 μg/ml immediately blocks synthesis, but does not decrease the puff size	Beermann (1966, 1967)
		Concentration 0.2 μg/ml reduces Balbiani rings, but does not decrease many small puffs in development	Clever and Romball (1966); Clever (1967)
		The granular component and the peripheral zone of the nucleolus reduced (1–2 μg/ml)	Stevens (1964); Perov and Chentsov (1971)
		On the inhibitor background the action of heat shock and ecdysterone induces puffs only up to 15–50% of normal size	Berendes (1968a)

		Induction of puffs by ecdysterone is blocked	Ashburner (1972c)
		"Puffs" formed in pericentromeric heterochromatic blocks in *Chironomus thummi*	Kiknadze (1965a–c, 1981a,b); Valeyeva *et al.* (1979, 1984, 1997); Kiknadze and Valeyeva (1980, 1983, 1991); Perov and Kiknadze (1980); (Kiknadze *et al.* (1983a)
		RNA polymerase II remains in the chromosomes despite inhibited RNA synthesis	Sass (1982)
		Fluorescence intensity of antibodies against DNA/RNA hybrids does not decrease	Alcover *et al.* (1982)
7-Amino-actinomycin D		Reduction of the nucleolus and Balbiani rings, and occurrence of giant "puffs" in the centromeric region of the chromosomes of C. *thummi* (see above)	Valeyeva *et al.* (1984); Valeyeva (1991)
α-Amanitin	RNA polymerase II inhibitor	Incorporation of [^{3}H]uridine into the chromosomes (1–50 μg/ml concentration); labeling of the nucleolus, synthesis of low molecular weight (4-5S) and ribosomal RNA not impaired	Beermann (1971); Wobus *et al.* (1971a); Ashburner (1972c); Egyhazi *et al.* (1972); Holt and Kuijpers (1972a,b); Serfling *et al.* (1972); Edstrom (1974); Majumdar *et al.* (1975); Law *et al.* (1976); Santelli *et al.* (1976); Sass (1980a,b); Chatterjee and Mukherjee (1982, 1984); Vlassova and Zhimulev (1988)
		Incomplete block of induction of the ecdysterone puffs, presumably because of poor permeability of inhibitor in cells	Ashburner (1972c)

(continues)

Table 3. *Continued*

Inhibitor	Mechanism of inhibition	Effect of inhibitor on transcription and morphology of chromosomes	Reference
		Incorporation of [^{3}H]UTP into isolated nuclei of salivary gland cells of *D. hydei* inhibited by 60%	Stein (1976)
		After 2 hr treatment there was no inhibition of [^{3}H]uridine incorporation in 0–45% of nuclei	Vlassova *et al.* (1987)
		When RNA synthesis inhibited, RNA polymerase II was not separated from the chromosome	Sass (1982)
		There are mutations of *D. melanogaster*, resistant to the effect of α-amanitin	Greenleaf *et al.* (1979); Phillips *et al.* (1982)
		Species of Drosophila mycrophage (*D. putrida, D. recens, D. tripunctata)* resistant to α-amanitin	Janike *et al.* (1983)
α-Amanitin with the addition of actinomycin D at low concentrations		Induction of puffs in the centromeric regions of the chromosomes of C. *thummi*. Treatment with α-amanitin alone does not induce these puffs	Valeyeva (1991)
Anisomycin	Protein synthesis inhibitor	After short treatment of C. *thummi* larvae (1 μg/ml, 3–6 hr) BR1 and BR2 puffing states remain unaffected	Diez *et al.* (1980)
Astrason pink		When larvae incubated (2.5 μg/ml) Balbiani rings decrease	Stockert (1990)
Benzamide		Inhibits in the chromosomes, not in the nucleolus	Sirlin and Jacob, (1964); Lakhotia and Mukherjee (1970a)
Berberine		When larvae incubated (25 μg/ml), Balbiani rings decrease	Stockert (1985, 1990)

Cycloheximide	Protein synthesis inhibitor	Treatments with 10 μg/ml on C. *pallidivittatus* larvae (3–24 hr) result in stimulation of the nonribosomal RNA synthesis. Ribosomal RNA synthesis is depressed	Diez *et al.* (1977)
		After short treatment of C. *thummi* larvae (10 μg/ml), 3–6 hr) BR1 and BR2 puffing states remain unaffected	Diez *et al.* (1980)
		Treatments of C. *thummi* larvae (20 μg/ml) inhibit RNA synthesis in chromosomes and nucleoli. New puffs are induced, in which [^{3}H]uridine does not incorporate	Valeyeva (1995)
5.6-Di-chloro-1-beta-D-ribofuranosil-benz-imidazole (DRB)	Inhibitor of transcription initiation	Incorporation of [^{3}H]uridine into the chromosomes and synthesis of hnRNA suppressed. Only partial inhibition of synthesis of low-molecular-weight RNA; incorporation of [^{3}H]uridine into the nucleolus not suppressed	Sirlin and Jacob (1964); Sirlin *et al.* (1966); Egyhazi *et al.* (1970, 1982); Egyhazi (1974, 1976a); Wiltenberg and Lubsen (1976); Lezzi *et al.* (1984); Holst and Egyhazi (1985)
		Phosphorylation of nonhistone and histone-like proteins is disturbed	Egyhazi and Pigon (1983); Egyhazi *et al.* (1984)
		Induction of puffs by ecdysterone is inhibited	Ashburner (1972c)
		Balbiani rings regress after 60 min treatment, their sizes restored in 60 min after the inhibitor has been washed off (Figure 32)	Egyhazi (1975); Andersson *et al.* (1982)
		As a result of inhibition of RNA synthesis, 5 nm fibrils of Balbiani ring got compacted into fibrils of 25 nm diameter	Andersson *et al.* (1982)

(continues)

Table 3. *Continued*

Inhibitor	Mechanism of inhibition	Effect of inhibitor on transcription and morphology of chromosomes	Reference
		When the inhibitor injected into the larvae of *C. tentans*, complete repression of Balbiani rings occurs in 44 of 72 nuclei. A dissociation of RNA polymerase II and the chromosomes takes place. Slight incorporation of [^{3}H]uridine retained along the entire length of the chromosome	Sass (1984)
		Treatment with the inhibitor produces distinct decrease in fluorescence of antibodies against DNA/RNA hybrids	Alcover *et al.* (1982)
		Strong inhibitory effect achieved after 2 hr incubation of *D. melanogaster* salivary glands with the inhibitor. However, antibodies against DNA/RNA hybrids detected in 16 regions of the chromosomes (84EF, 88D, 91D, 91F, 92A, 94C, 94EF, 43E, 46D, 47B, 47D, 50C, 51F, 53F, 57E, and 93D)	Vlassova and Zhimulev (1988)
		Combined treatment almost completely inhibits [^{3}H]uridine incorporation	Vlassova *et al.* (1987)
3′-Deoxy-adenosine (cordycepin)		Puff induction is inhibited by ecdysterone	Ashburner (1972c)
		[^{3}H]uridine incorporation into the nucleoli and the chromosomes almost completely inhibited in 80–85% of cells (Figure 33). All puffs and Balbiani rings regress	Diez (1973); Gopalan (1973a); Stockert and Diez (1979)
		Induction of puffs in response to change of the ionic composition of solution blocked	Gopalan (1973a)

		Strong inhibition of [^{3}H]uridine incorporation mainly into puffs. Incomplete reduction of label incorporation into the nucleoli	Majumdar *et al.* (1975)
Equinomycin		Decrease of Balbiani ring sizes	W. Beermann in: Stockert (1990)
Ethidium bromide		When larvae incubated (25 μg/ml), Balbiani rings decrease	Stockert (1990)
Ethionine		When larvae incubated (3 mg/ml) Balbiani rings decrease	Stockert (1990)
Orcein		When larvae incubated (25 μg/ml), Balbiani rings decrease	Stockert (1990)
Phosvitin	Possibly affects on RNA polymerase II	When injected into nucleus Balbiani rings regress	Egyhazi and Pigon (1986)
Polylysines and histones		[^{3}H]uridine incorporation inhibited by injection into larvae. Puff induction by ecdysterone prevented	Desai and Tencer, (1968)
Rifampicin	Binds to the subunit of *E. coli* RNA polymerase	Does not inhibit the induction of the ecdycterone puffs	Ashburner (1972c)
RNase	Digestion of RNA	Incubation of the salivary glands of *D. busckii* for 4–7 hr (3–5 mg/ml) produces: (1) inhibition of [^{3}H]amino acid incorporation; (2) disappearance of the nucleolus; (3) induction of about 40 puffs, some are the heat shock puffs	Ritossa and Borstel (1964); Ritossa *et al.* (1964, 1965)
		[^{3}H]uridine incorporation suppressed, the nucleolus completely destroyed	Prostakova (1968)
		Data on induction of a large number of puffs not confirmed on *C. plumosus* and *D. melanogaster*	Prostakova (1968); Ashburner (1970); I. F. Zhimulev (1970 unpublished observations)
Thioflavine T		When larvae incubated (2.5 μg/ml) Balbiani rings decrease	Stockert (1990)

occurrence of compaction: when salivary glands of *Drosophila* were treated with proflavin at a concentration effective to inhibition, the rate of puff regression was not retarded (Ashburner, 1972c).

In another series of experiments, when larval salivary glands of *C. tentans* were exposed to actinomycin D (0.1–2 mg/ml at 20°C), shrinkage of all the puffs occurred and RNA synthesis was inhibited in parallel. However, puffs did not shrink at 4°C, nor when exposed to both dinitrophenol (an ATP inhibitor) and actinomycin D. When the glands preincubated at a low temperature were thoroughly washed in a sucrose medium and were then transferred to the same medium at room temperature, shrinkage occurred later. Thus, actinomycin D binds to the chromosomes and inhibits RNA synthesis; but it does not cause shrinkage of puffs, which is presumably dependent on a process inhibited by low temperature and requiring ATP (Beermann, 1966, 1967).

It has been compellingly demonstrated that the interbands are transcriptionally active (see above). On this basis it might be expected that prolonged exposure of salivary glands to an inhibitor of RNA synthesis would lead to compaction of interband DNP and thereby to a considerable shortening of the chromosomes. However, even after RNA synthesis has been almost completely inhibited by DRB and α-amanitin, the active regions of the chromosomes did not become more tightly packed, nor did they convert into bands. Compact structures (minibands) were not discerned in the electron microscope, neither was shortening of the chromosomes (Vlassova *et al.*, 1987).

These data may be taken to mean that the compaction of the small puffs and the interbands is also not directly related to transcription inactivation.

C. Transcription and dosage compensation

1. The problem of dosage compensation

In *Drosophila*, the X chromosome: autosome ratio (the X : A ratio) is the primary determinant of the development of somatic sexual phenotype. 1X : 2A individuals are males; those with the 1 : 1 ratio (2X : 2A; 1X : 1A; 3X : 3A; 4X : 4A) are females; and those with an intermediate ratio (2X : 3A) are intersexual. In conformance with Bridges' balance theory of sex determination, 1X : 3A and 3X : 2A individuals are referred to as supermales and superfemales, respectively, 4X : 3A individuals are also females (see review: Baker and Belote, 1983).

Most known X-linked genes are equivalently expressed in males and females of *Drosophila*, despite unequal gene dosages, a single in the X of the male and two in the Xs of the female. This regulatory phenomenon has been first described in *D. melanogaster* by Bridges (1922), thoroughly studied by Muller, and termed "dosage compesation" (Muller *et al.*, 1931; Muller, 1932, 1950). The literature on this subject, including reviews, is voluminous (Muller, 1950; Stern, 1960; Gvozdev, 1968; Lucchesi, 1973, 1977, 1996; Schwartz, 1973;

Stewart and Merriam, 1980; Baker and Belote, 1983; Lucchesi and Manning, 1987; Belote, 1992; Nothiger, 1992; Henikoff and Meneely, 1993; Baker *et al.*, 1994; Kelley and Kuroda, 1995; Bashaw and Baker, 1996; Bone and Kuroda, 1996; Cline and Meyer, 1996; Willard and Salz, 1997). The main information bearing on this issue is given below.

Almost all the X-linked genes were found to be dosage compensated. But there are exceptions to this rule. One is the *bobbed* locus present on the X and Y chromosomes, such that there are two copies in each sex. The other noncompensated gene encodes larval serum protein: the LSP-1-α gene, whose two very closely lying copies (*LSP-1-*β and *LSP-1-*γ) are located on the autosomes. A third example of noncompensation is provided by the genes encoding yolk protein (YP) (Baker and Belote, 1983; see Baker *et al.*, 1994, for details).

The specific activities of the enzymes glucose-6-phosphate dehydrogenase and 6-phosphate dehydrogenase, which map to the X chromosome of *D. melanogaster*, are the same in the male and female X chromosomes (Kazazian *et al.*, 1965; Komma, 1966; Seecof *et al.*, 1969; Steele *et al.*, 1969; Gvozdev, 1970; Bowman and Simmons, 1973; Faizullin and Gvozdev, 1973; Lucchesi and Rawls, 1973a,b; Maroni and Plaut, 1973b; Stewart and Merriam, 1974; Williamson and Bentley, 1983; Birchler *et al.*, 1989). Similar results were obtained for tryptophan oxygenase (tryptophan pyrrolase) (Baillie and Chovnick, 1971; Tobler *et al.*, 1971). Males do not differ from females of *Drosophila* in the specific amount of protein in salivary gland secretion encoded by the *Sgs-4* gene (Korge, 1975, 1981).

The entire X chromosome seems to have the ability to compensate. When transferred to the autosome by insertional translocations, even small fragments of the X chromosome retain their level of dosage compensation (Muller and Kaplan, 1966; Seecof *et al.*, 1969; Korge, 1970a,b; Baillie and Chovnick, 1971; Tobler *et al.*, 1971; Bowman and Simmons, 1973; Lucchesi and Manning, 1987; Baker *et al.*, 1994). In contrast, the level of the expression of the autosomal gene does not alter when translocated to the X chromosome (Roehrdanz *et al.*, 1977).

As evidenced by P-element DNA transformation experiments, the *white*$^+$ and *Sgs-4*$^+$ genes (normally located on the X chromosome), can still dosage compensate in males when inserted together with the small flanking sequences contained in a transposon into a site on the autosomes (Hazelrigg *et al.*, 1984; Krumm *et al.*, 1985; see Baker *et al.*, 1994, for review).

Genetic analysis of the larval secretion *Sgs-4* gene revealed that two of its alleles in Karsnas and Samarkand stocks in *D. melanogaster* are not subjected to dosage compensation (Korge, 1981). Comparison of the base sequences of the 600 nucleotides adjacent to the 5′ upstream region in these stocks and dosage-compensated Oregon-R (wild-type) revealed nine neutral differences (Karsnas) and two base pair exchanges, C-to-T transitions, at positions -344 and -520 (Samarkand).

Base exchange at −344 produced a 50% decrease in the amounts of synthesized *Sgs-4* RNA. Some of the base exchanges in these stocks were found in the regions between −600 and −839. It is unclear which of the changes in the sequences respond to dosage effect (Furia *et al.*, 1984; Krumm *et al.*, 1985; Kaiser *et al.*, 1986; Hoffmann and Korge, 1987; see Baker *et al.*, 1994, for review).

It should be noted in this regard that the gene's capacity to compensate is also related to the site at which it is located in the chromosome. *LSP1α* mentioned above as a gene lacking dosage compensation acquires this ability when located to different site on the X chromosome (Ghosh *et al.*, 1989).

The autosomal *rosy* (Spradling and Rubin, 1983), *Adh* (Goldberg *et al.*, 1983; Laurie-Ahlberg and Stam, 1987), and *Ddc* (Scholnick *et al.*, 1983) genes become dosage compensated upon insertion in the X chromosome.

The transcriptional activities of the autosomal *Adh* gene of *D. melanogaster* and of the X-chromosomal *hsp82-neo* gene of *D. pseudoobscura,* inserted into P-element mediated transposon which was transposed at various sites in the *D. melanogaster* genome, were measured. It was found that both genes were fully compensated at their new sites in the euchromatic regions of the X. However, both were uncompensated when inserted at a site in β-heterochromatin near the base of the X and into the autosomes (Sass and Meselson, 1991). The transposable copia element continues showing dosage compensation when contained within the *white* transcriptional unit (Hiebert and Birchler, 1992). Thus, it may be inferred that the gene's ability to dosage compensate is strongly dependent on the molecular organization of the chromosome region nearest to the insertion point.

There are data indicating that the genes located in the autosomes can dosage compensate. In flies trisomic for the entire *2L* of *D. melanogaster,* the distally mapping structural genes for phosphoglycerate kinase, glycerophosphate dehydrogenase, and malate dehydrogenase produced the same level of activities as in the diploid control. The proximally mapping genes for alcohol dehydrogenase *(Adh)* and dopa decarboxylase *(Ddc)* failed to compensate; that is, they produced a 1.5-fold higher level of activities relative to diploid. Compensation seems to be effected through some other mechanisms than in the X chromosome (Devlin *et al.*, 1982, 1984).

In some species of *Drosophila,* the X chromosome consists of (or corresponds to) two (*D. pseudoobscura* and *D. americana*) or three *(D. miranda)* arms, containing the "standard" X chromosome and other chromosome arms. In *D. americana,* the two arms of the X chromosome consist of the A and B elements (the X and the second chromosome in *D. virilis*). In *D. pseudoobscura* this metacentric X chromosome derived by fusion of the original X with an autosome and a D-element corresponding to *3L* arm of *D. melanogaster* suggested that the genes of the left arm of the X chromosome (corresponding to the A element) might be able to dosage compensate, while those of the right arm (corresponding to the autosomes) might not (Muller, 1950). The time period since the autosomal arm differentiated into the sex chromosome during evolution was the primary determinant in establishment of dosage compensation. The level of activities of XR-linked glucose-6-phosphate dehydrogenase (G6PD) and esterase-5 (est-5) were indistinguishable in males and females of *D. pseudoobscura*. The gene for G6PD was located on the X and that for esterase-5 on chromosome 3L in *D. melanogaster*. The level of activity of the gene for isocitrate dehydrogenase (IDH) was completely compensated in males of *D. willistoni*. The gene IDH has been mapped to chromosome *3* in *D. melanogaster* (Abraham and Lucchesi, 1974).

In *D. miranda*, a second autosome fused with the Y chromosome, generating a neo-Y chomosome and a neo-X (X_2) (see for review Steinemann *et al.*, 1996).

2. Transcriptional activity of the sex chromosomes under dosage compensation

From the above observations it is apparent that mechanisms of dosage compensation must be sought among differences in the structural organization of the sex chromosomes between females and males.

It is long known that a single X chromosome in a male is very loose in its appearance and it can be as wide as two autosomes in the same nucleus in the polytene chromosomes of the salivary gland cells of *Drosophila*. The X chromosome is as thick as the autosomes in the nucleus in females (see for details: Zhimulev, 1992, 1996). In *Sciara ocellaris*, where females are XX and males are XO, the only male X salivary gland chromosome shows bloated appearance as well (Da Cunha *et al.*, 1994). The X chromosome of the male also differs from that of the female in protein composition (Rudkin, 1964; Chatterjee *et al.*, 1980; Ukil *et al.*, 1984, 1986). In *D. hydei*, the content of nonhistone proteins in the single X chromosome of males makes up about 75% of their total content in the two Xs in females (De Roda *et al.*, 1980). Autoradiographic study demonstrated that the single X chromosome of the male bound twice the amount of molecules of ^{3}H-labeled RNA polymerases of *E. coli* than did each of the two Xs of the female (Khesin and Leibowitch, 1974). In contrast, H1 histone labeled with FITC bound more intensely to fixed chromosomes of females than males (Leibovitch *et al.*, 1974, 1976; Leibovitch and Bogdanova, 1983).

The important regulatory loci controlling dosage compensation in *Drosophila* include *maleless (mle)*, and *male-specific lethals* (*msl-1*, *msl-2*, and *msl-3*). The normal products of these genes are necessary for male, but not female, viability. The mutations at these loci are male-specific lethals (Belote and Lucchesi, 1980a,b; Uchida *et al.*, 1981, reviews: Baker and Belote, 1983; Baker *et al.*, 1994; Bashaw and Baker, 1996). The *Sex lethal (Sxl)* gene plays an important role in controlling X-chromosome transcriptional activity by interacting with the male-specific lethals (e.g., Lucchesi and Skripsky, 1981).

Antibodies to their products associate with hundreds sites along the X chromosome in males and not in females (Kuroda *et al.*, 1991; Palmer *et al.*, 1993; Gorman *et al.*, 1995; Zhou *et al.*, 1995; Kelley *et al.*, 1995; Bashaw and Baker, 1995; Lyman *et al.*, 1997). All *msl* genes probably function in a multimeric complex (Bashaw and Baker, 1996).

The male X chromosome is enriched with histone H4 acetylated at lysine 16 (Turner *et al.*, 1992). The colocalization of H4Ac16 and MSL proteins in X chromosome of *D. melanogaster* and several other species suggests a link between MSL function and chromatin structure (Bone *et al.*, 1994; Bone and Kuroda, 1996).

In homozygous *mle* males, the morphology of the X chromosome is anomalous, acquiring the compaction degree of the autosomes and the X chromosomes in female (Belote and Lucchesi, 1980a). The X chromosome was half as wide in male larvae with the mutation Sxl^{MN1} as the X in control larvae (Lucchesi and Skripski, 1981). Antisera raised against the *D. melanogaster maleless*, *msl-1* proteins, and histone H4Ac16 generated a strong signal on polytene X chromosomes of *D. simulans* and *D. virilis*. In polytene chromosomes of *D. americana americana* only one arm was stained; it is homologous to element A, but not element B, in *D. pseudoobscura* in both arms of the X, in A and D elements (Bone and Kuroda, 1996). Similar data on location of H4Ac16 antibodies were described by Steinemann *et al.* (1996). In *D. miranda* anti-MLE, anti-MSL-1, and anti-H4Ac16 stain the XL plus XR chromosome arms and about 20 sites in X2 arm (Bone and Kuroda, 1996). Only partial staining of X2 chromosome was found when H4Ac16 antibodies were used. In this chromosome region of about 10% of the chromosome length at the distal tip show only background staining (Steinemann *et al.*, 1996). This nonstaining region also did not show higher RNA synthesis after [^{3}H]uridine incorporation (Strobel *et al.*, 1976).

The level of transcription of the X-linked genes in males enhances as decompaction degree increases. In autoradiographs prepared from the *D. melanogaster* polytene chromosomes of pulse-labeled salivary glands with [^{3}H]uridine (5 min), the number of grains was counted over the regions 1A-3B of the X and over the entire X chromosome, and also over the 3L segment in the same nuclei (61AF and 61A-68B). Next, the numbers of grains over the X chromosome segments were plotted against their numbers over the 3L segment for the two sexes. Comparisons showed no significant differences in the ratios between males and females and, consequently, the intensity of transcriptional activity in a single X chromosome of the male was twofold that in each of the two X chromosomes in the female (Mukherjee and Beermann, 1965).

This conclusion has been extensively supported by studies on the chromosomal basis of dosage compensation in *D. melanogaster* (Mukherjee, 1966; Kaplan and Plaut, 1968; Lakhotia and Mukherjee, 1969, 1972; Korge, 1970a, b; Holmquist, 1972; Ananiev *et al.*, 1974; Maroni *et al.*, 1974; Maroni and Lucchesi, 1980; Lakhotia *et al.*, 1981; Lucchesi and Skripsky, 1981; Chatterjee and Ghosh, 1985; Ghosh and Mukherjee, 1986), *D. hydei* (Mukherjee *et al.*, 1968), *D. virilis* and *D. americana* (Duttagupta and De, 1986), *Drosophila kikkawai*, *Drosophila bipectinata* (Lakhotia and Mukherjee, 1981), *Rhynchosciara angelae* (Casartelli *et al.*, 1969), and *S. ocellaris* (Da Cunha *et al.*, 1994). The discrete regions of the X chromosome are compacted independently of one another in *Drosophila*. There were considerable differences in RNA synthesis rates between males and females in only 4 of the 15 puffs in the X chromosome of *D. hydei* (Chatterjee and Mukherjee, 1971a,b). The male X chromosome fragments remained hyperactive when translocated to an autosome (Holmquist, 1972; Prasad *et al.*, 1981). An autosomal segment retained the level of transcriptional activity characteristic of the autosome when it was translocated to the male X chromosomes (Ghosh and Mukherjee, 1986).

Dosage compensation at the level of transcription was demonstrated in Northern blot hybridization using DNA from the G6PD (glucose-6-phosphate dehydrogenase) gene as a probe (Ganguly *et al.*, 1985). The rate of [^{3}H]uridine incorporation by the X chromosome in homozygous *mle* males was only 60–70% of that by the two X chromosomes in mutant females. The amount of RNA seen in Northern blots also decreased when DNA from the *Sgs-4* gene was used as a probe (Belote and Lucchesi, 1980; Breen and Lucchesi, 1986; Kaiser *et al.*, 1986).

The rate of [^{3}H]uridine incorporation into polytene X chromosomes was compared in larvae with different X:A ratios. In larvae with the X:A ratio varying from 0.33–0.5 (metamales and males) (Figure 34), one of the X chromosomes incorporated twice as much [^{3}H]uridine than found for females and metafemales whose X:A ratios were 1.0 and 1.5, respectively. In intersexes (0.66), the rate of label incorporation is intermediate between the two sexes (Maroni and Plaut, 1973a; Ananiev *et al.*, 1974; Lucchesi *et al.*, 1977).

Fixed squash preparations of the salivary gland chromosomes of *Drosophila* also retain the state of chromatin packing in the male X chromosome providing dosage compensation. When reaction mixtures consisting of *E. coli* RNA polymerase and four triphosphates (tritium-labeled UTP) were applied to squash preparations, autoradiographic measurements of transcriptional activities demonstrated, again, that the single chromosome of the male incorporates label

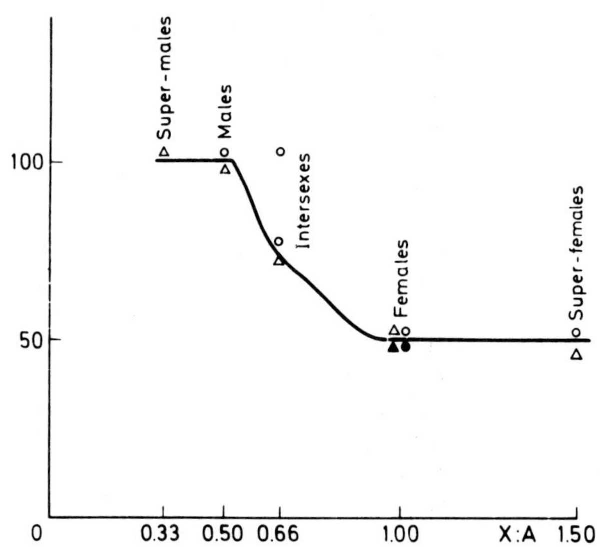

Figure 34. Dependence of the gene activity of the X chromosome on the ratio of X chromosome number to the autosome set number. Abscissa, ratio of sex chromosome numbers to autosome numbers; ordinate, silver grain number (open triangles) and 6 PGD activity (open circles) per X chromosome; filled figures show the same characteristics for triploid females. Transcriptional activity of the male X chromosome is accepted as 100% (reprinted by permission from Ananiev *et al.*, 1974 a,b).

twice as intensely as each of the two female X chromosomes (Leibovitch and Khesin, 1974; Khesin and Leibovitch, 1974; Leibovitch *et al.*, 1976; Khesin, 1980; Chatterjee *et al.*, 1981; Chatterjee, 1985a). Concominantly with this, transcriptional activity was the same in metafemales and females, the same in metamales and males, and its level was intermediate in intersexes (Leibovitch *et al.*, 1974, 1976; Khesin, 1980). According to other data (Gvozdev *et al.*, 1983), the levels of transcriptional activity are the same in intersexes and females. In males homozygous for the *mle* mutation, template activity of the X polytene chromosomal DNA was reduced by 30–40% in the *E. coli* RNA polymerase system (Chatterjee and Bhadra, 1985). The *transformer (tra)*, *tra2*, or *intersex (ix)* mutants affect somatic sex differentiation only in females. The null mutations at *double sex (dsx)* affect both XX and XY individuals causing them to develop as phenotypically intersex. All of these mutations do not play a role in the regulation of the template organization for dosage compensation in the polytene chromosomes measured as ^{3}H-UMP incorporation in *E. coli* RNA polymerase reaction mixture (Acharyya and Chatterjee, 1997).

In *Drosophila* species whose X chromosome consists of two arms, differences in the degree of the expression of dosage compensation were observed:

both arms were transcriptionally hyperactive in the males of *D. pseudoobscura* (Chatterjee *et al.*, 1974; Pierce and Lucchesi, 1980), *Drosophila persimilis* (Mutsuddi *et al.*, 1985), and *D. americana* (Duttagupta and De, 1986).

In *Drosophila miranda*, the left arm of the X chromosome (X^1) corresponding the original ancestral X of *Drosophila*, is normally dosage compensated. The activity of some genes is dosage compensated, but that of others is not in the right arm (X^2) corresponding to chromosome *3L* of *D. melanogaster* (Strobel *et al.*, 1978; Das *et al.*, 1982; Mutsuddi *et al.*, 1988; Krishnan *et al.*, 1991). Autoradiographic measurements of the rates of chromosomal RNA synthesis in the polytene X^1 and X^2 chromosomes of *D. miranda* demonstrated that the distal 10% of the X^2 was not dosage compensated and, in contrast, its middle section representing 30% of the length of the X showed dosage compensation. Apparently, the X^2 sex chromosome differentiation is still on its way to (evolutionary) completion (Strobel *et al.*, 1978).

Certain nucleotide sequences providing preferential association with proteins, the gene products of the *mle* type are of importance in mechanisms involved in establishment of dosage compensation. This role can be assigned to repeat types whose location is specific in the sex chromosomes—such as (a) a 372-bp repeat unit, confined to the X chromosome and distributed among many of its chromosomal sites in *Drosophila* (Waring and Pollack, 1987); and (b) the simple mono- and dinucleotide repeats (AC/GT type), whose total content in a single X chromosome is twice as great as in autosome samples.

There are indications that dosage compensation can occur in the autosomal genes (also see earlier): in trisomics for certain autosomes of *Drosophila*, a number of genes had levels of activities and rates of RNA synthesis close to those observed in diploids (Devlin *et al.*, 1982, 1984, 1985a,b; Ghosh, 1985; Ghosh and Mukherjee, 1987, 1988; Ghosh *et al.*, 1989).

The mechanism by which dosage is compensated is presumably different here. However, the level of chromatin template activity in fixed squash preparations of salivary gland chromosomes of trisomics for either *2L 3L*, when assayed with *E. coli* RNA polymerase, was 30–40% of that observed in diploids (Bhadra and Chatterjee, 1986). This is consistent with the conclusion that regulation is effected at the level of transcription.

A final word may be said about the hypothesis that the DNA amplification of the gene for the male X act as a regulatory mechanism in dosage compenastion. This hypothesis is based on the observation that a band in section F of the X chromosome occasionally is thicker in males than in females of *Drosophila arizonensis*. Band size in the two sexes was expressed as the ratio of the band whose thickness varied to the area of a marker neighboring band whose thickness did not vary. The ratio in males was threefold that in females (Bicudo, 1983).

IV. ORGANIZATION OF INTERBANDS

A. General characterization of interbands

The regions of polytene chromosomes lying between two bands (the chromomeres) are called interbands (the interchromomeres). The segment of the chromatid assigned to the interband are more decondensed and dispersed than those making up the bands. For this reason, when the stained chromosomes are viewed in light, phase-contrast, or electron microscopes, the interbands appear lighter than the bands. The precise identification of the interbands in polytene chromosomes has presented difficulties because they are not conspicious. In fact, the chromosome regions are composed of a series of thin bands whose sizes close to the resolving limit of the light microscope. The lightly staining spaces are very often mistakened for the interbands, although obviously, containing a series of true interbands. It is feasible to correctly identify the interbands only by electron microscopy provided that certain mapping rules are obeyed. However, analysis of this kind has not been performed in most cases, and the features observed for the small faintly staining regions containing both the bands and the interbands have been ascribed to the interbands.

With these inaccuracies in mind the following findings will be evaluated. According to data obtained for many dipteran species, the size of interbands varies in the 0.05–0.38 μm range, most frequently being 0.1–0.2 μm. The molecular size of the interband is 0.3–3.8 kb. The content of DNA in the interbands constitutes from 1 to 26% (3 to 5%, on the average) of its total content in the polytene chromosome.

The question of how DNA may be organized in interbands has been discussed ever since polytene chromosomes were discovered. Having observed that almost all [^{3}H]thymidine radioactivity is discontinuously distributed over the bands of the stretched chromosomes, Steffensen (1963a,b) came to the conclusion that DNA is probably absent in the interbands. According to hypothesis of Laird (1980), the DNA in the interbands would be reduced in polyteny (for detailed discussion see: Zhimulev, 1992, 1996). In fact, there is considerably less DNA in the interband, the DNP fibrils composing chromatids straighten out and lie far apart (Figure 35).

The numerous data obtained with extensive cloned chromosome regions containing a series of bands and interbands are not consistent with the former hypothesis (Figure 36). It is quite apparent that DNA in a chromosome cannot be discontinuous; it traverses the entire length of both the bands and interbands. Furthermore, there is evidence that the DNA consists of one uninterrupted molecule in the chromosomes: (1) the isolation of a DNA molecule of the size of the full chromatid from its one end to the other in gentle lysates (Kavenoff and Zimm, 1973; Kavenoff *et al.*, 1974); (2) the appearance of a

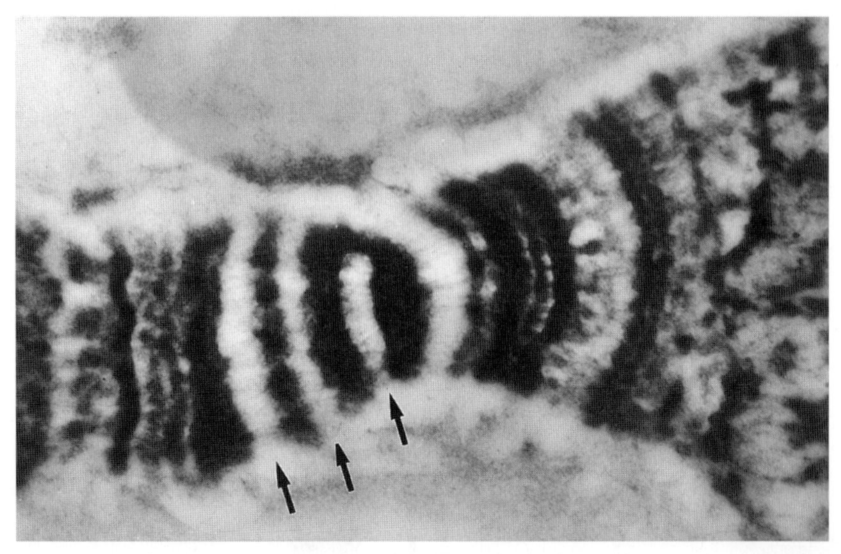

Figure 35. An electron microscope section of a salivary gland polytene chromosome of *D. melanogaster* Arrows point to the interbands (from V.F. Semeshin, unpublished observations).

bundle of chromatids of a length comparable with that of DNA in a single chromosome, when the polytene chromosome is stretched by a micromanipulator.

Cloning of a sizeable regions of the polytene chromosomes (see Figure 36) demonstrated that DNA is not underrepresented in the interband.

A specific structure has been assigned to the interband DNA enabling it to convert to the Z-form. The possible existence of Z-DNA in native preparations and the general conclusion that it may be a fixation artifact resulting from treatment with acetic acid were topics of a preceding review (Zhimulev, 1996). Furthermore, Z-DNA was not detected in the "true" interbands, but rather in the quite lengthy regions composed of thin loosened bands and interbands.

From their indirect immunofluorescence analysis, Donahue *et al.* (1986) made the conclusion that *Drosophila* histone H2A.2 preferentially binds to the interbands of polytene chromosomes. However, the acompanying phase contrast and UV fluorescence photographs show that only extensive loosened structures consisting of a series of thin bands alternating with interbands will label.

Compelling evidence that the interbands are active in transcription was provided in Section III,A. Taking into consideration the high constancy of

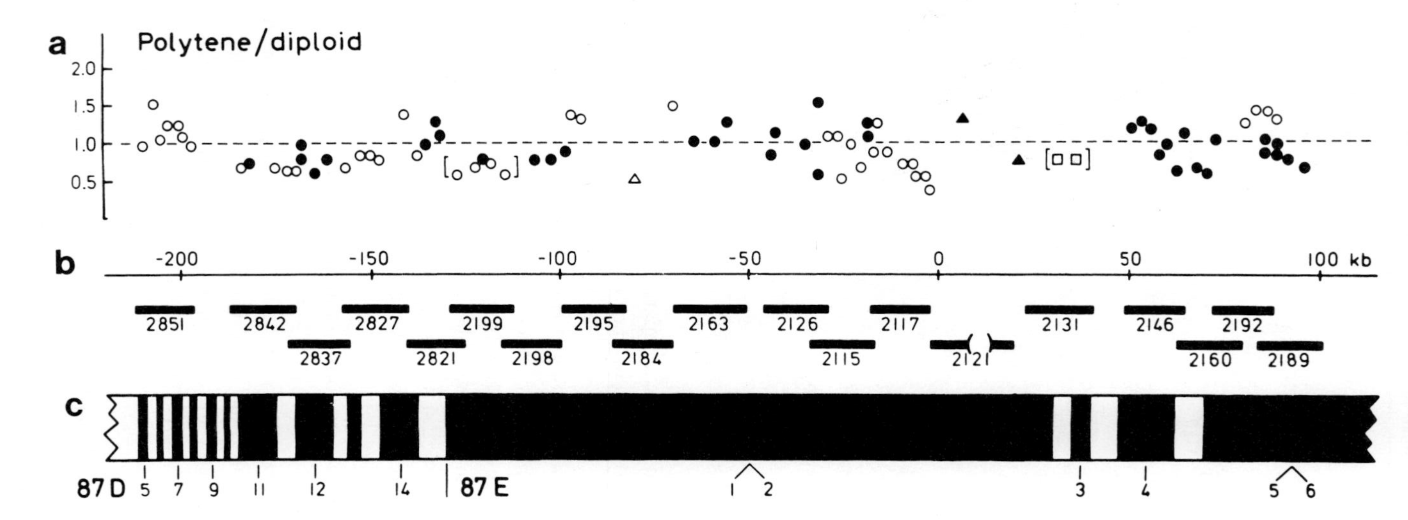

Figure 36. Relative level of polytenization in the bands and the interbands of the 87DE region of the third chromosome of *D. melanogaster*. (a) Ratio of representation of different DNA fragments in polytene chromosomes to their representation in diploid cells; (b) a physical map of the region (distance in kb); DNA clones are designated by short segments with numbers; (c) a cytological map of the region. Triangles, circles, and squares in a designate different restriction DNA fragments [reprinted with permission from *Nature* (Spierer and Spierer, 1984). Copyright (1984) Macmillan Magazines Limited].

band/interband patterns—i.e., essentially the same organization underlying the more or less decompacted segments in different chromosomes and at different stages of development—there is every reason to believe that the interbands are constantly transcriptionally active regions (Fujita, 1965; Speicer, 1974; Zhimulev and Belyaeva, 1975a,c).

B. Molecular and genetic organization of interbands

We are far from understanding the genetic function of the interbands, because the relevant information is scant.

In an attempt to explain why the number of genes needed for survival and the number of band/interbands in small regions of the polytene chromosomes are closely correlated (see Section V), Bgatov (1985; Bgatov *et al.*, 1986) suggested that the vital loci may be located in the interbands. However, electron microscopic localization of transposons inserted into polytene chromosomes did not support this hypothesis: in 19 *D. melanogaster* transformed lines transposon inserted namely into interbands and homozygotes for insertions remained viable (Semeshin *et al.*, 1986a,b, 1989a,b; 1994; 1997b).

Figure 14 (see earlier) shows how the interband between two bands at 31A and 31B is eliminated by a deletion or inactivation followed by compaction in Batumi-L strain. The strain is in a homozygous condition despite the fact that this interband is lacking.

The *Df(1)rst*^{2} deletion removes the 3C3-4-3C6-7 region, (i.e., two bands and two or three interbands). Homozygotes are viable, although expressing the phenotypes of the *rst* and *vt* mutations (Lindsley and Grell, 1968).

Indirect inferences concerning the location of the genes in the interbands can be based, in certain cases, on autoradiographs. Certain genes are repeated in tandem arrays in the genome, and they occupy thick bands in the chromosomes. The histone genes in 39DE and the 5S rRNA genes in 56F are good examples. Although cells very much require the products of these genes, puffs are not formed and [^{3}H]uridine is not incorporated at their location (Zhimulev, 1974; Zhimulev and Belyaeva, 1974, 1975b; Bonner and Pardue, 1977b; Bonner *et al.*, 1978). The occurrence of RNP-like particles, the products of transcription in the interband adjacent to the 5S rRNA cistron cluster in the 56F region in *D. hydei* (Alonso and Berendes, 1975), suggested that portion of the clustered 5S rRNA cistrons are located in the adjacent interbands. Although transcribed, they cannot form puffs because of their small size.

Poly(A)$^+$RNA isolated from the salivary gland polytene chromosomes of *D. virilis* were localized *in situ* with high resolving accuracy. The surface-spread polytene chromosomes had a linear extension three- to fourfold that obtained in conventional squash preparations. They were probed with biotinylated clone DNAs and processed for gold labeling. The autoradiographs were analyzed by electron microscopy (Figure 37). Of the 20 labeled loci, 10 or 11 were detected in the regions showing all the features of the interbands (Kress *et al.*, 1985; Kress, 1986). Since the labeled regions had not been mapped, the one appearing as a labeled interband might have contained a microband from which a "minipuff" has derived as a result of decompaction; poly(A)$^+$RNA hybridized precisely to the "minipuff."

In *C. thummi thummi, in situ* hybridization of a short (0.945-kb) DNA fragment containing the hemoglobin gene to polytene chromosomes revealed that its genome contains three labeling sites and hybridization is confined to the band–interband transitional zone in all the sites. Since certain important housekeeping genes are located in more than a single site, it was thought that the number of interbands housed the genes encoding entirely different proteins would be smaller than the number of interbands in the chromosomes (Whitmore, 1986). It should be recalled that mapping was precise only to the band–interband transition zone.

Based on thorough cytogenetic analysis, the X-linked *Notch* locus in *D. melanogaster* was localized to the 3C7 band. The recessive visible mutant

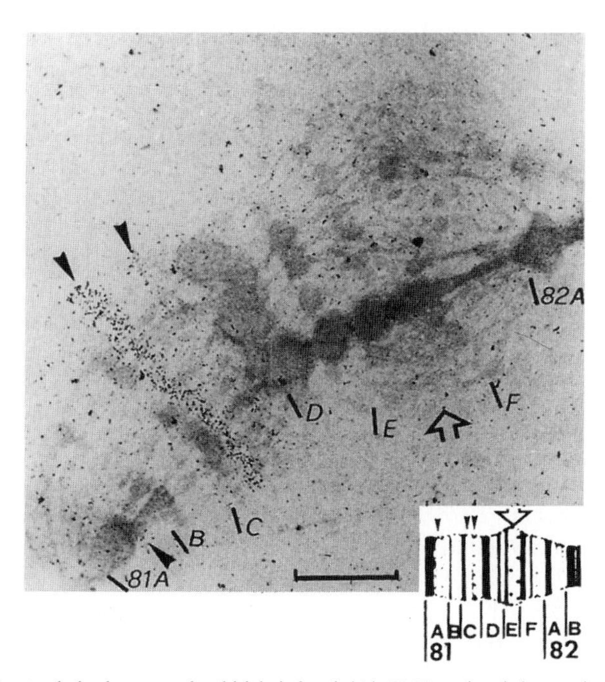

Figure 37. *In situ* hybridization of gold-labeled poly(A)$^+$RNA isolated from salivary glands to spread polytene chromosomes of *D. virilis*. Arrowheads point to labelled sites, open arrow indicates the unlabeled 81E puff (reprinted by permission from Kress *et al.*, 1985).

fa^swb at *Notch* was mapped to the leftmost end of the genetic map. The *fa^swb* mutant is associated with a cytological deletion of the interband between 3C5-6 and 3C7 under whose effect the two bands became fused (Figure 38) (Welshons and Keppy, 1975; Keppy and Welshons, 1977, 1980). A combination of high-resoluton hybridization with computer-based analysis allowed determination of the position of the hybridization signal relative to the band. When

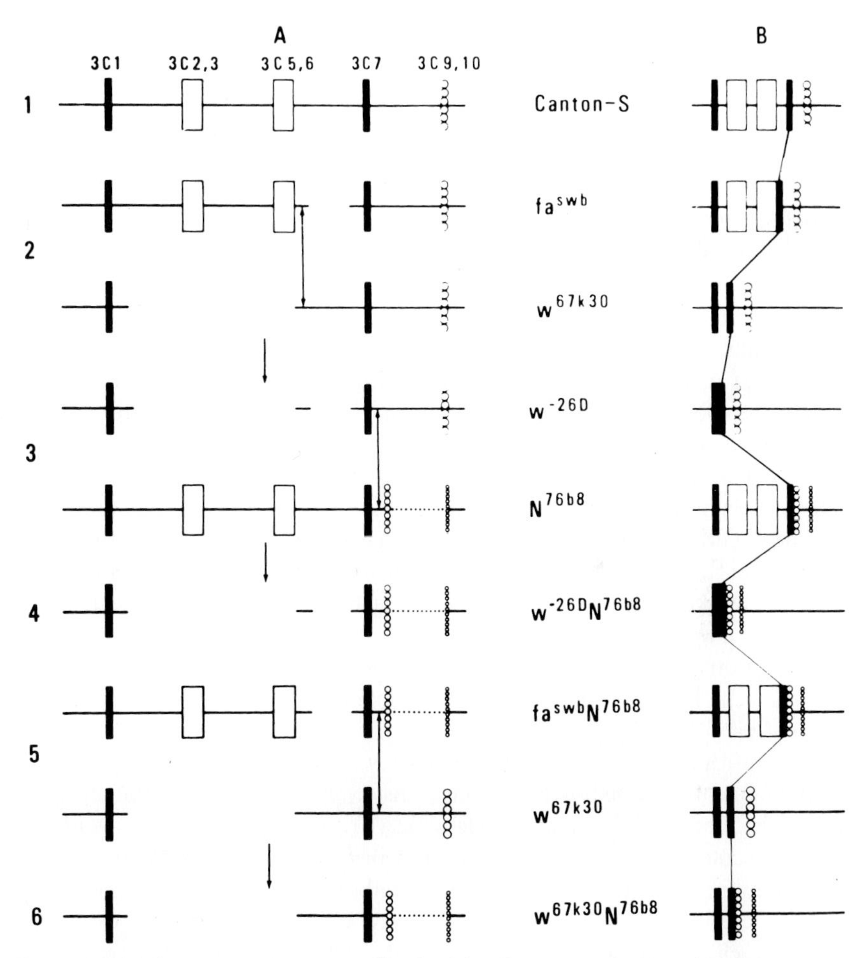

Figure 38. A schematic representation of bands of the 3C region in the normal X chromosome in *Canton-S* strain and in chromosomes with various rearrangements. (A) Disposition of the 3C1-3C9,10 bands in stretched chromosomes; (B) disposition of bands in unstretched chromosomes (reprinted by permission from Keppy and Welshons, 1980).

chromosomes were hybridized *in situ*, the *Ah1.6* probe consistently hybridized to the band 3C7 or the interband between 3C5-6 and 3C7. Thus, the gene *Notch* is contained within 3C7, and the interband 5′ to *Notch* contains its regulatory sequences as defined by the deletion mutation fa^{swb} (Rykowski *et al.*, 1988). However, how much of the clone hybridized to band or to interband was not determined precisely. The general conclusion was that the regulatory sequence of the *Notch* gene maps its 5′ end in the interband and that the remaining portion of the gene is localized in the neighboring 3C7 band. According to data obtained in this laboratory, regulatory sequences mapped to the 5′ and 3′ ends of the w^+ gene lie in the interband, and the structural gene itself resides in the 3C2–3 band (Rykowski *et al.*, 1988).

The cytogenetic organization of Balbiani ring 2 (BR2) in *C. tentans* is somewhat reminiscent of the situation above. The revelant gene, whose activation results in the formation of BR2 encoding a giant primary transcript of 37 kb, lies in the IV-3B10 band. This was demonstrated for Malpighian tubule chromosomes in which Balbiani rings are not formed and the usual bands and the interbands are localized at their sites (Derksen *et al.*, 1980; Sass, 1981).

When injected to *C. tentans* larvae, the drug DRB drastically inhibited chomosomal RNA synthesis resulting in a recondensation of the BRs and a recovery of a distinct banding pattern. Using a clone containing 152 bp of *Chironomus* DNA, *in situ* hybridization failed to localize the DRB-repressed gene to the band. It rather appeared that the gene partly resides in a neighboring faint interbandlike region (Sass, 1984). Different changes in the activities of puffs and Balbiani rings were found in *Chironomus* larva treated with dimethylsulfoxide (DMSO). However, all the puffs become inactivated after removal of DMSO and then they resumed transcriptional activity. Reactivation of the BR2 band was first observed in the brightest zone (the interband?) between the band 3B9/10, RNA polymerase II started to accumulate in the zone and showed label incorporation (Sass, 1981, 1982). Thus, it may be assumed that the 5′ end of the BR2 gene lies in the interband, and its main structural portion occupies the adjacent band.

With electron microscopic analysis of the chromosome regions carrying DNA fragments inserted by P-element transformation, studies on the organization of the interbands became more feasible. A new approach allowed avoidance of the difficulties arising in mapping and identifying the true interbands. In the case when transposon inserts into real interband, a new cytologically visible band appears (Figure 39). A genomic library was constructed from the transformed stock and probed with the insert DNA. The latter was used as a probe for screening the full-length native interband. The sequences of the interband nearest to the insertion site (497 bp distal and 792 bp proximal to it) were determined. The features of the isolated clone were as follows:

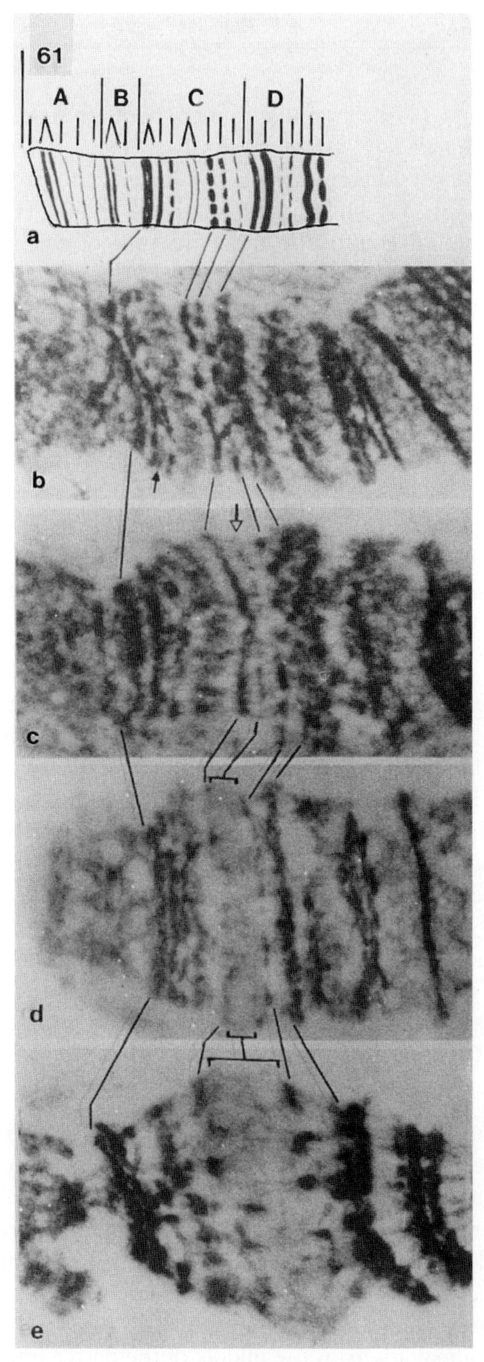

Figure 39. Electron microscopic mapping of the 61C region of chromosome 3L of *D. melanogaster* (a) Bridges' map; (b) the 61C region on the control strain without inserted transposon; (c) the band formed in the 61C region (arrow) as a result of compaction of transposon DNA; (d and e) induction of the heat shock puff during activation of the *hsp70* gene in the transposon (reprinted by permission from Semeshin *et al.*, 1989b).

1. It is unique.
2. It is A-T rich (53.4%), containing short 8- to 50-bp repeats showing 40–50% homology.
3. It contains motifs of different regulatory elements and two overlapping reading frames, one 555 bp long, the other 354 bp long (Figure 40). Of the motifs, one is homologous to the attachment site of topoisomerase II and the T-SAR-domain (the attachment region of the nuclear scaffold (Demakov *et al.*, 1991; 1993). It was demonstrated that the SARs are present at the 3′ and 5′ flanks of the gene, and in the promoter regions and at the boundaries between the transcriptionally active and inactive heterochromatin (Mirkovitch *et al.*, 1987).
4. Analysis of codon frequencies in open reading frames identified in the interband DNA (see Figure 40) and in 269 kb encoding proteins in *Drosophila* (Ashburner, 1989) revealed significant differences (Demakov *et al.*, 1993). This may be taken to mean that either information for protein synthesis is not encoded in these open reading frames or that they encode proteins using very rare codons. Similar characteristics were found for the interbands 3C1/3C2-3, 86B4/86B6, and 85D9/85D10 (Demakov *et al.*, 1997; Schwartz *et al.*, 1998).

Uncertainties in the interpretation of the above data derive from the failure to localize the part of the band to be considered. Figure 39 shows that a new band results from transposon DNA insertion. Although the P element inserts into the genome in any orientation, its DNA is transcribed only in one direction (O'Hare and Rubin, 1983; Karess and Rubin, 1984). Since the P-element promoter is weak and it can be transcribed actively throughout development (Laski *et al.*, 1986; O'Kane and Gehring, 1987), its DNA can be in a decompacted state. Insertion of the P element into the interband splits it because the P-element material is decondensed. When a P-element insertion occurs in the interband, the whole region increases in length (Semeshin *et al.*, 1989a,b).

Taking all this into consideration, it is obvious that the pattern of the 61C7 and 61C8 bands and the interband (see Figure 39) can result from one of the following events:

1. The transposon may be inserted in the 61C7 band/interband transition zone. If so, the 1.7-kb fragment at the 5′ end of the P element may become decompacted and give rise to a new pseudointerband. Then a new band composed of transposon material would follow and the native 61C7/8 interband was made to move away from its "own" 61C7 band by the transposon. In such a case, the interband sequence would be to the right of the insertion site (Figure 39).
2. The transposon may insert in the middle of the native interband. The interband would simply be extended; the increment would be equal to the length of the decompacted part of the P element at the 5′ end.

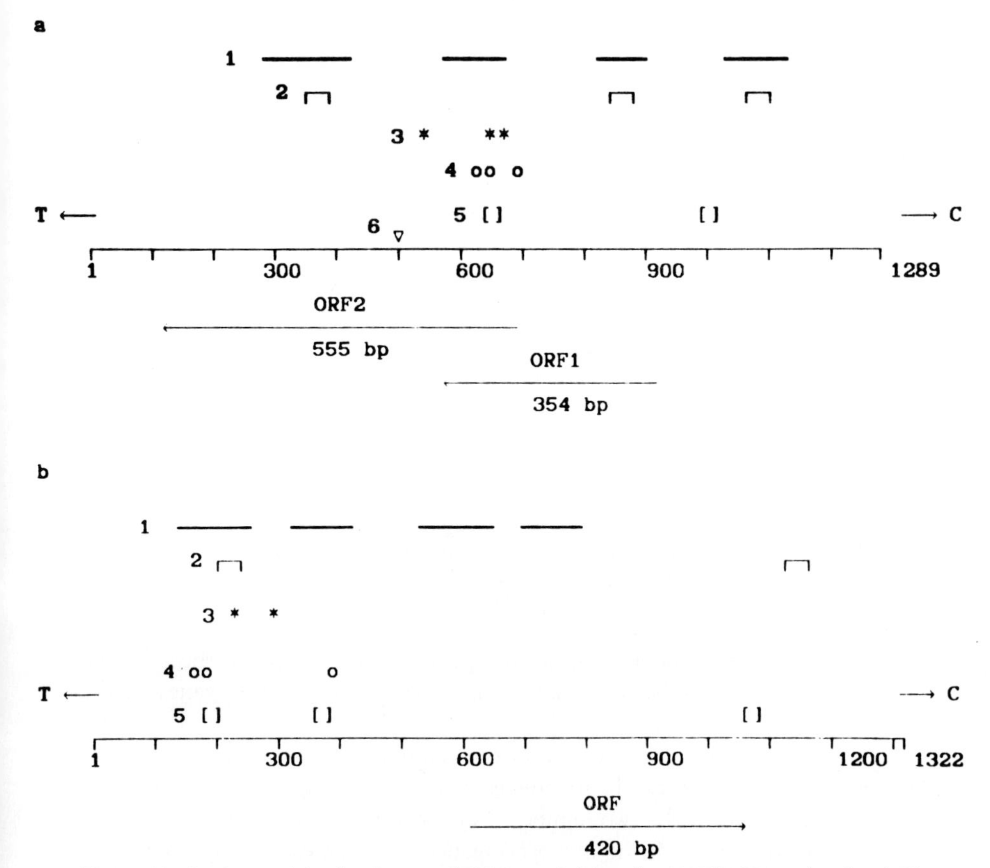

Figure 40. A scheme of the distribution of DNA motifs (a) in the 61C7/61C8 interband and (b) in a fragment removed by the *Df(1)fa*^{swb} deletion. 1, regions of direct and inverted repeats; 2, zDNA; 3, regions of topoisomerase II binding; 4, autonomously replicating sites; 5, regions of scaffold attachment; 6, insertion site. ORF, open reading frame. Arrows T and C point to the telomere and centromere, respectively. The data on the molecular organization of the interband DNA removed by the *fa*^{swb} deletion are from the research (Artavanis-Tsakonas *et al.*, 1983; Kidd *et al.*, 1986; Ramos *et al.*, 1989) and treated by S.A. Demakov (reprinted by permission from Demakov *et al.*, 1991, 1993).

This method did not make it possible to define at what distance from the insertion site the interband DNA ends and the material of the nearest neighboring bands (61C7 and 61C8) starts. For this reason, it was accepted that the interband size varies from 0.3–3.8 kb, and analysis was restricted to 1289 bp immediately adjacent to the insertion.

It was difficult to interpret the above results. However, a comparison of the organization of four interbands from different regions (Figure 40) revealed striking similarities: short repeats, fragments capable of converting into the Z-form of DNA, recognition sites for topoisomerase II, autonomous replication sites (ARS), and T-rich sequences, characteristic of the nuclear template associated DNA regions (the T-SAR-boxes) (Demakov *et al.*, 1997).

Probably interbands may have some controlling functions. When insertions of P-transposon obtained with help of an enhancer trapping method occur into interbands, *lacZ* reporter gene expresses in various organs (Semeshin *et al.*, 1997b).

In some cases the interband can completely consist of material of mobile element. Insertion of the P element splits the 10A1-2 band, and a new interband, containing P-element DNA, appears between two splitted parts (see Figure 20).

V. MOLECULAR AND GENETIC ORGANIZATION OF THE CHROMOMERES OF POLYTENE CHROMOSOMES

One approach to an understanding the genetic organization of the polytene chromosome bands is based on comparisons of the number of genes in the genome with the number of bands in polytene chromosomes—or, such comparisons as made in distinct regions of the genome. The distribution density of the genes in the region can be roughly estimated and compared with that of the bands by determining the two numbers (review: Zhimulev, 1994a).

Band number in the *Drosophila* genome can be estimated at 3.5 thousand, because about 1.5 thousand bands Bridges drew in his map as "doublets" are now generally recognized as an artifact (reviewed by Zhimulev, 1992, 1996). The problem of the correct gene number shall be eventually resolved at the molecular level when the entire *Drosophila* genome is deciphered.

A. Relation of band and gene numbers in chromosome regions

1. Calculations of the number of genes in the *Drosophila* genome

In *Drosophila,* estimates of the number of genes were mainly based on the frequencies of spontaneous and induced mutations since the 1920s (Lefevre, 1974c). Using this criterion, the number of genes on the X chromosomes was predicted to be in the range of 500–1280 (Muller and Altenberg, 1919; Muller, 1929; Gowen and Gay, 1933; Demerec, 1934; Lea, 1947).

In 1937, Alikhanyan (1937) and Shapiro (1937) developed a method allowing recovery of genes in small chromosome segments. Lethal mutations were induced in the X chromosome regions covered by short duplications. All males carrying lethals in the other chromosomes not covered by duplications died and only those carrying mutations in the cytological intervals of the duplication were selected. The number of bands in the duplication can be counted and compared with that of detected genes. This method was called "genetic exhaustion of a region" (Alikhanyan, 1937; Shapiro, 1937) and, later, "saturation with mutations." Alikhanyan (1937) extended his data on the number of genes in a small chromosome region to the entire X chromosome and thus obtained a value of 1231.

The "saturation method" has been extensively used since the 1960s and 1970s (Judd, 1962; Lyfschytz, 1967; Kaufman *et al.*, 1969), and saturation studies of many chromosome regions were performed in *Drosophila* (see Section V,A,2). Analysis of the obtained data led to the conclusion that total gene number of *Drosophila* is five to seven thousand (Judd *et al.*, 1972; Bishop, 1974; Hochman, 1974; Lewin, 1974 1975; Garcia-Bellido and Ripoll, 1978; Young and Judd, 1978; Spradling and Rubin, 1981). A good commentary of attempts to estimate gene number in *Drosophila* as 5000, was given by Olenov (1974) (see Section II,B).

Attempts were made to estimate the number of genes on the basis of data on chromosomal rearrangements (Herskowitz, 1950). Having found that most X-ray-induced mutations at the *white* locus are associated with chromosomal rearrangements, Herskowitz (1950) obtained an estimate of 726 loci by dividing the observed number of chromosomal rearrangements in the whole chromosome by the number of mutations induced at *white;* he thought that this value corresponded to the number of genes in the X chromosome. Lefevre (1974b) presented cytogenetic data supporting the one band–one gene hypothesis. He suggested that rearrangements affecting the same locus may be differently expressed in the phenotype—lethal, visibly mutant, or nonmutant. Rearrangements affecting the same band, but containing different functions, would complement one another. However, the rearrangements induced in the same bands (1A5-6, 1B3, 3C2, 3C5, 3F1, 8D8-9, 15F1-2, and 20A3) all showed the same phenotype. From these data it was inferred that the number of genes can approximately correspond to the number of bands; that is, to about five thousand on Bridges' map. It should be noted that only the genes required for survival (whose mutations are lethal) were taken into account in calculations yielding an estimate of 5000 genes in the *Drosophila.* However, the entire genome has genes other than the essential ones. For example, enzyme "null alleles" were detected in 13 of the 25 examined loci of *Drosophila;* individuals homozygous for these alleles were not lethal (O'Brien, 1973). In surveys of populations of *Drosophila* from North Carolina and Great Britain, enzyme-null alleles were recovered at 13 of the 25 autosomal loci. Individuals homozygous or hemizygous for 12 of the 13 null alleles were viable and fertile (Voelker *et al.*, 1980; Burkhart *et al.*, 1984). Also, at least 30 loci affecting flight ability were identified in addition to the lethals usually induced in the X chromosome (Homyk and Sheppard, 1977; Koana and Hotta, 1978; Homyk *et al.*, 1980), and an early study of another "nonvital" phenotype identified about 50 loci controlling female fertility (Mohler, 1977). Many other genes are known to be involved in the fly's biological clock, behavioral responses, taste sensation, vision, and other functions that are inessential for viability in the laboratory. Mutations in many genes are phenotypically undetectable by elementary screens, and they are identifiable only as DNA-fragment complementary to the RNA transcripts (see Section V,B).

Cytogenetic analysis of more than 1500 lethal X chromosomes demonstrated that the vital loci are nonuniformly distributed in the different regions of the X chromosome. The distribution was not correlated with either the number of bands and DNA content. Thus, the vital loci were numerous in the 3AB region; in contrast, only two were found in 3CD, a region of equal length. The number of vital loci was large in the 1B, 1F-2A, 10A, 11A, and 19EF regions and other ones; those with a small number of vital loci included 6EF and 10B-10E (Lefevre, 1981). Attempts were made to reveal the nonessential DNA sequences whose rearrangement breakpoints are not associated with phenotypically observable effect and are nonlethal (Young and Judd, 1978; Zhimulev *et al.*, 1981c, 1983). Difficulties in defining whether the breakpoint is in the true essential locus, in the single gene in a repeat cluster, or in the spacing between the genes limit the interpretation of these data.

Approaches other than those based on mutations that are associated with chromosome rearrangement breakpoints yielded different calculated estimates of gene number in *Drosophila*. Zhimulev and Belyaeva (1975a) emphasized that the total genome length is 280 centimorgans when expressed as recombinational units. Recombinational size of genes, such as *vermilion* and *rosy* were determined; their products are known, and they are 5×10^{-3} and 9×10^{-3} centimorgans respectively. The total number of genes can reach 30,000–50,000, if the genes are densely packed (i.e., >2 or 3 per band). Gilbert *et al.* (1984) reasoned that—if the total number of *Drosophila* genes is 5 to 10 thousand and if they were randomly distributed—there would be an average of one gene for every 15–30 kb. However, there are many lines of evidence indicating that gene density is high in the DNA flanking many cloned genes: clusters of transcript units contain up to one transcript every 2–4 kb (see Section V,B). Extrapolation over the entire genome gives approximately 30–50 thousand genes.

Based on measurements of renaturation kinetics of denaturated DNA, the estimated nucleotide sequence diversity is equal to about 120,000 different genes in *D. hydei*, sufficient to specify 50–100 different cistrons in a band

(Laird, 1971). Spacing between genes, introns, and repetitive nucleotide sequences are not taken into account in these estimates, which look like overestimates.

The kinetics of hybridization of cDNA with its template poly(A)$^+$RNA supported the notion that the number of genes considerably exceeds the number of bands. The degree of sequence complexity of poly(A)$^+$RNA from *Drosophila* cell culture suggested that there exist at least 7000 diverse transcripts with a molecular weight of 4×10^5 (Levy and McCarthy, 1975; Levy *et al.*, 1976; Izquierdo and Bishop, 1979).

According to Turner and Laird's data (1973) more than 30% of the unique sequences in the *Drosophila* genome are represented by transcripts in pupae. This may be taken to mean that about 23 thousand genes are transcribed only at this developmental stage.

Analysis of the molecular composition of the genome of the cellular slime mold *Dictyostelium discoideum* yielded interesting comparative results. The unique, nonrepetitive portion of the genome is of 3×10^4 kb (Firtel and Bonner, 1972), while the mRNA precursor is 1.0–1.1 kb (Firtel and Lodish, 1973). Thus, there may be about 30 thousand genes in the *D. discoideum* genome. About 56% of the unique DNA is transcribed during the 26-hr developmental cycle (16 to 17 thousand RNA transcripts of ~1 kb each) (Firtel, 1972). About 20% of the unique sequences are represented in cellular RNA (5.7 thousand of RNA transcripts) at all developmental stages. It is suggested that these nucleotides may be used for housekeeping functions (Firtel, 1972). Their number closely corresponds to that of the interbands in polytene chromosomes.

Subsequently, it became apparent that nonpolyadenylated RNAs represent approximately 70% of sequence diversity in polysomal RNA. Measurements of poly(A)$^-$RNA appeared warranted. Calculated estimates based on analysis of the kinetics of hybridization of mRNAs from prepupae of *D. melanogaster* corresponded to approximately 17 thousand transcripts of average size of 1250 bp. Thus, the calculated number exceeds the number of polytene chromosome bands by about three times (Zimmerman *et al.*, 1980).

Current estimates on the number of genes vary in broad limits. Data on the genes now identified in *Drosophila* lead to the number 6898 (Gelbart *et al.*, 1994). Lewin (1994) suggests that the total number is about 8000. Nusslein-Volhard (1994) admits existence of about 6000 essential genes, of which 5000 mutate to lethality and about 1000 to sterility. According to Miklos and Rubin (1996) the gene number is closer to 12,000, in part based on estimates of the number of transcription units. So, estimates have ranged from 3,000 to 5,000 genes (Lewin, 1974), to those given by Zhimulev and Belyaeva (1975a,c) that the number of genes in *Drosophila* "is not less than 20,000–30,000."

This value is almost one order of magnitude more than the number of

Table 4. Analysis of Data on Genetic Saturation of the Chromosome Regions in *Drosophila*

Chromosome region	Complementation groups	Number of bands[a]	Maximum and minimum ratio of the number of cistrons to that of bands	Notes
1B1-14, the X chromosome	14 or 15 complementation groups of lethals, *M(1)Bld* lethal, the loci of the visible *yellow*, *achaete*, *scute*, *silver* (Lefevre, 1974a, 1976b)	14(1) (Lindsley and Grell, 1968); 8–9 (Berendes, 1970); 10 (King, 1975)	19–20/8–14 = 2.5–1.3	The region is "unsaturated" because only 20 mutations giving rise to 15 complementation groups recovered (Lefevre, 1974a, 1976b)
1B10-1D6-E1, the X chromosome	10 genes, including *tw*, *su(s)*, *brc*, *M(1)B* (Voelker *et al.*, 1989)	15(4) (Lindsley and Zimm, 1992)	10/11–15 = 0.9–0.7	
1E1-2-2A1-2, the X chromosome	10 complementation groups of lethals, 3 single lethals loci of mutations with visible manifestation, *su(w^a)*, *sta* (Rayle, 1972; Lefevre, 1976b)	11(4) (Lindsley and Grell, 1968); 7–8 (Berendes, 1970); 11 (King, 1975)	12–15/7–11 = 2.1–1.1	1. The region may be regarded as unsaturated because of presence of three single lethals 2. Cytological limit of *Df(1)sta* used by Rayle is wider: 1E1-2-2B3-4 (Belyaeva *et al.*, 1982) 3. 10 complementation groups were detected in the 1E1-2-2A1-2 regions in independent experiments (Belyaeva *et al.*, 1980a,b; 1982)
2B3-5-2B7-8 (the region of the 2B5-6 bands and the edges of neighboring 2B3-5 and, possibly, 2B7-8 bands), the X chromosome	Three complementation groups of visible and lethal *sta*, *dor*, and *swi* mutations and *BR-C* (the *ecs* locus) (Belyaeva *et al.*, 1980a,b; Aizenzon *et al.*, 1980; Melnick *et al.*, 1993)	2–4(2) (Lindsley and Grell, 1968); 2–4 (King, 1975); 2–3 (Belyaeva *et al.*, 1980a,b)	4/2–3 = 2.0–1.3	

2B17-2C1 (region between *Dpy*^{2}*Y67g19.1* and T(1;3)(?) breakpoints)	Three complementation groups of lethal mutations (Schalet, 1983)	1 (Lindsley and Zimm, 1992)	3 : 1	
2C1-2-2F6, the X chromosome	17 complementation groups, *ph*, *fs(1)pcx*, *pn*, *fs(1)K10*, *kz* genes (Perrimon *et al.*, 1984, 1985)	29(6) Lindsley and Zimm, 1992)	22/23–29 = 1.0–0.76	
2D3-2F5, the X chromosome	8 complementation groups of lethals, 3 single lethals, the locus of mutation with visible expression of *prune* (Gvozdev *et al.*, 1973–1977; Alatortsev, 1997), the locus affecting female fertility *mel(1)R1* (Romans *et al.*, 1976)	12(2) (Lindsley and Grell 1968); 7 (Berendes, 1970); 16 (King, 1975)	12–13/7–16 = 1.9–0.8	1. Presence of 16 bands (King, 1975) not documented with the photographs 2. The *prune* locus most likely mapped to the region of the 2E1-2 band or near it (Lefevre, 1976a; Ilyina *et al.*, 1980) 3. *mel(1)R1* not tested for allelism to other mutations 4. All mutations not tested for allelism to other mutations (Perrimon *et al.*, 1984, 1985)
2F1-3A4, the X chromosome	11 lethals classified into six complementation groups (Nishida *et al.*, 1988)	10(2) (Lindsley and Zimm, 1992)	~0,7	

(*continues*)

Table 4. *Continued*

Chromosome region	Complementation groups	Number of bands[a]	Maximum and minimum ratio of the number of cistrons to that of bands	Notes
3A1-10, the X chromosome	6 lethal complementation groups, the *gt*, *tko*, *z* loci (Judd *et al.*, 1972; Shannon *et al.*, 1972a,b; Judd and Young, 1974; Lim and Snyder, 1974; Lui and Lim, 1975; Judd, 1975, 1976). Non-lethal *T(1;Y)p22* break (Young and Judd, 1978)	10(1) (Lindsley and Grell, 1968); 7 (Berendes, 1970); 8–9 (Beermann, 1972); 10–13 (King, 1975)	9/7–13 = 1.3–0.7	1. Presence of 13 bands (King, 1975) not documented by photographs 2. Increased intensity of *in situ* hybridization of total and complementary RNA to the 3A1-4 region (Gvozdev *et al.*, 1980) evidences for the presence of repeats in the region
3A2-8, the X chromosome	19 lethals (Bhakta and Mukherjee, 1984, 1991)			Allelism to B. Judd's lethal mutations was not identified
3B1-4, the X chromosome	7 complementation groups of nonlethal mutations (Judd, 1976), the *per*, locus controlling the emergence of adults (Konopka and Benzer, 1971), 2 loci of female sterility mutations (Judd and Young, 1974; Young and Judd, 1978), the *paralog* locus controlling the differenciation of oocytes in females and interaction with neighboring loci (Thierry-Mieg, 1982)	4(1) (Lindsley and Grell, 1968; 5 (Berendes, 1970); 5 (Beermann, 1972); 6 (King, 1975)	11/4–6 = 2.8–1.8	Presence of six bands not documented

3C1-3C7, the X chromosome	The loci of lethal mutations *Notch* and *1(1)3C3*, the loci of the visible *white*, *roughest*, and *verticals* mutations (Gersh, 1965; Lefevre and Green, 1972; Lefevre, 1974a)	7(2) (Lindsley and Grell, 1968); 4 (Berendes, 1970); 5–7 (King, 1975); 4 (Lefevre and Green, 1972)		1. According to the standard map (Lindsley and Grell, 1968), 14 loci of mutations located between *w* and *spl*, induced by various authors and not tested for allelism (see also: Lefevre and Green, 1972). 2. The region between the 3C2-3 and 3C5-6 bands is the region of intercalary heterochromatin (Beermann, 1972a; Zhimulev *et al.*, 1982b); 3. *Df(1)Su(z)J93* removing *rst* and *vt* but not removing *w* and *N* acts as a dominant suppressor of z^1, it also exhibits *rst* and *vt*, reduced viability and female sterility (Judd, 1995).
3C8-3D3, the X chromosome	No loci of visible or lethal mutations (Lefevre, 1974a), the *Sgs-4* locus, encoding the synthesis of polypeptides of salivary gland secretion (Korge, 1977a)	8(2) (Lindsley and Grell, 1968); 7 (King, 1975)		
3D4, the X chromosome	Two nonessential genes are identified within the band: *sam*, affecting male fertility and *dnc*, controlling learning capacity, female fertility, and cAMP synthesis by specific phosphodiesterase (Salz *et al.*, 1982)	1 (Lindsley and Grell, 1968)	2 : 1	1. Vital genes not identified because homozygotes for band deficiency survive 2. *sam* located proximally to *dnc* at a distance of 0.24 centimorgans 3. See Section V,B,6

(continues)

Table 4. *Continued*

Chromosome region	Complementation groups	Number of bands[a]	Maximum and minimum ratio of the number of cistrons to that of bands	Notes
The 4C5-6 band, the X chromosome	The *lac*, *Qd*, *bi*, *omb*, *rb* genes (Banga *et al.*, 1986)	2(1) (Lindsley and Zimm, 1992)	5/1–2 = 5–2.5	
6E1-7B1-2, the X chromosome	9 complementation groups of lethal mutations, the *cm* locus controlling eye color and the *Sxl*. The loci nonrandomly distributed: in the 6E1-6F4 region 6 complementation groups	10(4) (Lindsley and Zimm, 1992)	6/6–10 = 1.0–0.6	
	In the 6F4-5-7A1-2 one complementation group	9(4) (Lindsley and Zimm, 1992)	1/5–9 = 0.2–0.1	
	In the 7A3-7B1-2 four complementation groups (Nicklas and Cline, 1983; Sheen *et al.*, 1993)	8(3) (Lindsley and Zimm, 1992)	4/5–8 = 0.8–0.5	
The 7F1 band and adjacent interband (King *et al.*, 1986)	The *pt*, *otu*, *1(1)B4* genes and the gene for chorion protein (Spradling, 1980, 1981; Mohler and Carroll, 1984; King *et al.*, 1986)	1	4:1	

The 9F12-10A7 region, the X chromosome	7 loci of lethal mutations, loci controlling: eye color—*v*, sensitivity to ultraviolet light—*sev*, female fertility—*fs(1)BP2*, male fertility—*ms(1)BP6*; the *slm* locus, whose mutations delay development (Zhimulev *et al.*, 1981c, 1983, 1987a,b) The loci unevenly distributed in the region; three loci located in both 9F12 and 10A1-2 bands, in the 10A3-10A4-5 region represented by two bands there are also three loci. One gene identified in each of the 9F13, 10A6, and 10A7 regions (Zhimulev *et al.*, 1987a,b).	7 (Semeshin *et al.*, 1979; Zhimulev *et et al.*, 1981f)	12/7 = 1.7
The 10A8-9-10B17 region, the X chromosome	10 loci of lethal mutations, two loci of visible mutations: *tu(1)Sz*^{ts} and *ny*, the *hfs* locus whose decrease in dose produces female sterility (Lefevre, 1969; Zhimulev *et al.*, 1981c, 1987a,b; Dybas *et al.*, 1983; Geer *et al.*, 1983; Zhimulev, 1994b).	15 (Zhimulev *et al.*, 1981f)	13/15 ~ 1:1

(continues)

Table 4. *Continued*

Chromosome region	Complementation groups	Number of bands[a]	Maximum and minimum ratio of the number of cistrons to that of bands	Notes
	The loci unevenly distributed in the regions: three loci *hfs*, *1(1)L1*, and *1(1)EM16* located in the regions of two bands: 10A8-9 and 10A10-11. No genes identified in the 10B1-2 and 10B3 bands, 10 loci in the 10B4-10B17 region in 10 bands.			
The 10B17-10E1-2, region, the X chromosome	12 loci, including the structural gene encoding RNA polymerase II. Two genes and 6 DNA fragments encoding transcripts located in the 10C1-2 region (Voelker *et al.*, 1985)	18(7) (Lindsley and Zimm, 1992)	12/18–11 = 0.7–1.1	
The 14C1-2-14C6-7 region, X chromosome	6 essential genes and *non-A*, *mei41* (Banga *et al.*, 1995)	7(3) (Lindsley and Zimm, 1992)	8/7–4 = 1,1–2	
The 15A1 band (cytological location of the *rudimentary* locus (Lefevre, 1976a; Segraves *et al.*, 1983), the X chromosome	The locus encodes 4 enzyme activities of the primary steps of the biosynthesis of pyrimidines (Judd, 1976; Jarry, 1979)	1 (Lindsley and Grell, 1968)	1 : 1	Four enzyme activities associated with one polypeptide (Rawls and Fristrom, 1975; Segraves *et al.*, 1984; Freund and Jarry, 1987; Scott, 1987)
15A-F region, the X chromosome	17 genes (McKim *et al.*, 1996)			

The 17 region, the X chromosome	17 genes of which 71% of importance to oogenesis and embryogenesis (Eberl *et al.*, 1992)			Obviously this region is not saturated
The 18F4-5-19D3 region, the X chromosome	4 loci of mutations with visible expression and 5 loci of lethal mutations (Schalet and Lefevre, 1976)	20(8) (Lindsley and Grell, 1968)	9/12–20 = 0.8–0.5	At least, three essential loci in the region mutated only once
The 19E1-2-20A1-2, region, the X chromosome	14 loci of lethal mutations, 3 loci controlling male fertility, the *Pas, mal, mell, runt, R-9-29, R-9-28, EC 255, lf, vao, unc, lfl, eo* genes (Lifschytz and Falk, 1968a,b, 1969; Schalet and Finnerty, 1968; Falk, 1970; Lifschytz, 1971, 1973, 1978; Schalet and Lefevre, 1976; Lifschytz and Yacobovitz, 1978; Yamamoto and Miklos, 1984; Schalet, 1986; Miklos *et al.*, 1987; Baird *et al.*, 1990), the *flamenco* gene (Prud'homme *et al.*, 1995) In the 19F region six bands contain 20 complementation groups (Falk, 1970)	16(7) (Lindsley and Grell, 1968)		1. Little information on allelism of mutations induced by different authors 2. The 19E1-4 and 20A1-2 regions are sites of intercalary heterochromatin
The 21C6-D1-22A6-22B1 region, 2L chromosome	18 complementation groups of lethal mutations, the *ds, S/ast, LSP1β* genes (Roberts *et al.*, 1985)	21(4) (Lindsley and Grell, 1968)	18/17–21 = 1.0–0.9	

(continues)

Table 4. *Continued*

Chromosome region	Complementation groups	Number of bands[a]	Maximum and minimum ratio of the number of cistrons to that of bands	Notes
The 24D4-25F2, 2L chromosome	42 loci (112 lethal and visible mutations) (Szidonya and Reuter, 1988)	60(17) (Lindsley and Zimm, 1992) 48 (Semeshin and Szidonya, 1985)	42/43-60 = 1.0–0.7	
The 25A-27D region, 2L chromosome	18 complementation groups of lethal mutations, the *dp*, *M(2)S1*, *tkv*, *1 Gpdh*, spd^{fg} loci (Kotarski *et al.*, 1983)			The region obviously not saturated
	In 25E—seven essential genes and *clot*	2 (Kotarski *et al.*, 1983)	8/2 = 4	Cytological map given by Kotarski *et al.* (1983) does not correspond to standard map (Lindsley and Zimm, 1992)
	In 25F—five essential genes and a Gpdh	3 (Kotarski *et al.*, 1983)	6/3 = 2	
	In 26A—four essential genes	1 (Kotarski *et al.*, 1983)	4/1 = 4	
The 25D-F region, 2L chromosome	5 loci of lethal mutations, *hel* and *nlam* genes (Eberl *et al.*, 1997)			
The 31A-32A region, 2L chromosome	52 complementation groups	43(33) (Lindsley and Zimm, 1992)	52/33–43 = 1,6–1,2	
34D1-2-35B10, 2L chromosome	21 complementation groups of lethal and 9 groups of visible mutations (O'Donnel *et al.*, 1977; Woodruff and Ashburner, 1979a,b; Ashburner *et al.*, 1982a,b)	33(9) (Lindsley and Grell 1968)	30/24–33 = 1.3–1.0	1. Region not saturated because of 27 lethal loci 11 had a single mutation each 2. The 34E-35A region is the site of intercalary heterochromatin
The 35A1-B5 region, 2L chromosome	11 genes (Roote *et al.*, 1996)	9(2) (Lindsley and Zimm, 1992)	11/9–7 = 1,2–1,6	

The 36A6-7-36F1-2, region, 2L chromosome	10 complementation groups of lethal mutations, the *Myosin*, *pillow*, *dorsal*, *Bicaudal D*, *quail*, *reduced ocelli*, *nina D*, *kelch* (Steward and Nusslein-Volhard, 1986)	37(11) (Lindsley and Zimm, 1992)	19/26–37 = 0.7–0.5	
The 37B9-C1-37D1-2, 2L chromosome	12 complementation groups of lethal mutations and the *hook* locus (Marsh and Wright, 1979; Wright *et al.*, 1981; Wright and Beermann, 1981)	8–12 (distal break of the deletion determined with an error of 4 bands)	13/8–12 = 1.6–1.1	
	12, and possibly, all the 13 loci are located in a narrower region—37B10-C4 (Wright *et al.*, 1981)	7(2) (Lindsley and Zimm, 1992)	12(13)/5–7 = 2.6–1.7	
	10 complementation groups in the 37B13-37C1-3 region (Gilbert *et al.*, 1984)	2–3 (Lindsley and Zimm, 1992)	10/2–3 = 5–3.3	
	Presumably, only 1 locus is mapped in the 37C5-7 region (Wright *et al.*, 1981)	3 (Lindsley and Zimm, 1992)	1/3 = 0.33	
	No loci identified in the 37D1-2 region (Wright *et al.*, 1981)	2(1) (Lindsley and Zimm, 1992)	0	The 37D1-2 band is late replicating which is evidence of the presence of intercalary heterochromatin
The 39A-39F region, 2L chromosome	A cluster of the histone genes and 13–14 complementation groups of lethal mutations (Siegel, 1981)			

(continues)

Table 4. *Continued*

Chromosome region	Complementation groups	Number of bands[a]	Maximum and minimum ratio of the number of cistrons to that of bands	Notes
The 42A10-12-42B1 region, 2R-chromosome	Twenty mutations for 4 loci, including *EcR* (Bender, 1996)	7–9 bands		Region is not saturated
The 44F-45C5-6 region, 2R chromosome	The *rad(2)201*[G1] and 85 mutations induced by α-radiation, 18 complementation groups (Konev *et al.*, 1991a,b; 1994a,b)			Region is not saturated
The 49D3-4-50A1-2 region, 2R chromosome	17 complementation groups of lethal mutations and the *vestigital* gene (Lasko and Pardue, 1988)			Region is not saturated
The 50A region, 2R chromosome	3 genes: *1(2)P*, *cnn*, *drk* (Heuer *et al.*, 1995)	15(3)		Region is not saturated
The 57B17-57F6, 2R chromosome	18 complementation groups of mutations including *tud* and *Pu* (O'Donnell *et al.*, 1989)	41(9) (Lindsley and Zimm, 1992)	16/41–32 = 0.4–0.5	Region is not saturated
The 59F6-7, 2R chromosome	*1(2)tid* gene (Gundacker *et al.*, 1993; Kurzik-Dumke *et al.*, 1995)	2 (Lindsley and Zimm, 1992)	0.5	
The 62B3-62D3-4 region, 2R chromosome	14 genes (Sliter *et al.*, 1989; Wang *et al.*, 1994)	31 (Sorsa *et al.*, 1984)		Region is not saturated
The 63B region, 3L chromosome	11 complementation groups (Wohlwill and Bonner, 1991)	14(1) (Lindsley and Zimm, 1992)	11/14–13 = 0.9–0.8	
The 63E8-64A region, 3L chromosome	20 complementation groups (Harrison *et al.*, 1995)	21(4) (Lindsley and Zimm, 1992)	20/17–21 = 1.2–1.0	

The 67A2-D11-13 region, 3L chromosome	24 essential loci (Leicht and Bonner, 1986, 1988)	45(4) (Lindsley and Zimm, 1992)	24/34–45 = 0.7–0.5	
The 67B1-2 bands, 3L chromosome	7 transcripts within the 67B heat shock puff (Ayme and Tissieres, 1985) There are deletions produced in this region (Leicht and Bonner, 1986)	2 (Semeshin *et al.*, 1985a)	7/2 = 3.5	See Section V,B,2
The 68A2-3-68B1-3(C1) region, 3L chromosome	13 complementation groups of lethal mutations, the gene controlling the xanthine dehydrogenase *(lxd)*, the *SOD (superoxide dismutase)* gene (Campbell *et al.*, 1984, 1985; Crosby and Meyerowitz, 1986a)	9–11(2-3) (Lindsley and Zimm, 1992)	15/6–11 = 2.5–1.4	
The 68B1-3-68C5-6 region, 3L chromosome	Three protein secretion genes: *Sgs3, -7, and -8* (Akam *et al.*, 1978; Meyerowitz and Hogness, 1982; Crosby and Meyerowitz, 1986)	7–10(3) (Lindsley and Zimm, 1992)	3/4–7 = 0.8–0.4	
The 68C5-6-68C10-11 region, 3L chromosome	Three complementation groups of lethal mutations, the *rotated* gene (Crosby and Meyerowitz, 1986)	4–6(1) (Lindsley and Zimm, 1992)	4/3–6 = 1.3–0.7	
The 68C8-69B5 region, 3L chromosome	37 complementation groups of lethal mutations (Hoogwerf *et al.*, 1988)	35(8) (Lindsley and Zimm, 1992)	37/27–35 = 1.4–1.0	
The 70A2-3-70A5-6 region, 3L chromosome	Three complementation groups of lethal mutations (Zhimulev and Feldman, 1982)	2–3(1) (Lindsley and Grell, 1968)	3/1–3 = 3–1	1. The 70A1-5 region is the site of intercalary heterochromatin (Zhimulev *et al.*, 1982b)

(continues)

Table 4. *Continued*

Chromosome region	Complementation groups	Number of bands[a]	Maximum and minimum ratio of the number of cistrons to that of bands	Notes
				2. The region presumably enriched with repeats (Spradling *et al.*, 1975)
The 73A3-4-73B1-3 region, 3L chromosome	16 genes (McKeown *et al.*, 1987)	10(2) (Lindsley and Zimm, 1992)	16/8–10 = 2–1.6	
The 84A1-84B1-2 region, 3R chromosome	Five independent complementation groups of partly overlapping mutations, the tRNA genes *Met-2*, and *Lys-5* (Lewis *et al.*, 1980a,c)	8(3) (Lindsley and Zimm, 1992)	7/5–8 = 1.4–0.9	1. The 84A region is the site of intercalary heterochromatin 2. Independent groups are composed of 1–3 mutations
The 84B3-4-84D3-5 region, 3R chromosome	16 genes (11 essential and 5 nonessential) (Cavener *et al.*, 1986b)	16(4) (Lindsley and Zimm, 1992)	16/12–16 = 1.3–1.0	
The 84D11-12-84D16 region, 3R chromosome	25 essential genes, three valine tRNA genes (Lewis *et al.*, 1980b; Baker *et al.*, 1991)	27(6) (Lindsley and Zimm, 1992)	28/21–27 = 1.3–1.0	
The 86F1-2-87B15 region, 3R chromosome	25 complementation groups of lethal mutations (Gausz *et al.*, 1981)	36(8) (Lindsley and Grell 1968)	25/28–38 = 1.0–0.7	No mutations identified in the region of the 87A7 heat-shock puff
The 87C1-9 region, 3R chromosome	Four complementation groups of lethals and the *kar* locus located in the narrow 87C4-5-87C9 interval. No complementation group found in 87C1-3 region. Heat shock puff develops from 87C1 band (Gausz *et al.*, 1979)	6 (Lindsley Grell, 1968)	5/6 ~ 1.0	

The 87D2-4-87E12-F1 region, 3R chromosome	21 complementation groups of lethal and visible mutations (Chovnick *et al.*, 1978; Hilliker *et al.*, 1980, 1981b) The genes unevenly distributed in the region. Three loci *B16-1*, C9A, and *Ace* mapped within the 87E1-2 band. Four independently transcribed DNA sequences also located there (Spierer *et al.*, 1983) The *Act 87E* and *1(3)m112* genes located in the 87E9-12 region (Manseau *et al.*, 1988) Five transcripts synthesised on the 87E5-6 band sequences (Spierer *et al.*, 1983)	24–28(2–4) (Lindsley and Grell, 1968)	21/22–28 = 1.0–0.8	Three complementation groups were composed of a single lethal in the group (i.e., region was not saturated) See Section V,B,2
The 88F8-89B10 region, 3R chromosome	11 genes (Nelson and Szauter, 1992)	24(5) (Lindsley and Zimm, 1992)	11/19–24 = 0.6–0.5	
The 93C6-7-93F8 region, 3R chromosome	20 complementation groups of visible (including *ebony*) and lethal mutations. No mutations detected within the 93D heat shock puff (Scalenghe and Ritossa, 1976; Fortebraccio *et al.*, 1977; Mohler and Pardue, 1982a,b, 1984; Caizzi *et al.*, 1987)	29(7) (Lindsley and Zimm, 1992)	20/22–29 = 0.9–0.7	
The 96A17-19 region, 3R chromosome	7 complementation groups, including *ash2* gene (Adamson and Shearn, 1996)	3(1) (Lindsley and Zimm, 1992)	7/2–3 = 3.5–2.3	

(continues)

Table 4. *Continued*

Chromosome region	Complementation groups	Number of bands[a]	Maximum and minimum ratio of the number of cistrons to that of bands	Notes
The 97D2-9, 3R chromosome	Nine complementation groups (4 essential, 3 semilethals, 1-visible and 1-male fertility (Knibb *et al.*, 1993)	8(1) (Lindsley and Zimm, 1992)	9/7–8 = 1.3–1.1	
The 99D3-99E2-3 region, chromosome 3R	11 complementation groups of lethal mutations and presumably the ribosomal protein gene *rp49* (Vaslet *et al.*, 1980; Warmke *et al.*, 1989)	8–9(2) (Lindsley and Zimm, 1992)	12/6–9 = 2.0–1.3	
The fourth chromosome	37 complementation groups of lethal mutations, six loci of visible mutations, several (not precisely known) the so-called "minor" genes affecting the number of sternopleural bristles, sensitivity to ether, penetrance of the witty locus (Hochman *et al.*, 1964; Hochman, 1971–1976)	50(17) (Lindsley and Grell, 1968) 137 (Slizynsky, 1944)	43/33–137 = ?	1. Band number of 137 not at all documented 2. The chromosome presumably not saturated, because only a single mutation in each of the 4 loci is known 3. Gene number in the fourth chromosome estimated as 67–75 (Hochman, 1974) 4. There are three or four bands containing A-T-rich repeats in the chromosome (see review: Zhimulev, 1993, 1997)

Note. Minimal band number is obtained accepting "doublets" as single bands.
[a]The number of doublets are in parenthesis.

bands in *D. melanogaster* polytene chromosomes—maximally 5113 (Heino *et al.*, 1994), even including "doublets." However, according to calculation of Zhimulev (1996), there are about only 3500 bands in polytene chromosomes of *Drosophila* (see above).

2. Genetic analysis of polytene chromosomes regions

In view of the numerous genetic saturation studies of chromosome regions, it is valuable to examine Table 4. Two conclusions are immediately apparent: first, there is no region in which the number of identified genes exceeds by several-fold the number of bands; second, even the one-to-one relation of bands and genes seldom holds true. In fact, there occurs departures tenfold above and below the 1 : 1 relationship. In interpretating the gene number findings, the following should be kept in mind:

1. Gene mapping with the precision to a particular cytological structure, namely the band or the interband, is impractical using chromosomal rearrangements. The mapping precision of deletions cannot exceed 1 band plus 1, or 2, adjacent interbands or 1 interband plus edges of 1 or 2 neighboring band(s) (see Figure 41). It should be noted that electron microscopy does not much increase the resolving capacity of the method (Semeshin *et al.*, 1979b; Zhimulev *et al.*, 1981f). The only strategy allowing to make correct inferences concerning the localization of the gene in the band would involve oppositely directed deletions, each removing only a part of the band (see Figure 41i). However, only a few bands (e.g., 10A1-2) have been studied with this approach (see Section V,B,2).

   Hence, in mutational saturation of chromosome regions the detectable genes are not necessarily associated with the bands; they can be located in the interbands, too. In fact, their numbers are the same in the region (Lifschytz, 1971; Zhimulev *et al.*, 1981a).

   Taken together, the data do not provide evidence for genetic organization of the bands. They relate mainly to correlations between the number of genes and the number of bands/interbands in a region.
2. When a deletion surrounding a saturated site occurs in a region consisting of very thin bands (and this is most frequently the case because the majority of bands are thin), additional difficulties arise because of imprecise localization of the breakpoints of the deletion. As a result, the estimated gene-to-band ratio can vary widely depending on the number in the denominator (the number of bands). Thus, a consequence of ambiguous mapping of the suturation site in the 37C region, may be a variation of the gene to band ratio in the 2 : 1 to 12 : 1 range (Wright *et al.*, 1981).

   Another hypothetical source of errors in estimates of band number

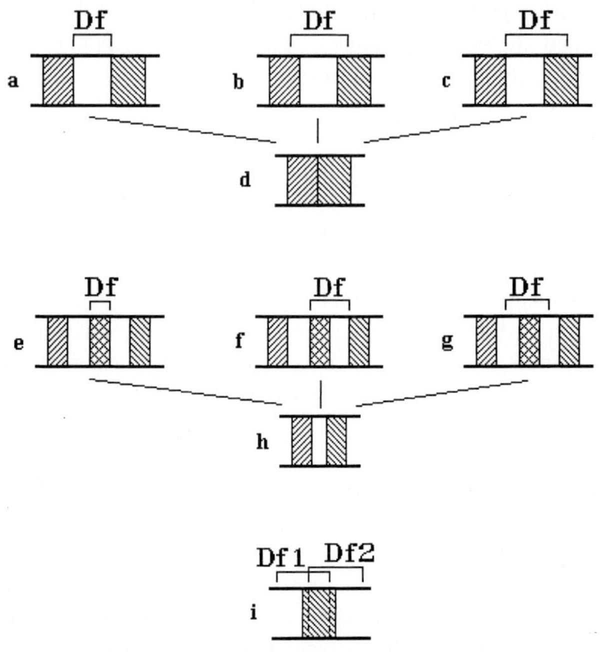

Figure 41. Resolution of mapping the deletions ends. Removal of (a) only the interband, (b) the interband and edges of both neighboring bands, (c) the interband and most of the neighboring band results in (d) the same banding pattern. Similarly, (h) identical banding pattern is produced, when only (e) the band, (f) the band and the neighboring interband, (g) the band and most of two neighboring interbands are removed. Two overlapping deletions (each removing only a part of the band) allow to map precisely the gene within the band (i) (reprinted by permission from Belyaeva *et al.*, 1976; and Zhimulev *et al.*, 1981a. Informational content of polytene chromosome bands and puffs. *CRC Crit. Rev. Biochem.* **11,** 303–340. Copyright CRC Press, Boca Raton, Florida).

may be incomplete polytenization. It was demonstrated that a series of middle repetitive sequences, including those which are transcribed, underreplicate (Woodcock and Sibatani, 1975). This suggested that the DNA of a part of the genes, hence, the bands, may be lost by the polytene chromosome; yet, the genes in question are present in the genome of germ line cells, and this may explain why the number of genes exceeds the number of bands (Woodcock and Sibatani, 1975). It remains to be determined how serious these doubts are.

3. Regions were saturated using mostly lethal mutations, which, as discussed above, do not allow to judge the complete genetic composition of the region.
4. It is not easy to decide when the real maximum of mutational saturation is reached in a region; that is, when the genes have been mutated at least

once and thereby made identifiable. It is believed that a region is saturated with mutations when several mutations have been generated at a locus. Data of this kind have frequently been treated mathematically, with the frequency distribution of loci approximately following the Poisson distribution. This analysis allows determinations of the total gene number in a segment of the genome: the difference between these values and the number of genes experimentally indentified will also be the number of the so-far undetected genes (Judd *et al.*, 1972; Barrett, 1980; Irick, 1983). However, the Poisson distribution is applicable only, if the following conditions are met: (a) the probability of a mutational event is very small; (b) all the genes are equal in size; (c) all the genes are equally mutable by a given mutagen; and (d) there is no locus–mutagen interaction (Barrett, 1980; Hilliker *et al.*, 1981a). These conditions can hardly be met in a real experiment. The 3AB region is a case in point (see Table 4). From earlier work, it first appeared that all the genes have been identified in the regions. The conclusion was based on the results of intense mutational saturation (Figure 42) and mathematical treatment (Judd *et al.*, 1972). Using another mutagen, three lethal mutations of a hitherto unknown essential locus were unexpectedly recovered (Liu and Lim, 1975); two other genes were also recovered in the course of selection of a new mutation that causes female fertility (Young and Judd, 1978).

From consideration of the above data, one must conclude that establishment of band–gene relationships in chromosome regions is insufficient for judging the genetic organization of the band. At this time, no information from the presented experimental data support the one-to-one relation of bands and genes (see Table 4).

It is appropriate to turn to a statement Lefevre and Watkins (1986, p. 869) made overviewing the results of their extensive studies on the organization of the *Drosophila* X chromosome: "We conclude that the one-band, one-gene hypothesis, in its literal sense, is not true; furthermore, it is difficult to support, even approximately. The question of the total gene number in *Drosophila* will, no doubt, eventually be solved by molecular analysis not by statistical analysis of mutation data or saturation studies."

3. Distribution of genes in the DNA molecule

Analysis of cloned DNAs containing sizeable coding regions has provided a different approach to the problem of the molecular and genetic organization of the bands. Accepting that a medium-sized band is 30 kb long and that each band is associated with a single gene locus, there would be one gene every 30 kb. However, already in the 1980s, based on the data on the organization of the

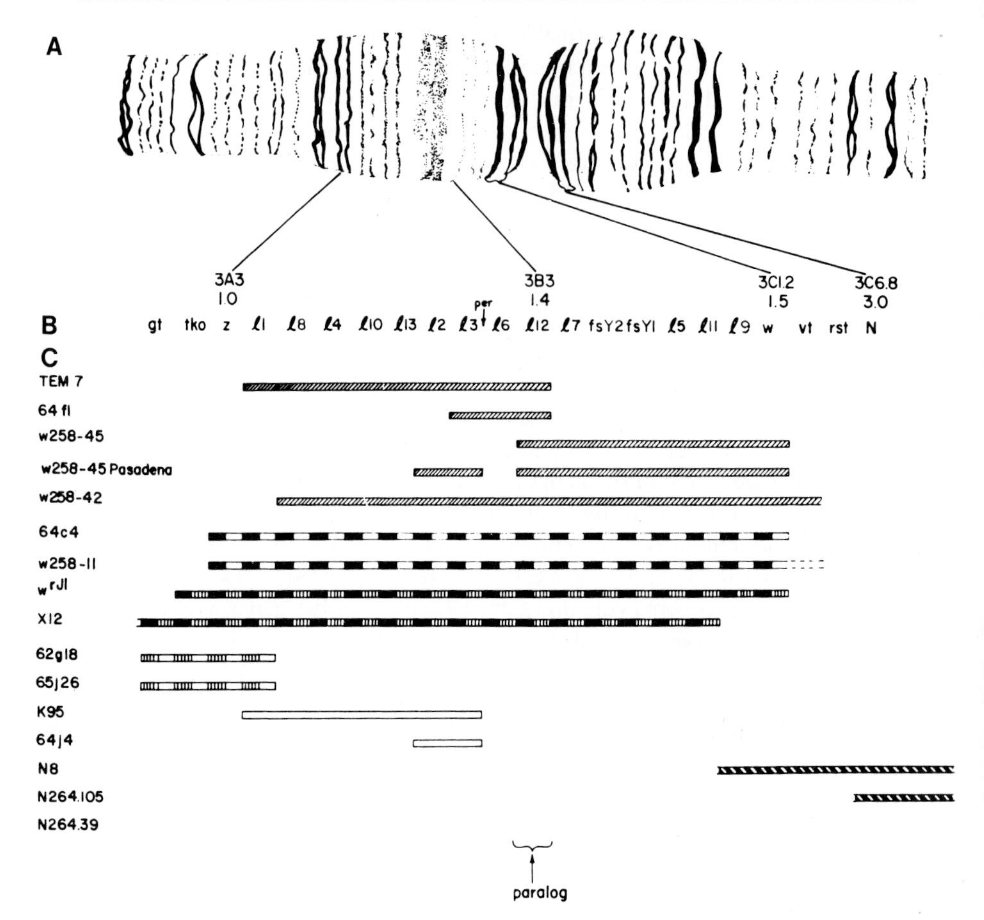

Figure 42. (A–C) Localization of genes in the *z-w* interval and (D) in *Ddc* gene cluster. (A) A cytological map of the region; (B) genes, the *per* gene is shown above the general list, the *paralog* gene is given at the very bottom of the figure; (C) extension of deletions and duplications; (D) *hk-TW1* interval (from Thierry-Mieg, 1982; Stathakis *et al.*, 1995).

rRNA, histone, and heat-shock protein genes, it was recognized that the two genes are closer to one another than previously thought, with densities up to one gene every 1–5 kb (Zhimulev, 1982; Gilbert *et al.*, 1984; Zhimulev and Belyaeva, 1985; Korge, 1987). One possibility is that the genes are located in several very thin bands lying close together and not larger than 1–5 kb; another possibility is that the two neighboring genes reside in a larger band.

With the advent of methods of molecular cloning of extensive DNA

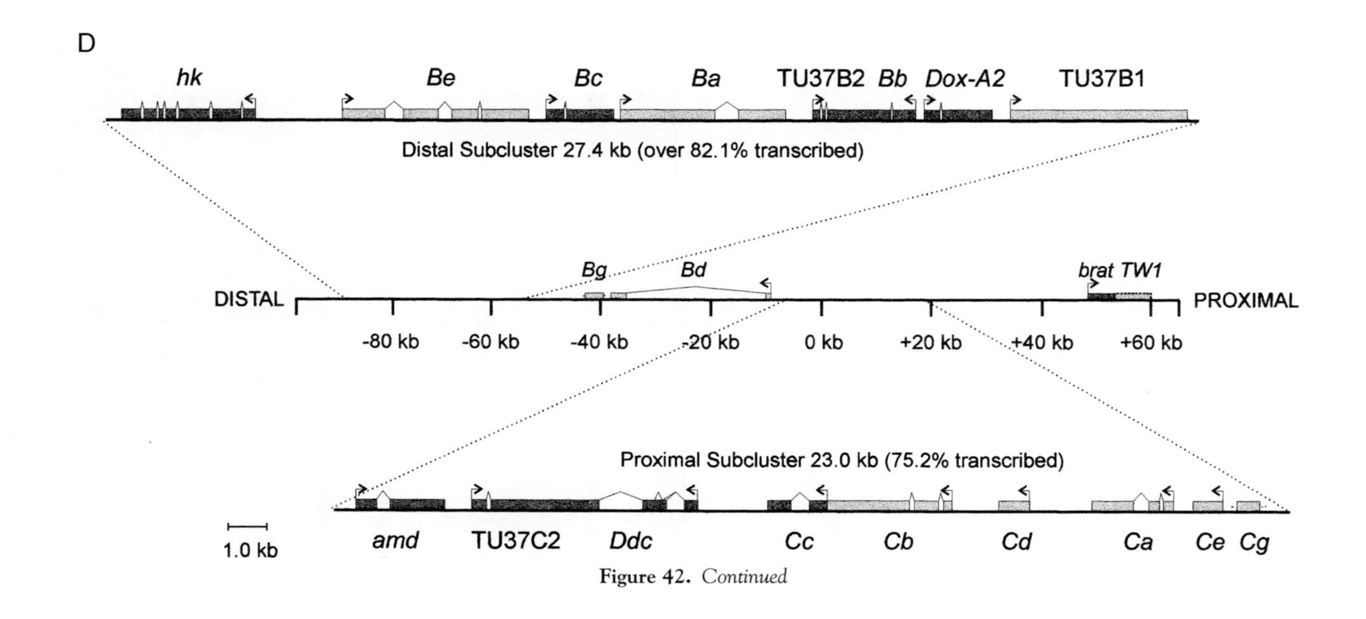

Figure 42. *Continued*

fragments and with mapping of cloned DNAs containing RNA coding regions, it became feasible to determine the densities of independently transcribed DNA fragments and to compare their distributions with those of the bands and the interbands on the DNA maps. This approach surely does not deepen insight into the organization of a particular band or interband. It does, however, establish correlations between the alternating coding and noncoding DNA fragments and the alternating bands and interbands.

Korge (1987) has called attention to the fact that the size of mature RNA can vary in a wide range: from 0.12 kb for 5S rRNA genes (Artavanis-Tsakonas *et al.*, 1977) to 7.3 kb for the *rudimentary* gene (Segraves *et al.*, 1983), 8.2 kb for the *sev* gene (Benerjee *et al.*, 1987), and 10.5 kb for the *Notch* gene (Grimwade *et al.*, 1985). The length of genomic DNA from which mRNA is transcribed varies more widely: from several thousands of base pairs to 103 kb for the *Antennapedia* gene (Scott *et al.*, 1983; Schneuwly *et al.*, 1986), 70 kb for the Ultrabithorax gene (Bender *et al.*, 1983a; review: Scott, 1987). However, genes of such sizes are quite exceptional. The other genes are, as a rule, much smaller (Table 5). These data on gene density cannot be regarded as averages for the whole genome because information about maximal density was included in the Table. The examples clearly show that the position of the genes and the chromomeres along the DNA molecule in many regions of the chromosomes are not correlated.

B. Molecular-genetic organization of individual bands

The number of genes would be about fivefold larger than the number of chromomeres in the *Drosophila* genome (see Section V,A,1); accordingly, the number of genes in the chromomere may vary from one (the informationally simple bands) to several tens (the informationally complex bands).

Different experimental studies provided data supporting the idea that there may exist of genetically complex bands.

1. When a band splits, the interband appears between the halves (see Section II,C); this indicates that the band contained at least three different DNA fragments before it split: two remained inactive at the endpoints and a third between the two became activated later.
2. Puff formation from a part of the band may be explained in this way. For example, the formation of the band from the proximal part of the 64E1-2 band, the distal part of the 71F1-2 band in *Drosophila* (Semeshin *et al.*, 1985a,b), or the formation of Balbiani ring from a part of the dense band called the centromere in *Acricotopus lucidus* (Staiber, 1982).
3. Studies of the bands from an evolutionary perspective has revealed their complex organization. For example, a number of related species of *D. virilis*,

Table 5. Genetic Content of Short DNA Fragments from the Dipteran genome[a]

Chromosome region	Gene (transcript) number	Total DNA fragment length[b]	Transcript number per a band[c]
	8 genes in subtelomeric region of the X chromosome from *l(1)EC1* to *ac* (Karpen and Spradling, 1990; Spradling *et al.*, 1992)	290(36,2)	*d*
1B1-4	Seven transcripts of the AS-C complex in two bands, 1B1-2 and 1B3-4 (Campuzano *et al.*, 1985; Cline, 1988; Gonzales *et al.*, 1989; Torres and Sanchez, 1989)	110(15.7)	7/2 = 3.5
1B9/10	3 genes in 8 kb (Duffy *et al.*, 1996)	8(2.7)	3/1 = 3.0
1B10-14	Six genes (Voelker *et al.*, 1989)	24(4.0)	6/4 = 1.5
1DE	Five transcripts, including the *su(w[a])* gene (Zachar *et al.*, 1987)	14(2.8)	*d*
2E	Three transcripts in the region of the *prune* locus (Frolov and Alatortsev, 1993)	~10(3)	*d*
	Six transcripts in 33 kb (Frolov and Alatortsev, 1994)	33(5,5)	*d*
3C	6 genes: *Sgs-4*, *Pig-1*, *ng-1* to *ng-4* in 11 kb fragment (see D'Avino *et al.*, 1995b for references)	11(1.8)	*d*
4B3-4-4C1	Three genes, including *mei-9* and *norpA* (Sekelsky *et al.*, 1995)	~35(~12)	*d*
6E1-3	Four partly overlapping transcripts (Watanabe and Kankel, 1990)	24(6.0)	4/3 = 1.3
7F1-2	Two chorionic protein genes, S36-1 and S38-1, and four transcripts (Spradling, 1980, 1981; Spradling and Rubin, 1981, Orr-Weaver, 1991)	12-16(2.0-2.7)	6/1 = 6
8F-9A	Two yolk protein genes separated by a region of 1 kb (Barnett *et al.*, 1980; Shepherd *et al.*, 1985; Bownes, 1994)	4.5(2.25)	*d*
10A1-2	22 transcripts (Kozlova *et al.*, 1997)	180(8,1)	22/1 = 22
10C1-3	Seven transcripts, including the RNA-polymerase II gene (Voelker *et al.*, 1985)	30(4.3)	7/1-3 = 7.0-2.3
14B-15B	45 RNA transcribing regions in 0 to 18 hr embryos (Surdej *et al.*, 1990)	800(17.8)	45/57(22) = 0.8-1.3
14D	5 genes (Hong and Ganetzky, 1996)	80(6)	5/4 = 1.25
14F2-3-15A1-2	Six independent transcripts and the *rudimentary* gene of 12.5 kl (Rawls *et al.*, 1986)	90(12.9)	*d*
15E	Seven transcripts around the *mei-218* region (McKim *et al.*, 1996)	48(6,9)	*d*

(continues)

Table 5. *Continued*

Chromosome region	Gene (transcript) number	Total DNA fragment length[b]	Transcript number per a band[c]
19F1-2	18 size classes of poly(A)$^+$ RNA species and part of tRNA gene cluster (Russel *et al.*, 1992)	50(2.8)	at least 18
23E-24B	Two genes, *odd* and *sob* (Hart *et al.*, 1996)	16.5(8)	*d*
24CD	Two *slp* genes (Grossniklaus *et al.*, 1992)	14(7.0)	*d*
25D1-E1	7 genes and transcripts (George and Terracol, 1997)	120(17.1)	7/7-9 = 1-0.8
26A	11 transcripts (Knipple *et al.*, 1991)	70(6.4)	11/7-9 = 1.6-1.2
28C1-4,5 (four bands)	Six transcripts, including the uratoxidase gene, five transcripts located in a narrower interval (18 kb) (Friedman *et al.*, 1991)	38(6.3)	6/4 = 1.5
28DE region	TRF and two other transcripts (Crowley *et al.*, 1993)	8(~3)	*d*
37B10-37D1	20 genes (Wright, 1987; Stathakis *et al.*, 1995)	near 160 (8.0)	20/8-12 = 2.5-1.7
42A	16 tRNA genes (Figure 43) (Yen *et al.*, 1977; Yen and Davidson, 1980; Kubli, 1982)	50(3.1)	*d*
44D	Two gene clusters at a distance of 11 kb apart. The first cluster contains 5 cuticalar protein genes (Lcp) two of them are transcribed in one direction, two in the opposite, the fifth is, possibly, a pseudogene (Snyder *et al.*, 1981a,b, 1982; Snyder and Davidson, 1983; Pritchard and Schaeffer, 1997)	7.9(1.58)	*d*
	These four genes are similarly arranged in *D. persimilis*, *D. pseudoobscura*, and *D. miranda* (Steinemann and Steinemann, 1990)		
	The other cluster contains three genes homologous to each other (55–60%), but not to the cuticular protein genes (Snyder and Davidson, 1983)	8(2.6)	*d*
48EF	Three genes: *Deb-A*, *Deb-B*, *Deb-C* (Vincent *et al.*, 1990)	4(1.3)	*d*
54A1-B1	Two α-amylase genes of 1650 bp transcribed in opposite directions (Gemmill *et al.*, 1983, 1986; Levy *et al.*, 1985; Boer and Hickey, 1986; Benkel *et al.*, 1987)	6(3)	*d*

Table 5. *Continued*

Chromosome region	Gene (transcript) number	Total DNA fragment length[b]	Transcript number per a band[c]
	The α-amylase gene is duplicated in *D. erecta*, *D. simulans*, *D. mauritiana*, *D. yakuba*, and *D. teissieri* (Dainou *et al.*, 1987; Bally-Cuif *et al.*, 1990; Hickey *et al.*, 1991; Shibata and Yanazaki 1995)		
	Three α-*Amy* genes in *D. pseudoobscura* mapped to the band in the region 73A (Aquadro *et al.*, 1991)	11-12 (3.6-4.0)	3/1 = 3
	Four α-*Amy* genes organized as two independent pairs in *D. ananassae* (DaLage *et al.*, 1992)	*d*	*d*
64A10-11 band	Three genes *rfc40*, *Rop*, and *Ras2* (Harrison *et al.*, 1995)	10.2(3.4)	3/1 = 3
65A5-6	A cluster of 12 cuticle protein genes (Charles *et al.*, 1997)	36(3)	12/1 = 12
66D11-15	Two chorion protein genes separated by a spacer of 1–2 kb and two transcripts (Spradling and Rubin, 1981)	12(3.0)	*d*
The 67B1-2 band (the heat-shock puff)	Seven genes including those that encode proteins *hsp 22*, *hsp 23*, *hsp 26*, *hsp 28* (Corces *et al.*, 1980; Craig and McCarthy, 1980; Wadsworth *et al.*, 1980; Voellmy *et al.*, 1981; Ireland *et al.*, 1982; Sirotkin and Davidson, 1982; Zimmerman *et al.*, 1983; Ayme and Tissieres, 1985)	12(1.7)	7/1-2 = 7-3.5
The 68C4 band	The genes encoding salivary gland secretion proteins *Sgs-3*, *Sgs-7*, and *Sgs 8* (Meyerowitz and Hogness, 1982; Crowley *et al.*, 1983, 1984; Meyerowitz and Martin, 1984; Meyerowitz *et al.*, 1985, 1987; Crosby and Meyerowitz, 1986b; Hoffmann *et al.*, 1991)	5(1.7)	3/1 = 3
	The genes similarly located in *D. simulans*, *D. erecta*, *D. yacuba*, and *D. teissieri* (Meyerowitz and Martin, 1984)		
The 69A1-3 bands	Esterase-6 and Esterase P genes (Balakirev and Ayala, 1996)	~3,5(1,7)	2/2-3 = 0.7-1
The 70C2 band	Three transcripts (Swaroop *et al.*, 1986)	31(10.3)	3/1 = 3

(continues)

Table 5. *Continued*

Chromosome region	Gene (transcript) number	Total DNA fragment length[b]	Transcript number per a band[c]
71CD	Nine different transcripts, including the *mex 1* gene, three genes overlapping each other and covering 4.5 kb: *Eip 28/29*, *gonadal*, *z600* (Cherbas *et al.*, 1986; Schultz *et al.*, 1986, 1991; Schultz and Butler, 1989; Cherbas, 1993)	18(2.0)	*d*
71EF	Eleven genes (Restifo and Guild, 1986; Wright *et al.*, 1996)	14(1.3)	*d*
73A	8 genes are in one-two bands (Butler *et al.*, 1986; McKeown *et al.*, 1987)	25(3.1)	*d*
	Four genes, including *argos* and *scarlet* (Wemmer and Klamb, 1995)	40(10)	*d*
73D2-7	Five genes, including *sina* and *Rh 4* (Carthew and Rubin, 1990)	35(7)	5/5 = 1.0
84AB	18 genes in the region of the *Antp-Complex* (Kaufmann *et al.*, 1990; Cribbs *et al.*, 1992; Gorman and Kaufman, 1995; Kapoun and Kaufman, 1995)	340(18.8)	*d*
	zen and *bcd* genes of *Antp-Complex* in *D. subobscura* occupy 5 kb (Terol *et al.*, 1995)	5(2,5)	*d*
84C1-2-84C8	Nine genes (Cavener *et al.*, 1986a)	140(15.6)	9/6-8 = 1.5-1.1
The 85D9/85D10 region	The band appeared as result of transformation with PlArB transposon containing *lacZ*, *Adh*, *ry*, and a gene of ampicillin resistance (Semeshin *et al.*, 1994)	19(4.75)	4/1 = 4.0
The 87A7 band (the heat-shock puff)	Two *hsp 70* genes encoding heat-shock proteins (Livak *et al.*, 1978; Moran *et al.*, 1979; Ish-Horowicz and Pinchin, 1980; Gausz *et al.*, 1981; Udvardy *et al.*, 1985)	8(4)	2/1 = 2
The 87C1 band (the heat shock puff)	Three *hsp 70* genes encoding heat-shock proteins and repetitive α- and β-sequences (Livak *et al.*, 1978; Moran *et al.*, 1979; Ish-Horowicz and Pinchin, 1980; Gausz *et al.*, 1981; Leigh Brown, 1981)	55(78.3)	3/1 = 3
The 87D5-87E5-6 region (14 bands)	43 transcripts (Gausz *et al.*, 1986)	315(7.3)	43/14 = 3.1

Table 5. *Continued*

Chromosome region	Gene (transcript) number	Total DNA fragment length[b]	Transcript number per a band[c]
The 89E1-4, *BX-complex*	Transcripts of homeobox-containing *Abd-B*, *abd-A* and *Ubx*, noncoding *bxd* and *iab-4*, *Glucose transporter-like* gene (*GIU*) (Lewis *et al.*, 1995)	300(50)	6/1-4 = 6-1.5
The 96F8-9-96F11-13 region	15 transcripts (Knust *et al.*, 1992; Schrons *et al.*, 1992)	65(4.3)	15/2-6 = 7.5-2.5
98F3-10	8 genes of *yema* cluster (Ait-Ahmed *et al.*, 1987, 1992)	39(4,9)	*d*
The 99C region	12 transcripts (Shaffer and MacInture, 1990)	35(2.9)	*d*
99D	Three genes: *sry a*, *sry b*, and *sry* A, and *rp* 49 ribosomal protein gene, the *janus* A and *janus* B genes. The latter two overlap each other (Vincent *et al.*, 1985; Payre *et al.*, 1989; Yanicostas *et al.*, 1989; Yanicostas and Lepesant, 1990)	8.4(1.4)	*d*
99E	Four expressing genes encoding cecropin, two pseudogenes, and *Anp* gene. Another cluster of cecropin genes is located in the neighboring 4 kb DNA (Kylstein *et al.*, 1990; Tryselius *et al.*, 1992; Clark and Wang, 1997)	7,0(1,0)	7/4-5 = 1.8-1.4
100F	*mod* and *Map205* genes (Garcino *et al.*, 1992)	about 20(10)	*d*
The DNA fragment in the region of the ribosomal protein locus	Seven genes (Vaslet *et al.*, 1980)	18(2.6)	*d*
	Ten actively transcribing genes (the "*wpp*" locus) (Hager and Miller, 1991)	42(4.2)	*d*
D. grimshawi	Three vitellogenin genes, *v1*, *v2*, and *v3*. The *v1* and *v2* genes are transcribed in opposite directions (Hatzopoulos and Kambysellis, 1987)	8(2.7)	*d*
D. miranda	5 genes of cuticular proteins-Lcp in Y chromosome	7,9(1,58)	*d*
	4 genes of Lcp in X2 chromosome (Steinemann and Steinemann, 1992; Steinmann *et al.*, 1996a,b)	6,7(1,69)	*d*

(continues)

Table 5. *Continued*

Chromosome region	Gene (transcript) number	Total DNA fragment length[b]	Transcript number per a band[c]
Species of *Drosophila montium* subgroup	*Four β-tubulin genes in D. auraria* are located in the same polytene chromosome band (Scouras *et al.*, 1994). In other 11 species these genes can be organized in a cluster, or separately (Drosopoulou and Scouras, 1995)	~24(6)	6/1 = 6
D. mulleri	Two alcohol dehydrogenase genes (ADH) and ADH-like gene (possibly, a pseudogen) (Posakony *et al.*, 1985)	7(2.3)	*d*
D. virilis	Two genes, *Egp-1* and *Egp-2* (the 55E puff) (Kress *et al.*, 1990; Thuroff *et al.*, 1992)	5(2.5)	*d*
	Two protein secretion genes *Lgp1* and *Lgp 3* (the 16A puff) (Kress *et al.*, 1990; Swida *et al.*, 1990)	2.9(1.45)	*d*
Chironomus tentans	Balbiani ring 2 gene occupies about 10% of 3B10 chromomere (Derkesen *et al.*, 1980)	*d*	*d*
C. thummi	Cluster of 5 hemoglobin genes in a single band	7(1.4)	5/1 = 5
	Cluster of 6 to 7 hemoglobin genes in a single band	13(~2.0)	6-7/1 = 6-7
	Location of each cluster in a single band was found in 13 other *Chironomus* species		

[a]*Drosophila melanogaster*, if not specified otherwise.
[b]DNA length in kilobases per gene in parentheses.
[c]Band number in the indicated region according to Bridges' revised map ("doublet" number is in parentheses); the data are not given if there is no precise location of the fragment in the region.
[d]No exact calculations.

on the one hand, and *D. americana americana*, *D. a. texana*, and *D. novamexicana*, on the other, differ in the presence of a cluster of 5S ribosomal RNA genes in a certain chromosome region showing identical banding pattern (Cohen, 1976; Wimber and Wimber, 1977). These bands in *D. virilis* obviously contain, apart from the 5S rRNA genes, some other sequences that are also present in the three other species of *Drosophila*.

The α- and β-serum proteins of the hemolymph genes are located in one of the bands in *D. pseudoobscura* which appears to be genetically complex (Dudley, 1983). In other species these genes are located in different regions (Brock and Roberts, 1983).

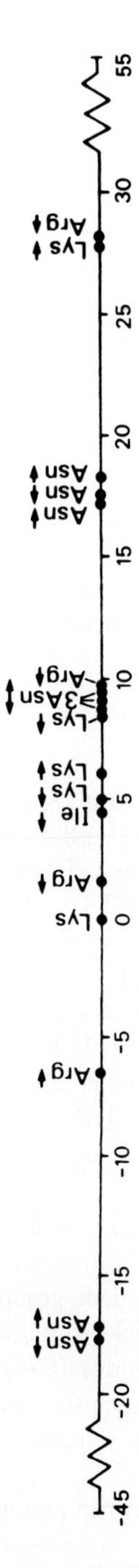

Figure 43. Localization of the t-RNA genes in the DNA of 42A region of *D. melanogaster* 2R chromosome (from Kubli, 1982).

4. The so-called heterozygous (heteromorphic, asymmetrical) bands occur frequently. Their condition is manifest as substantial difference in content and qualitative composition of DNA (for example, in the presence of certain repeats) between homologous bands in individuals caught from different populations or belonging to different subspecies, even within a species.
5. Insertions of many various mobile elements provide good examples of the informational complexity of the bands (review: Zhimulev *et al.*, 1981a).
6. Determination of the sizes of heterogeneous nuclear RNA transcripts from cultured *Drosophila* cells indicated that the modal class was of 1.4×10^6 daltons; in the case when the entire band was transcribed, the modal class would be 10×10^6 Daltons. It was concluded that the molecule of heterogeneous nuclear RNA would be too small to be the transcript of all the DNA of an average band (Lengyel and Penman, 1975). Consequently, other functions, too, may be encoded in the band. Along with these general notions, there is extensive information about individual bands allowing classification according to the degree of their informational complexity.

1. The polygene bands

In *D. melanogaster*, the genes encoding 5S rRNA compose a block of 160–200 identical sequences 385 bp long and with total cluster length of 60–80 kb. The repeat unit consists of a coding (~135 bp) region and a spacer (~250 bp) region (Quincey, 1971; Procunier and Tartof, 1976; Tschudi and Pirrotta, 1980; Spradling and Rubin, 1981; DeLotto *et al.*, 1982).

A cluster of genes coding for 5S rRNA was cytologically localized by *in situ* hybridization in the middle of a group of four dense bands in the 56F region (Wimber and Steffensen, 1970, 1973; Steffensen and Wimber, 1971, 1972; Prensky *et al.*, 1973; Grigliatti *et al.*, 1974; Procunier and Tartof, 1975; Hershey *et al.*, 1977; Procunier and Dunn, 1978). Furthermore, four to five tRNA genes were identified at the 3′ end of the cluster in a region 2.5 kb long (DeLotto *et al.*, 1982).

From a comparison of the DNA length spanned by these genes with that of the average band, it was concluded that at least one band in these region is polygenic; single genes or gene groups within a cluster are transcribed independent of one another: the deleted parts do not hinder the activity of its remaining parts (Procunier and Dunn, 1978). Clones of the 5S rRNA genes isolated from different parts of the cluster are normally transcribed in the 5S rRNA molecule when injected into *Xenopus* oocytes (Brown and Gurdon, 1978; DeLotto *et al.*, 1978).

Thus, at least one band in the region, where the 5S rRNA genes are located in *Drosophila,* has a complex organization, containing many copies of a gene or other sorts of genes. The sequences located in the band are not transcriptional units and they are functionally independent.

In the other studied species, such as *Drosophila* of the *virilis* group, *Drosophila funebris*, *D. hydei*, *C. tentans*, *C. thummi*, *Glyptotendipes barbipes*, and *Lucilia cuprina*, one to three regions—where the 5S rRNA gene clusters are located—were identified (Wen *et al.*, 1974; Alonso and Berendes, 1975; Wieslander *et al.*, 1975; Wieslander, 1975; Baumlein and Wobus, 1976; Cohen, 1976; Wimber and Wimber, 1977; Renkawitz-Pohl, 1978; Kress *et al.*, 1985; Bedo and Webb, 1990). This provides evidence for the polygenic organization of the bands in which the 5S rRNA genes are located in these species.

The 18S and 28S ribosomal RNA genes are similarly arranged in *Drosophila* and other eukaryotic organisms. The sequences of the 18S (1.56 kb) and 28S (3.6 kb) RNA of the transcribed and nontranscribed spacers compose a unit of about 11 kb (Figure 44). The DNA that codes for the 18S and 28S

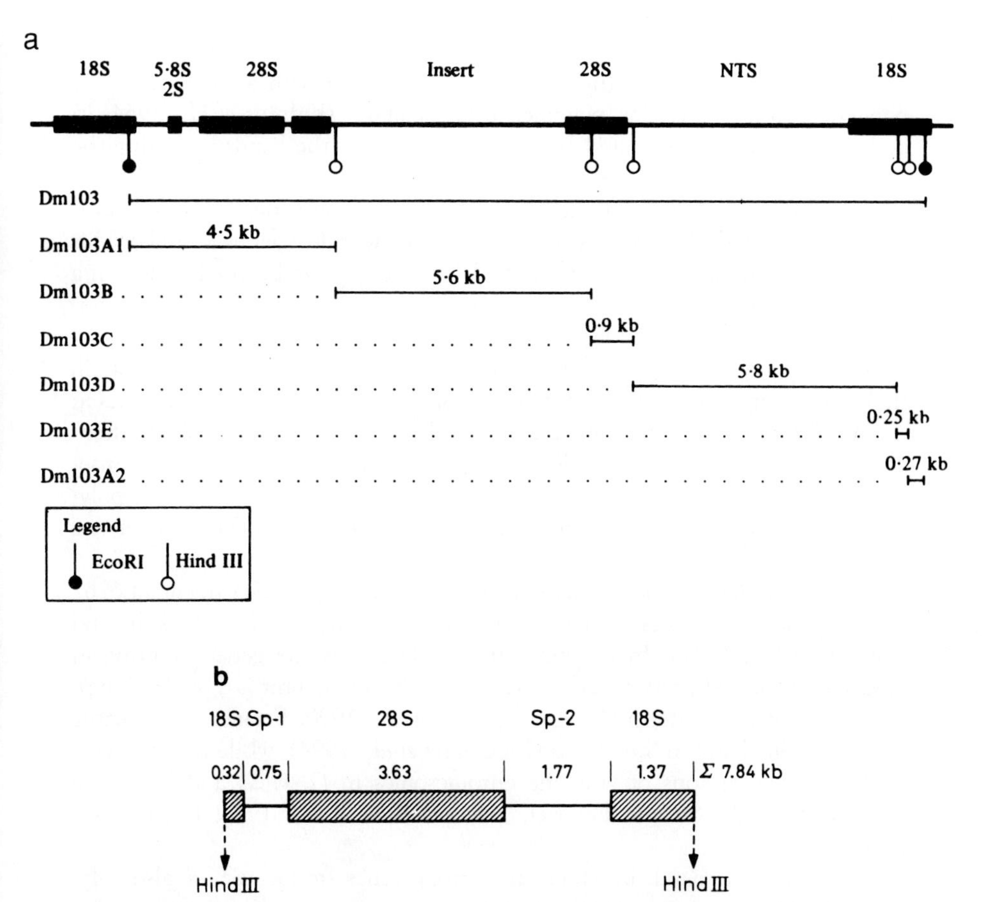

Figure 44. Organization of cistrons of ribosomal RNA in (a) *D. melanogaster* and (b) *C. tentans* (reprinted by permission from Peacock *et al.*, 1981; Degelmann *et al.*, 1979).

ribosomal RNA genes occurs in hundreds of copies and forms the nucleolar organizer (see Section VIII,B). These genes are organized in a similar manner in other dipteran species (Degelman *et al.*, 1979; Renkawitz *et al.*, 1979; Long and Dawid, 1980).

The structural organization of the chromosome is much perturbed in the region of the nucleolar organizer, so much so that it is not always feasible to identify the band from which it has originated. The nucleolar organizer has the appearance of a band in some species (see Section VIII,A); it appears as a large double band in *C. tentans* (Beermann, 1960; Pelling and Beermann, 1966).

The association between the nucleolar organizer and a single large band is clearly seen in larvae of *Acricotopus lucidus* heterozygous for the activity of the nucleolus (Panitz, 1960a). In *Glyptotendipes barbipes*, the block of ribosomal genes is located in two thin adjacent bands (Wen *et al.*, 1974).

The formation of the nucleolus is associated with a small region of polytene chromosomes, for example, in *C. plumosus* (Kalnins *et al.*, 1964a,b) and *C. thummi* (Kiknadze and Belyaeva, 1967); thus, the bands harboring the 18S and 28S rRNA genes are also polygenic.

The total length of the cluster of the histone genes in *Drosophila* is approximately 500 kb (100 repeat units 5 kb long with five genes in each). This is enough to provide the formation of 15–17 average bands. Single genes in a repeat unit are transcribed in the opposite directions (Figure 45). This is evidence that they function independent of one another.

Sea urchin 9S [^{3}H]RNA containing information about most, if not all, the histone genes of *Drosophila* hybridize *in situ* to the polytene chromosomes in the 39DE region of chromosome *2L* (Birnstiel *et al.*, 1973; Wimber and Steffensen, 1973; Pardue *et al.*, 1977; Serfling *et al.*, 1979; Fitch *et al.*, 1990). Consequently, at least the bands located in the middle of a labeled region are polygenic and contain a series of repetitive genes transcribed independent of one another.

In *C. thummi thummi*, five histone genes occupy a DNA fragment 6262 bp long and a gene cluster located in five bands in arm D (Hankeln and Schmidt, 1990, 1991). In *D. hydei*, the clustered histone gene occurring in tandemly organized units reiterated 120–140 times (the unit is 5153 bp long), occupy the band in the 50A region (Fitch *et al.*, 1990; Kremer and Hennig 1990) and one band in *S. ocellaris* (Da Cunha *et al.*, 1994), while this cluster is located in two regions of polytene chromosomes in *D. virilis* and *Drosophila hawaiiensis* (Anderson and Lengyel, 1984; Domier *et al.*, 1986; Fitch *et al.*, 1990).

Thus, the bands in which the histone genes are located are also polygenic in other species.

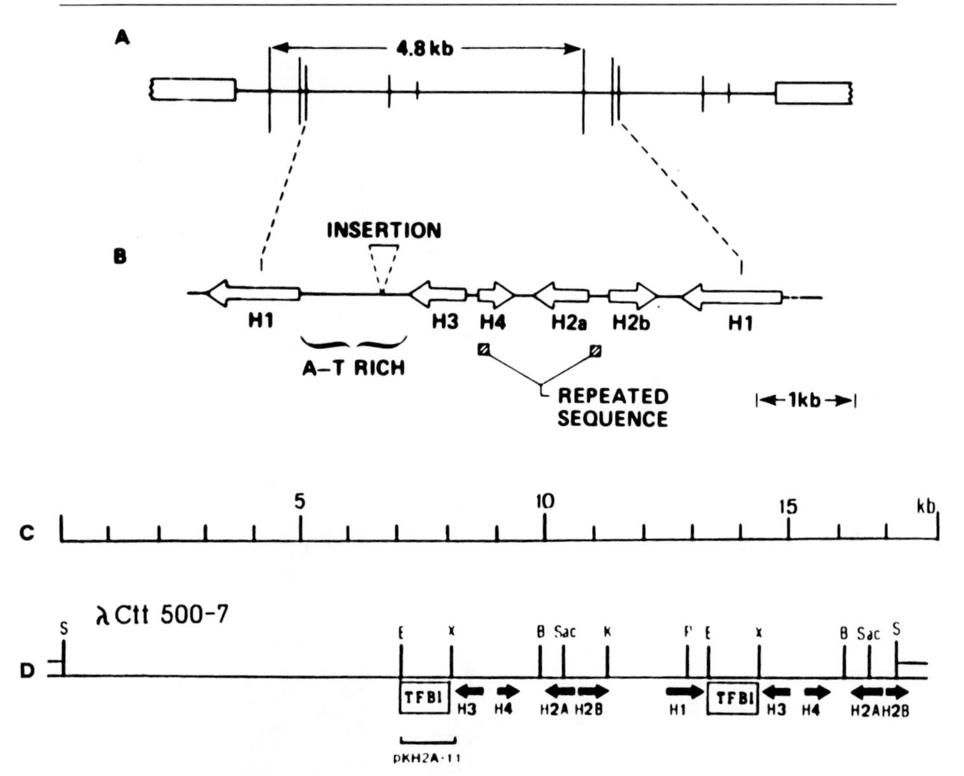

Figure 45. A scheme of the organization of the histone genes in the repetitive unit of (A and B) *D. melanogaster* and (C and D) *Chironomus thummi thummi*. (A) A restriction map of the cDm500 clone. Vertical lines of different height designate from left to right: *Bgl*II, *Bam*HI, *Hin*dIII, *Hpa*I, *Sst*I; (B) localization of the genes, arrows indicate the direction of transcription; (C) distance on the map; (D) restriction sites and localization of the histone genes, TFB1–mobile element (A and B from Lifton *et al.*, 1978; C and D from Hankeln and Schmidt, 1990).

2. Bands containing several genes (the oligogene bands)

The large single 10A1-2 band, as shown by experiments using differently directed overlapping deletions removing different parts of the band, contain two genes, one essential for survival and the *vermilion* gene (Lefevre, 1969, 1971; Lefevre and Wiedenheft, 1974).

Later, as a result of intensive studies of a small fragment, EM maps of this region were built (Semeshin *et al.*, 1979b; Zhimulev *et al.*, 1981f); many chromosomal rearrangements were generated and mapped (Zhimulev *et al.*, 1980a,b,d, 1981c,d, 1982a,d, 1987a,b). The region is saturated with various mutations, including morphological and lethal ones and others causing female

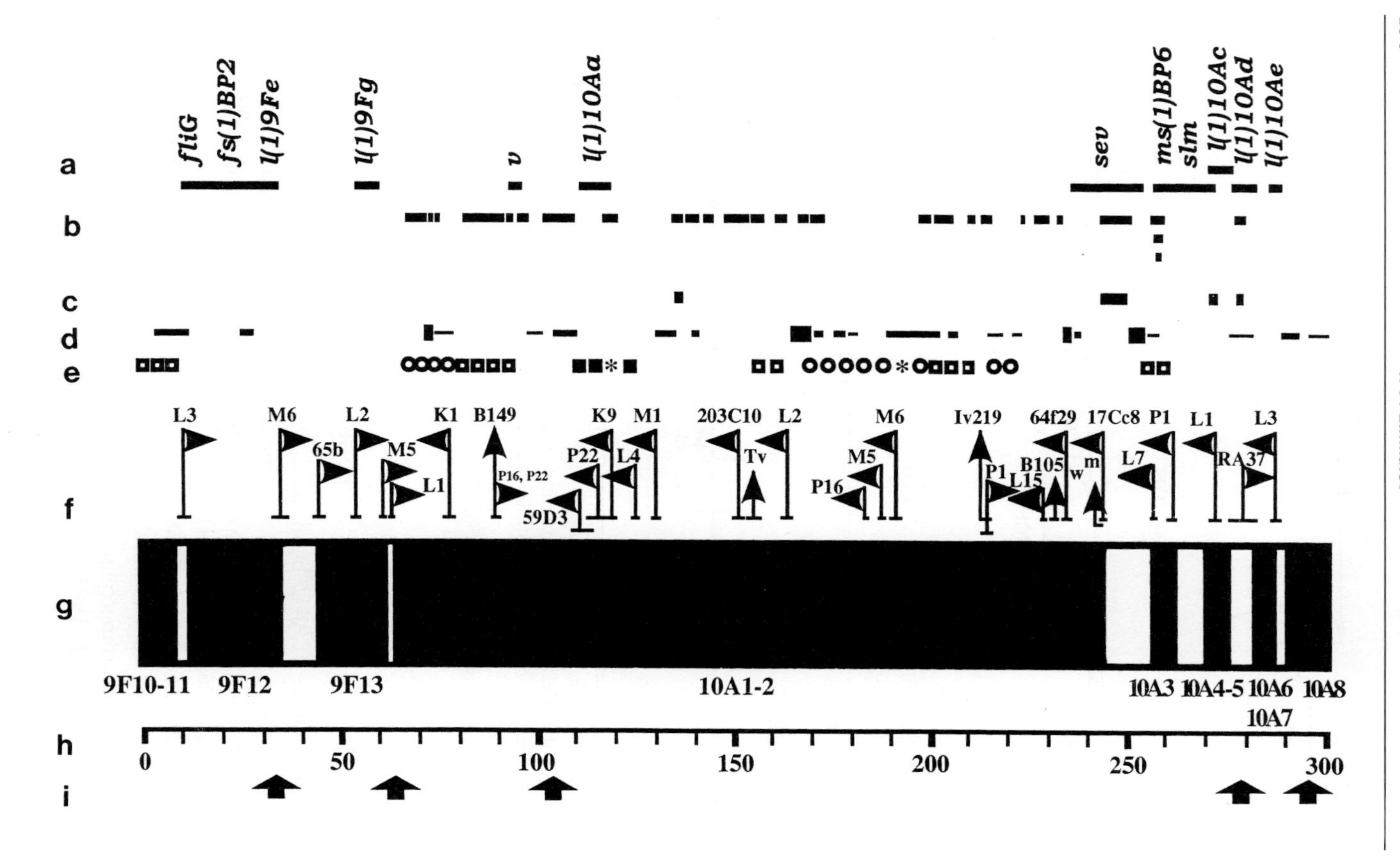
a
b
c
d
e
f
g
h
i
fliG
fs(1)BP2
l(1)9Fe
l(1)9Fg
v
l(1)10Aa
sev
ms(1)BP6
slm
l(1)10Ac
l(1)10Ad
l(1)10Ae
L3
M6
65b
L2
M5
L1
K1
B149
P16, P22
59D3
P22
K9
L4
M1
203C10
Tv
L2
P16
M5
M6
Iv219
P1
L15
B105
64f29
w
m
17Cc8
L7
P1
L1
RA37
L3
9F10-11
9F12
9F13
10A1-2
10A3
10A4-5
10A6
10A7
10A8
0
50
100
150
200
250
300

or male fertility (Belyaeva *et al.*, 1976; Zhimulev and Ilyina, 1980; Janca *et al.*, 1986; Pokholkova *et al.*, 1982; Geer *et al.*, 1983; Pokholkova and Zhimulev, 1984; Eberl and Hilliker, 1988; Pokholkova and Solov'yva, 1989; Pokholkova *et al.*, 1991). The 9F12-10A7 region containing seven bands was completely cloned (Kokoza *et al.*, 1990; Kozlova *et al.*, 1994).

Five cytogenetic zones were identified within the 10A1-2 band by complementation analysis and cytogenetic mapping. Three of these contain genes detectable by mutagenesis: *vermilion (v)*, *l(1)10Aa*, and *sev*. The others are zones of "silent" DNA spanning about 70% of the band length on the genetic map. The first silent zone spans from the distal edge of the band to the *v* gene. The second, largest zone is situated between *l(1)10Aa* and *sev*. The genes they may contain were not identified (Zhimulev and Belyaeva, 1985; Zhimulev *et al.*, 1987b).

Molecular genetic analysis (Figure 46) demonstrated that the 10A1-2 band is about 190 kb in size (Kozlova *et al.*, 1994). The *v* gene spans approximately 2.5 kb of genomic DNA; it contains six exons from which 1306 bp RNA is transcribed (Searles and Voelker, 1986; Walker *et al.*, 1986; Searles *et al.*, 1990). The v^{1}, v^{2}, and v^{k} mutations are caused by insertion of the Dm412 mobile element into the first exon; the v^{36f} mutation is a *roo* insertion into the fourth exon; v^{H2a} results from P-element insertion into the sixth; and v^{48a} is a 200-bp deletion in the third and fourth exons (Searles *et al.*, 1990).

The transcription products of the *v* gene accumulate at the end of the embryonic period, in larvae and adults, as demonstrated by Northern blotting analysis. Transformants carrying a construct containing promoter *v* and *lacZ* express *v* in fat larval bodies, and the expression is dependent on the nucleotides located between +19 and +36 from the transcription start, and it can occur without the TATA domain (Fridell and Searles, 1992).

Figure 46. A molecular and cytogenetic map of the 9F12-10A7 region of the X chromosome of *D. melanogaster*. (a) The loci detected in genetic experiments and mapped between the breakpoints of chromosomal rearrangements (Zhimulev *et al.*, 1987a,b); (b) *vermilion* transcript and 4-kb *Eco*RI transcript (according to Searles and Voelker, 1986; Walker *et al.*, 1986); nine transcripts (according to Hafen *et al.*, 1987; Banerjee *et al.*, 1987); other transcripts (according to Kozlova *et al.*, 1994, 1997); (c) cDNA clones; (d) fragments containing AC-repeats; (e) K1 repeats (open squares), K2 repeats (dark squares), repeats sharing homology with pericentric heterochromatin (asterisks and circles) (data from Kokoza *et al.*, 1997); (f) mapping of chromosomal rearrangements on physical (DNA) and cytological maps of the region. Arrows point to direction of deletions, arrows directed upward point to the breakpoint position of translocations and inversions; (g) banding pattern according to the data of electron microscopy. Band sizes are proportional to the extension of DNA comprised by the band; (h) a physical map of DNA, the distance is given in kb; (i) break points of "evolutionary rearrangements" (from Kokoza *et al.*, 1990).

The Dm412 insertions considerably lower the transciption level (Pret and Searles, 1991). The *su(s)* gene is involved in splicing of the mobile element from the exon (Fridell *et al.*, 1990).

In *D. melanogaster*, the *sev* gene spans a distance of about 16 kb in genomic DNA, and it encodes an 8.5-kb transcript (Banerjee *et al.*, 1987; Hafen *et al.*, 1987; Bowtell *et al.*, 1988). The *D. virilis sev* gene spans 19 kb in genomic DNA. The *sev* genes of both *Drosophila* species contain 11–12 exons (Michael *et al.*, 1990).

Twenty-three transcriptionally active fragments were found in the limits of the 10A1-2 band. These fragments each hybridized with at least one cDNA preparation and were distinct in the sense of being expressed in distinct developmental pattern (Figure 46).

One of the transcriptionally active fragments residing on the DNA map in position 133–138 kb (Fig. 46b,c) was used as a probe to screen a cDNA library. One of the isolated clones, c901, was sequenced and found to be 1277 bp with one long ORF, containing two in-frame AUG codons. The putative polypeptide encoded by c901 shows clear homology with the products of the genes *Delta* and *Serrate*, the products of which are thought to represent ligands for multifunctional *Notch* receptors and play a role in intercellular signaling. The c901 putative polypeptide has 7 complete EGF-like repeats and a region showing homology to the DSL (Delta, Serrate, lag-2) domain (Kozlova *et al.*, 1997). Analysis of tanscriptionally active fragments showed that there are four distinct transcriptionally active fragments in a distally located "silent zone" (coordinates 60–90). The middle part of the 10A1-2 band (coordinates 130–230), within the limits of the proximal "silent" zone, include 15 transcriptionally active transcripts (Kozlova *et al.*, 1997). Finally, repeated DNA of different types were found in the band (Kokoza *et al.*, 1997; see Figure 46).

The results of numerous experiments demonstrated that the genes located in the 10A1-2 band are functionally unrelated. More than that, the band is not associated with the adjacent interbands, thereby the hypothesis of "the cytogenetic unit" (band plus interband) is discredited. The results are summarized below.

1. Assuming that the band is the transcription unit bearing structural sequences, while the sequences required for transcription initiation are in the interband at the left and in the left silent DNA, the discontinuities of these two parts caused by the *T(1;Y)B149* translocation (Figure 47) would give rise to a phenotype characteristic of the mutations of the more rightward loci. The regulatory sequences would be concentrated in one part of the chromosome, while the structural part would be concentrated in its other part (Figure 47a). Therefore, males carrying the translocation would express at least one of the phenotypes: *v*, *sev*, or lethality. However, v^+ and sev^+ are viable. This means that the lefthand silent zone and the lefthand interband

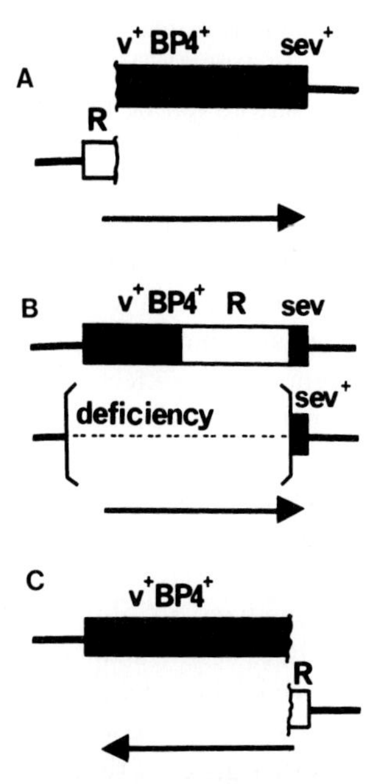

Figure 47. A scheme illustrating the absence of functional dependence between DNA fragments located in the 10A1-2 band. Arrows point to the putative direction of transcription; (a) the *T(1;Y)B149* translocation separates the distal zone of silent DNA and the adjacent distal interband from the other part of the band; (b) the *Df(1)sbr*^{K8} deletion removes two zones of silent DNA, the v^+ and *l(1)BP4*$^+$ loci; (c) the *T(1;2)lv-219* translocation separates the proximal part of the 10A1-2 band from the v^+ and *l(1)BP4*$^+$ loci. *l(1)BP4*$^+$ is a synonym of *l(1)10Aa* (reprinted by permission from Zhimulev *et al.*, 1981c).

and all the sequences to the right of the *T(1;Y)B149* breakpoint are transcribed independent of one another.

Considerations of this sort would lead to refutation of the hypothesis that all the sequences to the left of *sev* are functionally related to provide its expression. In such a case, the regulatory zone (R) would be more leftward, and heterozygotes for any deletion with a breakpoint to the left (see Figure 47b) and the *sev* mutation would have the *sev* phenotype. However, even removal of the greater part of the 10A1-2 band from the left does not affect gene expression. Thus, in terms of transcription from left to right, all the zones into which the band is divisible, as well as the adjacent interband, would be transcribed independently. Assuming that transcription starts with the right interband (or the endpoint of the band), removal of the right part by the *T(1;2)l-v219* translocation (see Figure 47c) would give rise to the mutant phenotype at the v^+ and *l(1)10Aa* loci, and this is also not the case (Zhimulev *et al.*, 1981c).

2. In *D. virilis*, *in situ* hybridization of the clones of DNA, contained in the *D.*

melanogaster 10A1-2 band, demonstrated that they hybridize with two different bands (Umbetova *et al.*, 1988; Kokoza *et al.*, 1990). It is clear that the v^+, $BP4^+$, and sev^+ genes need not be located in a single band to provide normal function, and this additionally supports the conclusion that the 10A1-2 band does not compose a functional unit.

3. The "foci" of action of these three genes (i.e., the organs where they function) were localized by genetic mosaic analysis and proved to be in distinct regions of the embryonic blastoderm: the *v* gene is in the regions that give rise to the fat body, Malpighian tubules, and eyes (Nissani, 1975; Bgatov *et al.*, 1984); the *sev* gene seems only to affect the cells of eye ommatidia (Harris *et al.*, 1976), in which a 288-kDa giant protein, the gene product is synthesized (Michael *et al.*, 1990); in *l(1)10Aa,* the focus (or foci) may be in the blastoderm region that gives rise to the central nervous system (Bgatov *et al.*, 1983, 1984). Thus, the *l(1)10Aa, sev,* and *v* mutations have entirely different foci of expression.
4. The three genes located in the band are also functionally unrelated as judged by the results of analyses of the cellular metabolic pathways in which their products may be involved. The *v* gene, whose product is tryptophan oxygenase, catalyzes the first step of the conversion of tryptophan into brown eye pigment. If the other genes of the 10A1-2 band were involved in this metabolic pathway, they would necessarily control some of the sequential steps. However, the genes controlling the enzymes catalyzing these steps have been thoroughly studied and mapped to distinct regions of the chromosome outside the 10A1-2 band (Lindsley and Grell, 1968; Dickinson and Sullivan, 1975; Kamyshev *et al.*, 1988; Lindsley and Zimm, 1992). The 23 transcriptionally active fragments are distinct in terms of their variable relative expression levels at different stages of embryogenesis and in different tissues, suggesting that each fragment contains at least one distinct transcription unit (Kozlova *et al.*, 1997).

Based on these data, it may be concluded that the 10A1-2 band is informationally complex and that it contains many functionally unrelated genes. Moreover, it contains repeated DNAs of several families: K1, K2, AC and GCT (Kokoza *et al.*, 1997).

The very accurate gene localization in the 9F12-10A7 on the cytogenetic (Zhimulev *et al.*, 1987a,b) and molecular (Kozlova *et al.*, 1994) maps made it possible to estimate the fit between the molecular and recombinational distances between the genes. The total distance between the very remote *fliG* and *l(1)10Ae* genes is 0.89 centimorgans or about 290 kb. A straight line connects the position of the other genes on both maps (Figure 48). According to these data, 0.01 centimorgans correspond to 3.5 kb. This is close to 3.8 kb/0.01 centimorgans estimated for the long region of the X chromosome between the *w* and *f* genes (Rudkin, 1965; Lefevre, 1971).

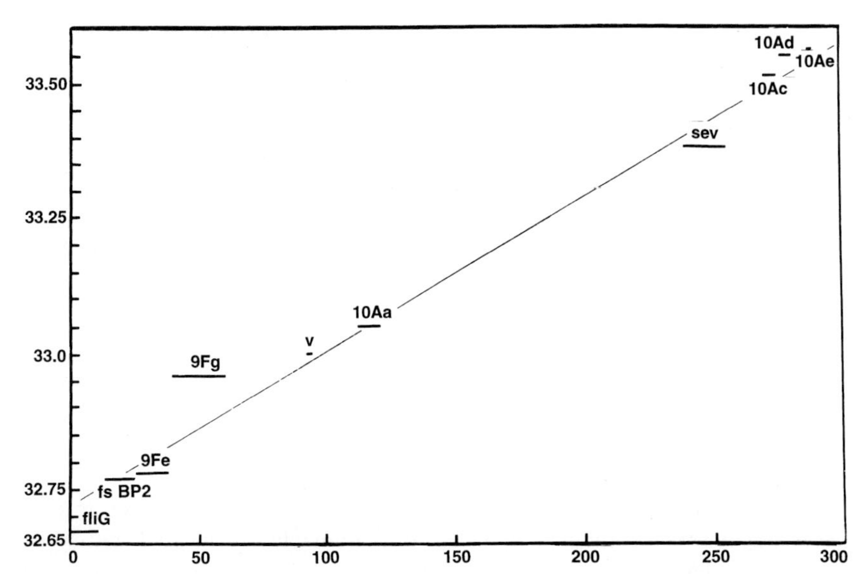

Figure 48. Correspondence between positions of the genes on molecular (abscissa) and genetic (ordinate) maps in the 9F12-10A7 region. Distances are given in kilobase pairs (abscissa) and map units (ordinate) (from Kozlova *et al.*, 1994).

The suggestion (Zhimulev *et al.*, 1981c, 1983) that the silent zone between the *l(1)10Aa* and *sev* genes may be a hot spot for the formation of chromosomal rearrangements appeared implausible after the band size was accurately determined and the rearrangements mapped (Kozlova *et al.*, 1994). The rearrangement breakpoints within the 10A1-2 band are apparently randomly located (see Figure 46).

Three genes controlling different functions were identified in the small 9F12 band: *l(1)9Fe* (viability), *fliG* (flight ability), and *fs(1)BP2* (female fertility) (Figure 46). This band is about 30 kb in size (Kozlova *et al.*, 1994).

As experiments with mosaics demonstrated, mutations of the *l(1)9Fe* gene are cell lethals; that is, the gene product is required for survival of the cells (Bgatov and Zhimulev, 1982, 1985; Bgatov *et al.*, 1986).

The *fliG* locus has two foci on the blastoderm map, the flight function being associated with wing determinants. The lethality focus is near the head capsule and first leg pair. Mutation of the *fs(1)BP2* gene causes female sterility due to underdevelopment of the ovaries (Bgatov and Zhimulev, 1985).

Seven genes (Figure 49) were identified in a 12-kb DNA fragment mapped *in situ* to the 67B heat-shock puff (see Table 5). According to electron microscopic data, this puff originates from two thin bands 67B1 and 67B2 (Semeshin *et al.*, 1984, 1985a). Either one or both of these bands obviously contain several genes.

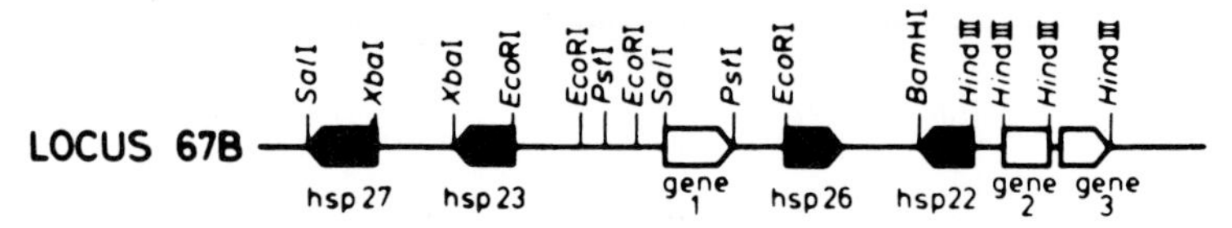

Figure 49. Organization of the 67B heat shock puff in *D. melanogaster*. Direction of transcription is indicated by arrows. The genes of the four small heat-shock proteins are depicted by black arrows [from Ayme and Tissieres (1985). Locus 67B of *Drosphila melanogaster* contains seven, not four, closely related heat-shock genes. *EMBO J.* **4**, 2949–2954, by permission of Oxford University Press].

By using *in situ* hybridization of cloned segments and deletion mapping, a detailed map covering the 87D5-87E5-6 regions (a walk extending over to 14 bands), was built for the *rosy-Ace* region of *D. melanogaster* chromosomal DNA (Bender *et al.*, 1983b; Hall *et al.*, 1983; Spierer *et al.*, 1983; Bossy *et al.*, 1984; Spierer, 1984; Gausz *et al.*, 1986; Nagoshi and Gelbart, 1987).

Two to three genes were identified by mutagenesis and seven to eight DNA fragments encoding poly(A)$^+$RNA presumably are located within the large 87E1-2 band. Five fragments encoding transcripts were detected in the left (proximal) part of the 87E5-6 band (Figure 50).

The *BX-C* is a complex that includes several genes and cis-regulating regons controlling the segmental pattern of the body of larvae and adults. When all the complex is deleted, the transformed segments resemble the normal second thoracic segments of the adult (reviews: Bender *et al.*, 1983a, 1985; Hogness *et al.*, 1985; Scott, 1987; O'Connor *et al.*, 1988; Lawrence, 1992; Hendrickson and Sakonju, 1995). The locus spans about 300 kb (Figure 51), it was mapped to the *Df(3R)P9* deletion, which removes 89E1-2-89E3-4 bands (Lewis, 1978, 1985; Sanches-Herrero *et al.*, 1985) or actually a single band (see for discussion Zhimulev *et al.*, 1982b).

There are various transcription units in *BX-C:* those producing protein-coding transcripts: *Ubx*, *abd-A*, *Abd-B*; and those producing noncoding transcripts: *bxd* and *iab-4*. Twelve cis-regulatory regions were found in *BX-C*, which are involved in controlling the development of organs and body structures (see Martin *et al.*, 1995, for references).

Mapping the transcripts, ORFs, and putative genes shown in Figure 51 gives strong support to the theory that the 89E1-4 band is really oligogenic.

The genes of the *Achaete-scute*-complex control the development of cuticular sensory organs (bristles) in *Drosophila*. The loss-of-function mutations *achaete (ac)* and *scute (sc)* suppress formation of bristles at correct positions; in contrast, gain-of-function mutations produce extra bristles in displaced positions (Serebrovsky and Dubinin, 1930; Agol, 1931; Dubinin, 1932, 1933; Garcia-Bellido, 1979; Ghysen and Dambly-Chaudiere, 1988).

A region comprising about 110 kDa of DNA (Figure 52), where the

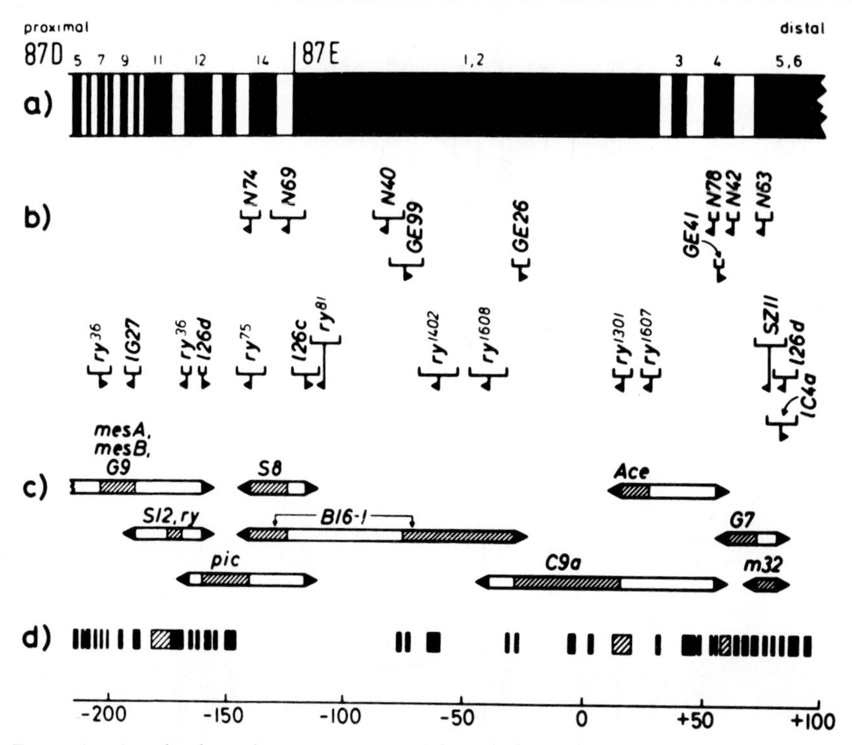

Figure 50. A molecular and cytogenetic map of the 87DE region. (a) banding pattern; (b) localization of breakpoints (brackets) and direction (arrows) of deletions; (c) localization of the genes detected by genetic methods; (d) localization of DNA fragments complementary to the transcripts. Bellow, a physical map of DNA (distances in kb) (from Gausz *et al.*, 1986).

genes of the *AS-C* complex and the adjacent *yellow* gene are located, was cloned (Campuzano *et al.*, 1985; Alonso and Cabrera, 1988; Gonzalez *et al.*, 1989; Campuzano and Modolell, 1992). Numerous chromosomal rearrangements affecting the *AS-C* were assigned positions on cytological and molecular levels (Muller and Prokofyeva, 1935; Garcia-Bellido, 1979; Campuzano *et al.*, 1985). On the molecular map, the cloned sequence was genetically defined by the y^{3P} (at + 73 kb) and the y^4 (+ 71 kb) insertions and on the proximal end by the breakpoint of the *260-1* deletion (− 36 kb) (Campuzano *et al.*, 1985). On the cytological map (Figure 53), these rearrangements were defined by two doublets 1B1-2 and 1B3-4; this indicated that seven genes are located in two bands. The genes of the complex are transcribed in opposite orientation (Figure 52); consequently, the band cannot make up a transcription unit. In *D. subobscura y*, *ac* and *sc* are also located in the same band of the 2B region (Botella *et al.*, 1996).

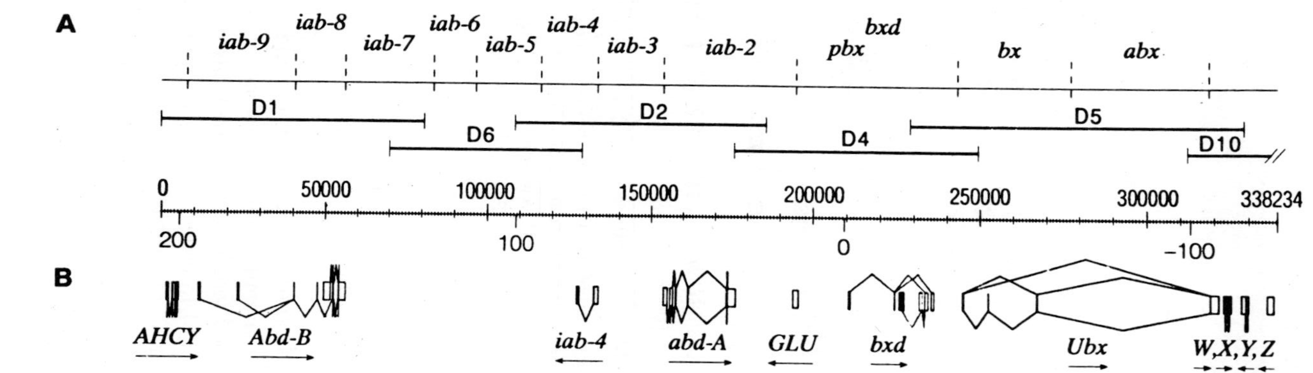

Figure 51. Genetic and molecular maps of the *Bithorax* complex. (A) Boundaries of cis-regulatory regions, delineated by hatched vertical lines, D1-D10-P1 bacteriophage BX-C insertions; the numbers below and above the line refer to the map positions in kilobases (from Hendrickson and Sakonju, 1995). (B) The transcription map based on intron–exon mapping, cDNA, and computer analysis. Transcript units include homeobox-containing *Abd-B*, *abd-A*, and *Ubx*; noncoding *bxd* and *iab-4*; and the open reading frames and predicted candidate genes. Arrows indicate the direction of transcription. The predicted genes labeled W, X, Y, and Z refer to two LDL receptor "a" repeats; serine protease-like, chaperonin-containing t-complex γ-subunit-like protein, and no-on A-like transient, respectively (from Martin *et al.*, 1995).

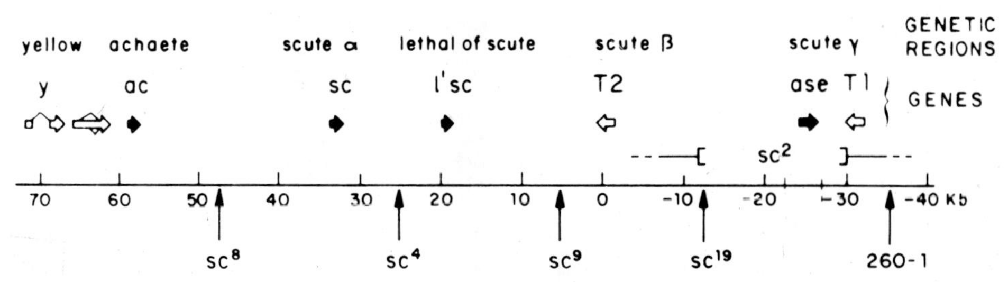

Figure 52. A scheme of the *achaete-scute* complex organization in *D. melanogaster* [from Gonzalez *et al.* (1989). Molecular analysis of the asense gene, a member of the achaete-scute complex of *Drosophila melanogaster*, and its novel role in optic lobe development. *EMBO J.* **8,** 3553–3562, by permission of Oxford University Press].

The dopa decarboxylase gene in *Drosophila* was mapped to the 37C1-2 band by *in situ* hybridization (Hirsh and Davidson, 1981). Another eight genes were localized in the flanking stretch of 25 kb (Figure 42d) (Eveleth and Marsh, 1986a,b; Marsh *et al.*, 1986; Spencer *et al.*, 1986a; Black *et al.*, 1987; Stathakis *et al.*, 1995). The *Cs* and the *Ddc* genes are transcribed in opposite orientation, and there is an 88-bp overlap between the 3' end of the *Ddc* gene and its 3′-adjacent gene (Spencer *et al.*, 1986b). The closely linked *amd* and *Ddc* genes in *Drosophila* share extensive homology (Eveleth and Marsh, 1986b). This is presumably because 14 of the 18 genes identified in the 37BC regions function in

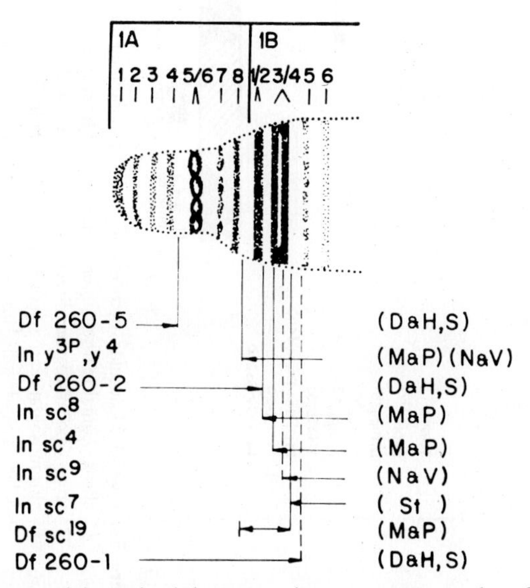

Figure 53. Localization of the ends of chromosomal rearrangements in the *achaete-scute* region of the X chromosome of *D. melanogaster* (from Garcia-Bellido, 1979).

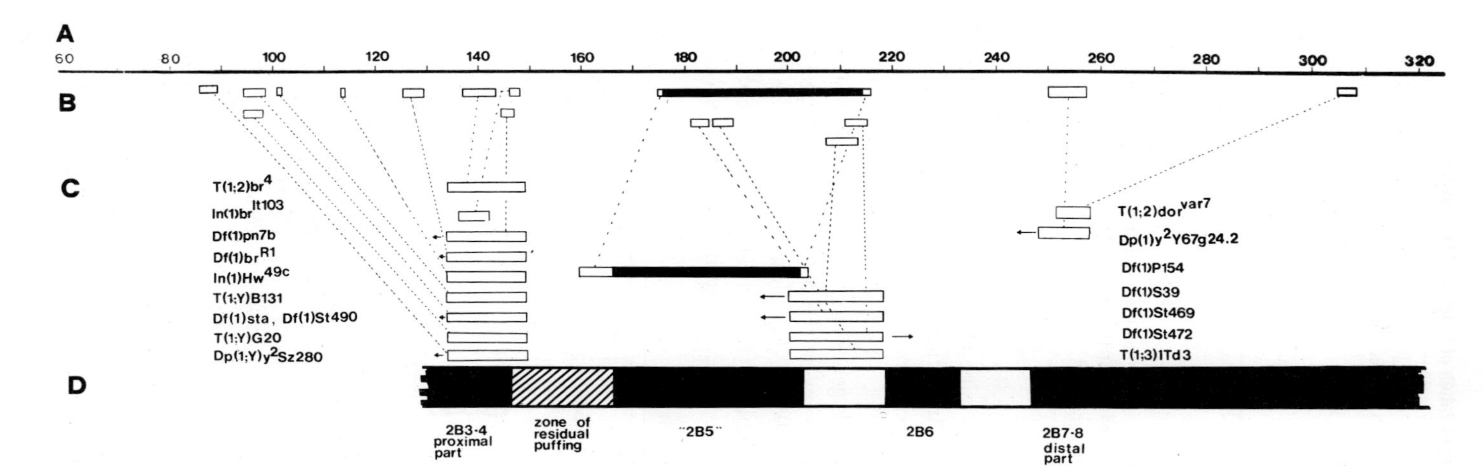

Figure 54. (A–D) Molecular–cytogenetic and (E) complementation maps of the 2B3-5-2B6 region. (A) A physical map of DNA; (B and C) mapped chromosomal rearrangements on the (B) physical and (C) cytogenetic maps. The limits of the localization of regions are indicated by open triangles; arrows point to the direction of deletions and duplications; (D) a cytological map of the region; (E) horizontal lines depict mutations and chromosomal rearrangements. Overlapping lines designate no complementation. The number of hypomorphic mutations noncomplementary only to deletions and "long" alleles are in a rectangle. Asterisks designate mutations induced by the P element, deletions are marked with triangles. Mutations and rearrangements affecting female fertility are underlined by thick lines. (A–D) Reprinted by permission after Belyaeva *et al.*, 1987, Protopopov *et al.*, 1991; (E) reprinted by permission from Mazina *et al.*, 1991.

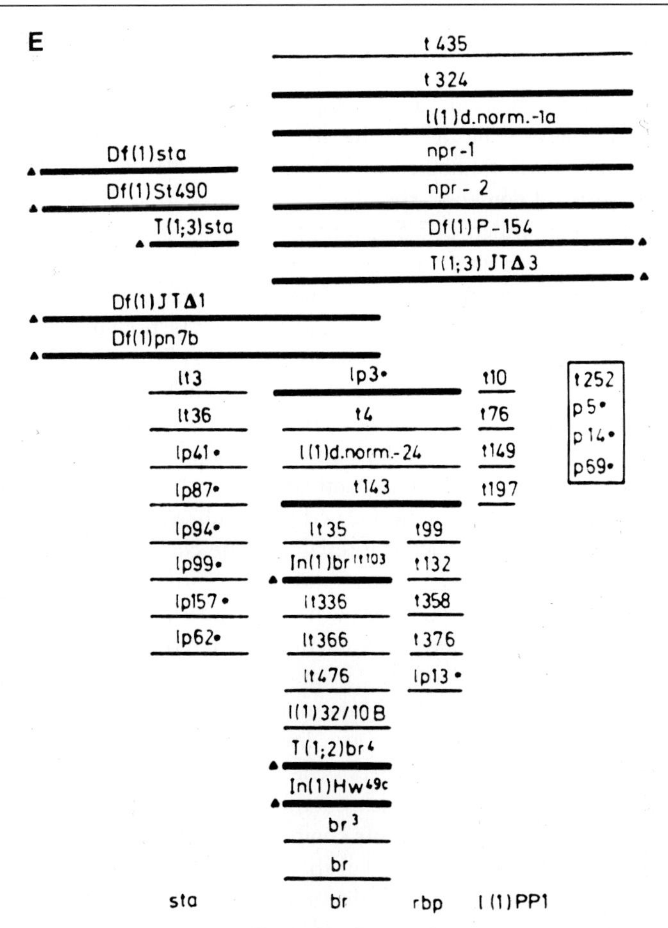

Figure 54. *Continued*

catecholamine metabolism and/or in the associated formation, sclerotization, and pigmentation of cuticle (review: Wright, 1987). It is appropriate to conclude by recalling that the 25-kb fragment may be associated with the 37C1-2 band with the stipulation that the resolution capacity of *in situ* hybridization does not allow a decision as to whether all the genes reside in this band or part may lie in the adjacent interbands.

In C. *thummi* (and 12 other species) there are two clusters of hemoglobin gene, five and six to seven genes in each. The gene cluster containing monomeric hemoglobin proteins III and IV (five genes) in all 13 species is located in a single band in chromosome arm E. The hemoglobin gene cluster containing the genes for the dimeric Hb VIIB proteins (six to seven genes) are located, again, in a single band in chromosome arm D (Schmidt *et al.*, 1988).

An EM study of puff formation in the 2B of the X chromosome of *D. melanogaster* demonstrated that one of the two bands in the 2B1-4 region splits and the ecdysterone-induced puff develops from the proximal part of the band named 2B5 (Belyaeva *et al.*, 1987). Its distal part remains compacted, reminiscent of the single band numbered 2B3-4 on Bridges' (1938) map. When inactive, the 2B3-4 band composes the entire material, so that 2B5 does not exist. This band has been called 2B3-5 in order to remain consistent with Bridges' nomenclature (Belyaeva *et al.*, 1987).

Genetic saturation experiments and molecular cloning of the 2B3-5-2B7-8 region (Figure 54) permitted localization of *ecs* (synonym is *BR-C*) to the right part of the 2B3-5 band. The *ecs* locus controls cell sensitivity to ecdysterone; the function of the distal part of the band is unclear—there is reason to suppose that it composes the *sta* gene. Thus, there is evidence indicating that the organization of the band is informationally complex and also that at least two of its parts may be autonomously activated.

The *dor* and *swi* genes are located in the chromosome region containing the 2B6 band and the adjacent interbands. The complementation maps for *swi* (Belyaeva *et al.*, 1989), *dor* (Rayle and Hoar, 1969; Bishoff and Lucchesi, 1971; Garen and Gehring, 1972; Belyaeva *et al.*, 1989), and *swi* are simple and nonoverlapping. In independent experiments several lethals were isolated which affected puparium formation and were located near the tip of the X chromosome (Rayle, 1967b; Kiss *et al.*, 1976). Mutations of the *npr* series appeared to be alleles of the *ecs* mutations and previously known *br* visible mutations (Belyaeva *et al.*, 1980a,b). Subsequently, the *ecs* gene was renamed *Broad-Complex* (BR-C). The *swi* was renamed *hfw* (Lindsley and Zimm, 1992).

The *BR-C* locus is genetically complex, containing several complementation groups. Mutations of a group fail to complement any of the other mutations of the group; for example, *br*, *rbp*, and *l(1)pp1*. These mutations do not complement those overlapping them; *t143*, for example (see Figure 54) and the "longer" mutations fail to complement any one of the alleles at the locus (*t324* and others shown in Figure 54) (Belyaeva *et al.*, 1978, 1980–1982, 1989; Aizenzon *et al.*, 1980, 1982; Ilyina *et al.*, 1980; Kiss *et al.*, 1980, 1988; Zhimulev *et al.*, 1980b, 1982c; Aizenzon and Belyaeva, 1982; Solovyova and Belyaeva, 1989; Mazina *et al.*, 1990, 1991; Balasov and Bgatov, 1992). The strongest mutations at the *ecs* locus (*lt324* or *lt435*; Figure 54) show the same phenotype, and developmental failure at the end of the third-larval instar is not due to ecdysterone deficiency: the puparium is not formed (Kiss *et al.*, 1976a,b), and the ecdysterone puffs are not induced (Belyaeva *et al.*, 1981); the larval and pupal organs do not undergo metamorphosis even when hormone is available (Kiss and Molnar, 1980; Kiss *et al.*, 1980, 1988; Fristrom *et al.*, 1981; Restifo and White, 1991, 1992).

Homozygotes for the weaker alleles at *BR-C* locus survive, but adults have mutant phenotypes, such as *br* or *rbp* (Aizenzon *et al.*, 1980). Heterozygotes for seven rearrangements and six point mutations are viable, though sterile (Mazina *et al.*, 1990, 1991). Disturbed fertility is not associated with defects of a minor chorion protein (called *s70*) mapped to the region of the 2B3-5 puff (Yannoni and Petri, 1984; Peterson and Petri, 1986). Subsequently, the *s70* gene was more accurately localized to the more distal cytological interval 2A1-2-2B3-4 (Mazina and Dubrovsky, 1990; Mazina *et al.*, 1991). Another mutation, *de12*, causing sterility (presumably associated with the absence of the egg filaments), is induced by insertion of the *gypsy* mobile element (Orr *et al.*, 1989; Huang and Orr, 1992).

Transplantation of the gonads isolated from BR-C heterozygous larvae to larvae carrying the *Fs(1)k1237* dominant mutation causing female sterility demonstrated that, when associated with the host's gonad ducts, the donor gonads retained their ability to function normally. This was taken to mean that female sterility in BR-C mutants results from disturbance of the somatic, not the germline, cells, possibly due to defects of the genital imaginal disks (Mazina and Korochkina, 1991).

The molecular and genetic organization of the 2B3-5-2B6 region became the subject of intensive research. The DNA from this region was cloned, and the rearrangement breakpoints and insertions of mobile elements were mapped within its confines (Chao and Guild, 1986; Galceran *et al.*, 1986, 1990; Belyaeva *et al.*, 1987; Protopopov *et al.*,1988, 1991; Sampedro *et al.*, 1989; Solovyova, 1992).

Mapping of the ends of chromosomal rearrangements on the physical map allowed definition of the limits of the genes in a region. The *BR-C* locus at the proximal site is defined by the *Df(1)St472* breakpoints (between 210.5 and 214.5 kb on the map; see Figure 54). The gene can be placed to the right of the *Df(1)sta* and *Df(1)St490* breakpoints (between 94 and 98.5 kb; see Figure 54). This appears accurate, because heterozygotes for these deletions and a number of *BR-C* mutants are sterile (Belyaeva *et al.*, 1987; Mazina *et al.*, 1991).

The *de12* is a mutation causing sterility. It is induced by insertion of the *gypsy* element and maps to position 169.5 kb (Huang and Orr, 1992). P-element insertions were mapped at positions from 168 to 172 kb, one insertion was mapped to the 200 to 204-kb region (Solovyova, 1992). Thus, the total length of the gene is not less than 115 kb.

Further information concerning the function of *BR-C* gene in the active puff will be given in Section VI,E.

The *dor* and *hfw (swi)* genes, which are associated with the 2B6 band, lie approximately in the 40-kb range in the 214- to 254-kb region (see Figure 54) (Belyaeva *et al.*, 1987, 1989; Shestopal *et al.*, 1997), and, finally, *sta* is in about 50 kb to the left of *BR-C* (Melnick *et al.*, 1993). The *dor* locus was cloned and characterized. The 3.1-kb *dor* transcript was detected by Northern hybridization at all stages of development and is expressed in third-instar larvae salivary gland cells. The 3180-bp *dor* cDNA predicts an ~115-kDa protein that contains a cysteine and histidine-rich, zinc-finger-like motif. The protein sequence reveals 23% identity to the *Saccharomyces cerevisiae* PEP3 protein (Shestopal *et al.*, 1997). The PEP3 protein is a protein membrane of vacuole. Mutants of *pep3* gene show weak abnormalities of endocytosis (Preston *et al.*, 1991; Robinson *et al.*, 1991). In *Drosophila* the DOR protein interacts with HOOK protein and possibly takes part in endocytosis (Kramer and Phistry, 1996).

There is reason to suppose that the juxtaposition of *ecs*, *dor*, and *swi* might have become fixed in the course of evolution. In fact, it was demonstrated by *in situ* hybridization of clones of DNA from the 2B3-5 puff of *D. melanogaster* that all three genes are also located in the region of the same puff in *Drosophila funebris*, *D. hydei*, *Drosophila repleta*, *Drosophila mercatorum*, *Drosophila paranaensis*, and *D. virilis*. The puff is the most distal subtelomeric. In *Drosophila kanekoi*, it lies more proximally. In *D. pseudoobscura*, it is in the proximal portion of the X chromosome (Kokoza *et al.*, 1991, 1992).

A cluster of several homeotic and other genes (Dunkan and Kaufman, 1975; Kaufman, 1978; Kaufman *et al.*, 1980; Lewis *et al.*, 1980a,b,c; Wakimoto *et al.*, 1980; Denell *et al.*, 1981; Wakimoto and Kaufman, 1981; Hazelrigg and Kaufman, 1983; Cribbs *et al.*, 1992) was mapped to the 84AB region (Figure 55). The entire *ANT* complex controls the development of the head and anterior thoracic segments (Gehring, 1985; Abbott and Kaufman, 1986; Scott, 1987). Using gene cloning (Scott *et al.*, 1983; Gehring, 1985; Mlodzik *et al.*, 1988), a physical map correlating with the genetic map was constructed for this region. It is about 300 kb long and contains eight transcript units of which one, *Antennapedia*, spans over 100 kb (Figure 56). *Antp* has two promoters and, as a consequence, two transcription units, one of 36 kb, the other of 103 kb. Both have the same exons grouped within 13 kb at the 3′ end. The length of the open reading frame is about 1.1 kb; that is, it constitutes about 1% of the length of the transcription unit (Garber *et al.*, 1983; Gehring, 1985; Laughon *et al.*, 1985, 1986; Schneuwly *et al.*, 1986; Scott, 1987). The homeodomains are encoded within the 3′ end of four genes (*Dfd*, *Scr*, *ftz*, and *Antp*) (Gehring, 1985; Levine *et al.*, 1985). Thus, a cluster of functionally associated genes is located in the 84A-B1-2 region, eight transcripts, two tRNA genes per 1–2 bands.

The structural organization of the *ANT* complex is generally the same in *D. virilis* and *D. subobscura* (Hooper *et al.*, 1992). In nine *Drosophila* species of the *obscura* group, the probe homologous to *D. melanogaster* DNA hybridized *in situ* to the DNA in the group of bands of the E element of the chromosomes of Muller's nomenclature (Terol *et al.*, 1991).

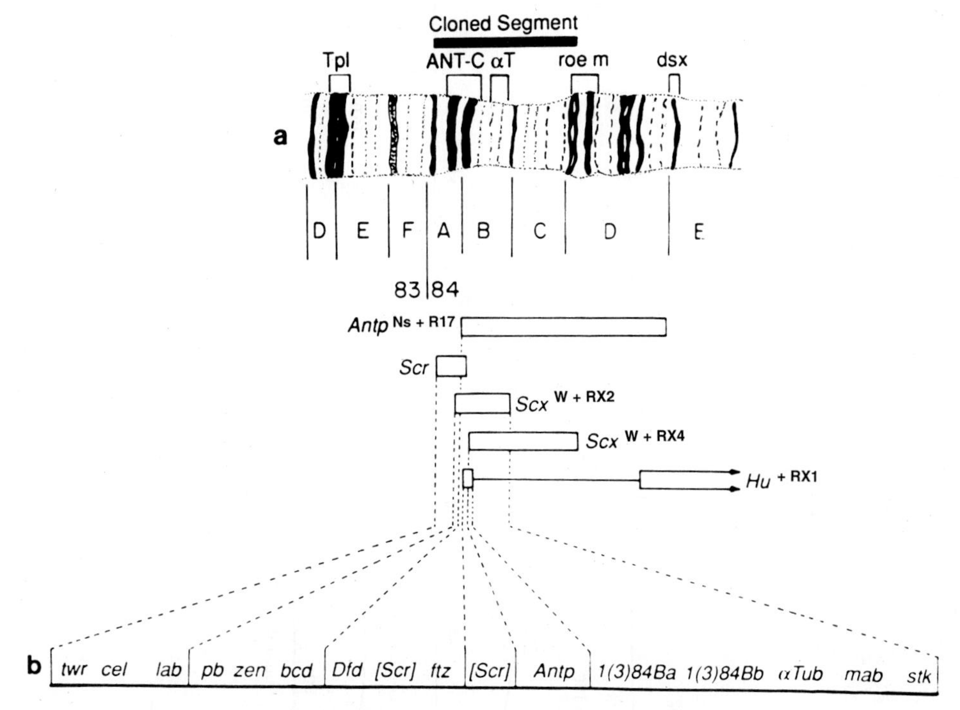

Figure 55. A cytogenetic map of the 84AB region. (a) Drawing of the portion of the third chromosome on which the ANT-C resides. Open bars below the chromosome show the extent of deletions; (b) genetic loci (from Lewis *et al.*, 1980b; Kaufman *et al.*, 1990).

3. The "simple" bands

It may be imagined that there exist bands containing DNA in amounts sufficient for a single gene (i.e., the bands may be simple in terms of information). However, there is, so far, no convincing evidence that such bands have been identified.

The organization of the 3AB region containing a series of very thin bands (Figure 42) strongly supports the "one band–one gene" hypothesis. The 3C2-3, 3C7, and 10B1-2 bands are among those that can be viewed as informationally simple. These bands lie in regions intensely saturated with mutations (Beermann, 1972; Zhimulev *et al.*, 1987a,b).

The *white (w)* gene maps at the distal end of the 3C2-3 band (approximately 10% of band length); *w* does not change its location when broken or transferred by inversion of the bulk of the band (Sorsa *et al.*, 1973). The *w* gene occupies about 14 kb (Levis *et al.*, 1982; O'Hare *et al.*, 1983, 1984).

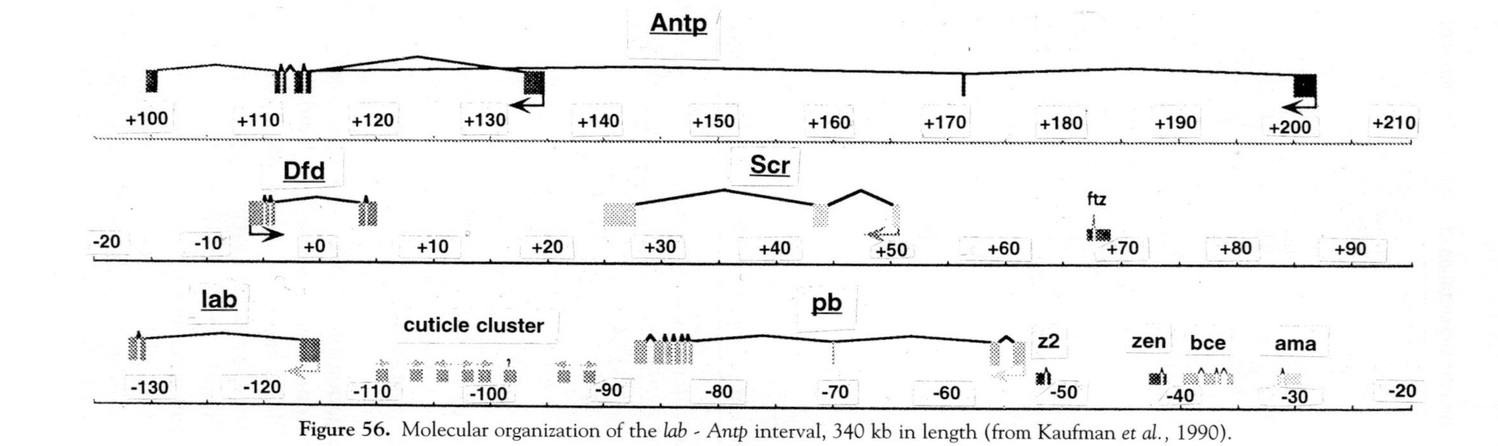

Figure 56. Molecular organization of the *lab* - *Antp* interval, 340 kb in length (from Kaufman *et al.*, 1990).

Analysis of P-element-mediated DNA transformation demonstrated that a DNA fragment 11.7 kb long composes the complete sequence of the w^+ gene (Gehring *et al.*, 1984; Hazelrigg *et al.*, 1984). There is an indication, albeit not well documented, that the regulatory portion of the w^+ gene lies in the distal 3C1/3C2-3 interband (Rykowsky *et al.*, 1988).

The 3C2-3 band is rather large and there are no accurate estimates of its DNA content; but if we accept that *white* is located on the distal part and *rst* on proximal part of the 3C2-3 band (Lindsley and Zimm, 1992), the band can compose about 180 kb of DNA (Davis and Judd, 1995). Consequently, about 160 kb is not asociated with the function of the w^+ gene. There is at least one gene between *white* and *roughest* (Davis and Judd, 1995).

The function of the remaining part of the band is unclear. No other genes were detected in the 3C2-3 band, although the region has been intensely studied (Beermann, 1972a). It may be assumed that either this DNA performs no function or it is silent (review: Zhimulev and Belyaeva, 1985); or perhaps it contains genes whose mutations are phenotypically not expressed and, hence, undetectable.

Gene *PP2* in *D. melanogaster* is about 120 kb long. Electronmicroscopic *in situ* hybridization of 5′ and 3′ flanking fragments of this gene showed their location to the left and to the right of the band 54B1-2. So, about 100 kb of the gene *PP2* completely occupy the whole band (Semeshin *et al.*, 1997a).

An RNA interval of about 37 kb long is transcribed from the DNA of the *Notch* gene, and the 10.4 kb poly(A)$^+$RNA is the mature processing product of a much larger primary transcript spanning this region (Artavanis-Tsakonas *et al.*, 1983; Kidd *et al.*, 1983, 1986; Grimwade *et al.*, 1985; Wharton *et al.*, 1985; Ramos *et al.*, 1989). Thus, most of the gene extends over a great part of the 3C7 band, although its size has not been accurately determined. The stretch of the gene from the 5′ end presumably lies in the distal 3C5-6/3C7 interband. The *Df(1)fa^{sbw}* deletion also carries away the distal 3C7 interband when it removes about 800 bp from the 5′ end of the gene (Fig. 57; also see Section IV,B). It may be assumed that the *Notch* gene spreads out over the entire band and the interband; the structural portion of the gene lies in the 3C7 bands, while the regulatory sequences are at the 5′ end, in the interbands.

Saturation experiments of 9F-10B regions did not reveal genes in the 10B1-2 and 10B3 bands (Geer *et al.*, 1983). Subsequently, a single *tu(1)Sz^{ts}* gene was mapped; this may not be the only gene in such a large band (10B1-2). The genes in this region may escape detection possibly because they contain DNA repeats, coding or not coding certain functions. The properties of intercalary heterochromatin in the 10B1-2 band indirectly support this assumption (Zhimulev *et al.*, 1982b). This band might not perform any genetic functions and contain only genetically silent DNA. The other large bands (10A1-2 and 87E1-2) are presumably partly composed of such DNA (Zhimulev and Belyaeva, 1985).

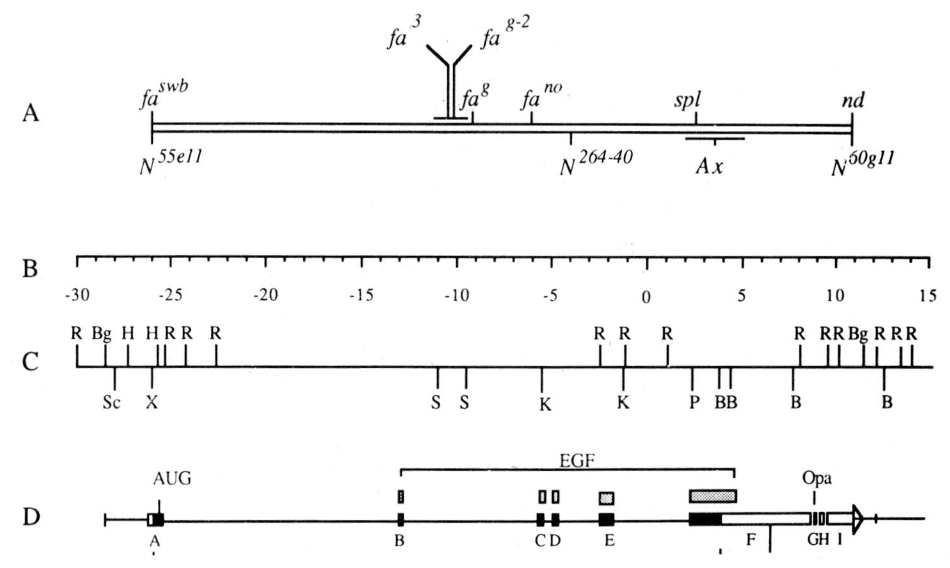

Figure 57. (A) Genetic and (B–D) molecular organization of the *Notch* locus of *D. melanogaster*. (A) Localization of mutations on the genetic map; (B and C) a restriction map: B, *Bam*HI; Bg, *Bgl*II; K, *Kpn*I; P, PstI; R, *Eco*RI; S, *Sal*GI; Sc, *Sac*I; X, *Xho*I. (D) A map of exons; AUG, translation start, EGF and Opa domains (from Ramos *et al.*, 1989).

4. The "artificial" bands

The bands resulting from genetic manipulations (chromosomal rearrangements or transformation) and not occurring in native polytene chromosomes may be called "artificial."

The visible mutant fa^{sbw} at the Notch locus results from the 800 bp $Df(1)fa^{sbw}$ deletion between the 3C5-6 and 3C7 bands. 3C7 becomes fused to 3C5-6 as a result (Keppy and Welshons, 1980). The cytology of the new band does not differ in any respect from that of the other bands, including 3C5-6 and 3C7, before they became fused. This may mean that compaction and banding pattern are independent; that is, the compacted regions, which are separated from the decompacted, can unite after the removal of the latter.

As Keppy and Welshons (1980) stress, the artificial construction of such bands implies that the existence of naturally occurring compound bands must be considered a reality. The 10A1-2 band is the case. DNA composing this band in *D. melanogaster* hybridizes *in situ* with two different bands in *D. virilis* located in the 2B and 5A regions (Kokoza *et al.*, 1990).

The transformation method provides an original approach to study the organization of the bands. It allows testing of the cytological consequences of sequences with known molecular genetic features as single copy into a genome. The insertion sites of the construct are identified by *in situ* hybridization; then, two possibilities are distinguished by electron microscopy: the arisen morphological structure is either a band or an interband.

Various DNA fragments are used as transforming constructs: the mobile P element, marker genes (the *rosy*, *Adh*, and *E. coli* β-galactosidase genes), and the expressing gene of DNA plasmids (Figure 58) (Semeshin *et al.*, 1986a,b, 1989a,b, 1994, 1997; Semeshin, 1990). According to the EM data, a new band is formed in each 19 of the 25 studied transformed regions (Figure 39). The sequences in such an artificial band remain functionally active: the marker genes function normally, the heat-shock puffs are readily activated, and the puffs are concomitantly formed.

These experiments have led to the following conclusions:

1. Since a single band is formed from functionally diverse DNAs, all the DNA becomes completely compacted without showing a band–interband pattern. This means that gene disposition in the two structures is not the necessary condition, as in the *Notch* locus, and the bands are not functionally associated with the interbands; thus, there is no reason to claim that the "band plus interband cytogenetic unit" exists (see Section II,B). If in the normal

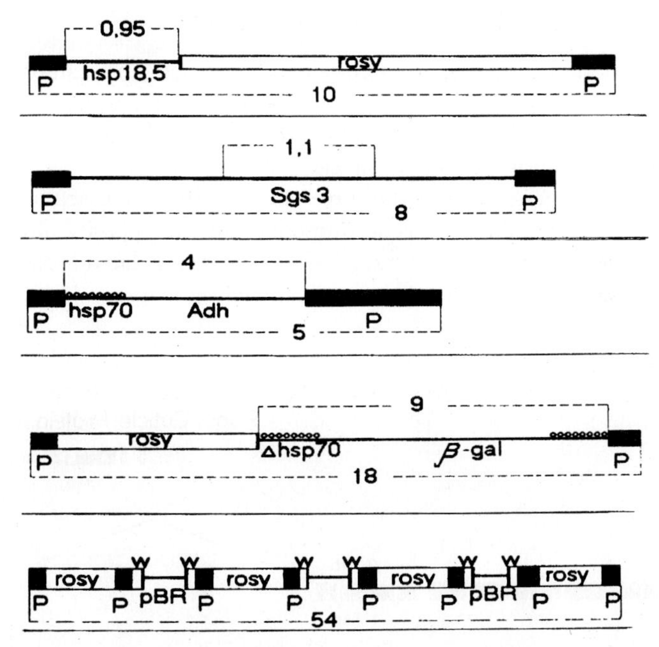

Figure 58. Genetic composition of insertions in transformed lines. Total length of the insert is indicated by the number at the dashed line (kb); the genes are *w*, *rosy*, *hsp18.5*, *hsp70*, *Sgs-3*, *Adh*, and *beta-gal*; P, sequences of the P element; pBR, fragments of the pBR322 plasmid (from Semeshin, 1990).

polytene chromosome, one of the genes, *rosy*, for example, were associated with the adjacent interband, and the association was required for function, then the interband would arise next to the compacted DNA of the *rosy* gene within the transforming construct. This line of reasoning may be applied to other genes and, if so, transformation would give rise to a series of bands and interbands at the insertion site, whose number would correspond to the number of genetic units in the insertion. However, there has never been eidence that this is so.

2. The assortment of genes derived from the various bands and interbands within the insertion is compacted into a single band; therefore, no specific mechanism provides compaction. The major requirement for compaction seems to be DNA in an inactive state. After induction of the heat-shock puffs or transcription of a P element, for example, there is decompaction of the DNA of certain genes in an insertion, the transformed band splits, and a puff develops. This means that different genes in the band can be activated independent of one another, and the promoter zones do not need to be located in the predecompacted regions, the interbands, for activation to occur (cf. Paul, 1972).

5. The bands containing overlapping genes

The organization of certain genes is quite unusual: other genes can be nested in their introns (reviews: O'Hare, 1986; Richards, 1987; Zhimulev, 1994a). Even if a single gene into which another gene has inserted is located in a band, the band is certainly informationally complex.

The *Gart* gene encodes the three purine pathway enzymatic activities; it contains six exons. Another gene transcribed in the opposite direction lies within the largest intron (Figure 59). Both genes are regulated independent of one another: *Gart* is active throughout development, the inserted gene is mainly

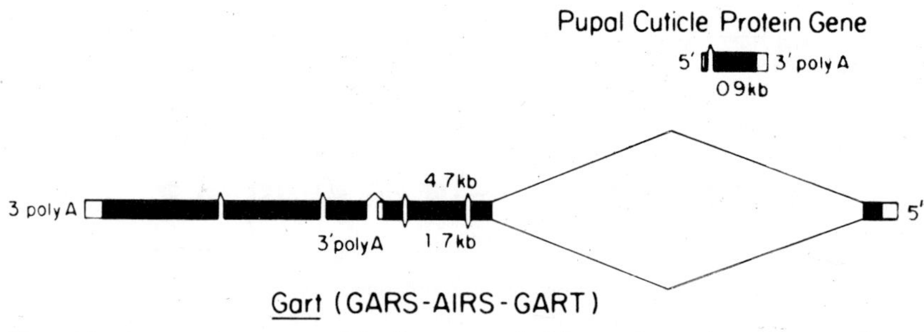

Figure 59. An exon–intron map of the *Gart* gene and of the pupal cuticular protein gene located in its intron (from Henikoff *et al.*, 1986, © Cell Press).

active during the 3-hour period in the abdominal epidermal cells of prepupae (Henikoff *et al.*, 1986).

The mutation of the gene exerting a pleiotropic effect on development maps to the 60-kb range from five transcripts are read through (Figure 60). Two tRNA genes are mapped within the complex (Gelbart *et al.*, 1985; Segal and Gelbart, 1985; Johnston *et al.*, 1990).

The *sina* gene, whose mutation arrests the development of the seventh central rhabdomere, is located in the large intron (about 9 kb) of the *Rh4* gene controlling the formation of photoactive pigment in the rhabdomere of the eyes of adults. The central seventh rhabdomere does not develop as a result of mutation (Montell *et al.*, 1987; Carthew and Rubin, 1990). The genes are transcribed in the opposite direction (Figure 61).

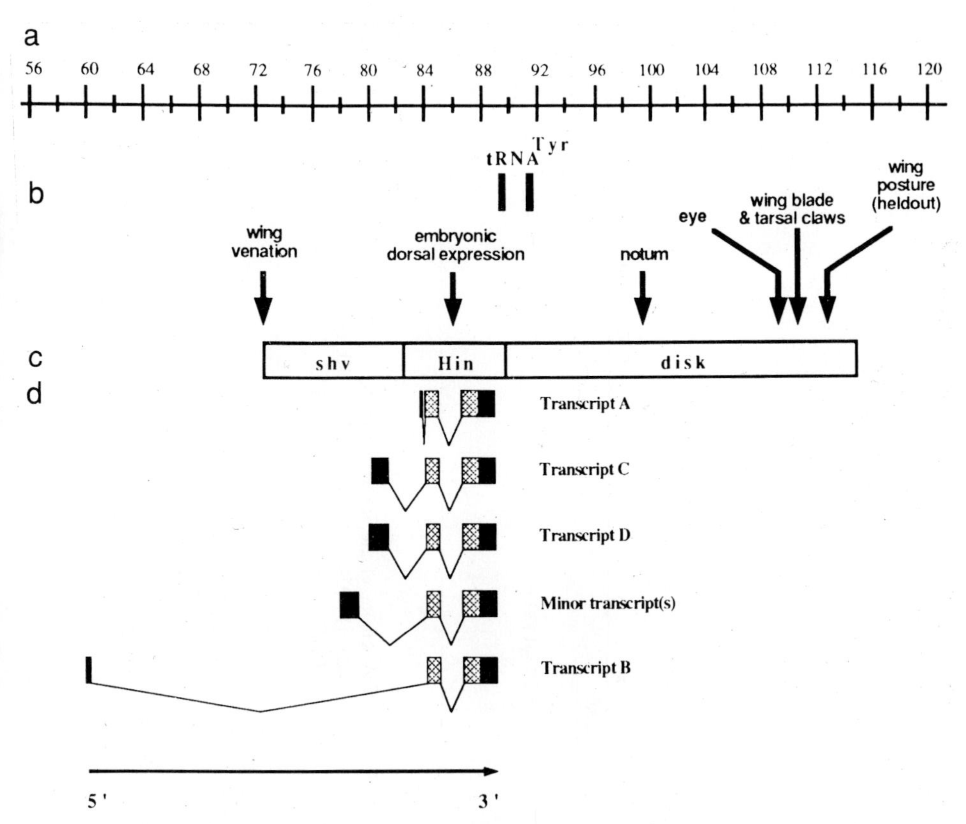

Figure 60. A scheme of the *dpp* gene in the 22F1-2 band of *D. melanogaster*. (a) A physical map of the region (distance in kb); (b) localization of the tyrosine t-RNA genes; (c) localization of three domains giving rise to different phenotypes; and (d) a scheme of the organization of transcripts (from Johnston *et al.*, 1990).

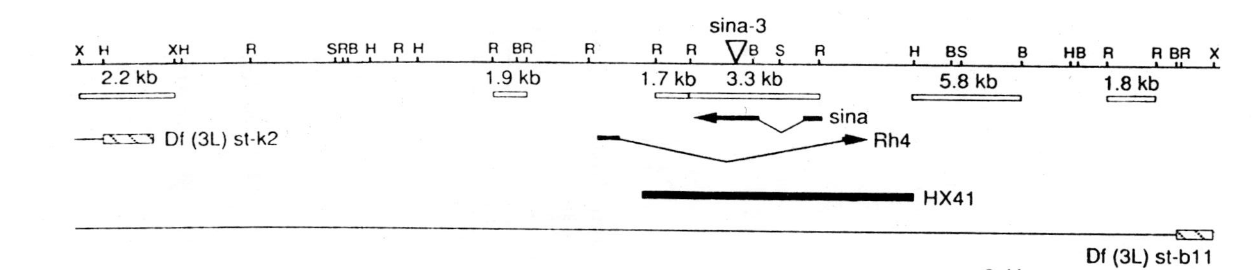

Figure 61. A scheme of the disposition of the *sina* and *Rh4* genes and fragments encoding RNA The top line shows the restriction map (X, XbaI; H, *Hind*III; R, *Eco*RI; S, *Sal*GI; and B, *Bam*HI). Open rectangles indicate fragments encoding RNA; HX41, the fragment used for transformation; Df, extension of deletions (from Carthew and Rubin, 1990, © Cell Press).

Many other genes are presumably organized in this complex fashion. The *Adh* and *Adh*[r] genes are probably nested in the *outspread* locus of *Drosophila* (McNabb *et al.*, 1996). Analysis of the complex complementation pattern of the mutation of the *Passover* locus (located in the X-chromosomal 19E3 region), showed that cis-interaction of two of its domains is required for the normal function of the gene. It was suggested that the gene is separated into two parts by at least one other gene residing in one of its introns (Baird *et al.*, 1990). A similar hypothesis was offered to explain the organization of the *Punch* gene (McLean *et al.*, 1990).

6. Genes covering several bands

Cytogenetic evidence localized the structural gene coding for cyclic AMP-phosphodiesterase (the *dunce* or *dnc* gene) at chromomere 3D4; another gene affecting female and male fertility *(sam)* was placed in this chromomere. The two *sam* alleles are separated by crossingover (0.02 ± 0.01 cM); the distance between the two *dnc* alleles is 0.04 ± 0.01 cM; and that between the *sam* and *dunce* is 0.08 ± 0.02 cM (Kiger and Golanty, 1979; Salz *et al.*, 1982; Salz and Kiger, 1984; Henkel-Tigges and Davis, 1990).

The organizational pattern of the *dnc* and the *sam* genes proved to be much more complex as judged by mapping on the cloned DNA and molecular characterization of the *dnc* and the *sam* region (Figure 62). It was found that *dunce* covers approximately 130 kb of genomic DNA, and that the other three genes *Pig1*, *Sgs-4, indnc*, are located in its giant intron (79 kb); also several other genes are distal to exons 1 and 2; one of these is *ng-1* (Davis and Davidson, 1984; Chen *et al.*, 1986, 1987; Digilio *et al.*, 1990; Furia *et al.*, 1990, 1992; Davis and Dauwalder, 1991; Qiu *et al.*, 1991). The *Sgs-4* gene maps to the 3C11–12 puff (Korge, 1977a; McGinnis *et al.*, 1980; Muskavitch and Hogness, 1980). Thus, the *dunce* gene with the inserted other three genes span a chromosome region containing six bands and interbands.

Transcriptions of the *Pig1*, *Sgs-4*, and *ng-1* genes—located in the 3C11-12 puff—are not under common hormonal control (Furia *et al.*, 1992).

The *E74* gene is located in the 74E region of chromosome *3L*; the 74E puff is the result of the activity of *E74* (see Section VI,E,1). Two bands, 74E1-2 and 74E3-4, become decompacted when the puff is formed (Semeshin *et al.*, 1985a).

Sh is a gene complex that encodes a family of potassium channel proteins in the neuromuscular system of *Drosophila*. The genetic complex is divisible into a haplolethal (HL in Figure 63), a viable (V), and a maternal effect (ME) region. The HL region is limited by the breakpoints of *T(X;Y)V7* and *T(X;Y)W32* and the ML region is limited by *T(X;Y)B55* and *T(X;Y)JC153* (Tanouye *et al.*, 1981; Baumann *et al.*, 1987; Kamb *et al.*, 1987; Ferrus *et al.*,

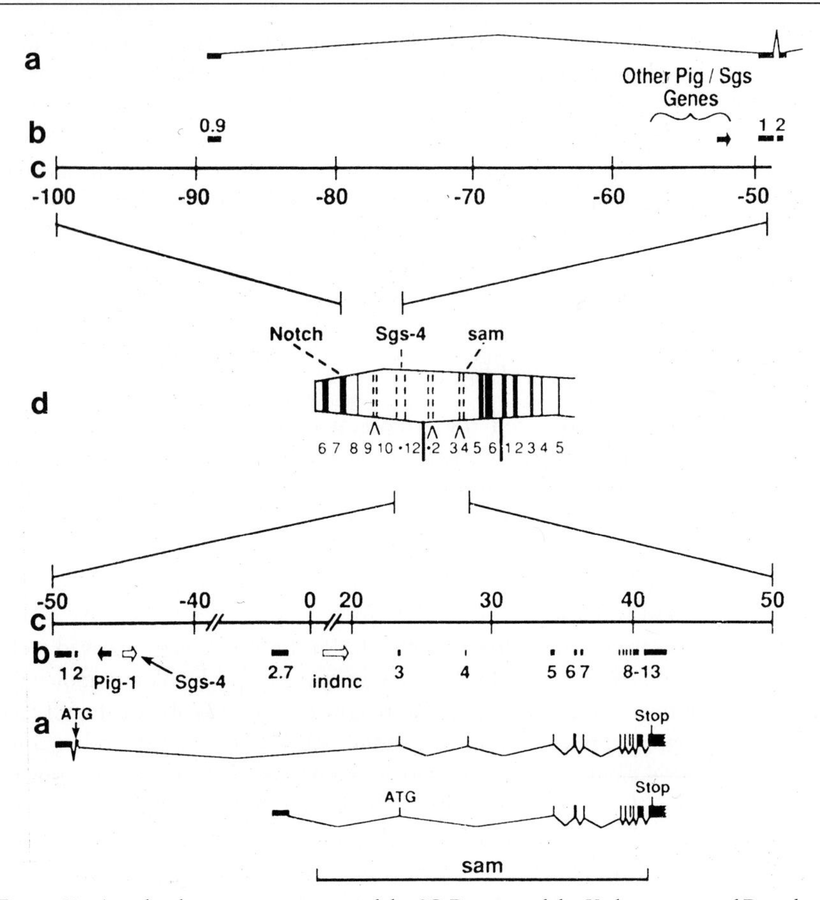

Figure 62. A molecular–cytogenetic map of the 3C-D region of the X chromosome of *D. melanogaster*. (a) A map of the *dunce* locus; (b) a map of transcripts in the 3CD region; (c) a physical map of the region; and (d) a cytological map of the region (reprinted from *Trends Genet.*, **7**(7), Davis, R. L., and Dauwalder, B., The *Drosophila dunce* locus: Learning and memory genes in the fly, 224–229, copyright 1990, with permission from Elsevier Science).

1990). Thus, the *Sh* locus spans 5–12 bands (the number depends on the localization accuracy of the rearrangements) on the cytological map and about 350 kb on the molecular map (Figure 63). The DNA region that was suggested to encode at least a part of the *Sh* transcript unit encompasses about 96 kb (+49 to +139 bp) and is separated by an intron 85 kb in length. The RNA of the coding fragment, supposedly of another gene (Kamb *et al.*, 1987), lies in the intron. A family of proteins, showing extensive homology to invertebrate troponin, is encoded by the HL region (Barbas *et al.*, 1991).

It is unclear from the examples just given how the patterns of the bands and the interbands may be formed from the DNA contained by a single gene.

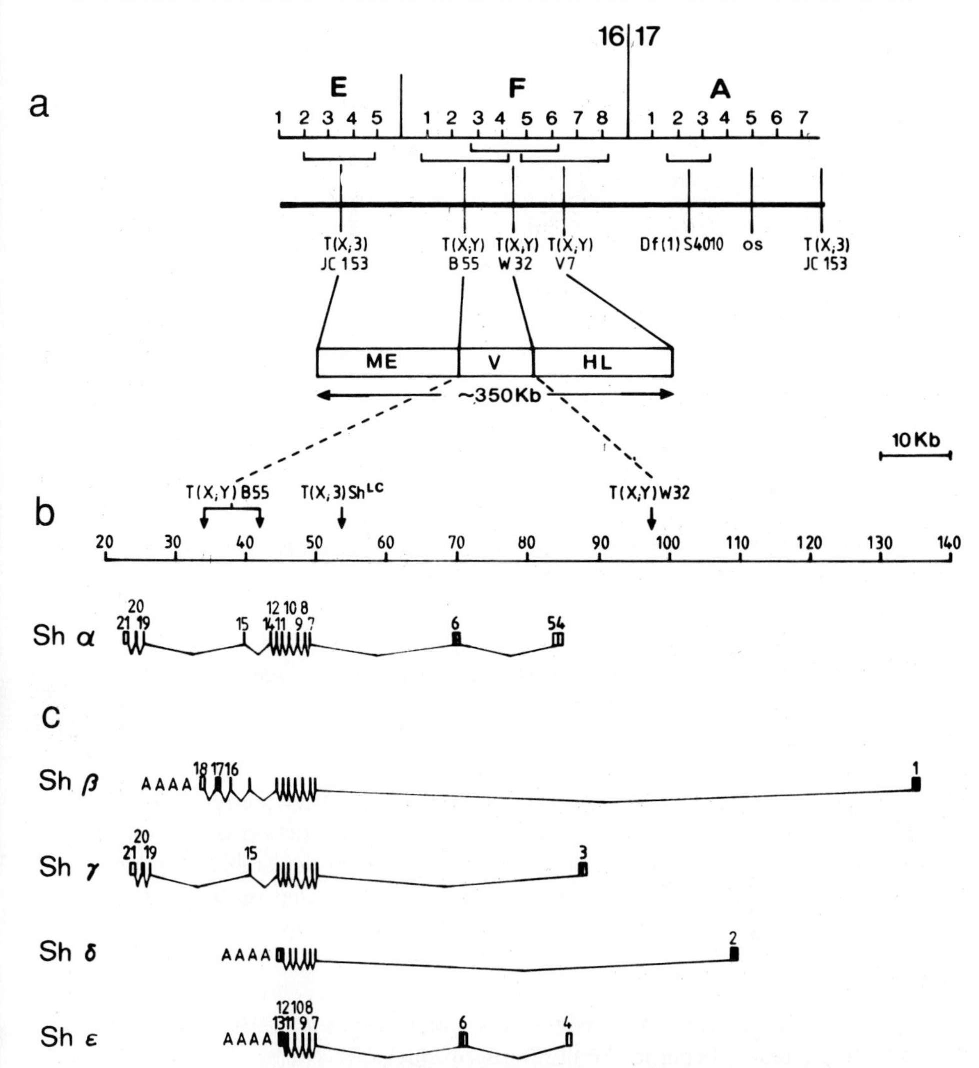

Figure 63. A molecular–cytogenetic map of the 16E-17A region of the X chromosome of *D. melanogaster*. (a) Localization of the breakpoints of chromosomal rearrangements on the cytological map of the 16E-17A; (b) a physical map of DNA; and (c) transcripts [from Pongs *et al.* (1988). Shaker encodes a family of putative potassium channel proteins in the nervous system of *Drosophila*. *EMBO J.* **7,** 1087–1096, by permission of Oxford University Press].

Findings discussed in this section show that there is no reason to consider the cytological band of polytene chromosome as a structure specially organized for location and function of a gene. The patterns of these results can be considered in the context of band organization as interpreted by Zhimulev and Belyaeva (1975a,c): a band consists of a DNA fragment containing genes, which are simply not active at this developmental stage or in this tissue; DNA of all genes (one or many) not active at this stage is compacted and is separated from other compacted DNA by active (decompacted) DNA sequences.

C. Bands, genes, and loops

The DNA fiber of the chromosome is much longer than that of the visible chromosome. The total packaging ratio of DNA in mitotic chromosome of *Drosophila* is about six to seven thousand. The packing ratio of DNA in bands can vary widely, from 30 to 380, but more commonly in the range of 50 to 100. The packaging of DNA molecules during the formation of a 30-nm superbead (nucleomere) chromatin fiber compresses the length of the DNA approximately 30- to 40-fold (see review: Zhimulev, 1992, 1996).

Various models have been suggested for DNA packaging at higher levels of organization. The loop model will be considered because it is amply substantiated by experimental data. The fiber is folded into 30- to 100-kb loops fastened at their bases by nonhistone proteins. In the compact mitotic chromosome, the neigboring loops are held together by protein–protein or protein–DNA interactions that form an internal protein scaffold. The major component of the scaffold is topoisomerase II. The scaffold is attached to specific DNA regions called the scaffold-associated regions (the SARs) (see reviews: Bode *et al.*, 1995; Razin *et al.*, 1995).

The compound lithium 3′-5′-diiodosalicylate (LIS) removes histones and other proteins at low concentrations of salts. These conditions are mild enough to spare specific scaffold–DNA interactions. After the removal of LIS the extracted nuclei are digested to completion by restriction nucleases, they preserve the SAR-site DNAs and thereby make them detectable (Laemmli *et al.*, 1978; Mirkovitch *et al.*, 1987; Jackson, 1991). Based on analysis of 18 SAR-associated regions from the *Drosophila* genome, the conclusions are as follows:

1. The SAR sites can be mapped to restriction fragments from 0.6–1.0 kb in size.
2. The SAR sites occur in the nontranscribed regions.
3. The distance between two adjacent sites varies from 4.5 to 112 kb.
4. No changes are observed in the position of the SAR sites upon induction of transcription.
5. The position of the SAR sites is the same in different cell types and also in the interphase and metaphase chromosomes (Mirkovitch *et al.*, 1987).

Are the bands, the loops, and the genes interrelated? Mirkovitch *et al.* (1987) believe that the highly transcribed genes occur in the small 4- to 17-kb loops; those with less abundant transcriptional activity would be found in the larger loops (50 kb or larger). The *hsp70*, the *actin* 5C gene, histone gene cluster, *Adh*, *Sgs-4*, *ftz*, and other genes were referred to the first class; the genes in the DNA fragment 320 kb long in the 87DE region (right arm of chromosome 3) were referred to the second class.

1. Histone genes

A SAR region was found in the nontranscribed spacer between the repetititive units of the histone gene cluster (Mirkovitch *et al.*, 1987) (Figure 64). One of the classes of *in vivo* topoisomerase II cleavage sites is mapped in nucleosome linker sites in the SAR of the histone gene loop between H1 and H3 genes (Kas and Laemmli, 1992). Because repetition of the SAR-site DNA is equal to the number of repetitive units, it may be concluded that there would be many loops in the DNA composing the polytene bands occupied by these genes.

2. *Adh*, *Sgs-4*, and *ftz* genes

Each of these developmentally regulated genes resides in a distinct loop and the SAR sites are located at the 5′ end of each gene at a distance of not more than 4.5 kb (Figure 65).

3. Genes of the 87DE region

At least 10 bands and interbands in the 87D region form the giant loop DNA longer than 78 kb. No less than eight genes are localized in it (Figure 65). The loop in which *Su(var)*, *G7*, and *m32* are localized forms two to three bands.

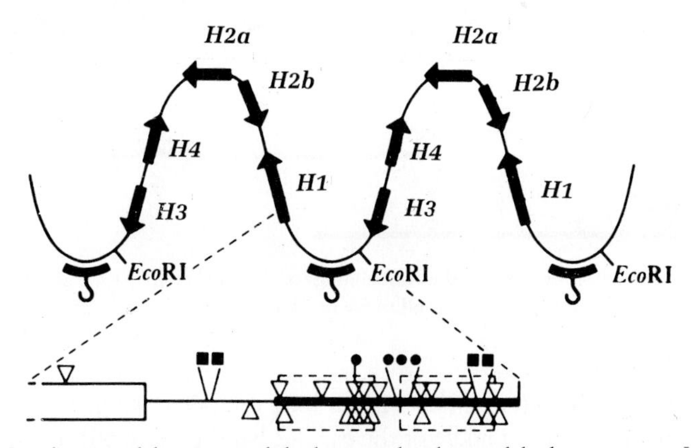

Figure 64. Localization of the genes and the loops in the cluster of the histone genes. H1 and H3, the histone genes; *Eco*RI, the restriction site. The arc with the hook designates the DNA fragment containing the SAR site sequences with 70% homology with the consensus of topoisomerase II binding site (from Mirkovitch *et al.*, 1987).

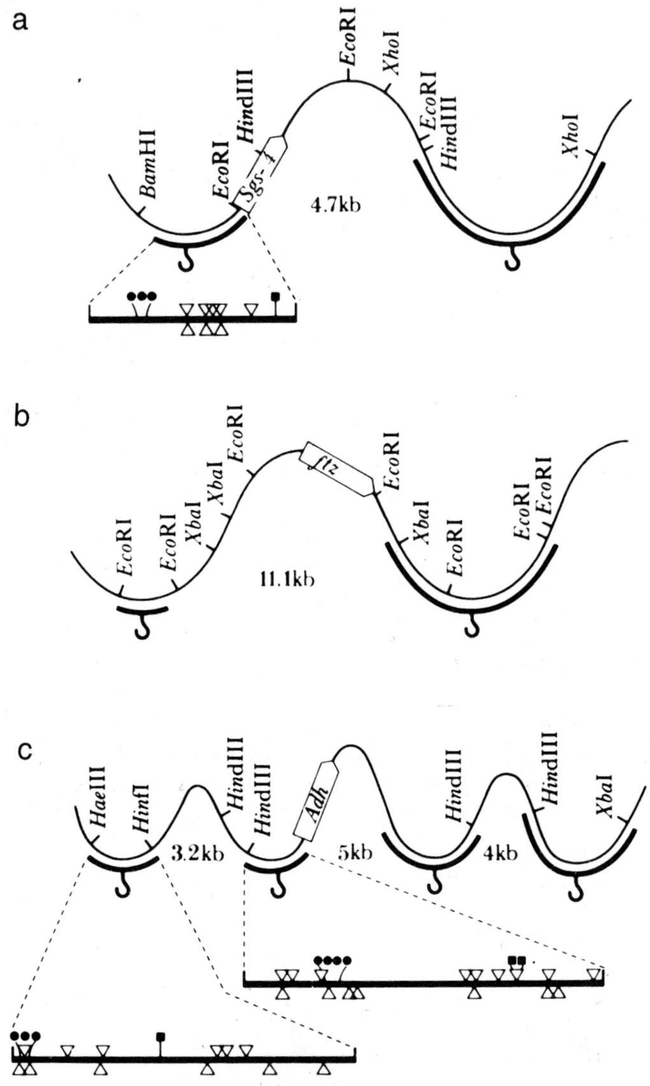

Figure 65. Localization of loops in the surroundings of the (a) *Sgs-4*, (b) *ftz*, and (c) *Adh* genes. For designations see the legend to Figure 64 (from Mirkovitch *et al.*, 1987).

Another loop composed of DNA 112 kb long contains two genes: its start and end were mapped to the middle of the 87E1-2 band (Figure 66). Finally, mainly two loops (the *S8* and *Ace* genes) form the edge fragments of the 87E1-2 band, the adjacent interbands, and parts of the more remote bands (Mirkovitch *et al.*, 1986, 1987).

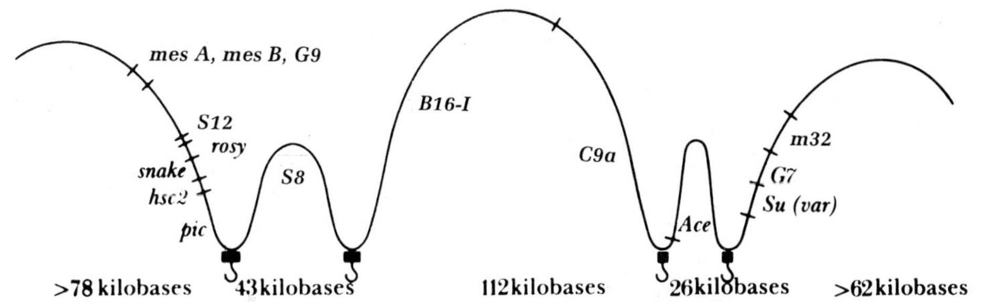

Figure 66. Distribution of the genes and the loops in the 320-kb-long DNA fragment located in the 87D5-87E5-6 region. Gene symbols are given along the loops (from Mirkovitch *et al.*, 1987).

4. 14B-15B region

On a long 800-kb DNA continuum from the 14B-15B region on the X chromosome of *Drosophila*, 86 scaffold-associated fragments were detected in the chromatin of 0- to 19-hr embryos. There are up to four to five transcribed regions in some loops and none in other loops. The total number of RNA synthesizing fragments in 0- to 18-hr embryos is 45 (Surdej *et al.*, 1990b). There are data on the distribution of cDNA fragments transcribed from larval, pupal, and adult poly(A)$^+$RNAs (Surdej *et al.*, 1990a). The total number of genes was not determined. However, it was concluded that—contrary to the data obtained by genetic analysis, according to which there are a few genes in the region—the fragment of the *Drosophila* genome is densely packed with the transcribed DNA. There are 57 bands on Bridges' map, 22 of which are "doublets"; thus, the real number of bands is about 35–40. Therefore, it may be expected that there is one band per two to three loops in the 14B-15B region.

Using a new approach to the chromosomal DNA loop excision, by topoisomerase II-mediated DNA cleavage at matrix attachment sites, the loop anchorage sites in this region were mapped. Most of the loop anchorage sites mapped by this method colocalized with weak SARs (Figure 67). Eleven anchorage sites delimiting 10 DNA loops ranging in sizes from 20 to 90 kb were found in this region.

It is suggested that SARs may constitute an essential but not sufficient element of DNA loop anchorage sites (Iarovaia *et al.*, 1996). Data were also obtained on the molecular organization of the SAR loci. Experiments designed to identify middle repeats (members of 22 families) did not reveal the presence of repetitive DNA in 22 of 86 SAR-containing fragments. Under more stringent conditions of DNA–DNA hybridization, there was no affinity for middle repeats (Surdej *et al.*, 1991). All the 27 sequences able to replicate autonomously (ARS) were found in the SAR-containing DNA fragments (Brun *et al.*, 1990).

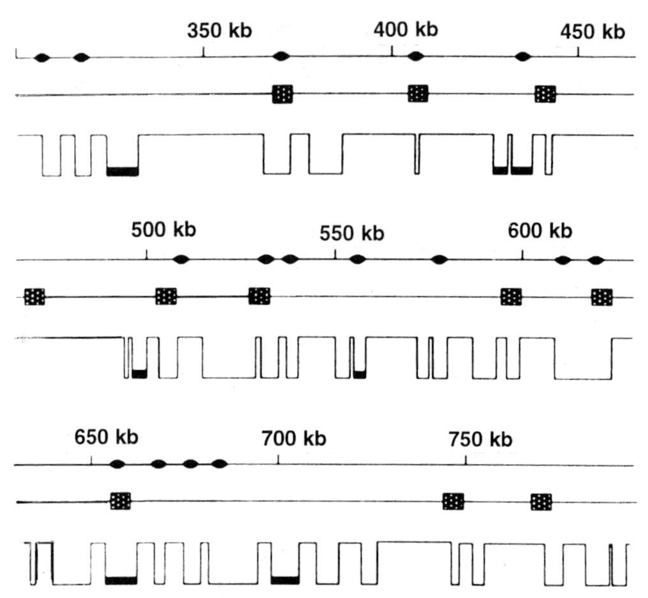

Figure 67. Position of DNA loop anchorage sites mapped by topoisomerase II-mediated DNA cleavage (shaded rectangles on the line below the scale) and position of SARs (lower portion of the broken bottom line). The upper and lower parts of the broken lines represent the SAR-free regions and regions bearing SARs of different strengths, respectively; the strong SARs being shown by thick lines. Solid ovals on the scale line show the positions of yeast ARSs (reprinted with permission from Iarovaia *et al.*, 1996).

The ensemble of these data allow us to regard the bands, the genes, and the loops as distributed independently of one another in the genome.

VI. PUFFS

When the genes located in the band are activated, chromosome texture in the region becomes diffuse and decompacted; it increases transversally, and a swelling into the puffed structure is produced. The largest puffs with distinctive disposition of the chromatids in the active region, Balbiani rings, occur, as a rule, in Chironomides, Simuliides, Culicides, and, occasionally, in certain *Drosophila* species. The DNA puffs in which DNA is amplified, and then RNA is synthesized occur in Sciarides. The usual puffs occur in all species of animals and plants having polytene chromosomes of the classical type. The question whether polytene chromosomes can puff in *Infusoria* has been previously discussed (Zhimulev, 1992, 1996).

The notion that the puffs of polytene chromosomes are indicators of gene activity was generally accepted. Grounds for its criticism were tenuous. Harris (1968) believed that the puff is not a satisfactory model for genetic regulation in normal diploid, moreover haploid, cells. His doubt stemmed from the erroneous conclusions that development proceeds normally in the absence of puffs (Goodman *et al.*, 1967) and that it is impossible to correlate puffing with metabolic patterns of the cell.

Thomson (1969) believed that in *Calliphora*, for example, the salivary gland and the bristle-forming cells differentiate last and they synthesize a very small number of specific proteins. The polytene chromosome cells of the fat body, which is one of the principal biosynthetic organs in larva, is very diffuse so that the puffs escape detection. In *Drosophila* as well, the fat body cells show a normal banding pattern, and they puff (Richards, 1980a, 1982).

Various aspects of the puffing phenomenon have been reviewed in detail (Beermann, 1956, 1962–1964, 1966; Gall, 1963; Mechelke, 1963; Beermann and Clever, 1964; Clever, 1964d, 1966a, 1968; Pavan, 1964, 1965a; Kiknadze, 1965b, 1966, 1970, 1971b, 1972, 1981a,b; Beermann and Clever, 1966; Kroeger and Lezzi, 1966; Kroeger, 1968; Laufer, 1968; Berendes, 1969, 1972a, 1973; Berendes and Beermann, 1969; Pavan and da Cunha, 1969; Ashburner, 1970a, 1972a, 1974, 1975, 1977; Mala, 1970; Nagel and Rensing, 1972; Edstrom, 1974; Pavan *et al.*, 1975; Rensing *et al.*, 1975; Ashburner and Berendes, 1978; Ashburner and Bonner, 1979; Zegarelli-Schmidt and Goodman, 1981; Zhimulev *et al.*, 1981a, 1994a; Bautz and Kabisch, 1983; Richards, 1985, 1986, 1997; Korge, 1987; Sorsa, 1988; Lepesant and Richards, 1989; Lezzi and Richards, 1989; Kress, 1996; Lezzi, 1996; Russell and Ashburner, 1996).

Because these kinds of studies have been mainly performed on salivary gland polytene chromosomes, and also because puffing appears to be chiefly under hormonal control, the anatomical and morphological features of the salivary glands and the hormonal systems of larvae will be characterized below.

A. Developmental features of dipteran insects

1. Endocrine system in insects and effect of hormones on development

The developmental cycles of all insects undergoing complete metamorphosis are generally similar. They include the embryonic stage; three or four larval stages; then, the animals proceed to an intermediate stage (puparium formation), which begins with eversion of the anterior spiracles of the trachea, the first steps of cuticle sclerotization, and ends with pupal molt and transition to the pupal stage in *Drosophila* (Figure 68). While this is occurring, a great part of the larval organs are lysed, and organs of the adult are formed from the imaginal

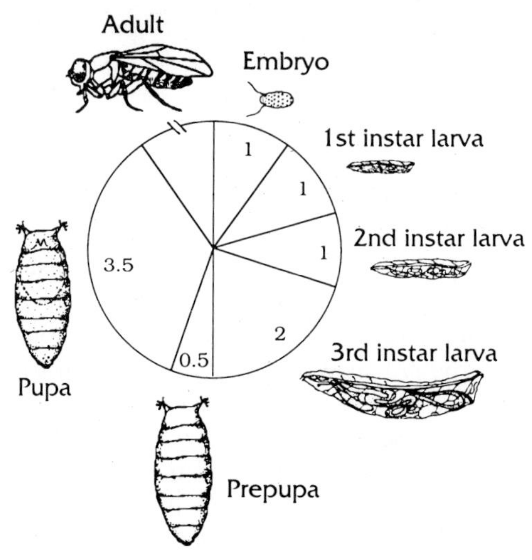

Figure 68. Developmental cycle in *D. melanogaster*. The numbers in the sectors of the circle designate the duration of the corresponding cycle in days (reprinted from *Trends Genet.*, 8(4), Andres, A. J., and Thummel, C. S., Hormones, puffs and flies: The molecular control of metamorphosis by ecdysone 132–138, copyright 1992 with permission from Elsevier Science).

disks. Metamorphosis is then completed (see reviews: Bodenstein, 1950; Shamshurin, 1972; Kiknadze *et al.*, 1975; Poluektova *et al.*, 1975a; Richards, 1981b; Sehnal, 1985; Richards and Hoffmann, 1986).

At the end of each successive larval instar, epidermis separates from the "old" cuticle, which the insect splits; it emerges and hardens its new cuticle. Molting and pupariation are under the control of three hormonal factors, as evidenced by the following results of ligature application and organ transplantation (review: Richards, 1981b).

1. If a ligature is applied behind the thoracic gland before the release of hormone, only the anterior portion molts. If the ligature is applied just prior to molt, after release of hormone, the process of the separation of epidermis from cuticle spreads over the anterior and posterior portions of the larva (Fraenkel, 1935).
2. If the brain is removed in a way that spares the prothoracic gland, molting is prevented provided that brain extirpation is early.
3. If the brain from a molting larva is implanted into a larva with removed brain, molting is induced in the latter.
4. If the brain and the prothoracic gland are implanted together into the posterior of a ligatured larva (see item 1), molting is induced.

The data indicate that the brain releases a hormone (ecdysiotropin) stimulating the synthesis of another hormone (ecdysterone) in the prothoracic gland (reviews: Karlson 1966, 1975; Karlson and Bode, 1969; Richards, 1981b; Steel and Davey, 1985). The secretion of the third juvenile hormone is associated with a small gland (the corpus allatum, CA), located behind the brain. When an active CA was transplanted from a young larva into an older larva, the latter did not undergo metamorphosis, a supernumerary instar formed instead. A young larva, whose CA was removed, molted into a small premature adult (review: Richards, 1981b). A scheme for the interaction of the three hormones and their effects on development is given in Figure 69. The type of molting depends on the balance between the concentrations of ecdysterone and juvenile hormone at each molt.

Although the existence of ecdysiotropin had already been postulated in the 1920s, its chemical structure is obscure. It is thought to be polypeptide with a molecular weight of 5 to 20 thousand daltons (review: Richards, 1981b; Steel and Davey, 1985).

The hormone secreted by the prothoracic gland was purified and isolated from hemolymph of *Bombyx mori* (Butenandt and Karlson, 1954). There are numerous analogs of the hormone differing in the number and position of hydroxyl groups. The prothoracic gland releases α-ecdysone, which hydroxylates at position 20 (Figure 70) under the effect of mitochondrial cytochromes and becomes about 20–200 times more biologically active than α-ecdysone (reviews: Burdette and Bullock, 1963; Richards and Ashburner, 1984). The hy-

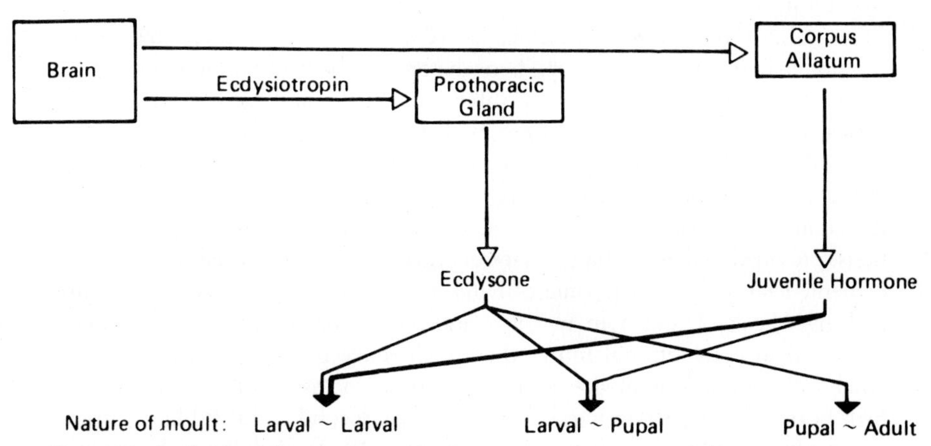

Figure 69. A scheme of the interaction of ecdysiotropin, ecdysterone, and the juvenile hormone (from Richards, 1981b).

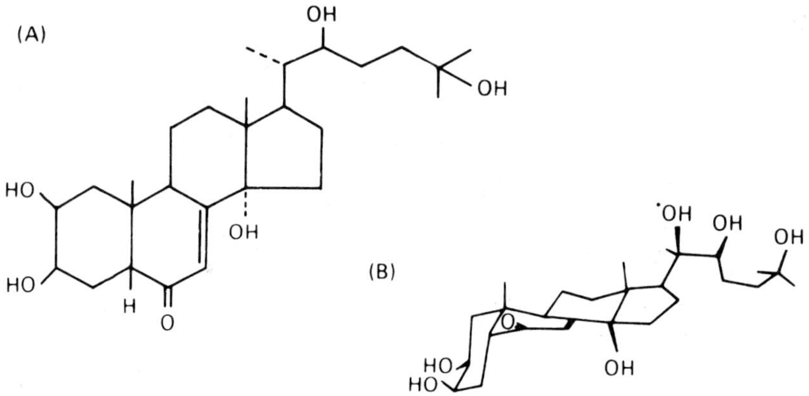

Figure 70. Structural formula of (a) α-ecdysone and (b) stereochemical formula of 20-OH-ecdysone. Position C_{20} is marked by a dot in b (from Richards, 1981b).

droxylated analog is named 20-hydroxyecdysone, 20-OH-ecdysone, β-ecdysone, or ecdysterone (Goodman *et al.*, 1978; Richards, 1981b; Richards and Ashburner, 1984; Steel and Davey, 1985; Koolman, 1989). α-Ecdysone is presumably the short-lived precursor of β-ecdysone (King and Siddal, 1969).

Of all the ecdysone analogs, ecdysterone binds most actively to the receptors of the hormone (Richards, 1981b). According to data obtained with imaginal disk cells (Fristrom *et al.*, 1973), salivary gland cells (Ashburner, 1973) and cells cultured from embryos (Cherbas *et al.*, 1977), the ecdysterone begins to exert its action at a concentration of 10^{-8} M and higher.

Ligature experiments (see Figure 103, below) demonstrated that a sharp rise in ecdysterone titer occurs 6 hr before the formation of puparium in *D. melanogaster* (Becker, 1962a; Ashburner, 1967b) and *D. hydei* (Berendes, 1965a) or 4.5 hr before its formation in *D. virilis* (Kress, 1974).

Summarizing the results of radioimmune assays of ecdysterone titers during *Drosophila* development, Richards (1981a) commented upon 11 peaks on the composite curve (Figure 71). Peak 1, which was detected in midembryogenesis, provided evidence that ecdysterone may possibly be stored during development. Peak 2 fell at the time of hatching (i.e., during a process resembling molting). Peaks 3 and 5 in mid-first and second larval instars could not be explained in conventional terms; they might have resulted from insufficicent synchronization of larval development or from inclusion of individuals with high hormone content during molt between the second and third larval instars (see peaks 4 and 6). According to Berreur *et al.* (1979), peak 7 may be a switch to programing pupariation. In Richards's view, this is effected by a rise in hormone titer before puparium formation (peak 8). There is agreement in the

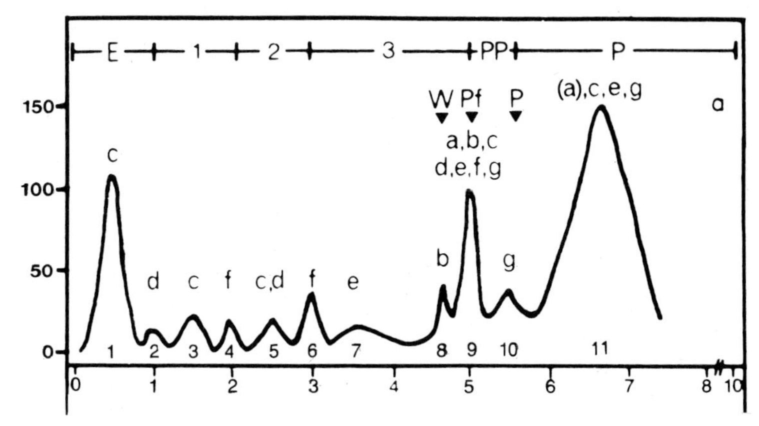

Figure 71. Changes in ecdysterone titer during development of *Drosophila*. Abscissa, time of development in days (25°C), ordinate, ecdysterone titer (relative units); 1–11, concentration peaks; E, 1, 2, 3, PP, and P, embryonic, three larval, prepupal, and pupal stages of development, respectively; W, beginning of the migration period of larvae; Pf, puparium formation. (a–g) Peak titres of ecdysterone according to the data of a (Borst *et al.*, 1974), b (De Reggi *et al.*, (1975), c (Hodgetts *et al.*, 1977; Kraminsky *et al.*, 1980), d (Garen *et al.*, 1977), e (Berreur *et al.*, 1979), f (Maroy *et al.*, 1980), and g (Klose *et al.*, 1980) (reprinted from *Mol. Cell Endocrinol.* **21,** Richards, The radioimmune assay of ecdysteroid titres in *Drosophila melanogaster*, 181–197, Copyright 1981, with permission from Elsevier Science).

litrature with respect to variations in ecdysterone titers in development in the final peak 9 (see Figure 71); this peak is thought to be directly associated with puparium formation. The high peak (11) in the mid-pupal period is presumably associated with histolysis and the formation of adult organs. An increase in hormone titer occurred in pupae just prior to cuticle deposition and before melanization of the bristles of adults (Bainbridge and Bownes, 1988).

From the mid-pupal period, ecdysterone titer sharply falls (see Figure 71). This is presumably associated with histolysis of the prothoracic gland during metamorphosis (Wigglesworth, 1955). However, in a number of experimental papers and reviews the very presence of ecdysterone in the body of females was questioned (Garen *et al.*, 1977; Richards, 1981b; Handler, 1982; Richards and Ashburner, 1984; Bownes, 1989). Doubts arose because vitellogenin, one of the major components of egg cytoplasm, is induced by ecdysterone; also, hormone induces the synthesis of the protein even in males of *Drosophila*.

In the mosquito of the *Aedes* genus, after a blood meal, ecdysterone titer in hemolymph of females rapidly rises; its maximum is reached after 36 hr. The hormone, in turn, induces the synthesis and secretion of vitellogenin later deposited in the egg (review: Bownes, 1989).

As measured by a radioimmunoassay, ecdysterone concentration is 24.0–37.6 pg/mg of body weight at the age of 1 day, 16.4–20.8 at the age of 5 days, and 10.3–11.8 at the age of 10 days in females of *Drosophila*. The hormone concentration is 3–6 pg/mg of body weight in males of *Drosophila* (Walker *et al.*, 1987). The source providing ecdysterone in adults is unproven. It is thought that ecdysterone is an ovarian hormone and that the ovaries are the major secretory organs of hormone (Garen *et al.*, 1977; Handler, 1982).

Radioimmune analysis of ecdysterone titers during postembryonic development in *C. thummi* demonstrated that the concentration of hormone is 150 ng/g of body weight in third instar larvae and 450 ng/g before pupation (Valentin *et al.*, 1978). Changes in ecdysterone titers were found in the salivary glands of metamorphosing *Rhynchosciara americana* larvae (Stocker *et al.*, 1984). In *Calliphora erythrocephala*, ecdysone concentration increased just before puparium formation and reached its maximum 5–6 hr after it (Shaaya and Karlson, 1965).

A compound characterized by its ability to inhibit the metamorphosis of larvae and prepupae was first isolated from males of *Hyalophora cecropia* in 1956. It has been named juvenile hormone I (JHI); there are other similar compounds (Figure 72), JHO, JHII, and JHIII. The hormone is completely synthesized in the CA. Measurements of JHIII titers in *D. melanogaster* during development revealed that they are highest in larvae crawling up the culture vessel before pupariation, and also in adults. Larvae and pupae were found to have the lowest JHIII titer (Sliter *et al.*, 1987). There are several reviews on the biosynthesis and physiological activity of the juvenile hormone (Richards, 1981b; Richards and Ashburner, 1984; Steel and Davey, 1985).

In *D. melanogaster*, mutations were identified that affect normal devel-

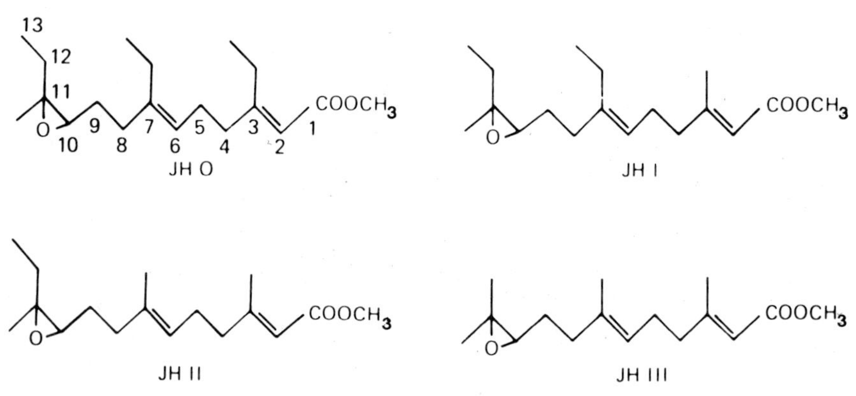

Figure 72. The chemical structure of the juvenile hormones JHO, JHI, JHII and JHIII (from Richards, 1981b).

opment by impairing hormonal functions (review: Ashburner, 1979). The characteristic features of some of the mutations are given below.

a. l(2)gl: lethal (2) giant larvae

Homozygotes for this mutation form puparium with a great delay compared to normal larvae or do not form it at all (Hadorn, 1937; Aggarwal and King, 1969; Korochkina, 1972; Richards, 1976a; Korochkina and Nazarova, 1977). The prothoracic glands are much reduced in size in mutants (Scharrer and Hadorn, 1938). Puparium formation of *l(2)gl* can be accelerated by transplantation of ring glands from normal larvae to mutants (Hadorn, 1937) by injection of an extract from *Calliphora erythrocephala* or *Bombyx mori* enriched with ecdysterone (Karlson and Hanser, 1952).

b. ecd1: lethal (3) ecdysone l^{ts}

After a temperature shift to 29°C early in the third instar, *ecd 1* larvae fail to pupariate; they remain as viable larvae for as long as 3 weeks. The ecdysterone titer is 3–10% of the normal in such larvae (Garen *et al.*, 1977; Lepesant *et al.*, 1978; Klose *et al.*, 1980; Redfern and Bownes, 1983; Berreur *et al.*, 1984). In mutant larvae, the metabolic rate of ecdysterone, particularly its conversion into 3-dehydroecdysone and 3-dehydro-20-oxyhydroecdysone, is decreased at the restrictive temperature (Somm-Martin *et al.*, 1988). The mutations are autonomously expressed in the ring gland (Henrich *et al.*, 1987).

c. *l(1)su(f)*^{ts67g}: *lethal (1) suppressor of forked*

Mutant larvae do not pupate at restrictive temperature (30°C) and die after 1–3 weeks at restrictive temperature. This is associated with an ecdysterone titer of 3–10% of normal (Dudick *et al.*, 1974; Klose *et al.*, 1980; Hannson *et al.*, 1981). After treatment of larvae with exogenous 20-OH-ecdysone (additive to food of larvae), pseudopupae are formed (Hannson and Lambertsson, 1983, 1984).

d. gt: *giant*

The third-larval instar is extended by 4 days at 25°C in *gt* homozygotes. Radioimmune assays demonstrated that an increase in hormone titer is associated with puparium formation; it is delayed by 4 days and the titer is lower than in wild-type larvae. Mutant mid-third instar larvae fed ecdysterone pupariate at the same rate as wild-type larvae. Development proceeds at the same rate as in wild-type larvae after puparium formation (Schwartz *et al.*, 1984).

e. dor: deep orange

This mutation affects eye pigmentation and causes female sterility; a synthetic lethal results from interaction of the *dor* and *rosy* mutations. Larvae homozygous for the lethal alleles at the *dor* locus fail to pupariate (Aizenzon *et al.*, 1982; Biyasheva, 1987; see also Lindsley and Zimm, 1992). The state of puffing suggests that ecdysterone is not present in hemolymph (see Section VI,D).

f. *l(1)HM4O*

The mutation produces a delay of 24–100 hr in puparium formation and has cell lethal effects at the pupal stage. Injection of ecdysterone increases the rate of puparium formation (Biyasheva *et al.*, 1991).

g. *sta: stubarista*

Homozygotes for these mutations die as larvae at the first instar, but the majority die at the third instar during the larval–pupal transformation. Some larvae pupariate, and flies emerge in 2–5% of cases. Injection of hormone to larvae stimulates puparium formation (Biyasheva *et al.*, 1991).

h. *grg: lethal (1) giant ring gland*

Third-instar larvae do not pupariate; they die after 10 days. Ecdysterone titer in homozygotes is 3–10% of the wild-type (Klose *et al.*, 1980).

i. $L(3)3^{DTS}$

This is a dominant temperature-sensitive mutation. Homozygotes for the mutation die as embryos. When the larvae are grown at the restrictive temperature of 29°C, the puparium is not formed. The ring gland of $L(3)3^{DTS}/+$ larvae is grossly hypertrophied at 29°C because the prothoracic cells are increased in size. Hormone titer was determined to be 5% of normal by the radioimmune method. Puparium is formed when the hormone is injected into larvae (Holden and Ashburner, 1978; Holden *et al.*, 1986). According to Walker *et al.* (1987), the titer of the hormone is a half of the normal in adult females, as a consequence of which they presumably are sterile.

j. *ft: l(2)ft*

Some of the nonpupariating larvae homozygous for the lethal mutation at this locus—$l(2)ft^{fd}$ and $l(2)ft^{al3}$—form puparium 10 days after egg laying (at 25°C). They possess the normal set of the ecdysterone induced puffs characteristic of this stage. The puff spectrum in the nonpupariating larvae corresponds to the ecdysterone stage. When salivary glands were incubated in a hormone-containing medium, the puff set corresponded to PS5-7 after 4 hr (at 22°C). The inability to pupariate may be due to a delay in ecdysterone secretion (Zhimulev and Szidonya, 1991).

k. $Df(2R)bw^5$

The mutation affects the histology of endocrine organs and development and differentiation of imaginal disks (Ashburner, 1979).

l. *ey^D: eyeless^D*

Homozygotes mainly die at the pupal stage and less often at the prepupal stage (Lindsley and Zimm, 1992). The puparium is formed after injection of ecdysterone (Arking *et al.*, 1975).

m. *l(3)tl: lethal (3) tumorous larvae*

Third-instar larvae fail to pupariate, remaining viable for a month. Salivary gland cells isolated from larvae do not contain the ecdysterone-induced puffs, indicating that hormone titer is not sufficiently high. When the salivary glands are incubated *in vitro* with ecdysterone, puffs are formed (Zhimulev *et al.*, 1976; Zhimulev, 1992).

n. *ecs: ecdysone sensitivity (BR-C: Broad Complex)*

The puparium is not formed in lethal mutants. In mosaics observed at the end of the third-larval instar, ecs^{+}/ecs cuticle cells are pigmented; they harden (pupariate) when exposed to ecdysterone. Fragments of *ecs/0* cuticle remain larval (white, not hardened); that is, they do not respond normally to ecdysterone despite its presence in hemolymph (Kiss *et al.*, 1976a,b, 1978). The gene is located in the early ecdysterone puff 2B3-5, and it controls the development of the entire cascade of gene activation induced by hormone (Belyaeva *et al.*, 1981, 1989).

o. *halfway: swi (singed wings)*

The *hfw* mutation disturbes development at the end of the third larval instar. The cuticle becomes pigmented and hardens only in the anterior portion of the larva. Pupal molt does not occur (Rayle, 1967). swi^{t467} mutants show the same phenotype; they presumably are alleles of *hfw* (Belyaeva *et al.*, 1980; Aizenzon *et al.*, 1982). The data suggest that the swi^{t467} mutation disrupts the chain of responses in the cascade of ecdysterone-induced gene activities (Belyaeva and Zhimulev, 1982).

p. *dre4*

Homozygotes for the *dre4* mutation fail to molt; their larval life is extended to the third instar. An upshift of temperature (31°C) at various times during development causes nonpupariation in the temperature sensitive mutants. The mutants showed no evidence of the differentiation of adult structures and flies do not emerge. The developmental disturbances in *dre4* mutants are associated with elimination of ecdysterone peaks (Sliter and Gilbert, 1992).

Factors in addition to mutations can cause developmental arrest. The gene of the insect baculovirus encodes an ecdysterone UDP-glycosyl-transferase (egt). Interaction of egt with its ecdysterone host can block molting in viral infected larvae (O'Reilly and Miller, 1989).

2. Anatomical and physiological characteristics of the salivary glands of dipteran larvae

Despite the diversity of cell types (Figures 73–75; also see Figs. 80, 84) the salivary glands have anatomical and physiological features in common, such as division into sections, secretory activity, and hormonal control of physiological processes.

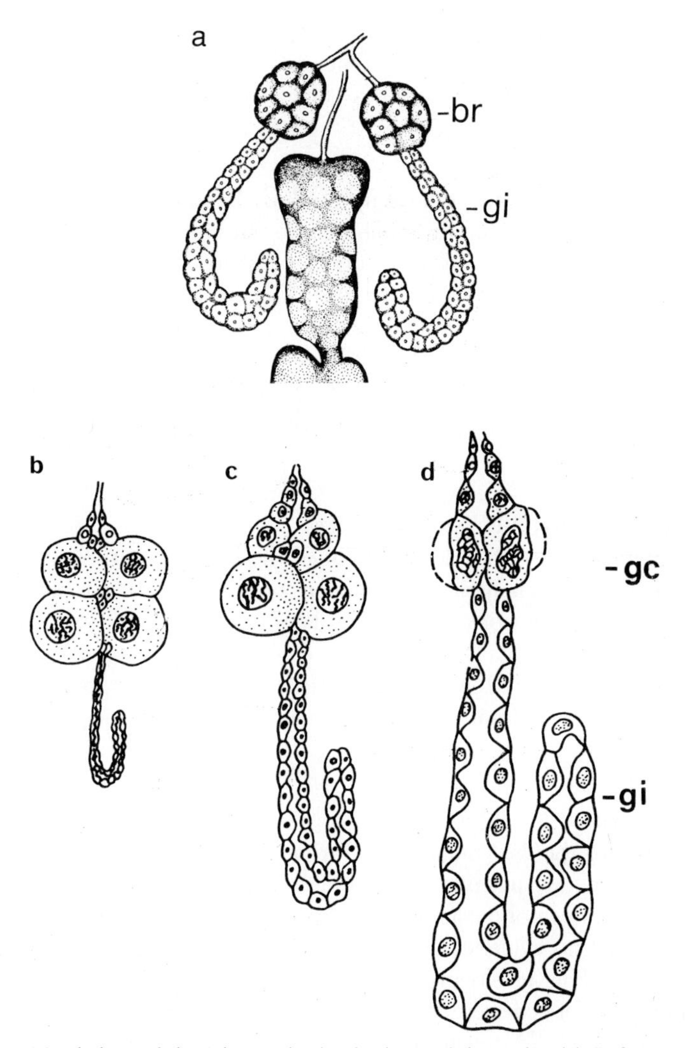

Figure 73. Morphology of the salivary glands of a larva of the midge (a) *Dichaetia pusilla* and (b–d) *Harmandia loewi*. (b, c, and d) Salivary glands of growing larvae 0.3–0.8, 0.8–1.8, and 1.8–3 mm long, respectively. br, Basal reservoir; gi, the gland itself; gc, giant cells (a, from Mamayev, 1968; b–d, from Novozhilov and Slepyan, 1976).

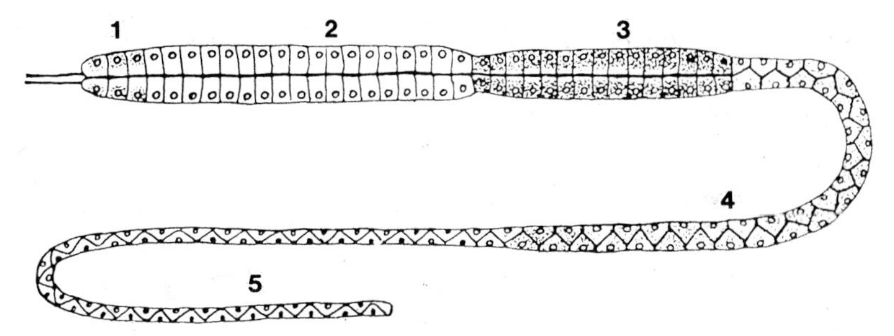

Figure 74. A drawing of one of the larval salivary glands of *Bradysia elegans* (Sciaridae). 1–5, parts of the salivary gland (from da Cunha *et al.*, 1973).

In the larvae of the gall midges, differentiation into segments is more prominent. The basal reservoir comprises two or several cells (see Figure 73); the salivary gland cells are giant in the former case. The gland proper consists of 40–130 cells.

The developmental changes in all the sections of the salivary glands conform to a definite pattern. In young larvae, the cells of the gland proper are underdeveloped compared to those of the basal reservoir. The ratio of cell sizes changes in the two sections throughout the growth period of larvae (Figure 73): the basal reservoir disintegrates and the gland proper becomes giant. It is believed that digestive enzymes, or poisonous substances (in predacious species), are synthesized in the cells of the basal reservoir. The gland proper produces a silky secretion, which is possibly used to form the cocoon during pupation (White, 1948; Henderson, 1967; Mamayev, 1968; Zhimulev and Lychev, 1972b; Novozhilov and Slepyan, 1976; Grinchuk, 1977).

In the representatives of the Sciaridae family, the salivary glands are very long, up to 1 cm in *Sciara coprophila* (Gabrusewycz-Garcia, 1975); they consist of about 300 cells lying in two rows on the same plane. Based on morphological criteria, the cells are readily divided into five sections (see Figure 74).

In the majority of the sciarid species, the proximal portion of the gland stains with the "PAS-reaction" (which occurs after application of the periodic acid–Schiff reagent), because it contains PAS-positive granules in abundance. Cell of this section release glue secretion throughout the entire larval life. This glue converts into a silky secretion used to form the cocoon. Cells from the distal section are more often PAS-negative; they contain protein granules up to 1.3–1.5 μm in diameter (Doyle and Metz, 1935; Metz, 1935, 1941; Jacob and Jurand, 1963; Phillips, 1965; Phillips and Swift, 1965; da Cunha *et al.*, 1969a, 1973a,b; da Cunha and Pavan, 1969; Pavan and da Cunha, 1969; Been and Rasch, 1972; Bianchi and Terra, 1975; Gabrusewycz-Garcia, 1975; Okretic *et al.*, 1977; Amabis, 1983; Laicine *et al.*, 1984).

Salivary gland proteins of the fourth instar larvae from *Rhynchosciara*

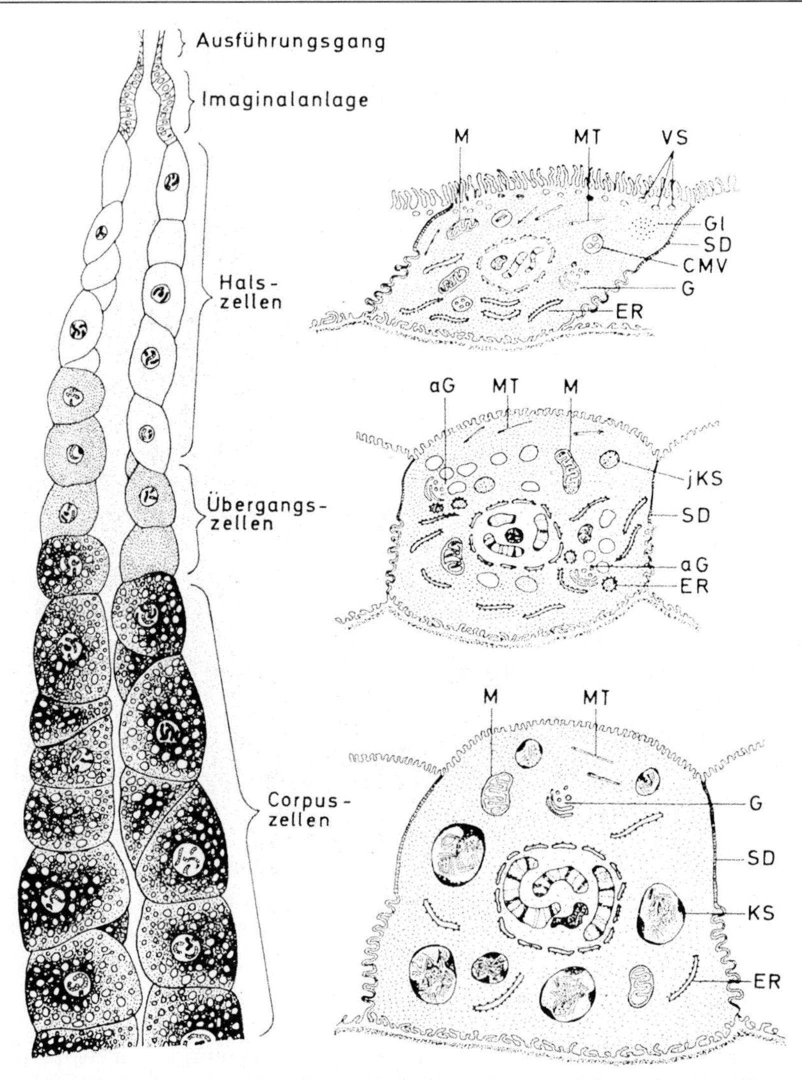

Figure 75. Proximal portion of the salivary gland of *D. melanogaster*. M, mitochondria; ER, ergastoplasm; G, Golgi's complex; MT, microtubule; CMV, vacuolar body; Gl, glycogen deposition; SD, desmosomes; jKS, aG, and VS, secretory granules (reprinted by permission from Gaudecker, 1972).

americana were resolved by SDS–polyacrylamide electrophoresis. The secretory product contains nine polypeptide bands, of which two major ones, P4 and P5, had molecular weights of 90 and 72 kDa, respectively; in several older larvae, there were additional bands P1 (100 kDa) and P8 (29 kDa) besides P4 and P5 (Winter *et al.*, 1977a,b, 1980). This secretion presumably is used to spin the communal cocoon in which the larvae complete their metamorphosis (Guaraciaba and Toledo, 1967; Terra and Bianchi, 1974, 1975; Bianchi and Terra, 1975). The chemical composition of the cocoon expressed as a percentage of its dry weight is as follows: protein, 38%; calcium carbonate, 43.5%, and carbohydrates, 10.6%. The cocoon is made up of specific silky filaments produced by the salivary glands and of material with high content of calcium carbonate deposited on the cocoon wall. Calcium carbonate is abundant in the Malpighian tubules, notably before pupation, and they presumably secrete it. Some proteins of the cocoon are homologous to those of hemolymph and possibly transported from it; the carbohydrate component may be produced by the fat body (Bianchi *et al.*, 1973; Bianchi and Terra, 1975; Terra *et al.*, 1975). Several secretion fractions were identified in *Bradysia* (Perkowska, 1963a).

The anatomy and functions of the salivary glands in various *Drosophilidae* species have been described in an extensive series of primary papers and reviews (e.g., Ross, 1939; Bodenstein, 1943, 1950; Ashburner, 1970a; Tulchin and Rhodin, 1970; Gaudecker, 1972; Lane *et al.*, 1972; Korge, 1977b, 1980; Berendes and Ashburner, 1978; Campos-Ortega and Hartenstein, 1985; Panzer *et al.*, 1992).

The salivary gland of Drosophilidae is a paired organ lying in the anterior portion of the larval body (Figure 75). Each gland contains, on the average, 128 cells; the number varies depending on culture conditions (Makino, 1938; Hadorn and Faulhaber, 1962; Altmann, 1966; Chaudhuri and Mukherjee, 1969; Berendes and Ashburner, 1978); the highest average number of cells is observed at 18°C (Altmann, 1966). The two glands of an individual are, as a rule, morphologically identical. There are exceptions to the rule. The left and right parts differ most in *Drosophila victoria* (Wharton, 1943) and in *Drosophila lebanonensis* (Ward, 1949; Eeken, 1977). In *D. melanogaster*, the anterior–posterior extent of the salivary gland primordium is established by the action of the homeotic gene *Sex combs reduced (Scr)*, but the dorsal–ventral extent is regulated by action of genes *decapentaplegic* and *dorsal* (Panzer *et al.*, 1992)

The function of the salivary glands before mid-third-larval instar are obscure. During late embryonic development, certain substances in the salivary gland lumen stain heavily with hematoxylin (Sonnenblick, 1950); this secretion may facilitates the crawling of the larva out of the egg membrane (Berendes and Ashburner, 1978). There are many indications that the salivary glands may be engaged in digestive processes during the second larval and most of the third-larval instars (Ross, 1939; Bodenstein, 1950; Fraenkel, 1952). There is no evidence supporting this possibility. In fact, enzymic activities thought to be responsible for digestion are detected neither in gastric cells nor in secretion (reviewed by Berendes and Ashburner, 1978). Lezzi and Richards (1989) believe that the salivary glands of *Drosophila* and *Chironomus* larvae do not produce saliva. Inferences about the activities of the salivary glands at these developmental stages are based on the presence of small granules beneath the apical plasma membrane and their release into the gland lumen (Berendes and Ash-

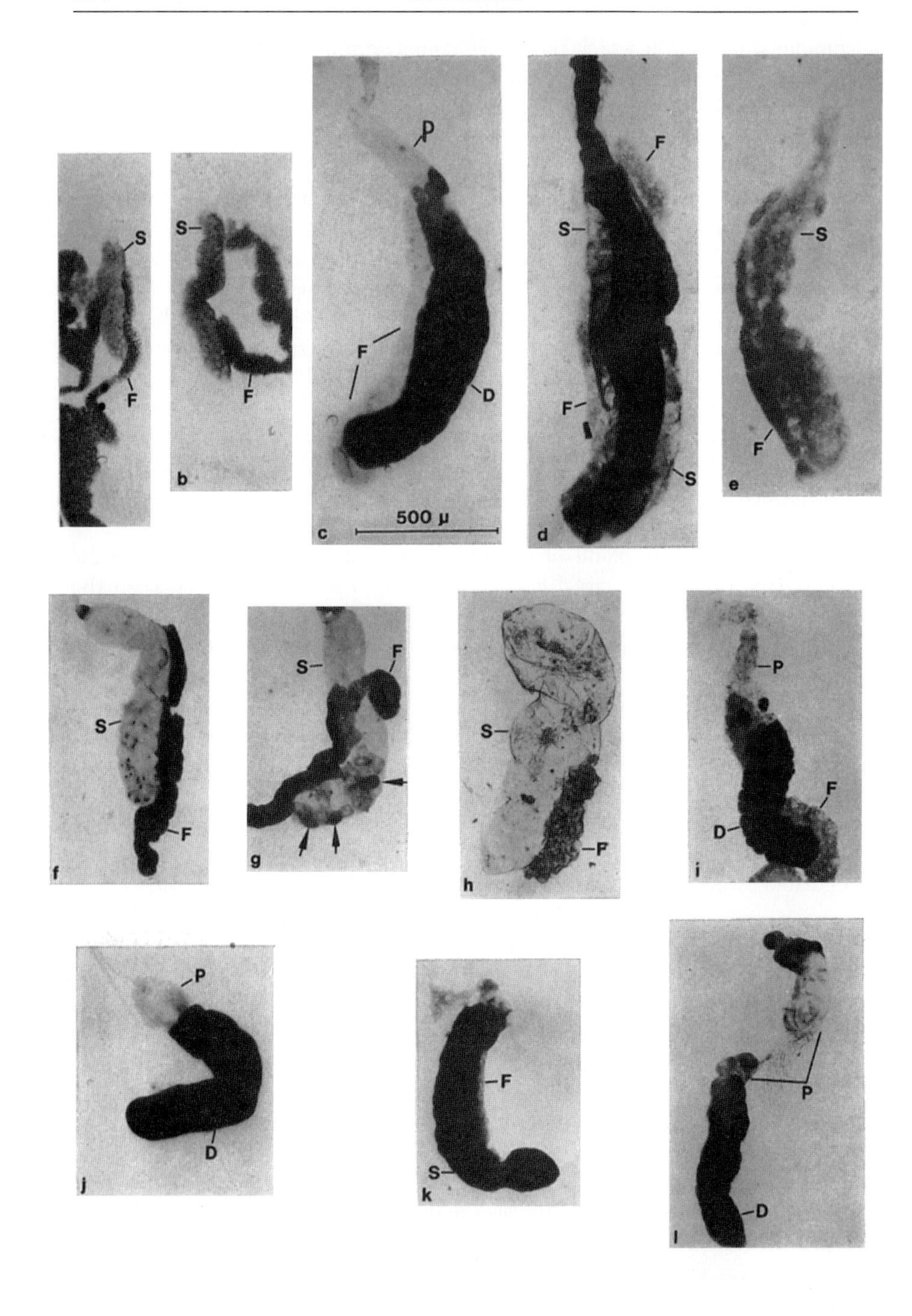
S
F
S–
F
b
P
F
D
c
500 μ
F
S–
F–
S
d
–S
F
e
S–
F
f
S–
F
g
S–
–F
h
–P
F
D–
i
P
D
j
–F
S–
k
P
–D
l

burner, 1978; Thomopoulos, 1988). Evidence for the external digestion of amylose, cellulose, and chitins was taken to mean that the salivary glands possess a digestive function at these developmental stages (Gregg *et al.*, 1990).

The most important function of the salivary glands at the end of the third-larval instar is synthesis of glycoprotein (described as mucopolysaccharide in the early papers) secretion, which is released at the time the larvae evert their anterior spiracles (the beginning of puparium formation) and which serves to attach the puparium to the substrate (Fraenkel, 1952; Fraenkel and Brookes, 1953).

There are various procedures for detecting secretion using the PAS reagent, which gives specific staining for glycoproteins, thereby providing visualization of large granules of high electron density in the electron microscope. The granules are of irregular shape; they are also identified in the cytoplasm of squash preparations provided that fixation in an ethanol–propionic acid mixture was sufficiently long at a temperature of about 0°C (Zhimulev, 1974). Isolated secretion can be characterized electrophoretically.

Twenty-four hours before the onset of puparium formation, the salivary glands start staining with the PAS reaction (Figure 76), and specific cytoplasmic granules up to 3 μm in diameter are seen in the light and electron microscopes (Figures 77 and 78). Secretory granules are formed in the Golgi's complex; then the membranes surrounding the granules coalesce with the membrane of the apical part of the cell and the granules are discharged into the gland lumen (see Figure 78). Secretion formation in *D. melanogaster* and related species has been fully reviewed (Ross, 1939; Painter, 1945; Hsu, 1948; Bodenstein, 1950; Gay, 1956; Kaufmann and Gay, 1958; Swift, 1962; Rizki, 1967; Smith and Witkus, 1970; Vidal *et al.*, 1971; Gaudecker, 1972; Lane *et al.*, 1972; Gaudecker and Schmale, 1974; Zhimulev, 1974, 1975; Kolesnikov and Zhimulev, 1975a; Korge, 1975, 1977b, 1980; Zhimulev and Kolesnikov, 1975a; Thomopoulos and Kastritsis, 1979; Karakin, 1985).

In *l(3)tl* mutants, whose salivary glands are not stained with the PAS reaction (Zhimulev and Kolesnikov, 1975), secreted material is nevertheless synthesized (Zhimulev, 1973a).

Figure 76. Dynamics of accumulation of PAS-positive material in the larval salivary glands of *D. melanogaster*, (a–e) laboratory stock Batumi-L, (f–h) in *l(3)tl* mutants, (i) *l(2)gl* mutants, (j and l) after culturing *in vivo*; (k) in larvae with a ligature separating the salivary gland from the source of ecdysterone. (a) 80, (b) 90, (c) 100 and (d) 120 hr after egg laying; (e) a 2-hr prepupa; (f) PAS-negative staining at 168 hr; (g) single cells (indicated by arrows), PAS-positive stained at 168 hr; (h) 17 days after egg laying; (i) 168 hr; (j) salivary gland isolated from larvae at the stage of the second larval molt incubated in the bodies of adult females for 3 days; (k) salivary glands were transplanted together with fat body; (l) the same, glands without fat body. S, salivary gland; F, fat body; P and D, proximal and distal portions of salivary gland (reprinted by permission from Zhimulev and Kolesnikov, 1975a).

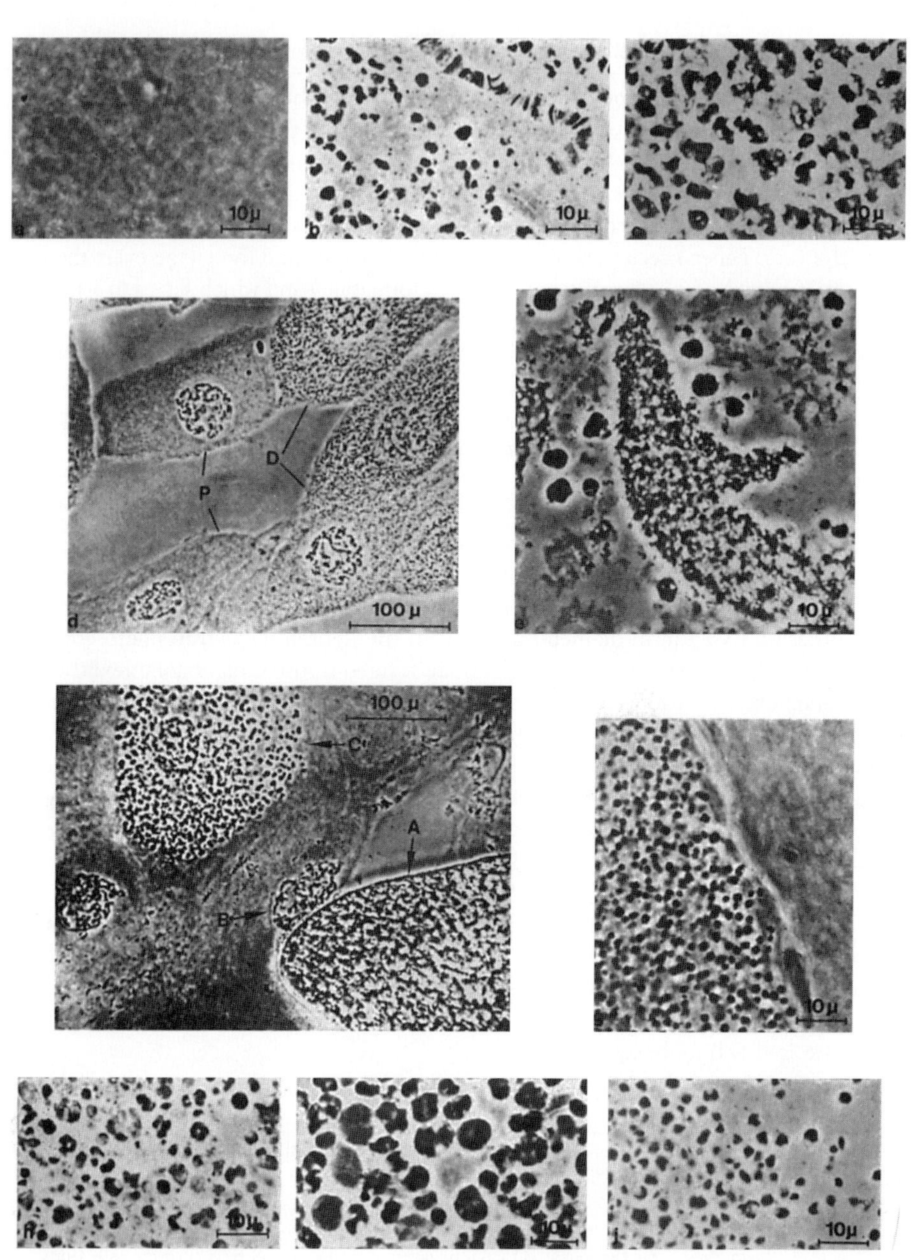

Figure 77. (a–f) Glycoprotein granules in cell cytoplasm on squashed preparations of the larval salivary glands of Batumi-L strain during normal development; (g) in the *l(3)tl*; (h) *l(2)gl* mutants; (i) in larvae with ligatures, and (j) after transplantation of salivary glands into the abdomen of an adult females. (a) 90, (b) 95, (c) 100, and (d) 100 hr, boundary between cells of the proximal (P) and distal (D) portions; (e) secretory granules from salivary gland duct; (f) salivary gland of a 120 hr larva before spiracle eversion. Arrows indicate (A) secretion strand in salivary gland duct, (B) cells with and (C) without glycoprotein granules (reprinted by permission from Zhimulev and Kolesnikov, 1975a).

The PAS-positive granules and strands of secretion were detected at the end of the third-larval instar in *Drosophila flavorepleta* (Wiener *et al.*, 1964), *Drosophila repleta* (McMaster-Kaye and Taylor, 1959; McMaster-Kaye, 1962), *Drosophila robusta* (Lesher, 1951, 1952), *D. hydei* (Berendes and Bruyn, 1963; Berendes, 1965a,c, 1972b; Poels, 1970, 1972; Poels *et al.*, 1971), *D. pseudoobscura* (Harrod and Kastritsis, 1972a,b), *Drosophila lebanonensis* (Eeken, 1977), and *Drosophila auraria* (Thomopoulos *et al.*, 1989, 1991, 1992). In *Drosophila simulans*, *Drosophila yakuba*, *Drosophila eugracilis*, *Drosophila atripex*, *Drosophila bipectinata*, and *Drosophila malerkotliana* the average size of granules varies in the 2.1–4.5 μm range (Thomopoulos and Kastritsis, 1979). It is surprising that the glands of larvae belonging to species of the *Drosophila suzukii* group stain very weakly with the PAS reaction (*D. suzukii*, *Drosophila lucipennis*), or not all (*Drosophila rajasekari*). Secretory granules are nevertheless synthesized in salivary gland cytoplasm, although in smaller amounts (1.7–2.47 μm) (Thomopoulos and Kastritsis, 1979).

In *D. virilis*, synthesis of secretion starts at the end of the third larval instar (Bodenstein, 1943, Swift, 1962), during the 32- to 40-hr time span after the second larval molt; the amount of secretion continuously increasing up to 70 h (Onishchenko *et al.*, 1976). According to Kress (1974, 1982), the secretion of larval glue proteins *LGP1* and *LGP2* starts 120 hr after egg laying and reaches its maximum rate about 20 hr later at the time when the migrating period starts. Analysis of *LGP1* mRNA was detectable in the salivary glands of late third instar larvae at the age of 113–140 hr (Swida *et al.*, 1990).

Shortly before the larvae start to evert their anterior spiracles and the onset of puparium formation, secretion is released into the lumen, where it lies like a tight cable, producing swelling of the gland (Lane *et al.*, 1972; Zhimulev and Kolesnikov, 1975a; Kolesnikov and Kokoza, 1985). At the time of spiracle eversion, the larva contracts, and secretion is discharged from the gland lumen. It hardens some minutes later and serves as an adhesive of the pupae to the substrate. The larval salivary glands begin histolysis shortly after the secretion has been extruded (Ross, 1939; Bodenstein, 1950). However, according to other data (Mitchell *et al.*, 1977), the salivary gland continues functioning as a secretory organ throughout and beyond the prepupal stages.

After secretion has been discharged from the lumen, the salivary glands are more or less electron lucent; they again become opaque after 3 hr, and secretion released by cells almost to the end of the prepupal stage make their first appearance in the proximal portion of the lumen, then in the whole gland (Korge, 1977b).

Drosophila gibberosa larvae do not attach themselves to the substrate before pupariation. For this reason, the larval salivary glands contain only a small amount of protein. It continues to increase after puparium formation, and there is up to 20 μg of secretion per gland in 20-hr prepupa; the amount of secretion decreases before puparium formation (Shirk *et al.*, 1988).

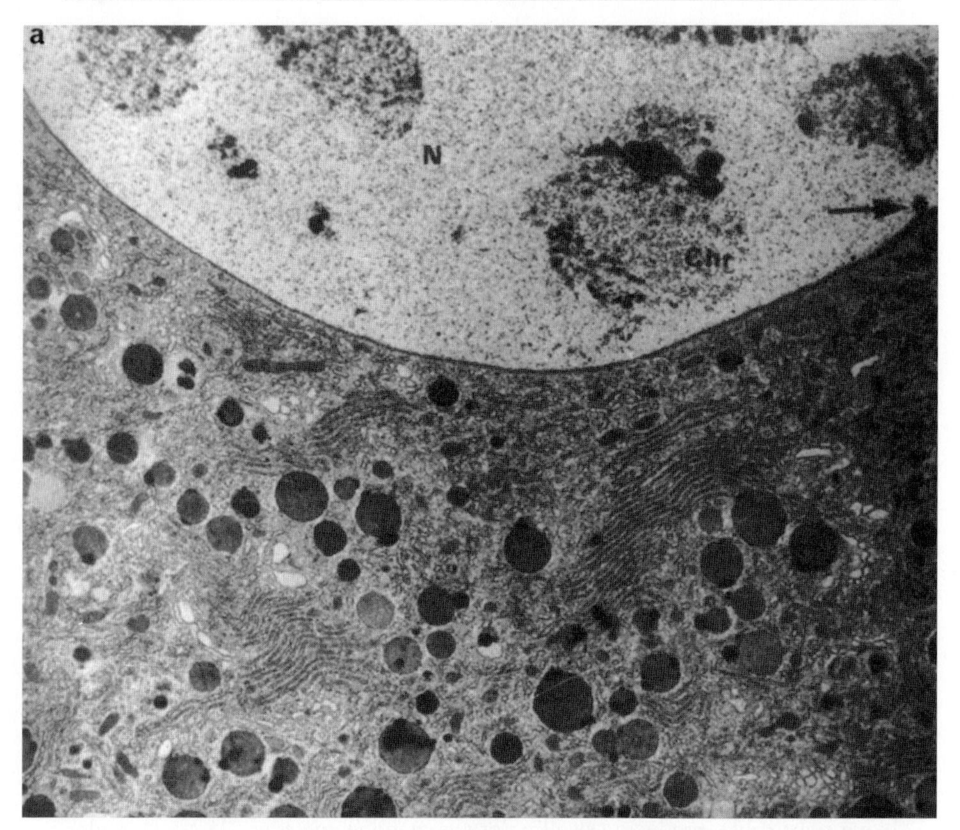

Figure 78. Secretory granules (arrows) in cell cytoplasm of the salivary glands of larvae of (a) *D. pseudoobscura* and (b and c) *D. melanogaster*. (a) Small granules in cytoplasm, (b) bigger granules near salivary gland lumen; and (c) glycoprotein flowing into salivary gland duct (L) (a, from Harrod and Kastritsis 1972a; b and c, reprinted with permission from Lane *et al.*, 1972).

Since the pioneering biochemical studies on glycoprotein secretion of larvae (Blumel and Kirby, 1948; Kodani, 1948; Perkowska, 1963b), the secretion in the gland lumen was isolated; this was facilitated by the fact that it stretches like a tight cord and constitutes up to 30% of the protein of the salivary gland, which is equivalent to about 4 μg (Kodani, 1948), 1 μg (Zhimulev and Kolesnikov, 1975a; Korge, 1977b), and 2–3 μg (Kokoza and Karakin, 1985). Treatment of the salivary glands with 95% ethanol hardens the secretion; it becomes opaque (in contrast to the cellular background) and separable by needles. Secretion becomes soluble again after transfer to water (Kodani, 1948).

Kodani (1948) described secretion as a mucoprotein containing the amino sugar glucosamine and the proteinaceous moiety represented by 15 amino

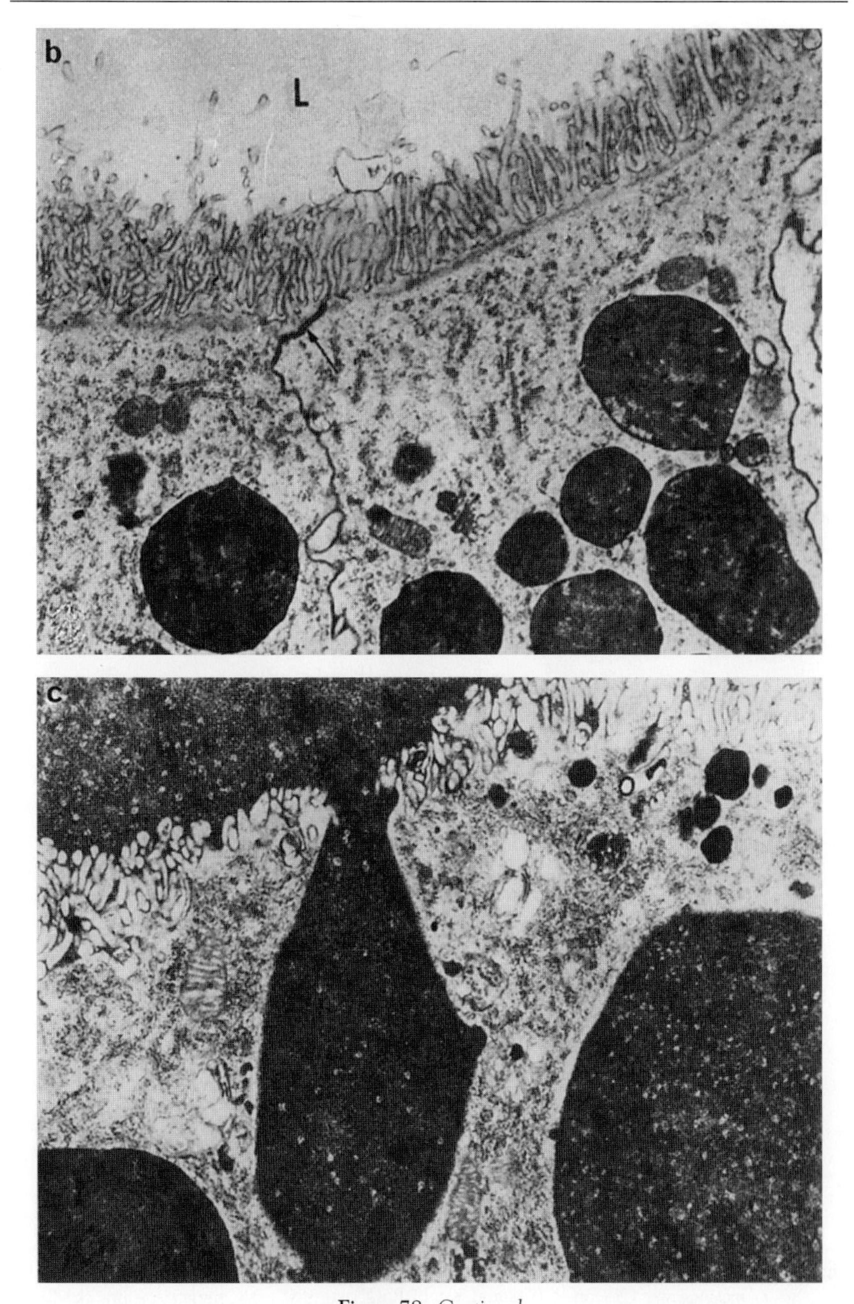

Figure 78. *Continued*

acids. In confirmation of his results, it was demonstrated that the secretion consists of 30% sugars and 70% amino acids (Ashburner and Blumenthal, in Ashburner, 1970a). In *D. virilis*, secretion is composed of 35.5% protein, 45.5% reducing sugars, and 19% amino sugars (Perkowska, 1963b).

PAGE electrophoresis of secretory proteins in the presence of 8 M urea allowed detection of five fractions in *D. melanogaster* (Korge, 1975, 1977a,b, 1980; Akam *et al.*, 1978; Kokoza *et al.*, 1980; Velissariou and Ashburner, 1980, 1981; Kokoza and Karakin, 1981).

Several fractions were additionally identified in concentrated or gradient gels (Kokoza *et al.*, 1980; Korge, 1981; Kokoza and Karakin, 1982). Figure 79 presents an illustrative electrophoregram of glue protein secretion; the nomenclature of the known fractions is given in Table 6. SDS–PAGE electrophoresis revealed eight fractions with apparent molecular weights varying from 8.5 to 360 kDa (Beckendorf and Kafatos, 1976; Korge, 1977b; Kokoza *et al.*, 1980; Gautam, 1983, 1984; Hansson and Lambertson, 1984). One secretory protein class, *Sgs-3*, yields four fractions in two-dimensional electrophoresis (Kokoza *et al.*, 1980, 1982; Kokoza and Karakin, 1985). The data for the correspondence of fractions according to the authors are given out in Table 7.

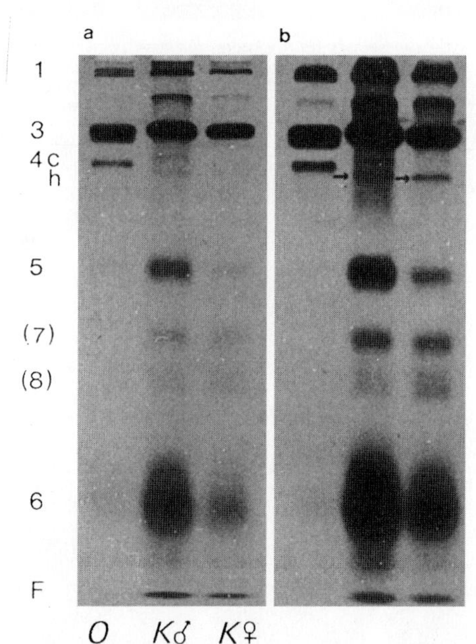

Figure 79. Electrophoretic separation of secretory proteins of the larval salivary glands of *D. melanogaster* Oregon-R (O) and Kochi-R (K) strains. (a) Coumassie stain; (b) fluorography after incorporation of [^{14}C]proline. Numbers at the left designate fraction number; subfractions are designated by letters. F, front of electrophoresis (reprinted by permission from Korge, 1981).

Table 6. Nomenclature of Salivary Gland Secretion Proteins of the Larvae of *Drosophila melanogaster*[a]

Numeration of secretion proteins based on the data of			
Kokoza and Karakin (1985)	Kokoza *et al.* (1980)	Korge (1981)	Velissariou and Ashburner (1981)
Sgs-1	1	1	1
Sgs-6	—	—	6
Sgs-3	3	3	3
Sgs-4	4	4	4
Sgs-5	5	5	5
Sgs-10	7a	7	—
Sgs-9	7b	8	—
Sgs-8	6a	6a	—
Sgs-7	6b	6b	—

[a]After Kokoza and Karakin (1985).

The carbohydrate moiety of secretion is visualized after staining of the electrophoregram with the PAS procedure (Beckendorf and Kafatos, 1976; Kolesnikov and Kiknadze, 1976; Korge, 1977b; Kokoza *et al.*, 1980), *Sgs-3*, *Sgs-7*, and *Sgs-8* were found to be the most glycosylated fractions (Kokoza *et al.*, 1980; Kokoza and Karakin, 1981). Protein fractions are phosphorylated (Khabibullaev *et al.*, 1986). Presumably, the above described secretion in the *l(3)tl*

Table 7. Molecular weights (kDa) of the secretion proteins of *D. melanogaster* according to the data of electrophoresis in polyacrylamide gel in the presence of sodium dodecylsulphate (after Kokoza and Karakin, 1985)

Beckendorf and Kafatos (1976)	Korge (1977b)	Kokoza *et al.* (1980)
P1 (130–160)	A (360)	Sgs-3 (180)
P2a (120)	B (160)	Sgs-3 (140)
P2b (105)	—	Sgs-3 (85)
P3 (91)	C (100)	Sgs-4 (80)
—	—	Sgs-3 (65)
P4 (20.5)	D (23)	Sgs-5 (20)
P5 (15)	E (12)	Sgs-9, Sgs-10 (13–15)
P6 (8.5)	—	Sgs-7, Sgs-8 (10)

mutant (Zhimulev, 1973a) may be an entirely proteinaceous component devoid of carbohydrates.

The antigenic profile of secretion contains five major fractions (Karakin *et al.*, 1977a,b; 1983) similar in antigenic properties (Karakin *et al.*, 1983; Kokoza and Karakin, 1985). Immunochemical similarity presumably results from homology of polypeptide sequences. Comparisons of the peptide maps of the secretory proteins *Sgs-4*, *Sgs-5*, *Sgs-7*, *Sgs-7b*, *Sgs-6a*, and *Sgs-6b* (for nomenclature see Table 6) revealed that, of the total number of 6–14 peptides in each *Sgs* protein, from 3 to 6 (after treatment with trypsin) or 5 to 7 peptides (after treatment with chemotrypsin) are also identified in the other *Sgs* proteins (Shcherbakov *et al.*, 1985).

Comparative analysis revealed high homology between the nucleotide sequences of the secretory protein *Lgp-1* and *Lgp-3* genes in *D. virilis* (Swida *et al.*, 1990).

In summary, the salivary gland secretion of *D. melanogaster* contains no more than 10 fractions. A smaller number (2–5) of "major" fractions was detected in glycoprotein secretion of the salivary glands in *Drosophila varians*, *Drosophila bipectinata*, *Drosophila malerkotliana*, *D. pseudoobscura*, *Drosophila serrata*, *Drosophila auraria*, *Drosophila quadraria*, *Drosophila triauraria*, *Drosophila takahashii*, and *Drosophila equinoxialis* (Manousis, 1985; Manousis and Kastritsis, 1987). Two to three major fractions were identified in the secretion isolated from *D. virilis* salivary glands. The synthesis of *Lgp-1* occurs in three steps. A protein fraction with a molecular weight of 138 kDa is modified by glycosylation. Glycosylation occurs at threonine residues. *Lgp-2* has a molecular weight of 15 kDa, and it is weakly glycosylated. The gene encoding the *Lgp-3* fraction is 300 bp in length (Kress, 1979, 1982, 1986; Kress and Swida, 1990; Swida *et al.*, 1990). According to other data, the secretion of this species contains five to eight fractions (Kolesnikov and Aimanova, 1985; Aimanova *et al.*, 1987a,b).

In members of the *nasuta* subgroup of *Drosophila* species, the total number of fractions identified in secretion was 11, of which 6 major fractions were glycosylated (Ramesh and Kalisch, 1987, 1988a,b; Kalisch and Ramesh, 1988, 1997a; Zajonz *et al.*, 1996a,b,d). In eight representatives of the species of the *D. nasuta* subgroup, a close number of fractions, which are species specific, were detected (Ramesh and Kalisch, 1989). No less than 10 fractions were found in *Drosophila gibberosa* (Shirk *et al.*, 1988) and in several other species (Zajonz *et al.*, 1996b; Kalisch and Ramesh, 1997b,c).

In view of all these results, the data that larval secretion in *D. melanogaster* contains 69 protein fractions, of which 40 are PAS-positive in *D. melanogaster* larvae (Kolesnikov, 1977, 1990; also see Kolesnikov *et al.*, 1975, 1976a,b; Kolesnikov and Kiknadze, 1976) may be ascribed either to inadequacy of PAAG electrophoresis, using 8M urea or acetic acid, which might have subjected polypeptides to degradation or to inaccurate isolation of proteins from residual gland cells. Whatever the explanation may be, these results may be regarded as artifactual.

Information about differentiation of the salivary gland cells into proximal and distal portions according to their capacity to synthesize glycoprotein secretion was given above. The differences between the two portions proved to be more consequential.

The *D. melanogaster* genome was transformed with 1453-bp fragments carrying the regulatory part of gene P25 *Bombyx mori* coding for silk protein fused with the *E. coli* β-galactozidaze (*lacZ*) coding sequence. The fragments with *p25-lacZ* constructs were used to make transgenics. In them, only cells of the anterior region of the salivary glands expressed the *lacZ* reporter, downstream from the promoter gene, irrespective of larval age. It was concluded that, although *Drosophila* salivary glands do not produce silk, their anterior regions retain trans-acting factors capable of recognizing cis-regulatory elements of a *Bombyx* silk protein gene (Bello and Couble, 1990).

Experiments with hybridization of DNA from the *Egp-1* and *Egp-2* genes, isolated from the 55E puff of *D. virilis*, to RNA on salivary gland preparations demonstrated that the RNA of these genes is synthesized in all the cells of the gland throughout larval development. However, transcripts disappear from cytoplasm of the distal gland from mid-third instar, remaining only in cells of the proximal gland; they disappear from that part, too, before puparium formation (Thuroff *et al.*, 1992). Within the nucleus glue mRNA of the glue protein genes *Egp-1*, *Egp-2*, *Lgp-1*, *Lgp-2*, and *Lgp-3* accumulates in primary compartment developing at a single chromosome site. Then transcripts become homogenously dispersed throughout the whole nucleus, a clear border being formed by the nuclear envelope.

After some delay cytoplasmic export of transcripts appears to start simultaneously at all points of the nuclear envelope (Thuroff *et al.*, 1992; Kress, 1996).

A transgenic strain of *D. melanogaster* was obtained using a fusion gene construct containing the *lac-Z* sequences and 532 bases upstream and the gene *Egp-1* and the 300-bp-downstream sequences. All the cells expressed the fusion gene at the early larval stage and only the proximal cells during the third-larval instar (Thuroff *et al.*, 1992).

In *D. hydei*, larval salivary glands transplanted into the abdomen of adults ceased to differentiate into the distal and proximal portions as judged by their synthethic capacities. In fact, all the salivary gland cells began to synthesize salivary gland secretion (Berendes and Holt, 1965).

Studies of intracellular and extracellular electric phenomena in the salivary gland cells of Diptera (Loewenstein and Kanno, 1962, 1964; Bozhkova *et al.*, 1970a,b; Kislov and Veprintsev, 1970; Cohen, 1977; Cohen *et al.*, 1982; also see reviews by Berendes, Ashburner, 1978; Zhimulev, 1992) revealed that there is virtually intracellular electrical coupling in the undifferentiated salivary

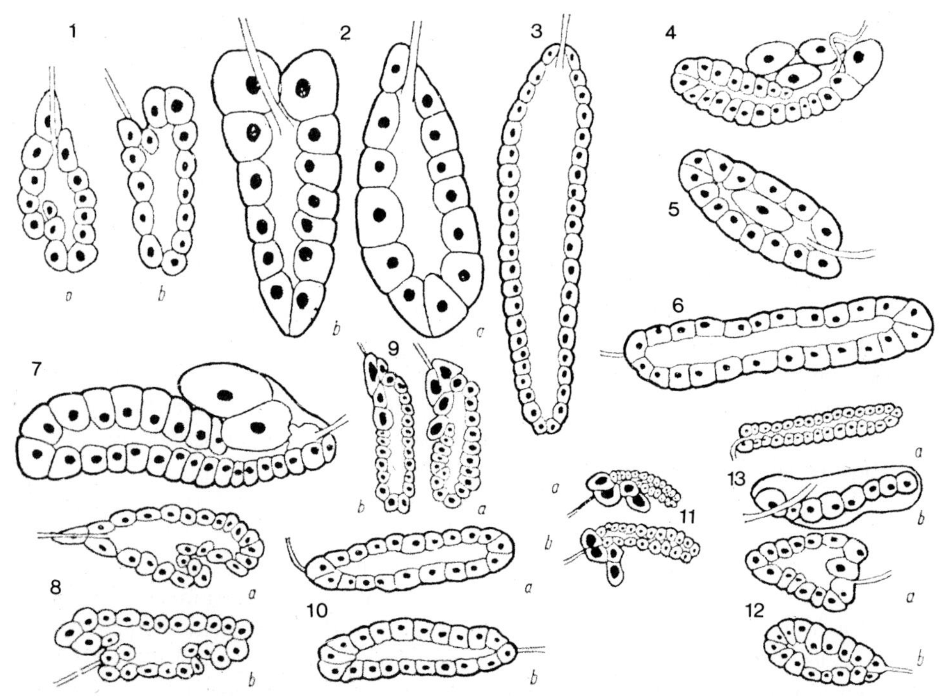

Figure 80. Morphology of the salivary glands in larvae of various species of *Chironomids*. (a) Right gland; (b) left gland. 1, *Cryptochironomus* gr. *anomalus;* 2, *Cryptochironomus* gr. *conjungens;* 3, *Cryptochironomus demeijerei;* 4, *Cryptochironomus* gr. *defectus;* 5, *Cryptochironomus* gr. *fuscimanus;* 6, *Cryptochironomus monstrosus;* 7, *Cryptochironomus nigridens;* 8, *Cryptochironomus* gr. *pororostratus;* 9, *Cryptochironomus redeckei;* 10, *Cryptochironomus rolli;* 11, *Cryptochironomus supplicens;* 12, *Cryptochironomus* gr. *viridulus;* 13, *Cryptochironomus* gr. *vulneratus;* 14, *Prodiamesa olivacea;* 15, *Syndiamesa* gr. *nivosa;* 16, *Cricotopus dizonias;* 17, Orthocladiinae gen. l. *macrocera;* 18, *Clinotanypus nervosus;* 19, *Pelopia punctipennis;* 20, *P. villipennis;* 21, *Cryptochironomus obseptans;* 22, *Demeijerea rufipes* (1–20 from Konstantinov and Nesterova, 1971; 21–22 from Kurazhkovskaya, 1969).

glands at the beginning of the third instar. Electrical cell coupling appears and increases at the mid-third-instar period in cells synthesizing glycoprotein secretion (Bozhkova *et al.*, 1970a,b).

Differentiation of the cytoplasm was described in some mutants. Slizynski (1963, 1964a,b) reported that a "vacuoli-like body" is present in the distal salivary gland cells of a *fat* (*ft*) mutant in *D. melanogaster* at the end of the second instar.

This so-called vacuolar body is very similar to the nucleolus, but it is larger and bound by a membrane. The number of vacuoles varies from 0 to 3 per cell; their average number is 0.6 per cell. The vacuoles do not stain by the Feulgen reaction, nor with methyl green-pyronine; they are PAS-negative, they

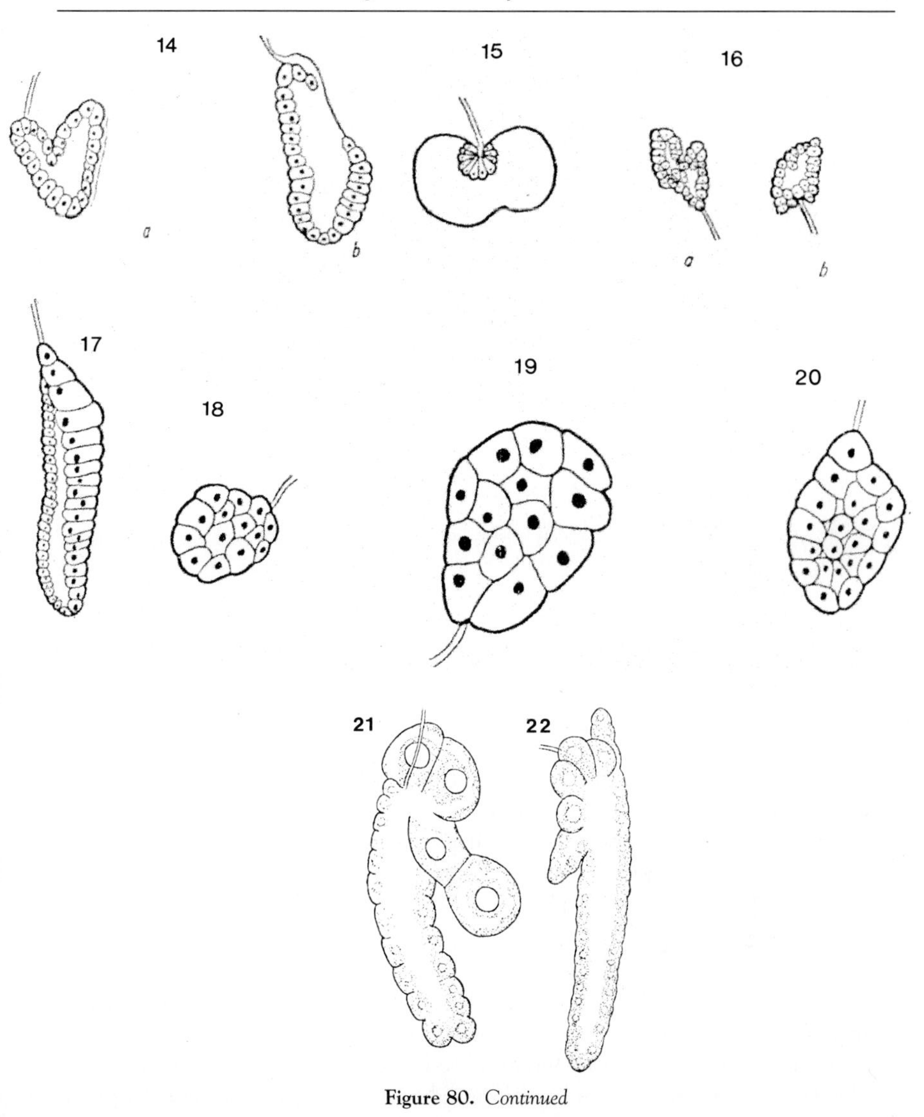

Figure 80. *Continued*

do not stain with alcian blue or toluidine blue; however, they stain with the Millon reaction, bromphenol blue, and modifications of Sudan black reaction. It was inferred that the so-called "vacuoles" are lipoprotein in nature (Chaudhuri, 1969; Chaudhuri and Mukherjee, 1969).

Vacuoles were not detected in homozygotes for the lethal alleles at the *ft* locus (Zhimulev and Szidonya, 1991).

A comparison of the structure of the salivary glands in Chironomidae

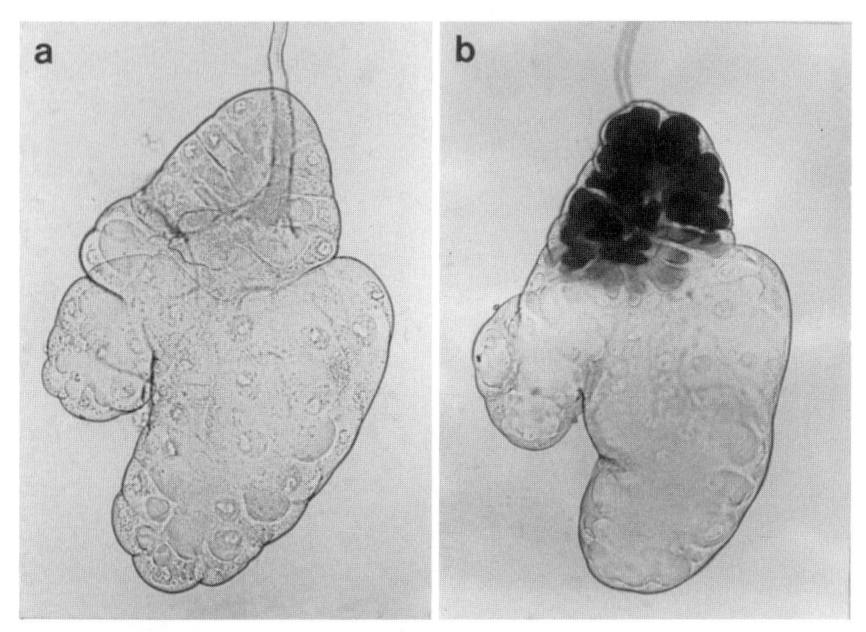

Figure 81. Differentiation of the salivary glands of (a and b) *Acricotopus lucidus*. (a) Larva; (b) prepupa (from Baudisch, 1963b).

species reveals remarkable morphological differences (Figures 80 and 81): the average number of cells per gland may vary betwen species: from ~13 in *Tanytarsus* gr. *lobatifrons* to ~61 in *Glyptotendipes glaucus* (Konstantinov and Nesterova, 1971). The two glands in a larva can be symmetric or asymmetric (Figure 80; also see Staiber and Behnke, 1985). In certain species, the gland, at least outwardly, seems to not differentiate into two parts; in other species, the two parts differ in their cell size (Figure 81).

Rambousek (1912) suggested that the giant formations in the salivary gland nuclei of *Chironomus* larvae must be regarded as chromosomes unusually changed to perform a secretory function, to release a secretion by which detritus is glued. Many subsequent studies addressed both the organization of the physiological functions of the glands and relations between gland activity and polytene chromosomes (Tanzer, 1922; Parat and Painleve, 1924a,b; Kolmer and Freischmann, 1928; Parat, 1928; Kryukova, 1929; Beams and Goldsmith, 1930; Gatenby, 1932; Walsche, 1947; Yoshimatsu, 1962; Yoshimatsu and Uehara, 1968).

The main product of the salivary glands is glycoprotein, synthesized in the cytoplasm and secreted as granules mainly in the Golgi complex. The granules migrate to the cell apex; the lumen releases its contents into its surroundings (Beermann, 1961; Yoshimatsu, 1962; Yoshimatsu and Uehawa,

1968; Kato *et al.*, 1963; Kato and Sirlin, 1963; Kroeger, 1964; Jacob and Jurand, 1965; Clever, 1969; Kloetzel and Laufer, 1968–1970; Kiknadze *et al.*, 1975; Grossbach, 1977; Agapova and Kiknadze, 1985a,b).

The rate of the release of secretion of the salivary glands can be considerably increased by treating larvae with pilocarpine, which causes a compression of the salivary gland and a release of secretion from its lumen (Beermann, 1961; Clever, 1969; Valeyeva, 1975; Mahr *et al.*, 1980; Meyer *et al.*, 1983; Filippova, 1985). In contrast, atropine blocks the secretion process (Grossbach, 1977).

The various cells of a gland can differ in their secretory functions; this appears to be reflected in the morphological differences between the cells (Figure 81). Although similar in appearance, cells can differentiate to perform different functions. Based on morphological features, the salivary glands of *Acricotopus lucidus* can be divided into the anterior, main, and lateral lobes (Figure 81); large amounts of β-carotinoids accumulate in the anterior lobe during prepupal development. The main and lateral lobes differ in high hydroxyproline content from the anterior lobe (Mechelke, 1953, 1958; Baudisch, 1960, 1961, 1963a–c, 1967, 1972, 1977; Baudisch and Panitz, 1968; Baudisch and Hermann, 1972). When viewed by electron microscopy, the cells of the main and lateral lobes appears as a fibrous structure, while those of the anterior lobe are globular (Dobel, 1968).

In many chironomid species, belonging to the *Cryptochironomus*, *Trichocladius*, and *Zavrelia* genera (Beermann, 1952b, 1956, 1959, 1961), four to six specialized cells (Sonderzellen) are distinguished in the anterior part of the gland near the opening of the gland duct. Special gland cells are abundant in the secretion surrounding them. Various inclusions, reminiscent of autophage vacuoles, were observed in the special lobe cells by electron microscopy (Kloetzel and Laufer, 1969). Along with electron-dense material that is also present in the main lobe cells, the special lobe cells contain granules filled with material of low electron density (Agapova and Kiknadze, 1978).

The lobe cells differ in other respects. In the C. *thummi thummi*, [^{14}C]lysine is incorporated more intensely into the specialized cells than into all the other gland cells (Sebeleva, 1972). Staining of the salivary glands with various cytochemical methods (alcian blue at different pHs; and various pretreatments with the Hale reagent, Bismark brown, and the PAS reagent, among others) demonstated that the secretion of the special lobe and of the main gland is very different. The difference is mainly due to the carbohydrate component: the concentration of neutral and, presumably, acid polysaccharides is presumably higher in the special lobe. The amino acid spectra of proteins are the same in both lobes (Sebeleva and Kiknadze, 1977; Kiknadze *et al.*, 1979; Sebeleva *et al.*, 1981, 1985).

In C. *thummi*, the weight of the secretion of the special lobe is about 0.6 μg (which amounts to about 7% of the total weight of gland secretion).

This secretion is actively extruded from the gland during the light day time and this makes its chemical analysis quite difficult (Sebeleva *et al.*, 1985). Secretion was isolated from the special lobe and sp3 and an additional fraction with a molecular weight of 160 kDa, was isolated; this fraction was missing in the secretion of the main lobe (Kolesnikov *et al.*, 1981; Sebeleva *et al.*, 1981).

Antibodies against a hybrid protein containing fragments of *E. coli* β-galactosidase and a low-molecular-weight (67-kDa) protein encoded by Balbiani ring 6, specific to the special lobe, were raised. Analysis of these antibodies revealed similarities between the protein secretion of the special and main lobes (Bogachev *et al.*, 1989, 1990).

The main lobe of the salivary gland is a machinery extremely active in protein synthesis: the cell cytoplasm contains up to 25 ng RNA (the data for *C. thummi* were provided by Zakharenko *et al.*, 1978), thereby ensuring the intense processes of translation. The salivary glands of *C. thummi* contain 9–10 μg (Kiknadze *et al.*, 1979) or 7–14 μg (Sebeleva *et al.*, 1985) of protein; its content is 15–30 μg in *C. tentans* (Darrow and Clever, 1970). It is thought that up to half of this amount (Darrow and Clever, 1970; Grossbach, 1977; Kiknadze *et al.*, 1979; Rydlander and Edstrom, 1980; Serfling *et al.*, 1983) and up to 80% of labeled amino acids are incorporated into proteins (Wobus *et al.*, 1970; Grossbach, 1977).

The rate of the synthesis of protein secretion is high. There is evidence indicating that protein is synthesized in an amount equal to its weight for 12 hr in the salivary glands of *C. tentans* (Beermann, 1972); according to other evidence, up to 50 μg of secretion can be synthesized for 5 hr in the larvae of *C. thummi* (Serfling *et al.*, 1983).

Giant polysomes contain RNA templates (Clever and Storbeck, 1970; Franke *et al.*, 1982; Kiseleva, 1989). The average number of ribosomes in the giant polysomes of *C. tentans* is about 100 (Franke *et al.*, 1982) and from 79 to 146 in *C. thummi;* its average number is 97 (Kiseleva and Masich, 1991). Protein is presumably synthesized on relatively long-lived mRNA templates. Glandular protein synthesis proceeds at a similar rate as before the addtition of actinomycin D and after RNA synthesis has been inhibited by the drug for 48 hr (Clever, 1969; Clever *et al.*, 1969a,b; Doyle and Laufer, 1969b; Clever and Storbeck, 1970; Wobus *et al.*, 1972).

Meaningful biochemical studies on salivary gland secretion in chironomids became feasible only after more efficient methods facilitating its collection and analysis were developed (review: Grossbach, 1977). Two methods are currently being used: (1) the isolated gland is ruptured, and secretion flows freely into saline; (2) the secretion is directly precipitated into the gland with cold ethanol or an ethanol–acetic acid (3:1) mixture; the secretion strand is isolated when removing debris of glands by needles (Grossbach, 1969; Wobus *et al.*, 1970; Kopantzev *et al.*, 1979, 1985a,b; Rydlander and Edstrom, 1980).

A detailed biochemical characterization of the secretion in Chiromomus can be found in the review of Kopanzev *et al.* (1985b). Some of the relevant information is given below.

Salivary gland secretion contains proteins and carbohydrates—which are covalently bound to proteins; for example, neutral sugars—whose weight is up 2% of the total weight of the secretion (Grossbach, 1969, 1977) or an even greater percentage (Hertner *et al.*, 1983). An unusually high content of certain amino acids was found in the protein components of the secretion of *C. tentans*, *C. thummi* and *Chironomus pallidivittatus*; the content of lysine, proline, serine, and glutamic acid amounts to 8–24% (Grossbach, 1969, 1977; Serfling *et al.*, 1983). In most *Chironomus* species, 5 to 6 fractions of secretory protein constituting 95% of total secretion can be seen in the electrophoregrams. The remaining 5% were represented by minor fractions. They may be numerous. There are, for example, up to 11 fractions in *C. plumosus* (Grossbach, 1968, 1969, 1974, 1977; Wobus *et al.*, 1970–1972; Beermann, 1972; Pankow *et al.*, 1976; Hardy and Pelling, 1980; Rydlander and Edstrom, 1980). In *C. thummi*, the molecular weights of the fractions are: sp2, 220 kDa; sp4, 180 kDa; sp5, 35 kDa; sp6, 18 kDa; and sp7, 16 kDa (Serfling, 1983; Kopantzev *et al.*, 1985a,b); according to other data (Kolesnikov *et al.*, 1981), their molecular weights are, respectively: 180, 150, 35, 18, and 15 kDa.

In *C. tentans*, the secretion represented by the sp1 fraction is separated by various modifications of electrophoresis into two to three, even four, fractions with molecular weights from 800 to 1400 kDa (Grossbach, 1968, 1969, 1974, 1977; Edstrom *et al.*, 1980; Hardy and Pelling, 1980; Hertner *et al.*, 1980; Rydlander and Edstrom, 1980; Rydlander *et al.*, 1980; Galler *et al.*, 1984; Kao and Case, 1985, 1986; Case, 1986).

Subsequently, two giant polypeptides similar in size were detected in the secretion of *C. thummi* (Kolesnikov *et al.*, 1981; Serfling *et al.*, 1983; Cortes *et al.*, 1990).

The secretory polypeptides have been repetitively cleaved into fractions with multiple molecular weights. This was observed after electrophoretic separation of soluble and undegraded secretion was exposed to endogenous or exogenous (papain) proteases (Hertner and Mahr, 1979; Hertner *et al.*, 1980). The observations suggested that the high-molecular-weight fractions of the secretion may contain repetitive fragments about 200 amino acid residues in length (Hertner *et al.*, 1980; Grond *et al.*, 1987).

There is reason to believe that the secretory proteins of *Chironomus* are related. Analysis of peptide maps demonstrated that limited proteolysis of secretory proteins sp-2, sp-4, sp-5, sp-7, and sp-8 yields a set of peptides with similar, if not the same, molecular weights (Kopantzev *et al.*, 1985a,b). Fractions sp-1a and sp-1b were found to be immunochemically similar (Rydlander and Edstrom, 1980; Rydlander, 1984); furthermore, antisera raised against these fractions occasionally precipitated proteins with a lower molecular weight of 165 kDa (Hertner *et al.*, 1983). There are, presumably, homologous regions in the secretory protein with molecular weights of 220 and 185 kDa (Brumley *et al.*, 1992).

A comparison of amino acid sequences in the constant regions of secretory polypeptides in three species of *Chironomus* (*C. tentans*, *C. thummi*, and *C. pallidivittatus*) disclosed considerable similarities. In fact, 7 of the 30 amino acid residues were completely constant (Hamodrakas and Kafatos, 1984; Wieslander *et al.*, 1984).

From the mid-1960s, the question of whether the secretory proteins are directly synthesized in the salivary gland or partly released from hemolymph has been discussed. Using antibodies raised against collected secretion of *C. thummi*, it was found that all the antibodies bind to both the secreted and hemolymph proteins (Laufer and Nakase, 1964, 1965b; Laufer, 1965, 1968; Laufer and Goldsmith, 1965; Doyle and Laufer, 1968, 1969a,b; Schin and Laufer, 1974). This finding suggested that the salivary glands may be organs transporting protein from hemolymph (Laufer, 1968) or the adjacent fat body (Sebeleva *et al.*, 1985). Some support was provided by the results obtained with cloning the genes encoding the secretory proteins in *D. virilis*. Two genes localized in puff 16A in *D. virilis* are weakly expressed in salivary gland as well as in midgut and fat body cells (Kress *et al.*, 1990). The objection to this idea was that the antiserum against salivary gland secretion of *C. tentans* reacts extremely weakly with larval hemolymph (Grossbach, 1977).

Kopantzev *et al.* (1985b) thought that Laufer was in error mainly because he did not use in this experiment pure secretion isolated from the gland lumen—rather, secretory material collected from the substrate and dissolved in saline. Because the major secretory fractions dissolved poorly in the usual buffer solutions, the content of minor water soluble proteins of hemolymph was possibly high in the material Laufer used for immunization. In more direct experiments, antiserum against homogenate of larvae without salivary glands did not cross react with secretory proteins (Kopantzev *et al.*, 1979, 1985b).

Experiments were performed to test the possibility that secretory proteins may be transported across the fat body. High-molecular-weight material was not detected in that organ (Sebeleva *et al.*, 1986). Thus, there is every reason to believe that the secretory proteins are organ specific.

The salivary gland secretion of *Phormia regina* acts as an adhesive to attach the prepupal case to a substrate (Fraenkel, 1952; Fraenkel and Brookes, 1953; Moorefield and Fraenkel, 1954).

The biochemical composition of the secretion is as follows: carbohydrate, glucosamine, two unknown components, amino acid residues (glycine, alanine, serine, threonine, valine, leucine, tyrosine, proline, aspargic and glutamic acids, arginine, lysine, and cysteine), and free amino acid lysine (Moorefield and Fraenkel, 1954).

The larval salivary gland is 6–7 mm long and 0.8 mm wide in another representative, *Calliphora erythrocephala*. Proteins were isolated by *in vitro* incubation in Ringer solution and they separated into three major and four to five minor fractions (Price, 1974).

In *Sarcophaga bullata*, the activity of the salivary glands is detected at the prepupal stage. About 28 hr after puparium formation, the salivary glands swell, gradually increase in size, and their contents become yellow brown. At about 6 hr or so, (2 hr before pupal molt) the glands empty their contents again (Whitten, 1975).

In the melon fly *Dacus cucurbitae*, the salivary glands are PAS-negative during the entire 168 hr of larval development; later, the larvae stop feeding and abandon their culture medium. PAS-positive material appears in the cytoplasm. Secretion begins to be released at a time when the salivary gland chromosomes cease incorporating [^{3}H]thymidine. At the age of 240 hr, the salivary glands show bloating with abundant secretions filling the lumen (Gopalan and Dass, 1972a).

The salivary glands of the third- to fourth-instar larvae of *Simulium nudifroms* are about 5 mm long and 0.4 mm wide. The gland has the appearance of a wide duct filled with secretions. Electron-dense secretory granules are seen in the cell cytoplasm in cross sections. Larger membrane-bound granules occupy the apical part of the cells. A release of the contents of the granules into the glandular duct is evident in sections (McGregor and Mackie, 1967).

The larval salivary glands of females of *Anopheles stephensi* consist of two cylindrical lateral lobes; each lobe is divided by a thin constriction into a distal and a proximal portions. The glands also contain a third medial lobe similar to the distal and proximal portions of the lateral lobe. Gigantic masses of secretion accumulated in an enormous vacuole are seen in the cells. The secretion cycle is presumably long in these cells.

The whole gland resembles the proximal portion of the lateral lobes in males (Wright, 1969).

The structure of the salivary glands of *Aedes aegypti* was found to be similar. The secretion contains polysaccharides and proteins in this species (Orr *et al.*, 1961).

3. Hormonal regulation of physiological functions in salivary gland cells of the larvae of *Diptera*

At the end of the larval and the beginning of prepupal development, glycoprotein synthesis continues concomitant with the rise in ecdysterone titer. Hormone is released by the cells, and the salivary glands undergo histolysis after the stage of puparium formation. In *D. virilis*, the salivary glands transplanted into the abdomen of adult females were cultured for 14 days. The glands became flat and wide under the pressure of lucid aqueous fluid accumulated in the gland duct. When one- to four-ring glands were transplanted at the same time, salivary gland cells showed signs of histolysis, notably opaqueness (Bodenstein, 1943).

Although leaving some issues unclear, such as secretion characterization, these suggest that at least histolysis is under the ecdysterone control. For this reason, it was of interest to analyze the findings on the endocrine regulation of the various processes occurring in the gland during the larval–pupal transition and during larval molt in *Chironomus*.

In *C. thummi*, during the molt that occurs from the third to the fourth instars, the salivary glands cease releasing secretion, which bloats all the internal cisternae of the cells. According to EM data, the number of secretion granules near the Golgi zone decreases, and DNA replication virtually ceases. All these processes are rapidly recovered after molt is completed. The most significant changes occur during the larval–pupal transition (Kiknadze *et al.*, 1975; Agapova and Kiknadze, 1979).

Information about hormonal induction of the synthesis of glycoprotein secretion in *Drosophila* is contraversial. Secretion is synthesized in the *l(2)gl* mutants (Zhimulev, 1975; Zhimulev and Kolesnikov, 1975a; Richards, 1976a), whose hormone titer decreases sharply (see Section VI,A,1). When the anterior parts of embryos of *Drosophila* or salivary glands of early third-instar larvae (just after the second molt)—that is, long before the onset of secretion synthesis—were transplanted into the abdomen of adult females, secretion was also synthesized and the gland stained with the PAS-reagent (Berendes and Holt, 1965; Ashburner and Garcia-Bellido, 1973; Zhimulev, 1975; Zhimulev and Kolesnikov, 1975a).

In individuals homozygotes for the *l(3)tl* ecdysterone-deficient mutation, the salivary gland cells were mainly PAS-negative; nevertheless, secretion of an indefinite type was synthesized (Zhimulev, 1973a; Zhimulev and Kolesnikov, 1975).

Contradictory data were obtained on the role of ecdysterone in the initiaton of the synthesis of protein secretion in other ecdysterone-deficient mutants. In Northern blot hybridization of larvae homozygous for the *l(1)su(f)*^{ts67g} mutation—developing at a restrictive temperature (30°C), starting from the second-larval instar—*Sgs-3-*, *Sgs-4-*, *Sgs-7-*, and *Sgs-8*-encoded RNAs were not detected; however, the mRNAs were detected in the nuclei, when ecdysterone was added to the food (Hansson and Lambertsson, 1983, 1984). The amount of *Sgs-3* RNA was highest 1 hr after the beginning of hormonal induction; induction was inhibited in the presence of cycloheximide (Hansson and Lambertsson, 1989).

The initiation of expression of the *Sgs-3* gene inserted into the genome by transformation also requires ecdysterone (P. H. Mathers, in Meyerowitz *et al.*, 1987). In the *l(1)npr-1* mutant (the allele at the *ecs* or the *BR-C* locus), the cells are insensitive to this hormone and, as a consequence, they do not express the *Sgs-3* gene (Crowley and Meyerowitz, 1984). Secretions are synthesized and released into the gland lumen in homozygotes for the *BR-C lt435* allele of this gene (Biyasheva, 1987). The induction of the synthesis of the secretory protein may be associated with a small peak of hormone concentration (peak 7 in Figure 81) at the beginning of the third-larval instar.

It was convincingly demonstrated that the inhibition of secretion synthesis is under ecdysterone control (Berendes and Holt, 1965; Poels *et al.*, 1971; Poels, 1972). *In vivo* treatment of the salivary glands of *D. melanogaster* with hormone caused (within 15 min) a sharp decrease in the rate of RNA transcription from the genes encoding the *Sgs-3* and *Sgs-4* proteins. The hormone-induced inhibition may be sensitive (*Sgs-3*) or insensitive (*Sgs-4*) to cycloheximide (Crowley and Meyerowitz, 1984; Meyerowitz *et al.*, 1987; Hansson and Lambertsson, 1989).

As a study of the glucosamines metabolism in *D. virilis* salivary glands demonstrated, the glands capacity to form amino sugars is under the control of enzymes glutamine-fructose-6-phosphate aminotransferase and UDP-*N*-acetylglucosamine pyrrophosphorylase. Their activities sharply decreased before puparium formation (Kress and Enghofer, 1975; Enghofer *et al.*, 1978; Enghofer and Kress, 1980; Kress, 1981), and the ecdysterone titer increased in parallel. There was a concomitant drastic decrease in the rate of incorporation of

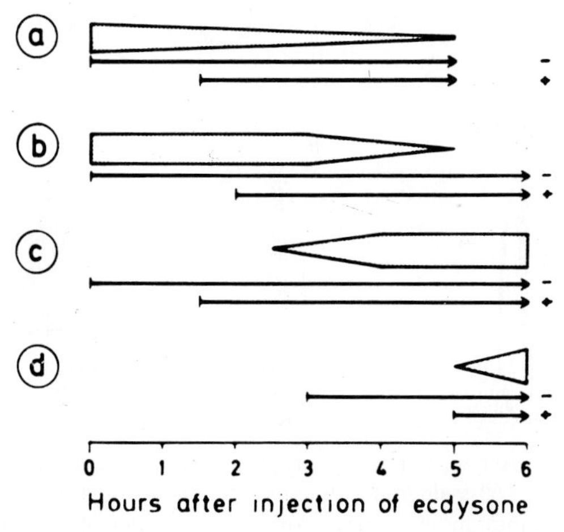

Figure 82. Changes in the activity of (a) glutamine-fructose-6-phosphate aminotransferase; (b) intensity of [^{14}C]glucose incorporation into the glycoprotein fraction of *Lgp 1*; (c) size of salivary gland duct; and (d) synthesis of prepupal proteins of salivary gland after injection of ecdysterone into larvae of *D. virilis* (reprinted by permission from Kress, 1981). Figures under horizontal line designate hours after injection of ecdysterone into larvae (0.02 μg/larva). Thickness of figures reflects the intensity of the process. The length of the arrow under the figure designates the time elapsed after injection of actinomycin D (60 ng/larva); a minus indicates depression of hormone-induced function under the effect of actinomycin D; a plus indicates normal performance of function despite the presence of antibiotic.

[14C]glucose into glycoprotein of the salivary glands and secretion (Kress, 1981, 1982, 1986; Kress and Swida, 1990).

Incorporation of [14C]glucose or [14C]proline was completely inhibited 5 hr after injection of α-ecdysterone (Figure 82). When hormone was injected together with actinomycin D, the synthesis of secretory proteins was not arrested even after 8 hr. However, when antibiotic was injected 2 hr after hormone, it ceased 8 hr after the treatment (Kress, 1977, 1979).

The appearance of ecdysterone arrested the transcription of a number of other genes active throughout entire development; for example, that of the *Egp-1* gene in *D. virilis* localized in the tissue-specific 55E intermolt puff (Thuroff *et al.*, 1992) and under the control of glycosamines (Kress, 1973). Following

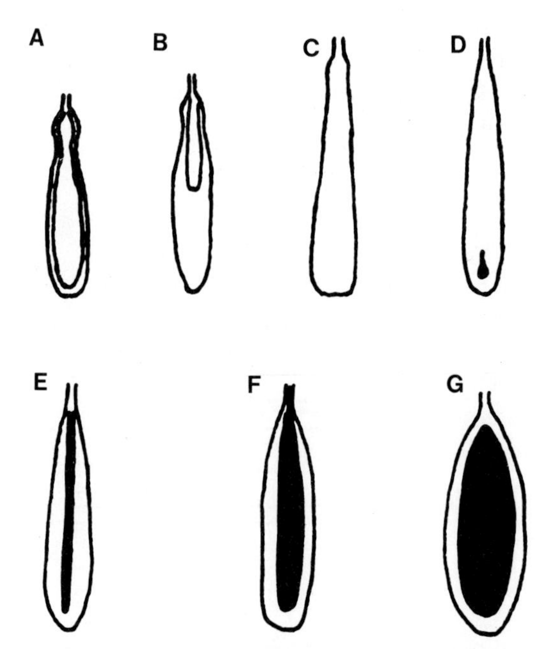

Figure 83. Change in morphology of the salivary gland during the second half of third-larval instar of *Drosophila* (A) midthird-larval instar, gland lumen is completely open; (B) the last third of the third instar, starting from the distal portion of gland, the cells are brought close together plugging gland cavity; (C) cells are opposed along the entire length of the gland and filled with glycoprotein secretion, the puff set in the chromosomes corresponds to stage PS1; (D) appearance of a small cavity in the most distal portion of the gland due to secretion of glycoproteins, the set of ecdysterone puffs corresponds to stage PS2-3; (E) a thin secretion branch lies along the gland length in the duct, PS4-5; (F) the thickness of the secretion strand corresponds to one third of that of the salivary gland, PS6-7; (G) the gland appears as a swollen sac filled with secretion; developmental stage—just before discharge of secretion into surroundings and spiracle eversion, PS9-11 (original drawing made according to G. Richards, personal communication, 1979, and data by Zhimulev *et al.*, 1981e).

the injection of larvae of C. *thummi* (fifth phase) with ecdysterone the Golgi complex was found to be inactivated. This may be evidence that the synthesis of secretion then becomes inactivated (Kiknadze *et al.*, 1984).

Secretion release from cells into the gland lumen is a hormonally controlled function. During development (Figure 83), the enfoldment of the secretory process and, consequently, the lumen's state and the gland's morphology are closely correlated with the degree of the induction of the ecdysterone puffs (the so-called puff stages; see Section VI,C,1).

Salivary glands at the mid-third-larval stage may be implanted into the abdomen of adults up to 40–50 days; secretion strands are then not seen in the gland lumen (Berendes and Holt, 1965; Aizenzon, 1979). The injection of ecdysterone into the abdomen may result in the appearance of secretion in the lumen already after 9–10 hr (Zhimulev, 1975; Aizenzon, 1979). According to other data, secretion can be also released into the duct without injection of hormone (Bodenstein, 1943; Korge, 1977b). These differences may be due to the variations in the amount of ecdysterone injected into the abdomen of females. These amounts would be too small to induce the ecdysterone puffs, yet large enough to induce secretion.

When the anterior part of 10- to 20-day-old embryos was transplanted into the abdomen of adults, secretion filled the salivary gland lumen of 10-day-old transplants (Korge, 1980). This occurred possibly because the ecdysterone producing prothoracic gland was transplanted together with salivary glands. When ecdysterone was injected into early third-instar-larvae of *D. hydei*, secretion was released into the lumen, while the cell cytoplasm became PAS-negative. The time elapsed from injection to secretion release was related to the age of the exposed larval and hormone concentration. The interval varied from 8 in 120 hr larvae to 4 in 150 hr larvae (Poels, 1970; Poels *et al.*, 1971). Secretion was released into the lumen after hormone injection into 168 hr larvae homozygous for the *l(2)gl* mutation or mid-third-instar larvae of the wild-type. In contrast, no release of secretion was observed in the part of the salivary gland separated by a ligature from the ring gland (Zhimulev, 1975).

Experiments in which the salivary glands of *D. hydei* or *D. melanogaster* were cultured *in vitro* demonstrated that the secretion is synthesized by the gland cells, although not released into the lumen (Poels, 1970; Boyd and Ashburner, 1977). Five hours after the addition of hormone, the secretion appeared in the lumen (Ashburner, 1972b; Poels, 1972). The secretion process requires hormone-induced synthesis of RNA and proteins, because pretreatment of the glands with actinomycin D or cycloheximide greatly inhibits secretion (Poels, 1972).

The salivary gland cells of *D. melanogaster* begin to release secretion 3.5 hr after exposure to hormone and cease releasing it 8 hr after the exposure. To completely evacuate the glue from the gland, the hormone must be in contact with the cells for about 3 hr; thereafter the hormone can be washed off:

β-ecdysone is about 170 times more active than α-ecdysone. Inhibitors of RNA and protein synthesis prevent secretion like in *D. hydei* (Boyd and Ashburner, 1977).

In *D. virilis*, secretion is discharged by the cells at a high concentration of hormone (Kress, 1974). When injected, α-ecdysone induced the discharge of glycoproteins into the lumen. When hormone was injected together with actinomycin D or cycloheximide, it did not serve its inducive function. When injected 1.5–2 hr before inhibitors, α-ecdysterone induced release of the secretion (Kress, 1977).

The results obtained with ecs^{lt435} mutants were unexpected. Homozygotes lost their sensitivity to ecdysterone. Judging by staining with the PAS reagent, no secretion was released in the lumen of 100% of the 33 glands studied in 120- to 140-hr larvae; but in 100% of 15 glands studied in 160-hr larvae secretion was in the lumen (Biyasheva, 1987). Injection of ecdysterone into the fourth-instar larvae of *C. thummi* resulted in massive extrusion of the secretion into the gland reservoir and from the gland (Kiknadze *et al.*, 1984; Filippova, 1985).

The first data indicating that histolysis in the salivary gland may be under hormonal control were obtained by Bodenstein (1943). Later on, Clever (1965a) suggested that the ecdystrone-induced puffs may be involved in the activation of lysosomes leading to histolysis. There is no doubt now that the disintegration of the larval organs is induced by ecdysterone on the background of a low concentration of the juvenile hormone (review: Lockshin, 1985).

In many experiments the salivary gland isolated from early third-instar larvae and transplanted into the abdomen of adult females remained viable to 50 days. However, up to 60% of the glands underwent lysis 24 hr after the injection of ecdysterone and 69–93.5% 48 hr after it. After injection of hormone together with actinomycin D, less than 20% of the salivary glands lysed. Transplantation experiments with the salivary glands of zero-hour prepupae, which had been maintained in larval ecdysterone-containing hemolymph for the previous 6 hr, demonstrated that 87% of the glands lysed after 4 days (Aizenzon and Zhimulev, 1975) or earlier (in an experiment) after 18–24 hr in an *in vivo* system of transplants from 4- to 8-hr prepupae (Staub, 1969). The salivary glands can be incubated for a long time *in vitro* in artificial media. Added ecdysterone quite rapidly leads to histolysis. The larval organs of homozygotes for the *ecs(BR-C)* lethal mutation did not undergo histolysis because thay were not sensitive to hormone (Kiss and Molnar, 1980).

Viability of the salivary glands of *Rhynchosciara angelae* in an *in vitro* system depends on the developmental stage of the larvae from which the glands were isolated. The ecdysterone-induced 2B puff (see Section VI,F), chosen as a marker of the stage of development, is active for 5 days before pupation. When the glands were isolated from larvae in which the puff had been active only for the first two to three days, the glands could be incubated for several days.

Incubation of glands, whose puff has passed the peak of development, were less successful because of their more frequent death. When peak activity had passed, and the puff had already become inactivated, the glands became rapidly inviable *in vitro* because they underwent histolysis (Simoes and Cestari, 1981).

In *D. melanogaster* homozygous for rbp^4 and rbp^5 (BR-C mutations) larval salivary glands do not histolyze and can be found even in late pupae (Restifo and White, 1991, 1992; Zhimulev *et al.*, 1995a).

Recent studies confirmed that salivary gland histolysis is stage-specific steroid-triggered programmed cell death response. Stage-specific induction of the *Drosophila* death genes *reaper (rpr)* and *head involution defective (hid)* immediately precedes the destruction of the salivary gland. The *diap-2* anti-cell death gene is repressed in larval salivary glands as *rpr* and *hid* are induced. This gene is repressed by ecdysterone *in vitro* under the same conditions that induce *rpr* expression (Jiang *et al.*, 1997).

There is indirect evidence for the role of ecdysone. In Sciarides larvae infected with microsporidia, ecdysterone-induced puffs do not develop. As a result, these cells do not undergo lysis during metamorphosis, and they are detected even in the body of imago (Pavan *et al.*, 1975).

Ecdysterone induces histolysis and, consequently, it must induce the activity of lysosome enzymes (Wattiaux, 1969). Indeed the activities of lysosome enzymes increase at the end of the last larval instar—the beginning of the prepupal period in dipteran insects. Acid phosphatase, which is detectable by histochemical staining of lysosomes in the salivary gland chromosomes (Misch, 1962; Rasch and Gawlik, 1964; Clever, 1965; Schin and Clever, 1965, 1968a; Anastasia-Sawicki, 1974; Gaudecker and Schmale, 1974; Jurand and Pavan, 1975), passes into the cytoplasm when undergoing lysis (Clever, 1965; Schin and Clever, 1965, 1968a,b).

The activities of many lysosome enzymes increase at the end of the last instar (the beginning of the prepupal period); for example, those of acid phosphatase (Barker and Alexander, 1958; Aizenzon *et al.*, 1975; Filippova, 1985), alkaline phosphatase in *C. thummi* (Filippova, 1985), but not in the house fly (Barker and Alexander, 1958); those of proteases (pH 3.5) (Rodems *et al.*, 1969; Clever, 1972), acid hydrolases (Laufer and Schin, 1971; Schin and Laufer, 1972, 1973), DNases (Laufer *et al.*, 1963, 1965, 1968; Laufer and Nakase, 1965a; Laufer, 1968; Boyd and Logan, 1972), β-glucoranidase (Vature and Sawant, 1971), as well of total lysosome activity (Hegdekar and Smallman, 1967).

In *Drosophila* treatment of transplanted salivary glands with ecdysterone induces an increase in the specific activity of acid phosphatase, which is sensitive to actinomycin D and cycloheximide (Aizenzon *et al.*, 1975). Injection of ecdysterone into the larvae of *C. thummi* (at the fifth phase of the fourth instar) results in about a twofold increase in the activity of acid phosphatase (Filippova, 1985).

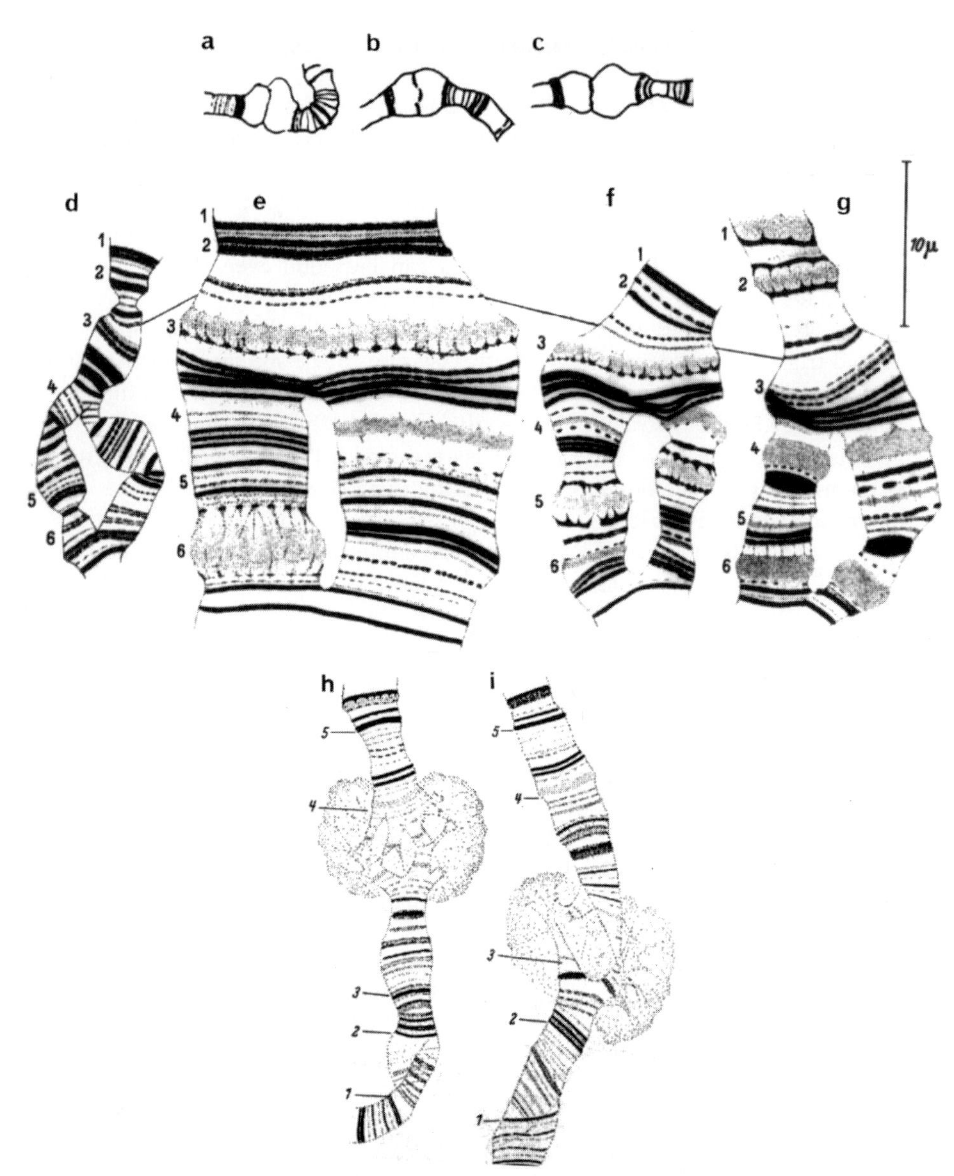

Figure 84. First depictions of the puffs of polytene chromosomes. (a–c) Drawings of puffs in the 74-75 regions of chromosome 3L of *D. melanogaster;* (d–g) variation in localization of six puffs (designated by numbers 1–6) in region 14 of the third chromosome of *Chironomus tentans* in (d) midgut, (e) salivary gland, (f) Malpighian tubules and (h) rectum cells. Pairing is disrupted in the region of the heterozygous inversion. (h and i) Localization of Balbiani rings in the fragment of the first chromosome of *Trichocladius vitripennis* in cells of (h) the main lobe of the salivary gland and (i) in the special cells with granules in cytoplasm. Homologous regions in the chromosome are designated by numbers 1–5 (a–c, from Painter, 1935; d–g, reprinted by permission from Beermann, 1952a; h and i, reprinted by permission from Beermann, 1952b).

From these observations it is evident that many processes in the salivary gland cells enfold under ecdysterone control. This hormone induces many changes in the genetic activity of the puffs.

B. Morphological aspects of puffing

1. First descriptions of puffs

Painter (1935) defined the series of relatively achromatic segments (Figure 84) as the salient features of the third chromosome of *D. melanogaster* (Painter, 1935). These are two more-or-less swollen segments (which proved to be in the 74–75 regions of chromosome 3), limited by dense bands at the tips and separated by a dense band one from another. "These two segments show a striking variation in diameter although within an individual I am under the impression that the form is fairly constant. In some larvae they may be as much as three times the average diameter of the element" (Painter, 1935, p. 305). There are no thicknenings in the given regions in the other larvae; they contain the usual bands. Bridges (1935, 1938) called some of the swellings, for example, in the X-chromosomal 2B region, "puffs."

Working with the chromosomes of *Sciara ocellaris*, Poulson and Metz (1938) established that the giant swellings (bulbs) characteristic of this species are formed from bands. The authors reached the conclusion that all the nuclei have the same set of swellings in all the salivary gland cells of a larva and the location of the swellings vary widely among larvae.

At the time it was unclear what functions the puffs may perform, and they ceased to be a matter of general concern. By the end of the 1940s and beginning of the 1950s, the genetic factors in polytene chromosomes started to be mainly associated with the bands, and this raised the question of whether the bands may be involved in the differential activity of the genes during development.

Studying the morphology of a small fragment (about 40 bands) in the third chromosome of several organs in *C. tentans*, Beermann (1952a) identified six regions in which puffs formed at the site of the bands (Figure 84). The location of the puffs is invariable in the chromosomes of an organ; however, it is different in the different organs of a larva—puffing is in some way correlated with the state of cell differentiation. Beermann concluded that: "Puffing obviously indicates changes, most probably increases, in the activity of gene loci. Hence, nuclear differentiation in cells of different function would be characterized by different specific patterns of activation along the chromosome" (Beermann, 1956, p. 222).

Breuer and Pavan (1952) concomitantly detected that bulb formation in polytene chromosomes of *Rhynchosciara angelae* conforms to a definite pattern: first, the bulbs appear and disappear only in larvae of a certain age; second, the bulbs are present only in salivary gland cells, and not in the chromosomes of Malpighian tubules and gut; instead, the usual bands are formed. The authors came to the conclusion that different cells of an organism have the same genes, but certain genes are "more stimulated" than others in some tissues because swellings are formed.

It was then found that certain specific functions of the cell and the presence of Balbiani rings are correlated. In *Trichocladius vitripennis*, six "special" cells differ from the bulk of the salivary gland cells in granular cytoplasm (see Section VI,A,2). Two regions of the chromosome, whose cells of the main portion show a normal banding pattern, are represented by Balbiani rings in granular cells (Figure 84). In contrast, another locus is represented by the bands in the "granular cells" and by a Balbiani ring in the cells of the main portion (Beermann, 1952b).

In *Acricotopus lucidus*, the entire anterior lobe functionally differs from the other cells (Figure 81). Two specific Balbiani rings and a giant puff were identified in the lobe (Mechelke, 1953).

Thus, by the middle of the 1950s, the concept crystallized that structural changes in the chromosomes, namely loosening of band material and the formation of a specific puff-like swelling in it, are morphological manifestations of gene activation and that these changes correlate with a definite state of differentiation (Beermann, 1952a,b, 1956; Breuer and Pavan, 1952, 1955; Mechelke, 1953).

2. Chromosome material involved in puffing

It has been commonly believed that the puff originates from the decompacted material of the single band (Beermann, 1956, 1962; Pelling, 1966; Clever, 1968; Shcherbakov, 1968; Ashburner, 1970a; Kiknadze, 1971b, 1972; Pelling, 1972). However, from the earliest descriptions of the puffs, the idea that many bands may be included in the region of the active puff gained increasing support. Thus, Poulson and Metz (1938) observed that up to 30 bands are included in each puff in *S. ocellaris*. One of the puffs originates from "more than 20 bands" in *Chironomus plumosus* (Belyanina, 1977). There are numerous examples of multibanded puffs: they occur in *Acricotopus lucidus* (Panits, 1965), *C. thummi* (Kiknadze, 1972), in *Drosophila adiastola* and *D. heedi* (Raikov, 1973), *D. melanogaster* (Burnet and Hartmann-Goldstein, 1971), *D. pseudoobscura* (Stocker and Kastritsis, 1972) (see also review: Kroeger and Lezzi, 1966). Such puffs form in polytene chromosomes in the ovarian nurse cells of *Calliphora erythrocephala* (Ribbert, 1979).

Puffing extends over to the adjacent bands during the formation of the large puffs in the foot pad cells of *S. bullata* (Bultmann and Clever, 1969). In *S. coprophila*, when the DNA puffs reach their maximum sizes, several adjacent bands decondense and the puffs become spherical (Gabrusewicz-Garcia, 1975).

The formation of the complex 2-48-BC in *D. hydei* was most thoroughly studied by Berendes and colleagues (Berendes, 1968a, 1969, 1972a; Berendes *et al.*, 1974). In this region, there are five thin bands (the thickest is B5) which are limited by two groups of dense bands. The five bands are discernible neither in the light nor in the electron microscope after 3 min of heat shock; only separate clumps of condensed chromatin scattered throughout the entire zone of the puff are seen. Uncoiling is over after the next 2–3 min, and the puff is represented by parallel stretched fibrils, which are covered by RNP granules, 30–40 nm in diameter, after another 5 min. In the region of the uncoiled B5 band, aggregates of these granules start to form at this time. The aggregates are seen dispersed in the entire puff after 30 min.

These and other observations raised the question whether it is reasonable to regard the multibanded puffs as a single functional entity. In fact, the possibility could not be ruled out that different independently arising puffs could be located at one site. An important, although not decisive criterion of the integrity of the complex puff, may be obtained by analysis of the sequence of the development and regression of the puffs. This would allow clarification of puffing coordination in single bands.

Beermann (1962, 1967) has paid great attention to this issue. The thorough studies on development in Balbiani rings in the representatives of various *Chironomus* species demonstrated that, with the exception of the two to three nearest adjacent bands at either side, the rest of the bands remains unpuffed. Beermann introduced two notions, the "appearance zone," which is the locus at which puffing starts, (Entstehungsort), and the "activity zone" (Aktivitatsort), which encompasses all the fully developed puffs. Beermann emphasized that in *Chironomus* the development of even giant puffs, such as Balbiani rings, starts with the decondensation of the single band, and a long active region composed of several bands results from their secondary involvement in puffing. However, he accepted the possible existence of the complex puffs that are sequentially involved. Mechelke (1961) defined the puff maximum as the zone from which the puffs with maximal diameter in the fully developed puff starts to expand. He described the migration of puff maxima along the chromosome length: the formation of Balbiani ring starts at one tip of the group of 20 bands, then puffing spreads over to reach the other tip of the chromosome region (Figure 85). He explained this movement by the sequential organization of the genes in the chromosome conforming to the pattern of the organization of the histidine–tryptophan operon complex in *Salmonella*.

In the course of development of *D. hydei*, movement of puff maxima was described in many puffs; X-chromosomal maxima were observed in the 7D-8C region: in 8A, 8B, and 8C and occasionally in the 7D puffs (Berendes, 1965a). A similar movement of puff maxima was found in the pulvilli cells of *Sarcophaga bullata* (Whitten, 1969), in the bristle-forming cells of *Calliphora*

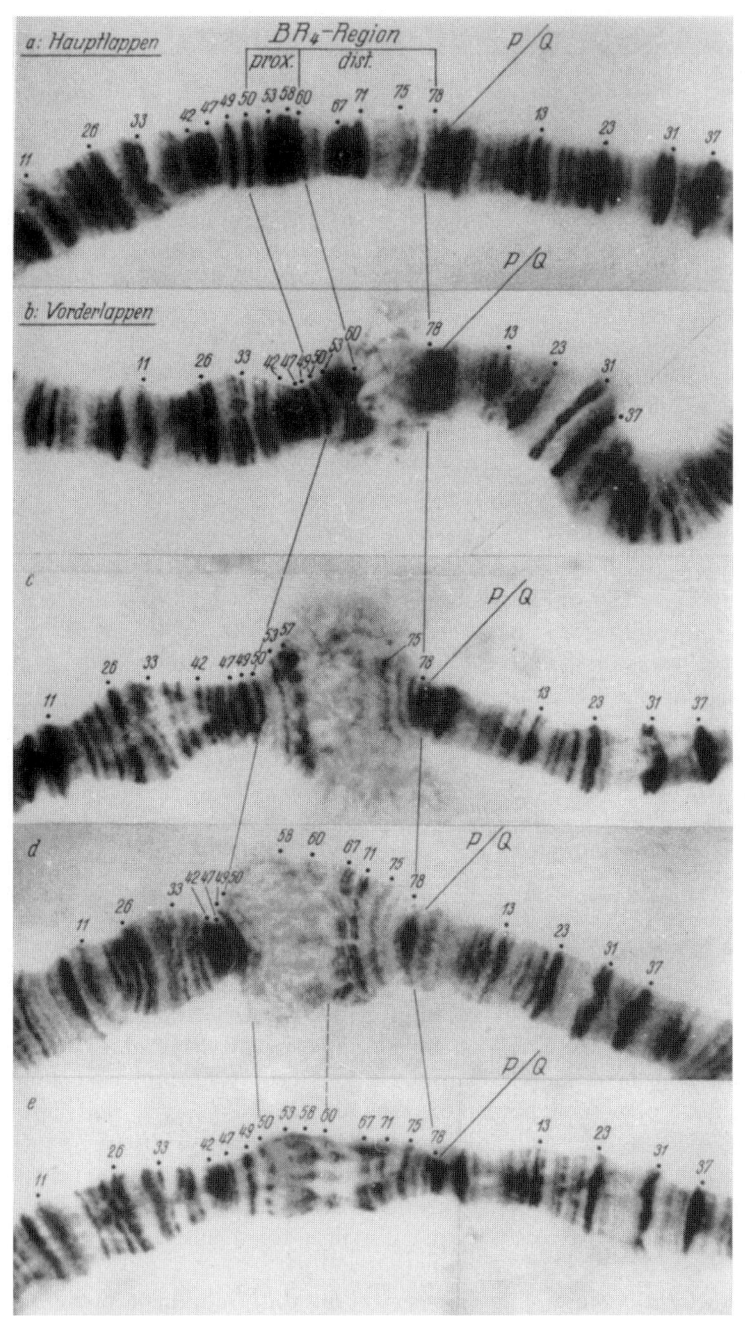

Figure 85. Movement of puff maximum during the formation of Balbiani ring 4 in the 50–78 region of the *Acricotopus lucidus* second chromosome. (a) A chromosome region of the main lobe of the gland where the Balbiani ring is not expressed; (b–e) sequential steps of band activation and movement of the zone of the active puff in cells of the anterior lobe (reprinted by permission from Mechelke, 1961).

erythrocephala (Ribbert, 1972), and in the salivary gland cells of *D. melanogaster* (Zhimulev, 1974a; Korge, 1977a; Semeshin *et al.*, 1985a).

The existence of multibanded puffs and movement of puff maxima allowed the conclusion that the large puff can accomodate many bands. This raised the question as to whether the entire material of the puff is active, and, if so, to what extent is it activated in a coordinated fashion? Some investigators held the view that the band material from which the puff commences in the large puffs is transcriptionally active (Pelling, 1971), while the other part of the puff is "passively uncoiled." Passive uncoiling implies decompaction of certain chromosome regions in the developing puff, loosening of its DNA, and its transcription start. Changes in morphology and structure of the regions nearby a functional region (loss of distinct banding pattern and loosening of texture) could not be attributed to transcription; rather, it could perhaps be the consequence of loosening of DNP (deoxyribonucleoprotein) fibrils by RNA and proteins required for its synthesis which have accumulated in the active zone of the puff (Lychev, 1968). A similar explanation for the movement of puff maxima was offered by Burnet and Hartmann-Goldstein (1971). Supporting the idea of passive uncoiling, Lychev referred to the descriptive studies of more than 120 puffs in *Drosophila* according to which increase in puff size as their development proceeds always correlates with a unidirectional shift in puffing with the large dense bands acting as natural restrictors. However, if the less-tightly packed light regions of the chromosome were at the other end of the puff, puffing could spread also in this direction. Based on morphological observations, it could not be determined how many bands are passively involved in puffing and transcribed (Kroeger and Lezzi, 1966). This remained an open question according to Ashburner and Berendes (1978).

There was some difference in opinion as to localization of the actively transcribed regions in the puff. A line of evidence indicated that the transcription region passes like a narrow strip along puff maximum (Berendes, 1968a, 1969). It was Pelling's (1972) view that the transcribed part of the puff itself is confined to the material of the band from which the puff originates. However, there was no subsequent evidence that this is the case. Silver grain density was uniformly distributed over the puffs after pulse labeling with tritiated uridine for 1 min. These were the largest puffs characteristic of normal development (63E, 71CE, 82F), and also the 63B, 87A, and 87B heat-shock puffs (all of these being regions of chromosome 3 in *D. melanogaster*) (Belyaeva and Zhimulev, 1976).

The major evidence obtained so far indicates that the complex puffs originate from series of puffs that are adjacent, although functionally independent. The following facts confirm this conclusion:

1. Panitz (1964) studied in detail regression in Balbiani Ring 4 (BR4) under normal conditions and in transplants of the salivary glands of fourth-instar larvae into prepupae with regressing BR4. This BR4 in *A. lucidus* was be-

lieved by Mechelke (1961) to be the discovery of moving puffing maxima. However, in the transplantation experiments, there was no real movement of puffing maximum: puffing regression in the proximal zone was associated only with partial activity in the distal zone; the 58–60 unpuffed bands were interposed between the two zones, giving the appearance of two separate puffs in this region. Therefore, even though the activity of certain bands requires the activation of other bands, it cannot be claimed that the activity of each band depends on the activation of the adjacent band. Independently behaving areas are detected in this complex puff, and this makes the idea that BR4 is a functionally integrated entity doubtful. Later developmental studies on puff activity led to the same conclusion (Staiber and Behnke, 1985).

An EM analysis of the 50CD region of *D. melanogaster's* second chromosome, in which movement of puff maximum was observed (Zhimulev, 1974a), detected that five puffs are formed in it; of these, three adjacent ones sequentially replaced each another. The puffs are separated by the unpuffed bands from one another at certain stages of development; that is, the puffs can be thought of as functioning independently (Semeshin *et al.*, 1985a).

The 3C8-D1 puff of the *D. melanogaster* X-chromosome is periodically active; the puff is developed in third-instar larvae due to decompaction of the bands in the distal end of the puff; then it disappears before puparium formation and appears again in 5- to 9-hr prepupae; the puff then involves the more proximal bands of the 3D region. The *Sgs-4* RNA is transcribed from the puff during the first period of its activity. The second peak of its activity is not associated with synthesis of this protein (Korge, 1977a). This provides evidence that the puff is informationally complex.

2. One of the *T(1;3)sc*$^{260\text{-}15}$ (Ashburner, 1972a) translocation breakpoints is within the 71CE largest puff of *Drosophila*. It is a typical puff, whose bands are activated and inactivated synchronously (or asynchronously) and are engaged in transcription (Zhimulev and Belyaeva, 1975b). When separated, both parts of this puff are able to puff in a coordinated and independent manner (Ashburner, 1972a). H. Bauer (in Panitz, 1964) identified a translocation separating a Balbiani ring into two equivalent parts in *Smittia*. Other cases are known (although they are less informative) where the large puff is separated by chromosomal rearrangement from another puff: the 74EF and 75B large puffs are separated by *T(2;3)C65* (Ashburner, 1972a); or chromosome material is separated by translocations from the heat-shock puffs (Ellgaard and Brosseau, 1969). These rearrangements actually do not perturb the structure of the puff itself.
3. In certain cases, the puff can include the unpuffed bands; for example, in *Drosophila*, the 2B3-5 puff is such that the material of the band, intermingling with its own material and that of the puff, penetrates into the puff

> (Bridges, 1935, 1938; Belyaeva *et al.*, 1980, 1987). Because these bands are believed to represent intercalary heterochromatin, the puffs were occasionally called "heterochromatic" (review: Zhimulev, 1993). Another such puff was detected in the 85F region of *Drosophila* (Baricheva *et al.*, 1987); a Balbiani ring was identified in *D. auraria* (Scouras, 1984), as were a number of putatively heterochromatic puffs in *Simulium nolleri* (Shcherbakov, 1968) and, possibly, a DNA puff in *S. ocellaris* (Perondini, 1971). Material deeply staining with orcein is seen in the puffs.

Transformation data evidence that the peculiar morphology of the 2B3-5 puff is due to neighboring heterochromatin. When a cDNA fragment of the BR-C (*ecs*) gene, whose activation leads to the formation of the 2B3-5 puff contained by a transposon, inserts into the 76C region, the puff forming during induction already does not appear heterochromatic (E. S. Belyaeva, C. Bayer, J. Fristrom, and I. F. Zhimulev, 1994, unpublished observations).

Puffs, whose bands decompact at the same time, are of interest with regard to their functional complexity (multibanded structure) (Figure 86). The EM data give good reason to believe that the bands are functionally coordinated and that the DNA of the different bands in the formed puffs controls a single function; for example, only one gene was detected in each of the 74EF and 75B puffs (see Section VI,E,1), despite the apparent involvement of several bands in puffing (see Figure 86). Based EM observations on puff development in 63 regions of *Drosophila's* polytene chromosomes (Semeshin *et al.*, 1982, 1984, 1985a,b, 1986, 1989; Baricheva and Semeshin, 1985; Baricheva *et al.*, 1987; Semeshin, 1990), the complex puffs originating from 2 to 4 bands were demonstrated to comprise almost half of the studied puffs (Table 8). The other half of the studied regions comprise puffs originated from a single band (Figure 87).

Although the "one band–one puff" rule is generally accepted, it must be viewed with reservation, because supporting evidence, particularly electron-microscopic, is scanty. Reference is usually made to Keyl (1965), who described the b3,11 puff in the third chromosome of *C. thummi* (see Figure 18, above), which forms only from a massive band: puffing starts in the distal region, less commonly in its proximal part; puffing separates a part of the band which appears as a loosened mass, then puffing spreads over to the rest of the band region. When reaching its full development, the puff does not extend beyond the band, which uniformly but not completely uncoils: although the band is bulky, the diameter of the puff is only twice greater than that of the chromosome. Keyl regards this band as an example of a genetic system composed of interrelated, simultaneously acting functional entities.

Examples of development of the puffs forming from a single band in *C. thummi* have been described by Kiknadze (1972). Thus, the 2-7A3 puff originates from a single thin band and the 2-19B5 puff from a thick one. Kiknadze emphasized that the tips of the band remained uncoiled in both cases,

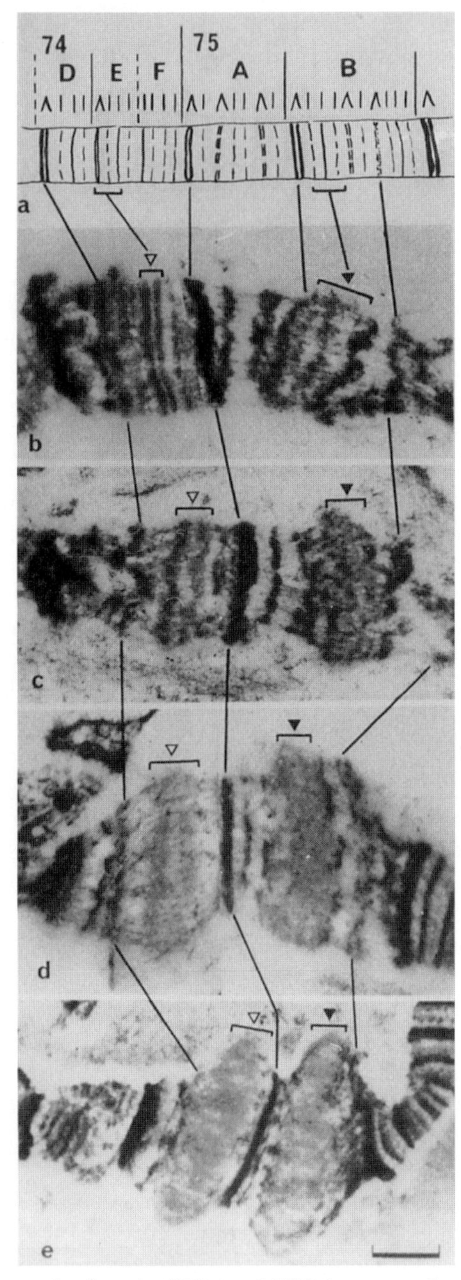

Figure 86. Development of puffs in the 74E1-4 and 75B3-5 regions with concomitant decompaction of several bands (in brackets with triangles). (a) Bridges' map; (b) PS1; (c) PS2; (d) PS3-4; (e) PS-6. Bar = 1 μm (reprinted by permission from Semeshin *et al.*, 1985a).

Table 8. Development of Puffs in *D. melanogaster*[a]

Number of chromomeres involved	Type of chromomere decompaction	Chromosome location	Number of puffs
One chromomere	Decompaction of the whole chromomere	2B11; 3C11-12; 21D3; 29F3; 35A1′; 35B3; 50C23; 55C3; 58E3-4; 62C2′; 63B9; 63E2-3; 63F7; 68C3; 70C5; 70C6; 76C3; 79F1′; 90B7; 95D11	20
	Puffing of part of band	1E3-4; 35B1-2; 64E1-2; 93D6-7	4
	Splitting the band and puffing of its parts	2B3-5; 2C1-2	3
Two chromomeres	Simultaneous decompaction of both bands	3D1,2; 25B4,5; 25C5,6; 50C9,10; 50C11,12; 50C17-18,20-21; 56E1-2,4-5; 60B12,13; 62E3,3′; 62B4,5; 63D1,2; 63F5,5′; 67B1,2; 71C2,3; 71D3,4; 71E3,4; 72E3,4; 72D10,11; 74E1-2,4; 85D1,2; 85F3,4-5; 99F4,5	21
	Decompaction of whole band and a part of neighboring	22B3,4-5; 71E5,F1-2	2
	Decompaction of neighboring parts of two bands	1C1,2-3; 7A1-2,4-5	2
Three chromomeres	Simulataneous decompaction of all three bands	21E3′,4,4′; 33B8-9,10,11-12; 50F4-5,6-7,8; 55E2,3,4; 75B3,4,5; 78C4,5-6,6′; 82F5,6,7	7
Four chromomeres	Simultaneous decompaction of all four bands	87A6,7,8,9	1
	Decompaction of three bands and partial decompaction of one band	87B13,14,15,C1	1

[a]According to EM analysis of Semeshin (1990).

posing a morphological restriction to the separation of the puff from the interband spacing. However, these regions of the band tips presumably are also partly decondensed and, in the case of maximum puffing, they undergo changes—which makes it difficult to accept that puffing is restricted to the material of a single band; however, the central part of the puff is surely more uncoiled than its tips and the adjacent bands. For this reason, these puffs are referred to as "multibanded."

Puffs originate from a part of the band in a number of cases: a part of a band become puffed when other part is in a compacted state (Figure 88; cf. Figure 17). This indicates informational complexity of the chromomeres and

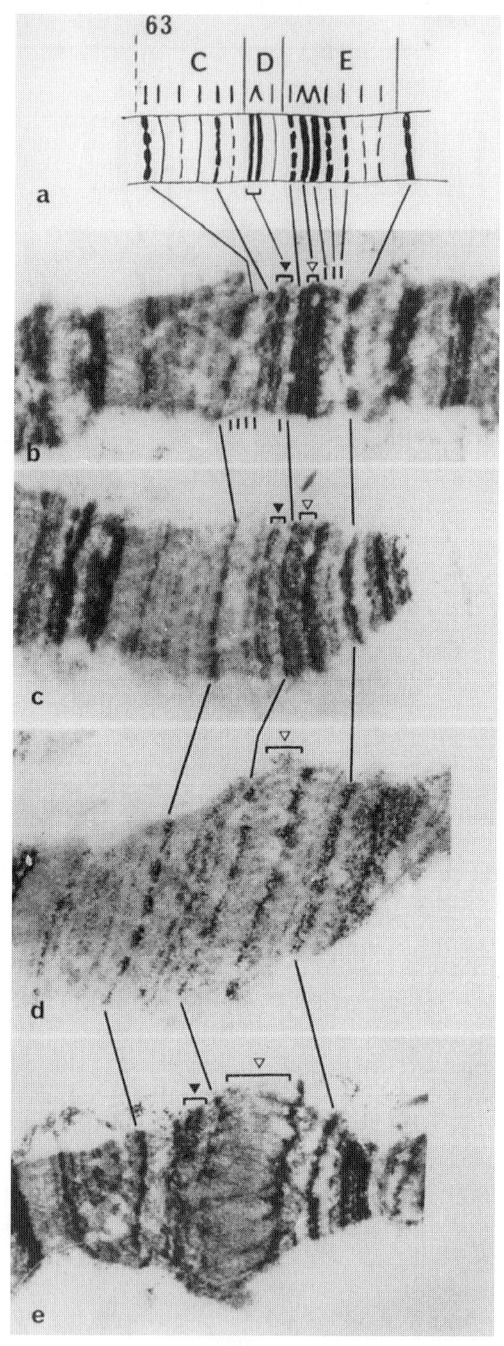

Figure 87. Development of a puff from a single 63E2-3 band. (a) Bridges' map; (b) PS1; (c) PS8; (d and e) PS10 (reprinted by permission from Semeshin *et al.*, 1985b).

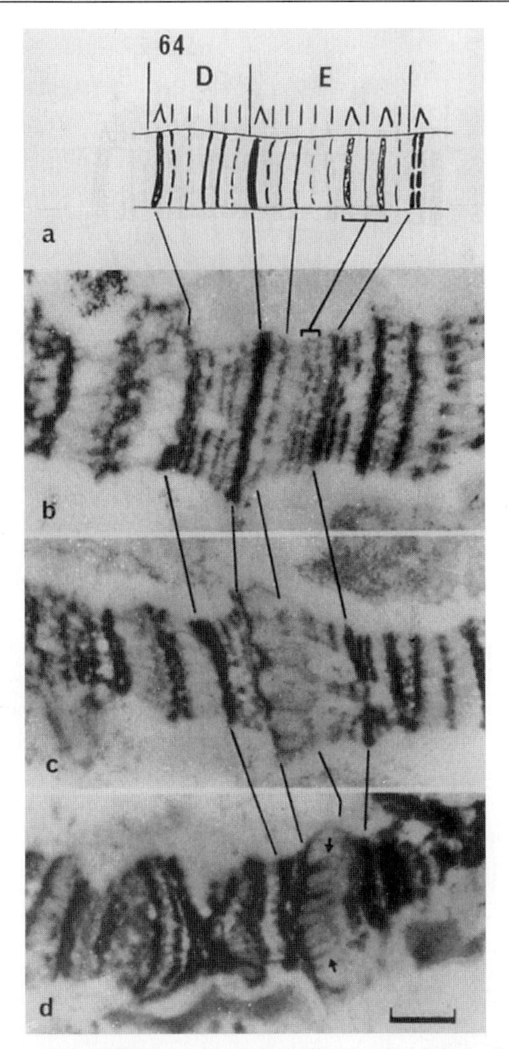

Figure 88. Development of a puff from a part of the 64E1-2 band. (a) Bridges' map; (b) prepupa, 5 hr; (c and d) heat shock, 3 and 5 min, respectively. Arrows point to fibrils in the puff. Bar = 1 μm (reprinted by permission from Semeshin *et al.*, 1985b).

the ability of their various functional units to be expressed independently of the other units. These features become more conspicuous when band puffing is caused by transformation (Semeshin *et al.*, 1986, 1989). Insertion of a transposon containing known DNA sequences in a mobile P element (Figures 89a–89c) led to the formation of a new band at the insertion site (Figures 89e–89g). When the puff resulted from activation of simple sequences—for example, the *Sgs* and *Adh* genes with their promoters—the entire band material

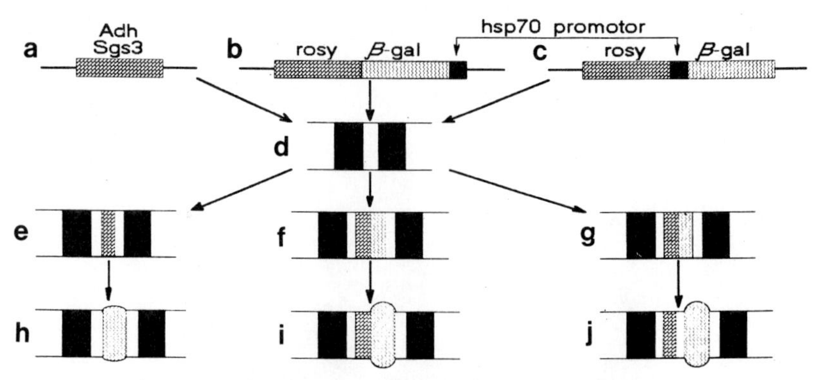

Figure 89. A schematic representation of bands and puffs in the transformed regions. (a–c) Constructs of DNA fragment; (d) initial state of the region; (e–g) formation of new bands; (h–j) their puffing (from Semeshin, 1990).

puffed (Figure 89h,i). When a transposon includes two genes (*rosy* and *E. coli lacZ*, or *rosy* and *hsp*) in inactive state (Figures 89b,c), a single band is derived from it. Change in morphology associated with puff activation under the effect of heat shock depends on where the *hsp* promoter is located: on the tip of the band (Figure 89i) or in its middle (Figure 89g). In the former case, after a brief heat shock, starting from the right tip, puffing propagates from the right-hand part of the band leftward towards its center (Figures 89f,i). In the latter case, the band splits into two parts (Figures 89g,j) with concomitant formation of an interband, separating the two parts, which therefore arise as a result of activation of the *hsp* promoter. A large puff is formed from the part of the band that has split off, while its left part remains in a compacted state.

3. Puff size and methods for its estimation

The developmental cycle of the puff includes the following steps: one band or a group of deeply staining bands become diffuse; their staining intensity decreases (the bands appear grey, when viewed in the phase contrast microscope); and, concomitantly, the outlines of the band become blurred; the swelling increases in size (Figure 87). Decompaction reaches its maximum in a fully developed puff (Sorsa, 1969); the puff loses stainability; it becomes "white" or transparent when stained or viewed in the phase-contrast microscope. The same steps are followed in the reverse order during puff regression. The swelling diminishes first; then the material becomes completely compacted, giving rise to the band.

Different points of view were expressed concerning the factors determining the size of the puffs when it has reached its maximum development.

Some authors took the view that puff size depends on the DNA amount in the band from which the puff originates (Mechelke, 1959; Pavan *et al.*, 1975); others believe that it depends on the amount of synthesized RNA (Pelling, 1964; Kroeger, Lezzi, 1966). On Pelling's view (Pelling, 1972), transcription rate is considerably higher in the giant puffs (Balbiani rings) than in the usual smaller puffs. This is supported by the calculated number of molecules of RNA-polymerase II: 120 per chromatid in Balbiani rings and 2 in the usual puff.

Having summarized the data on the effects of various factors on puff size, Belyaeva (1982) came to the conclusion that two conditions must be fulfilled to provide the formation of the large puff: high transcription rate and transcription of a very long region. Consequently, the activity of the unique gene 1–2 kb in size cannot be manifested as puff formation; instead a small puff, even an interband, was formed. This assumption was confirmed by the results of analysis of the puffs in the transformed regions of polytene chromosomes: the genes approximately 1–2 kb long form structures similar to the interbands or the micropuffs (the *Sgs-3* and *hsp18.5* genes); sequences about 3 kb in length (the *Adh* gene, for example) give rise to average-sized puffs; DNA fragments 9–11 kb long (the *lacZ* gene) give rise to the large puffs (Semeshin *et al.*, 1986, 1989; Semeshin, 1990).

A positive correlation was established between transversal and longitudinal sizes of the puffs at the same locus: a greater length of the puff corresponds to its greater diameter (Poluektova, 1970a).

No differences were found in autosomal puff size between males and females of *D. hydei* and *D. melanogaster* (Berendes, 1965a; Ashburner, 1967b; Chatterjee and Mukherjee, 1971a,b). Studies on 10 autosomal puffs in females and males of *D. hydei* demonstrated that puff size and absolute incorporation of [^{3}H]uridine were the same in seven regions in the two sexes; grain number decreased in these regions in the male chromosome (Chatterjee, 1991). Sex differences in the puff size in the X chromosome were observed in *D. virilis:* the puffs were longer in males than females (Poluektova, 1975a).

There is information concerning the linear specificity of puff size. In males of *D. melanogaster*, the 2B puff is longer in the proximal direction, and it is ellipsoid in the strain with a duplication in the 3A-4C region: all the bands up to 3A12 were found to be puffed. The shape of the puff is recovered in recombinants in which a duplication has been removed (Rasmuson, 1967).

Transversal size and puff number are used as criteria for determining the total activity of the puffs in the nucleus. The following approaches are used to estimate puff size:

1. The visual method. The puffs in a preparation are arranged on a time scale reflecting the developmental stages of the large puff (Figure 90): loosening of the band is followed by an increase in its diameter. The ratio of the

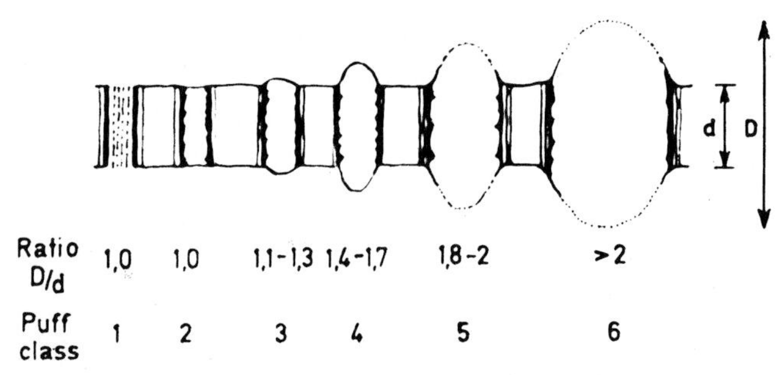

Figure 90. A scale for estimating puff size (from Belyaeva *et al.*, 1974; and reprinted by permission from Zhimulev, 1974a,b).

maximum puff diameter to the thickness of the unpuffed chromosome region is visually assessed, quantitatively expressed, and converted into "scores" (Figure 90). This method makes possible rapid assessments of large chromosome samples and analysis of many puffs (Berendes, 1956a; Lychev, 1965, 1968; Ashburner, 1967b; Lychev and Medvedev, 1967; Bultmann, 1969; Zhimulev and Lychev, 1970; Burnet and Hartmann-Goldstein, 1971; Sorsa and Pfeifer, 1972). The accuracy and reproducibility of the method are good. Analysis of a sample of 200 puffs, repeated after 2 weeks, revealed differences in 1 score in 43 cases (Berendes, 1965a). In the other experiments, repeated estimates of puff size agreed with the previously obtained in approximately 80% of cases. The differences were small in the other cases, 1 score on the average (Lychev and Medvedev, 1967; Zhimulev and Lychev, 1970). The difference between repeated estimates are ± 0.3 scores, on the average, 1–2 scores in the extreme cases (Belyaeva and Zhimulev, 1974).

2. Puff measurement is a more accurate, although slower, method based on measurement of the diameter of puffing maximum and relating it to the diameter of the thinner part of the chromosome (Lychev, 1968); or to a particular marker-band—a black compact band not undergoing puffing during the studied span of time during development. Then the obtained values are statistically treated, with mean values being calculated (Ashburner, 1971). These are different approaches to estimate the relative activity of Balbiani rings. First, there is a morphological scale, based on the ratio of the puffed part to chromosome diameter (Lezzi and Gilbert, 1969; Yamamoto, 1979; Beermann, 1971, 1973; Diez, 1973; Colman and Stockert, 1975; Diez *et al.*, 1980; Mahr *et al.*, 1980; Nelson and Daneholt, 1981). Second, the diameter of the chromosome is measured with a micrometer (Sass, 1981). Third, the area of Balbiani ring is measured and related to the area of the chromosome where it is located (Pankow *et al.*, 1976; Lezzi *et al.*, 1981; Meyer *et al.*, 1983). Fourth, the ratio of the area of the outline of Balbiani ring (Figure 91) to chromosome diameter is determined (Stockert, 1990).

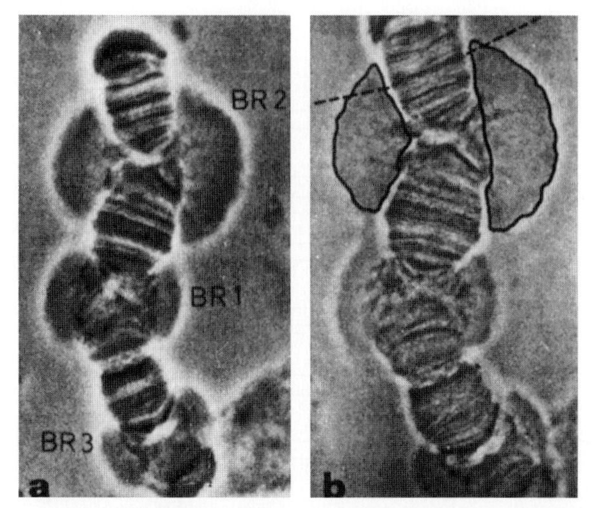

Figure 91. Estimation of Balbiani ring activity. (a) Fourth chromosome of C. *pallidivittatus;* (b) the same chromosome. Outline of active part of the Balbiani ring is made by continuous line. Marker band is marked by dashed lines. For determination of the BR activity, ratio of BR square to diameter of the band is estimated (from Stockert, 1990).

C. Activity of puffs during development

1. Puffing of salivary gland chromosomes

In species of the dipteran order, larval development takes several days during which time dramatic and readily observable changes occur in puffing. In *D. melanogaster*, which is the most thoroughly studied species, puffs during the last 24 hr of larval and during 12 hr of prepupal development have been described in detail. All the puffs are divisible into the large puffs, i.e., the swelling proper, and the small expansions not associated with an increase in chromosome diameter; the bands in such a region are, however, loose. The puffs scored on a scale from classes 1 and 2 are shown in Figure 90.

At particular developmental stages each larva or prepupa possesses a characteristic set of large puffs. In the largest young larvae, whose polytene chromosomes are already large enough to permit their analysis (about 100–110 hr after egg laying or somewhat after the mid-third larval instar), several large intermolt puffs are present in the chromosomes (68C, Figure 92). At the very end of larval development, about 6 hr before puparium formation, these puffs become inactivated on the background of a sharp increase in ecdysterone titer, and the large puffs start to form; the early puffs (74EF and 75B in Figure 92) appear fast, followed by the late puffing (78C, 66B, 71CE: Figure

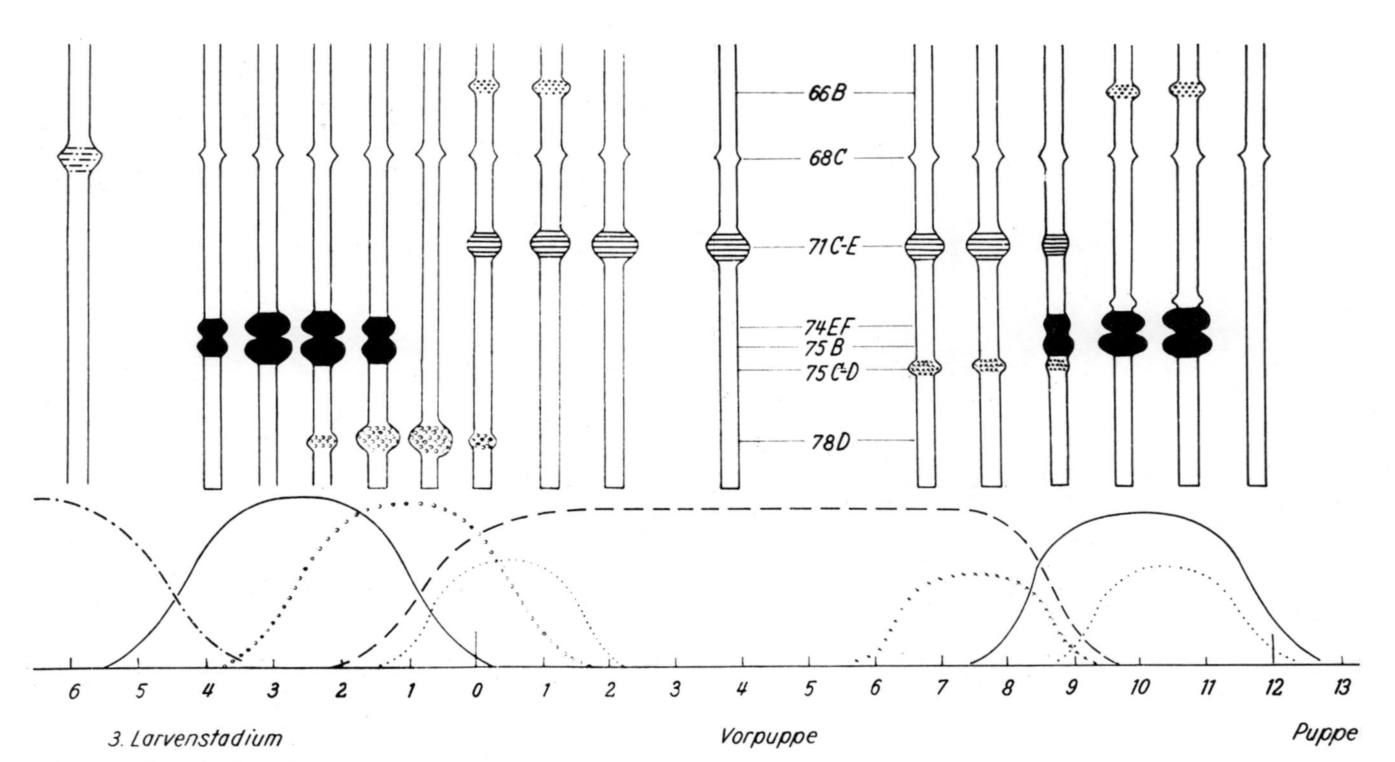

Figure 92. Change in the puff activity in chromosome 3L of *D. melanogaster* at the end of third-larval instar and prepupal development. 66B–78D (78C) designation of puffs. The puffs on the chromosomes and the curves for changes in their activity are depicted by the same hatching. Abscissa, time of development in hours; 6–0, before puparium formation; 0–12, after puparium formation (reprinted by permission from Becker, 1959).

92). In zero-hour prepupae, the activity of the puffs is highest; that is, their number and size of the puffs are highest. During the next 2–3 hr, their activity sharply falls, and several mid-prepupal puffs appear (75CD in Figure 92). At the end of the prepupal stage a wave of the late and the early puffs passes (Figure 92). The total number of the large puffs, whose activities change during this period of development, is about 120 (Becker, 1959, 1962a,b; Lychev, 1965, 1968; Ashburner, 1967b, 1969b,c,f, 1972a, 1975, 1978; Burnet and Hartmann-Goldstein, 1971; Sorsa and Pfeifer, 1972; Belyaeva *et al.*, 1974; Zhimulev, 1974; Korochkina and Nazarova, 1977; Loreto *et al.*, 1988).

The activity of certain puffs replaces that of the other puffs in an ordered sequence. The developmental stage characterized by a certain puffing pattern is called the "puff stage" (PS) and assigned a corresponding order number, according to Becker (1959), who first described these stages. Developmental schedules for the substitution of one puff stage by the next were tabulated (reviews: Ashburner, 1972a, 1975; Ashburner and Berendes, 1978), (Table 9; Figure 93): 10–11 of the stages were assigned to the end of the larval instar, another 10 to the prepupal period.

The spectra of polypeptides synthesized in salivary glands also change in a fashion which is related to the puff stages (Figure 94). About 50 stage-specific fractions were detected by one-dimensional electrophoresis. Some fractions, which were active at the PS1 stage, disappeared; others appeared at the end of the third instar, when the ecdysterone titer was rising. Then polypeptides specific to mid- and late prepupae appeared (Zhimulev *et al.*, 1981e; Dubrovsky and Zhimulev, 1985, 1988; Dubrovsky *et al.*, 1985). Twenty-six polypeptides were detected by two-dimensional electrophoresis (Poeting *et al.*, 1982).

Painter (1934) noted early that there is little variation in the puffs in the various gland cells. He observed that the type of swelling is quite constant in an individual. There was no instance when a puff was missing at a stage when it ought to have formed together with the other characteristic puffs (Becker, 1959); there is little variation in size of puffs within glands in *C. tentans* (Clever, 1962a). In *D. hydei*, puffing is identical in various gland cells, although differences in sizes of puffs can reach two scores (Berendes, 1965a). In 13 2R chromosomes of a single gland of *D. melanogaster*, differences in size of puff formed by the same region reached two scores in 35% of cases; it did not exceed one score in all the other cases (Lychev and Medvedev, 1967). Of 29 loci of 14 3L chromosomes, the differences were one score for 16 regions, two scores for 10 regions, and three scores for one region, and there were no differences in scores for two regions (Zhimulev and Lychev, 1970).

In zero-hour prepupae of *D. melanogaster*, the frequencies and mean sizes of the 63E1-5 and 71CE puffs were not different among salivary gland cells (Belyaeva and Zhimulev, 1973, 1974). There were no differences in puffing between the sister glands of a larva (Becker, 1959; Zhimulev and Lychev, 1970,

Table 9. Puff Stages (PS) of Salivary Gland Chromosomes of the Third-Instar Larvae and Prepupae of *Drosophila melanogaster*[a]

Puff region	Puff stages (PS)																				
	1	2	3	4	5	6	7	8	9	10	11	12	13	14	15	16	17	18	19	20	21
X chromosome																					
2B5-6 (2B3-5)[1]	+	++	++	++	++	++	++	++	++	(+)	–	(+)	(+)	(+)	(+)	–	–	(+)	++	++	++
2B11-13 (2C1-2)[1]	–	–	(+)	+	++	++	++	++	++	+	(+)	(+)	(+)	–	–	–	–	–	+	+	–
2EF	–	–	–	+	+	+	+	+	+	+	++	++	+	+	+	–	–	–	+	+	++
3C	++	+	–	–	–	–	–	–	–	–	–	–	–	(+)	+	++	++	++	++	++	–
16C	–	–	–	–	–	–	–	–	+	+	+	+	–	–	–	–	–	–	–	–	(+)
2L chromosome																					
21C	(+)	(+)	–	–	–	–	–	–	–	–	–	–	–	–	–	–	–	–	(+)	(+)	(+)
21F	–	–	–	–	–	+	+	+	++	++	++	–	–	–	–	–	–	–	–	–	–
22A	–	–	–	–	–	–	–	–	–	–	–	–	–	–	–	–	–	–	++	+	–
22B4-5	–	(+)	+	+	+	+	–	–	–	–	–	–	–	–	–	–	–	–	–	–	+
22C	–	–	+	–	–	+	+	+	++	++	++	–	–	–	–	–	–	–	–	–	–
23E	–	+	+	+	+	+	+	++	++	++	++	+	+	–	–	–	–	++	++	++	++
25AC	++	+	–	–	–	–	–	–	–	–	–	–	–	–	–	–	–	–	–	–	–
34A (345-6)[2]	+	+	+	+	+	+	+	–	–	–	+	++	++	++	++	++	+	+	+	+	+
2R chromosoma																					
42A (42A4-18)[2]	+	+	+	+	+	+	+	+	+	–	–	–	–	–	–	–	–	+	+	–	–
46A	–	–	–	–	–	+	+	+	+	+	–	–	–	+	+	+	+	+	–	–	–
46F (46FG1-9)[2]	–	–	–	–	+	+	+	+	+	+	+	++	++	++	+	+	+	+	+	–	–
47A (47A9-19)[2]	–	+	+	+	+	+	+	++	++	++	+	+	+	+	+	+	+	+	+	++	++
48B	–	–	–	–	–	+	+	+	+	+	(+)	–	–	–	–	–	–	–	–	–	(+)
52A	–	–	–	–	–	–	–	–	–	–	–	–	–	–	–	+	+	+	+	–	–
55E	–	–	–	–	+	+	+	+	+	+	+	+	+	+	–	–	–	–	–	–	–
58BC	–	–	–	+	+	+	+	–	–	–	–	–	–	–	–	–	–	–	–	–	–
60B	–	–	–	–	–	–	–	–	–	–	–	+	+	++	++	+	+	+	+	–	–

									3L chromosome												
62E	−	−	−	+	+	++	++	++	++	++	+	−	−	−	−	−	−	−	−	++	++
63E (63E1-3)[3]	−	−	−	−	−	−	−	+	+	++	++	+	+	−	−	+	+	++	+	−	−
63F (63F3)[2]	−	+	+	+	++	++	+	−	−	−	−	−	−	−	(+)	−	−	−	−	−	−
66B	−	−	−	−	−	−	−	−	+	+	++	++	(+)	(+)	(+)	(+)	(+)	+	++	++	++
68C	++	+	−	−	−	−	−	−	−	−	−	−	−	−	−	−	−	−	−	−	−
69A	(+)	−	−	−	−	−	−	−	−	−	−	−	−	−	−	−	−	++	−	−	−
71DE	−	+	+	+	+	+	+	+	++	++	++	++	++	++	++	++	+	+	(+)	(+)	(+)
71F1-2[4]	−	−	−	−	−	−	−	−	−	++	++	++	++	++	++	++	+	+	(+)	−	−
74EF (74E1-4)[4]	−	+	++	++	++	++	++	+	+	−	−	−	−	−	−	−	−	−	++	−	−
75B	−	+	++	++	++	++	++	+	+	−	−	+	+	+	+	+	+	+	++	+	+
75CD	−	−	−	−	−	−	−	−	−	−	−	−	−	−	−	−	−	++	+	−	−
78D (78C)[3]	−	−	−	−	−	+	++	++	++	++	+	−	−	−	−	−	−	−	−	−	−
									3R chromosome												
82F	−	−	−	−	−	−	−	−	++	++	++	+	−	−	−	−	−	+	−	−	−
85D (85D1-2)[4]	−	−	−	−	−	−	−	−	−	−	+	++	++	++	++	++	++	+	−	−	−
85F	+	+	+	+	+	+	+	++	++	++	++	++	++	++	++	++	++	++	++	++	++
90BC	++	+	+	+	+	+	+	−	−	−	−	+	++	++	++	+	+	−	−	−	−
91D	−	−	−	−	−	−	−	−	−	+	++	++	+	−	−	−	−	−	−	−	−
93F (93F9-10)[2]	−	−	−	−	−	−	−	−	−	−	−	−	−	−	−	−	−	−	++	++	++
95F	−	−	−	−	−	+	+	+	+	+	−	−	−	−	−	−	−	(+)	−	−	−
98F	−	−	−	−	−	−	+	+	++	++	++	−	−	−	−	−	−	−	−	++	++
100DE	−	−	−	−	−	−	−	−	−	−	−	−	−	−	+	++	++	++	−	−	−

[a]After Ashburner and Berendes (1978). Correct name is given in parenthesis according to:
1, Belyaeva *et al.* (1980a); 2, Belyaeva (1982); 3, Semeshin *et al.* (1985b); and 4, Semeshin *et al.* (1985a).

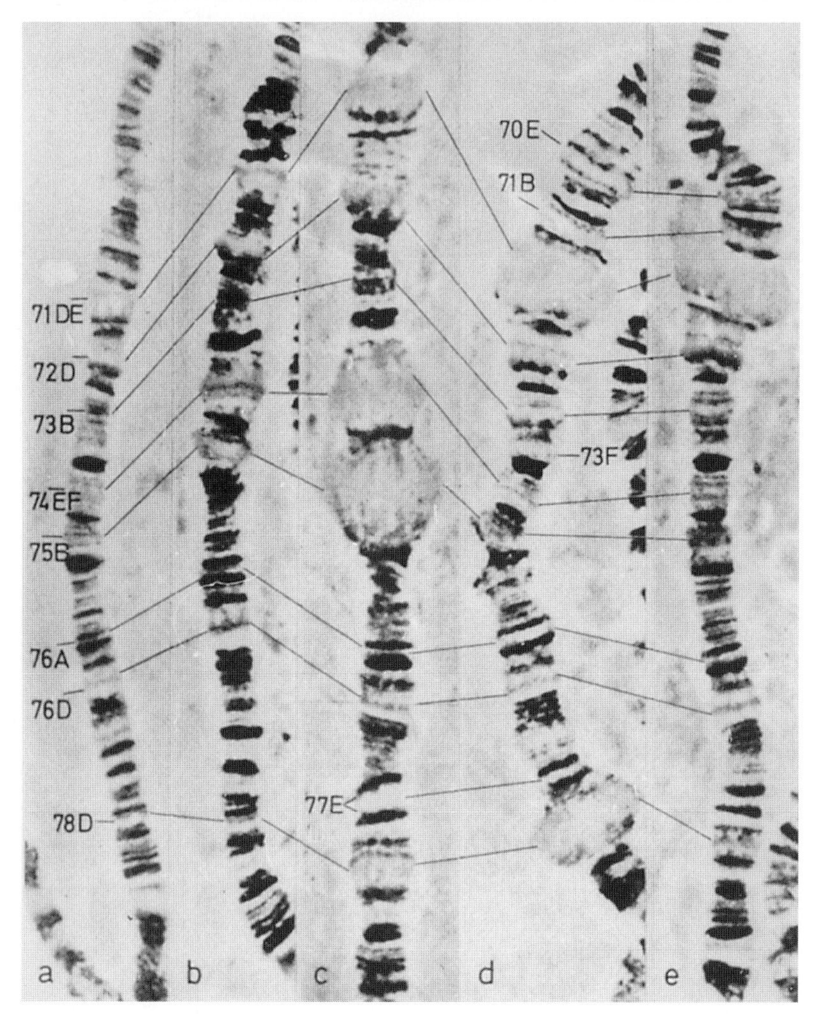

Figure 93. Changes in puffing in the proximal region of chromosome 3L of *D. melanogaster*. (a) PS1; (b) PS2-3; (c) PS6; (d) PS10 (zero-hour prepupa); (e) PS14 (4-hour prepupa) (reprinted by permission from Ashburner, 1972a).

1972a), with a few exceptions (Berendes, 1965a; Ashburner, 1967b). Puffing patterns were identical even in species of *Drosophila*, showing considerable differences in size of salivary glands (H.D. Berendes, in Ashburner, 1970a).

It is more difficult to estimate variations between individuals. The puffing stages from PS1-PS2 at the end of the third larval instar are substituted during the last 6 hr before puparium formation. However, even when the most

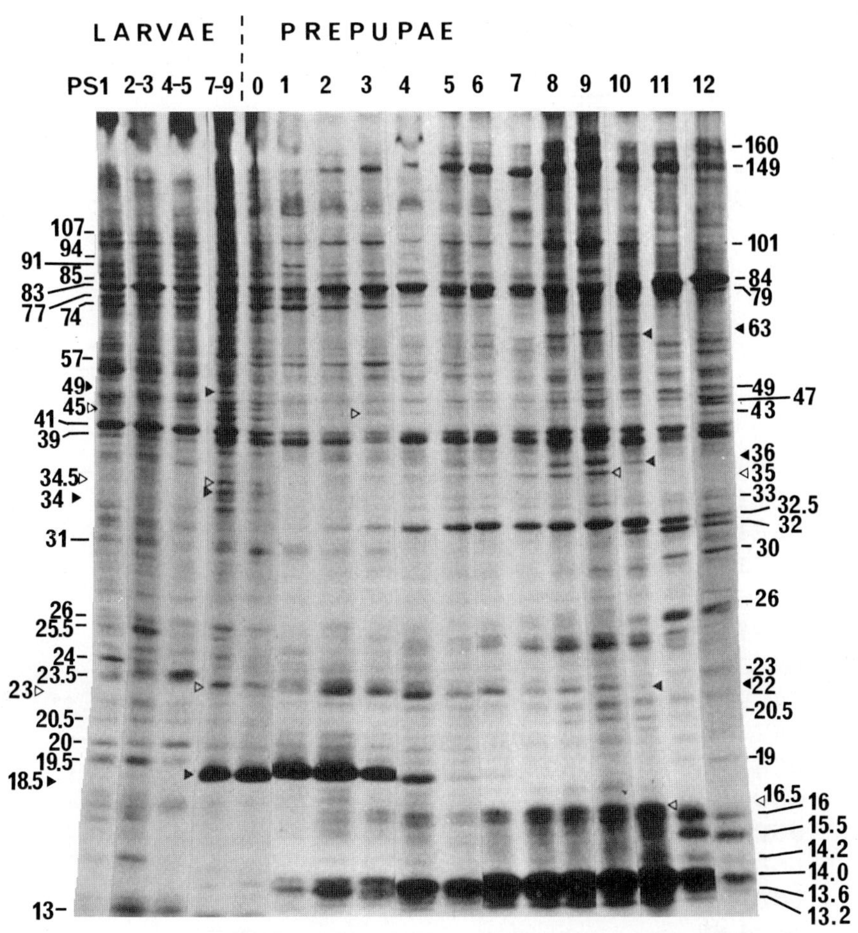

Figure 94. Pattern of polypeptide synthesis in the salivary glands of the *Drosophila* larvae at stages PS1-PS7-9 and in prepupae aged 0–12 hr. Proteins were labelled with [^{35}S]methionine and separated in polyacrylamide-SDS gradient gel. Numbers designate apparent molecular weights of polypeptides (reprinted by permission from Zhimulev *et al.*, 1981e).

accurate method was used to synchronously select developing larvae at the time of the second larval molt, variation in the time of puparium formation reached 10–12 hr in the population (Gaudecker, 1972; Zhimulev, 1974b; Zhimulev and Kolesnikov, 1975b). For this reason, samples of larvae can include both individuals at the PS1 stage, those whose larval wave of puffs has already passed (PS10-11), and individuals at all the intermediate stages.

Nevertheless, the available data evidence that the puff set at each

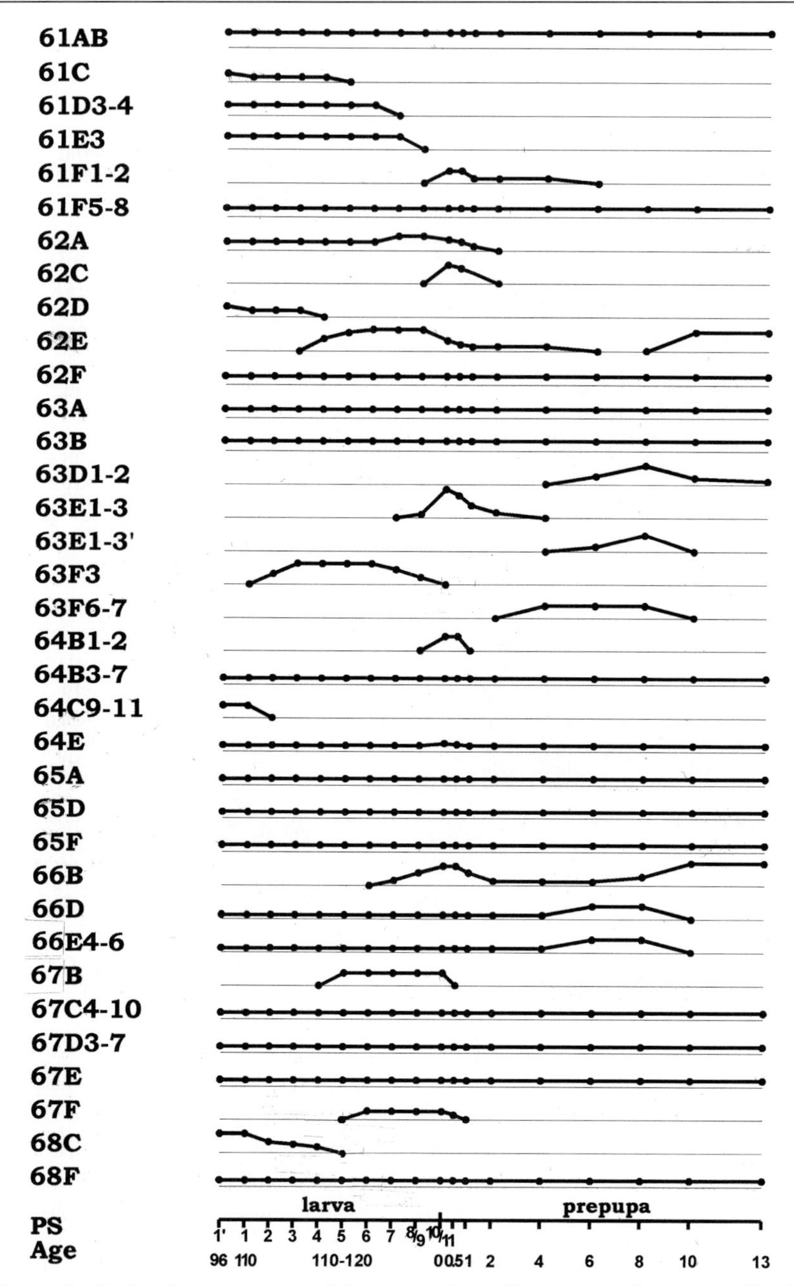

Figure 95. Graphical representation of changes in the puff activity in chromosome 3L of *D. melanogaster*. Abscissa, 1′–10/11 puffs stages; 0.5–13 hr after puparium formation (from Belyaeva, 1982).

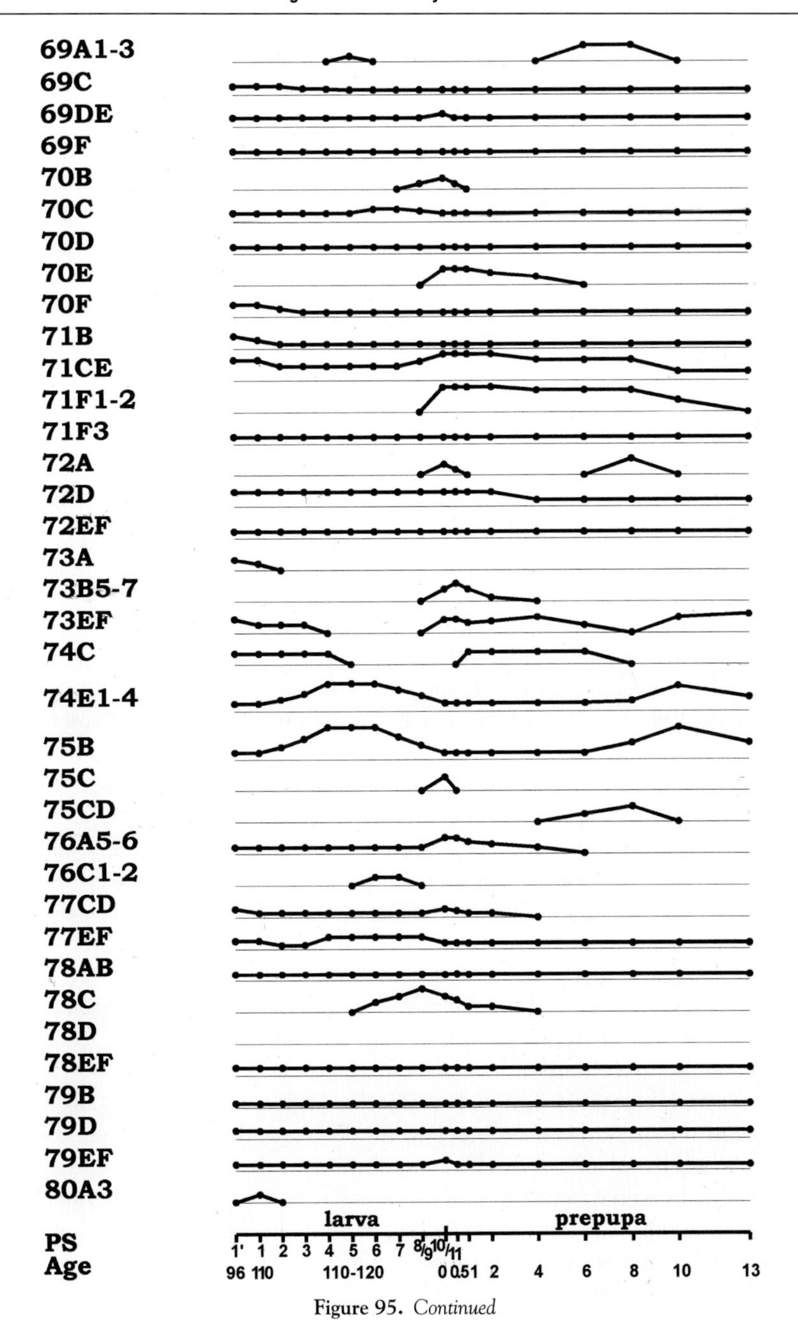

Figure 95. *Continued*

developmental stage is extremely constant. It is quite apparent that each PS has particular characteristic puff set, and that these sets are not "littered" by other puffs (Table 9). Independent samples of third-instar larvae did not differ in the frequencies and sizes of puffs (Lychev and Medvedev, 1967; Zhimulev and Lychev, 1970).

Constancy of puffing pattern is brought into prominence when a PS is related to stage-specific morphological markers of a larva. For example, samples of synchronously developing individuals can be chosen at the stage of anterior spiracle eversion (zero-hour prepupa), which takes a few seconds to complete (Poluektova, 1969; Burdette and Carver, 1971; Sorsa and Pfeifer, 1972). The results of a special study demonstrated that, in zero-hour prepupae of *D. melanogaster*, the frequency of presence of the two large puffs at 63E1–5 and 71CE is 100%; while the variation in size expressed as the ratio of the diameter of the puff to that of the adjacent band was 1.3–3.0, being the same in larvae developing under normal and very overcrowded conditions, as well as on food on which one generation of *Drosophila* was reared (Belyaeva and Zhimulev, 1973; 1974).

Particular PSs consistently correspond to particular morphological types reflecting the state of the salivary gland and of its duct (Figure 83).

Finally, no qualitative differences in the enfoldment of the puff stages were found between males and females of *D. melanogaster* and *D. simulans* (Ashburner, 1967b, 1969b,c). No sex differences in the puffs were observed in *D. hydei* (Berendes, 1965a).

In addition to large swellings, many regions show different degrees of compaction; for example, such regions occur in 65 of 120 subdivisions of the map in the 2R chromosome of *Drosophila* (Burnet and Hartmann-Goldstein, 1971). The total number of the large and small puffs, whose number is about 340 in *D. melanogaster* (Belyaeva and Zhimulev, 1973, 1974; Belyaeva *et al.*, 1974; Zhimulev, 1974a,b; Belyaeva, 1982), corresponds to the number of discrete RNA-synthesizing regions and hybridization DNA/RNA sites. The changes in the activity of the small puffs possibly escape detection because they are too small. The morphology of the small puffs does not change; hence, they are invariably active during development (Figure 95). This constancy may be due to continuous transcriptional activity in region of the small puff and to changes in the activities of different bands decompacting in the small puffs, so that different bands puff at each time point, even though the general appearance of the decompacting chromosome material in the puff remains unaltered.

The substitution of puff sets, particularly during interlarval molts, does not proceed in species with longer developmental cycles; in Chironomides and Sciarides, for example, as dramatically as in Drosophilidae, and their puffing stages have not been described. The frequency of the occurrence of a puff was occasionally used as a criterion of its activity (Clever, 1962a,b, 1963a, 1964d;

Kiknadze, 1965b, 1972). Nevertheless, puffing activity sharply increases at the end of the fourth-larval instar before puparium formation in *Chironomus* as in *Drosophilidae* (Figure 96); and the small puffs were also described so that the total number of puffs in *Chironomus* corresponds to that identified in *Drosophila*, namely about 340. Comparisons of puffing patterns in the chromosomes of third- and fourth-instar larvae of *C. thummi* demonstrated that more than 50% of the puffs remained unaltered during the studied developmental period (Kiknadze, 1976, 1978, 1981a,b).

Comparisons of puffing patterns during larval molt and metamorphoses are of interest. In *C. tentans*, the set of puffs changing during larval molt and metamorphosis are in the main similar (Figure 96).

Similar results were obtained for *C. thummi*. The characteristic features of puffing during molt include retention of the larval puffs, which usually regress during metamorphosis, and the less-prominent metamorphosis puffs (Kiknadze, 1976, 1978, 1981a,b). The results obtained for *Drosophila* were different. In *D. melanogaster*, when larvae pass from the second to the third instars during a molt, the salivary glands are very small, polyteny level is low, and the chromosomes are virtually not amenable to analysis. There are indications that several large puffs are present during larval molt (Becker, 1959).

The data obtained for *D. gibberosa* are remarkable. In this species, salivary gland polytene chromosomes are at relatively early developmental stages of sufficient size so that the puffs can be readily analyzed. It was found that the metamorphosis puffs, and all the other large puffs, are not formed during the second larval molt (Roberts and MacPhail, 1985).

A relation between change in the activity of nuclei and photoperiodicity was established. Nuclear and nucleolar volumes changed in a regular fashion during 24 hr in salivary gland cells of *Drosophila*. All the rhythms were

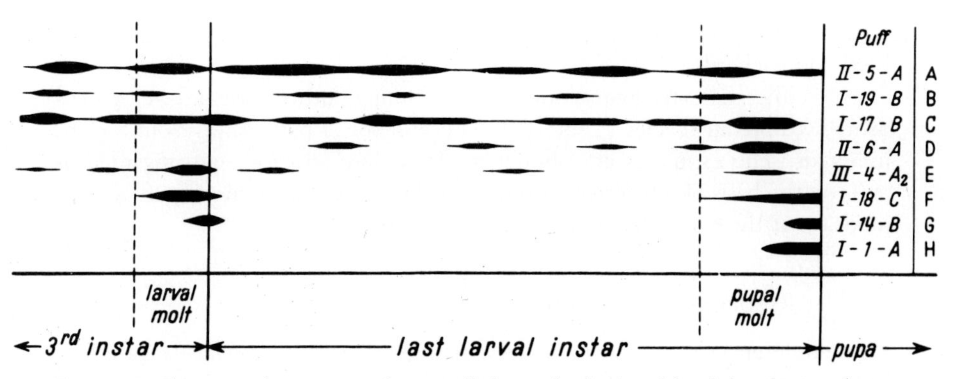

Figure 96. Changes in the activities of some puffs during the third- and fourth-larval instars during larval and pupal molts (from Clever, 1964d).

bimodal with a period of 12 hr. The 74EF/75B metamorphosis puffs most frequently appeared at a certain time of the day, followed by peaks of puparium formation 6–8 hr later (Rensing and Hardeland, 1967; Rensing, 1971). These effects were reproducible in an *in vitro* system (Rensing, 1969). *D. lebanonensis* showed a very strict circadian rhythm with regard to puparium formation (Rensing and Hardeland, 1967). Larvae collected from the same batch of eggs laid during a 2-hr period formed puparium during the first 6 hr of the dark period on day 7. When they failed to pupate during this time, there also followed a new peak of puparium formation during the first 6 hr of the dark period, later, on day 8 (Berendes and Thijssen, 1971).

The puffs characteristic of the stage of puparium formation appeared after injection of ecdysterone into the larvae of *D. lebanonensis;* however, they did not form puparium for about another 20 hr; that is, puff induction in salivary gland cells and cuticle sclerotization were uncoupled (Eeken, 1974).

Under short daylight conditions, development was delayed in mid-fourth-larval instar which was extended to 150 days. Diapause followed (Ineichen, 1978; Ineichen *et al.*, 1979). The ecdysterone level was then lowest (Valentin *et al.*, 1978). *C. tentans* lacked the metamorphosis puffs, including the early ecdysterone I-18C puff and Balbiani ring 1 (Clever, 1962b; Ineichen, 1978; Meyer *et al.*, 1983). Balbiani ring 2 and the juvenile hormone induced puff I-19A remained active (Ineichen, 1978).

The morphology and puff inducibility in the I-18C region changed in diapausing larvae during 24 hr. The puff was usually decondensed during the dark part of the day, and it was maximally condensed during its light part. However, the puff decondensed for a short time, 3 hr after the beginning of the light period, while it transiently condensed 3 hr after the beginning of the dark period. The sizes of BR2 and total chromosome length changed according to another schedule.

The hormone ecdysterone ceased to induce the I-18C puff during the condensation period in a part of the chromosomes (Lezzi *et al.*, 1989a,b, 1991; Lezzi and Richards, 1989).

There is extensive information on puffing during normal development in many dipteran species (Table 10). The number of puffs varies from tens to hundreds. The estimates are often biased, because structures are regarded as puffs on the basis of subjective morphological criteria. Furthermore, puff activity increases at the end of larval instar in all species.

2. Tissue specificity of puffing

Tissue differences in puffing have been first described in *C. tentans* (Figure 84). Later it was reported that Balbiani rings in different parts of the salivary glands in *Chironomus* are specific (Beermann, 1961) and that the physiological changes

Table 10. Puffing in the Salivary Gland Chromosomes of Different Representatives of Diptera

Species	Information on puffs	References
Acricotopus lucidus	305 puffs (299 properly puffs, the nucleolus, 5 Balbiani rings) described on the basis of decompaction of bands, 349 puffs—on the basis of RNA-specific staining and incorporation of [^{3}H]uridine	Staiber and Behnke (1985)
Chironomus plumosus	119 puffs	Maksimova and Ilyinskaya (1983)
	In the fourth instar larvae in decondensed (puffed) state are 55% of the area of the Y chromosome of summer larvae; 29% in autumn and 74%; in winter larvae	Maksimova (1983)
	30–50 and 40–60 binding sites of antibodies against DNA/RNA hybrids in short and long 3rd polytene chromosomes, respectively	Vlassova *et al.* (1991)
Chironomus sp.	11 puffs in first and second chromosomes	Helling and Ruqi (1985)
C. *tentans*	The number of puffs increases during the larval molt, it decreases during the fourth larval instar. Then there is a sharp increase of puff activity during pupation (see Figure. 96)	Clever (1961); (1962a,b); (1963a,b); (1964a–d); (1966a,b)
	Approximately 10% of all bands are in a puffed state	Beermann (1963b)
	277 regions discretely incorporating [^{3}H]uridine in the course of development or staining with toluidine blue	Pelling (1964)
C. *thummi* (described as C. *dorsalis*)	In the first and second chromosomes of larvae and prepupae, 31 RNA-containing regions are detected by staining with pyronine. In pupae, a single Balbiani ring remains	Kiknadze and Filatova (1963)
	A hundred puffs at the beginning of the fourth larval instar (approximately 160) at the end of it. Approximately 10 puffs remain active in late pupae	Kiknadze (1965b)
C. *thummi*	Approximately 336 puffs at the third and fourth larval instars and in the course of metamorphosis. The sizes of approximately 40% of puffs change markedly	Kiknadze (1972, 1976, 1978, 1981a,b, 1985); Kiknadze *et al.* (1981)
	Changes of puff sizes described in development	Kroeger (1964)

continues

Table 10. *Continued*

Species	Information on puffs	References
	Of 180 puffs described, 22 show the most marked changes during metamorphosis	Kiknadze *et al.* (1972)
Dacus cucurbitae	246 puffs described, during 120 hr of larval development	Gopalan and Dass (1972b)
Drosophila (Hawaiian species)	Short descriptions of puffs in several species	Raikow (1973); Jeffery (1976)
D. ananassae	48 puffs whose activities increase in prepupae described	Moriwaki and Ito (1969)
D. auraria	Developmental changes of 69 puffs, including 2 Balbiani rings	Scouras and Kastritsis (1984); Mavragani-Tsipidou and Kastritsis (1986)
D. bicornuta	Developmental changes in Balbiani ring 1 and 37 puffs	Mavragani-Tsipidou *et al.* (1992b); Mavragani-Tsipidou and Scouras (1992)
D. busckii	Several puffs described	Kroeger (1960)
D. cubana	247 puffs described in the third chromosome of the third-instar larvae and 1 to 6-hr prepupae	Valente and Cordeiro (1991)
D. diplacantha	5 big intermoult, 8 early and 1 late ecdysterone puffs and Balbiani rings	Mavragani-Tsipidou *et al.* (1994)
D. gibberosa	Puffing described in larvae from the end of the second larval instar to 28 to 32-hr prepupae. Almost no puffs during the second molt and the first 24 hr after it. Puffing activity sharply increases in 80 hr after molt with the beginning of the migration period of larvae and reaches maximum at the time of puparium formation. This is followed by a new peak of puffing approximately in 16 to 28-hr prepupae. Totally, 86 loci puffed	Roberts and MacPhail (1985)
D. guanche	151 puffs at the end of the third larval instar–prepupal stage (0–28 hr) of development	Molto *et al.* (1987b, 1988)
D. hydei	Changes described in 148 puffs of larvae, prepupae and early pupae	Berendes (1965a)
D. lebanonensis casteeli	89 puffs detected in the middle of the third larval instar. Five new puffs appear and 7 inactivated before puparium formation	Berendes and Thijssen (1971)

D. melanogaster	Changes in approximately 120 large puffs at the end of the third larval instar and 0 to 12 hr prepupae described	Becker (1959, 1962a,b); Lychev (1965, 1968); Ashburner (1967b, 1969b,c,f, 1972a, 1975, 1978); Burnet and Hartmann-Goldstein (1971); Sorsa and Pfeifer (1972); Korochkina and Nazarova (1977, 1978); Dennhofer and Muller (1978); Dennhofer (1979); Loreto *et al.* (1988)
	Approximately 340 puffs, including small ones	Belyaeva *et al.* (1974); Zhimulev, (1974a,b); Belyaeva (1982)
	In zero-hour prepupae approximately 400 sites of discrete incorporation of [^{3}H]uridine	Zhimulev and Belyaeva (1974, 1975b)
	355 binding sites of antibodies against DNA/RNA hybrids in zero-hour prepupae	Vlassova *et al.* (1984, 1985a,b); Umbetova *et al.* (1994)
D. pseudoobscura	Of 176 puffs described in larvae, prepupae and pupae 111 change during development. Activity peaks appear before puparium formation and during pupation. A small peak in middle prepupae	Stocker and Kastritsis (1972, 1973)
D. seguyi	2 intermoult, 6 early and 3 late ecdysterone puffs	Mavragani-Tsipidou *et al.* (1994)
D. serrata	Changes in the activity of 10 large puffs, including a Balbiani ring described	Mavragani-Tsipidou *et al.* (1990a)
D. subobscura	106 puffs described, 34% of which are stage-specific in larvae and prepupae	Frutos and Lattorre (1982a,b); Latorre *et al.* (1983); Pascual *et al.* (1983); Molto *et al.* (1987b)
	Approximately 140 puffs in H271 strain and only 96 in H271 inbred strain	Frutos *et al.* (1984b)
	166 puffs in prepupae	Pascual *et al.* (1985)
D. virilis	At the end of the third larval instar 360 bands activated judging by the results of staining with methyl green and pyronine. 108 sites showed permanent activity, 138 are intermolt, 34 early induced, and 82 late induced	Kress (1972, 1993)

continues

Table 10. *Continued*

Species	Information on puffs	References
	142 puffs described from the middle of the third larval instar to histolysis of salivary glands (48 hr), 43% of the puffs continuously active, 47.1% of the puffs arose and became inactive during this period, 9.9% of the puffs "pulsed"	Poluektova (1975a)
D. vulcana	Balbiani ring 1, 2 intermoult puffs, 1 postintermoult puff, 9 early ecdysterone puffs, 3 late puffs	Pardali *et al.* (1996)
D. willistoni	277 puffs identified in the third instar larvae - prepupae (1–6 hr) in the third chromosome	Valente and Cordeiro (1991)
Rhynchosciara angelae	In the cells of the S1 region of the salivary gland 51 large puffs and hundred of small ones described, approximately 10–20 DNA puffs	Breuer (1967); Guaraciaba and Toledo (1968); Pavan *et al.* (1975)

in salivary glands are correlated. Since then, many cells, tissues and organs with fairly large polytene chromosomes had puffs amenable to analysis and comparison with those in the other organs were detected. In the first studies it was already noted that the peak activity of the puffs in giant foot-pad cells is preceded by a period when these cells secrete cuticle (reviewed by Zhimulev, 1992, 1996).

There were detailed studies of the puffs in the giant cells of polytene chromosomes of foot-pad cells of *Sarcophaga bullata*. A large set of the most prominent puffs of the genome was the subject of Whitten's (1969) investigation. According to other data, there are approximately 600 puffs in polytene chromosomes of the type; however, it proved to be feasible to study in detail only 71 puffs in chromosome B (Bultmann and Clever, 1969; Clever *et al.*, 1969). There is a general agreement between the conclusion drawn from the two studies. Chromosomes are observable from day 3 of pupal development, when chromosomes still increase in size, to cell death, when flies emerge.

An overall scheme for the puffing schedule is given in Figure 97. Each puff makes its appearance in an ordered sequence. The majority appears only during a short interval. The number of puffs present throughout development is very small. Certain puffs appear twice. Two peaks of activity are readily distinguishable; one appears 6–7 days after white puparium, the other 10 days after it. Approximately 50% of the potentially active puffs appear on days 5–6. These "early" puffs disappear a day later. A set of the "late" puffs appear on days 7–10 and disappear half a day later before the first pulvilli cells begin the break down. However, several puffs remain; they continue to incorporate [^{3}H]uridine, even after some cells have broken down (Bultmann and Clever, 1969). In another species, *Parasarcophaga ruficornis*, similar patterns of changes were found for 44 studied puffs (Kaul *et al.*, 1981, 1983).

These data indicate that there is a cyclicity in changes of chromosome activity, which may be related to specific changes in functions of giant cells: labeling of cuticle with leucine and glucose is heaviest after the first peak of puff activity and sclerotization of cuticle starts after the second peak (review: Zhimulev, 1992, 1996).

Each fly has 24 giant pulvilli cells. Comparisons of puff sets in different cells of a pupa revealed that their activities are highly synchronous, with some variations. For example, the puffs develop somewhat earlier in cells of the anterior than posterior foot and earlier in inner than outer foot-pad cells (Bultmann and Clever, 1969; Clever *et al.*, 1969).

Puffing in the process of development of bristle-forming cells has been studied in *Calliphora erythrocephala* and *Sarcophaga barbata*. Puffs are numerous in these cells; there are about 600 puffs in all the chromosomes in *S. barbata;* for this reason, a few samples have been studied in detail (about 100 puffs) in each species (Ribbert, 1972; Trepte, 1976, 1977). Viewed broadly, regularity in

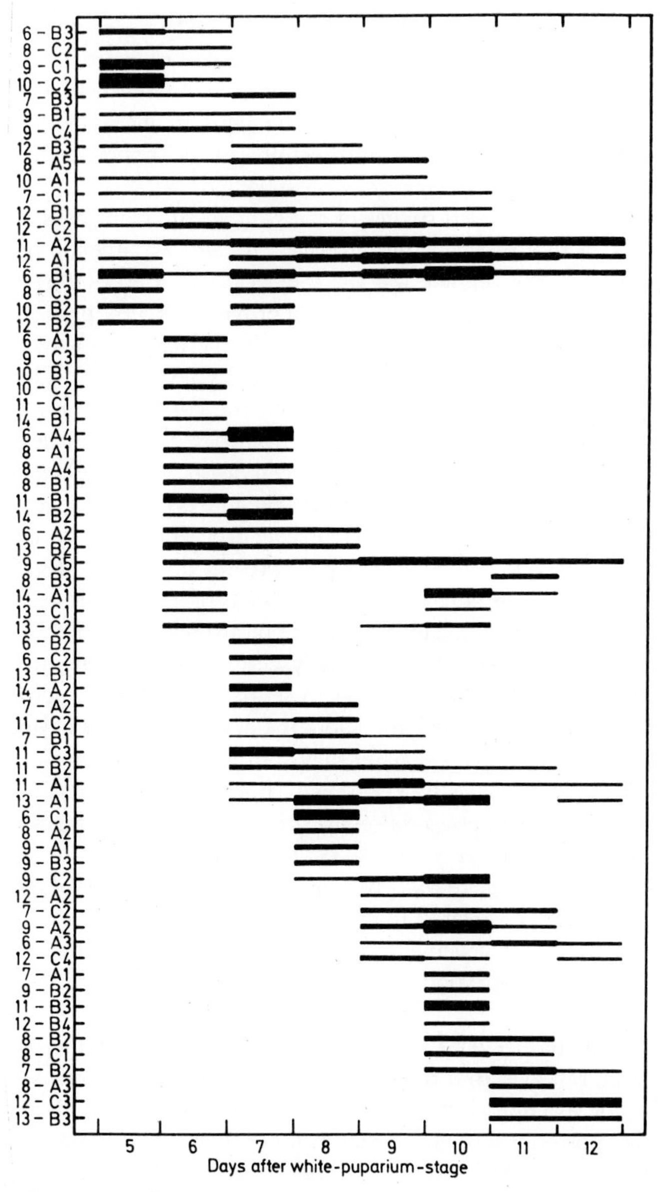

Figure 97. Changes in puffing patterns in the B-chromosome of the giant foot pad cells of *Sarcophaga bullata*. Width of line represents mean size of puffs named on ordinate. Abscissa, days after formation of white puparium (reprinted by permission from Bultmann and Clever, 1969; Clever *et al.*, 1969a).

changes of puffing patterns are the same in the two species. All the puffs appear and disappear in an ordered sequence; most puffs are active only for a short time. In *C. erythrocephala* the puffs reach their maximum activity for several parameters: the number of observed, newly formed puffs or those reaching maximum. Then, puff activity slowly decreases to day 8 and rapidly after this day. The peak of puff activity is presumably induced by ecdysterone, because its maximal titer is on days 4–5 in *Calliphora* (Ribbert, 1972).

There are comparative data on puffing activity in trichogen and tormogen thoracic cells, in the bristle-forming cells of scutellum, and the fifth abdominal tergite, as well as in trichogen and giant foot pad cells.

Comparisons of the puffs in scutellar trichogen cells and the anterior thoracic cells in *C. erythrocephala* demonstrated that, when puff sets are identical in the scutellar cells, some puffs are replaced by others somewhat earlier than in the thoracic cells; this generally correlates with more rapid development of scutellar bristles (Ribbert, 1972). In *S. barbata,* 105 puffs were detected in trichogen cells and 102 puffs in tormogen cells of the fourth chromosome during development. The puffing patterns of these sister cells were very similar. In 49 identified loci, there were slight differences in the synchrony of puff timing (Trepte, 1976). Comparisons of the puffs of trichogen cells on scutellum and fifth abdominal tergites also revealed that their sets are similar; however, their appearance in abdominal cells was delayed for a day or so (Figure 98). Processes, such as growth of trichogen cells and bristles, sclerotization of cuticle, and attainment of maximum sizes by the nucleus were also delayed for a day (Trepte, 1980).

According to preliminary data in *S. bullata*, comparisons of the activity of 11 puffs in giant foot-pad and trichogen cells demonstrated that 4 puffs were active in the chromosomes of both cell types in pupae at the age of 4–5 days, 3 puffs were active in trichogen cells, and only 4 in pulvilli cells (Whitten, 1969).

The bursicone, which controls hardening of cuticle in adult, induces three new puffs in the chromosomes of tormogen cells in newly emerged flies (Trepte, 1976).

The secondary polytene chromosomes in the ovarian nurse cells of *C. erythrocephala* (reviewed by Zhimulev, 1992, 1996) have characteristic regions containing puffs and Balbiani rings. The features by which these chromosomes are recognized are by the conspicuous swellings adjacent to the centromeric regions of each of the five long chromosomes called "kinetochorial bulbs" (Bier, 1960; Ribbert, 1979). They possibly correspond to the kinetochorial baskets described for the trichogen chromosomes in this species (Ribbert, 1967). While the latter show no evidence of enhanced RNA synthesis, kinetochorial bulbs label very heavily after short (3–20 min) autoradiographic exposure to [^{3}H]uridine. Their diameter can be 3.2- to 5-fold that of the nonpuffing region of the chromosome; they are not Balbiani rings, because their shape is not torus-

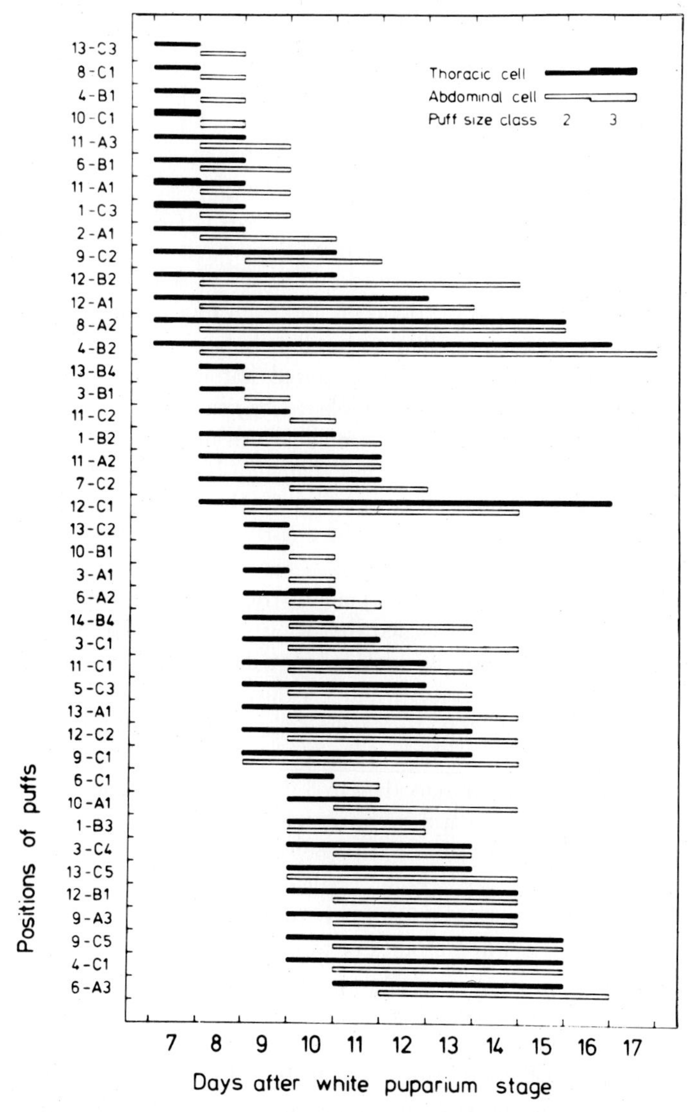

Figure 98. Changes in the activities of chromosome IV puffs of *Sarcophaga barbata* in the thoracic and trichogen cells. Width of bar reflects puff size. Abscissa, days after white puparium formation (from Trepte, 1980).

like. Ribbert (1979) suggested that such swellings result from the activation of a series of closely located loci interrupted by the inactive regions of the chromosomes. These regions have the appearance of dotted bands traversing the "kinetochorial bulbs".

The puffs are generally more developed in nurse cell chromosomes than in somatic (trichogen) cell chromosomes in this species. However no qualitative comparisons were made (Ribbert, 1979).

Developmental changes in puffing activity in *C. erythrocephala* can be studied from the second to the fifth stages of oogenesis, when degeneration of nurse cells becomes visible. At this stage, RNA in the nurse chamber is synthesized in considerable amounts (making up more than 95% of the amount stored in the mature oocyte). In addition to large amounts of nucleoside-triphosphates (0.4 μkg), the mature oocyte contains about 2.5 μg of macromolecular RNA (with a molecular weight exceeding 20-thousand daltons). More than 2% of this RNA is polyadenylated (Kirchhoff, 1981). One would expect that this RNA synthetic activity might correlate with puffing of polytene chromosomes, since nurse cells are the only sources of RNA for the growing oocytes. However, chromosome analysis demonstrated that puffing pattern did not change throughout the period when the secondary polytene chromosomes amenable to analysis and about 95% of the macromolecular RNA was synthesized. The two obvious exceptions were that the Balbiani ring was maximally active at the end of oogenesis, and there were transient puffs in the subterminal block of heterochromatin of the second chromosome in the middle of oogenesis.

It was found that the temperature at which flies were exposed after secondary polytene chromosomes have already developed is influential. After transfer from 14° to 21°C, the size of the puff somewhat increased, but the puff did not change qualitatively (Ribbert, 1979).

In *Anopheles superpictus*, polytene chromosome banding patterns of the ovarian nurse cells and larval salivary glands are so different that homologous bands can be only tentatively designated on chromosome maps. This was attributed to the inconsistent localization of the puffs in the chromosomes of the two organs. Fourteen characteristic puffs were detected in the chromosomes of the nurse cells, many of them were mapped to the centromeric regions of the chromosomes (Coluzzi *et al.*, 1970; Coluzzi and Kitzmiller, 1975).

Two puffs and a Balbiani ring, which incorporate [^{3}H]uridine most intensely, were described in chromosome *3L* of *Anopheles stephensi* nurse cells. Polytene chromosomes showed a most distinct banding pattern 12–36 hr after a blood meal. Analysis of preparations made at 2-hr intervals during this period did not reveal changes in the morphology of the puffs of chromosome *3L*. New puffs did not appear.

There are two puffs at the proximal end of the X chromosome at the stage of their maximum activity. When expanded, these puffs resemble the "kinetochorial bulbs" of *Calliphora* (Redfern, 1981a). The general conclusion is that puffing does not change appreciably during development in the two species.

In the *fs(2)B* mutant of *D. melanogaster*, only the small puffs continuously active in the salivary gland chromosomes can be seen in the secondary polytene chromosomes; they show a clear-cut banding pattern. Puff diameter never exceeded chromosome diameter. The only large puff exceptional in this respect was in the 3F region of the *X* chromosome, which was seen in a part of the cells (H. Gyurkovics, 1983, personal communication). Several puffs were identified in the nurse cells of the *otu* mutant of *D. melanogaster* in regions 3CD, 7E, 8C, 11B, 22F, 39E, 61A, 79D, and one that looked like a Balbiani ring: 47A (N.I. Mal'ceva, I.F. Zhimulev, V.F. Semeshin, and D.E. Koryakov, 1995, unpublished observations).

Two transcriptional features characteristic of the secondary polytene chromosomes from ovarian nurse cells are of interest:

1. Their more intensive incorporation of [^{3}H]uridine compared to somatic cells. Provided that the conditions of [^{3}H]uridine labeling experiments are identical, trichogen cell preparations of the chromosomes of *Calliphora erythrocephala* must be exposed 8–10 times longer than ovarian nurse cell preparations to obtain the same grain counts (Ribbert, 1979). Similar differences were observed in labeling intensity of chromosomes of the salivary glands of larvae and ovarian nurse cells of *Anopheles stephensi* (Redfern, 1981a).
2. In *Calliphora erythrocephala,* nurse cell polytene chromosomes differ from trichogen cell polytene chromosomes in their pattern of [^{3}H]uridine incorporation. Trichogen cell chromosomes show a very continuous label distribution, while all the regions of the secondary polytene chromosomes label almost with equally high intensity (Bier *et al.*, 1969; Ribbert and Bier, 1969). This may be attributed to the fact that the autoradiographs were obtained from chromosomes with incompletely paired oligotene fibrils; as a consequence, small longitudinal displacement of these fibrils, still occurring in the secondary polytene chromosomes, resulted in homogeneous grain density (Ribbert, 1979). However, there seems to be another plausible explanation. As Ribbert and Bier (1969) emphasized, after pulse labeling, [^{3}H]uridine incorporation was not restricted to the puffs and Balbiani rings, the usual band–interband regions of the chromosome did show label incorporation. Like the nurse cells of *Anopheles stephensi*, almost each chromosome region incorporated [^{3}H]uridine, although it did not puff (Redfern, 1981a). Thus, the smaller puffs perform the major role in transcriptional activity.

The problems remain as to how puffing changes in cells not subjected to histolysis during metamorphosis—such as Malpighian tubules—and how the adult retains the same organ the larva once had. In *D. hydei,* comparisons of puffing patterns in larval and adult Malpighian tubules demonstrated that 12 puffs, which had been active in larvae, were missing in adults. Puffs specific to adults were not found (Berendes, 1966a). In *C. thummi*, developmental changes in puffing activity were followed in pupae. Puffing activity was low in the middle of pupal development, it sharply decreased before adult molt (Buetti, 1968). In *D. gibberosa*, the rapid and cyclic changes in the puffs in Malpighian tubules during the larva–prepupal transition were replaced by a relatively invariable pattern in young, middle-aged, and old adults (Roberts, 1983). The I-19A specific puff, which is induced by the juvenile hormone in larval salivary glands, appears in Malpighian tubule cells within 1 hr after its topical application to adults of *C. tentans*. The higher was the concentration of applied hormone, the faster was the inducible puff inactivated (Holderegger and Lezzi, 1972; Lezzi and Gilbert, 1972).

Information concerning tissue-specificity of puffing activity is given in Table 11 (also see Fig. 99).

3. Genetic control of puffing

A remarkable stability underlies the patterns of puffing activity that is characteristic of a species. Thus, 82–83 puffs were detected in two different strains of *D. melanogaster*, Oregon-R and *vg6*; the timing of activity was identical in 53 puffs (64%) in Oregon-R and *vg6*; 10 puffs (12%) differed in their mean sizes;

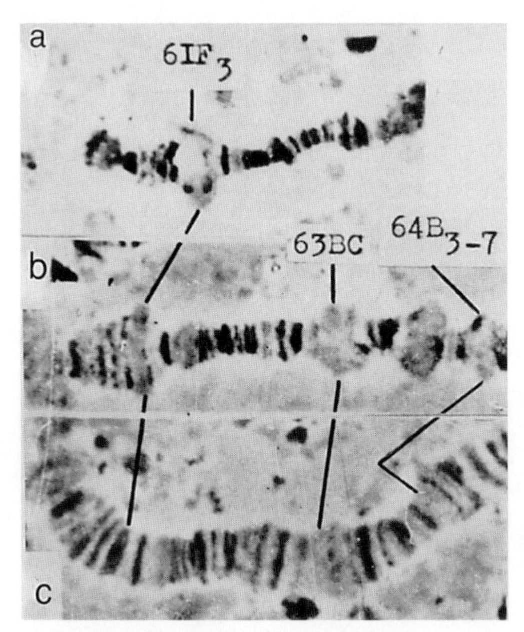

Figure 99. Difference in puff activity in the 61F3 region between cells of the (a and b) proximal and (c) distal parts of salivary glands of *D. melanogaster* (from Zhimulev, 1973b).

Table 11. Tissue Differences in Puffs and Balbiani Rings in Representatives of Diptera

Species	Compared organs or cells	Specificity of puffing	Reference
Acricotopus lucidus	Main, lateral and anterior lobes of the salivary gland	Balbiani rings 3 and 4 active only in the anterior lobe	Mechelke (1953, 1958)
		The II/0-17 and I/D-9 puffs specific to the main and lateral lobes.	Panitz (1965)
		11 puffs and 3 Balbiani rings (BR3, BR4, and BR7) specific to the anterior lobe, 25 puffs in the anterior lobe larger than in the other lobes, 8 puffs (including BR1 and BR2), specific to the main and lateral lobes, 14 puffs larger in size in the main and lateral lobes, 188 puffs are the same in different lobes	Staiber and Behnke (1985)
	Salivary glands and Malpighian tubules of larvae	204 puffs in Malpighian tubules versus 349 in salivary glands, 5 puffs more active in cells of Malpighian tubules	Staiber and Behnke (1985)
	Salivary gland, internal variability	Cells (6 in 2000 preparations) detected at the boundary of the anterior and main lobes of the gland with intermediate set of Balbiani rings. These cells presumably resulted from fusion of cells of different lobes	Staiber (1986)
Anopheles stephensi	Salivary gland of larvae and ovarian nurse cells of adults	Salivary glands specific Balbiani ring (3D4), nurse cells specific 6BC and 5C4-5 puffs	Redfern (1981a)
Calliphora erythrocephala	Ovarian nurse cells of adults and trichogen cells of pupae	Direct comparison of puffing impossible because of incomparable chromosome banding pattern of two cell types	Ribbert (1979)
Chironomus tentans	Salivary glands, Malpighian tubules, rectum, midgut	Variation in occurrence of some puffs, Balbiani rings identified only in salivary gland chromosomes	Beermann (1952a, 1956)

	Malpighian tubules in female and male adults	A large puff, virtually Balbiani ring in male, a small puff in female	Lezzi *et al*. (1961)
Cryptochironomus	Cells of the "special lobe" and the remaining part of the salivary glands	Presence of specific Balbiani rings in "special cells"	Bauer (1953); Beermann (1956, 1961)
C. gr. *defectus* (karyotipe I), C. *obreptans*, C. *ussouriensis*	Cells of main and special lobes of salivary gland	One of the Balbiani rings is only in main lobe cells	Istomina *et al*. (1992)
Drosophila auraria	Salivary glands, Malpighian tubules, fat body	Balbiani rings specific to salivary glands only. In all, 73 sites puffed in the chromosomes of fat body, 56 in those of midgut; of those 17 never occurred in salivary gland	Mavragani-Tsipidou and Scouras (1991)
D. gibberosa	A pair of usual (C) Malpighian tubules and a pair of rings (R)	In late larvae puffing almost identical in tubules of both types. Puffs are larger in the C tubules	Roberts (1971)
	Larval salivary glands, Malpighian tubules, fat bodies near the salivary glands and in the anterior part of the body	Puffing pattern in fat body of two types almost identical, and they differ markedly from those in Malpighian tubules and salivary glands	Roberts (1972)
	Salivary glands, midgut	In midgut cells 71 puffs described (compared to 86 in salivary glands) starting from the second larval molt till the puparium formation. Homologous puffs appear in stomach cells earlier than in salivary gland cells	Roberts (1988)
	Salivary glands, fat bodies, midgut, Malpighian tubules	Late larval–early prepupal puffing is enhanced in salivary glands relative to gut, fat bodies and Malpighian tubules	Roberts and Jakobsen (1991)
D. hydei	Proximal and distal parts of the salivary glands	Of 110 puffs only 47B is specific to the distal part, 38 puffs differ in size	Berendes (1965a)

continues

Table 11. *Continued*

Species	Compared organs or cells	Specificity of puffing	Reference
	Salivary glands, Malpighian tubules, stomach of larvae	76 puffs common to all the three organs, 34 puffs occur in one or two organs. Most differences due to the earlier puff occurrence in a gastric cells than in the other ones	Berendes (1965a,b, 1966a)
D. lebanonensis	Sister salivary glands are different in size and morphology	No differences in puffing detected	Berendes and Thijssen (1971)
D. melanogaster	Proximal and distal parts of salivary glands	The 15BC puff larger in size in the proximal part	Becker (1959)
		61F3 puff specific to the proximal part only (see Figure 99), 27E1-2—to the distal part. Differences in size in 18 regions (including 15B)	Zhimulev (1973b)
	Salivary glands and fat body of larvae and prepupae	The number of puffs in chromosome 3L in cells of the two types is approximately the same, the large puffs are common, the puffs of the end of the third instar appear earlier in fat body cells	Richards (1980b, 1982, 1985)
	Salivary gland and other organs	The 68C puff expressed only in salivary gland	Meyerowitz and Hogness (1982)
	Salivary gland, hindgut and midgut, prothoracic gland	Specific puffs for hindgut: 93A and 96B; midgut: 64B, 65A and 67C; prothoracic gland: 21BC and 22A	Hochstrasser (1987)
D. melanogaster, heterozygote DTS^{-3}/e^{11} (29°C)	Ring and salivary glands of larvae	Of 24 puffs in 31 compared chromosome in 9 no differences found, 6 present in ring glands only; 9 in salivary glands only. The large puff in the 102C region of the fourth chromosome detected in ring gland	Holden and Ashburner (1978)
D. melanogaster, stock *otu*	Ovarian nurse cells and salivary gland of larvae	The 32B, 68C and 75B puffs present	Sinha *et al.* (1987)
		No puffs detected	Heino (1989)

		No puffs characteristics to salivary glands were found in nurse cells. Puffs specific to nurse cells: 3CD, 7E, 8C, 11B, 22F, 39E, 61A, 79D. The 47A looks like Balbiani ring in the nurse cells.	Mal'ceva *et al*. (1995); Mal'ceva, N. I., Zhimulev, I. F., Semeshin, V. F., and Koryakov, D. E. (1995), unpublished observations
		Puff inducible by ecdysterone in salivary gland cells cannot be induced in nurse cells.	Zhimulev and Mal'ceva (1997)
		Heat shock puffs are very small in nurse cells in comparison with salivary gland cells	Heino *et al*. (1995); Zhimulev and Mal'ceva (1997)
D. virilis	Salivary gland, midgut, fat body	The genes mapped to the 16A (secretion protein) and 55E (the gene active in larvae before the ecdysone stage) puffs expressed in salivary glands only	Kress *et al*. (1990)
Melanagromyza obtusa	Salivary glands and midgut	Of 49 puffs studied 12 was common for two organs; 5 specific for salivary glands, 12 for gut	Gupta and Singh (1983)
Rhynchosciara angelae	Salivary glands and Malpighian tubules	Seven DNA-puffs of the C chromosome present in salivary glands and absent in Malpighian tubules	Breuer and Pavan (1954, 1955)
Sarcophaga barbata	Trichogen and tormogen cells of pupae	In the fourth chromosome 105 puffs identified in trichogen and 102 puffs in sister tormogen cells. Many differences observed in puff size	Trepte (1976)
	Bristle-forming cells on the thorax and abdomen	Puffs mainly identical, ontogenetic puff changes occurred in abdominal cells a day later than in thoracic cells	Trepte (1980)
S. bullata	Foot pad cells on the mesothorax and prothorax, internal and external cells on the leg	Activity differences of the 14-A2 puffs	Bultmann and Clever (1969)

continues

Table 11. *Continued*

Species	Compared organs or cells	Specificity of puffing	Reference
Sciara coprophila	The proximal and distal parts of the gland	Of 18 DNA puffs 12 of the same size in both parts, three puffs (III-2B, III-15B, and IV-12A) specific to the proximal part and two puffs (IV-15B and II-11A) to the distal part. Size differences observed for a single puff	Gabrusewycz-Garcia (1975); see also: Crouse (1968); Stocker and Pavan (1974, 1977); Perondini and Dessen (1985)
Stictochironomus sp.	Cells of the main special lobes	An additional Balbiani ring in special lobe	Istomina and Kiknadze (1993)
Trichocladius vitripennis	Cells of the main and the special lobes of salivary glands	Two Balbiani rings active only in six granular cells of the special lobe; one in cells of the main lobe	Beermann (1952b)

16 (19%) showed timing differences; and 4 (5%) differed both in timing of their activity and their mean puff size (Ashburner, 1969d,e). Only five specific puffs (23C, 42C, 46D, 61F and 47F) were detected in Parvoo-wild strain. However, they all were detected later in a laboratory stock of Batumi-L (Zhimulev, 1974a).

Comparisons of the sizes and frequencies of puffs in the X chromosomes of zero-hour prepupae from a natural population (Alma-Ata), three laboratory strains (Batumi-L, Canton-C, and Oregon-R), and two inbred strains (HA and BA) did not reveal qualitative differences in puffing. The differences in puff size between the stocks did not exceed a score of 0.6 (Belyaeva and Zhimulev, 1973, 1974). Two lines of *D. virilis,* numbers 101 and 9, differed only in the size of the puffs at 3-C6 (Poluektova, 1975b). In three strains of *D. subobscura* differing by inversions in the X chromosome, no differences in puffing patterns were found in zero-hour prepupa (Latorre *et al.*, 1984).

In *C. thummi,* with prevailing puff number (336), two additional puffs were found. The frequency of the IV-Dc puff varied in the range of 2.1 to 6.2% (the Novosibirsk population) to 58.9% (the Germany population) and that of IIC1h-j puff in the 1.2 to 3.0% range (Kiknadze *et al.*, 1988).

Puffing patterns are also quite stable, except for their derangements caused by mutations. Thus, in the *giant* mutants of *Drosophila,* despite considerable developmental delay, puffing generally did not differ from normal (Becker, 1959). In homozygotes for certain mutant genes dying at the end of the third larval instar, puffing remains normal to their death, as, for example, in *lethal (2) histolytic* and *lethal (2) giant disks* (Ashburner, 1970d, 1972a). The metamorphosis puffs are absent from the mutants *l(2)gl, halfway,* and dor^{l} (Becker, 1959; Ashburner, 1972a, for details see Section VI,D). The observation that the specific 24DE puff (Slizynski, 1964) occurs only in *fat* strain was not confirmed, since this puff is usually seen in normal strain (Zhimulev, 1974a,b; Ashburner, 1975). Detailed comparisons of 77 puffs in *fat* and Oregon-R wild-type demonstrated that only 7 are present in the control and 12 are more active in *fat*. The 24DE puff was not entered in this list (Chaudhuri and Mukherjee, 1970).

Data on wide variations in puffing within populations of *D. melanogaster* came as a surprise against a background of the information just given. Thus, there are 201 puffs in larvae of the population of Dilijan, their number is 298 in that of the Nikitski botanical garden, and it is 273 in that of Magarach. From 16 to 25%, on the average, of the puffs are variable (Kiknadze, 1965; Kiknadze *et al.*, 1965, 1972, p. 151). The 4AB and 17E puffs described as specific to the *yellow* mutant in the above mentioned papers usually occur in normal strains (Belyaeva and Zhimulev, 1973, 1974).

The results obtained for the effect of crosses between closely related individuals (inbreeding, selection for expression of a particular character) were contradictory. There is a line of evidence indicating that there is no effect; according to another line it is very strong (Table 12).

Table 12. Occurrence of Heterozygosity for the Active Regions of the Polytene Chromosomes

Species	Characteristics of heterozygosity	References
Acricotopus lucidus	Tissue-specific heterozygosity for the nucleolus: the nucleolus is symmetrical in the anterior part of the gland, the nucleoli are heterozygous in the central and lateral lobes (see Figure 166). The nuclei of Malpighian tubule cells have also symmetrical and asymmetrical nucleoli	Panitz (1960a)
Boophthora erythrocephala	Heterozygosity for the nucleolus occasionally observed	Petrukhina (1968)
Bradysia tritici	Heterozygous puffs	Perondini and Dessen (1969)
Chironomus	Heterozygosity for the large puff detected in hybrids between Palestinian species of *Chironomus*	Goldschmidt (1942)
	Heterozygosity for puffs in larvae of Chinese populations (Figure 101)	Hsu and Liu (1948)
	Heterozygosity for Balbiani ring in the third chromosome	Beermann (1959)
C. *pallidivittatus*	Heterozygosity for Balbiani ring 4 does not depend on pairing of homologs	Beermann (1961)
	Heterozygous Balbiani ring 6 (the third chromosome) arises after 7 days of induction with galactose	Beermann (1973)
C. *plumosus*	Heterozygosity for Balbiani ring	Nesterova (1967)
	In some cases, heterosygosity for Balbiani ring 2	Bukhteeva (1974)
	Asynchronous puff development in the different homologs	Maksimova (1974)
	In one larva (0.77%) of the Estonian population, heterozygosity for a puff of chromosome 1. Heterozygosity for a puff in the fourth chromosome in 4 larvae of the Volgograd population (4.93%) and one larva of the Estonian (0.77%) population	Belyanina (1977)
C. *tentans*	Two of the three Balbiani rings and one puff are occasionally heterozygous	Beermann (1952a, 1956)
	On two homologs the 14C2 puffs differ in size, intensity of [^{3}H]uridine incorporation and the amount of bound antibodies against RNA polymerase II (see Figure 101)	Beermann (1952a); Pelling (1964); Sass and Bautz (1982b)
Hybrids of C. *tentans*/ C. *pallidivittatus*	Heterozygosity for Balbiani ring 4 (present in the *pallidivittatus* homolog) in special cells (SZ)	Beermann (1961)
C. *tepperi*	Heterozygosity for the nucleolus in the 16B region results from deletion of the nucleolar organizer	Martin (1974); Eigenbrod (1978); Lentzios and Stocker (1979)

C. thummi (described as *C. dorsalis*)	In pupae in one homolog the nucleolus inactivated earlier than in the other	Kiknadze (1965b)
C. thummi	Many puffs induced with oxytetracycline are heterozygous	Serfling *et al.* (1969)
	A heterozygous puff in the 4-Dc region (additional Balbiani ring), 8% of larvae have a homozygous puff in all gland cells, the puff is heterozygous in 50% of larvae	Kiknadze and Panova (1972)
Cnetha dacotensis	Heterozygous nucleoli. High incidence of B chromosomes is associated with nonexpression of the nucleolar organizer	Procunier (1982)
Cnetha djafarovi	Heterozygosity for additional nucleoli	Kachvoryan (1990)
	Heterozygosity for a puff in the 18–19 region of the second chromosome	
Cryptochironomus gr. *defectus*	Heterozygosity for the main nucleolus detected in 14.1% of larvae, 22.7% are heterozygous for the additional nucleoli. Heterozygosity for puffs	Miseiko and Popova (1970a,b); Istomina, *et al.* (1992)
Hybrids of *Drosophila gymnobasis* × *D. silvarentis*	Heterozygosity for puff size on the end of the X chromosome	Raikow (1973)
D. lummei	A heterozygous puff in the region of the β-esterase genes in hybrids from crosses of strains with normal gene expression and null allele strains	Polyakova *et al.* (1990)
D. melanogaster	The 2B13-17(2C1-2) puff in the *halfway* mutant functions asynchronously in the different homologs	Rayle (1967)
	Heterozygosity for the 3C puff. When the homologs synapse, the puff develops in both, and proteins are synthesized specific to both homologs. In the case of asynapsis, the puff is on one homolog	Korge (1975)
Hybrids of *D. melanogaster* × *D. simulans*	The 63E puff is of much larger size in the former species, the 46A puff is missing in *D. simulans*	Ashburner (1969c)
	Considerable differences in incorporation of ^{3}H-uridine in the 4DE, 5CD, 7AB, 8D regions of the X chromosome and in some regions of the autosomes	Chatterjee and Ghosh (1985)
D. pseudoobscura	In the *sal* mutants puffing is reduced in the adjacent region. In *sal*/*sal*$^{+}$ heterozygotes, the mutation is not manifested when homologs synapse, puff develops according to the genotype when homologs asynapse	Levine and van Valen (1962); Gersh (1972a,b, 1973)

continues

Table 12. *Continued*

Species	Characteristics of heterozygosity	References
Hybrids of *D. virilis* × *D. texana*	In zero-hour prepupae the 3C4 and 3BC puffs occur in *D. virilis* and are missing *D. texana*. The 3C4 puff heterozygous in hybrids, it occurs on the homolog of *D. virilis*. At the 3B3 locus in the unpaired homologs, the puff heterozygous more frequently in offspring from *D. virilis* × *D. texana*	Poluektova (1969, 1970a,b)
Endochironomus tendens	Some heterozygous puffs	Michailova (1992)
Eusimulium aureum	The additional nucleolus active in a number of individuals	Dunbar (1959)
Harmandia loewi	Asynchronous reduction of homologous puffs in the course of cesation of functioning of salivary gland cells	Zhimulev and Lychev (1972c)
Odagmia maxima	Heterozygous additional nucleoli	Ralcheva (1977)
O. obreptens	Heterozygous puffs (till 31% of larvae)	Ralcheva (1977)
Odagmia ornata	Heterozygosity for additional nucleoli	Shcherbakov (1966a)
	Heterozygous puffs	Shcherbakov (1968)
Parasarcophaga misera, *P. ruficornis*	The heterozygous puff in the 7BC region in foot pad polytene chromosomes	Tewari and Agrawal (1984)
Phaseolus coccineus	Sharply asymmetrical incorporation of [^{3}H]uridine into the nucleolar organizer of the polytene chromosomes of haustorium	Cionini *et al.* (1982)
Procecidochares utilis	Heterozygous puffs on unpaired homologs	Bush and Taylor (1969)
Prosimulium hirtipes	Heterozygosity for Balbiani ring in one nucleus (?)	Grinchuk (1967)
P. inflatum	Nucleolus active in the X chromosome. Putative nucleolar organizer in the Y chromosome represented by a dense heterochromatic band	Basrur (1959)
P. nigripes	Heterozygous nucleoli	Ralcheva (1977)
Psychodidae (representatives of family)	Heterozygosity for puffs and nucleoli	Pavan *et al.* (1968)
Sciara ocellaris	Heterozygosity for puffs at the end of the second chromosome. Loosening of two bands and their conversion into a puff mistaken for a deletion by the author	Rohm (1947)

	Heterozygous DNA and RNA puffs	Pavan and Perondini (1967); Pavan *et al.* (1968); Perondini and Dessen (1969, 1983)
	Asymmetry in the size of homologous puff on different homologs results from the size difference of the puffed band	Pavan and da Cunha, (1969a)
	The 12B3 puff functions asynchronously in the different homologs	Perondini and Dessen (1969)
S. pauciseta II	Two heterozygous RNA puffs and a single DNA puff	Gabrusewycz-Garcia (1971)
	The larger DNA puff develops from the larger homologous band	
Simulium nolleri	Three types of karyotypes identified: 1. Puff present on both homologs at identical loci 2. Puff observed in only one of the homologous chromosomes 3. Puff missing in both homologs	Shcherbakov (1966b, 1968)
S. ornatipes	Heterozygosity for the nucleolus	Bedo (1978)
Telmatoscopus sp.	Heterozygous puffs	Amabis and Simoes (1972)
Tetisimulium condici	Heterozygous nucleoli in 8–9% of larvae in Armenian populations	Kachvoryan (1989)
Urophora cardui	Asynchrony in the regression of several puffs on unpaired homologs	Mainx (1976)
Wilhelmia equina	Heterozygous puffs in single instances	Grinchuk (1968)
	Heterozygosity for the nucleolus detected in several nuclei	Grinchuk (1969)
	Heterozygosity for the nucleolus	Bedo (1978)

Patterns of puffing activity are also very similar in closely related species. Comparisons in X chromosome puffing pattern of *D. melanogaster* and *D. simulans* revealed slight differences in puffs. The 7B1-3 puff is absent from *D. melanogaster;* it is active in *D. simulans* (Ashburner, 1969b). The 4F1-4 puff is active in the mid-third-larval instar in *D. melanogaster*, whereas it is active in zero-hour prepupae in *D. simulans* (Ashburner, 1969b; Zhimulev, 1974a,b). About 50% of the autosomal puffs that these two sibling species have in common were identical in regard to mean size and timing activity. The remaining puffs differed in their timing activity or their mean size. Puffing pattern of *D. melanogaster* differed qualitatively in the presence of the specific 46A puff. The 63E puff was much smaller in *D. simulans* (Figure 100) than in *D. melanogaster* (Ashburner, 1969c, e).

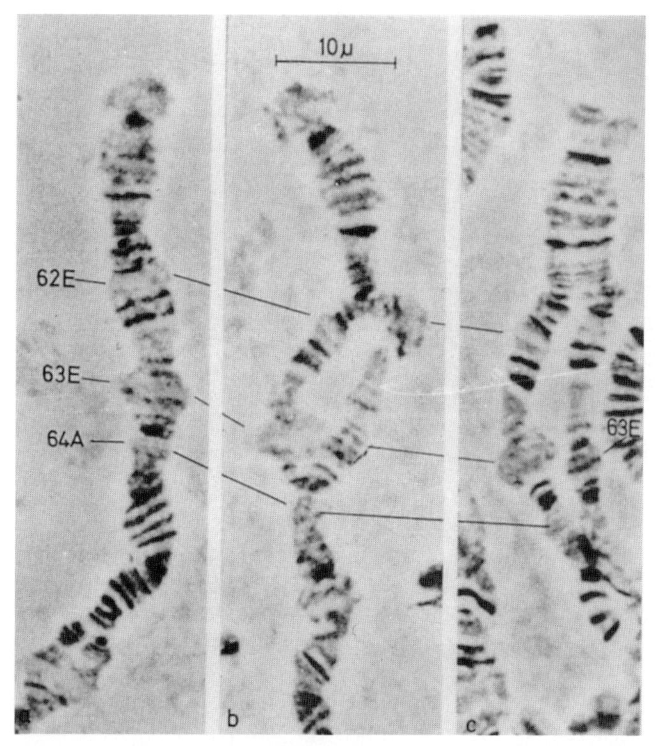

Figure 100. Dependence of the size of the 63E puff in *D. simulans* x *D. melanogaster* hybrids on synapsis of homologous chromosomes. (a) The large 63E puff in the chromosomes of hybrids with pairing homologs; (b and c) the large puff on *D. melanogaster* homolog, the small puff on *D. simulans* homolog (reprinted by permission from Ashburner, 1969c, 1970d).

The puffs on chromosome 3L in 4 closely related species of *D. melanogaster, D. yakuba, D. teissieri,* and *D. simulans* showed almost identical activities, and a general timetable for the changes in the large puffs was established (Ashburner and Lemeunier, 1972). The puffs of *D. melanogaster* can be recognized in the chromosomes of *Drosophila takahashii* (Ashburner, 1972a; Ashburner and Berendes, 1978).

The polytene chromosome puffing patterns of *Droshophila guanche* were compared with the puffing patterns of the closely related *D. subobscura*. The number detected in *D. guanche* was 150; it was 166 in *D. subobscura*. Of these, 20 puffs occurred only in the former and 7 only in the latter species (Molto *et al.*, 1988).

The number of puffs and presumably micropuffs on the third chromosome is close in two other related, "almost homosequential," *Drosophila* species: 277 in *Drosophila willistoni* and 247 in *D. cubana* (Valente and Cordeiro, 1991).

Single puffs can be "turned on" or "turned off" quite frequently—for example, during the formation of chromosomal rearrangements or in the vicinity of breakpoints.

In *Acricotopus lucidus* a small inversion was induced by irradiation. It breaks the block of heterochromatin, whose fragment induced an additional Balbiani ring when transposed at a new position (Staiber, 1982). A case was described where a Balbiani ring is present in an inverted chromosome fragment (Mechelke, 1960). A Balbiani ring forms exactly in the same way in *Chironomus nuditarsis* (Pulver and Fischer, 1980). In a population of the midge *Simulium nolleri*, a translocation associated with the apearance of an additional nucleolus was detected (Shcherbakov, 1965). Differences in puffing are produced by the existence of the E12 inversion in *D. subobscura* (Frutos *et al.*, 1987). Some specificity in puffing can arise in regions remote from those containing inversions (Latorre *et al.*, 1988).

The formation of the large puff in the 10C3 region is possibly related to genetic sex determination in *Culex pipiens*. Males are heterozygous for the region: one heterozygous chromosome has the "euchromatic" band; the other has the "heterochromatic" puff. Females are homozygous for the puff (Dennhofer, 1975; Sweeny *et al.*, 1987).

Numerous cases are known in which variation in the activity of single regions lead to puff heterozygosity (Figure 101). Asynchronous puffing of homologous chromosomes may be a cause of puff heterozygosity. The band on one chromosome commences and terminates puffing earlier than the band on the other chromosome. This results in the formation of an asymmetric puff. Asymmetry may be due to differences in sizes of homologous bands from which the puff originates (Pavan and da Cunha, 1969).

Changes in transcription rate are correlated in homologous puffs of different sizes. In the salivary gland chromosomes of *C. tentans*, the 14C2 puff

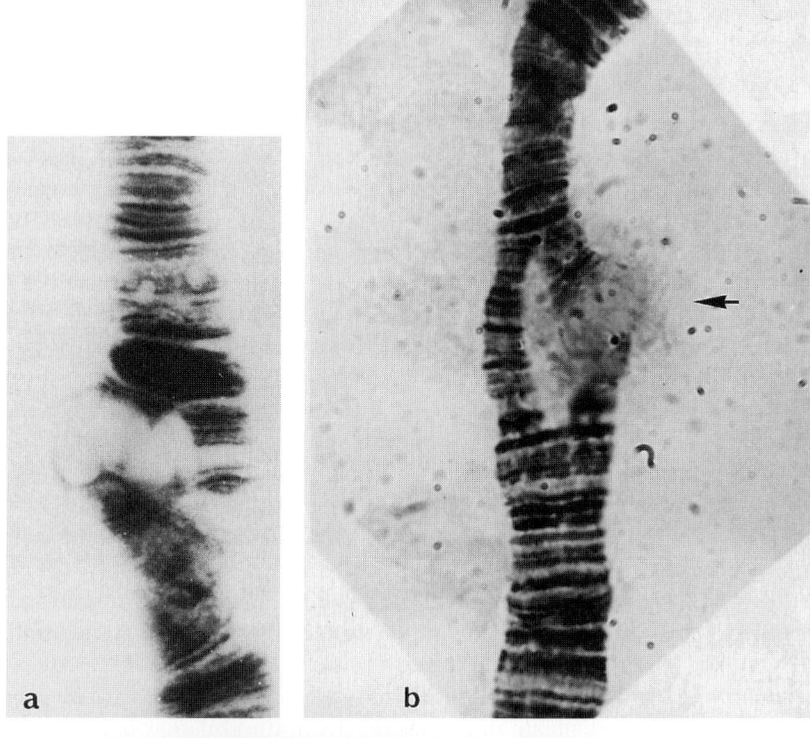

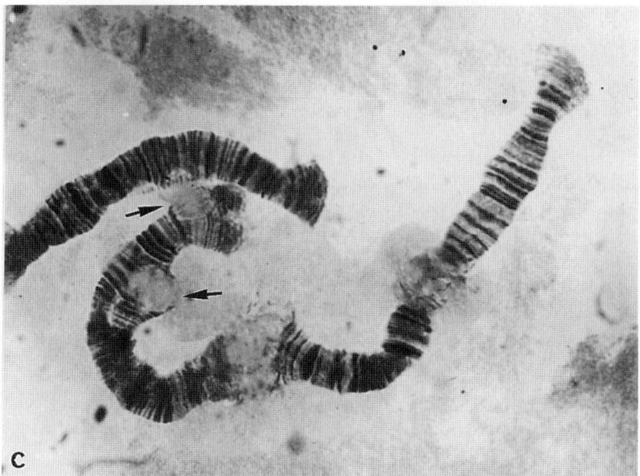

Figure 101. Heterozygous puffs. (a) A heterozygous tetracycline-induced puff in *Chironomus thummi* (reprinted by permission from Serfling *et al.*, 1969); (b and c) *Cryptochironomus* gr. *defectus* (arrows) (from Miseiko and Popova, 1970a,b); (d–f) heterozygosity for the 14C2 puff in *C. tentans*, (d) the chromosomes under phase contrast, (e) incorporation of [^{3}H]uridine, (f) localization of antibodies against RNA polymerase II (reprinted by permission from Sass and Bautz, 1982b); (g and h) the heterozygous puffs in *Chironomus* sp. Bar = 10 μm. (from Hsu and Liu, 1948).

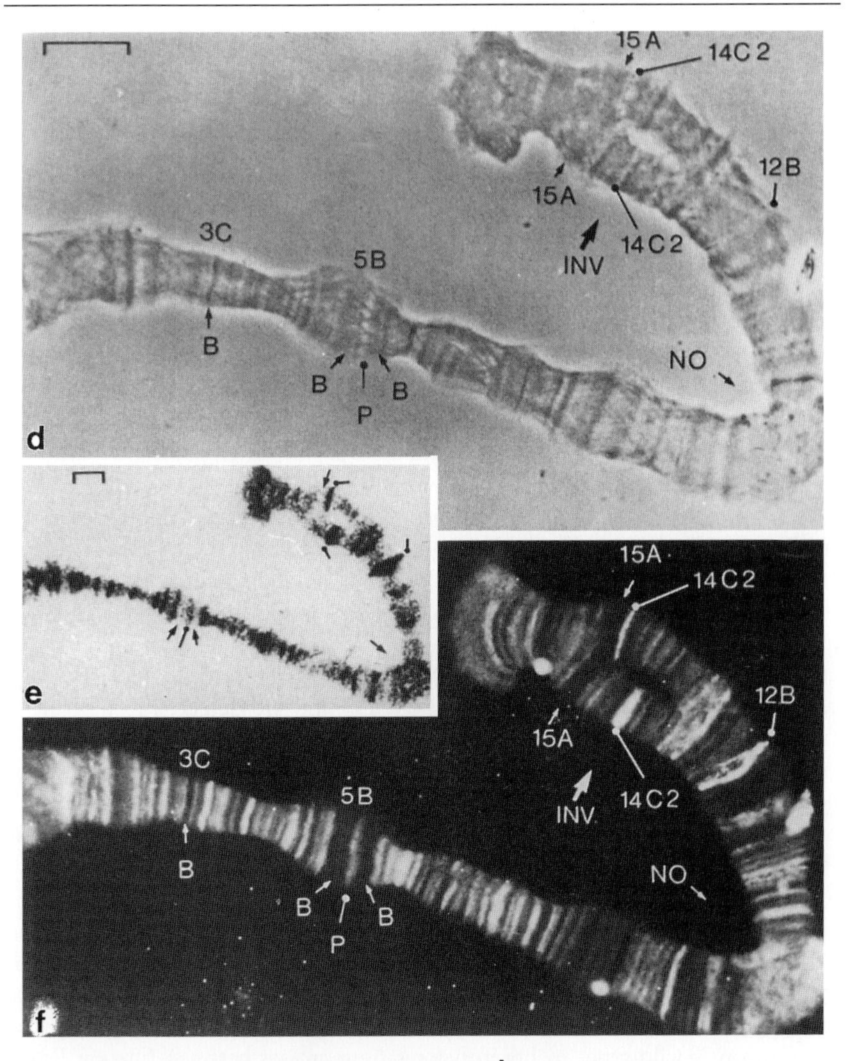

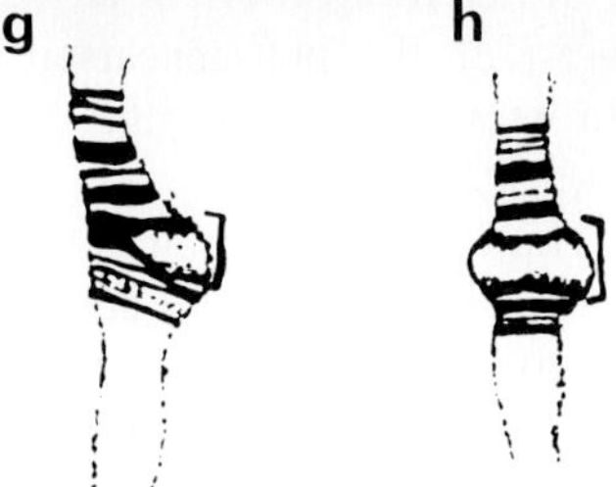

Figure 101. *Continued*

is heterozygous for size (Beermann, 1952a; Figure 84: puff 6). It was found that homologous chromosomes differ according to their sizes in the rate of [^{3}H]uridine incorporation (Pelling, 1964; Sass and Bautz, 1982b), staining with azure (Pelling, 1964), and binding of antibodies against RNA polymerase II (Figure 101). Shcherbakov (1966b, 1968) associated the presence of heterozygous puffs with mutations, in particular with the "puff organizer" which determines the start of despiralization and its pattern in the chromonemes. Panitz (1965) believed that local heterochromatization of a locus, corresponding to the puff of one of the homologous chromosomes, underlies the appearance of puff heterozygosity. Of particular interest was Ashburner's findings (1967a, 1969a,d; 1970b, 1974b) of heterozygous puffs, whose expression depends on the synaptic condition of the homologous chromosomes. The 63E1-3 puff is very large in zero-hour prepupae of *D. melanogaster* (Figure 100); it is small in *D. simulans*. In hybrids, the puff is large if homologs pair; when pairing is disrupted, the puffs are of the size characteristic of parental species.

All the heterozygous puffs can be assigned to three major classes according to their behavior in heterozygous condition (Ashburner, 1970b,d, 1974b; Korochkin and Belyaeva, 1972):

(1) Class I Mutations causing the formation of the heteromorphic puff in heterozygotes irrespective of the synaptic state of both homologous chromosomes. Such mutations frequently occur in representatives of the dipteran order in natural populations (Table 12) and are detected as a result of crosses between related species of the *Drosophila* genus: the 46A puff in the *D. melanogaster–D. simulans* cross (Ashburner, 1969a, 1970b), the puffs in hybrids between *Drosophila* species of the *willistoni* group (Cordeiro, cited after Ashburner, 1974b), and presumably the 3C4 puff in the *D. virilis–D. texana* hybrid (Poluektova, 1969, 1970a,b).

(2) Class II These are mutations causing puffing of two homologous bands in homozygotes irrespective of the synaptic state of both homologous chromosomes. A good example is the 22B4-5 puff identified in many genetic strains and active at well-defined developmental stages in *D. melanogaster* (Ashburner, 1972a). However, it does not occur in the inbred Cambridge stock of the Oregon-R wild-type. In heterozygotes from crosses between Oregon-R and the strain with the 22B4-5 puff, this puff appears at the appropriate time and on both homologous chromosomes irrespective of their synaptic condition (Ashburner, 1969a,c; 1970b, 1974b).

(3) Class III These are from mutations causing puffing of both homologous chromosomes, when the parental chromosomes are synapsed, and puffing of one homolog in the case of asynapsis. Such mutations were detected in two regions of the chromosomes of *D. melanogaster*, 22B8-9 on the second chromosome and 64C on the third. The homomorphic puff forms at both loci in interstrain heterozygotes when the corresponding chromosome regions are synapsed. The puff forms only on the homologous chromosome from the puff active parent when the homologs are asynapsed (Ashburner, 1967a, 1969d, 1970b, 1974b). Three small puffs at the end of the third chromosome of *D. pseudoobscura* are presumably also of class III. Wild-type larvae have these puffs; *sal* mutants do not. In heterozygous condition, *sal*/+, the puffs are active on the homolog from the normal strain when this asynapsis does not occur. These three puffs are inactivated in *sal* as result of a single mutation (Levine and Van Valen, 1962; Gersh, 1972a,b). The genetic site of mutations affecting the activity of the 22B8–9 and 64C puffs and the bands from which the puffs derive are very close together if not coincident (Ashburner, 1974b). This provides evidence indicating that, in at least two instances, the DNA fragments controlling puff activity are located in the puff or in the band itself. Experimental removal of the adjacent chromosome region by inversions or translocations from the puffs did not result in puff inactivation (Ellgaard and Brosseau, 1969; Ashburner, 1970b).

(4.) Class IV Such mutations can also be distinguished. The 2B13-17 puff in the mutant *halfway* of *D. melanogaster* can be assigned to this tentative class on the basis of asynchronous behavior (Rayle, 1967), and the puffs active on both homologous chromosomes and at different times in *S. ocellaris* can also be referred to this class on the basis of their polymorphism (Perondini and Dessen, 1969).

Ashburner (1974b) offered two explanations for mutations of class I. They may either result from deletions of the structural gene itself or be mutations of a region, which is not a part of the structural gene in the sense that it is a functional rearrangement for the structural gene, although not transcribed. For example, it may be the factor which controls puffing in the region.

Mutations of classes II and III presumably are not mutations of the structural genes located on the puffs: the genes are transcribed, and this results in puff formation. It may be conceded that the mutation has occurred at a single locus controlling puff activity (class II). To explain the behavior of mutant loci of class III (Figure 102), it was assumed that the regulating element of gene (R) and the structural part (S) are contiguous. This is a good reason why puff formation on homologous chromosomes is synapsis conditioned. The explanation also implies that the transfer of the signal from the regulatory to the S structural gene is chromosome limited, and it cannot be transferred through the nuclear sap (Ashburner, 1970d, 1974b).

4. Factors affecting puffing activity

The puffing of polytene chromosomes can be influenced by (1) injection of the tested substance into larvae; (2) culturing salivary glands *in vitro* in an medium

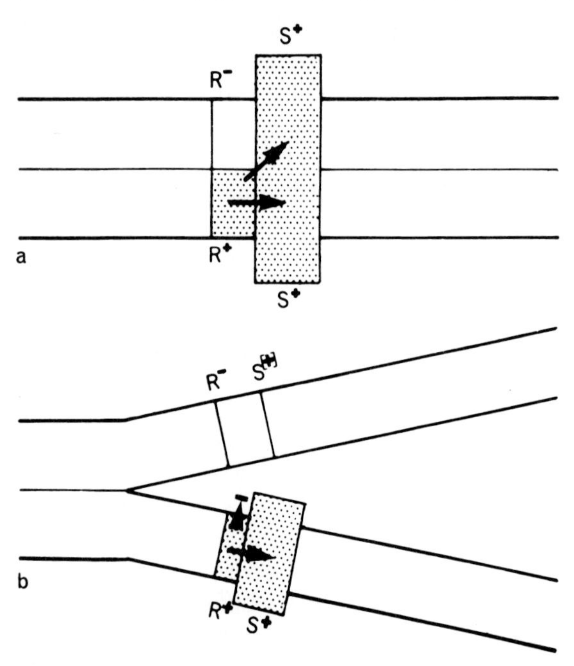

Figure 102. Scheme illustrating synapsis-dependent heterozygosity of puffs (from Ashburner, 1970b,d; 1974b). (a) Homologs synapse and information transmitted from the intact regulator (R) region activates alleles of the structural gene (S) on both homologs; (b) homologs are asynapsed. R, the locus is efficient for only the cis-S-locus; S, locus in the other homolog remains inactive.

containing the substance; or (3) addition of the substance to the medium in which the larvae are cultured.

The hormone ecdysterone and heat shock were shown to have very strong effects on puffing activity. Their effects have been described in detail in Sections VI,D and VI,G. The data on the effects of other factors are summarized in Table 13.

Because there is a continual dramatic change in puffing activity in the salivary glands during development and at any of its stage, it is rather difficult to distinguish the puffs normally seen in development from the experimentally induced.

According to Panitz (1972), all experimentally induced changes in puffing can be reduced to: (1) alteration in puff size; (2) induction of puffs normally never seen during development or at any of its stages; or (3) inactivation of the puffs. Distinctions are very difficult to make whether there is an

Table 13. Effect of Various Factors on Puff Activities

Acting factor	Type of action	Species	Result of action
Actinomycin D	Incubation of larvae	Chironomusthummi	"Puffs" formed in pericentric heterochromatic blocks. No RNA synthesis was found in these puffs (Kiknadze, 1965a–c, 1981a,b; Valeyeva *et al.*, 1979, 1984, 1997; Kiknadze and Valeyeva, 1980, 1983, 1991; Perov and Kiknadze, 1980; Kiknadze *et al.*, 1983a; see Zhimulev, 1997 for review)
Alcohols: ethanol, glycerol	Maintenance of larvae in solution	*C. thummi*	Effect is much weaker than of galactose treatment: BR1 are larger than BR2 in 30–35% of cells after 2 days treatment. Methanol and isopropanol are not efficient at all (Cortes *et al.*, 1990)
Aluminum lactate	Incubation of salivary gland for 2 hr (13.2 μg/ml)	*Simulium vittatum*	Of 9 ecdysterone-induced puffs, puffing completely inhibited in 7 (Sanderson *et al.*, 1982)
α-amino acids (histidine, methionine, tryptophan, D-tryptophan, tyrosine)	Incubation of salivary glands (0.0022–0.06 M)	*D. melanogaster*	A new 68D puff appeared under the effect of L- and D-tryptophan detected, the 50F puff is induced under the effect of L-methione (Fedoroff and Milkman, 1964, 1965). This conclusion not confirmed by Ashburner (1970a)
Amino acids (all but tyrosine)	Incubation of salivary glands (0.08 M)	*Chironomus thummi*	D- and L-tryptophan induce 7 new puffs, the same puffs of the smaller size arise under the effect of L-phenylalanine. The remaining amino acids are not active in puff induction (Gopalan, 1973b)
Anasthetic agents: diethyl ether, chloroform	Treatment of larvae with an air flow containing anaesthetic agent: 2-5 min: ether; several seconds: chloroform	*Trichosia pubescens*	In salivary gland cells, the 15E large puff induced, reaching maximum size in 60–100 min and inactivated in 200 min. Several smaller puffs also arise. Puffs not induced in midgut and Malpighian tubule cells (Amabis and Janczur, 1978)
Aneuploidy		*D. melanogaster*	Large X-chromosomal aneuploidy may cause an abnormal puffing pattern of the male X chromosome (Chatterjee, 1992)

continues

Table 13. *Continued*

Acting factor	Type of action	Species	Result of action
Aurantine (a mixture of actinomycins C1, C2, and D)	Additive to larvae incubation medium (0.01–50 μg/ml)	*C. dorsalis* (correct: *C. thummi*)	Puffs induced in centromeric regions (see "Actinomycin" above)
Benzamide	Incubation of salivary glands (1.3 mg/ml)	*D. melanogaster*	The 93D puff sharply increases in size, RNA synthesis inhibited in the other regions (Lakhotia and Mukherjee, 1970, 1980).
Berberine sulphate	Incubation of larvae in a solution (10–25 μg/ml, for several hours)	*Chironomus tentans*, *C. pallidivittatus*	Balbiani rings reduced or completely disappeared (Stocker, 1985, 1990)
5-Bromdeoxyuridine	Addition to food (10^{-2}–10^{-3} M)	*D. virilis*	Activity changes of 16 puffs, 6 of which never observed during normal development (Poluektova *et al.*, 1975, 1976)
			The stage-specific puffs mainly change (Mitrofanov and Poluektova, 1977; Poluektova and Mitrofanov, 1977)
Cadmium chloride	Addition to food (0.02–50 mg/liter)	*D. melanogaster*	No abnormalities in puffing observed (Sorsa and Pfeifer, 1973b)
[^{3}H]Carcinogenic hydrocarbons	—	*D. hydei*	Incorporation into polytene chromosomes (Emmerich and Schmialek, 1966)
Chromium chloride	Addition to food (2.5×10^{-5} M)	*D. melanogaster*	Four new puffs appeared in third-instar larvae (Nikiforova *et al.*, 1970) (see text for comments)
cAMP	Treatment of salivary glands (10–500 μg/ml)	*D. melanogaster*	[^{3}H]Uridine incorporation and puffs size studied after pretreatment with cAMP (Chatterjee and Mukherjee, 1992). Drawback of the study: larvae age was not determined, absolute grain number compared in different preparations
	Treatment of salivary glands	*D. melanogaster*	Decrease of activity of the 2B puff and increase of activity of 3C (Rensing and Hardeland, 1972)
		D. melanogaster, *D. hydei*	No effect detected (Leenders *et al.*, 1970; M. Ashburner, in: Ashburner and Berendes, 1978)

	Incubation of salivary glands in solution (10^{-2}–10^{-6} M, 1 hr)	*D. melanogaster*	Induction of the 3CD, 32CD, 34A, 47A, 61B, 61CD, 88F, 98F, 100C, 100E puffs (Felten, 1978)
Cobalt chloride	Addition to food (1.02×10^{-3} M)	*D. melanogaster*	Four new puffs detected in third-instar larvae (Rapoport *et al.*, 1971b) (see text for comments)
Colchicine	Treatment of larvae with solution (0.01–1.0%)	*D. virilis*	The 2Aj-p giant puff (the homologous 93D in *D. melanogaster)* induced after 24–48 hr of treatment (Gubenko and Baricheva, 1979; Gubenko, 1989)
	Incubation of salivary glands	*D. melanogaster*	The 93D puff specifically induced. Incorporation of [^{3}H]uridine into the other chromosome regions inhibited (Lakhotia and Mukherjee, 1984)
			The 93D puff induced. Other microtubulin poisons (cold shock, chloral hydrate, diamid, podophyllotoxin, vinblastin, griseofulvin, nocodazole) do not induce this puff (Singh and Lakhotia, 1984)
	Additive to food		13% of puffs "activated" and 7% "inhibited" in Oh prepupae. According to other data, 18.54% "activated" in females and 14.50% in males, 16.93% and 9.16% "inactivated," respectively. The 93D puff decreases in females, it increases in males (Drozdovskaya, 1988a–c)
Cycloheximide	Treatment of larvae (10 μg/ml, 15–20 hr)	*Chironomus tentans*	Induction of several puffs (Clever, 1967)
	Treatment of larvae	*C. thummi, C. strenzkei*	No puffs induced (Serfling *et al.*, 1969)
	Injection into larvae	*D. melanogaster*	Induction of new puffs (Ashburner, 1972a)
	Incubation of salivary glands (100 μg/ml, 3–4 hr)		The 63E, 47B, and the 50CD puffs induced (Schoon and Rensing, 1973)
	Incubation of larvae (20 μg/ml)	*C. thummi*	Puffs formed in pericentromeric regions of three large chromosomes (Valeyeva, 1995)

continues

Table 13. *Continued*

Acting factor	Type of action	Species	Result of action
Cycloheximide (with or without emetin)	Incubation of larvae (0.29 μM cycloheximide and 1.8 μM emetin, 5 days)	C. *tentans*	Differential inactivation of secretion protein synthesis and regression of Balbiani rings 1 and 2 (Almeida and Edstrom, 1983).
Cytoplasm of the eggs of *D. melanogaster*	Incubation of cell nuclei	*D. busckii*	No changes found in 3 of the 7 regions studied, three puffs disappeared, one newly appeared (Kroeger, 1960)
Dicyandiamide	Incubation of salivary glands (0.1 M)	*D. melanogaster*	Differences between control and experiment not large (Mukherjee, 1968)
Digitonin	Incubation of salivary glands in solution (0.01%, 20 min)	*D. melanogaster*	2 hr after treatment with digitonin, activities of 35 puffs change: the previous puffs, activated (6), new puffs (29). The new puffs include all the heat shock puffs (Myohara and Okada, 1988a–c)
Dimethylsulfoxide (DMSO)	Treatment of larvae with solution (10%, 2–3 hr 18°C)	*Chironomus tentans*	BR3 is sharply induced as the result of treatment. After removal of DMSO, all the puffs and BR rapidly collapse and the banding pattern is restored; BR is newly activated and puffs are induced, giant new puffs appear, which were not previously observed. Salivary glands do not respond *in vitro* to DMSO. Similar effect of DMSO also observed for the following species C. *pallidivittatus*, C. *thummi*, C. *melanotus*, C. *plumosus*, C. *annularius*, *Glyptotendipes barbipes*, *Prodiamesa olivacea* (Sass, 1980a, 1981)
Ethanol	Treatment of larvae (100 mM)	C. *pallidivittatus*	Balbiani ring 6 induced (Edstrom *et al.*, 1982)
		C. *thummi*	Moderate inducibility of BR1 and repressivity of BR2 (Cortes *et al.*, 1990)
Extracts from freshly laid eggs and first instar larvae	Incubation of salivary glands 1–2 hr	*D. melanogaster*	Of 7 glands in 1 gland 60B, 63C and 73B new puffs detected (Tokimitsu, 1968)

Galactose		C. *thummi*	BR2 repressed and BR1 activated. When larvae are first treated with galactose, then with inhibitors of protein synthesis, reactivation of BR2 occurs (Santa-Cruz *et al.*, 1978; Diez *et al.*, 1980, 1990; Cortes *et al.*, 1990)
			Structures resembling Balbiani rings arise in one of the strains, in the first and second chromosomes, possibly, homologous BR6 in C. *tentans* (Santa-Cruz and Diez, 1979)
Galactose + glycerol		C. *thummi*	Inducive effect of galactose considerably enhanced (Cortes *et al.*, 1990)
Gibberellic acid	Incubation of larvae in a solution (0.5–1.4 μg/ml)	*Acricotopus lucidus*	BR1, BR2, and BR6 regressed in the central and lateral lobes. The sizes of BR3 and BR4 in the anterior lobe did not change. Upon transfer to fresh medium BR activity restored (Panitz, 1967; Baudisch and Panitz, 1968)
	Injection into larvae		BR2 rapidly regreses (Panitz, 1972)
	Addition to food	*Sciara coprophila*	No changes in puffing pattern (Rash and Lewis, 1968)
Gibberellic acid-3	Injection of solution at a concentration of 3 μg per individual	*Drosophila hydei*	In 10h the 4-72B puff and the 2-21B appear, 78B and 77BC puffs decrease in size in 15–20% of larvae (Alonso, 1971)
Glycerol	Treatment of larvae (50 mM, 6 days)	C. *pallidivittatus*	Balbiani ring 6 induced (Edstrom *et al.*, 1982)
Glucosamine	Injection to larvae	D. *virilis*	The 55E puff inhibited in 5h (Kress, 1973)
Glutamic amino acid	Addition into food	D. *melanogaster*	In zero-hour prepupae, of 98 studied puffs, 20 were larger and 18 smaller compared to control (Dennhofer, 1978)
Heat-shock-producing factors	—	—	See Section VI,G
Hormones	—	—	See Section VI,D
Hydroquinone	Treatment of larvae with a solution (0.1–1.0 mg/l)	C. *thummi*	"New puffs induced" (Simakov *et al.*, 1990)
Hydroxyurea	Injection	*Bradysia hygida*	Inhibition of all the DNA puffs (Sauaia *et al.*, 1971)

continues

Table 13. *Continued*

Acting factor	Type of action	Species	Result of action
		Rhynchosciara angelae	Incorporation of [^{3}H]thymidine into DNA puffs inhibited, no effect on RNA puffs (Machado-Santelli and Basile, 1978)
Inbreeding	36–41 genetions of inbreeding (brother × sister),	*D. melanogaster*	Occurrence frequency of 28 puffs decreased in the third instar larvae (Lychev, 1965)
	100 generations of inbreeding according to the brother × sister scheme		No deviations from control in 66 regions of the X chromosomes of Oh prepupae (Belyaeva and Zhimulev, 1973, 1974)
—	288 generations of inbreeding	*D. subobscura*	138 puffs in 0–2.5 hr normal prepupae and 96 puffs in inbred prepupae are found (Frutos *et al.*, 1984a)
Increased temperature (29–30°C)	Prolonged effect on development of larvae	*D. melanogaster*	In zero-hour prepupae, "activity" of some puffs increased, decreased of other puffs (Drozdovskaya, 1975)
Incubation of salivary glands *in vitro*	Incubation in saline	*D. melanogaster*	The 67C puff induced after 90 min of incubation (Becker, 1959)
		C. tentans	Changes in puffing detected after 2 hr incubation of salivary glands in some media (Clever, 1965a,b)
	Incubation in Poels' medium	*D. hydei*	Appearance of the 95D and 97A puffs (Poels, 1972)
	Incubation in synthetic media	*Rhynchosciara angelae*	A giant puff in heterochromatin of the X chromosome induced in contaminated cultures. In "clean" cultures this puff arises after 15–20 days of incubation (Simoes and Cestari, 1969)
	Salt solution	*D. melanogaster*	The 50F, 57D, 84E puffs induced, rarer the 4C, 5C, 26A, 76D (Ashburner, 1970c)
			Appearance of the 63BC, 67B, 87A, 87C, and 93D heat-shock puffs (Nagel and Rensing, 1971)
	Artificial mediums		After incubation of prepupae salivary glands in three artificial media, no puff deviations compare to the *in vivo* system. Differences observed after the gland incubation of PS8-11 larvae (Dennhofer and Muller, 1978)

Grace medium		Up to 20 puffs formed after 12 hr incubation of salivary glands in Grace medium. (5B, 18C, 26A, 29CD, 30EF, 47AD, 50CD, 57D, 62C, 66BC, 73A, 78E, 79E, 84E, 85B, 91D, 94C, 94E, 95AB, 99B) (Ashburner, 1972b) Puffs appear according to a definite timetable. Appearance of the 57D puff is inhibited with cyclogeximide (Ashburner, 1972b, 1974a)
Grace medium	*D. auraria*	10 puffs induced in fat body cells and 9 in midgut cells (Mavragani-Tsipidou and Scouras, 1991)
C50 medium	*D. virilis*	6 puffs induced (Poluektova *et al.*, 1980a)
		Presence of yeast extract in medium blocks development of the 5B1 and 34A puffs (Poluektova *et al.*, 1985)
Incubation of salivary glands in medium C46P (6 hr)	*D. melanogaster*	During 6 hr of incubation 10 puffs induced, 4 of which (18D, 47B, 66B and 73A) induced also in Grace medium (see Ashburner, 1972b)
		The sets of environmental puffs are identical in *dor* and *gt* strains (Biyasheva *et al.*, 1985)
	D. virilis	Induction or repression of some puffs in certain strains. Puffing in most regions is blocked during incubation of glands in C50K medium adapted by embryonic cells (Poluektova *et al.* 1985)
	D. hydei	Of 69 puffs 12 did not change in 7 days after implantation, the others 27 varied to some degree (Berendes and Holt, 1965)
Robb's medium	*D. melanogaster*	After 24 hr incubation of salivary glands and nurse cells 47D, 50C, puffs appear (Zhimulev and Mal'ceva, 1997)
		New 60D puff appears in salivary glands.
		After 6 hr incubation without ecdysterone and 6 hr with ecdysterone, 46F and 47D puffs are induced in salivary glands only (Zhimulev and Mal'ceva, 1997)

continues

Table 13. *Continued*

Acting factor	Type of action	Species	Result of action
Incubation of salivary glands in imago abdomen		*D. melanogaster*	Formation of the 78E puff (Staub, 1969; Ashburner and Garcia-Bellido, 1973)
			The 68B, 79C, and the 63BC heat-shock puffs irregularly formed (Staub, 1969)
			After 2 days incubation of the third instar larvae salivary glands, the 4F1-3, 25D, 39B, 47D, 47A, 50C5-11, 50D, 60E1, 71C, 72C, 78E, 82D, 82F, 88EF, 93D, 101F puffs identified (I. F. Zhimulev and E. S. Belyaeva, 1972, unpublished data)
		D. virilis	Changes in puffing observed in 35 regions. Of these, only 2 puffs newly appear. Appearance of puffs depending on female age detected in 8 regions. When transplanted to male abdomens changes of 26 puffs observed (Poluektova *et al.*, 1978a)
Irradiation with gamma rays	900 R at a dose of 5250 R/min	*D. melanogaster*	5–6 hr embryos irradiated. Puffs of chromosome 2R analysed in third-instar larvae (15 irradiated, 33 control). No differences in occurrence frequency of puffs were identified (Lychev, 1968)
	9cgR	*Chironomus thummi*	In F2-F4 larvae some new puffs appeared (Gunderina and Aimanova, 1998)
K, Mg ions	50 mM KCl/50 mM $MgCl_2$	*C. thummi*	Puffs induced in 7 regions (Gopalan, 1973a)
Lead nitrate	Incubation of salivary gland (1–3 hr)	*Chironomus* sp.	Inactivation of several Balbiani rings, induction of several puffs (Rathore and Swarup, 1982a)
Low temperature (+2°C)	Prolonged effect on development of larvae (more than 1 month)	*C. plumosus*	The IB16 puff formed, which is also induced by treatment of larvae with high temperature (60°C) Maksimova and Il'inskaya, 1983)
Lysine-rich histone of the thymus	Injection	*C. thummi*	No changes in puffs and BRs, incorporation of [^{3}H]uridine inhibited (Desai and Tencer, 1968)

		D. melanogaster	Combined injection with ecdysterone prevents the appearance of the hormone-induced puffs (Desai and Tencer, 1968)
Magnetic field	Treatment of salivary gland	*Sciara coprophila*	Increase of [^{3}H]uridine incorporation into polytene chromosomes (Goodman *et al.*, 1983, 1987a)
Menthol	Incubation of larvae in a saturated aqueous solution	*C. plumosus*	As a result of treatment of larvae to produce narcosis, a set of large puffs appears in all the studied larvae (Il'inskaya and Maksimova, 1983)
Methoxyethyl mercuric acetate, methooxyethylmercuric chloride, mercuric chloride, methylmercuric chloride	Addition to food (0.02–10 mg/l)	*D. melanogaster*	In zero-hour prepupae, larval puffs do not regress (3B, 74EF, 75B, 78D). Some puffs missing (3A, 4A, 28A, 46F, 47A, 47F, 61C, 62C, 73B) (Sorsa and Pfeifer, 1973a)
Microsporidian infection		*Rhynchosciara angelae*	Induction of a giant puff in the region of centromeric heterochromatin of the X chromosome (Diaz and Pavan, 1965)
Neomycin sulfate	Incubation of salivary gland	*Chironomus* sp.	Induction of new puffs (Rathore and Swarup, 1980)
Oxytetracycline	Addition to incubation medium of larvae (2 mg/ml, 20 hr)	*Chironomus thummi*	The first puffs appear in 2 hr of treatment. In all, more than 30 new puffs induced. The nucleoli and BRs are simultaneously reduced. (Serfling *et al.*, 1969)
		C. strenzkei	No puffs formed (Serfling *et al.*, 1969)
Pactamycine	Treatment of larvae	*Chironomus thummi*	Puff induction (Clever, 1967)
	Incubation of salivary glands (100 μg/ml, 3–4 hr)	*D. melanogaster*	The 63E puff induced (Schoon and Rensing, 1973)
Paraamino benzoic acid	Addition to food (0.005%)	*D. melanogaster*	"Activation" of 22% of puffs (Drozdovskaya, 1989), according to others (Drozdovskaya and Rapoport, 1991), the activities of maximally 29.6% of puffs change, 18.51% are activated, 11.11% are inhibited
Pilocarpine	Maintenance of larvae in solution	*Chironomus*	Treatment with pilocarpine produces release of secretion from the gland (Beermann, 1961, 1973; Clever, 1969; Valeyeva, 1975; Firling and Kobilka, 1979; Mahr *et al.*, 1980; Meyer *et al.*, 1980, 1983; Olins *et al.*, 1980; Stockert, 1990)

continues

Table 13. *Continued*

Acting factor	Type of action	Species	Result of action
			During restoration of secretory activity, the size of two puffs increases, as well as that of 4BR1 and 4BR2 in C. *thummi* (Valeyeva, 1975)
			In C. *tentans* BR1 and BR2 are sharply activated. [^{3}H]uridine incorporation increases 15-fold (Mahr *et al.*, 1980; Meyer *et al.*, 1980, 1983)
Proflavin			"Puffs" induced in centromeric heterochromatin (E. Serfling, cited after: Ashburner, 1970a)
Puromycin	Injection into larvae	C. *tentans*	Induction of several puffs (Clever, 1967)
	Treatment of larvae	C. *thummi*, C. *strenzkei*	No puff induction observed (Serfling *et al.*, 1969)
	Injection into larvae	D. *melanogaster*	Induction of several new puffs (Ashburner, 1972a)
Raus sarcoma *virus*	Addition to food	D. *melanogaster*	No reliable differences in puffing pattern (Burdette and Yoon, 1967)
			Decreased size of 5 studied puffs (2B, 71, 72, 74, 75) (Burdette, 1969a)
Ribonuclease	Incubation of salivary glands (3–5 mg/ml)	D. *busckii*	The nucleolus disappears after 1 hr incubation. 90% of nuclei have approximately 40 induced puffs after 7 hr incubation (Ritossa and Borstel, 1964; Ritossa *et al.*, 1964, 1965)
		D. *melanogaster*	Above data not confirmed (Paul, J. S., 1965, cited after: Ashburner, 1970a, p. 48)
	Incubation of salivary glands (0.1-5 mg/ml)	*Tendipes plumosus*	No puff induction detected (Prostakova, 1968)
	Incubation of salivary glands (0.5%)	D. *virilis*	Induction of new puffs not described (Smith, 1969)
	Incubation of salivary glands (5–20 mg/ml)	D. *melanogaster*	No new puffs detected after 4–7 hr of incubation (Zhimulev, 1970; not published)

Ribonucleotide absence in nutritive medium	Maintenance of larvae on medium	*D. melanogaster*	In the control group containing abundant yeast, the 48F puff induced, on an RNA-deficient medium the 50E, 54A, 55D puffs induced (Burnet and Hartmann-Goldstein, 1971)
Sacharose, tregalose		*C. thummi*	The same effect as exerted by galactose (Cortes *et al.*, 1990)
Selection for acceleration (475 generations) and deceleration (330 generations) of development		*D. melanogaster*	In zero-hour prepupae of the strain selected for deceleration development (fly emergence on day 13) many puffs have smaller compared to control (Loreto *et al.*, 1988)
Sodium rodanid	Addition to food (2×10^{-4} M)	*D. melanogaster*	24 new puffs detected in the third-instar larvae (Sakharova *et al.*, 1971) (see text for comments)
Sodium arsenate		*D. melanogaster*	In zero-hour prepupae 20 new puffs appear (Drozdovskaya and Rapoport, 1974) (see text for comments)
Sodium perchlorate	Additive to food (2.5×10^{-5} M)	*D. melanogaster*	New puffs appear in third-instar larvae (Rapoport *et al.*, 1970) (see text for comments)
Sodium tetraborate	Addition to food (17.5×10^{-4} M)	*D. melanogaster*	Numerous changes in puffs of larvae (Rapoport *et al.*, 1971a) (see text for comments), 21 new puff (Drozdovskaya, 1971). In zero-hour prepupae, 14 new puff formed (Drozdovskaya, 1974) The listed puffs are known for normal development
Solutions with increased salt concentration			See Section VI,E,3
Sugars D(+)maltose D(+)lactose D(+)glucose D(+)galactose D(−)fructose D(−)mannose D(+)sacharose	Incubation of larvae in solution (1%)	*C. pallidivittatus*	After 48 hr incubation with maltose, glucose, and galactose, BR1 becomes larger than BR2 (see Figure 154); BR6 arises. The increase of BR4 is characteristic for "special cells" (Beermann, 1973; Botella and Edstrom, 1991)

continues

Table 13. *Continued*

Acting factor	Type of action	Species	Result of action
		C. *tentans*	BR2 decreases and BR1 increases, BR6 does not arise (Beermann, 1973)
			Under the effect of galactose the number of 75S RNA molecules decreases in BR2 and increases in BR1 (Nelson and Daneholt, 1981)
Thioacetamide	Addition to silt (0.0005–1.0%)	*Chironomus thummi*	Inactivation in 8 chromosome regions, sharp increase in activity in the other regions, appearance of some Balbiani rings (Kiknadze and Filatova, 1963)
Thiroxine	Incubation of salivary glands (0.0006 mg/ml, 60 min)	*Chironomus* sp.	Induction of new puffs in the third chromosome and regression of puffs in chromosomes 1, 2, and 4 (Rathore and Swarup, 1982b)
Triphenylstibine $Sb(C_6H_5)_3$	Addition to food	D. *melanogaster*	Both "activation" and "depression" of puffs observed in all the chromosomes of "early prepupae" (Drozdovskaya *et al.*, 1977c)
Trypsin	Injection	C. *thummi*	New puffs do not form, but incorporation of [^{3}H]uridine increases twofold (Desai and Tencer, 1968)
Urea		D. *melanogaster*	In zero-hour prepupae 6–52% of puffs stimulated; 0–20%, repressed (Drozdovskaya *et al.*, 1977b)
Urethan (ethyl carbamate)		D. *melanogaster*	In prepupae decrease of activity or disappearance of 13–60% of puffs in different chromosomes (Drozdovskaya, 1977a)

inducive effect on puffing patterns. Complications arise when estimates are made on the basis of the frequency of the occurrence of the puffs, especially the small ones.

Another reason for complications in estimating experimentally induced puffing is asynchronous development of larvae. Larvae at the end of the third *(Drosophila)* or fourth (*Chironomus*, Sciaridae) instar are mostly used in the experiments (i.e., several hours before the start of pupation), when the chromosomes reach their maximum sizes and are most amenable to analysis. However, the most dramatic and rapid changes in puffing occur during the last 5–6 hr of development in *Drosophila*. Thus, a factor not directly acting on the genome of salivary gland cells can somewhat decelerate or increase the developmental rate of larvae in the experimental groups. This fact must be taken into account in studies of chronic effects (addition of a substance to food) and mutant genotypes. The experimental and control samples may differ in developmental age, thereby causing statistically significant differences in the puff sets of their chromosomes. Although extremely large, the differences do not reflect the changes induced by the tested factor. Consequently, the results of experiments in which the puffs were analyzed at the end of the third-larval instar can be meaningful only when the puffs never observed during normal development are induced. Synchronously developing larvae must be compared virtually in all the experiments, irrespective of the mode of exposure, to exclude unacceptably large differences between puffing patterns in a sample. Even with the most accurate synchronization methods, the larval sample used to analyze the puffs are composed of individuals at incomparable ages (see above). Thus, a group can include both individuals, activated before puparium formation and whose puff wave has not yet started to pass and individuals whose puff wave has already passed.

Therefore, the only way to obviate this difficulty is to include individuals homogeneous with respect to chronological and developmental age in the experimental and control samples; for example, zero-hour prepupae, whose differences in age do not exceed a few minutes. The onset of this stage is readily recorded, when the anterior spiracles evert for 1–2 sec in a larva, which has crawled out of food and stuck to the glass tube. A sample of zero-hour prepupae is homogenous in terms of developmental time, and this developmental stage is convenient for comparing puffing in individuals of different experimental groups (Poluektova, 1969; Burdette and Carver, 1971; Sorsa and Pfeifer, 1972; Belyaeva and Zhimulev, 1974).

Reliable and easily reproducible experiments include culturing of salivary glands and other organs in appropriate media. In such a case, the simplest control would be the sister salivary gland, because the glands do not differ in puffing. However, to obtain a homogeneous sample of salivary glands, in this case, it is also more expedient to use zero-hour prepupae or to set up a homogeneous group of salivary glands by selection based on morphological criteria (Fig. 83).

Another method for selection of synchronously developing larvae on the basis of morphological changes in imaginal disks was developed for Chironomids. With these criteria, the entire fourth-larval instar and prepupal stage (12–13 days) were divided into nine phases, with each phase taking 1 day (review: Kiknadze *et al.*, 1975).

In all these studies, without exception (see Table 13), only the largest puffs at the end of the third-larval instar and Balbiani rings in Chironomids, which do not represent all the transcriptional activity of the chromosomes, are included in analysis. It is unclear how the small puffs and the interbands respond. There is reason to expect changes precisely in this group of active regions, not in the pattern of the stage-specific puffs, which is mainly characteristic of the already-differentiated cells. Thus viewed, the data appear incomplete.

Compounds strongly affecting the physiology of insects are used in most studies. For example, 5-bromodeoxyuridine is lethal at high concentrations and semilethal at low ones (about 20% of larvae survive) (Poluektova *et al.*, 1975b); treatment with certain organic mercurial compounds is teratogenic (Sorsa and Pfeifer, 1973a). Tetraborate induces a decrease in the size of eyes leading to their obliteration (Drozdovskaya, 1974). Sodium arsenite induces the formation of melanotic tumors (Drozdovskaya, Rapoport, 1974).

Information presented in Table 13 evidences that puff-inducing activity has been demonstrated for amino acids, benzamide, diethyl ether, chloroform, bromodeoxyuridine, intracellular infections, adult hemolymph, gibberellic acid, digitonin, colchicine, menthol, oxytetracycline, pactamycine, puromycine, sugars (galactose, saccharose, tregalose), alcohols, various incubation media and cycloheximide. Benzamide inhibits RNA synthesis in most chromosome regions, except the "special heat shock puff" at 93D; berberine sulphate, gibberellic acid (in Chironomus), oxyhydrourea are able to inactivate puffs; glucosamine inhibits the specific 55E puff in *D. virilis*.

It is difficult to unambiguously estimate the effect of many substances because the criteria for "activation" or "inactivation" of the puffs are not clear-cut. For this reason, the puffing data recorded include an increase or a decrease in the frequencies of the puffs and a change in their sizes without reference to estimates of their natural variability. Furthermore, it is often difficult to interpret the results, because photographs of the induced or repressed puffs are lacking.

There were other explanations of uncertainties arising in estimation of the results of experimental modification of puff activity (Belyaeva and Zhimulev, 1974; Ashburner and Berendes, 1978). Thus, it was reported that certain chemical compounds, morphogens, have specific effects on puffs (Rapoport *et al.*, 1970, 1971a,b; Sakharova *et al.*, 1971), and the genes responsible for puffing in response to an administered morphogen were localized cytologically. The data were ambiguous and did not allow this conclusion. The authors described specific induction—an increase in puff size by one, more rarely by two, scores; for example, in the 1C, 4C, 4F, 5C, 8B, 10E, and 11E regions—without providing data on variation in the puff sizes. It is known that puff size can differ by 1–2 scores in the control samples. Taking this variation into account, an increase by 1–2 scores in an experiment is unlikely to be significant.

The very low frequency of many puffs (20–30%) also came as a surprise. However, it is known that the frequency of puffs reaches 100% in well-synchronized samples. Larvae at different stages of the third instar were presumably included in the experimental and control series. In fact, the list of the control puffs contained those characteristic of the "old" and "young" larvae (3C, 2B, 8D, and 16DE). As a result, the difference in the age of single individuals was not less than 6 hr in the sample. Asynchrony in development of the studied individuals (without lacking estimates of puff variability) influenced these results. For example, the 21C, 61AB, 64B, 98E, and 99EF puffs referred to either the specific rhodonite-induced puffs (Sakharova *et al.*, 1971) or to the control puffs (Rapoport *et al.*, 1971a); the 25BC, 33E, 34A, 69C, and 70C puffs referred to either the specific boron-induced puffs (Rapoport *et al.*, 1971a) or to the control puffs (Rapoport *et al.*, 1970a).

D. Hormonal control of puffing

There is extensive literature concerning the hormonal control of puff activity (see reviews by Clever, 1963c; Beermann and Clever, 1964, 1966; Kroeger and Lezzi, 1966; Kroeger, 1968; Ashburner, 1970a, 1972e, 1974b, 1980, 1990; Lezzi and Gilbert, 1972; Ashburner *et al.*, 1974; Ashburner and Richards, 1976b; Richards, 1976e, 1980b, 1985, 1992, 1997; Ashburner and Berendes, 1978; Kress, 1981; Kramerov and Gvozdev, 1984; Pongs, 1984a,b, 1988; Richards and Ashburner, 1984; Richards and Lepesant, 1984; Cherbas *et al.*, 1986a; Korge, 1986; Koolman, 1989; Segraves and Richards, 1990; Thummel, 1990, 1996; Andres and Thummel, 1992; Zhimulev, 1994a; Lezzi, 1996; Russel and Ashburner, 1996).

1. The role of ecdysterone in the control of puff activity

Becker (1959) was the first to provide rigorous proof that hormones induce puffing. He demonstrated that homozygotes for the *l(2)gl* mutation, whose ecdysterone-releasing ring gland is smaller than normal, lack the metamorphosis puffs. Slight chromosome swellings were concomitantly detected at 3C11-12, 71CE, and 68C; that is, in the regions of the intermolt puffs. Becker (1959) thought that "inadequate hormonal situation" was one of the reasons why the metamorphosis puffs were absent from these homozygotes.

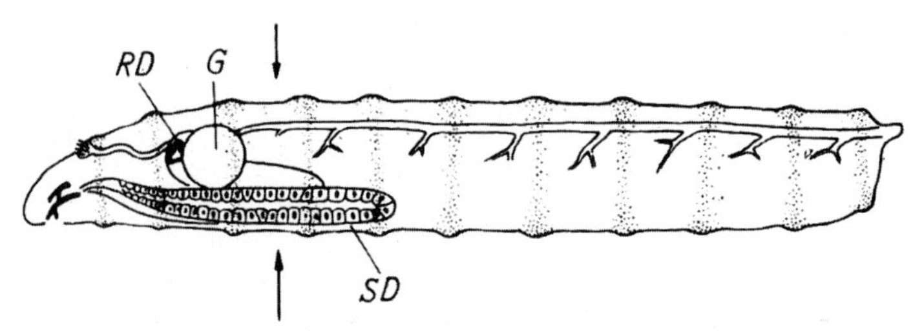

Figure 103. Scheme of ligature application (indicated by arrows) to larvae of *D. melanogaster*. RD, G, and SD: ring gland, brain, and salivary gland, respectively (reprinted by permission from Becker, 1962a).

In subsequent ligature experiments, the posterior region of the salivary gland was separated from the ring gland, the source of the hormone (Figure 103). It was found, first, that the metamorphosis hormone is released about 6 hr before puparium formation and, second, that the metamorphosis puffs are formed only from the part of the salivary gland that remained immersed in ecdysterone-containing hemolymph (Figure 104). In salivary glands from third-instar larvae transplanted into the abdomens of larvae at the end of this instar, when they already possessed the metamorphosis puffs, the puffs also developed in implanted gland cells (Becker, 1962a,b, 1965).

Injection of ecdysterone into *C. tentans* larvae demonstrated that hormone directly induces the metamorphosis puffs (Figure 105): the puff I-18C appeared, and another, I-19A, became inactivated within 2 hr. Such an alternation of puff activity is also characteristic of normal metamorphosis (Clever and Karlson, 1960).

These data suggested a new model for the effect of the vertebrate steroid hormones in which they enter the cell nucleus inducing the synthesis of new enzyme molecules (Karlson, 1963,1965, 1967, 1975; Karlson and Sekeris, 1966).

Figure 104. Formation of puffs depending on hormonal status in hemolymph (reprinted by permission from Becker, 1962a). (A) As a consequence of ligature application, puparium is formed only in the anterior half of the larva, the metamorphosis puffs: 74EF, 75B, and 78D (78C) are present only in the anterior part of salivary gland (the chromosome at the top) and are absent in the posterior, "larval" part (the chromosome at the bottom); (B) puparium has formed in both parts of the larva, and the same puffs are present in cells of both parts of the salivary gland; (C) dependence between time of ligature application and puparium formation. Abscissa, time in hours between ligature application and puparium formation. Ordinate, proportion of individuals (percentage) forming puparium only in the anterior (open circles) or both (closed circles) parts of the larva. Solid triangles and squares indicate the number of individuals in which the metamorphosis puffs were only in the anterior (triangles) or both (squares) parts of salivary glands.

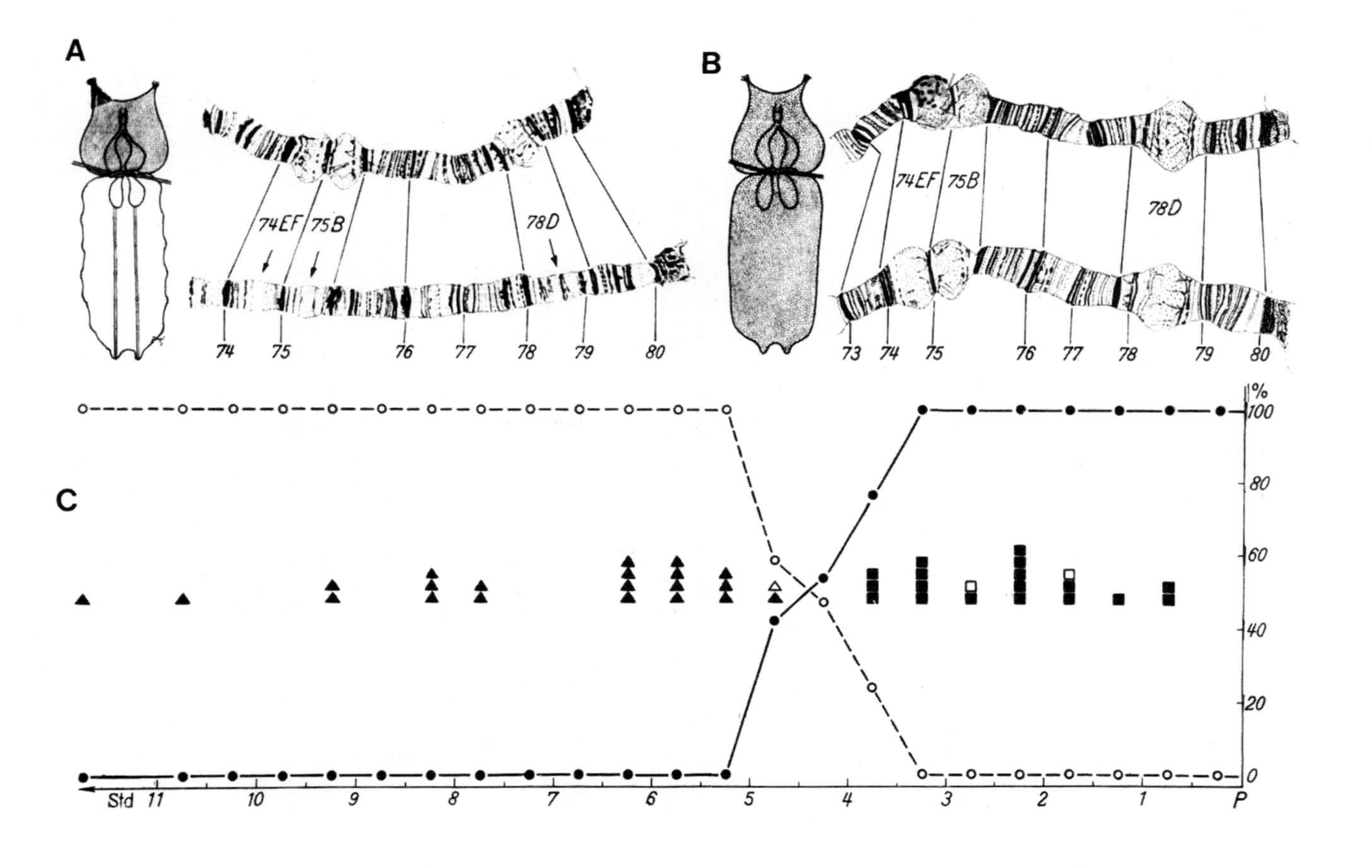
A
74EF
75B
78D
74
75
76
77
78
79
80
B
74EF
75B
78D
73
74
75
76
77
78
79
80
C
%
100
80
60
40
20
0
Std 11
10
9
8
7
6
5
4
3
2
1
P

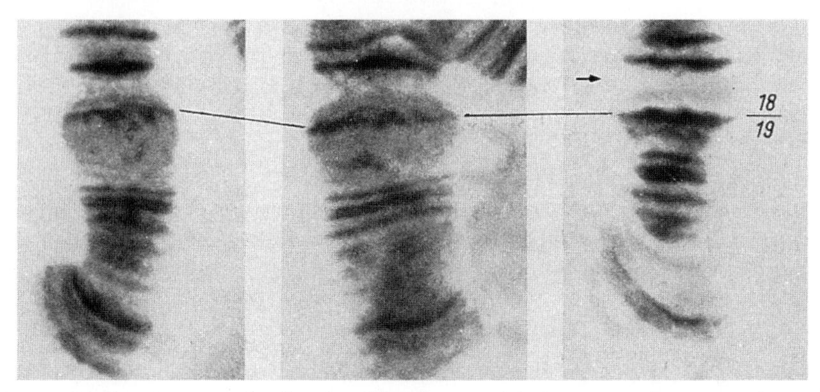

Figure 105. First depiction of puffs induced by injection of the hormone ecdysterone to larvae of *C. tentans*. The induced I-18C puff is marked with an arrow (from Clever and Karlson, 1960).

Subsequent studies by Clever (1961–1967) provided definite evidence for the dependence of changes in puffing on the presence of ecdysterone in the course of metamorphosis. Since then, much has been learned about the induction of the metamorphosis puffs in various dipteran species: in *Acricotopus lucidus* (Panitz, 1960b, 1964; Panitz *et al.*, 1972), *D. melanogaster* (Burdette, 1964, 1969a; Burdette and Anderson, 1965b; Burdette and Kobayashi, 1969), *D. hydei* (Berendes, 1967a), *Drosophila lebanonensis* (Berendes and Thijssen, 1971; Eeken, 1974), *D. pseudoobscura* (Stocker and Kastritsis, 1973), *D. virilis* (Kress, 1972; Poluektova *et al.*, 1980a), *Simulium vittatum* (Sanderson *et al.*, 1982), *D. auraria* (Mavragani-Tsipidou and Kastritsis, 1986), *D. bicornuta* (Mavragani-Tsipidou *et al.*, 1992a) in sciarids (see Section VI,F), *C. thummi* (Kroeger, 1964; Kiknadze *et al.*, 1972; Korochkina *et al.*, 1972), and in Hawaiian *Drosophila* (Raikow, 1973).

Injection of ecdysterone into mid-third-instar larvae of *D. hydei* produced considerable changes in puffing identical to those occurring during the last 6 hr of normal development: 15–20 min after the injection, 3 puffs formed anew, and 18 puffs showed an increase in activity. Five new "late" puffs appeared 4–6 hr later (Berendes, 1967a).

A medium called Grace's was found to be satisfactory for short-term culture of larval salivary glands. Ashburner (1972b, 1974b) applied this medium to treat the glands with ecdysterone to reproduce *in vitro* the puffing cascade in late third-instar larvae of *D. melanogaster*. The puffs occurred in the same sequence as during normal development.

Not only the activities of puffs, but also those of nucleoli and Balbiani rings are under the control of ecdysterone. In *Sciara ocellaris*, the size of the nucleolus and intensity of [^{3}H]uridine incorporation into ribosomal RNA gradually decreases at the end of the fourth instar. Analysis of nucleolar behavior in ligated larvae supported the following conclusion: in larvae ligated prior to the critical stage, the nucleolus continues to remain expanded, unlike in normal larva. In larvae ligated after the critical stage, the nucleolus regresses at a rate similar to that of nonligated normal larvae (Dessen and Perondini, 1973, 1976, 1985c). In *D. hydei*, rapid inhibition of rRNA synthesis occurred when salivary glands were cultured in an ecdysterone-containing solution (Hughes and Kambysellis, 1970). A normal electrophoretic RNA profile was observed in this species despite a high ecdysterone titer (Alonso, 1973a).

In *Acricotopus lucidus*, this hormone inactivates Balbiani rings 3 and 4 in a manner similar to that observed during metamorphosis. This was demonstrated by the culture of isolated salivary glands in prepupal or pupal hemolymph, by salivary gland transplantation experiments and also by studies of responses following hormone injection (Panitz, 1960b, 1964; Panitz *et al.*, 1972b).

In *C. tentans*, the injection of ecdysterone stimulates Balbiani rings 1 and 2 in young larvae (Lezzi and Gilbert, 1969), but according to other observations stimulates only Balbiani ring 1 (Meyer *et al.*, 1983; Lezzi and Richards, 1989). In *C. thummi*, the sizes of Balbiani rings 1 and 2 increase considerably after the injection of ecdysterone or hydrocortisone (Korochkina *et al.*, 1972), although the ability of hydrocortisone to induce puff activity is questionable (Ashburner, 1974b). According to other data, only Balbiani ring 2 positively correlates with an increase in ecdysterone titer in this species (Kiknadze, 1981; Lezzi and Richards, 1989).

Balbiani rings in certain species of the *montium* group *(D. auraria, Drosophila bicornuta, and Drosophila serrata)* respond quite similarly to ecdysterone: the onset of their activation is associated with darkening of the anterior spiracles of late-instar larvae (Scouras and Kastritsis, 1984; Mavragani-Tsipidou and Kastritsis, 1986; Mavragani-Tsipidou *et al.*, 1990a, 1992b).

2. Cascade of changes in puff activity under effect of ecdysterone

With respect to their patterns of response to ecdysterone, the puffs may be divided into three classes: the intermolt and the early and the late puffs. The same classes are observed during normal development (Ashburner *et al.*, 1974; Ashburner and Richards, 1976b).

The intermolt puffs are active in the middle of the last larval instar. Between molts, for example, the puffs at 25AC and 68C (Figure 106) are rapidly inactivated by hormone (see also Belyaeva *et al.*, 1981; Biyasheva *et al.*, 1981; Strachnuk *et al.*, 1991). The intermolt puffs were not detected in the fat body and midgut cells in *D. auraria* (Mavragani-Tsipidou and Scouras, 1991).

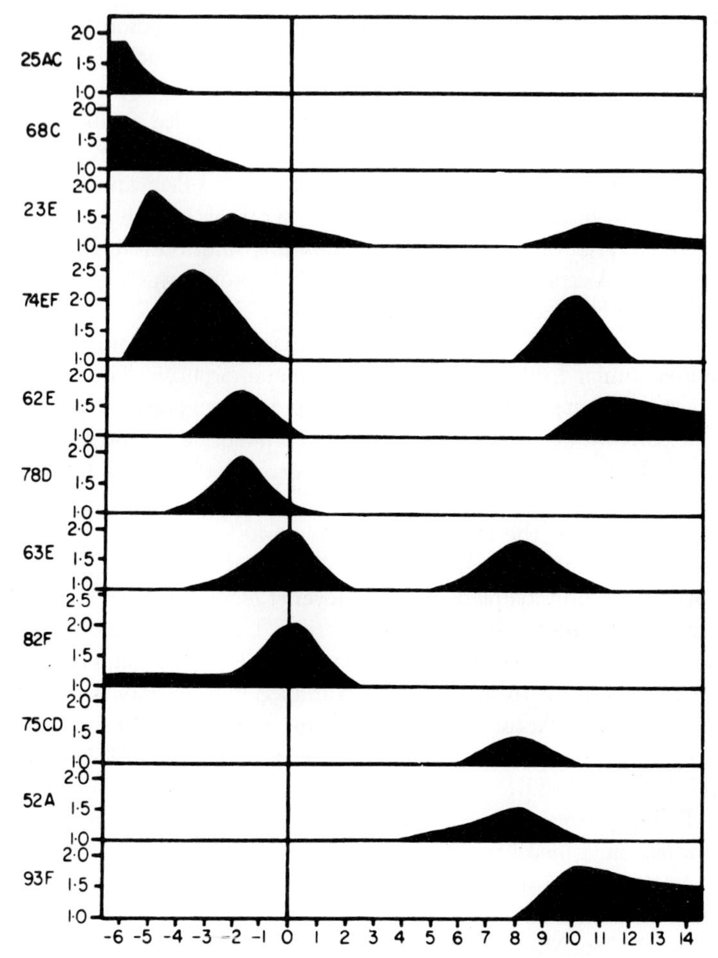

Figure 106. Changes in the activity of certain puffs during the last hours of larval and prepupal development of *D. melanogaster*. Abscissa, time in hours before (minus) and after puparium formation (0–14) (at 25°C); ordinate, relative puff size (from Ashburner and Richards, 1976b).

The early ecdysterone puffs, among the first activated at the onset of metamorphosis during development (23E and 74EF; Figure 106), are induced within 30 min by hormone in an *in vitro* system. The puffs reach their maximum sizes within 2 to 4 hr (Figure 107b). In fat body cells of *Drosophila*, the 74EF and 75B early puffs were induced almost immediately and they attained their maximum sizes within 30 min (Richards, 1982).The late ecdysterone puffs (62E,

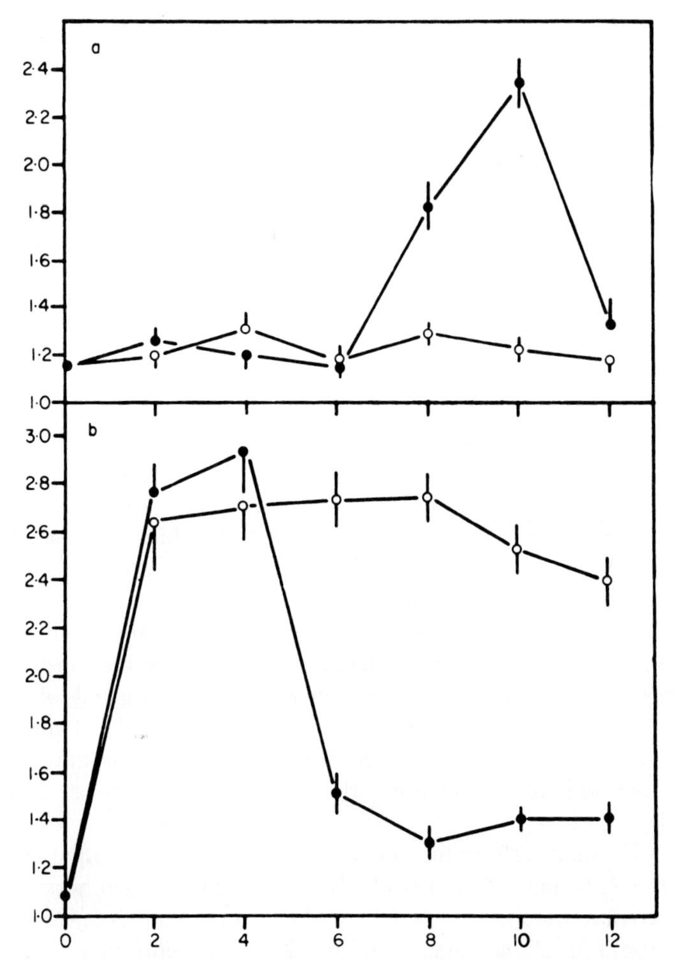

Figure 107. Development of (a) the late 82F and (b) early 75B puffs under the effect of ecdysterone alone or ecdysterone together with cycloheximide, the inhibitor of protein synthesis. Solid circles, treatment with ecdysterone (2.5×10^{-6} M); open circles, ecdysterone + cycloheximide (7×10^{-4} M). Abscissa, incubation time of salivary glands at 20°C; ordinate, ratio of the puff diameter to that of the nonpuffed band (from Ashburner and Richards, 1976b).

78D, 63E, and 82F; Figure 106) are induced after a delay of some hours (Figure 107a).

The existence of the early and late ecdysterone puffs revealed by Clever (1964a,d, 1965a) was confirmed later in additional investigations of hormonal induction in *D. hydei* (Berendes, 1967a), *D. lebanonensis* (Berendes and

Thijssen, 1971), *D. melanogaster* (Ashburner, 1972b), *D. pseudoobscura* (Stocker and Kastritsis, 1973), *Rhynchosciara hollaenderi* (Stocker and Pavan, 1974), *D. virilis* (Kress, 1979), *D. gibberosa* (Roberts, 1983), *Trichosia pubescens* (Amabis and Amabis, 1984), *D. auraria* (Mavragani-Tsipidou and Kastritsis, 1986), *D. virilis* (Poluektova *et al.*, 1986, 1987b), and *D. seguyi* plus *D. diplacanta* (Mavragani-Tsipidou *et al.*, 1994). The late ecdysterone puffs were not detected in the midgut chromosomes of *D. auraria* (Mavragani-Tsipidou and Scouras, 1991).

The early and late puffs differ in many functional features:

1. Rapidly reacting puffs in *D. melanogaster* respond to ecdysterone concentration over a wide range of at least 600 (Ashburner, 1973). The same range of effective concentrations was found for *D. hydei*. The concentrations of hormone most effective in puff induction were 33×10^{-2} to 33×10^{-4} Cu/μl (Berendes, 1967a). In *D. melanogaster*, a 50% induction of the 74EF and 75B early puffs was achieved at a concentration of hormone of 1×10^{-7} M. According to Ashburner's (1973) calculations, hormone concentrations vary between 0.8×10^{-7} M and 3.1×10^{-7} M in white prepupae of other dipteran species (mainly Calliphoridae). In *D. hydei* (Leenders *et al.*, 1970) the hormone's concentration in vivo is 5×10^{-6} M.

   Features such as a puff's ability to rapidly respond to ecdysterone in various insects were exploited in development of the "express method" for determining the hormomal status of larvae. The salivary glands of the larvae of *D. virilis* were transplanted into abdomens of nymphs of grasshoppers and larvae of beetles. Judgements concerning the hormonal status of the host were based on the response of the early puffs (Poluektova *et al.*, 1987a).

   Hormone at very high concentrations can partly inhibit induction. In *D. hydei*, the puffs did not reach their optimal sizes when hormone concentration was above 33×10^{-2} Cu/μl as a result of its injection (Berendes, 1967). In *D. melanogaster*, hormone at a concentration of 10^{-4}–10^{-3} M did not exert an inhibitory effect (Figure 108) (Ashburner, 1973; Belyaeva *et al.*, 1981), but it does so at a concentration of 10^{-2} M (Belyaeva *et al.*, 1981; Biyasheva *et al.*, 1981). The maximal sizes of the early puffs decrease with a decrease in hormone concentration: the puffs also become almost inactive more rapidly (Figure 109). The variation of hormone concentrations effective in inducing puffing of the late puffs is much narrower than that observed for the early puffs: they are not induced at all at concentrations below 5×10^{-8} M, and they are maximally induced at a concentration of 2×10^{-7} M or higher.
2. Effects of the inhibitors of protein synthesis. When ecdysterone was administered 1 hr after the injection of puromycin into *C. tentans* larvae, the early puffs formed normally; thus, prior protein synthesis was not required for

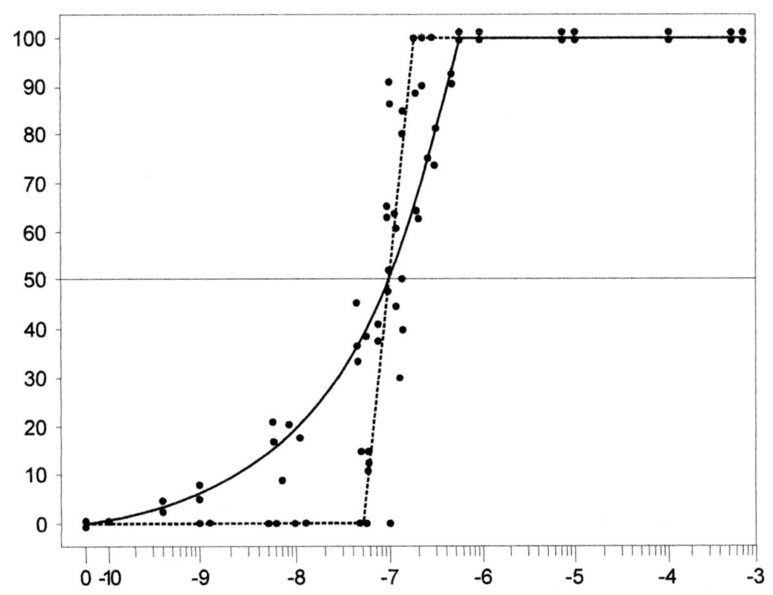

Figure 108. Dose–effect curve describing the dependence of the size of the early (continuous line) and late (discontinuous line) puffs on ecdysterone titer. Abscissa, logarithm (10) of ecdysterone titer; ordinate, puff diameter as percentage of maximum development (from Ashburner, 1973).

their induction. However, the late puffs were not induced after inhibition of protein synthesis. This indicated that the newly synthesized protein product was a necessary requirement for the induction of the late puffs (Clever, 1964a, 1965a). Clever suggested that the entire wave of the ecdysone puffs is a cascade regulatory circuit with the products of the early puffs inducing the late puffs (Clever, 1966a; Clever and Romball, 1966). This conclusion was extended to the puffing patterns of *Drosophila*. Two of the three early puffs (74EF and 75B) were induced when protein synthesis was inhibited. However, normal inactivation of the early puffs did not occur (Figure 107). The development of the third early puff 23E was blocked by all the tested inhibitors of protein synthesis (cycloheximide, puromycin, and anisomycin). All the late puffs were not induced (Figure 107), as in Clever's experiments (Ashburner, 1974a). In *D. virilis*, cycloheximide also inhibited the induction by ecdysterone of the late puffs when it was added to the medium no later than 2.5 hr after hormone (Kress, 1979). The intermolt puffs responded differently to inhibition: the inhibitor slowed the regression of 68C; antibiotics were without effect on the regression of the puff 25AC (Ashburner, 1974a). In *D. virilis*, inactivation by ecdysterone of the inter-

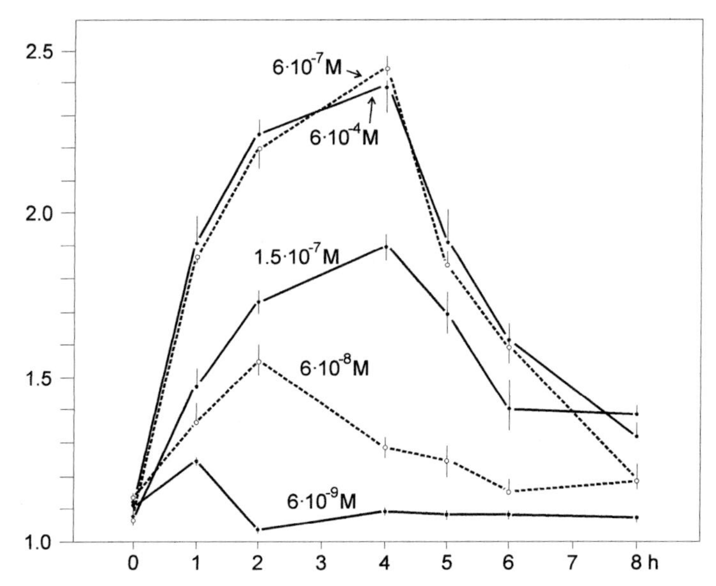

Figure 109. Processes of induction and regression of the 74EF early puff in relation to ecdysterone titer in incubation medium (from Ashburner, 1973). Abscissa and ordinate, the same as in the legend to Figure 107.

molt puff 97E was inhibited by cycloheximide when it was added to the medium not later than 1 hr after hormone (Kress, 1979).

3. Consequences of hormone withdrawal. Becker (1959) isolated salivary glands from *Drosophila* larvae at different stages of puff development and transferred them to an incubation medium. The glands were washed from hormone. In puff stage 6 (PS6) glands, all the ecdysterone puffs regressed after 1 hr of incubation, some late puffs appeared (71CE). The late puffs did not develop, when salivary glands were explanted before PS6. Similar results were obtained for the explanted salivary glands of *Sciara coprophila*. In the absence of hormone, the extent of puffing depended on the stage at which the salivary glands were explanted (Cannon, 1965).

In *Drosophila*, two of the three studied intermolt puffs (25AC and 68C) did not show reinduction on withdrawal of ecdysterone. The third puff (3C) regressed during the first hour of incubation irrespective of whether hormone was present in medium. The puffs recovered their normal size 5–6 hr after hormone withdrawal (Ashburner and Richards, 1976b; Bonner and Pardue, 1977a).

Removal of hormone from the incubation medium at any step of incubation caused the early puffs to rapidly regress (Figure 110). The regression of the puffs was not inhibited by cycloheximide. These puffs could be reinduced by exposure to hormones; however, their sizes were smaller. This indicated that as the time of cell exposure to hormone increases, the puffs became progressively more sensitive to its effect (Ashburner and Richards, 1976b; Bonner and Pardue, 1977a).

The late puffs can be divided into two groups on the basis of their response to hormone washout. The "early-late" puffs 62E and 78C appear somewhat earlier than the "late-late" 22C, 63E, and 82F puffs, which form at the very end ot the puff wave; that is, in zero-hour prepupa (Figure 106). The early-late puffs become inactivated on hormone washout, and they are reinduced after readdition of hormone. The "late" puffs are induced when larval glands are incubated for 8–10 hr (Figure 107). All three puff types are very rapidly induced upon removal of hormone. Reexposure to hormone inhibits these puffs (Figure 111).

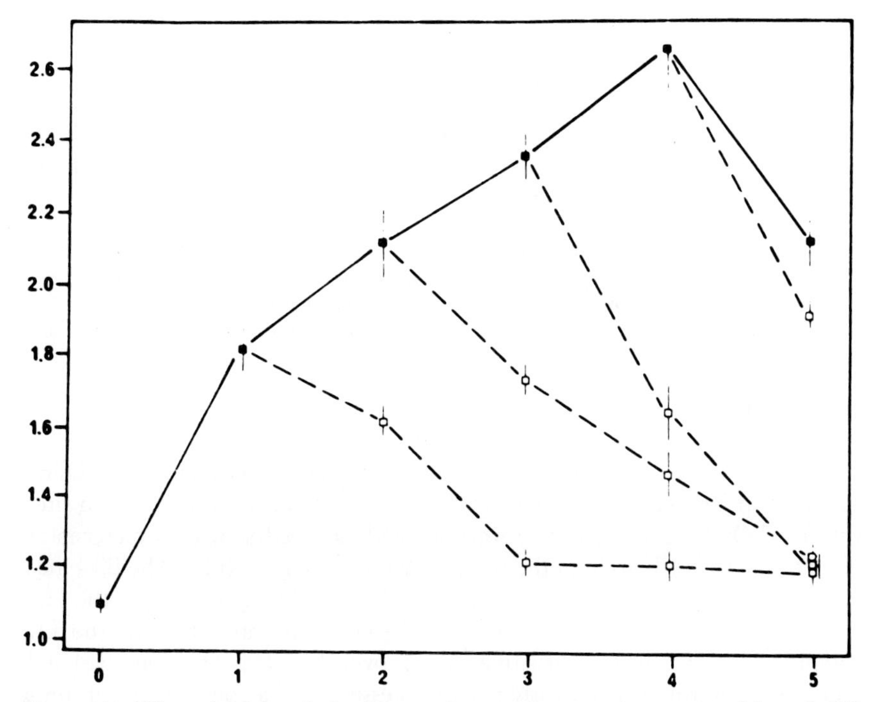

Figure 110. Effects of withdrawal of ecdysterone from incubation medium on size of the early 74E puff in *Drosophila* (from Ashburner and Richards, 1976a). Continuous line indicates a change in puff size in medium with hormone; a dashed line indicates when ecdysterone is washed off. Abscissa and ordinate, the same as in the legend to Figure 107.

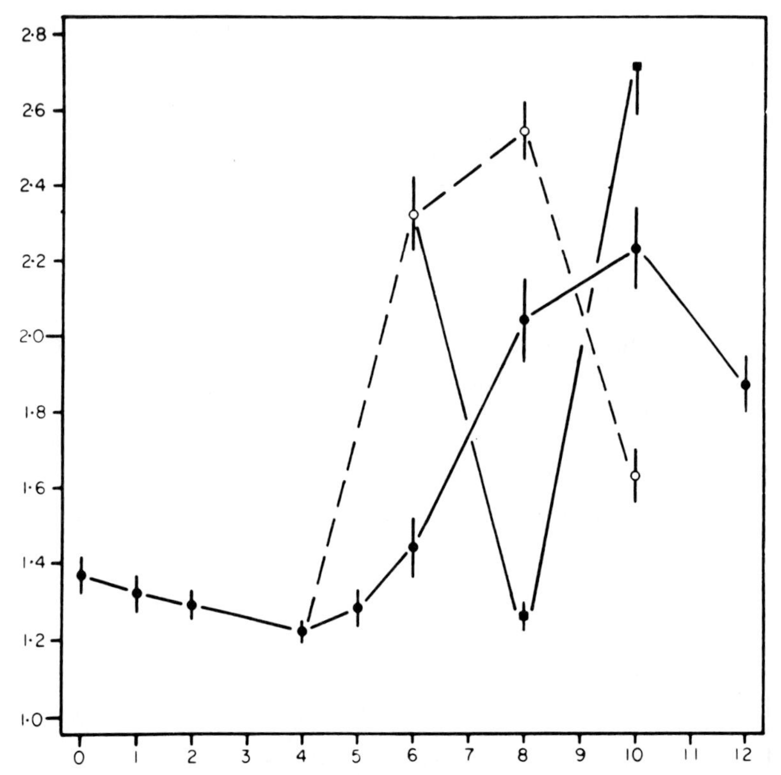

Figure 111. Precocious induction of the late 82F puff as a result of ecdysterone withdrawal after 4 hr incubation with hormone. Continuous line indicates incubation of salivary glands in a medium with the hormone; dashed line indicates incubation without hormone. Abscissa and ordinate, the same as in the legend to Figure 107 (from Ashburner and Richards, 1976b).

Ashburner *et al.* (1974) have proposed a model to explain how puffing activity can be controlled at both the early and late ecdysterone-induced sites (Figure 112): the first step is the reversible binding of ecdysone (E) by a receptor protein molecule (R) to form the receptor protein complex (ER). The ER binds to the puff sites: it induces the early puffs (Figure 112) and represses the late puffs ("-"). The product (P) of early ecdysone puffs interacts with the ER complex, thereby exerting a positive control over the late puffs (inductive) and a negative control over the early puffs (repressive). As a consequence of inhibition of the synthesis of the product P by cycloheximide, the early puffs are not inactivated at the appropriate time, and the late puffs are not activated. The ER complex disassociates on ecdysone washout, and induction of the late

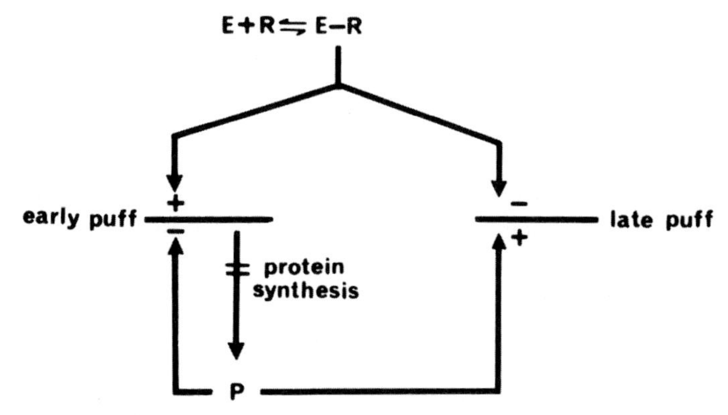

Figure 112. A model of the induction of the early and late puffs with ecdysterone (from Ashburner *et al.*, 1974). See text for details.

puffs immediately commences, because P has been already synthesized by the early puffs. Disassociation of ecdysterone from its receptor results in rapid inactivation of the early puffs. Considering all of the experimental evidence, this model explains well hormonal induction in an *in vitro* system.

Hormone also induces the gene activity in imaginal disk cells (Natzle *et al.*, 1986; Sin, 1987; Natzle, 1993). The locations of the genes are: 66C, 63E, 84E(93E), 70A, 64B, and 65B (Natzle *et al.*, 1986). However, imaginal disk RNA does not hybridize *in situ* to the sites of the ecdysterone puffs in salivary gland polytene chromosomes (Bonner and Pardue, 1976a).

The idea that the early and the late puffs are interrelated finds direct and indirect support in the following observations:

1. The induction of the DNA puff 2B in *Rhynchosciara* (see Section VI,F) is detectable within 3 hr after injection of total salivary gland RNA at the same developmental stage as the RNA puffs, into cytoplasm of cells. In controls, in which RNA was isolated from cells at the same developmental stage as the host cell, the puffs were not induced. When the injected material was treated with RNase, induction did not occur (Graessmann *et al.*, 1973).
2. Treatment of salivary glands with detergents that destroy the cell membrane increases cell permeability to high-molecular-weight substances; the cells become permeable (Myohara and Okada, 1990). The early puffs could be induced when permeabilized PS1 salivary glands were incubated in a PS8-9 gland homogenate supplemented with ecdysterone and ribonucleoside triphosphates (Myohara and Okada, 1987, 1988a).

3. Genotypes that were duplicated (three gene doses) or deficient (one dose) for the 74EF and 75B regions were constructed by combinations of translocations between 3L and the Y chromosomes of *D. melanogaster*. The early puffs were active for less time in the duplication genotype, and the other early puffs were active longer in the deficient genotype than in salivary gland from triploid larvae. Many, but not all, of the late puffs responded more rapidly to ecdysterone in triploids for the 74EF and 75B puffs, while the response to hormone of the same late puffs was delayed in deficient genotypes (Walker and Ashburner, 1981).

The wave of puffs induced by ecdysterone in salivary gland cells in an *in vitro* system never passes beyond the stage characteristic of 3- to 4-hr prepupa; that is, the puffs 75CD and 93F puffs are not induced, and the 74E, 62E, and 63E puffs are not reinduced (Figure 106).

The midpupal puffs 52A, 52C, 63E, 69A, and 75CD arise 4–5 hr after puparium formation, reaching their maximum sizes at 8 hr during normal development. These puffs are inhibited by ecdysterone when salivary glands of zero-hour prepupae are incubated under *in vitro* conditions (Richards, 1976b). Moreover, only 2–3 hr incubation of salivary glands of zero-hour prepupae in hormone-free medium or in medium containing it at a concentration not exceeding 5×10^{-9} M (Richards, 1976c) resulted in their induction.

Induction of the mid-prepupal puff in an ecdysterone-free medium is dependent on inhibitors of protein synthesis. The culture of salivary glands of zero-hour prepupae in an ecdysterone-free medium containing cycloheximide does not result in the formation of these puffs (Richards, 1976c). When larval salivary glands were incubated with ecdysterone, the 75CD puff arose only after 6 hr of hormone exposure, followed by 4 hr incubation in a hormone-free medium (Richards, 1976c).

Thus, RNA and proteins synthesized during the preceding larval wave of the ecdysterone-induced puffs are required for the induction of the mid-prepupal puffs. There are data indicating that the puffs can be induced when an extract from PS10/-11 prepupae is added. This evidences that the presence of a specific agent is required to provide the induction of the mid-prepupal puffs (Dennhofer, 1979).

The late prepupal puffs, active in 8- to 12-hr prepupae, were induced by ecdysterone when salivary glands of 8-hr prepupae were cultured with hormone *in vitro* (Richards, 1976d). However, cells must become competent to react to hormone more early: these puffs are inducible by hormone only in 6- to 8-hr prepupae; the glands must pass through the mid prepupal period, when the hormone titer is low (Richards, 1976b,c). Competence to react to hormone is also acquired when salivary glands of 0-, 2-, and 4- hr prepupae are cultured in a medium without hormone for 3.3 and 2 hr, respectively. Actinomycin D or cycloheximide added to the medium block the ability to become competent to repond the exposure to hormone (Richards, 1976c).

Thus, the late prepupal puffs also require RNA and proteins to be synthesized during the passage of the larval wave. An attempt was made to model the complete puffing sequence in larval salivary glands from PS1 to the end of prepupal development (initial exposure to hormone for 6 hr, hormone washout for 3 hr, followed by ecdysterone exposure for 2–4 hr at 25°C). As a result, it was demonstrated that the 93F puff formed only when the initial exposure to hormone was longer than 5 hr; that is, when all the wave of the larval ecdysterone puffs had passed (Richards, 1976d).

From the results presented in this section it follows that the changes in puffing observed at the end of larval–prepupal development are cascade type, consisting of several steps in which the activities of the preceding puffs induce those of the next puffs.

3. Effect of mutations on puffing

l(2)gl

Various mutations of the *l(2)gl* gene arrest development at different stages. In homozygotes for certain mutations, growing larvae fail to form puparium and they remain as third instar larvae to day 14. Homozygotes for the other alleles form puparium (Becker, 1959; Korochkina, 1977).

In homozygotes for "larval allele mutations," the ecdysterone puffs are not observed; the intermolt puffs 3C11-12, 71CE, 68C, and, occasionally, the heat-shock puffs are active (Becker, 1959). These observations were subsequently confirmed (Ashburner, 1972a; Richards, 1976a, 1985; Korochkina and Nazarova, 1977, 1978).

The larval puffs form normally in "prepupal lethal" mutants (Korochkina and Nazarova, 1977, 1978).

When salivary glands of homozygotes for the larval allele $l(2)gl^4$ were kept in Grace's medium containing ecdysterone (2.37×10^{-6} M), the intermolt puff 25AC regressed; the 23E, 74E, and 75B puffs developed and regressed in a 6-hr period, as occurs in the salivary gland cells of wild type. The late ecdysterone puffs 22C, 63E, and 82F were normally activated (Richards, 1976a).

l(3)tl

Homozygotes for this mutation do not form puparium at the end of the third larval instar, and they live for 30 days (Rodman, 1964; Rodman and Kopac, 1964; Kobel and Breugel, 1967; Breugel and Bos, 1969; Zhimulev *et al.*, 1974). The chromosome bands shorten and fuse in a complex manner during this period (see Section II,C). In 144-hr larvae, chromosome 3L lacks the 68C intermolt puff and the ecdysterone-induced puffs. The small puffs (33B, 33E,

34D, 35F, 55B, 58DE, 88D, 88E, 91CE, and 93D) are found in the other chromosomes. Larvae aged 192 and 384 hr were injected with ecdysterone. An ecdysterone puff pattern characteristic of PS10-11 was observed 6 hr after treatment in all the larvae of the first group. Some puff sets were incomplete; for example, the puffs 66B, 85D1-2, or 85F1-6 were not found when a puff was observed at 63E region. Injection of hormone is mainly without effect in larvae aged 384 hr. Of the 10 studied larvae, the small puffs 2B, 74E, and 75B were observed only in 1 larva (Zhimulev *et al.*, 1976).

Giant

Larvae homozygous for *giant* lethal mutations develop with a delay of 4 days to the stage of puparium formation. They show a low penetrance of the phenotype, reaching only 40% in certain heteroallelic combinations (gt^{E6}/gt^{13Z} or gt^{1}/gt^{13Z}) (Kaufman, 1972; Richards, 1985). These heterozygotes stop feeding and begin to wander on day 5, some pupate; others return to the food and start to wander 2–3 days later (as do giant homozygotes). The chromosomes undergo an additional replication on days 7–8 (Richards, 1985). Puffing is normal in these larvae despite delayed development (Becker, 1959; Richards, 1985).

DTS-3

At the permissive temperature for this dominant temperature-sensitive mutation (22°C), polytene chromosomes and puffs do not differ from normal. At the restrictive temperature of 29°C, the changes occurring during metamorphosis are not observed in unpupated larvae aged 6–14 days. The loci of the early puffs (23E, 74E, and 75B) show puffing activity in about 20% of larvae; the early-late puffs (62E and 78C) appear. The intermolt puffs 3C, 68C, and 90B (but not 25AC) are active in the 6- to 8-day larvae heterozygous for *DTS-3*. Certain puffs not seen during normal development were observed; for example, at 25D and 97D. In larvae grown at the restrictive temperature, the normal larval puffs were induced after 6 hr of *in vitro* culture in the presence of ecdysterone. The set of puffs induced in giant glands of mutant larvae cultured in the presence of hormone was similar to the *in vivo* pattern. Certain puffs (63E, 67A, 65D, and 73F/74A) characteristic of the ring gland regress under the influence of hormone; and puffs that are also induced in salivary gland chromosomes appear (2B, 23E, 62E, 63F, 74E, 75B, and 78C) (Holden and Ashburner, 1978).

dor

Homozygotes for a lethal allele, *l(1)7*, of the *dor* gene die after a prolonged third-instar life. Transplantation of normal ring glands did not result in formation of pupae. Thus, failure to pupariate is not due to deficiency in ecdysterone

(Goodwin, 1955); *l(1)7* homozygotes lack the late ecdysterone puffs (Rayle, 1967a,b). Detailed studies demonstrated that larvae homozygous for the dor^{lt187} mutation die toward the end of the third instar; only the intermolt puffs appear in these homozygotes. All the puffs that are characteristic of the larval puff sequence were induced by the culture in C46 medium containing ecdysterone (Biyasheva *et al.*, 1981, 1985, 1986). Transcript level of the *hsp27* ecdysone-induced puff is severely reduced in the dor^{22} mutant (Huet *et al.*, 1996).

halfway(swi)

Judging by Rayle's descriptions (Rayle, 1967a,b), the puff 2B13-17 (i.e., 2C1-2, see Table 9) is absent in hemizygous males. The swi^{t467} mutation, apparently at the *halfway* (*hfw*) locus, occasionally expresses the *hfw* phenotype; puparium forms only in the anterior part of the pupa (Aizenzon *et al.*, 1982); this may be why the mutation was referred to as a *hfw* allele (Lindsley and Zimm, 1992). Almost all larvae homozygous for the *t467* mutation that survive to the end of the third instar form puparium and undergo larval molt; however, few adults emerge. The temperature-sensitive period of the *t219* mutation was placed at the very end of the third instar and the pupal period (Aizenzon *et al.*, 1982). The size of the body, salivary glands, and polytene chromosomes are normal in larvae homozygous for the lethal allele. Secretion is released into the salivary gland lumen at the end of the third larval instar. However, the duration of larval development depends on culture conditions. Puparium is formed after a short lag under optimal conditions; it is delayed up to 2 days in somewhat overcrowded cultures (Belyaeva and Zhimulev, 1982). Puffing in homozygotes proceeds quite normally from PS1-PS8. Drastic changes commence from PS8: more than 30 puffs, whose appearance or considerably increase in size is characteristic of the period, are either reduced or absent from mutants (Figure 113). The mid-prepupal puffs (2- to 8-hr prepupae) in general develop normally; however, the size of the late prepupal ecdysterone puffs is smaller than normal. The culture of salivary glands in an ecdysterone-containing medium resulted in the normal appearance of the early and the early-late puffs. However, the late puffs formed neither after the culture of PS1 glands for 6 hr, nor after that of PS6 glands for 2.5 to 3.5 hr. Injection of ecdysterone into PS1 larvae was also without effect. Nevertheless, the late puffs were still competent to respond to hormone exposure. After the larval glands were washed free of hormone, these puffs were induced to sizes observed in controls (Figure 113). On the basis of these experimental results, the conclusions are as follows: (1) mutations of the *swi* gene decrease the effectiveness of swi^{+} gene product in induction of the late puffs; (2) the *swi* locus presumably controls most of late puffs; and (3) different signals are required to cause the regression of the early puffs and the late to be induced. The *swi* gene is concerned only with the latter process (Belyaeva and Zhimulev, 1982).

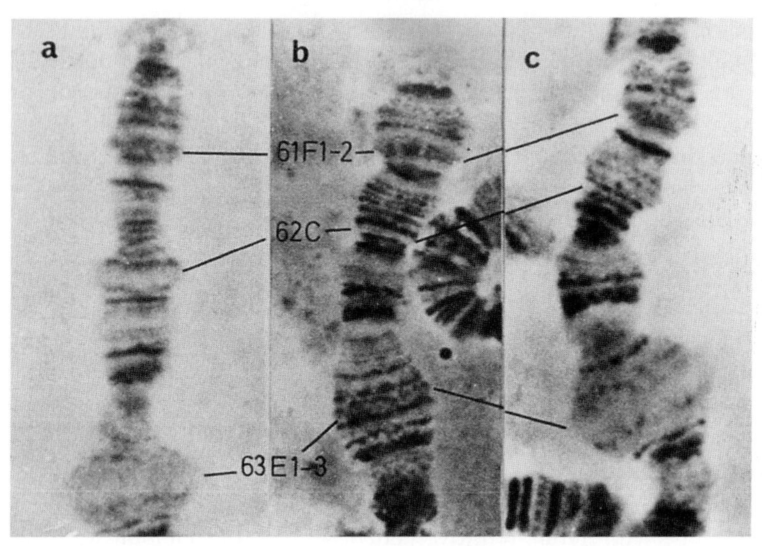

Figure 113. The 62C and 63E1-3 late puffs in zero-hour prepupae of (a) control *yellow* strain and (b) y, hfw^{t467}, as well as of (c) y, hfw^{t467} salivary glands isolated from larvae and incubated for 2 hr without hormone (reprinted by permission from Belyaeva and Zhimulev, 1982).

EcR

Five mutations were isolated which mapped to the EcR gene (Bender, 1996).

E74

Mutations in *E74A* and *E74B* are lethal during prepupal and pupal development. A large number of puffs normally active at puparium formation were significally reduced in sizes in *E74A* mutant: 21F, 22C, 62F, 63E, 71E, 72D, 82F, and 83E. The 62E "early-late" puff was also reduced. Two other early late puffs, 78C and 46E, and the 23E early puff were not affected. The 85D puff appeared to be larger at PS11 than in normal strain. In the *E74B* mutant, most of the late puffs were unchanged, with only 63E and 82F showing smaller sizes (Fletcher *et al.*, 1995).

The following conclusions concerning the effects of the above mutations at various steps of hormonal induction appear justified (Figure 114). Several mutations arrest development at the larval/pupal transition: *dor*, *l(2)gl*, *l(3)tl*, and *DTS-3*. Homozygotes show a preecdysterone puffing pattern readily induced on addition of exogenous hormone. These mutations seem to affect the process at the step of biosynthesis of hormone and its binding to the receptor. The *ecs* mutation at the BR-C locus acts at the step of the ecdysterone wave of puffing (Figure 114). The *swi* (*hfw*) locus disrupts the transmission of information between the early and late puffs (Figure 114).

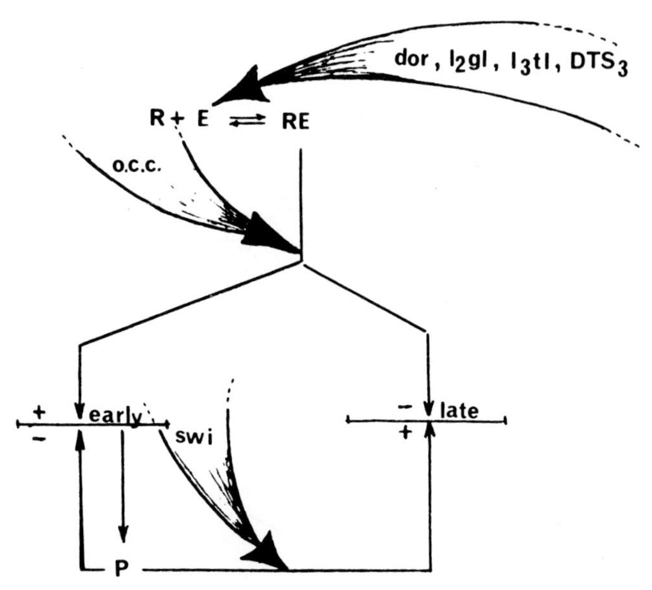

Figure 114. A scheme of the hormonal regulation of puffing with added data on genetic control of separate steps of induction. Arrows point to steps controlled by the indicated loci, the *dor*, *l(2)gl*, *l(3)tl*, *DTS-3*, *swi (hfw)*, and *o.c.c.* (*ecs* or *BR-C*) genes. The other designations are the same as in the legend to Figure 112 (reprinted by permission from Belyaeva *et al.*, 1982).

4. Ecdysterone analogs and vertebrate steroid hormones

Many ecdysterone analogs are known. Figure 115 presents the results of experiments in which the early ecdysterone puff 74E was induced by several different analogs. To obtain these results, PS1 larval glands were cultured in Grace's medium containing an analog for up to 6 h. Virtually all analogs are able to induce puffs, but they differ in their effectiveness in puff induction. The compounds depicted in Figure 115 have a steroid nucleus and side chains of different lengths. Activities of analogs without these chains (IX) or with a carbon chain (X) are high. If the chain has only one hydroxyl group (VI and VIII) or it has an isomerased hydroxyl group at position C22 (VII), the activities are intermediate; whereas compounds with two or more hydroxyl groups in the side chain (I, II, III, IV, and V) are very active, even when a part of the side chain has the shape of lacton ring (IV). The most active compounds have hydroxyl groups at position C20 or C22. The usual α-ecdysone (II) without hydroxyl group at position C20 is considerably less active than 20hydroxyecdysone (Ashburner, 1971, 1974b). Similar results were obtained when the ecdysterone analogs ecdysterone or 2β-, 3β-dihydroxy, or 5β-cholestenone were injected into larvae

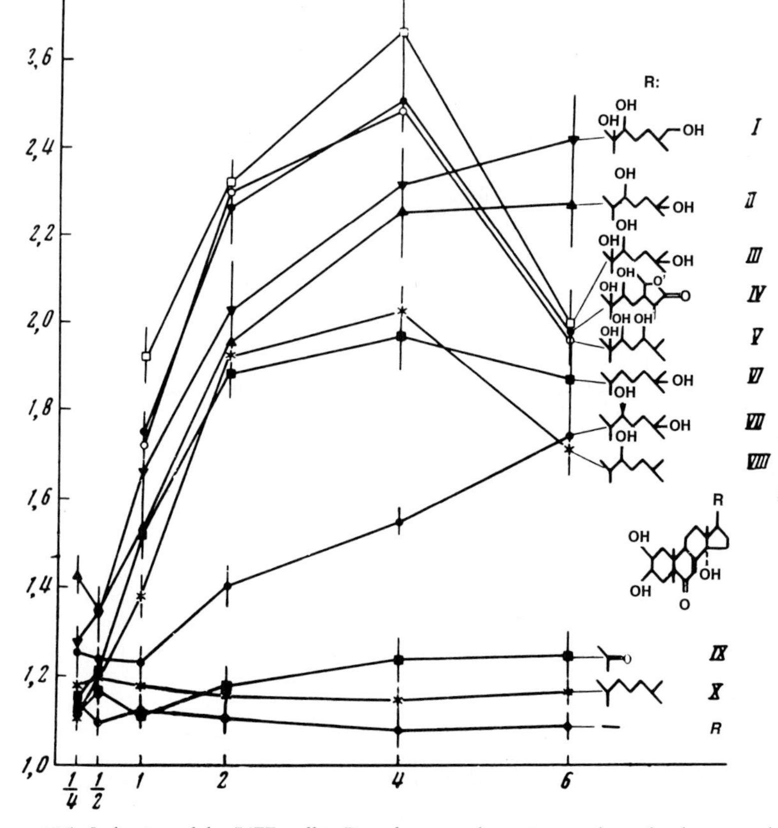

Figure 115. Induction of the 74EF puff in *D. melanogaster* by various analogs of ecdysterone during *in vitro* cultivating of salivary glands. Abscissa and ordinate, the same as in the legend to Figure 107 [reprinted with permission from *Nature* (Ashburner, 1971). Copyright (1971) Macmillan Magazines Limited; Ashburner, 1974b]. I, 25-deoxy-20,26-dihydroxy-α-ecdysone (inocosterone; 3.55×10^{-4} M); II, α-ecdysone (3.66×10^{-4} M); III, 20-hydroxy-α-ecdysone (β-ecdysone, 3.55×10^{-4} M); IV, ciasteron (3.43×10^{-4} M); V, 20,24-dihydroxy-α-ecdysone (iterasterone, 3.55×10^{-4} M); VI, 22-deoxy-α-ecdysone (3.8×10^{-4} M); VII, 22-iso-α-ecdysone (3.66×10^{-4} M); VIII, 25-deoxy-α-ecdysone (3.80×10^{-4} M); IX, 2-β-3-β-14-α-trioxyhydro-5-β-pregy-7en-6,20-dion (poststerone, 4.7×10^{-4} M); X, 22,25-deoxy-α-ecdysone (3.91×10^{-4} M); R, control, without ecdysone.

of *D. hydei:* observations made 10 min and 4 hr after induction demonstrated the series of changes known to occur during 6 hr before puparium formation (Berendes, 1968b).

In *C. tentans*, β-ecdysone initially induces only the early IV-2B puff, while α-ecdysone induces only the I-18C puff in *in vitro* experiments. Injection of α-ecdysone into larvae induces the puff I-18C within 15 min and the puff IV-2B with a lag which is longer than 30 min after the injection. The induction pattern was reversed for these puffs when β-ecdysterone was injected (Clever *et al.*, 1973). In *Acricotopus lucidus*, ecdysterone and inocosterone were much more effective in induction of Balbiani rings than α-ecdysone (Panitz *et al.*, 1972b).

As demonstrated for *in vitro* cultured glands of *D. melanogaster*, the early larval and the prepupal puffs responded much weaker to α- than to β-ecdysone (Ashburner, 1973; Richards, 1976d); the last named author estimated the difference as about 200-fold.

In another study, Richards (1978a) compared the activities of α-, β-ecdysone, and of their two analogs—3-dehydro-β- and 3-dehydro-α-ecdysone—with respect to their induction of the early and the late larval puffs, repression of the mid-prepupal ones. β-Ecdysone ranked first in its involvement in all the processes; 3-dehydro-β-ecdysone, α-, and 3-dehydro-α-ecdysone ranked next (Figure 116). Richards did not detect qualitative differences, in contrast to Clever, who did (Clever *et al.*, 1973). The differences in the biological activities of the analogs are, presumably, due to their different activities in the formation of the ecdysone–receptor complex (Ashburner *et al*, 1974; Richards, 1978a).

The results obtained with puff inducton by vertebrate hormones cortisone and hydrocortisone were discrepant. Some authors indicated that the hormones can stimulate puffing processes (Gilbert and Pistei, 1966; Kiknadze *et al.*, 1972; Korochkina *et al.*, 1972); others found that they can suppress puff formation (Goodman *et al.*, 1967; Goodman, 1969). According to the data of additional investigators, the hormones are without effect (Midelford and Rasch, 1968; Rasch and Lewis, 1968; Sang, 1968; Smith *et al.*, 1968). According to Korochkina *et al.* (1972)'s studies of *C. thummi*, treatment with hydrocortisone induced an increase in the sizes of most of the studied puffs.

Ashburner (1974b) found that other steroids, such as the vertebrate hormone hydrocortisone, do not induce puffing in an *in vitro* system. Hydrocortisone does not induce the early puffs even when present in a 100-fold excess (Ashburner and Berendes, 1978). The authors accounted for these observations by the fusion of the α- and β-rings at position 5a in the molecule of vertebrate steroid hormones. As a result, the molecule "flattens," acquiring a shape much different from the sharply "bent" molecule of ecdysterone, whose characteristic feature is fusion of the α- and β-rings at position cis-5b.

5. Juvenile hormone

The exact mechanisms of action of juvenile hormone are not clear. It is only known that there is an antagonistic interaction of juvenile hormone and ecdysterone. Arguments in favor of this notion stem from observations that juvenile

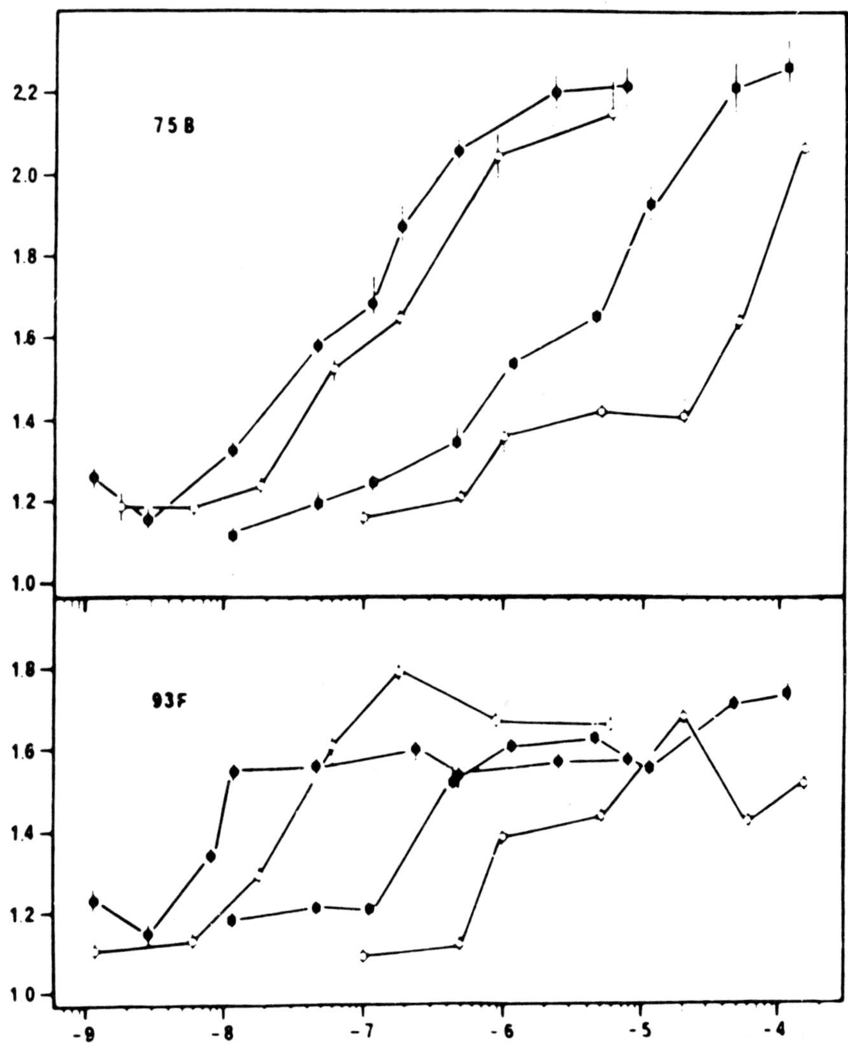

Figure 116. Dose–effect curves describing dependence of the size of the 75B and 93F late prepupal puffs on the concentration of ecdysone analogs (reprinted from *J. Insect Physiol.* **24,** Richards, The relative biological activities of a- and b-ecdysone and their 3 dehydro derivatives in the chromosome puffing assay, 329–335, Copyright 1978, with permission from Elsevier Science). Salivary glands of 6-hr prepupae incubated in media containing analogs, 2–4 hr. Solid circles, β-ecdysone; open circles, 3-dehydro-β-ecdysone; solid squares, α-ecdysone; open squares, 3-dehydro-α-ecdysone. Abscissa and ordinate, the same as in the legend to Figure 108.

hormone blocks ecdysterone-stimulated proliferation in cell culture or evagination of imaginal disks in culture. There are numerous data indicating that juvenile hormone is not a direct inhibitor of the effect of ecdysterone, because it stimulates larval processes by arresting the enfoldment of metamorphosis (see review by Richards and Ashburner, 1984).

The data on the effects of juvenile hormone on puffing are not numerous and are sometimes contradictory. In *C. thummi*, the sizes of 7–14 puffs observed during prepupal larval development considerably decreased after 6 hr of exposure to hormone. However, the size of one of the Balbiani rings (IVB), which is active in larvae and inactive in prepupae during normal development, increased under the influence of juvenile hormone (Laufer and Greenwood, 1969; Laufer and Calvet, 1972). When fed a synthetic analog of juvenile hormone, the larvae of *Chironomus* stopped developing at the prepupal stage. Studies of puffs in region 91 in these prepupae demonstrated that the sizes of nine puffs, which were quite conspicuous in the controls, were greatly reduced in treated individuals. An increase in the sizes of Balbiani rings, which are characteristic of the earlier developmental stages, was observed (Laufer and Holt, 1970). Local application of substances possessing juvenile hormone activity induced a decrease in Balbiani ring I and the appearance of the puff I-19A in salivary gland and Malpighian tubule cells. This puff is repressed by ecdysterone during normal development (Lezzi and Gilbert, 1969, 1972; Holderegger and Lezzi, 1972). The larval puff is also induced by juvenile hormone in prepupae of *C. thummi* (Lezzi and Frigg, 1971).

It was found that the response of puffs to juvenile hormone depends on its concentration. For example, the formation of the puff I-19A was induced when isolated salivary glands of *C. tentans* prepupae were incubated in a medium containing β-ecdysone and C18-cecropin juvenile hormone at a concentrations of approximately 3.4×10^{-7} M. The ecdysterone-dependent puff was repressed when glands were incubated in a medium containing a high dose of juvenile hormone, whereas both puffs were repressed when its doze was 10^{-4} M (Lezzi, 1974, 1975; Lezzi and Wyss, 1976).

The results of studies in *Drosophila* were also contradictory. According to Rensing (1975), juvenile hormone (JH-farnesyl-methyl-ether) causes the 3C intermolt puff to form (Rensing, 1975). A more significant disruption of chromosome puffing pattern was described by Poluektova *et al.* (1980b,c, 1981a, 1986). When salivary glands of *D. virilis* were cultured in medium C50 containing both juvenile hormone and ecdysterone, the activity of 70 puffs altered. Fifty-seven chromosome regions respond to ecdysterone alone and 41 to juvenile hormone alone (Poluektova *et al.*, 1980b,c, 1981a). Addition of a mixture of juvenile hormone and ecdysterone to the medium in which salivary glands were cultured induced a decrease in the sizes of the early puffs without affecting their possible induction (Poluektova *et al.*, 1986).

In *D. melanogaster,* injection or topical application of third-instar larvae with *Cecropia* juvenile hormone I (methyl-12,14-dehomojuvenate) had no effect on puffing in zero-hour prepupae (Ashburner and Berendes, 1978).

The puffing cascade in *D. melanogaster,* in which the induction by ecdysterone of the metamorphosis puffs in the presence of juvenile hormone is followed by ecdysterone withdrawal and reinduction of the prepupal puffs, is schematically represented in Figure 117. Clearly, juvenile hormone does not affect the early and the late larval puffs. However, the sizes of the ecdysterone prepupal puffs are reduced (Figure 117). When salivary glands of 8-hr prepupae

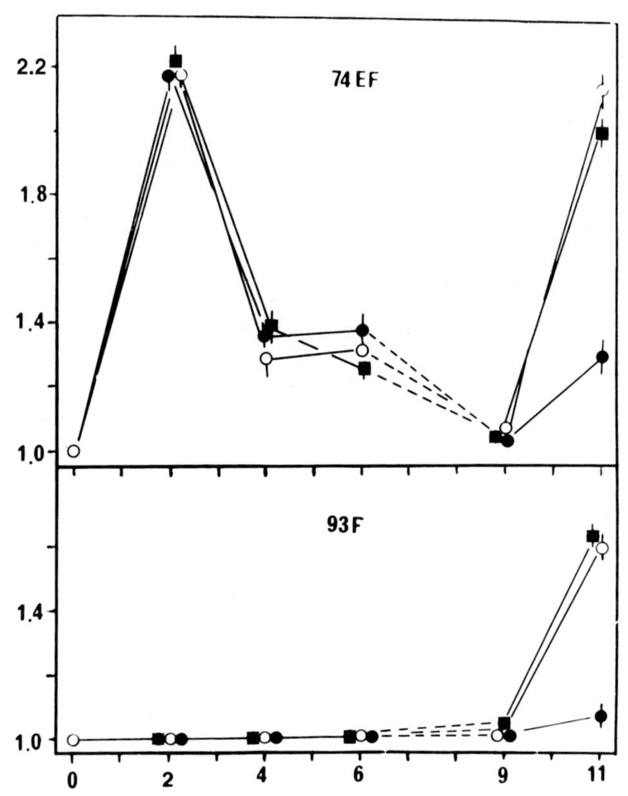

Figure 117. Inhibitory effect of the analog of the juvenile hormone (ZR 515) on the induction of the 74EF and 93F late prepupal puffs (from Richards, 1978b). Salivary glands of larvae at the PS1 stage were incubated for 6 hr with ecdysterone and the juvenile hormone (continuous line), then they were incubated in a medium without hormone for 3 hr (discontinuous line), thereafter ecdysterone was added again (continuous line). Solid square, control (without JH); solid circle, all media contained ZR 515 (2.9×10^{-5} M); open square, media contained methylepoxyhexadecanoate. Abscissa and ordinate, the same as in the legend to Figure 107.

were cultured in the presence of both ecdysterone and juvenile hormone, induction of the prepupal puffs was not inhibited.

E. Molecular and genetic characterization of hormonal induction

1. Molecular organization of the ecdysterone puffs

More than 50 ecdysteroid-regulated genes in *D. melanogaster* have been described in the literature. Four critical changes in ecdysteroid-regulated gene activity during third-instar larval development were found: at 78–88, ca. 100, 106–108, and 114 hr after egg laying. After puparium formation, ecdysterone induces successive waves of transcription (Andres *et al.*, 1993). Some of the genes are located in polytene chromosome puffs.

a. The 2B3-5 puff

The 2B1-10 region, which contains an early ecdysterone-inducible puff, was saturated with mutations of which those at the *ecs* (ecdysone sensitivity), *swi* (singed wings), and *dor* (deep orange) loci cause disturbances in the normal puffing sequence (Belyaeva *et al.*, 1981, 1989; Belyaeva and Zhimlev, 1982; Zhimulev *et al.*, 1982c; Biyasheva *et al.*, 1985).

Several lethals were isolated which affected puparium formation and were located near the tip of the X chromosome (Kiss *et al.*, 1976). Mutations of the *npr* (nonpupariating) series were mapped genetically very close to the *ecs* gene and appeared to be alleles of the *ecs* mutations as well as the previously identified *br* visible mutations (Belyaeva *et al.*, 1980a,b). Subsequently, this gene, which controls sensitiviy of *Drosophila* cells to ecdysterone, was renamed BR-C (broad complex), as introduced in an earlier section. The BR-C locus coincides with the cytological limits of the 2B3-5 early ecdysterone-inducible puff (Belyaeva *et al.*, 1987). The puff forms from a part of the 2B3-4 band, which splits into two new bands during activation; the 2B3-5 puff originates from the proximal part of the band (Figure 54). Localization of rearrangement breakpoints made possible the determination of the size of the largest DNA fragment involved in the formation of the puff. The fragment extends from the *In(1)br*^{lt103} breakpoint to the distal *Df(1)St472* breakpoint; thus, it covers about 65 kb of DNA. Both rearangements do not affect puff size. When DNA is removed by the *Df(1)P154* deletion, the size of the puff decreases. The *T(1;3)ITd*^{3} translocation that breaks within the 2B3-5 region forms a small puff from the distal part of the region. Because *In(1)br*^{lt103}, *Df(1)P154*, and *T(1;3)ITd*^{3} express the phenotype of the BR-C mutations, this was evidence for a good correspondence between the limits of the gene and the puff (Belyaeva *et al.*, 1987; Protopopov *et al.*, 1991). However, the boundaries of the gene presumably stretch beyond the puff. The region of the genetic map, where female fertility factors are located, lies in the region more distal to the breakpoints of *Df(1)sta* or *Df(1)St490;* that is, more distal than 95 kb on the map (Figure 54). The *In(1)br*^{lt103} inversion maps between positions designated 146.5 and 148 kb. This indicates that the region of the gene between these rearrangements (approximately 50 kb) is not involved in the formation of the puff.

The BR-C was cloned (Chao and Guild, 1986, Belyaeva *et al.*, 1987; Protopopov *et al.*, 1991). It was found to encode a family of DNA-binding proteins that share a common (core) amino-terminus, fused by alternative splicing to one of four C2H2-type zinc-finger domains, called Z1, Z2, Z3, and Z4 (Di Bello *et al.*, 1991; Bayer *et al.*, 1996, 1997; Sandstrom *et al.*, 1997). The core domain contains a 113-amino-acid sequence that is conserved in the transcriptional regulators encoded by the *tramtrack* gene as well as a large number of other proteins, primarily zinc-finger proteins (see Bayer *et al.*, 1996, for details). The relevant zinc-finger encoding exons, the core domain, and others are shown in Figure 118b.

Some of the BR-C mutations have been located on physical map (Galceran *et al.*, 1986, 1990; Belyaeva *et al.*, 1987; Izquierdo *et al.*, 1988; Orr *et al.*, 1989; Protopopv *et al.*, 1991; Solovyova, 1992) (Figure 118). Mutations resulting in the *br* phenotype are located in the interval between the breakpoint of *In(1)Hw*^{49c} and the mutation *br*^{28} (Figure 118c). The latter mutation is associated with a P-element insertion into zinc-finger domain- 2. Therefore, the *br* function maps between 110 and 200 kb on the map. Mutations essential for survival and shown in Figure 118g (except *In(1)Hw*^{49c}) are lethal when homozygous or heterozygous with each other. *In(1)Hw*^{49c} breaks DNA to the left of the first exon of the gene. All the other rearrangements that disrupt the continuity of DNA between the first exon and last Z3 domain are homozygous lethal. However their behavior in heterozygous combination with *In(1)Hw*^{49c} is peculiar. Heterozygotes between the inversion and rearrangements breaking the locus between the first and second exons (*br*^{26}–*br*^{6}) are almost completely viable. Heterozygotes for this inversion and P-element-induced mutations *P3* and *P13* inserted in the exon to upstream of the core are also viable. Heterozygotes of *In(1)Hw*^{49c} with rearrangements damaging the core and Zn-finger exons (*npr7* and *Df154*) are completely lethal. This may mean that full lethality results only from disruption of core and 2-exons in one chromosome and some DNA sequence at the 5′ end of the gene on the other chromosome. These data indicate that the function of viability is distributed along the gene to about the same extent as the *br* function.

The *rbp* phenotype (Fig. 118f), absence of visible phenotypes in hypomorphs (Fig. 118h), and *br* function of *In(1)Hw*^{49c} are associated with insertions of mobile elements in exon 4; these are situated distal to the core (P-element insertions in Figure 118h), in an intron (*rbp*^{tn}), or upstream of the transcribed part of the gene [as in *In(1)Hw*^{49c}]. Mutations resulting in sterility are located at both ends of the physical map. *Tp(1;3)sta* is sterile as a homozygote. Females

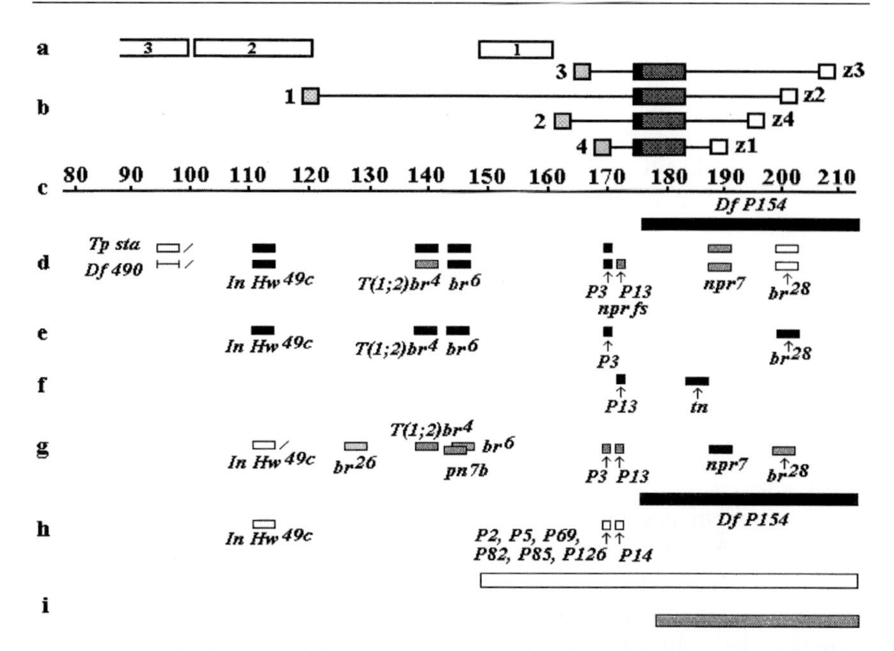

Figure 118. Molecular map of the *BR-C* gene (from Zhimulev *et al.*, 1995a,b). (a) Predicted regulatory elements, 1–3 (see text for details); (b) intron–exon map of transcripts, z1-z4 zinc finger domains (from DiBello *et al.*, 1991; Bayer *et al.*, 1996); (c) physical map, in kilobases (Chao and Guild, 1986; Belyaeva *et al.*, 1987); (d) localization of female sterility function in heterozygotes when *Tp(1;3)sta* or *Df(1)St490* is in one chromosome and any of the mutations listed to the right are in other chromosome. Solid rectangles indicate full sterility; shadowed ones, partial sterility; open rectangles, the fertile combinations (from Mazina *et al.*, 1991). Data on gypsy element location (npr^{fs}) are taken from Orr *et al.* (1989) and Huang and Orr (1992); (e and f) *br* and *rbp* phenotypes respectively (from Belyaeva *et al.*, 1987, 1989; Protopopov *et al.*, 1991; Solovyova, 1992). *tonock* (rbp^{tn}) was isolated by C. Moran and recognized as a P-element insert in the intron upstream of the z1 exon and its *rbp* allele (J. Fristrom, personal communication); (g) localization of breaks resulting in reduced viability. Homozygotes for $In(1)Hw^{49c}$ have normal viability, heterozygotes between the inversion and breaks, shadowed or dark show normal viability or lethality, respectively. Shadowed break points are lethal as homozygotes; (h) hypomorphs (from Solovyova, 1992); (i) ecdysterone-induced puff develops normally from DNA fragment (open rectangular) or has reduced sizes (shadowed) in chromosomes of larvae homozygous for br^6 inversion or *Df(1)P154* deficiency.

heterozygous for the transposition or *Df(1)St490* and any of the other mutations situated to the right of them (Figure 118d) are sterile as well—except *Tp(1;3)sta*/npr^7 and *Df(1)St490*/npr^7, which are partially fertile. *Df(1)St490*/$T(1;2)br^4$ females are also partially fertile. Eggs in the ovaries of sterile females appear to have developed normally but are not laid. Females, heterozygous for *Df(1)St472*, delimiting the right zone of BR-C and any of the *sta* alleles, including *Df(1)sta* and *Df(1)St490*, have normal fertility. Therefore, on the physical map of the BR-C the fertility function maps between 95 and 210 kb. Because breaks of DNA in *Tp(1;3)sta* or *Df(1)St490* also affect fertility, some part of the locus must be situated to the left of these rearrangement break points. For a detailed description of female sterility caused by BR-C mutations, see Zhimulev *et al.* (1995a,b).

In conclusion, no clear correlation was found between location of mutations on the physical map of the DNA and their presence in different complementation groups.

Complex transcriptional organization of the BR-C implies a complex pattern of controlling elements in the gene. It has been shown that ecdysterone induces a normal puff in the 2B3-5 region of the salivary gland chromosomes of the *Df(1)pn7b*, *Df(1)br*^{26}, and *T(1;2)br*^{4} variants, as well as in females homozygous for inversions *br*^{6} and *Hw*^{49c} (see Figure 118g). These breakpoints have been mapped in positions between 113.7 and 148 kb. All DNA sequences necessary for ecdysterone-activated transcription in the salivary gland cells and for the 2B3-5 puff activation are present between breakpoints of the *br*6 inversion (146 and 5–148 kb) and the start of transcription (Figure 118b). Hence there must be a promoter element (element 1 in Figure 118a) that responds to hormonal signal in this fragment, as suggested by Belyaeva *et al.* (1987). This element (between 148 and 160 kb) permits activation of all zinc-finger RNAs and puff formation. Two promoters, in positions 165 and 167 kb, have been suggested by DiBello *et al.* (1991); Bayer *et al.* (1996) concluded that there is one promoter at site 165 kb.

An additional controlling element (2 in Figure 118a) is located between *Tp(1;3)sta* and *Df(1)St490* (between 99 and 120 kb in Figure 118a); it may also be required for normal BR-C gene function (Belyaeva *et al.*, 1987). Galceran *et al.* (1990) identified a "cis-acting long-distance element" positioned between 102 and 115 kb on the map. According to DiBello *et al.* (1991) and Bayer *et al.* (1996) this promoter is situated at 120 kb. A third element (3 in Figure 118a) appears to be required for complete gene function, because chromosome breakage by *Tp(1;3)sta* and *Df(1)St490* results in sterility. Its proximal border coincides with breakpoints of these rearrangements. The distal limits of this element have not been mapped (Figure 118a).

Transcripts were detected at all the embryonic, larval, and adult stages; for example, in the ovaries during development (Galceran *et al.*, 1990). Transcription from the BR-C locus in salivary glands starts 12 h before puparium formation. Transcripts are detected later at the stages of prepupal development and partly at the pupal stage (Karim and Thummel, 1992), or (according to other data) near the beginning of the third instar. In early third-instar larvae, the most abundant BR-C RNA is Z3, then the levels of Z1, Z3, and Z4 RNA increases between early and mid third instar. Z2 RNA is not abundant in this tissue (Huet *et al.*, 1993; von Kalm *et al.*, 1994). The transcript containing Z1 appears in the salivary glands of 94-hr larvae; the RNA's abundance increases by 114 hr and then decreases before puparium formation. Transcripts appear anew in 4–12 hr prepupae. The Z2 transcripts are detected at PS3-13, during the time interval from 84-hr larvae to 20-hr prepupae, with a reduction in abundance occurring at PS11 stage and in 5- to 7-hr prepupae (Huet *et al.*, 1993).

In the imaginal disks, ecdysterone-induced BR-C RNA isoform accumulation exhibits two divergent expression profiles. The Z2, Z3, and Z4 isoforms accumulate to high levels at the beginning of the ecdysterone induction and then, after several hours, quickly disappear. The Z1 isoform continues to accumulate, while the other declines (Bayer *et al.*, 1996).

Effect of specific developmental pathways correlate well with the restricted expression patterns of Z1 and Z2 BR-C protein isoforms, supporting their assigment to the *rbp* and *br* functions, respectively (reviews: Bayer *et al.*, 1996; Thummel, 1996). The presence of pairs of C_2H_2 zinc-fingers in the BR-C proteins suggests that they regulate gene activity by binding to DNA and modulating transcription (see for discussion Bayer *et al.*, 1995).

Using specific antibodies against Z1, Z2, and Z3 BR-C protein isoforms demonstrated that these proteins accumulate in the nuclei of all larval and imaginal tissues. Monoclonal antibodies directed against the BR-core domain bind to numerous sites of salivary gland polytene chromosomes (Emery *et al.*, 1994). BR-complex isoforms exhibit different expression profiles. Prepupal salivary glands destined for hystolysis during metamorphosis express Z1 isoforms, while imaginal disks shift from the synthesis of Z2 isoform to the synthesis of Z1. The prepupal central nervous system expresses all isoforms, with Z3 predominating (Emery *et al.*, 1994).

The BR-C proteins are required in many tissues as shown by the effect of mutations at the locus on morphogenesis of adult structures and gene activity in larval organs. The product of the 2B3-5 puff regulates transcription of a large set of more than 100 puffs in salivary gland cells. In the 120- to 130-hr and in the earlier larvae, homozygous for the *lt435* (*npr6*) mutations, the intermolt puffs are of normal sizes. The 3C and 25B puffs regress, and 68C remains active later. Of the ecdysterone puffs, only 2B3-5 is very active, the 75CD mid-prepupae puff is prematurely active, which is strange enough (Belyaeva *et al.*, 1981). In *in vitro* experiments, the ecdysterone puffs are not induced at any one hormone concentration in the 10,000 range. Only puffs 74EF and 75B were found (Figure 119).

The entire wave of the larval ecdysterone puffs occurred in the control strain *yellow* (for examples, see Fig. 120), after 6 hr of culture of salivary glands in a medium with hormone. The features of response of *lt435* (*npr6*) mutants to hormone are as follows: (1) The intermolt puffs 25B and 3C are completely

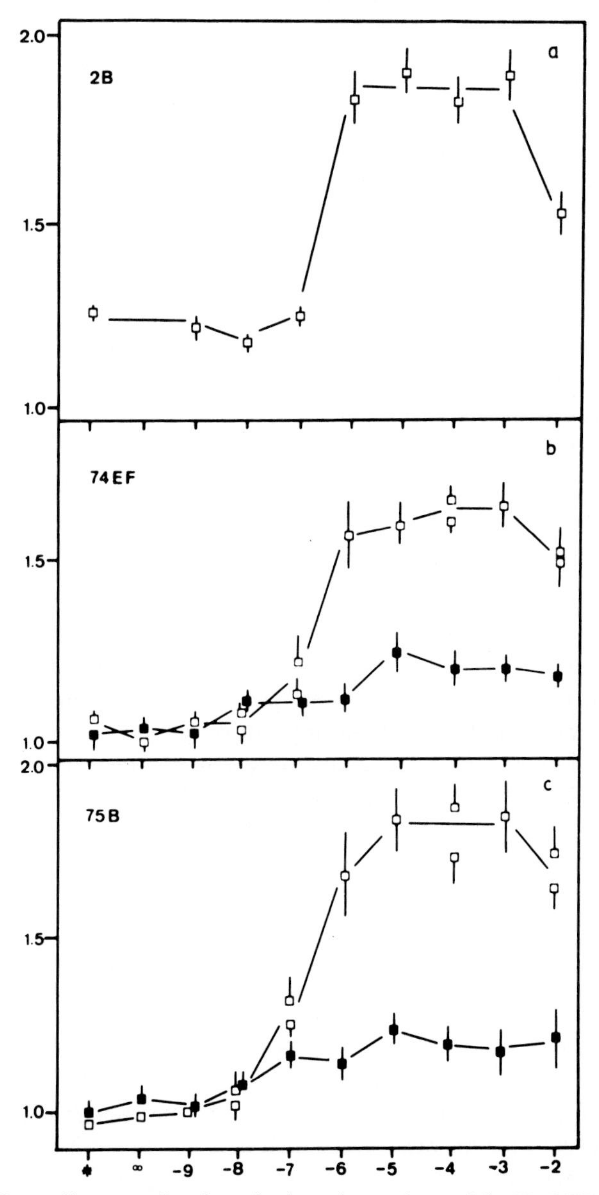

Figure 119. Dose–effect curve describing the dependence of sizes of the 2B, 74EF, and 75B puffs and $\log_{10}$ concentration of ecdysterone in the *lt435* mutant and control *yellow* strain (from Belyaeva *et al.*, 1981). The graphs show mean values of the relative sizes of puffs after 2 hr of incubation in a solution with hormone. Solid squares, *lt435* homozygotes; white squares, *yellow*. Abscissa and ordinate, the same as in the legend to Figure 108.

inactivated. Their inactivation is independent of the presence of ecdysterone. The 68C puff remains active (Figure 120). (2) The early ecdysterone puffs are only partly induced; for example 74E and 75B. (3) The 2B puff is induced, but not inactivated at the appropriate time. (4) The late puffs are not induced at all (Figure 120). The described abnormalities in induction of the ecdysterone puffs indicated that the cells of these mutants did not respond to ecdysterone, and that BR-C functions are required for the induction of all the later puffs (Belyaeva *et al.*, 1981; Biyasheva *et al.*, 1981).

Analysis of puffing changes *in vivo* in the *l(1)pplt10* mutants demonstrated abnormalities in the ecdysterone puffs: the size of the puffs (the 5 early and 16 late) at 21 loci was considerably decreased, and the intermolt puffs regressed less compared to normal. The next wave of the mid prepupal puffs passed normally. Here again, 21 ecdysterone puffs were underdeveloped in the late prepupal cycle (Zhimulev *et al.*, 1980c, 1982c).

Restifo and White (1992) showed that salivary glands of larvae homozygous for *rbp* mutations do not histolyze and can be found even in late pupae (72 hr after pupariation). Puffing patterns in *rbp*^{5} mutants are normal until PS7-8, but later many puffs do not develop normally: 2B3-5 does not inactivate and is still active even in 24-hr prepupa. Some of the latest larval puffs (66B, 82F) are active at zero-hour prepupae, but the majority of the puffs characteristic of this stage are underdeveloped. In homozygotes for *rbp*^{5} premature activation of 75CD occurs; puffs form at 46F and 93F9-10, where ecdysterone receptor genes are located (see Table 15). It looks as if *rbp*$^+$ represses these genes; but the mutant *rbp*^{5} product cannot exert that function, and the puffs are activated (E. S. Belyaeva and I. F. Zhimulev, 1996, unpublished observations). Z1 is sufficient to improve viability and rescue late puff formation in salivary gland chromosomes of the mutant carrying a Z1-encoding *rbp*^{5} transgene. Salivary gland histolysis is induced in these transgenes as well (E. S. Belyaeva and I. F. Zhimulev, 1996, unpublished observations; Balasov and Zhimulev, 1997). According to Karim *et al.* (1993), both *rbp* and *l(1)2Bc* (= *l(1)pplt10*) are required for glue gene induction in mid-third-instar larvae, glue gene repression in prepupae, and for the complete ecdysone induction of certain early mRNAs (E74, E75A, and BR-C), as well as efficient repression of most early mRNAs in prepupae.

Mosaics were obtained starting with *npr-1* heterozygotes (*npr-1* appears to be located with BR-C and is named *npr3* according to Lindsley and Zimm, 1992). *npr-1/0* (hemizygotes) tissue could respond to ecdysterone, and tanned puparial tissue (*npr-1/+*) was sharply separated by distinct boundaries from the white larval tissue (*npr*-1/0) (Kiss *et al.*, 1976). Tissues homozygous for BR-C mutations do not undergo metamorphosis *in vitro* (Kiss and Molnar, 1980).

Mutations at or deletions of the BR-C locus considerably reduce the synthesis of *Sgs* mRNAs: to 7% in *Sgs-4* (the 3C puff), 27% in *Sgs-5* (the 90BC

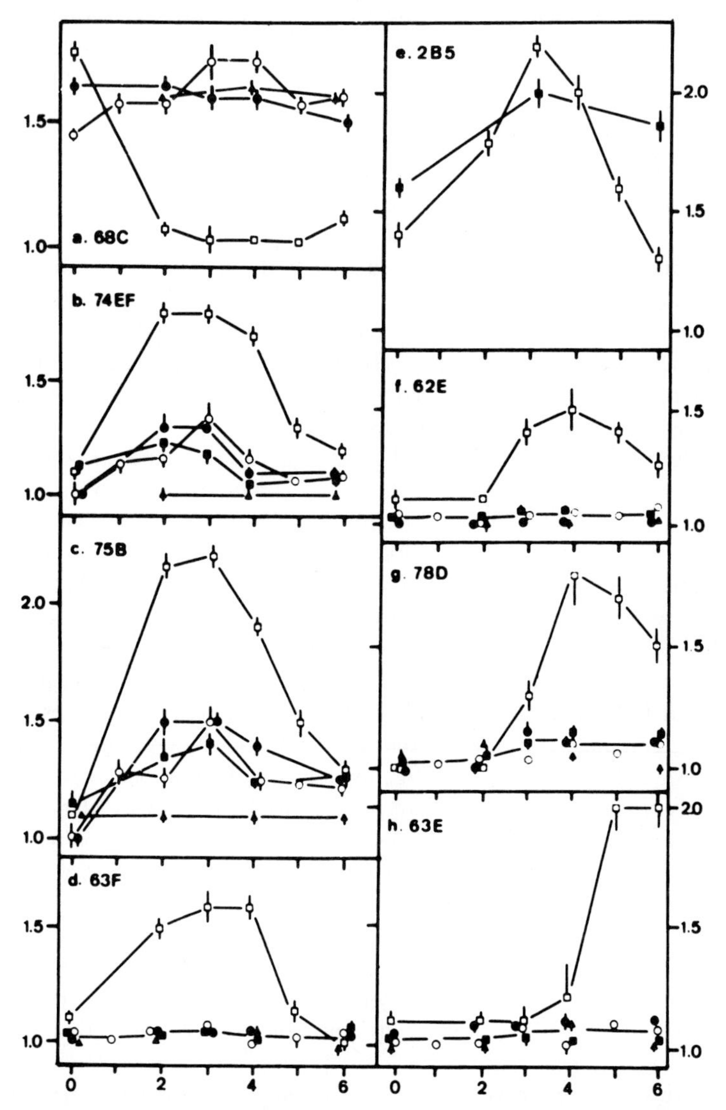

Figure 120. Change in the activity of the (a) 68C, (b) 74EF, (c) 75B, (d) 63F, (e) 2B, (f) 62E, (g) 78C, and (h) 63E puffs during incubation of salivary glands with ecdysterone. White squares, control *yellow* strain; white circles, y, *l(1)t435* males; black squares, y, *l(1)t435* females; black circles, males in which the 2B3-5 puff and the *ecs* locus are removed by the deletion; black triangles, males with the deletion of the puff, incubation without hormone. Abscissa and ordinate, the same as in the legend to Figure 107 (from Belyaeva *et al.*, 1981).

puff); and they completely inactivate the transcription of the *Sgs-3*, *Sgs-7*, and *Sgs-8* genes (the 68C puff) (Crowley *et al.*, 1984; Vijay Raghavan *et al.*, 1988; Galceran *et al.*, 1990; Guay and Guild, 1991). The product of the BR-C gene concomitantly binds to the enhancer regions of the *Sgs-4* gene (see Figure 121). The BR-C and E74 may function together to regulate ecdysterone-inducible target gene transcription: the *rbp5/E74*A^- allele combination causes a significant reduction in salivary gland glue (*Sgs-5*) transcript accumulation (Fletcher and Thummel, 1995).

In homozygotes for the lethal mutations of the BR-C genes, 24 to 26 ecdysterone-induced polypeptides are not synthesized in the salivary glands. When the BR-C locus is present in three doses, the amounts of synthesized proteins and *hsp23* and *hsp28* RNAs, whose structural genes are located in another ecdysterone-induced puff (67B) markedly increases, and about 15 ecdysterone-induced proteins appear 1 hr earlier in salivary glands *in vitro* (Dubrovsky and Zhimulev, 1985, 1988; Dubrovsky *et al.*, 1985, 1990). Deficiency of the BR-C as well as mutations of *npr* type reduce the expression of genes *hsp23* and *hsp27* by 95 to 99% (Dubrovsky *et al.*, 1994, 1996). According to other data, *hsp27* expression is moderately reduced in *brt435*. The br^5, rbp^5, or $l(1)2Bc^1$ variants had very little effect (Huet *et al.*, 1996).

The normal product of the *rbp* function of the *BR-C* complementation group is required for the expression of six genes mapping to the late ecdysterone puff 71F. Mutations in the *br*, *l(1)2Bc*, and *l(1)2Bd* groups have no effect on the expression of these genes (Restifo and Guild, 1986; Guay and Guild, 1991; Crossgrove *et al.*, 1996).

BR-C regulates activity of several genes in fat body (Antoniewski *et al.*, 1993, 1994, and references therein). The lethal BR-C mutations *lt435* and *t10* reduce the levels of transcription of the *FbP1* and *FbP6* genes expressed in the fat body (Lepesant *et al.*, 1986; Nelson *et al.*, 1991).

The BR-C proteins are required in many tissues, as shown by the effect of BR-C mutations on morphogenesis of adult structures (Sandstrom *et al.*, 1997) and expression of dopa decarboxylase (DDC) in the epidermis (Hodgetts *et al.*, 1995) and on the central nervous system (Restifo *et al.*, 1995).

The BR-C may be a target for transinteraction with other genes. Alleles at four loci have an effect on the expression of mutations of the *br* group: *stubble-stubloid* act as dominant enhancers; and *enhancer of broad* (E(br)), a recessive lethal, is a dominant enhancer of the br^1 allele. Two additional loci, *blistered* and *l(2)B485*, interact with *E(br)* and *Sb-sbd*. *blistered* (but not *l(2)B485*) interacts with br^1. All three loci (*blistered*, *l(2)B485*, and *E(br)*) are located in a very short genetic interval between the genetic markers *speck* and *irregular facets* in the 60C5-6-60E9-10 region of chromosome 3 (Beaton *et al.*, 1988; Gotwals and Fristrom, 1991).

b. The 2C1-3 puff

Detailed electron-microscope and cytological analyses have shown that correct name of the puff is 2C1-3 (Belyaeva *et al.*, 1980a), although several authors have named it 2B13-18. A fragment that maps to the 2C1-3 puff was isolated from an embryonic cDNA library using an oligonucleotide probe containing a DNA binding domain in steroid hormone receptors. The conceptual amino acid sequence of protein of the 2C1-3 puff shares at least four homologous regions with many vertebrate receptors. Region I includes two cysteine–cysteine zinc fingers, indicating DNA-binding activity; the II–IV regions are located in the carboxyl-terminal portion of the molecule that forms the binding site with hormones, and resembling similar sequences in vertebrate steroid/thyroid hormone receptors. It was called *ultraspiracle (usp)* gene (Henrich *et al.*, 1990; Oro *et al.*, 1990).

c. The 3C puff

Mapping of the electrophoretic variants of the salivary gland secretory proteins of *Sgs-4* demonstrated that their encoding gene is located on the X chromosome at position 3.0 on the genetic map and within the 3C11-12 band on the cytological map (Korge, 1975, 1977a, 1981; Kokoza *et al.*, 1980, 1982; Kokoza and Karakin, 1981; Kokoza, 1985). The DNA of this puff was cloned (McGinnis *et al.*, 1980; Muskavitch and Hogness, 1980, 1982). At least six genes were mapped within the 3C11-12 puff (Figure 121; cf. Figure 62): *Sgs-4*, *Pig-1*, and ng-1, -2, -3, and -4. (Muskavitch and Hogness, 1980; Hofman and Korge, 1987; Furia *et al.*, 1985, 1990; D'Avino *et al.*, 1995b). The three genes are located in the DNA fragment 11 kb in length; the distance between the *Pig-1*, *Sgs-4*, and *ng-1* genes is 813 bp and 6 kb, respectively. Although sharing some similarities, these genes are quite different with respect to timing of expression and the tissue sets where they express: *Sgs-4* RNA was detected only during the second half of the third instar, from 91 hr to the end of PS3 (Muskavitch and Hogness, 1980; Korge *et al.*, 1990), while transcriptional activity of *ng-1*, *ng-2*, and *ng-3* was expressed at high levels throughout the third instar until the end of this stage (Furia *et al.*, 1990; D'Avino *et al.*, 1995a,b). *Pig-1* RNA first starts to appear as early as in 12- to 18-hr embryos; it is observed throughout larval life and as late as in 5-day-old adults (Chen *et al.*, 1987; Hofmann and Korge, 1987; Barnett *et al.*, 1990). The expression of the *Sgs-4* gene is not limited to the salivary glands. Larvae transformed with a construct containing a regulatory fragment of the *Sgs-4* gene and the coding portion of the *Adh* gene express the *Sgs-4* promoter in the proventiculus of larvae as early as in the first instar. In the salivary glands of third-instar larvae, the increase in *Sgs-4* mRNA levels is several thousandfold (Barnett *et al.*, 1990). The length of *Sgs-4* mRNAs varies from 0.95 to 2.7 kb from one strain to another. Almost half of the gene consists of tandemly repeated sequences of 21 bp (Muskavitch and Hogness, 1980, 1982).

Five tissue-specific DNAse I-hypersensitive sites (see Figure 121) were

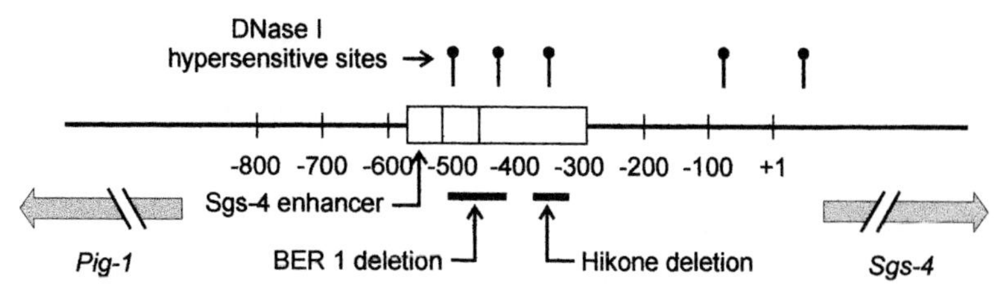

Figure 121. Disposition of the *Pig-1* and *Sgs-4* genes and the region between them (from Barnett *et al.*, 1990). The horizontal line represents the region 800 bp long between the *Pig-1* and *Sgs-4* genes, symbols of sites hypersensitive to DNase I are over it, short horizontal lines depict fragments removed by the deletions Berkeley and Hikone.

detected in the interval between −480 and +30 bp in the regulatory region of the *Sgs-4* gene (Shermoen and Beckendorf, 1982; McGinnis *et al.*, 1983). When the *Sgs-4* gene was transformed with cloned fragments containing the regulatory sequences lying between −838 bp 5′ and −130 bp 3′ remote to it, *Sgs-4* controlled expression was inferred to be at normal levels (McNabb and Beckendorf, 1986). Removal of the regulatory sequences by deletions significantly affects gene expression. The length of the deletion in Ber-1 is 95 bp, it extends from −496 to −392 (Figure 121). In homozygotes for the deletion, transcription level of the *Sgs-4* gene both in the salivary glands and the proventriculi makes up not more than 0.01% of wild-type. The Hikone wild-type strain contains a shorter deletion of 52 bp (Figure 121). The level of gene expression in both organs makes up 5–10% in homozygotes (Muskavitch and Hogness, 1982; Shermoen and Beckendorf, 1982; Barnett *et al.*, 1990). This region extending from −49 to −568 bp controls tissue- and stage-specificity (Shermoen *et al.*, 1987; Jongens *et al.*, 1988).

The regulatory sequences of the *Pig-1* and *Sgs-4* genes lie within the interval between the two genes (see Figure 121), and they presumably overlap. When five sequences were removed to −392 bp in transformant lines (Figure 121), *Pig-1* expression was at lower than normal levels. Thus, the stretch of DNA necessary for correct transcription of the *Sgs-4* gene is also required for normal developmental *Pig-1* expression. However, a certain deletion (*Ber-1*) causes a 5000-fold reduction in the level of expression of *Sgs-4*, but only a 4-fold one for *Pig-1* (McNabb and Beckendorf, 1986; Mougneau *et al.*, 1993). The Broad-Complex is required for proper timing of the switch from *Pig-1* to *Sgs-4* (also see below). BR-C proteins bind *in vitro* to elements regulating *Sgs-4*. rbp^+ function is necessary for induction of the glue gene; it acts through Broad-Complex binding sites (von Kalm *et al.*, 1994).

Data on the sizes of the regulatory region at the 5′-end of the *Sgs-4* gene that ensure normal induction of the puff are contradictory. In transformants with an inserted construct which includes the *Sgs-4* coding portion, 2 kb downstream and 840 bp of upstream material, the DNA of the inserted fragment expresses with stage and tissue specificity, although no puffs form. The sequences required for puff formation are contained within a larger (16–19 kb) DNA fragment (McNabb and Beckendorf, 1986). According to other data, a DNA fragment containing 840 bp 5′ and 2 kb of the coding region are sufficient to ensure puff formation and dosage compensation (Krumm *et al.*, 1985; Hofmann *et al.*, 1987; Korge *et al.*, 1990).

Korge (1977a, 1981) has shown that the *Sgs-4* protein is not synthesized in salivary gland cells, and the puff 3C11-12 is missing in certain strains. An *Sgs-4* allele *Sgs-4*^{K} was weakly expressed when made homozygous; however, *Sgs-4*^{K} showed enhanced expression when made heterozygous with a normal *Sgs-4* allele (from Oregon-R wild-type); its expression was enhanced nine-fold and four-fold when pairing in the neighborhood of the *Sgs-4*$^{+}$ gene is duplicated. No enhancement was observed when contact between the *Sgs-4*^{K} and *Sgs-4*$^{+}$ (Oregon-R) genes was disrupted by rearrangement of the *Sgs-4* alleles (Kornher and Brutlag, 1986).

Various proteins bind to the regulatory fragments of the *Sgs-4* gene. There are four binding sites of *secretion enhancer binding proteins (SEBP)* (Lehmann and Korge, 1995, 1996). The protein product of the BR-C gene binds to the region extending from −300 to −600 bp (Figure 121). The proteins containing the zinc fingers Z1 and Z2 bind to the region of the most proximal part of the promoter zone; the protein Z3 binds to its distal part (Beckendorf *et al.*, 1992). Despite *SEBP*s corresponding with Broad-Complex protein binding sites, the *SEBP-2* factor is encoded by *fork head (fkh)* gene (Lehmann and Korge, 1996). Antibodies against RNA polymerase II bind to the material of the 3C11-12 puff in the chromosomes of the normal strains, but not of strains with the *Ber-1* deletion (Figure 121) (Steiner *et al.*, 1984; Korge, 1987). Antibodies against topoisomerase I strongly bind to the puff in *Sgs-4*$^{+}$ strains, but much weaker in strains carrying loss-of-function mutations (Steiner *et al.*, 1984).

Nuclear proteins, tentatively designated as Bj6 and Bx42, bind to the material of the 3C11-12 puff when it is active. Bringing the *Sgs-4* gene into a new position by P-element-mediated transformation causes these two proteins to bind to the 2.5-kb fragment located upstream of the start of *Sgs-4* transcription. The fragment fails to bind to Bx42 protein in strains with a 52-bp deletion within it (Saumweber *et al.*, 1990).

Transcriptional activity of *ng-1*, *ng-2*, and *ng-3* is distinct from that of the *Sgs-4* gene located in the same 3C puff, as well as all other intermolt genes. The *ng* genes start to be transcribed at the beginning of the third instar, and *Sgs-4* at the early wandering stage. While null mutations of the BR-C strongly affect the level of the *Sgs-4* transcription, induction of *ng* does not depend on BR-C functions (D'Avino *et al.*, 1995a,b).

d. Puff 5A

Puff development at zero-hour prepupae in the 4F-5A region of the X chromosome is often named 4F9-10 (Ashburner, 1969b,f, 1975). However according to photographs in those papers, this puff develops in the 5A3-4 regions (Belyaeva *et al.*, 1974). The 5A puff is active during prepupal development from 0 to 10 hr after spiracle eversion (Zhimulev *et al.*, 1982c). One late gene was located in the "4F puff" (Wolfner, 1980). This gene is transcribed between 0 and 10 hr prepupae (Andres *et al.*, 1993).

e. 25B puff

The gene encoding the Sgs-1 protein maps to chromosome *2* in the 25A3-D2 region. The 25B intermolt puff is located in this region (Velissariou and Ashburner, 1980).

f. 35B2-5 puff

Low levels of RNA associated with a puff at 35B2-3 is detected by *in situ* DNA–RNA hybridization of an *Adh-* containing 2.7-kb DNA fragment. The puff forms in the region where the *Adh* gene is located. However, enzyme activity (alcohol dehydrogenase) was not detected in the salivary gland cells. In contrast, the puff is missing in midgut cells where *Adh* activity is high (Visa *et al.*, 1988). According to electron microscopic data, the puff in this region originates from the 35B3 band, while the 35B1-2 band remains compacted (Baricheva *et al.*, 1987). Therefore, it is not clear whether the formation of the puff in this region results from activation of the *Adh* gene. Puff formation and the *Adh* gene possibly involve the neighboring bands.

g. 39BC puff

The puff develops at PS8-11 (Zhimlev, 1974a). The *DHR39* ecdysterone receptor gene was provisionally mapped in the cytological limits of this puff (Ohno and Petkovitch, 1992; Huet *et al.*, 1995). Maximal transcriptional acivity is from PS6 to PS11 (Huet *et al.*, 1995) in the salivary glands and in 108-hr larvae–16-hr pupa (Horner *et al.*, 1995).

h. 42A puff

The *EcR* gene, encoding an ecdysterone receptor, maps within the region of this puff. The length of the gene is about 35 kb; ~6-kb RNA is transcribed from it (Figure 130). Hormone-binding and DNA-binding domains are contained in the protein molecule. This is a characteristic feature of all the steroid hormone receptors. The product of the *EcR* gene binds to the promoters of the ecdysterone-induced genes; for example, to the promoter of the ecdysterone-induced *hsp27* gene (see Section VI,H). The *EcR* transcripts are detected throughout development: transcriptional activity is high during embryonic development; trace amounts of *EcR* RNAs are detected during the first and second larval instars and there is a sharp enhancement in 96 to 120-hr larvae. Transcription remains at high levels in prepupae and 132- to 168-hr pupae (Koelle *et al.*, 1991; Huet *et al.*, 1993). Western blot analysis detected *EcR*-encoded proteins at each development stage in 108- to 120-hr larvae, prepupae, and pupae (Koelle *et al.*, 1991). Mutations of the *EcR* gene have been isolated (Bender, 1996; Bender *et al.*, 1997).

i. Puff 46F

The puff maps to the *DHR3* gene, encoding an ecdysone receptor. It is about 15 kb long (Koelle *et al.*, 1992). Transcripts become visible in salivary gland cells at 84 hr until PS6 and then in 5- to 15-hr prepupa (Huet *et al.*, 1993, 1995). In rbp^5 mutants the 46F puff appears at PS1 (E. S. Belyaeva and I. F. Zhimulev, 1996, unpublished observations). DHR3 represses the ecdysterone induction of early genes (Lam *et al.*, 1997; White *et al.*, 1997).

j. Puff 63F

Ubiquitin is a 76-amino-acid protein. It is found bound to histone H2A and occurs within the transcriptionally active regions. The ubiquitin gene maps within the early ecdysone puff 63F by *in situ* hybridization. It maps in this puff, in the region of two puffs separated by a compact band in *D. simulans*, *D. teissieri*, and *D. hydei*. The gene is transcribed, as evidenced by hybridization of DNA clones to the RNA synthesized in the puff. Nucleotide sequences of about 229 bp long, repeated over 15 times, were detected by restriction analysis within the cloned fragment. The sequence of ubiquitin is most conserved and it occurs in many higher eukaryotes (Izquierdo *et al.*, 1981, 1984; Arribas and Izquierdo, 1984; Arribas *et al.*, 1986). According to other data, the *E63-1* and *E63-2* genes are located in this puff. Ecdysterone induction of the *E63-1* is restricted to the salivary gland cells. The *E63-1* encodes a Ca^{2+}-binding protein related to calmodulin. The *E63-2* is 1254 bp in length (Andres and Thummel, 1995).

k. Puff 67B

The puff contains seven genes which are under control of both heat shock and ecdysterone (see Section VI,H for a detailed desciption). Increase in *hsp27* transcripts in the salivary glands depends on the increase in ecdysone titre characteristic of the end of the third larval instar. Level of the transcription in *ecd1* mutants at a restrictive temperature (29°C) is significantly lower then at 25°C (Huet *et al.*, 1996). During normal development transcription in salivary glands of the *hsp27* gene is demonstrated, beginning with 94-hr larva to 4-hr prepupa and then again at 7–8 hr and 10–13 hr (Huet *et al.*, 1996). Other developmental profiles were obtained by Dubrovsky *et al.* (1994). Nevertheless, expression of *hsp23*, *hsp26*, and *hsp27* were observed between PS2 to PS6, that is, at times when ecdysone-inducible puffs are active.

l. Puff 68C

The gene encoding *Sgs-3* protein was mapped between the *thread* and *hairy* loci on chromosome *3* at position 35.0 on the genetic map and to the 68A6-C11 region on the cytological map (Korge, 1975; Akam *et al.*, 1978; Kokoza *et al.*, 1982). Three transcribed sequences spanning 5 kb (Figure 122) were found in the region of the 68C puff. The lengths of the transcribed RNAs are 360, 320, and 1120 nucleotides. These RNAs are polyadenylated and found in polysomes (Meyerowitz and Hogness, 1982; Meyerowitz *et al.*, 1985a,b). The RNAs correspond to the *Sgs-8*, *Sgs-7*, and *Sgs-3* genes (Crowley *et al.*, 1983; Garfinkel *et al.*, 1983). P-element transformation experiments in which sequences of an intronless *Sgs-3* gene were inserted into the genome demonstrated that the absence of its intron does not affect the synthesis of *Sgs-3* RNA and protein (Mettling *et al.*, 1987).

Three secretory proteins share homologous regions: each has a 23-amino-acid terminus, which is not present in the mature form of the protein. Each also has a carboxyterminal site of about 50 amino acids that are cross homologous (Garfinkel *et al.*, 1983). The distinctive features of the *Sgs-3* polypeptide are: (1) a third control module, 234 amino acids in length; by contrast, the *Sgs-7* and *Sgs-8* polypeptides have only two modules each. The third control module consists of tandem repeats of the proline-threonine-threonine-threonine-lysine sequences; the molecule is composed of 40% threonine. (2) The *Sgs-3* protein is more glycosylated, presumably because of sugar attachment to threonine residues (Beckendorf and Kafatos, 1976; Crowley *et al.*, 1983; Garfinkel *et al.*, 1983); the length of the protein molecule can vary (Mettling *et al.*, 1985).

A DNA sequence about 5 kb long containing three genes completely overlaps the puff zone. Using *in situ* hybridization and high resolution microscopy, the most distal fragments were mapped to 68C4 and the most proximal

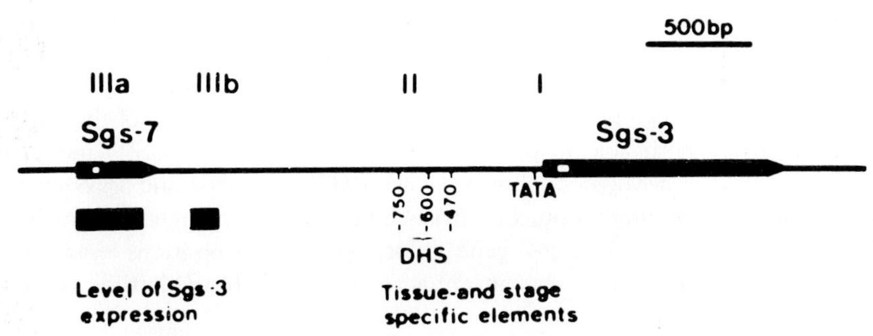

Figure 122. Organization of the *Sgs-7* and *Sgs-3* gene cluster in *D. melanogaster* (from Lezzi and Richards, 1989). I, proximal promoter, including the TATA-domain, sequences up to −130 bp suffice to provide stage- and tissue-specificity of the gene at a low level (Raghavan *et al.*, 1986); II, cluster of stage-specific sites of hypersensitivity to DNase I essential for normal expression of the gene (Ramain *et al.*, 1986); IIIa and IIIb, elements required for strong expression of the *Sgs-3* gene (Giangrande *et al.*, 1987); obviously, the coding sequences of the *Sgs-7* gene affect the expression of the *Sgs-3* gene.

to 68C7 (Kress *et al.*, 1985). The three genes lie in the middle of the sequence between the breakpoints of *Df(3L)vin*[3] and *In(3L)HR15*; that is, in a fragment of 20 kb (Meyerowitz *et al.*, 1985a,b).

The results of P-element-mediated transformation demonstrated that a fragment of 20 kb is sufficient for the expression of the cluster of three *Sgs* genes. Puffs form at the new sites of insertion only, when a fragment containing all the three glue protein genes is inserted into the genome (Richards *et al.*, 1983; Crosby and Meyerowitz, 1986b). All the three genes express at the same time, suggesting that they may have similar cis-acting elements (Meyerowitz and Hogness, 1982). As observed in P-element transformation studies, the *Sgs-3* gene is correctly expressed, provided that the transgene contains 2.27 kb of the *Sgs-3* sequence and some amount of DNA of the 3′ *Sgs-3* sequence (Richards *et al.*, 1983; Meyerowitz *et al.*, 1985a,b). A sequence from −2760 to −983 bp upstream from the *Sgs-3* gene contains two regions that modestly enhance expression (Giangrande *et al.*, 1987). Another element between positions −983 and −130 bp is responsible for a 10- to 20-fold increase in levels of expression over those observed for constructs covering only the proximal 130 base pairs (Bourouis and Richards, 1985a,b; Meyerowitz *et al.*, 1987; Ramain *et al.*, 1988). The tissue- and stage-specific regulatory elements located 130 bp upstream of the transcription start site ensure full-level expression by interacting with a remote enhancer-element (Bourouis and Richards, 1985a,b; Meyerowitz *et al.*, 1985a, 1987; Raghavan *et al.*, 1986; Crosby and Meyerowitz, 1986b; Giangrande *et al.*, 1987; M. Martin *et al.*, 1989a,b). A site located between −106 and −56, is sufficient for correct tissue- and stage-specific expression (Roark *et al.*, 1990). Two point mutations in the *Sgs-3* TATA sequence (TAAA or TAGA) reduce 50-fold transcription efficiency of the *Sgs-3* gene (Giangrande *et al.*, 1989). Within the 130 bp proximal fragment—a region stretching from −56 to −98 (M. Martin *et al.*, 1989a; Roark *et al.*, 1990)—where two groups of almost identical sequences are located, is important for full-level expression.

There are two regulatory elements in the upstream region of the *Sgs-3* gene which are both able to bind ecdysterone receptor (EcR/USP) and the product of the *fork head gene* (Lehmann *et al.*, 1997). The *Sgs-7* and *Sgs-8* genes are in the opposite orientation; the distance between their 5′ ends is 475 bp. The regulatory site of the *Sgs-7* gene is located between positions −43 and −211 bp. This element acts as an enhancer for *Sgs-3*, too (Mettling *et al.*, 1988).

The same regulatory motifs were found for five *Sgs*-genes (Table 14). The sequences in these elements are conserved in *D. simulans*, *D. erecta*, and *D. yakuba* (C. H. Martin *et al.*, 1988); and in *D. virilis* in the *Lgp-1* gene (Swida *et al.*, 1990). The consensus TTTG presumably controls tissue-specificity of expression (Jongens *et al.*, 1988).

The hormone ecdysterone is one of the trans-regulatory factors that affects the activity of the *Sgs* genes. Ecdysterone first stimulates (Hansson and

Table 14. Sequence Homology in Regulatory Regions of the *Sgs* Genes in *D. melanogaster*[a]

Genes	Sequences and their upstream positions	
Sgs-3	[−93]-TGTTTG (18 bp)	TCCATT-[−62]
	[−624]-TATTTG (28 bp)	TCCTTA-[−583]
Sgs-7/8	[−108]-GGTTTG (39 bp)	TCCATT-[−55]
	[−138]-TGTTTG (31 bp)	TCCAGT-[−95]
Sgs-4	[−324]-TATTTG (29 bp)	TCCTAC-[−282]
	[−458]-ATTTTG (31 bp)	TCGAGT-[−502]
	[−545]-TATTTG (31 bp)	TCCACA-[−501]
Sgs-5	[−119]-TGTTTG (39 bp)	TCAATA-[−67]
Consensus	TNTTTG (18–39 bp)	TCCANTA

Note. In square parenthesis positions of nucleotides following (left) or preceding (right) the homology region. Position of transcription initiation is −1.

[a]After Todo *et al.* (1992).

Lambertson, 1983) then inhibits the synthesis of secretion proteins. In *l(1)su(f)*^{ts67g} mutants, which are ecdysterone-deficient at restrictive temperatures, the puff 68C forms, but *Sgs*-RNA are not detected by Northern blot hybridization (Hansson *et al.*, 1981). It is not clear how secretion synthesis is provided in the other ecdysoneless mutants, such as *l(2)gl, l(3)tl, dor,* or in the *BR-C* mutants. In salivary glands isolated from mid-third-instar larvae and incubated *in vitro,* RNA synthesis in these genes continues for hours. However, a brief treatment with ecdysterone (15 min, 10^{-5} M) is sufficient to stop RNA synthesis (Crowley and Meyerowitz, 1984; Meyerowitz *et al.*, 1985).

The question was raised as to how the same hormone can induce and suppress the same gene. Two hypotheses were proposed to answer the question (Meyerowitz *et al.*, 1985a): (1) The initial induction of salivary gland cells by hormone results in activation of other genes, whose activities also affect later sensitivity to hormone in a reverse manner. (2) The turning on and off of the *Sgs* genes is possibly dependent on the different levels of hormone in the cell. The product of the *BR-C* gene is the second agent trans-regulating the *Sgs-3* and *Sgs-7* genes. In the *BR-C* mutants *lt435* and *npr-1*, the 68C puff is not inactivated (Belyaeva *et al.*, 1981; Crowley *et al.*, 1984). Despite the presence of the puff in *BR-C* mutants, Northern blot analysis did not reveal RNAs (Crowley *et al.*, 1984).

When the *Sgs-3* gene is active in DNA fragments containing the 5′ end, several DNAse-I hypersensitive sites—one in the region of the proximal promoter (−75 bp) and a cluster of sites in the distal regulatory region (−470, −600, −750)—were detected (Ramain *et al.*, 1986, 1988). GEBF-I (the glue enhancer binding factor), a possible product of the *BR-C* gene, binds to the

DNAse-I hypersensitive site at position −600. The factor is absent from mutants of the gene (Georgel *et al.*, 1991, 1992, 1993).

Based on comparisons of restriction maps of cloned DNA, RNA sizes and transcription direction at locations of the *Sgs* genes in representatives of closely related species of *Drosophila* of the *melanogaster* group (*D. melanogaster*, *D. simulans*, *D. erecta*, *D. yakuba*, and *D. teissieri*), it was concluded that their DNAs evolved rapidly. Moreover, the frequencies of nucleotide substitutions, insertions or deletions, inversions, and gene duplications were high. In contrast, in the neighboring distal 13-kb region, insertion/deletion events occurred 5–10 times less frequently (Meyerowitz and C. H. Martin, 1984; C. H. Martin and Meyerowitz, 1986, 1988; C. H. Martin *et al.*, 1988; Georgel *et al.*, 1992).

m. The region of the complex puff 71C3-4-D1-2

The genes encoding the ecdysone-induced proteins (Eip28/29) are located in the 71CE region that harbors a set of small puffs. However, it is not clear whether the formation of these puffs and *Eip28/29* activation in cell lines (called Kc) and also in proventricular, lymph gland, and hemolymph cells are related. Three short receptor binding regions were found on either side of the structural part of the gene: Prox and Dist (521 and 2295 bp downstream of the polyadenylation site), presumably required for gene response in cell culture, and a third region, upstream (-440) and not required for it. Each receptor-binding region functions as an *EcRE* site when located upstream from the promoter not responsive to ecdysterone. These regions contain the imperfect palindrome 5′-RG(GT)TCANTGA(CA)CY-3′. In the absence of hormone any one of the EcREs represses gene activity; when hormone is added, repression is abolished at the site, and additional activation is provided (Cherbas *et al.*, 1986, 1991).

n. The puff 71CE

The *Sgs-6* gene maps to chromosome 3 at position 42.0 in the 71C1-2-F3-5 region. The 71CE puff is located in this region (Velissariou and Ashburner, 1981).

o. The 71EF puff

In the 71EF region a complex puff develops. During late third-larval instar swelling develops in the 71E region; and at zero-hour prepupae, the neighboring band 71F1-2 unravels, forming the largest puff in *D. melanogaster's* genome: 71EF (Semeshin *et al.*, 1985a). Eleven developmentally regulated genes are nested in this puff (Figure 123). They are arranged in a cluster of 10 genes (*L71* genes) that are organized as five divergently transcribed gene pairs (Restifo and Guild, 1986a,b). The *L71-1* gene is localized by *in situ* hybridization in the 71E region; the *L71-10* is on the border of the 71F1-2 (Wright *et al.*, 1996). According to Wright *et al.* (1996), an ancestral L71 gene duplicated to form

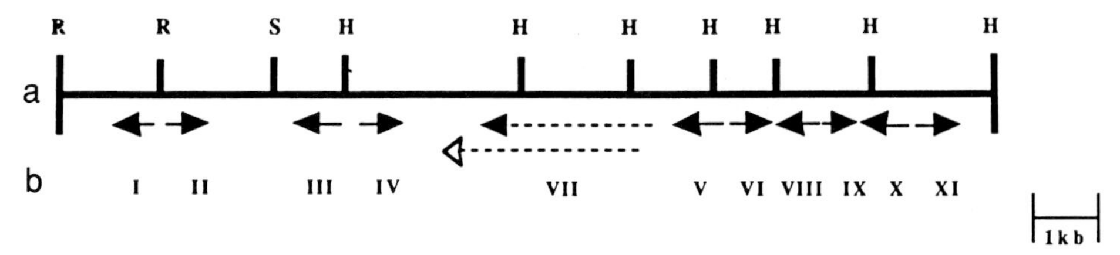

Figure 123. A molecular genetic map of the 71EF puff. (a) A restriction map (R, *Eco*RI; S, *Sal*I); (b) the I-XI genes, arrows indicate the direction of transcription (from Restifo and Guild, 1986a; Wright *et al.*, 1996).

the first gene pair, which was in turn duplicated to form the set of gene pairs. These genes (468 to 772 bp in length) encode small basic proteins with a conserved backbone of cysteine residues. The eleventh gene, *I71-7,* shows the regulatory and biochemical characteristics of the salivary gland intermolt glue proteins (Wright *et al.*, 1996). The *L71-1* to *L71-6* genes are expressed as late genes with transcript accumulation from 0- to 10-hr prepupa; *I71-7* (VII) is expressed with an intermolt accumulation profile (Restifo and Guild, 1986a; Guay and Guild, 1991). Northern blot analysis revealed typical late gene expression profiles for *L71-8* and *L71-9:* from 0- to 14-hr prepupa. The *L71-10* and *L71-11* genes show a similar but slightly delayed temporal pattern (Wright *et al.*, 1996). Expression of the cluster is under control of the BR-C and *E74A* genes (Guay and Guild, 1991; Urness and Thummel, 1995). The ETS domain DNA-binding protein encoded by the *E74A* early gene directly induces *L71-6* transcription (Urness and Thummel, 1995).

p. The 74E puff

The puff results from concomitant decompaction of the 74E1-2 and 74E3-4 bands (Semeshin *et al.*, 1985a). Cloning of this puff DNA (Burtis *et al.*, 1984, 1990; Moritz *et al.*, 1984; Thummel *et al.*, 1984; Jankneсht *et al.*, 1989) and subsequent molecular analysis demonstrated that two overlapping transcripts, *E74A* and *E74B*, are encoded in a region 60 kb long (Figure 124); however, each has its own transcription start site and a common 3′ terminal domain (Jankneсht *et al.*, 1989; Burtis *et al.*, 1990). Splicing of exons precedes polyadenylation during transcription (LeMaire and Thummel, 1990). Two proteins are translated from these transcripts. The TATA domain in the promoter zone maps at positions −17 and −24 in the *E74A* transcript unit. It was found that the sequence in an *in vitro* transcription system does not perform its function; (it does not provide accuracy of the start or transcription of its intensity. A sequence similar to a TATA domain is downstream from the start site of transcription at positions from +34 to +40 (Thummel, 1989). The structure of the *E74* gene seems highly conserved: the number and location of exons in

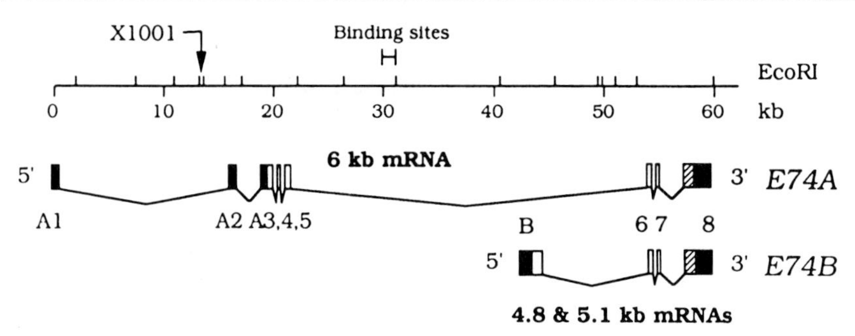

Figure 124. Structure of the *E74* gene (the 74EF puff). Solid rectangles, the 5′ and 3′ flanking regions not coding protein; open rectangles, protein coding sequences. Hatched part of exon 8 represents the fragment encoding the ETS-domain. Arrow points to the localization of the break of the *X1001* translocation; the "binding region," localization of three binding regions of proteins E74A and E74B (from Karim and Thummel, 1991).

D. virilis and *D. pseudoobscura* are, in general, the same as in *D. melanogaster* (Jones *et al.*, 1991). *E74* DNA hybridizes to the polytene chomosomes of six species of the *Drosophila* montium group (Drosopoulou *et al.*, 1997). In addition to *E74A* and *E74B*, four other transcripts (two at each flank) extending from 0.9 to 6.0 kb were detected in the relevant DNA fragment (Burtis *et al.*, 1990).

The mRNAs of the *E74* genes appear in cytoplasm at different times because of the threefold difference in their transcribed length (20 kb for *E74B* and 60 kb for *E74A*). The rate of *E74* transcript elongation is 1.1 kb/min, and this accounts for the 40 min difference in the transcription times of the two mRNAs (Thummel *et al.*, 1990; Thummel, 1992). Initiation of *E74B* transcription is sensitive to a lower titer of ecdysterone. As a result, the difference between the time of the appearance of the two transcripts reaches 9 hr. Thus, an invariant order in the synthesis of two proteins encoded by a single *E74* gene is an intrinsic feature of structural organization (Karim and Thummel, 1991, 1992). *E74A* transcripts are detected on Northern blots in 16- to 20-hr embryos (the signal is especially strong in 18-hr embryos). Weaker signals were obtained for 21- 24-hr embryos, as well as 36- to 48-hr and 60- to 72-hr larvae (i.e., before the first and the second larval molts). The signal is very strong in 108- to 120-hr larvae, and in 0- to 2-hr plus 10-hr prepupae. The gene is active throughout pupal development (reaching a maximum in 156- to 168-hr pupae) and in emerging adults. It is expressed in many larval organs (Thomas and Lengyel, 1986; Thummel *et al.*, 1990). Some differences in developmental profiles of *E74A* and *E74B* RNAs were found by Huet *et al.* (1993, 1995). *E74B* is transcribed from 86-hr larva to PS1 and then in 8- to 9-hr prepupa; *E74A* starts at PS2 and finishes at PS10, then in 1- to 4-hr and in 8- to 12-hr prepupa

(Huet *et al.*, 1993). The peak of protein appearance is delayed 2 hr after transcript abundance reaches its peak and 4 hr after puff development reaches its peak. Such a temporal uncoupling is not observed in 10-hr prepupae (Figure 125). The *E74B* gene is active in 18- to 21-hr embryos, and not during larval molts, the gene becomes active later from 96- to 108-hr to 192- to 204-hr pupae and it reaches maxima in 108- to 120-hr larvae and 180- to 204-hr pupae. The transcript is not detectable in adults (Thummel *et al.*, 1990).

The *E74* transcripts appear within 2 hr of the addition of ecdysterone to cultured salivary glands, and they disappear by 8 hr after it. When the glands are exposed to ecdysterone and cycloheximide, the *E74* transcripts appear later and they do not disappear by 8 hr of incubation; in contrast, they accumulate in great amounts presumably because transmission of the inactivation signals of the early puffs is blocked (Thummel *et al.*, 1990).

The *E74* gene encodes two proteins, E74A and E74B. Both have the same 85-amino-acid motif at their common C-terminal region. This is the ETS-domain which shows homology to the oncogene ETS of the avian erythroblastosis virus (Janknecht *et al.*, 1989; Burtis *et al.*, 1990; Karim *et al.*, 1990; Urness and Thummel, 1990). The ETS-domain is present in five additional *Drosophila* genes (Chen *et al.*, 1992).

E74 binding sites within the *Drosophila* genome were localized by staining the salivary gland polytene chromosomes with antibodies directed against

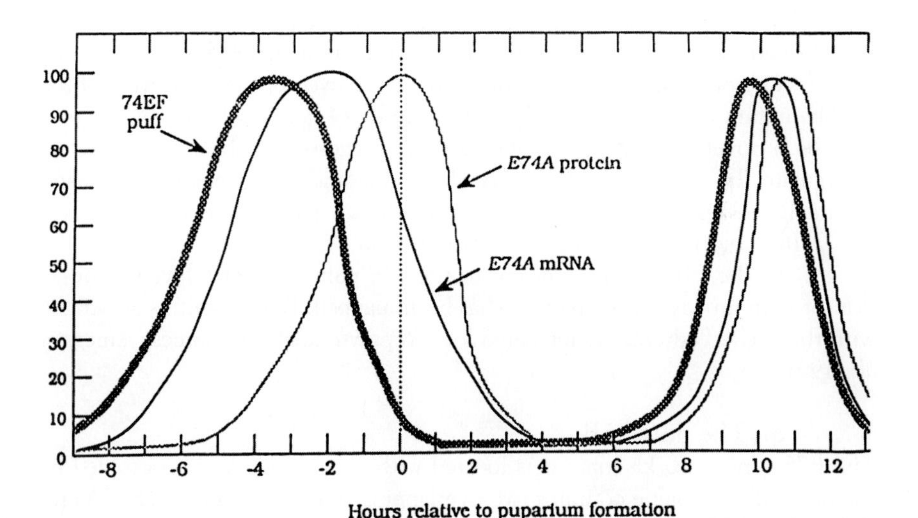

Figure 125. Temporal relationships between the processes of puff formation and accumulation of mRNA and proteins, the products of the 74EF puff. Abscissa, time before and after zero-hour puparium formation. Ordinate, the degree of the expression (percentage) (from Boyd *et al.*, 1991, *Development* **112,** 981–995, Company of Biologists Ltd.).

E74A protein. Approximately 70 sites in the polytene chromosomes stained with the antibodies (Urness and Thummel, 1990). Most of the sites correspond to the ecdysterone puffs. As shown in the figures, these sites include the 2C1-3 (referred to as 2B13-17 in the paper), 3A, 3C, 4F, 9E, 22B4-5, 27C, 42A, 43E, 47A, 47B, 50CD, 52C, 58BC, 62E, 63F, 64A, 71CE, 72D, 74EF, 75B, and 78C regions. Some puffs (2B3-5 and 26B) did not stain with antibodies (Urness and Thummel, 1990).

A 312-bp fragment that bounds the E74A protein was mapped in the middle of the *E74* gene, 11 kb upstream from the *E74B* transciption start site (Figure 124). It encompasses three sites sharing a common core with the consensus C/AGGAA. This supports the hypothesis (Ashburner *et al.*, 1974) that the products of the early puffs may regulate the activity of the early puffs (Urness and Thummel, 1990).

Recessive loss-of-function mutations specific to each transcription unit are predominantly lethal during prepupal development. *E74B* mutants are defective in puparium formation and head eversion and die as prepupae, while *E74A* mutants pupariate normally and die either as prepupae or pharate adults. Most puffs in salivary gland chromosomes are only modestly affected by the *E74B* mutations, whereas a subset of late puffs are submaximally induced in *E74A* mutant prepupae (Fletcher *et al.*, 1995). Among 30 primary ecdysone-regulated genes, expression of majority is unaffected by *E74* mutations. But *E74* is necessary for regulation of many ecdysone secondary-response genes; *E74B* is required for the maximal expression of glue genes, and *E74A* for maximal induction of late genes (puffs) (Fletcher and Thummel, 1995a). This defines *E74A* as one of regulators of late puff activity. Analysis of the phenoypes of the double-mutant animals reveals that BR-C and *E74* genes act together to produce both novel and synergistic effects, which suggests that these genes interact to regulate the expression of salivary gland glue and late genes (Fletcher and Tummel, 1995b). Ectopic expression of E74B can partially repress the *E78B* and *DHR38* orphan receptor genes (Fletcher *et al.*, 1997).

In the 74F region, adjacent to the 74E puff, another relevant gene—*Pep* (protein on edysterone puffs)—has been mapped. PEP protein is associated with the active ecdysterone regulated loci on polytene chromosomes (Amero *et al.*, 1991).

q. The puff 75B

The *E75* gene is 50 kb long; it is located within a 300-kb region occupied by the puff 75B. The gene contains two overlapping transcription units, *E75A* and *E75B*. The *E75A* unit encompasses the entire length of the gene; it has six exons and the mRNAs are of 4.9 and 5.7 kb (Figure 126). *E75B* is transcribed from 20 bp, and it forms two mRNAs, 5.2 and 6.0 kb in length. Transcription of these ecdysterone-inducible RNAs is initiated by two different promoters that are 30 kb apart. The transcripts are detected 2–6 hr after exposure to ecdysterone, and they disappear 8 hr later. When salivary glands were incubated in culture medium containing both ecdysterone and cycloheximide, the amount of RNA detected on Northern blots increased during entire incubation and reached its maximum by 8 hr (Segraves and Hogness, 1984, 1990; Feigl *et al.*, 1989). This was supposedly because the signals inactivating the late puffs were not delivered in the presence of antibiotic (see Figure 112). The *E75A* and *E75B* transcription units are induced at the same time the ecdysterone concentration reaches its maximum (Karim and Thummel, 1992). During development *E75B* transcript appears at PS1-5, PS7-9, PS12-15, and PS17-20 (Huet *et al.*, 1993, 1995). The *E75A* transcript appears at 94 hr of larval development, at PS3-10, (especially active at PS5-9), PS19-20 (Figure 131) (Huet *et al.*, 1993, 1995). Analysis of the E75 amino acid sequences demonstrated strong similarity to proteins of the steroid receptor superfamily (Feigl *et al.*, 1989; Segraves and Hogness, 1990; Segraves, 1991).

When the 75B regions are in a single dose, a considerable delay in increase in the activity of pyruvate kinase and arginine kinase in imaginal disks, which usually occurs in prepupae, was observed in larvae. This was presumably because the genes encoding these proteins are targets for reception of signals from the 75B puff (Collier and James, 1987).

r. The 75CD puff

The *β-FTZ-F1* gene (Figure 130), a member of the steroid receptor superfamily, maps within this puff. *β-FTZ-F1* is repressed by ecdysterone and therefore is active during the brief interval of low hormone titer in mid prepupae. Antibodies to the FTZ-F1 protein bind to 150 chromosome regions, including the 75CD puff (Lavorgna *et al.*, 1991, 1993; Woodard *et al.*, 1994). The 75C transcript appears at PS1–9 and again in 6-hr prepupa (PS16; see Figure 131). In *rbp*^{5} mutants this puff is prematurely induced at PS2 (E. S. Belyaeva and I. F. Zhimulev, 1996, unpublished observations). In a strong BR-C mutant, *l(1)t435*, this puff is always present (Belyaeva *et al.*, 1981). Ectopic expression of *FTZ-F1* after heat-shock *P-[F-F1]* transformed larvae leads to enhanced levels of BR-C, *E74*, and *E75* gene transcription and premature induction of stage-specific 93F puff (Woodard *et al.*, 1994).

s. The 78C puff

This puff in region 78 was referred to as 78D (e.g., Becker, 1959; Ashburner, 1967). Revision of the position for this region locates this puff at 78C4-6 (Semeshin *et al.*, 1985a). According to Ashburner and Richards (1976) it is an early late puff. It contains a single ecdysone-inducible gene consisting of two nested transcription units, *E78A* and *E78B* (Figure 130). The gene encodes the two members of the nuclear hormone superfamily. The *E78B* is directly induci-

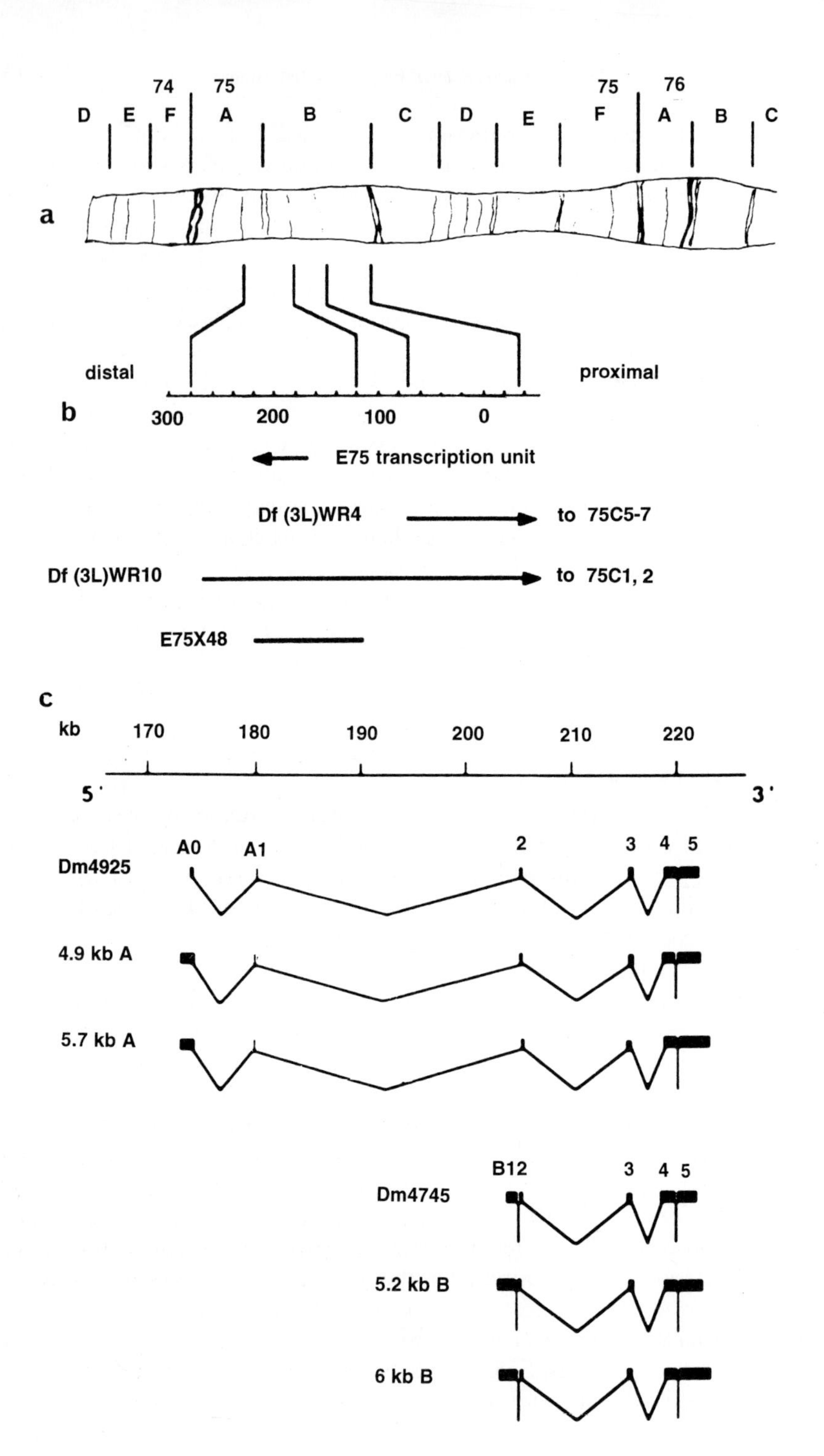
74
75
75
76
D
E
F
A
B
C
D
E
F
A
B
C
a
distal
proximal
b
300
200
100
0
E75 transcription unit
Df (3L)WR4
to 75C5-7
Df (3L)WR10
to 75C1, 2
E75X48
c
kb
170
180
190
200
210
220
5'
3'
A0
A1
2
3
4
5
Dm4925
4.9 kb A
5.7 kb A
B12
3
4
5
Dm4745
5.2 kb B
6 kb B

ble by ecdysone in late third-instar larvae and depends on ecdysone induced protein synthesis for its maximal level of expression (Stone and Thummel, 1993). E78B transcript appears at PS2-PS12 and again at PS16 (Huet *et al.*, 1995). The wild-type E78 functions are dispensable for normal development but required for the maximal activity of several late puffs, 63E and 82F. These puffs are reduced in size in *E78* mutants (Russell *et al.*, 1996).

t. The puff 90B3-8

The DNA from the *Sgs-5* gene hybridizes to the chromosome in the region of the 90B3-8 intermolt puff (Guild, 1984; Guild and Shore, 1984). The region of the salivary gland secretion gene *Sgs-5* contains 769 nucleotides and is divided into three exons by small introns. The protein-synthesizing region begins after a short (33 nucleotide) sequence of the untranslated RNA. The size of the *Sgs-5* protein is 163 amino acids. The *Sgs-5* gene has features characteristic of the other genes transcribed by RNA-polymerase II: the TATA (-25 bp) and the CAAT (-75 bp) domains, the polyadenylation sites. The sequence AA(AT)NTAAAGNNT (between -61 and -71 bp) usually found in the other Sgs genes was identified in the 5′ nontranscribed region of the *Sgs-5* gene (Shore and Guild, 1986). A secretion protein characterized in SDS-PAAG system was called P5 (Beckendorf and Kafatos, 1976). The locus of this fraction was genetically mapped between *bx* (3-58.8) and *sr* (3-62.0), i.e., within the interval in which the puff 90B3-8 occurs. It was suggested that both proteins are located in the same puff (Hoshizaki *et al.*, 1987).

u. The puff 93F

This puff is active in 9- to 14-hr prepupa (Figure 106). The *E93* gene located in the puff spans at least 55 kb of genomic DNA and encodes a 140-kDa protein. The *E93* mRNA is not induced by direct ecdysterone administration of late larval salivary glands but is induced 12 hr later by the prepupal ecdysteroid pulse. The *E93* transcripts are expressed in the gut, fat body, central nervous system, and imaginal disks. In rbp^5 mutants the 93F9-10 puff appears prematurely in larvae (E. S. Belyaeva and I. F. Zhimulev, 1996, unpublished observations). It has been proposed that *E93* may function as a stage-specific transcription factor in the prepupal salivary gland, acting in a regulatory hierarchy that results in the histolysis of the salivary gland (Baehrecke and Thummel, 1995). However, programmed cell death can be induced by culturing salivary glands of zero-hour prepupae in imago abdomen under conditions in which E93 is not induced. It may mean that everything necessary for the histolysis is transcribed

Figure 126. Molecular–cytogenetic structure of the 75B locus: (a and b) correspondence of cytogenetic and molecular maps and (c) transcript localization. (a) A cytological map of the region; (b) a molecular map and localization of rearrangements; (c) a map of exons in two transcripts (from Segraves and Hogness, 1990).

during larval wave of ecdysterone inducible puffs (Aizenzon and Zhimulev, 1975).

v. The I-18C puff of C. *tentans*

I-18C DNA was isolated by microdissection and microcloning (Hertner *et al.*, 1986). The ecdysteron-induced gene lies within a fragment 7.9 kb long (Amrein *et al.*, 1986). At least five transcripts are encoded by the gene (Figure 127): one of 1.8 kb with an open reading frame 417 bp in length, a 6.5-kb transcript, and 4.6-kb transcripts with open reading frames not longer than 270 bp. The 1.8-kb RNA is included in polysomes; the 4.6-kb RNA shows features characteristic of ribosome-free RNPs or monosomes (Amrein *et al.*, 1988; Lezzi *et al.*, 1989b; Dorsch-Hasler *et al.*, 1990).

w. The II-14A, II-9A, and IV2B puffs

These three puffs contain genes coding for receptor proteins homologous to *D. melanogaster* proteins *usp*, *DHR*, and *E75* respectively (Wegmann *et al.*, 1996; Lezzi, 1996).

x. The puff in region 10 of the X chromosome of *Drosophila nasuta*

The appearance of this giant puff at the preecdysterone stage in larvae temporally correlates with the appearance of salivary gland secretion fractions. The puff is enormous, possibly because the genes encoding secretion proteins are clustered (Ramesh and Kalish, 1988b).

y. *Drosophila lummei*

In this species, the α- and β-esterase genes map to the 2B2-2G5 regions, tentatively to the puff regions (Korochkin and Evgen'ev, 1982). The activities

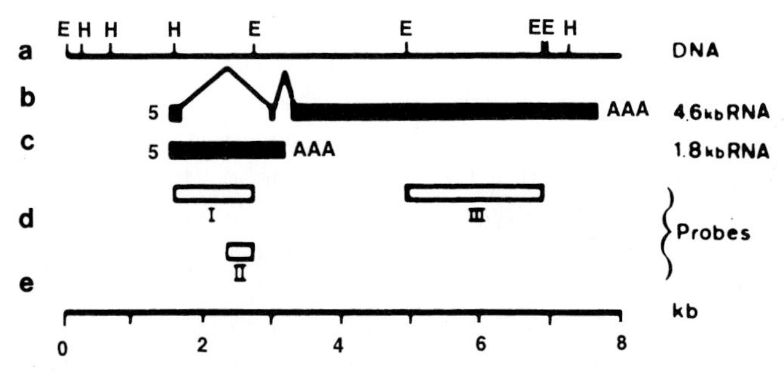

Figure 127. Organization of DNA in the I-18C ecdysterone puff of C. *tentans* (a) A restriction map of DNA; (b) 4.6-kb transcript; (c) 1.8-kb transcript; (d) probes; (e) scale in kilobases (reprinted from *Gene* **65**, Amrein *et al.*, Two transcripts of the same ecdysterone-controlled gene are differentially associated with ribosomes, 277–283, 1988 with kind permission of Elsevier Science – NL, Sara Burgerhartstraat 25, 1055 KV Amsterdam, The Netherlands).

of esterases and the puffs in the regions where these genes are located are correlated (Korochkin and Poluektova, 1983). In late third-instar larvae of *D. littoralis* (line 340), the ecdysterone puff forms in the 2G5 region, and the activity of β-esterase is high. In line 1012, a block of heterochromatin is located in this region; the activity of β-esterase is decreased; and the 2G5 puff is missing. A correlation of this kind was found for *D. lummei*. Constant puffing in the 2G1 region correlates with continuous activity of α-esterase (Yankulova *et al.*, 1986; Polyakova *et al.*, 1989). The puff 2G5 is absent from the line carrying the null allele of the β-esterase gene (Polyakova *et al.*, 1989, 1990). The data suggest that there may be a correlation between puff activation and enzyme appearance.

z. The 16A and 55E puffs in *Drosophila virilis*

The DNA from the intermolt puff 16A was isolated by microdissection. Two genes, *Lgp-1* and *Lgp-3*, were detected within it. Secretion protein RNAs are transcribed from these genes. Their transcription in salivary gland cells commences at the mid-third instar (at 118 hr), and it terminates by the time of puparium formation. Small amounts of these genes are transcribed in midgut and fat body cells (Kress *et al.*, 1990; Swida *et al.*, 1990). Comparison of the organization of the *Lgp-1* gene with that of the known *Sgs* genes of *D. melanogaster* revealed striking similarities between nucleotide sequences of *Lgp-1* exon-1 and the first exons of the *Sgs-3*, *Sgs-7*, and *Sgs-8* genes. Other fragments also show homology (Swida *et al.*, 1990). There is also a strong similarity in the organization of the promoter zones. In the *Lgp-1* gene from positions -318 to -281, a dyad of motifs ATTTGTCCAT-18-ATAACAAAT showing striking similarity to the element in the promoter gene *hsp23* of *D. melanogaster*—ATTTTCCAT-19-ATGGCAGAT—was found. The halves of this element in *D. virilis* and *D. melanogaster* are an imperfect reverse repeat. The halves act as independent ECR-binding sites. Another motif, AAGGGTTCA which is similar to the *hsp27* element of *D. melanogaster* (see Section VI,H), was detected at positions -303 to -294; that is, in the above mentioned dyad of *D. virilis*. Two motifs that are present in all the *Sgs* genes of *D. melanogaster* were detected in the proximal part of the 5′ nontranscribed promoter of the *Lgp-1* gene. The first motif, TATTTGCTC, is called "Prox1"; it lies between positions -122 and -113 (Figure 128); the second ATTT-4AATTG (Prox1′) is located at the same distance from Prox1 both in *Lgp-1* and *Sgs*, 19 bp downstream (Swida *et al.*, 1990; Kress and Swida, 1990). Removal of DNA fragments, containing either the proximal or the distal element reduced *Sgs-3* expression. Removal of both the proximal and the distal DNA elements abolished *Sgs-3* expression (M. Martin *et al.*, 1989a). The "Prox2" motif is characterized by the -CAGT-G-, identified in the nearest 20 bp upstream of the TATA domain in different genes (Figure 128). Analysis of the *Lgp-1* transcripts during the larval–pupal transi-

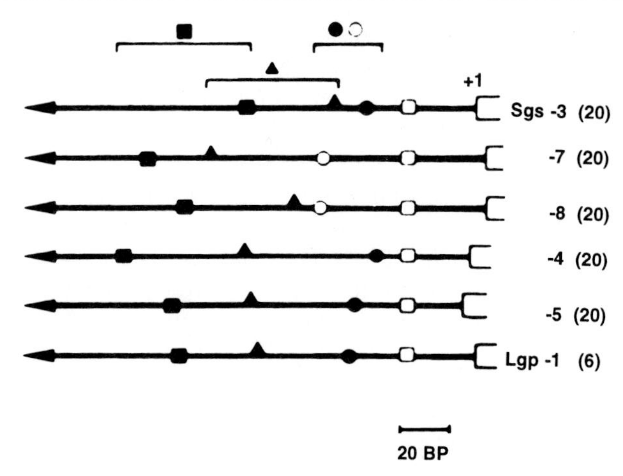

Figure 128. Similarity in disposition of conserved motifs of nucleotide sequences in the 5′-promoter zone of the *Sgs3-7* genes in *D. melanogaster* and *Lgp-1* in *D. virilis*. The transcribed part of the gene is designated by a rectangle with number +1, the non-transcribed promoter region is designated by a horizontal line. Open squares, the TATA-domain; solid circles, *Prox2* motif; open circles, element of symmetrical dyad motifs bound to *Prox2;* black triangle, *Prox1′* motif; black square, the *Prox1* motif (reprinted by permission from Kress and Swida, 1990).

tion periods demonstrated a progressive decrease in transcript length due to continuous shortening of the poly(A)$^+$ tracts (Swida *et al.*, 1990).

The 55E puff was called an intermolt one, because it is inactivated by ecdysterone at the end of the third larval instar. The DNA of this puff was isolated by microcloning. Two genes were identified in a 11-kb fragment. The genes are transcribed exclusively in the salivary gland cells during all three larval instars (Kress *et al.*, 1990).

2. Ecdysterone receptors

From the 1960s it has been known that [^{3}H]ecdysterone is mainly detected in cell nuclei (Karlson *et al.*,1964; Emmerich, 1969a,b; Weirich and Karlson, 1969; Claycomb *et al.*, 1971; Berendes, 1972a). The hormone binds to the chromosomes in the nuclei. Experiments were performed in which exogenous ecdysterone was crosslinked *in situ* to polytene chromosomes of *D. melanogaster* and *C. tentans* by photoirradiation. The crosslinked hormone was localized by immunofluorescent microscopy. Thus, hormone was crosslinked to its binding site prior to immunodetection. It was shown that the ecdysterone puffs are high-affinity binding sites for the hormone (Gronemeyer and Pongs, 1980; Gronemeyer *et al.*, 1981).

In the PS1-10 salivary glands of *Drosophila* larvae, there is a very strong correlation between the distribution of the puffing sites characteristic of a particular PS and ecdysterone binding to these sites. The hormone was detected in the intermolt puffs, and in the early and the late ecdysterone puffs (the two exceptions were the early puff at 23E and the intermolt puff at 25AC) throughout the active periods of the puffs (Dworniczak *et al.*, 1982, 1983). In the ecd^1 mutant, whose ecdysterone titer is sharply reduced, the hormone was not detected in the puffs (Dworniczak *et al.*, 1982).

Discussion of these data raised two questions (Belyaeva and Zhimulev, 1982): (1) Why does the chromocenter fluoresce brightly (Gronemeyer and Pongs, 1980)? and (2) Why does the number of fluorescent regions exceed by far that of the large ecdysterone puffs? The ecdysterone-binding regions are seen in the photographs accompanying the paper of Dworniczak *et al.* (1983). These regions include, for example, the 1E (Figure 1B), 2D and 4C (Figure 1E), 73F, and 56E1-2 (Figure 3A) regions, as well as many puffs in the 70C-75B (see their Figure 3C). Small puffs were described in all these regions (Belyaeva *et al.*, 1974; Zhimulev, 1974).

When injected into the isolated nuclei of the salivary gland cells, pure hormone did not induce the ecdysterone puffs (Alonso, 1972). Injection of hormone into the nucleus of the salivary glands of *D. hydei* resulted in a considerably smaller induction of the ecdysterone puff 4-78B than its injection into the cytoplasm (Brady *et al.*, 1974).

Ecdysterone complexed with protein is released from cells (Emmerich, 1970, 1972, 1975; Butterworth and Berendes, 1974; Maroy *et al.*, 1978; Schaltmann and Pongs, 1982; Lehmann and Koolman, 1988; Turberg *et al.*, 1988). The protein–receptor complexes and ecdysterone isolated from different cell types of *Drosophila* have similar sedimentation coefficients, 6 to 6.3 S. Partial purification of receptor proteins yielded a 130-kDa fraction with receptor activity (Pongs, 1989). Indirect evidence that ecdysterone binds to receptor molecules come from experiments in which the salivary glands were pretreated with N-ethyl-maleimide (NEM), which irreversibly alkylizes sulfhydryl reacting agents and thereby inactivates them. After this treatment, the induction of the 74E puff was inhibited and the extent of inhibition increased with increasing NEM concentrations. Pretreatment of the salivary glands with NEM for 2 to 3 min completely blocked the induction of the puff (Ashburner, 1972d; Rensing and Fischer, 1975).

Receptors are presumably synthesized in cells independently of the presence of hormone in culture medium. The salivary glands of first-instar larvae grown in adult abdomens normally react to ecdysterone by puffing at the responsive sites 74EF and 75B (Ashburner and Garcia-Bellido, 1973). However, nurse cells of adult *D. melanogaster*, mutant for *otu* gene, are not competent to react upon ecdysterone. *In vitro* cultivation of the nurse cells with ecdysterone for periods up to 24 hr, with hormone added after preincubation in pure medium, did not result in induction of any ecdysterone puff, probably because of absence of receptors (Zhimulev and Mal'ceva, 1997).

Cloning of genes encoding protein receptors, determination of amino acid sequences, and detection of characteristic domains in the protein molecules ushered in a new era in the study of receptors (see reviews by Evans, 1988; Segraves, 1991; Oro *et al.*, 1992; Richards, 1992, 1996; Mangelsdorf *et al.*, 1995; Thummel, 1995; Bender *et al.*, 1997; White *et al.*, 1997). Proteins that bind hormone show strong structural homologies. They have a DNA-binding domain 66–68 amino acids in length (Figure 129) and a domain (spanning approximately 220 amino acids) that binds to hormone or other ligands.

The steroid, thyroid, and retinoid hormone receptors, as well as the receptor-like molecules—whose ligands (the orphan receptors) are as yet not identified—have a similar organization. The number of identified orphan receptors is increasing. This suggests that additional steroid hormones may exist.

In *Drosophila*, 16 genes encoding proteins of the steroid receptor superfamily were identified (Table 15); 8 of these are either induced or repressed by the hormone at the level of transcription. *DHR38*, *DHR78*, and *DHR96* are mapped in the polytene chromosome regions 38E1-2, 78D4-5, and 96B12-14 (Fisk and Thummel, 1995; Sutherland *et al.*, 1995). None of these regions correspond to puffs in the salivary gland polytene chromosomes, and *DHR38* maps to a very condensed band (Pokholkova *et al.*, 1997).

In *Drosophila* the ecdysterone receptor consists of a heterodimer. One part of the receptor is encoded by the *EcR* gene (Koelle *et al.*, 1991) in three protein isoforms: ECR-A, *ECR-B1*, and ECR-B2 (Talbot *et al.*, 1993). All three ECR isoforms bind DNA in heterodimer with the protein → the product of the *ultraspiracle* gene: USP (Henrich *et al.*, 1990; Oro *et al.*, 1990; Shea *et al.*, 1990; Yao *et al.*, 1992, 1993; Thomas *et al.*, 1993; see for details, Thummel, 1995). All sites occupied by ECR antibodies in larval polytene chromosomes also stain with USP protein (Talbot, 1993; Yao *et al.*, 1993). DHR38 protein can compete *in vitro* against EcR for dimerization with USP (Sutherland *et al.*,

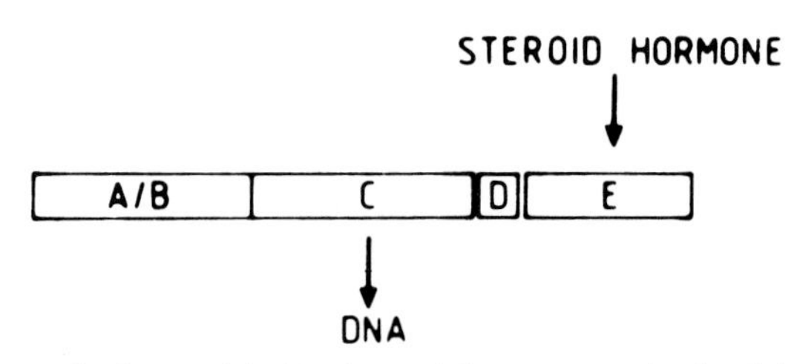

Figure 129. A general scheme of the structure of the protein molecule of the steroid hormone receptor. (A and B) The domain with unknown functions; (C) the DNA binding domain; (D) the hinge region; (E) the steroid hormone binding domain (from Pongs, 1988).

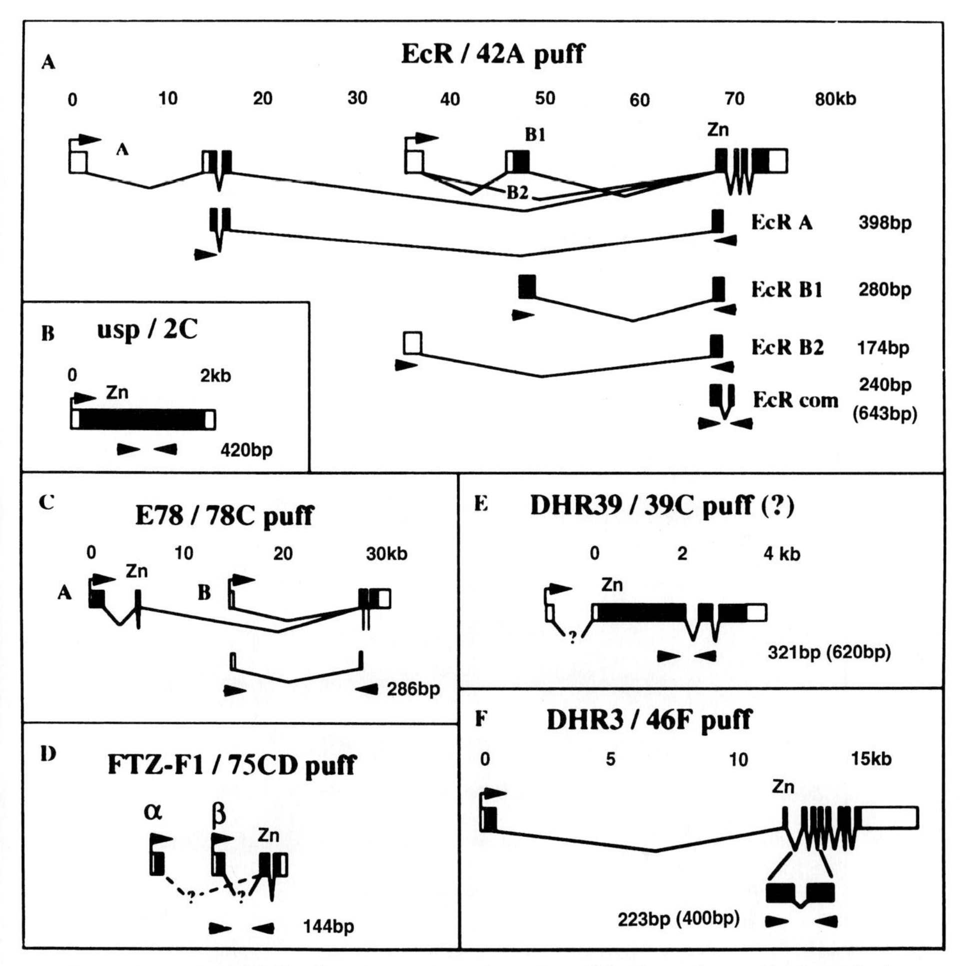

Figure 130. Structure of ecdysone receptor genes in *Drosophila*. For each gene (given with the corresponding chromosomal site and puff) the exons are depicted as box segments, the introns as lines. Coding regions are shown as solid boxes, nontranslated regions as open box segments (from review by Huet *et al.*, 1995, *Development* **121,** 1195–1204, Company of Biologists Ltd.).

1995), and DHR78 homodimer can bind as a homodimer to the *Eip 28/29 EcRE*, thus preventing the EcR-USP ecdysone receptor from activating the target promoter (Zelhof *et al.*, 1995).

Heterodimers of ECR and USP bind to the ecdysterone response element (EcREs) and to hormone, modulating transcription in these dimeric forms of the regulators. Riddihough and Pelham (1987) demonstrated that a 23-bp sequence from the *hsp27* gene promoter was necessary and sufficient to confer ecdysone inducibility on a heterologous reporter gene in a *Drosophila* cell trans-

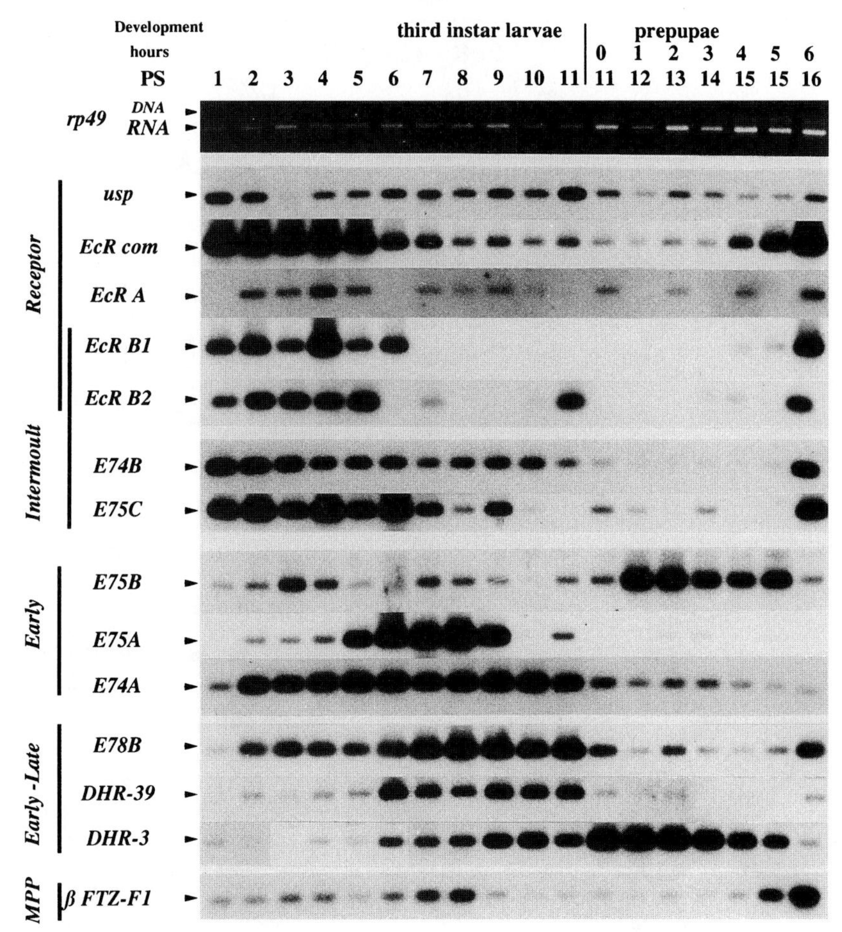

Figure 131. Transcripts profiles in salivary gland cells during the late larval ecdysone response from puff stage 1 to PS16. Each line derives from the same RT-PCR assay. With the exception of *rp49* and *E74* the molecules tested are members of the nuclear receptor family (from Huet *et al.*, 1995, *Development* **121,** 1195–1204, Company of Biologists Ltd.).

fection assay. It was later confirmed that this 23-bp element acts as an ecdysone response element (EcRE). The canonical EcRE at −537 in the *hsp27* gene has the following sequence: GGTTCAaTGCACT (see Cherbas *et al.*, 1991; Antoniewski *et al.*, 1993, 1996; Huet *et al.*, 1996 for references).

Two naturally occurring EcREs, from the ecdysone-inducible *hsp27* and *Eip28/29* promoters, are bound by EcR-USP (Yao *et al.*, 1992). Several ECR-USP binding sites have been identified upstream of the E75A promoter (Talbot, 1993). Ecdysterone receptors were identified in C. *tentans* (Imhof *et al.*, 1992,

Table 15. Characteristics of Proteins from the Superfamily of Nuclear Hormone Receptors of *Drosophila melanogaster* and *Chironomus tentans*[a]

Gene and symbol	Cytological localization	Homology of amino acids with hologous receptor of vertebrate (%)		Ecdysterone regulation[b]	Gene function
		DNA-binding domain	Hormone-binding domain		
		Drosophila melanogaster			
seven-up (svp)	87B	94	93		Embryonic lethal; rhabdomeres R3, R4, R1 and R6 are converted into R7
ultra-spiracle (usp)	2C1-3	86	49	−	The product constantly needed in the course of development
E75	75B	79	35	+	The early ecdysterone-inducible puff
E78	78C	72	35	+	The early-late ecdysterone puff
Ecdysone receptor, EcR	42A	81	34	+	The product binds to ecdysterone and activates ecdysterone-stimulated loci
Ftz-F1	75CD	88	33	+	Presumably binds and activates the genes *fushi tarazu*
tailless (tll)	100A	81	41		Controls development of the posterior part of the embryo
dHNF4	29E	90	67		
knirps (kni)	77E	—	—		Controls embryonic segmentation of the abdomen
knirps related (knri)	77E	—	—		Expression like in *kni*
embryonic-gonad (egon)	79B	—	—		Expression in embryonic gonads
DHR3	46F	75	20	+	
DHR 38	38E	88	54	−	
DHR 39	39C	62	17	+	
DHR 78	78D	74	26	+	
DHR 96	96B	4	28	+	
		Chironomus tentans			
—	2-17C	95[c]	75[c]		

[a]Data from Segraves (1991); Oro *et al.* (1992); Fick and Thummel (1995); Thummel (1995); Richards (1996).

[b]Response (+) or no apparent transcriptional response (−) upon ecdysterone in cultured larval organs

[c]Comparison with the corresponding domains of *EcR* protein of *D. melanogaster* (Imhof *et al.*, 1992).

1993; Wegmann *et al.*, 1995; Lezzi, 1996) and *C. thummi* (Deak and Laufer, 1995). Antisera against *C. tentans* EcR (cEcR) binds to approximately 50 transcriptionally active loci. Among them there are early ecdysteroid-inducible puffs sites, such as the locus containing the genes coding for the homolog of the E75 protein in *D. melanogaster*, as well as late ecdysterone puffs (a locus of a gene homologous to *usp* in *D. melanogaster*). Balbiani rings do not react with the antisera (Wegmann *et al.*, 1995). USP and ECR protein colocalize in polytene chromosomes of *C. tentans* (Lezzi, 1996).

3. Accessibility model for hormonal induction of puffs

In the 1960s, Kroeger put forward the hypothesis that ecdysterone may act at the level of cell membranes by preferentially altering their ion permeability. According to this model, hormone can affect the nuclear envelope. This leads to an accumulation of potassium ions in the nucleus and exerts an effect on puff activity in certain regions. The model is based on evidence indicating that the puff sets change in *C. tentans* after incubation of the salivary glands in solutions of different salts, that there is a mimicking of ecdysterone puffing at increasing K^+ concentrations (1.4% without Na^+ ions), and a regression to preecdysterone puffing in solutions at increasing Na^+ concentrations (0.6%, without K^+ ions). Other salts ($ZnCl_2$ and $MgCl_2$) can cause "rejuvenation" of puffing in some cases (Kroeger, 1963–1968; Kroeger and Lezzi, 1966; Lezzi, 1966, 1967a,b; Lezzi and Kroeger, 1966; Lezzi and Gilbert, 1970; Lezzi and Frigg, 1971; Lezzi and Robert, 1972; Kroeger *et al.*, 1973b; Kroeger and Muller, 1973; Kroeger and Trosch, 1974; Wuhrmann *et al.*, 1979; Wuhrmann and Lezzi, 1980a,b).

Because the structure of the nucleoprotein complexes largely depends on salt and ion composition of medium, the Kroeger hypothesis prompted a series of experiments (Berendes *et al.*, 1965; Clever, 1965a,b; Elgaard and Kessel, 1966; Rensing and Fischer, 1975) and raised discussions (Ashburner, 1970a; Ashburner and Cherbas, 1976; Kroeger, 1977). Taken together, the results of experiments designed to test this hypothesis were discrepant. In fact, the exposure of the salivary glands to potassium ions (Berendes *et al.*, 1965) yielded ambiguous interpretations: some investigators regarded the results as evidence for the Kroeger's hypothesis, others as evidence against it (see review by Ashburner, 1970a). In *D. hydei*, various changes in the ultrastructural organization of the salivary gland cells occur under the effect of potassium, sodium, magnesium, and calcium chlorides; however, no changes in puffing were observed (Ellgaard and Kessel, 1966). The notion that $ZnCl_2$ has an effect on puffing did not support the hypothesis in some studies (Clever, 1965b), but did in others (H. D. Berendes, cited after, Ashburner, 1970a).

Incubation of the salivary glands of late larvae of *D. melanogaster*, which had been stimulated by ecdysterone in artificial media containing substances affecting the ion permeability of membranes (trinactin, valinomycin, *N*-ethylmaleimid, tetraammonium chloride, tetrodoxin, and ouabaine), did not result in induction of the late puffs nor in the regression of the early (Rensing and Fischer, 1975). This was not surprising in view of the fact that the late puffs are not induced in the absence of ecdysterone.

Additional evidence was accumulating that ions are involved in the process of gene activation induced by hormone. In *C. thummi* the I-18C and IV-2B puffs are induced by both ecdysterone *in vivo* and by incubation of isolated nuclei in medium with high potassium content, and the intermolt puff I-19A is induced when sodium content is high (Lezzi, 1966, 1967a,b). This decompaction of the puff regions is not related to transcription (Lezzi and Robert, 1972). The presence of another inducer, ecdysterone, is a necessary requirement for adequate puff induction.

In diapausing larvae of *C. tentans*, there are periods when the I-18C region is supercompacted and not responsive to hormone (see Section VI,C,1). Studies on isolated chromosomes and nuclei, measurements of activities of intracellular ions, and use of ouabain (an inhibitor of active ion transport) all support the conclusion that changes in ion balance ensure the conditions required for decondensation of the material of the I-18C puff (Wuhrmann *et al.*, 1979; Wuhrmann and Lezzi, 1980a,b). The ion model for puff induction by

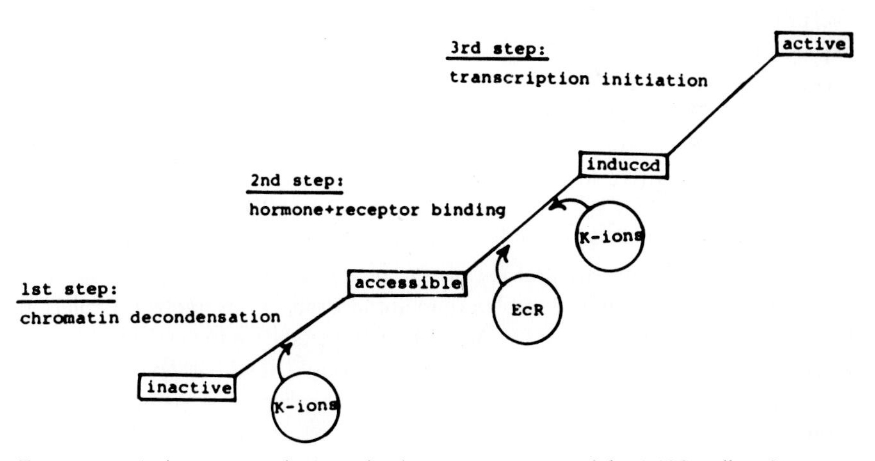

Figure 132. A three-step mechanism of ecdysterone activation of the 1-18C puff in *C. tentans*. The gene (rectangle) passes through four stages of activation. ERC, the ecdysone–receptor complex (from Lezzi and Richards, 1989).

hormones was developed on the basis of all these considerations (Figure 132). The first step is an K^+ ion-dependent decompactization of chromatin in the region, with the result that it becomes receptive to the action of the hormone receptor complex (EcR); the puff is induced, and transcription starts (Lezzi, 1980, 1996; Lezzi *et al.*, 1989b, 1991; Lezzi and Richards, 1989).

4. The scheme of cascade gene regulation under the effect of ecdysterone

The Ashburner model of cascade gene regulation (see Figure 112 above) faithfully reflects the relations between the formation of the early and the late puffs at the end of the third larval instar. The years elapsed from the time the model was developed have brought us to a clearer understanding of hormonal induction. Several new modifications of the model appear; these reflect hierarchial relationships between different waves of puffing (Woodard *et al.*, 1994; Huet *et al.*, 1995; Thummel, 1995). Among them, the modification developed by Richards (1992) looks promising: Ecdysterone is first released during the mid-third-larval instar. The 2B, 74E, and 42A puffs encoding receptor proteins and transcription factors controlling ecdysterone induction are concomitantly induced. The products of these genes are involved in the induction of the *Sgs* genes. Later on—6 hr before puparium formation, at the time when ecdysterone is released again—the *Sgs* genes are inactivated and, concomitantly with this, the wave of the late larval puffs is triggered (Figure 133). Hormone binds to the heterodimer receptor (R0 and EcR in Figure 132) and this complex binds to the promoter zone of the early ecdysterone-induced genes. As a result of the expression of the early genes, different proteins, including other protein receptors of ecdysterone (R1 in Figure 133), are transcribed. Then, a new complex, hormone plus two receptor molecules included in heteropolymer (EcR + R1 in Figure 133), act on the early-late puffs making them express, with the result that new receptors (R2 in Figure 133) are synthesized. When complexed with hormone and the EcR receptor, the new receptors induce the activity of the late puffs (Richards, 1992).

However, the temporal patterns of genes located in the puffs of different types appear to be more complex than anticipated. For example, the relation between the early puff 74E and some of the late puffs is readily established only when the puffing process itself is taken into account. Maximal development of the puff 74E precedes puparium formation by 4 hr; however, accumulation of protein product from this puff reaches maximum at the time of zero-hour prepupa (Figure 125). Thus, it is not clear how, if at all, the 74E product is involved in the activation of the puff in zero-hour prepupae; moreover, in the early-late puffs that are activated still earlier. A product of the puff 74E is obviously necessary for events occurring in the mid-prepupal period.

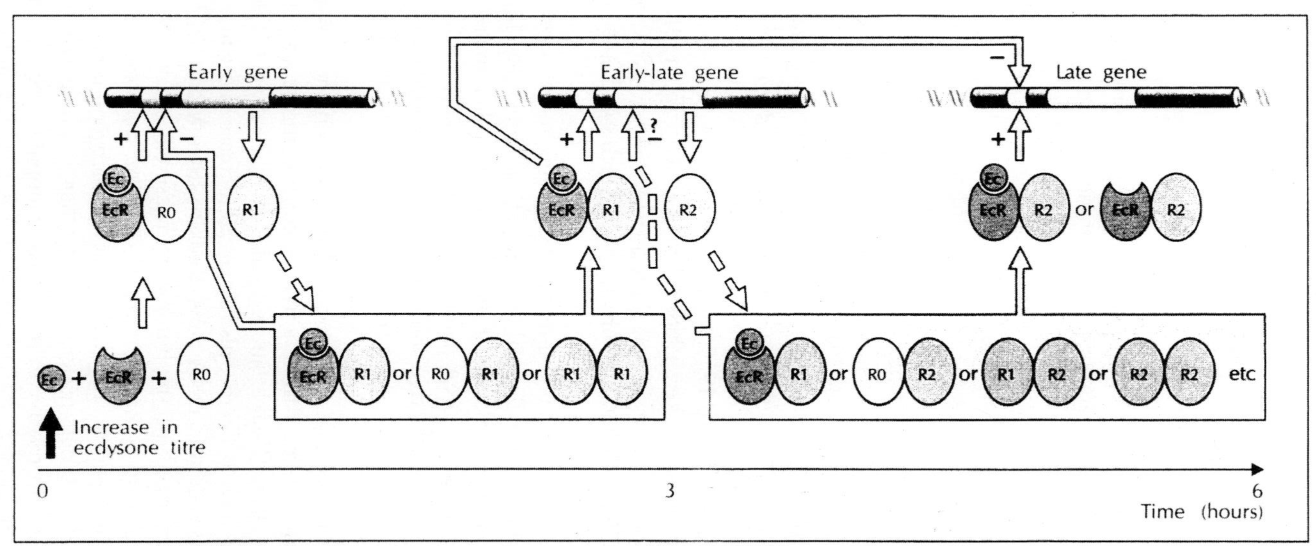

Figure 133. A model of ecdysterone activation of a gene cascade in the salivary gland cells of *Drosophila*. Ec, ecdysone; EcR, product of the *EcR* gene (protein receptor from the 42A puff); R0, receptor of the same group as *EcR*; R1 and R2, products of the early and the early–late puffs, respectively (from Richards, 1992).

F. The DNA puffs

Certain puffs in representatives of the Sciaridae family are loci involved in DNA amplification. They differ in this respect from the great majority of puffs. As a result of decompaction and, consequently, a decrease in DNA concentration in puff volume, the usual RNA puffs become light-staining (for example, with orcein) or transparent in appearance (when viewed in phase-contrast). In contrast the DNA puffs remain darkly stained because of the accumulation of extra DNA. However, their staining intensity can vary from very weak to very strong corresponding to the bands or heterochromatin (Figure 134; also see Rasch, 1970; Perondini and Dessen, 1988). According to the first descriptions, of DNA puffs during their formation a group of obviously euchromatic bands are transformed into a compact block of typically heterochromatic material, which subsequently develops into an enormous (bulb-like) structure. This process is reversible (Breuer and Pavan, 1954, 1955).

DNA synthesis that is excessive with respect to polytenization cycles occurs in the DNA puffs, as shown by chromosome staining (Breuer and Pavan, 1954, 1955) and by autoradiographical studies on the incorporation of [^{3}H]thymidine (Ficq and Pavan, 1957; Ficq *et al.*, 1958; Pavan, 1959a,b, 1965a; Swift, 1962; Mattingly and Parker, 1968b; Simoes *et al.*, 1970, 1981; Berendes and Lara, 1975; Machado-Santelli and Basile, 1973, 1975, 1978; Goodman *et al.*, 1976; Zegarelli-Schmidt and Goodman, 1981; Amabis and Amabis, 1984b; Lara *et al.*, 1991).

Studies on incorporation of 5-bromo-2′deoxyuridine demonstrated that DNA is synthesized preferentially in the DNA puffs (Figure 134). Injection of hydroxyurea into fourth-instar larvae of *Bradysia hygida* resulted in an inhibition of the development of all the DNA puffs. The inhibitory effect of hydroxyurea presumably is due to the selective inhibition of RNA synthesis (see below) during amplification (Sauaia *et al.*, 1971; Machado-Santelli and Basile, 1973, 1978).

There is a considerable increase in DNA mass in the regions of the DNA puff after termination of their activity compared to the stage preceding the formation of swelling (Rudkin and Corlette, 1957; Swift, 1962; Rasch and Rasch, 1967; Rasch, 1970). There is also a marked increase in dry mass concentration and protein content in DNA puffs (Rudkin, 1964; Vidal, 1977). Crouse and Keyl (1968) performed spectrophotometric measurements of DNA content of one of the three puffs on chromosome *II* of *Sciara coprophila*. During the studied developmental period the marker band underwent three replication rounds, while the DNA puff underwent five; that is, there was a fourfold increase in DNA content relative to the reference unpuffed regions. Hybridization experiments with cloned DNAs from the DNA puffs C3 and C8 of *Rhynchosciara americana* (Glover *et al.*, 1982; Lara *et al.*, 1985; Millar *et al.*, 1985), from the C4 puff of *B. hygida* (Paco-Larson *et al.*, 1990b), or from the II/9 puff of *S. coprophila* (DiBartolomeis and Gerbi, 1989; Wu *et al.*, 1990) were used to determine the degrees of extra replication. It was shown that the puffs pass (on average) three replication rounds and only one in the control regions. Because of its location in the DNA puff, the DNA of the mobile element is under the control of mechanisms regulating DNA amplification (Smith *et al.*, 1985). Further spectrophotometric measurements of DNA showed that there is a geometrical increase in the DNA content. This evidences that the DNA puff can undergo additional rounds of DNA replication (Crouse and Keyl, 1968).

Since their discovery, it has been suggested that the DNA puffs may be the sites of specific DNA amplification necessary for differentiation at developmental stages when excessive amounts of the gene product are required. The amplification product which was called "metabolic" DNA may either remain in the chromosome or be released from it (Pavan, 1959b, 1965a,b; Lara, 1987, 1990).

The extra synthesis of DNA in the DNA puff starts before or simultaneously with swelling the chromosome region and in association with RNA synthesis (Breuer and Pavan, 1955; Ficq *et al.*, 1958; Pavan, 1959; Pavan and da Cunha, 1969a). Of the 10 DNA puffs on the chromosomes of *Rhynchosciara angelae*, in 2 puffs DNA accumulated in the bands before puffing started; DNA synthesis took place at different steps of puff development in the other 8 puffs. This DNA presumably was not released from the puff; after puff regression, the band that had been previously puffed became much larger than it had been before puffing (Crouse, 1968; Guevara and Basile, 1973; Pavan *et al.*, 1975; Lara *et al.*, 1991).

Studies on [^{3}H]uridine incorporation have revealed that, concurrent with DNA amplification, RNA synthesis occurs in the puffs (Ficq *et al.*, 1958; Ficq and Pavan, 1961; Swift, 1962; Pavan, 1965b; Gabrusewycz-Garcia, 1968; Pavan and da Cunha, 1969a; Amabis *et al.*, 1977; Bonaldo *et al.*, 1979; Simoes and Cestari, 1981; Amabis and Amabis, 1984b). The DNA puffs are sites where RNA polymerase II is located (Santelli *et al.*, 1976). DNA/RNA hybrid antibodies (Busen *et al.*, 1982) and antibodies against the DNA binding Ku-antigen—which maps not only to the DNA puffs, but also to other transcriptionally active regions of the chromosomes—are detected in the DNA puffs (Amabis *et al.*, 1990). The content of methylated cytosine is increased in the fully expanded puffs, methylation is weak in the underdeveloped puffs (Wei *et al.*, 1981).

There was much earlier work on the transcriptional activity of the DNA puffs. Using nuclear RNA from the stage of puff formation for hybridization to DNA isolated from larvae before and after DNA puff formation (Meneghini *et al.*, 1971; Armelin and Marques, 1972; Balsamo *et al.*, 1973a), it was found that the hybridization level was higher to the DNA isolated after puff

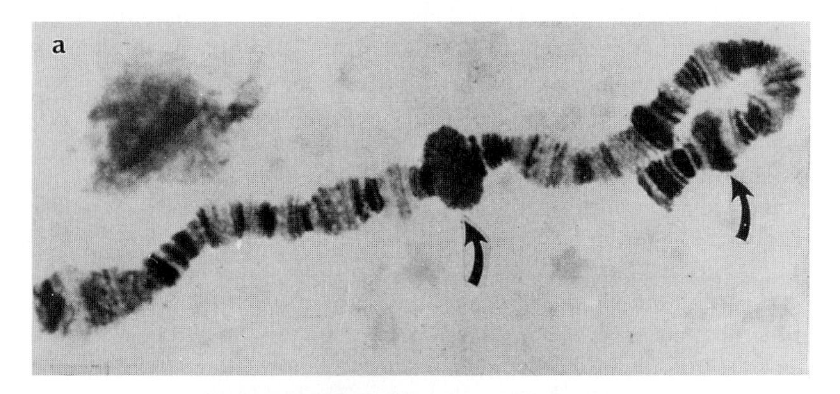

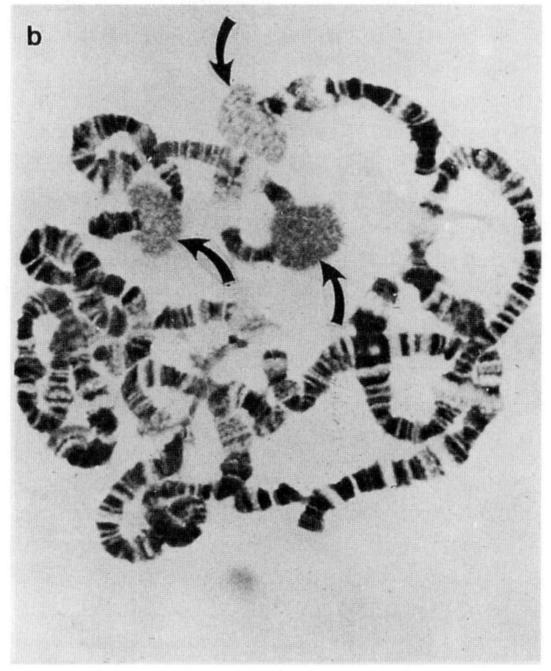

Figure 134. The DNA puffs. (a and b) Polytene chromosomes of *Sciara coprophila*, the DNA puffs are indicated by arrows; (c) incorporation of [^{3}H]thymidine into the DNA puff of *S. coprophila*; (d–f) incorporation of [^{3}H]uridine into the 2B DNA puff of *Rhynchosciara americana*; (g) incorporation of [^{3}H]5-bromodeoxyuridine into the chromosome of *Rhynchosciara hollaenderi*. C3, the DNA puff (a–c) from Puff development and RNA synthesis in *Sciara* salivary gland chromosomes in tissue culture, Cannon, G. B., *J. Cell. Comp. Physiol.*, copyright © 1965 Wiley-Liss, Inc. Reprinted by permission of Wiley-Liss, Inc., a subsidiary of John Wiley & Sons, Inc.; (d–f) reprinted by permission from Amabis *et al.*, 1977; (g) from Bradshaw and Papaconstantinou, 1970.

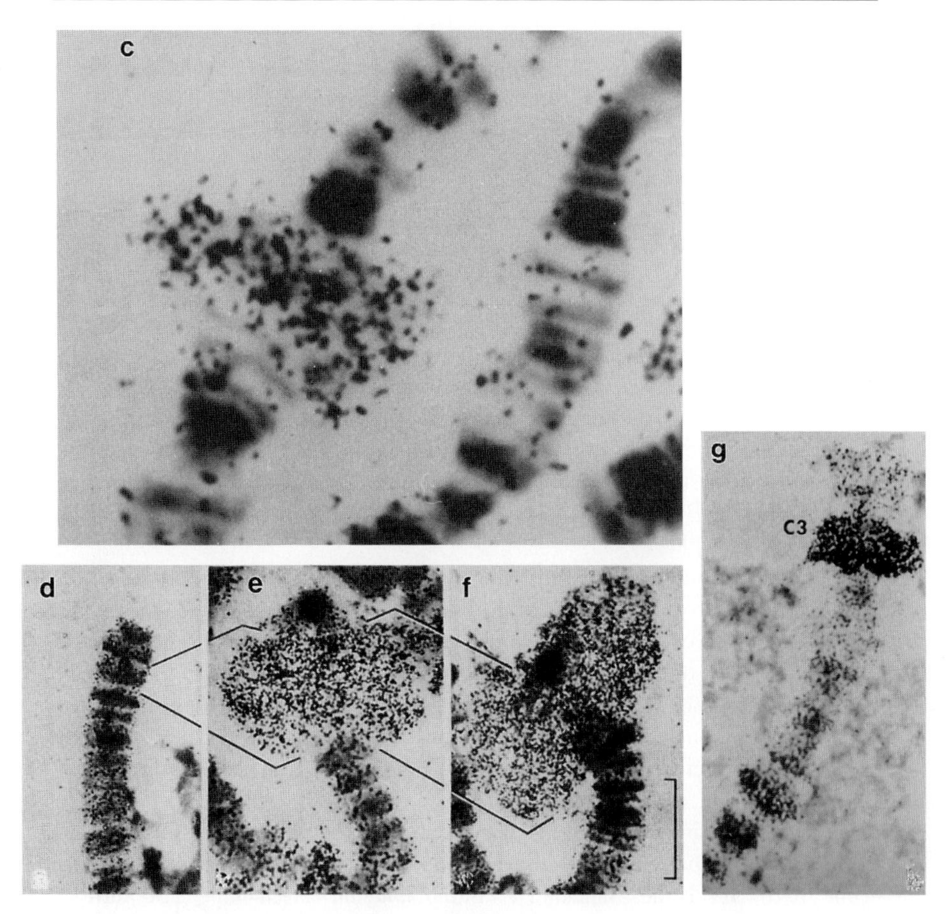

Figure 134. *Continued*

formation. This suggested that an intermediate fraction is the only repetitive DNA fraction synthesized during DNA puff formation. RNA able to hybridize to repetitive DNA was not detected, although nuclear RNA hybridized to DNA with various degrees of reiteration. It was concluded that the RNA from the DNA puff is confined to the cell nucleus (Balsamo *et al.*, 1973b) and that its transport to the cytoplasm remains to be proven (Edstrom, 1974).

These questions were resolved after the DNA from the DNA puffs was cloned according to one of the following schemes:

1. Clones were studied representative of genes active during the period when the DNA puffs are active. *In situ* hybridization confirmed that the assignment of clones to the DNA puffs was correct (Glover *et al.*, 1982).

2. Microlibraries of clones were generated by microdissection of predefined chromosome fragments. They were assayed for homology to eiter poly(A)$^+$ RNA or a cDNA library. Subsequently, the clones were also mapped *in situ* (Pirotta *et al.*, 1988; Paco-Larson *et al.*, 1990a). A 14S poly(A)$^+$ RNA was isolated by microdissection from the B2 DNA puff of *Rhynchosciara*. It also hybridizes *in situ* to the B2 DNA puff (Bonaldo *et al.*, 1979). The RNA from the puffs C3 and C8 hybridized on Northern blots to the DNA of these puffs (Zaha *et al.*, 1984).

Taken together, these data indicate that RNA is normally synthesized in the DNA puffs and that it is normally transported to the cytoplasm.

Molecular genetic analysis of the organization of the puff DNAs in *S. coprophila* demonstrated that two genes, II/9-1 and II/9-2, whose nucleotide sequences show 85% identity, are located in the puff II/9 in a region covering about 4 kb (DiBartolomeis and Gerbi, 1989). Two genes are located in each of the puffs B10 in *B. higida* (Fontes *et al.*, 1990) and two in C8 in *R. americana* (Osorio and Lara, 1990).

A possible mechanism of transport of RNA synthesized in the DNA puffs in the form of micronucleoli has been discussed (da Cunha *et al.*, 1969b; White, 1973). In *S. coprophila*, the micronucleoli or RNP globules are formed in the nuclei of the salivary gland cells (Gabrusewycz-Garcia and Kleinfeld, 1966). No excess of DNA is observed in the micronucleoli because it remains on the chromosomes during puff regression. In *Hybosciara fragilis*, some amounts of DNA are detected in the micronucleoli (da Cunha *et al.*, 1969b; da Cunha, 1972).

As already noted, the DNA puffs are found only in Dipteran insects of the Sciaridae family. A single instance was mentioned where the DNA puffs occur in the black flies *Simulium venistum/Simulium verecundum*. The authors relied on the observation that one of the puffs arising during the decompaction of six bands was strongly Feulgen-positive (Rothfels *et al.*, 1982). However, no evidence for DNA amplification was provided.

At least one puff at the end of chromosome A of *Sarcophaga bullata* synthesizes DNA (Whitten, 1965, 1969). The illustrating photograph suggests that this puff develops between the regions of intercalary heterochromatin; the result is that the compact and active chromatin intermingle, as in the case of the puff 2B of *D. melanogaster* (see Section VI,B,2). According to Bultmann and Clever (1969), all the puffs in foot-pad cells stain purple with azure B and incorporate [^{3}H]uridine. The incorporation of [^{3}H]thymidine was not preferential. Thus, there is no reason to believe that the DNA puffs are formed in foot pad cells.

The DNA puffs are specific to the cells of a single larval organ, the salivary gland; they are not detected in the cells of Malpighian tubules or mid-

or hindgut (Breuer and Pavan, 1955; Guevara and Basile, 1973; Pavan *et al.*, 1975). The number of salivary-gland puffs of this type is not large—about 10 in different species (Breuer and Pavan, 1955; Breuer, 1967; Guevara and Basile, 1973; Pavan *et al.*, 1975; Stocker and Pavan, 1977; Laicine *et al.*, 1984). Detailed studies on the DNA puffs in *S. coprophila*, *S. impatiens*, *S. pauciseta I* and *II*, and *S. prolifica* demonstrated the presence of 9 large and 9 small puffs in this species (Gabrusewycz- Garcia, 1964, 1971, 1975). According to other data 12 DNA puffs were detected in *S. ocellaris* (Perondini and Dessen, 1985).

The DNA puffs are concomitantly induced in all the cells of both the proximal and distal portions of the salivary gland (Cannon, 1965), although certain puffs are formed earlier in the distal part (Gabrusewycz-Garcia, 1975); puffs in different parts of the gland also show certain specificities (Table 11).

Characteristic stages are distinguished during the formation of the DNA puffs of *S. coprophila:* (1) the puffs are small; the increase in the diameter of the swelling relative to the unpuffed part of the chromosome is small. The puffs are compacted, they stain intensely with orcein or by Feulgen. (2) the puffs become larger; the intensities of their staining with orcein and by Feulgen, as well as incorporation of [^{3}H]thymidine, are comparable with those observed at stage 1. (3) the puffs reach their maximum sizes both in the lateral and longitudinal directions; neighboring bands become decompacted and spherical; staining with orcein and by Feulgen is very weak or diffuse; and intensity of [^{3}H]thymidine incorporation decreases. (4) The sizes of the puffs decrease; because of compaction of chromosome material the intensity of staining for DNA again increases, and that of [^{3}H]thymidine incorporation decreases (Gabrusewycz-Garcia, 1975).

Characteristic morphological changes in the DNA puffs also occur at the ultrastructural level. At the early and the late stages of development, when much of the puff material is in the compacted state, the DNA puffs resemble condensed bands. At the stage of maximal puff development, most of chromatin is diffuse; the DNA puffs resemble the usual RNA puffs. RNP particles, 23.8 nm in size, are detected in some of the DNA puffs; granule sizes and the variation in mean granule size is smaller than found in the RNA puffs granules (Gabrusewycz-Garcia and Garcia, 1974).

The DNA puffs appear only during the final stage of larval development when it is proceeding normally (Crouse, 1968; Mattingly and Parker, 1968a; Gabrusewycz-Garcia, 1975; Pavan *et al.*, 1975; Stocker and Pavan, 1977; Amabis, 1983a,b), that is, at the time when the titer of hormone ecdysterone is high (Stocker *et al.*, 1984).

It seemed doubtful whether DNA puffs can be under hormonal control: DNA puff formation was inhibited, while development and metamorphosis were on schedule in cortisone-fed larvae. It was even concluded that puffs are not essential for the normal development to occur in all the other Diptera (Goodman *et al.*, 1967; Goodman, 1969). However, these early findings proved not to be telling. According to Crouse (1968), injection of ecdysterone into larvae induced rapid development of the DNA puffs at their usual sites. Injection of ecdysterone together with cortisone usually resulted in the development of these puffs, while cortisone alone did not induce the puffs. Nevertheless, development was delayed in cortisone-fed larvae. Crouse believed that the DNA puffs would develop if the larvae had the chance to grow to the time of puparium formation; that is, cortisone does not inhibit the DNA puffs (they would be simply missing at this developmental stage); also note that cortisone is rather toxic, retarding development. It was shown further that exogenous cortisone acetate has no effect on the DNA puffs (Midelfort and Rasch, 1968).

The results of many experiments are known in which the DNA puffs were induced by injection of ecdysterone into larvae (Crouse, 1968; Gabrusewycz-Garcia and Margles, 1969; Stocker *et al.*, 1972, 1973; Stocker and Pavan, 1974; Berendes and Lara, 1975; Fresquez, 1979; Amabis and Amabis, 1984a; Dessen and Perondini, 1985b; Alvarenga *et al.*, 1991; Lara *et al.*, 1991; Gerbi *et al.*, 1993); incubation of salivary glands with hormone have been performed (Goodman *et al.*, 1976; Alvarenga *et al.*, 1979).

In ligature experiments, the anterior part of the larva was separated from the ring gland, the ecdysterone source, shortly before the formation of the DNA puffs in *Rhynchosciara angelae;* this prevented puff induction (Amabis and Cabral, 1970; Dessen and Perondini, 1973; Amabis *et al.*, 1977; Amabis and Amabis, 1984b). When salivary glands from larvae of *R. angelae* at the early stages of development lacking the DNA puffs were transplanted into larvae with the puffs, these were induced 4–5 days later in the transplants (Amabis and Simoes, 1971; Amabis, 1974). In contrast, transplantation of the salivary glands from larvae with developed puffs to younger larvae led to their rapid (within 5 hr) and complete regression (Simoes *et al.*, 1970).

Two genes (*II/9-1* and *II/9-2*)—in the DNA puff II/9A of *S. coprophila* coding for secretory proteins—were cloned (Gerbi *et al.*, 1993). Promoter sequences of the *II/9-1* gene resembles a consensus ecdysone response element (EcRE); these sequences confer ecdysone inducibility to a reporter gene introduced into *D. melanogaster* and bind EcR *in vivo* and *in vitro* (Bienz-Tadmor and Gerbi, 1990; Bienz-Tadmor *et al.*, 1991; Gerbi *et al.*, 1993). The putative origin of DNA replication lies further upstream of the transcription control region of gene *II/9-1*. Preliminary observations suggest that an EcR-binding sequence exists close to that putative origin (F. D. Urnov and S. A. Gerbi in Lezzi, 1996). In antiserum against the semiconserved D-domain of *Chironomus tentans* ecdysone receptor protein (cEcR) binds to DNA puff sites of *Trichosia pubescens* (Stocker *et al.*, 1997).

Induction of the DNA puff is indirect: (1) When larvae were treated with ecdysterone, the start of DNA replication and amplification in some DNA puffs was delayed up to 19–23 hr (Stocker and Pavan, 1974; Berendes and Lara,

1975; Fresques, 1979; Amabis and Amabis, 1984a). (2) Formation of the DNA puff B2 in *Rhynchosciara* is induced in young larvae by injection of total RNA isolated from the salivary glands during the period of development when the DNA puffs are active (Graessmann *et al.*, 1973). (3) Injection of actinomycin D after ecdysterone reduced the rate of [^{3}H]thymidine incorporation at the amplifying regions within 15 h (Berendes and Lara, 1975; Amabis and Amabis, 1984a). (4) The ecdysterone induction of DNA puffs is blocked by protein synthesis inhibitors (Lara *et al.*, 1991; Gerbi *et al.*, 1993); according to other data, ecdysterone induced DNA puff formation, when isolated salivary glands were incubated for up to 4 hr in medium containing cycloheximide (Goodman *et al.*, 1976).

Lara *et al.* (1991) envisaged a stepwise enfoldment of the process of induction of the DNA puffs: (1) an increase in the ecdysterone titer induces some of the early RNA puffs and inhibits the other puffs that are active during the midfourth instar; (2) the products of the activities of the early RNA puffs and ecdysterone induce DNA amplification and the DNA puffs; concomitantly with this, another group of the RNA puffs later in the sequence are induced; (3) the products of the late puffs suppress the activity of the DNA puffs and of the early RNA puffs.

The functions of the DNA puffs are presumably twofold: involvement in the preparation of cells to histolysis and encoding of secretion polypeptides.

Data indicating that infected cells in which the DNA puffs do not form are not lysed supported the idea that the DNA puffs are involved in histolysis (Pavan *et al.*, 1971, 1975). This does not appear convincing in view of the fact that injection produces a very complex syndrome severely affecting chromosome organization and cell physiology.

The finding of heavier polysomes at this stage of development is indirect evidence that the presence of the DNA puffs gives rise to changes in cell functions (Alliaga *et al.*, 1975). Fluorographic analysis of polypeptides of the salivary glands of *R. americana* revealed five fractions synthesized only during the period of development when the DNA puffs are active. The appearance of polypeptide 5 (with a molecular weight 21 kDa) and of a 16S poly(A)$^+$ RNA correlates with the apearance of the B2 puff, since the polypeptide, the transcript, and the puff all occur in the most proximal part of the salivary gland (Winter *et al.*, 1977b, 1980; Okretic *et al.*, 1977; Alvarenga *et al.*, 1991). Using translation of messages transcribed from the RNA puffs, in a system of lysates of rabbit reticulocytes, it was shown that poly(A)$^+$ RNA isolated from the cells of the proximal part of the gland is translated into polypeptides, including the 21-kDa protein (Toledo and Lara, 1978). These polypeptides were detected in the salivary gland secretion used to spin the common cocoon (Winter *et al.*, 1977a).

A temporal correlation was established between the appearance of specific polypeptides and the formation of the DNA puffs in the salivary glands of *B. hygida* (Laicine *et al.*, 1982, 1984) and *Trichosia pubescens* (Ferreira and Amabis, 1983). Molecular characterization of the C8 gene from the C8 late DNA puff in *R. americana* show that this gene encodes for secretory protein (Frydmann *et al.*, 1993).

G. The heat-shock puffs

During the early 1960s Ritossa (1962–1964b) made the observations that several new puffs are inducible by a brief exposure of larvae to high temperature (heat shock) or by treatment of salivary glands with 2,4-dinitrophenol, an inhibitor of oxidative phosphorylation in the polytene chromosomes of *Drosophila busckii* (Figure 135). The discovery stirred great interest and gave impetus to numerous studies. Based on the original results a new biological phenomenon

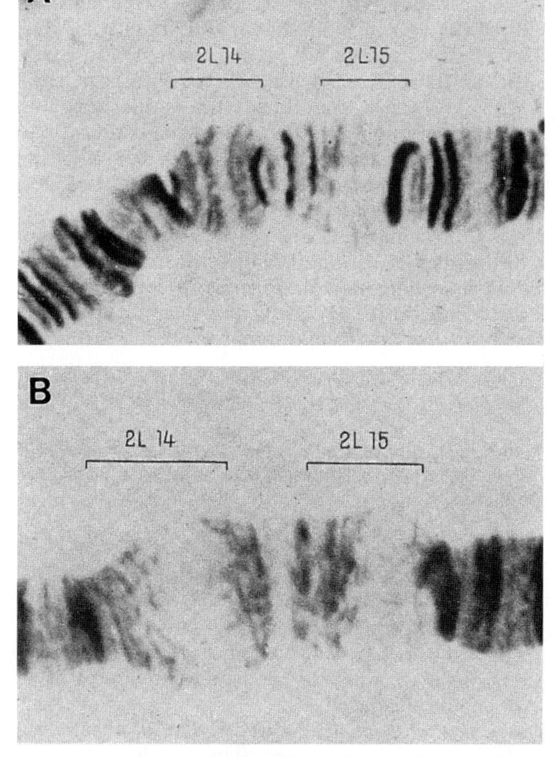

Figure 135. First depiction of the heat-shock puffs in polytene chromosomes of *Drosophila busckii*. The 2L14 and 2L15 regions in larvae at temperature of (a) 25°C and (b) after 30 min treatment at 30°C (from Ritossa, F. (1962). A new puffing pattern induced by temperature shock and DNP in *Drosophila*. *Experientia* **18**, 571–573).

was described, the syndrome of cell response to environmental stresses. The relevant literature is voluminous (see reviews by Berendes, 1973; Berendes *et al.*, 1974; Ashburner and Bonner, 1979; Rodionov *et al.*, 1981; Zegarelli-Schmidt and Goodman, 1981; Adams and Rinne, 1982; Lozovskaya *et al.*, 1982; Rensing *et al.*, 1982; Schlesinger *et al.*, 1982; Alahiotis, 1983; Craig, 1984; Lozovskaya and Evgen'ev, 1984; Neidhardt *et al.*, 1984; Nover, 1984, 1990; Nover *et al.*, 1984; Atkinson and Walden, 1985; Bienz, 1985; Bonner, 1985; Pelham, 1985, 1986, 1988; Burdon, 1986, 1988; Lindquist, 1986; Lakhotia, 1987; Pardue *et al.*, 1987; Piper, 1987; Lindquist and Craig, 1988; Morimoto *et al.*, 1990; Schlesinger, 1988; Sorger, 1991; Zhimulev, 1994a; Lezzi, 1996).

The main points concerning the induction and molecular genetic organization of the heat-shock puffs and genes will be summarized below and in Section VI,H.

The heat shock puffs (Figures 136 and 137) were found in all the studied species with polytene chromosomes. The number of puffs in the genome varies from one to nine. The optimal temperature for induction is 33–37°C (Table 16). The heat-shock puffs are induced not only by high temperature, but also by many chemical agents: uncouplers of oxidative phosphorylation, inhibitors of electron transport, substances which act as hydrogen acceptors (Table 17), and mitochondria appear to be the primary targets of heat shock (Leenders *et al.*, 1974; Sin, 1975; Sin and Leenders, 1975). Substances that decrease the concentration of ATP did not induce puffing (Leenders and Backer, 1972; Leenders and Berendes, 1972; Leenders *et al.*, 1974; Behnel and Rensing, 1975; Ellgaard and Maxwell, 1975; Koninkx, 1975; Koninkx *et al.*, 1975; Leenders, 1975; Sin and Leenders, 1975; Vossen *et al.*, 1977; Belew and Brady, 1981a).

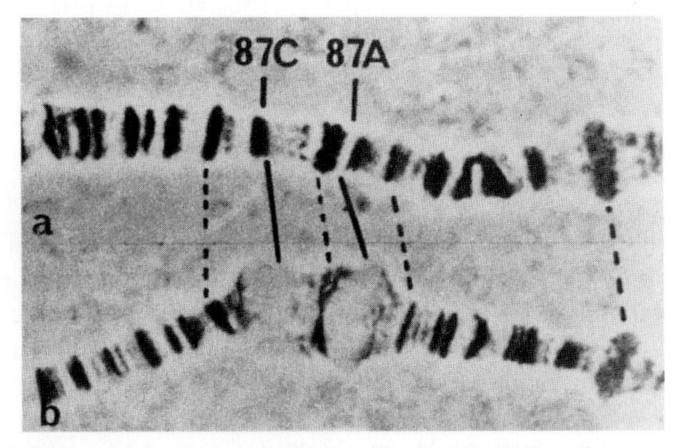

Figure 136. Induction of the 87A and 87C heat-shock puffs in *D. melanogaster*. (a) Control; (b) induced puffs (40 min at 37°C) (from Ashburner and Bonner, 1979, © Cell Press).

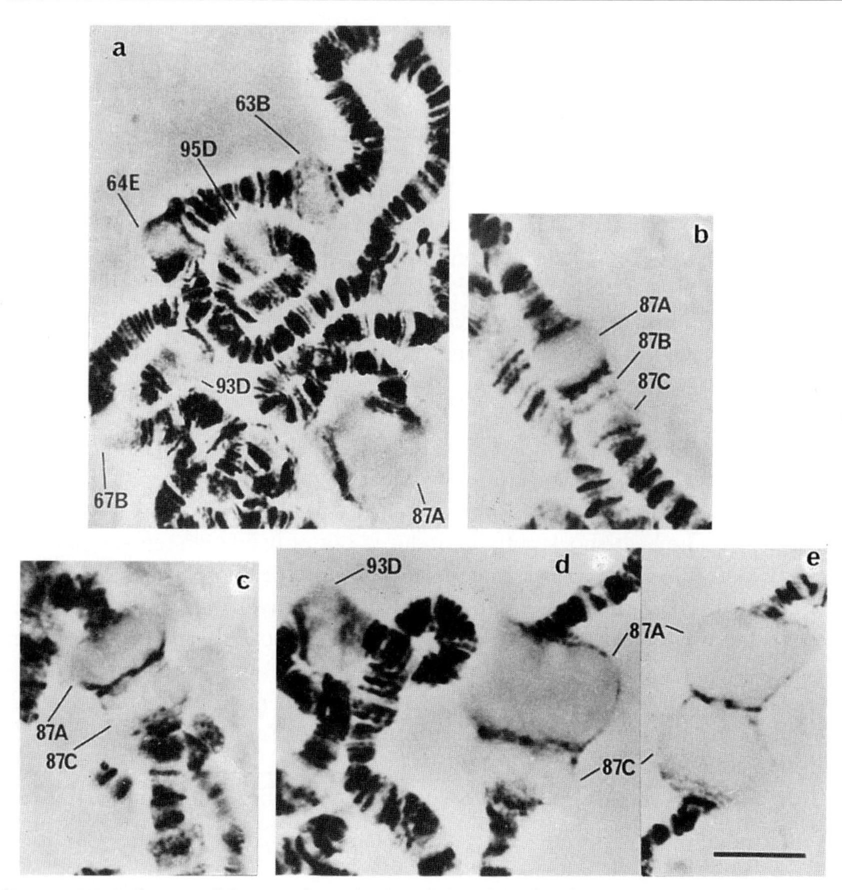

Figure 137. Induction of the giant heat-shock puffs by ethanol and ecdysterone in *D. melanogaster*. (a) General appearance; (b–e) 15–60 min after treatment. Bar = 10 μm (from Vlassova *et al.*, 1983).

However, exogenous ATP inhibits induction of the heat-shock puffs (Leenders and Berendes, 1972; Behnel and Rensing, 1975). In *D. hydei*, the induction of these puffs is also inhibited by the addition of malate and succinate to medium (Leenders and Berendes, 1972). In *D. melanogaster*, these substances do not block the induction of puffs (Behnel, 1978; Ashburner and Bonner, 1979).

The cells respond very rapidly to the inducer: puffs start to form within 1 min following the shift to the elevated temperature; the puffs reach their maximal sizes within 20–30 min following it; thereafter they regress for several

Table 16. Effect of Temperature Changes on Activity of Polytene Chromosomes[a]

Species	Treatment conditions	Puff induction	References
Anopheles stephensi	Treatment of ovarian nurse cells with temperatures of 37°, 39°, and 41°C	Six puffs induced at 39°C, only a single puff induced at 37°C. Transcription strongly suppressed at 41° C	Nath and Lakhotia (1989)
Chironomus	39–40°C	At first complete regression of puffs followed by their supernormal development	Yamamoto (1970)
C. plumosus	Transfer of larvae from 2–4° to 18°C	Large puffs induced in all the chromosomes	Il'inskaya (1981)
C. tentans	37°C, 30–60 min	Four to five new puffs induced in salivary glands and Malpighian tubules. The other puffs, including Balbiani rings, decrease	Sass (1980a); Lezzi *et al.* (1981, 1984); Lezzi (1984)
	42°C	The heat-shock puffs not induced. Balbiani rings do not regress	Sass (1980a); Lezzi *et al.*, (1981, 1984); Lezzi (1984)
	30°C	Transcription inhibition in the 1–18C puff	Lezzi *et al.* (1992)
	34–37°C	Transcription induction in this puff	Lezzi *et al.* (1992)
	37°C, 50–60 min	7 major heat-shock loci induced	Sass (1995)
C. thummi	15 min of heat shock in the 0–40°C range.	Puffs induced when temperature is raised or lowered on 5°C. Induction of the 4-A1b puff, regression of the nucleolus and Balbiani rings (I and II), variable changes in activities of 10 other puffs	Reznik *et al.* (1981, 1983, 1984)
	35°C	Induction of 7–8 puffs and Balbiani rings (T-BRIIIR), with lower frequency of BR in the second and the fourth chromosomes, decrease of [^{3}H]uridine incorporation into puffs and nucleolus	Santa-Cruz *et al.* (1981); Morcillo *et al.* (1981, 1982); Carmona *et al.* (1985)
	39°C, 5 min	Balbiani ring first decreases, then sharply increases	Santa-Cruz *et al.* (1981); Aller *et al.* (1982)
	35°C, 1 hr	Puff III-A3b, induced by heat shock not inhibited by DRB	Barettino *et al.* (1988b)

continues

Table 16. *Continued*

Species	Treatment conditions	Puff induction	References
C. *thummi piger*	35°C, 1–3 hr	Induction of telomeric BR in the fourth chromosome, transcription continues under heat shock in five puffs at least	Morcillo *et al.* (1988)
Drosophila americana	38–40°C, 40 min	Five puffs	Duttagupta and De (1987)
D. *ananassae*	37°C, 20 min	Nine puffs	Lakhotia and Singh (1982)
D. *arizonensis*	38–40°C, 40 min	Six puffs	Duttagupta and De (1987)
D. *auraria*	37°C, 10–60 min	Of 7 puffs induced 5 are active during normal development as well. Inactivation of Balbiani rings and [^{3}H]uridine incorporation into them	Scouras *et al.* (1986)
D. *bifurca*	35°C	Seven puffs	Berendes (1966c)
D. *busckii*	30°C, 30 min	At least 2 puffs in chromosome 2L	Ritossa (1962)
D. *buzzatti*	35°C	Seven puffs	Berendes (1966c)
D. *eohydei*	35°C	Seven puffs	Berendes (1966c)
D. *guanche*	31° and 37°C	Induction of five puffs. Stronger response of puffs at 37°C	Molto *et al.* (1987c)
D. *hamatofila*	35°C	Seven puffs	Berendes (1966c)
D. *hydei*	Transfer from 15° to 25°C	Puffs induction	Berendes and Holt, (1964)
	Transfer from 25° to 35°C, 1 hr	The heat-shock puffs induced in cells of stomach, midgut, Malpighian tubules, salivary glands	Holt and Berendes (1965)
	35°C	Seven puffs	Berendes (1966c)
	37°C, 2 hr	At least 4-81B puff	Koninkx (1976)
	—	The heat shock puffs induced in salivary glands after 3-week incubation in abdomen of adults followed by heat shock	Berendes and Holt (1965)
D. *kikkawai*	37°C, 20min	Six puffs	Lakhotia and Singh (1982)
D. *melanogaster*	37°C, 45 min	At least three puffs in chromosome 3R	Ritossa (1963)
	Incubation of salivary glands (37°C, 30–45 min)	At least 87A and 87C puffs induced	Ellgaard and Brosseau (1969)

Transfer from 25° to 37°C, 40 min	Nine puffs (33B, 63C, 64F, 67B, 70A, 87A, 87C, 93D, 95D)	Ashburner (1970c); Lakhotia and Singh (1982); Lapta and Shakhbazov (1984)
Treatment with temperature of 24–38°C range	Maximum induction of the 87A and 87C puffs at temperatures of 34–38°C	Ellgaard (1972)
37°C, 2.5 min	Along with the other puffs the small 86F3-6 puff induced	Zhimulev and Grafodatskaya (1974); Gausz *et al.* (1981); Semeshin *et al.* (1982)
37°C, 25 min	[^{3}H]uridine incorporation is inhibited into all the regions, except the nucleolus and regions: 2E1-2, 5C3-16, 10B3-16, 10EF, 18D3-13, 22E1-2, 24D3-11, 26A3-9, 37EF, 38F1-6, 43E3-18, 44DE, 45E1-2, 47B1-8, 47D1-4, 48E3-12, 49E1-15, 50C3-7, 50D1-7, 51D3-12, 56E1-6, 61C1-4, 63B, 64E3-13, 67AB, 70B1-7, 73D1-7, 78E1-2, 84E1-6, 85B6-9, 87A7-10, 87B10-15, 88E3-5, 93D1-10, 95D1-11	Belyaeva and Zhimulev (1976)
37°C, 45 min	Occurrence of puffs in cells varies from 86% (95D) to 99% (93D and 67B)	Zhimulev *et al.* (1976)
Incubation of ovarian follicle of adults at 4°C (1.5 hr), then in culture at 22°C	Synthesis of RNA hybridizing *in situ* at regions at least of three puffs: 87A, 87C, and 93D	Petri *et al.* (1977)
37°C	Sites of *in situ* hybridization of total cytoplasmic poly(A)$^+$RNA (from cell culture) labeled at 37°C (in addition to the heat-shock puffs): 2C, 3C, 7B, 11E, 16A, 18A, 30B, 32A, 51C, 52B, 55F, 58A, 61F, 62B, 62D, 63F, 64B, 79C, 79F, 82AC, 85B, 85D, 86C, 88E, 89A, 92D, 94D, 100D	Spradling *et al.* (1977)
Transfer from 10° to 24°C (30 min)	Only the 93D puff induced	Lakhotia and Singh (1985)

continues

Table 16. *Continued*

Species	Treatment conditions	Puff induction	References
	Transfer from 10° to 37°C	Sharp decrease of transcription, induction of the heat shock puffs, except 93D	Lakhotia and Singh (1985)
	Transfer from 24° to 10°C	The 93D puff increases, the other puffs not induced	Lakhotia and Singh (1985)
	Transfer from 16° to 37°C or 25–37°C	When transferred from 16° to 37°C the 87A7 puff in midgut cells is much smaller than 87C1. When transferred from 25°C to 37°C, reverse pattern	Hochstrasser (1987)
D. melanogaster, l(2)gl	37°C, 40 min	The 95D, 93D, 87B, 87A, 63BC, 67B puffs	Ashburner (1970c)
D. melanogaster, l(3)tl	In 384–408 hr larvae without heat treatment	The heat-shock puffs in some cells (1–34%)	Zhimulev *et al.* (1976)
	Heat shock 37°C, 45 min	In 144-hr larvae inducibility of puffs varies from 2% (95D) to 65% of cells (63BC). In 384-hr larvae; from 42% of cells (67AB) to 73% (93D)	
D. melanogaster, ebony or *black*	37°C, 1 hr	In homo- or heterozygotes for mutations, the 93D puff not induced, the other heat-shock puffs active	Lakhotia *et al.* (1990)
D. melanogaster, otu	37°C	Weak induction of heat-shock puffs	Heino *et al.* (1995); Zhimulev and Mal'ceva (1997)
D. melanogaster, rbp^4 and *rbp^5*	"normal" development	In 24-hr prepupal salivary glands heat shock appears	M. L. Balasov (1994); unpublished; Zhimulev *et al.* (1995a)
D. mercatorum	35°C	Seven puffs	Berendes (1966c)
D. mulleri	35°C	Seven puffs	Berendes (1966c)
	38–40°C, 40 min	Five puffs	Duttagupta and De (1987)
D. nasuta	37°C, 20 min	Six puffs	Lakhotia and Singh (1982)
D. neohydei	35°C	Seven puffs	Berendes (1966c)
D. nigrohydei	35°C	Seven puffs	Berendes (1966c)
D. pseudoobscura	35–37°C, 15–75 min	Four large heat-shock puffs	Pierce and Lucchesi (1980)

D. repleta	35°C	Seven puffs	Berendes (1965d, 1966c)
D. simulans	37°C, 30 min	The same puffs induced as in *D. melanogaster*. The 93D puff is of smaller size, total inhibition of transcription	Chowdhuri and Lakhotia (1986)
	37°C, 2.5 min	The same puffs and the small 86F3-6 puff	Semeshin *et al.* (1982)
D. subobscura	37°C, 20–45 min	Seven new puffs, activity increased in several puffs	Pascual and de Frutos (1984)
	31–37°C	Activity increase of 15 puffs	Pascual and Frutos (1985)
	31° and 37°C	Induction of 11 puffs	Pascual and Frutos (1986); Pascual *et al.* (1987)
	31°C	Induction of 6 puffs	Arbona and Frutos, (1987a,b)
	31° and 37°C	Five puffs, stronger puff response of at 31°C	Molto *et al.* (1987c)
	31–34° and 37°C	The 2C and 27A puffs activated only at 31°C and 34°C, the 15DE and 31C/D puffs only at 37°C. 10 total heat-shock puffs	Pascual and Frutos (1988)
D. virilis	Larvae or salivary glands *in vitro* at 37°C	Appearance of new puffs	Kenney and Hunter (1977)
	37°C, 30 min	Nine new puffs, in addition puff abnormalities detected in 47 regions	Poluektova *et al.* (1978b)
	Transfer from 25° to 37°C	Five new puffs induced; the same in *D. littoralis*, *D. texana*, *D. montana*, *D. novamexicana*	Gubenko and Baricheva (1979); Gubenko *et al.* (1991a,b)
	38–40°C	Five puffs	Duttagupta and De (1987)
D. vulcana	36°C	8 puffs	Pardali *et al.* (1996)
D. gr. *willistoni*	37°C, 40 min	From 3 to 5 puffs are induced in seven species; *hsp70* has been located by means of *in situ* hybridization	Bonorino *et al.* (1993a,b)
Glyptotendipes barbipes	Temperature shock	Puffs not incorporating [^{3}H]uridine induced in heterochromatic regions	Walter (1973)
Melanagromyza obtuza	39°C, 30 min	Three puffs in salivary gland cells; in midgut cells, five puffs	Singh and Gupta (1985)

continues

Table 16. *Continued*

Species	Treatment conditions	Puff induction	References
Parasarcophaga ruficornis	40°C, 30 min	The single large 12A puff induced in foot pad cells	Ranjan and Kaul (1988)
Sarcophaga bullata	37°C (20–180 min)	A giant puff formed in polytene chromosomes of foot pad cells. Maximum development in 20 min, sharp decrease of puff by 60 min. Synthesis of three proteins induced	Bultmann (1986a)
	The same + iodoacetamide (0.03 mM)	Puff induced	Bultmann (1986b)
	The same + iodoacetamide (0.07 mM)	Puff not induced	
	The same + fluoride (5 mM)	Puff induced	
	The same + fluoride (10 mM)	Puff not induced	
	The same + arsenite (0.05 mM)	Puff induced	
	The same + arsenite (0.1 mM)	Puff not induced	

[a]Salivary glands if not specially indicated.

Table 17. Agents Inducing the Heat-Shock Puffs

Inducing agents	Puff induction	Species	References
Ammonium chloride	0.01 M; at least 93D, 87A, and 87C induced	*D. melanogaster*	Scalenghe and Ritossa (1976)
Amytal	3.4×10^{-4} M; all puffs induced	*Drosophila hydei*	Leenders and Berendes (1972)
Anoxia	Puff not induced	*D. busckii, D. melanogaster*	Ritossa (1964a); Ashburner (1970c)
	120–180 min in an atmosphere of CO_2. Puffs not formed. [^{3}H]uridine incorporation not inhibited	*Chironomus thummi*	Barettino *et al.* (1988a)
	6 hr in an atmosphere of nitrogen, 25°C; induction of the giant heat-shock puffs	*Sarcophaga bullata*	Bultmann (1986b)
Anoxia + arsenite	0.1 mM; puffs not induced		Bultmann (1986b)
Antimycin A	All puffs induced except 81B	*Drosophila hydei*	Leenders and Berendes (1972)
Arsenate	10 mM; the heat-shock puff not induced	*Sarcophaga bullata*	Bultmann (1986b)
	10–20 mM at ph6.8; puff induced		Bultmann (1986b)
Arsenite	7.5×10^{-5} M; all puffs induced. Maximum reached in 2 hr. All puffs inactivated in 8 hr	*Drosophila hydei*	Leenders and Berendes (1972); Koninkx (1976); Vossen *et al.* (1977)
	Puffs not induced, [^{3}H]uridine incorporation inhibited	*Chironomus thummi*	Barettino *et al.* 1988a
Atractyloside	10^{-4} M, puffs not induced	*Drosophila hydei*	Leenders *et al.* (1974b)
		D. melanogaster	Behnel and Rensing (1975)
Azide	4 mM; puff induced	*Sarcophaga bullata*	Bultmann (1986b)
Azide + arsenate	10 mM; puff induced		Bultmann (1986b)
Azide + arsenite	10 mM; puff not induced		Bultmann (1986b)
Azide + iodoacetamide	0.1 mM; puff not induced		Bultmann (1986b)
Azide + fluoride	10 mM; puff not induced		Bultmann (1986b)
Azide + malonate	20 mM; puff induced		Bultmann (1986b)
Benzamide	Puffs not induced	*Chironomus*	B. B. Nath and S. C. Lakhotia, in Lakhotia (1987)

continues

Table 17. *Continued*

Inducing agents	Puff induction	Species	References
	1 mg/ml, 10 min; the 2C puff induced	*Drosophila ananassae*	Lakhotia and Singh (1982)
	1 mg/ml, 10 min; the 11BC puff induced	*D. kikkawai*	Lakhotia and Singh (1982)
	1.3 mg/ml, 10 min, incubation of salivary glands; the 93D puff induced, [^{3}H]uridine incorporation in other regions reduced	*D. melanogaster*	Lakhotia and Mukherjee A. S. (1970a); Lakhotia and Mukherjee, T. (1980)
	1 mg/ml, 10 min, 10°C; inhibition of RNA synthesis, including the 93D puff		Lakhotia and Singh (1985)
	1 mg/ml, 10 min; RNA synthesis reduced, including the 93D puff		Lakhotia and Singh (1985)
	1 mg/ml, 10 min; the 93D puff induced		Lakhotia and Singh (1982, 1985)
	1 mg/ml, 10 min; the 48A puff induced	*D. nasuta*	Lakhotia and Singh (1982)
	1 mg/ml, 10 min, 24°C; the 2-58C puff induced	*D. pseudoobscura*	Burma and Lakhotia (1984)
	1 mg/ml, 24°C, 10 min; the 93D puff induced	*D. simulans*	Chowdhuri and Lakhotia (1986)
	1 mg/ml, 10 min; incubation of organs: puff not induced neither in salivary glands, nor in midgut	*Melanagromyza obtusa*	Singh and Gupta (1985); Singh, (1988b)
	1 mg/ml, 30–120 min; puff not induced, strong inhibition of [^{3}H]uridine incorporation	*Chironomus thummi*	Barettino *et al.* (1988a)
Cadmium chloride	Treatment of cell culture (10 μM–1 mM) gives rise to synthesis of the heat shock proteins	*Drosophila melanogaster*	Courgeon *et al.* (1984)
Chloral hydrate	0.1 M, 24°C, 1 hr; a puff homologus to 93D of *D. melanogaster* induced	*D. ananassae*, *D. kikkawai*, *D. nasuta*	Mukherjee *et al.* (1982)
	0.1 M, 24°C, 1 hr; the 93D puff not induced	*D. melanogaster*	Mukherjee *et al.* (1982)
Chloramphenicol	2 mg/ml (treatment of larvae); sharp decrease of [^{3}H]uridine incorporation.	*D. melanogaster*	Behnel (1982)
	Puffs not induced. Sharp decrease of incorporation of [^{3}H]uridine	*Chironomus thummi*	Barettino *et al.* (1988a)

Coffeine	When salivary glands incubated, the heat-shock puffs induced (10^{-2} M, 40–90 min, 24°C). Maximum development of the 87A and 87C puffs, weaker development of 63C, 67B, 93D, 95D puffs	*Drosophila melanogaster*	Srivastava and Bangia (1985b)
Colchicine	The puffs not induced	*Chironomus*	B. B. Nath and S. C. Lakhotia, in Lakhotia (1987)
	100 μg/ml, 24°C, 40 min; induction of the 93D puff, RNA synthesis inhibited, including the 93D puff	*Drosophila melanogaster*	Lakhotia and Singh, (1982); Lakhotia and Mukherjee (1984); Lakhotia and Sharma (1995)
	Incubation of glands (100 μg/ml, 45 min, 24°C). The 2-58C puff induced	*D. pseudoobscura*	Burma and Lakhotia (1984)
	100 μg/ml, 24°C, 40 min; the 93D puff induced	*D. simulans*	Chowdhuri and Lakhotia (1986)
	Treatment of larvae in 0.1 μg/ml solution, the 20CD puff induced	*D. virilis, D. montana, D. littoralis, D. texana, D. novamexicana*	Gubenko and Baricheva (1979)
Cold shock	Transfer from 25°C to cold (4–8°C) for 1 hr. The 93D puff induced and RNA synthesis inhibited	*D. melanogaster*	Mukherjee *et al.* (1982); Singh and Lakhotia (1984)
Cyanide	10^{-6}–10^{-2} M, the puffs not induced	*D. hydei*	Leenders *et al.* (1974b)
	The same	*D. melanogaster*	Behnel and Rensing (1975)
Cycloheximide	No heat-shock puffs induced	*D. melanogaster*	Schoon and Rensing (1973); P. Burma and S. C. Lakhotia, in Lakhotia, (1987)
	Kc cell culture. Transcripts of the 93D puff detected		W. Bendena, in Lakhotia (1987)
Dicoumarol	10^{-3} M, all the puffs induced	*D. busckii*	Ritossa (1964a)
Digitonin	When salivary glands are treated, 35 puffs, including all nine heat-shock puffs, activated	*D. melanogaster*	Myohara and Okada, (1988b,c); Asaoka *et al.* (1994)
Dinactin	All puffs induced *in vitro* (10^{-6}–10^{-8} M)	*D. melanogaster*	Rensing (1973)

continues

Table 17. *Continued*

Inducing agents	Puff induction	Species	References
2.4-Dinitrophenol	Incubation of salivary glands (10^{-3} M, 30 min, 25°C), all the heat-shock puffs induced	*D. busckii*	Ritossa (1962)
	Incubation of salivary glands, all the puffs induced (10^{-3} M, 25°C)	*D. melanogaster*	Ritossa (1963)
	10^{-3} M; the heat-shock puffs induced	*D. hydei*	Leenders (1971); Leenders and Beckers (1972); Koninkx (1976)
	Maximum induction of the puffs after treatment with solution at a concentration of $10^{-4} - 10^{-3}$ M, all the puffs induced		Ellgaard (1972); Ellgaard and Maxwell (1975)
	The same	*D. hydei*	Leenders and Berendes (1972)
	10^{-3} M; induction of the puffs *in vitro*	*D. melanogaster*	Rensing (1973)
	Induction depends on pH, maximum size of the puffs at pH6.75, the puffs are not induced at pH5.9		Behnel and Seydewitz (1980)
	10^{-3} M, 24°C, 30 min; the heat shock puffs induced in salivary gland and midgut cells	*Melanagromyza obtusa*	Singh and Gupta (1985)
	The 0.02 mM; heat-shock puff induced	*Sarcophaga bullata*	Bultmann (1986b)
	0.1–1 mM, the puffs not induced, [^{3}H]uridine incorporation inhibited	*Chironomus thummi*	Barettino *et al.* (1988a)
Ethanol	5–10%, 60 min (treatment of larvae). All puffs induced	*Drosophila melanogaster*	Desrosiers and Tanguay (1985)
Ethanol + ecdysterone	7.4×10^{-6} M ecdysterone + 8% ethanol; incubation of salivary glands; the 63B, 64E, 67B, 87A, 87C, 93D, 95D giant heat shock puffs (see Figure 136)		Vlassova *et al.* (1983)
Fluoride	10^{-2} M; puffs not induced	*D. busckii*	Ritossa (1964a)
Fusidic acid	10 μg/ml, 63BC, 87A, 87C, and 93D are induced	*D. melanogaster*	Schoon and Rensing (1973)
2-heptyl-4-hydroxyquinoline *N*-oxide	0.25 mg/ml; all puffs induced except 81B	*D. hydei*	Leenders and Berendes (1972)

Homogenate of salivary glands, treated with heat shock (37°C, 60 min)	The 93D puff induced	*D. melanogaster*	Mukherjee and Lakhotia (1981; Lakhotia and Singh (1982); Singh and Lakhotia (1983)
	The 48C puff induced	*D. hydei*	Lakhotia and Singh (1982)
	The 48A puff induced, no puffs formed, when *D. hydei* salivary glands treated with *D. nasuta* salivary gland homogenate	*D. nasuta*	Lakhotia and Singh (1982)
	No puffs induced, no inhibition of [^{3}H]uridine incorporation detected	*Chironomus thummi*	Barettino *et al.* (1988a)
Hydrocortisone	3–15 mM, 120 min; induced at least 87A and 87C, 70-kDa polypeptide synthesized	*Drosophila melanogaster*	Caggese *et al.* (1983)
Hydrogen peroxide	10^{-3}–2.5×10^{-2} M, all the puffs induced		Compton and McCarthy (1978)
	The hsp 68, 70, and 23 polypeptides synthesized		Courgeon *et al.* (1988)
Hydroxylamine	0.01 M, at least 93D, 87A, 87C induced		Scalenghe and Ritossa (1976)
Incubation conditions *in vitro*	Induction of the 93D puff in "old culture medium"		Bonner and Pardue (1976b)
Incubation in adult abdomens	After 18 hr incubation the 63BC puff arises		Staub (1969)
Iodoacetate	10^{-3} M, the puffs not induced	*D. busckii*	Ritossa (1964a)
Magnezium chloride ($MgCl_2$)	96 mM; incubation of salivary glands induces the 2AB, 10BC, 47BC, 63BC, 87A, 87C, 90BC puffs, particularly 93D	*D. melanogaster*	Bhadra *et al.* (1992)
Menadione (vitamin K_3)	10^{-3} M, all the puffs induced	*D. hydei*	Leenders (1971); Leenders and Berendes (1972)
Methylene blue	10^{-2} M; all puffs induced, only in 50% of salivary glands		Leenders and Berendes (1972)
	10^{-3} M; puffs not induced	*D. melanogaster*	J. J. Bonner, in Ashburner and Bonner (1979)
Oligomycin	Saturated solution; puffs not induced	*D. hydei*	Leenders and Berendes (1972)

continues

Table 17. *Continued*

Inducing agents	Puff induction	Species	References
	$>10^{-5}$ M, induced at least 63C	*D. melanogaster*	Behnel and Rensing (1975)
Oligomycin + atractyloside	Saturated solution; all the puffs induced except 81B	*D. hydei*	Leenders and Berendes (1972)
Oligomycin + KCN	Saturated solution; all the heat-shock puffs induced, except 81B	*D. hydei*	Leenders and Berendes (1972)
Paracetamol (acetaminophen or *N*-acetylaminophenol)	After incubation of salivary glands (10^{-2} M, 60 min, 24°C) the 93D puff induced	*D. melanogaster*	Srivastava and Bangia (1985a)
Poisons destroying microtubules (griseofulvin, oncodazole, vinblastin, chloral hydrate, diamide, podophyllotoxin, nocodazole)	Puffs not induced	*D. melanogaster*	Singh and Lakhotia (1984)
Puromycin	Incubation of salivary glands in 10 μg/ml solution results in induction of 63BC, 87A, 87C, and 93D puffs		Schoon and Rensing (1973)
Recovery from anoxia	1 hr of anaerobiosis + 10 min on air, the heat-shock puffs formed	*D. busckii*	Ritossa (1964a)
	Incubation in an atmosphere of CO, CO_2 and N_2, 3 hr. Puffs arise 15 min after restoration of an oxygen atmosphere	*D. hydei*	Breugel (1965, 1966); Leenders and Knoppien (1973); Koninkx (1976)
	Puffs induced after 60 min exposure to CO_2 atmosphere	*D. melanogaster*	Burdette and Anderson (1965a)
	1 hr in N_2 atmosphere; after 20–25 min on air, the heat-shock puffs reach maximum size		Ashburner (1970c)

	All heat-shock puffs including small 86F3-6 puff appear in larvae maintained under a layer of saline (1 hr, 25°C) and 20 min after transfer to air.		Zhimulev and Grafodatskaya (1974)
	—		Mukherjee and Lakhotia (1982)
	Restoration after anoxia induces seven heat-shock puffs	*D. subobscura*	Arbona and Frutos (1987a,b)
	30–120 min in N_2 atmosphere; puffs not induced [3H]uridine incorporation inhibited.	*Chironomus thummi*	Barettino *et al*. (1988a)
	30–120 min in CO_2 atmosphere; the heat-shock puffs and the heat shock BR induced.		Barettino *et al*. (1988a)
Rotenon	Saturated solution, all the puffs induced	*D. hydei*	Leenders and Berendes (1972)
Selenium	After addition into medium (31 mg/kg) in 1 of 25 prepupae the 93D, 95D, 87A, 87C, 63C, 67F and 67B puffs were induced	*D. melanogaster*	Santos *et al*. (1993)
Sodium azide	3×10^{-3} M; all puffs induced	*D. busckii*	Ritossa (1964a)
Sodium cacodylate	10 μg/ml, 80 min, 24°C; all the heat-shock puffs induced except 93D	*D. melanogaster*	Lakhotia and Singh (1985)
	The same at 10°C; the puffs not induced		
Sodium salicylate	10^{-2} M, 30 min, 25°C; incubation of salivary gland, all the puffs induced	*D. busckii*	Ritossa, 1962
	Induction of 93D, 87A, and 87C at least	*D. melanogaster*	Ritossa (1963)
Suspension of mitochondria	After injection of mitochondrial extract subjected to heat shock into cytoplasm of salivary gland cells, the heat-shock puffs appear	*D. hydei*	Sin (1975)
Trinactin	In *in vitro* system, 10^{-3} M trinactin induces the 63B puff. There is no induction when ATP is added	*D. melanogaster*	Behnel and Rensing (1975)
	Puff size correlates with potassium concentration in the medium.		Behnel and Seydewitz (1980)
	Maximum puff size at pH 6.75, no puffs induced at pH 5.9		

continues

Table 17. *Continued*

Inducing agents	Puff induction	Species	References
Tyamphenicol	Induction of the 93D puff	*D. melanogaster*	Behnel (1982)
Tyrosine	The 2-48D puff induced, concomitantly, the activity of tyrosine-aminotransferase sharply increases	*D. hydei*	Leenders and Knoppien (1973); Brady and Belew (1981); Belew and Brady (1981a)
Uridine	10^{-2} M; at least 93D, 87A, and 87C induced	*D. melanogaster*	Schalenghe and Ritossa (1976)
Valinomycin	2×10^{-5} M; all puffs induced	*D. melanogaster*	Rensing (1973)
Vitamin B_6	Puffs not induced	*Chironomus*	B. B. Nath and S. C. Lakhotia, in Lakhotia (1987)
	The 2C puff induced	*Drosophila ananassae*	Lakhotia and Singh (1982)
	Injection 0.5 μl of 0.1 M, induction of the 48C puff	*D. hydei*	Leenders *et al.* (1973)
	10^{-7}–10^{-2} M, puff formed in 5 min, maximal development in 2 hr		Brady and Belew (1981)
	5×10^{-2} M; incubation of salivary glands, all puffs induced		Leenders *et al.* (1973)
	2×10^{-2} M; puffs not induced		Leenders *et al.* (1973)
Vitamin B_6 + oligomycin	2×10^{-2} M; only the 48C puff induced		Leenders *et al.* (1973)
	Saturated solution, puffs not induced	*D. melanogaster*	J. J. Bonner, in Ashburner and Bonner (1979)
	The 11BC puff induced	*D. kikkawai*	Lakhotia and Singh (1982)
	The 93D puff not induced	*D. melanogaster*	Lakhotia and Singh (1982)
	The 48A puff induced	*D. nasuta*	Lakhotia and Singh (1982)
	Treatment of larvae in solution of 6×10^{-2} M; the 20CD puff induced	*D. virilis*, *D. montana*, *D. texana*, *D. littoralis*, *D. novamexicana*	Gubenko and Baricheva (1979)

hours (Figure 138). The maximal puff size in *D. melanogaster* induced in the experiments depends on severity of temperature treatment (Figure 138; see also Berendes, 1968a; Lewis *et al.*, 1975; Ashburner and Bonner, 1979; Shakhbazov and Taglina, 1990). In the salivary gland cells of *D. hydei* treated with arsenite, the puffs reached their maximum sizes 2 hr after incubation and they were inactivated after 8 hr of incubation (Vossen *et al.*, 1977).

If heat shock is extended for more than 1 hr, the normally developing puffs, active at the time the temperature was raised, regress (see reviews by Ashburner and Bonner, 1979; Sass, 1995).

There are at least three types of responses at the level of RNA synthesis.

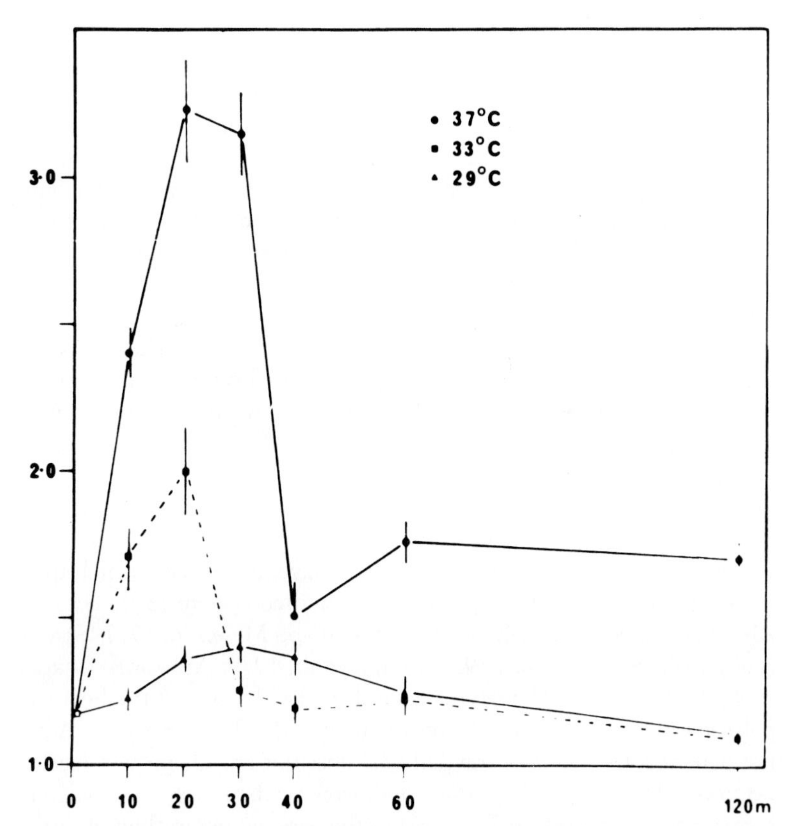

Figure 138. Changes in the size of the 87C1 heat-shock puff of *Drosophila* related to intensity and time of treatment. Abscissa, time of heat treatment (min); ordinate, puff sizes (from Lewis *et al.*, 1975).

1. RNA synthesis is induced in the heat-shock puffs. According to autoradiographic data, the puffs incorporate [^{3}H]uridine (Ritossa, 1964b; Berendes, 1968a; Tissieres *et al.*, 1974; Belyaeva and Zhimulev, 1976; Vossen *et al.*, 1977; Bonner and Kerby, 1982). *In situ* hybridizations showed that RNA synthesis occurs in the induced puffs when the cultured *Drosophila* cells are subjected to heat shock (McKenzie *et al.*, 1975; Spradling *et al.*, 1975, 1977; Bonner and Pardue, 1976b; Henikoff and Meselson, 1977; McKenzie and Meselson, 1977; Pardue *et al.*, 1977; Livak *et al.*, 1978; Lubsen *et al.*, 1978; Sondermeijer and Lubsen, 1979; Artavanis-Tsakonas *et al.*, 1979b). The inhibitors of RNA synthesis actinomycin D and α-amanitin block transcription and, concomitantly, the formation of the heat-shock puffs (Berendes, 1868a; Ellgaard and Clever, 1971; Holt and Kuijpers, 1972b; Compton and McCarthy, 1978; Law *et al.*, 1979; Rowe *et al.*, 1987). However, there are exceptions: DRB does not block the incorporation of [^{3}H]uridine into the heat-shock puff III-A^{3b} in *C. thummi* (Barettino *et al.*, 1988b). Telomeric Balbiani ring is not induced in response to a 2-hr pretreatment with actinomycin D (5 μg/ml), α-amanitin (1 μg/ml), or DRB (30 μg/ml). All three inhibitors completely inhibited RNA synthesis; nevertheless, Balbiani ring is induced (Barettino *et al.*, 1982). If novobiocin, an inhibitor of topoisomerase II, is added to incubation medium prior to puff induction, activation of the heat shock genes does not follow. If novobiocin is added after induction, transcription is rapidly turned off, but the chromatin organization is in an "active" configuration (Han *et al.*, 1985).
2. Synthesis of RNA in most genes that do not encode the heat-shock protein—with the exception of the histone genes (Spadoro *et al.*, 1986) or the mitochondrial genome genes (see review by Ashburner and Bonner, 1979)—is inhibited in the heat-shocked cells. RNA synthesis virtually ceases along the entire length of chromosome. The RNA from cells developing at the normal temperature (25°C) appears as a smear (Figure 139) on electrophrams. Thus it was shown that the sizes of the molecules vary widely. When cells are heat shocked, only several fractions, which hybridize *in situ* to the heat-shock puffs, are synthesized (Figure 139). These RNAs are detected in polysomes (Henikoff and Meselson, 1977; Spradling *et al.*, 1977; Mirault *et al.*, 1978; Moran *et al.*, 1978; Artavanis-Tsakonas *et al.*, 1979b). Table 18 presents data of the localization of the heat-shock polysome mRNAs. The RNAs of the small puffs 33B, 64F, and 70A are not detected; the RNA of the puff 93D does not encode protein (see Section V,H,2). According to autoradiographic data, only the nucleoli and several other regions (see Table 16) of the genome retain their ability to incorporate [^{3}H]uridine (Belyaeva and Zhimulev, 1976; Lubsen *et al.*, 1978; Evgen'ev *et al.*, 1979, 1985; Mitchell *et al.*, 1979; Mukherjee and Lakhotia, 1979; Lakhotia and Mukherjee, 1982; Evgen'ev *et al.*, 1985;

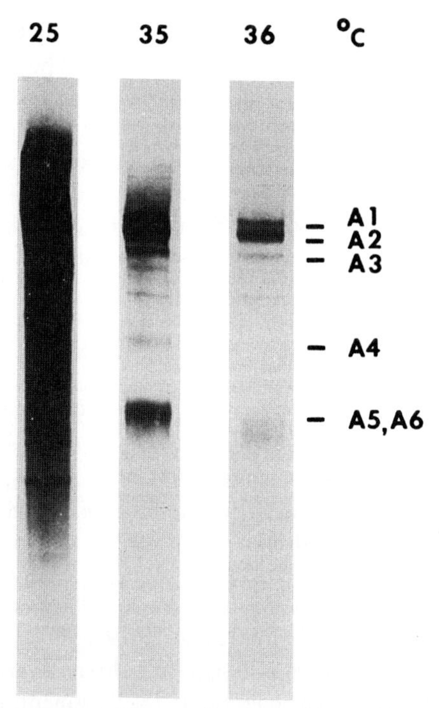

Figure 139. Cytoplasmic poly(A)$^+$ RNA in cell culture of *Drosophila* at different temperatures. A1–A6, fractions of heat-shock RNAs (from A. Spradling in Ashburner and Bonner, 1979, © Cell Press).

Choudhuri and Lakhotia, 1986; Nath and Lakhotia, 1989; Ghosh and Mukherjee, 1990). Based on these data, it may be concluded that RNA synthesis—effected by RNA polymerase II in the regions, with the exception of the heat-shock puffs—is inactivated. In fact, when the temperature is raised, antibodies against the enzyme are detected only in the induced puffs and in several other regions (Greenleaf *et al.*, 1978; Kramer *et al.*, 1980; Bonner and Kerby, 1982; Bonner, 1985; Weeks *et al.*, 1993).

Table 18. Puffs, RNA and Heat-Shock Proteins[a]

Puffs	33B	63BC	64F	67B	70A	87A/87C	87C	93D	95D
RNA fractions	—	A1	—	A5	—	A2	α-β-α (A4)	—	A3
Protein fractions (kDa)	—	82	—	27, 23, 26	—	70, 72	—	—	68

[a]See Ashburner and Bonner (1979) for details.

In parallel with these changes, the detectabiliy of RNA polymerase II in the cytoplasm increases (Jamrich *et al.*, 1977b). The transcription-dependent presence of U1 and U2 small nuclear ribonucleoproteins (snRNPs), which are components of the spliceosomes or necessary for splicing, have been immunolocalized to *C. tentans* heat-shock puffs (Sass and Pederson, 1984).

3. Normal processing of ribosomal 5S and 18 + 28S RNAs is disrupted (Ellgaard and Clever, 1971; Rubin and Hogness, 1975; Yost and Lindquist, 1986).

New polysomes containing newly synthesized mRNAs transcribed from the heat-shock genes appear within 10 min after the start of temperature treatment (reviewed by Ashburner and Bonner, 1979). Heat-shock proteins then appear on the electropherograms (Figure 140), and the synthesis of the other polypeptides concomitantly ceases (Tissieres *et al.*, 1974; Lewis *et al.*, 1975; McKenzie *et al.*, 1975; Koninkx, 1976; Mirault *et al.*, 1978; Mitchell and Lipps, 1978; Chomyn *et al.*, 1979; Mitchell *et al.*, 1979; Lindquist, 1980; Storti *et al.*, 1980; Tanguay and Vincent, 1981; Lakhotia and Mukherjee, 1982; Wadsworth, 1982; Ballinger and Pardue, 1985; Bultmann, 1986a; Morcillo *et al.*, 1988; Pascual and Frutos, 1988; Singh and Lakhotia, 1988; Nath and Lakhotia, 1989).

Cessation in protein synthesis results from a very rapid decay of the preexisting polysomes (McKenzie *et al.*, 1975; Biessmann *et al.*, 1978a,b; Sondermeijer and Lubsen, 1978). When cells are subjected to a strong heat shock, the RNA released from these polysomes is degraded; the RNA remains unaltered at the optimal temperature; however, it can be isolated and translated in an *in vitro* system (Spradling *et al.*, 1977; Mirault *et al.*, 1978). Normal protein synthesis is recovered within 1 hr after the shock irrespective of the presence of actinomycin D. This indicated that mRNAs retained their ability to function in translation when freed of ribosomes by heat shock (McKenzie, 1977; Storti *et al.*, 1980).

The ease with which the heat-inducible polypeptides can be detected made possible studies on the genomic response to agents stressing cells of organisms lacking polytene chromosomes. With this approach, it was demonstrated that the heat-shock syndrome develops in all species of animals, plants, and bacteria (see monograph by Schlesinger *et al.*, 1982).

A comparison of heat-shock proteins from *D. melanogaster* salivary glands by two-dimensional gel electrophoresis revealed 14 fractions (Buzin and Petersen, 1982). The majority corresponded to those revealed by one-dimensional gel electrophoresis: 83, 70, 68, 34, 26–28, and 21–23 kDa. The molecular weights of the small fractions are 44 and 66 kDa. Protein Hsp70 could be separated into seven groups slightly differing in isoelectric points. The Hsp proteins with molecular weights of 21–23 kDa were separated into eight components (Buzin and Petersen, 1982); Hsp72 and minor 38 and 74 kDa fractions isolated from polysomes were also identified (McKenzie and Meselson, 1977). mRNA was mapped *in situ* to the puffs: *hsp22-28* to 67B, *hsp68* to 95D, *hsp70* to 87A and 87C, and *hsp83* to 63BC (Table 18). The number of protein fractions is generally correlated to the number of puffs: from nine in *D. melanogaster*, to two to three in the representatives of Sarcophagidae (Tissieres *et al.*, 1974; Lewis *et al.*, 1975; Bultmann, 1986a; Joplin and Denlinger, 1990; Sass, 1995; Drosopoulou *et al.*, 1996; Konstantopoulou *et al.*, 1997).

The *hsp* genes are as a rule induced synchronously in the cells of the same or different organs. The puffs are induced at the same time in midgut and salivary gland cells of *D. busckii* (Ritossa, 1964a), in the cells of the foregut,

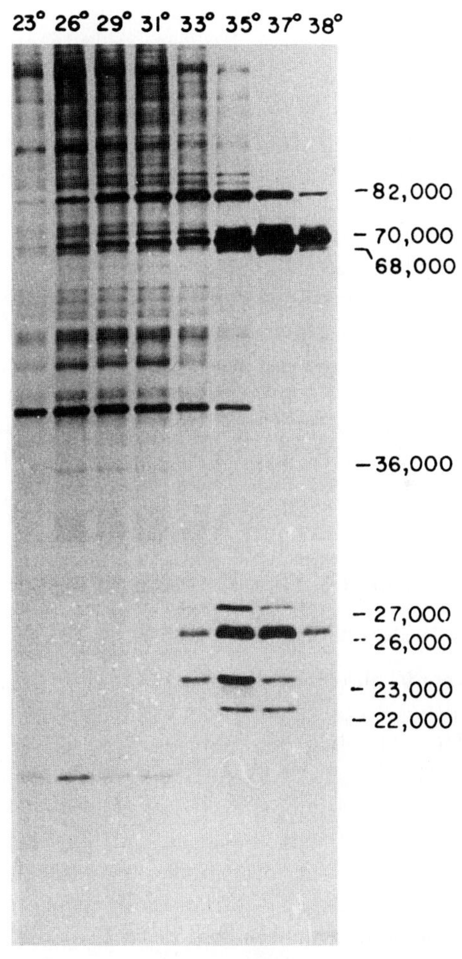

Figure 140. Protein synthesis in cell culture of *Drosophila* after heat shock (from S. L. McKenzie in Ashburner and Bonner, 1979, © Cell Press).

midgut, and Malpighian tubules of *D. hydei*, as well as in salivary glands transplanted into adult abdomens (Holt and Berendes, 1965). In salivary gland cells of *D. melanogaster*, synthesis of all the eight heat-shock proteins is induced within 10 min after the start of heat treatment; its maximum is reached after 1 hr; and 3 hr later its rate decreases by half (Tissieres *et al.*, 1974; Lewis *et al.*, 1975; Ashburner and Bonner, 1979). Nevertheless, the response is tissue specific (Sondermeijer and Lubsen, 1978). In *Melanagromyza obtusa*, three heat-shock puffs (also inducible by dinitrophenol) were found in salivary gland cells and five in midgut cells (Singh and Gupta, 1985). In *C. thummi*, the puffs responded asynchronously in some variants of heat shock (5 min at 39°C): the III-A3b puff formed within 30 min, the rest within 90 min (Carretero *et al.*, 1986). The 87A and 87C heat-shock loci were induced unequally in midgut cells depending on growth temperature (Hochstrasser, 1987, see table 16). In *Sarcophaga crassipalpis*, synthesis of two proteins, Hsp 65 and Hsp 72, is induced; Hsp 65 is expressed in the cells of the brain and integuments of third-instar larvae. Protein synthesis proceeds to completion in the two organs during pupation; only Hsp 72 is induced later. Induction of these two proteins is also tissue specific in adults (Joplin and Denlinger, 1990).

Certain other genes seem to be induced by stress, besides the heat-shock genes considered above. Thus, heat treatment (38°C) of *in vitro* cultured *Drosophila* cells results in the synthesis of short (320 bp) RNA molecules and their accumulation in the cytoplasm. This RNA inhibits translation in a lysate system of rabbit reticulocytes (Kawata *et al.*, 1988a,b). Nothing is known about their function.

There are indications that heat shock evokes a new round of DNA replication in 87A, 87C, and 93D puffs of *D. melanogaster* (Ghosh and Mukherjee, 1996).

H. Molecular organization of the heat shock puffs and genes

1. The heat-shock genes encoding polypeptides

The *hsp70* genes were localized to two puffs, 87A7 and 87C1, by *in situ* hybridization of *hsp70* mRNA to polytene chromosomes of *D. melanogaster* and *D. simulans* (Henikoff and Meselson, 1977; Spradling *et al.*, 1977; Lis *et al.*, 1978; Livak *et al.*, 1978; Schedl *et al.*, 1978; Artavanis-Tsakonas *et al.*, 1979a,b; Holmgren *et al.*, 1979). Electronmicroscopic data showed that these puffs are localized somewhat differently than was originally thought. In *D. melanogaster* and *D. simulans*, the puff 87A originates from four thin bands at 87A6–87A9. Four thin bands (87B13–87B15, 87C1) are also involved in the formation of the puff 87C in *D. melanogaster* and only three (87B13–87B15) in *D. simulans* (Semeshin *et al.*, 1982). Deletion of one or other puff does not affect the electrophoretic protein pattern, suggesting that the genes are repetitive (Ish-

Horowicz *et al.*, 1977, 1979a; Caggese *et al.*, 1979; Ish-Horowicz and Pinchin, 1980; Udvardy *et al.*, 1982).

The major repeated unit of the *hsp70* gene contains 2.7 kb DNA (the Z element) comprising the 2.4-kb protein (Z)-coding region and a 0.35 kb nontranscribed 5′ flanking *Znc* (Z-noncoding) region (Figure 141). Elements about 150 bp in size (Xa-c) lie upstream of the gene. They retained homology and are separated by DNA sequences of about 200–300 bp (the Y elements). The second Z element lies nearby, at about 1700 bp apart in an opposite orientation (Figure 141). By use of deletions (Costa *et al.*, 1977), it was shown that the *hsp70* genes are arranged in this manner in the 87A puff (Lis *et al.*, 1978; Schedl *et al.*, 1978; Artavanis-Tsakonas *et al.*, 1979a,b; Craig *et al.*, 1979; Ish-Horowicz *et al.*, 1979a,b; Mirault *et al.*, 1979; Moran *et al.*, 1979; Ish-

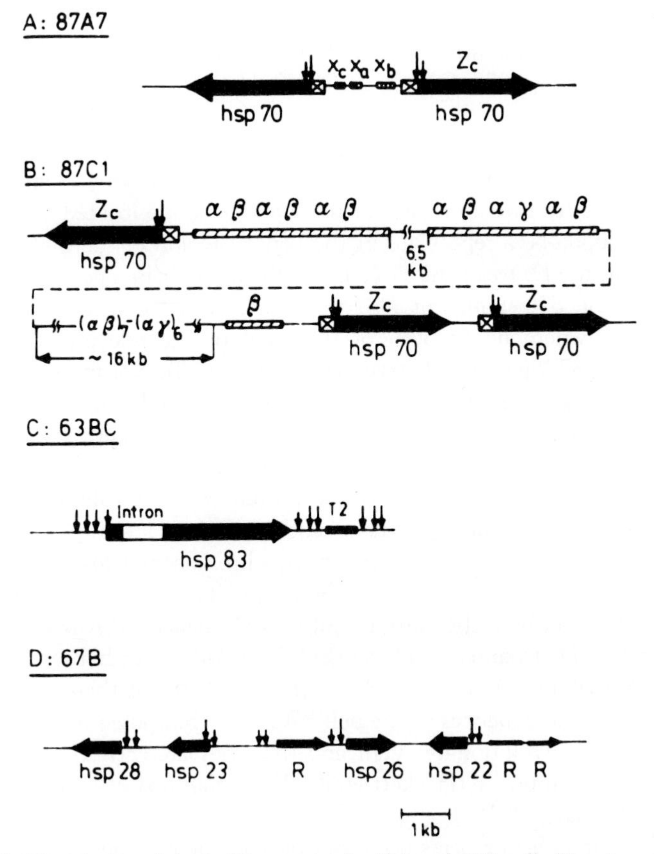

Figure 141. Organization of the *hsp70* genes in the 87A and 87C puffs. Direction of arrows Zc corresponds to the direction of transcription. Thin vertical arrows point to the DNase I hypersensitive sites. The X fragments in the scheme of the 87C puff are omitted. See text for details (from Nover *et al.*, 1984).

Horowicz and Pinchin, 1980; Goldschmidt-Clermont, 1980; Lis *et al.*, 1981b; Udvardy *et al.*, 1982). In *D. simulans*, *Drosophila mauritiana*, *Drosophila teissieri*, and *Drosophila yacuba*, the *hsp70* genes in the 87A puff are also arranged as pairs in an opposite orientation. The *hsp70* genes are also located in the 87C puff (Leigh Brown and Ish-Horowicz, 1980, 1981).

In *D. melanogaster*, the *hsp70* genes are presented in three to five copies in the 87C puff (Leigh Brown and Ish-Horowicz, 1980, 1981; Leigh Brown, 1981), separated by a DNA segment about 40 kb in length (see Figure 141) from which another heat-shock-inducible RNA is transcribed. No homology between this RNA and *hsp70* was found (Henikoff and Meselson, 1977; Lis *et al.*, 1978). The 40-kb segment contains a set of repeated units: the alpha (617 bp) and the beta (1.1 kb) repeats alternate; then, after a spacer 6.5 kb in size (see Figure 141), a block of repeat units alpha-beta-alpha-gamma-alpha-beta follows. The gamma repeat is of 733 bp in this unit. In all, at least 12 alpha-beta and 6 alpha-gamma units were identified in the 87C puff (Lis *et al.*, 1978, 1981a,c; Livak *et al.*, 1978; Craig *et al.*, 1979; Hackett and Lis, 1981; Mason *et al.*, 1982).

It was demonstrated by a high-resolution method (employing a biotin-labeled DNA probe) that the stretch of the 87C puff is readily subdivided into three parts. The hybridization sites of the *hsp70* gene are located at the flanks; those of the alpha-beta repeat are in the central region of the puff (Simon *et al.*, 1985a). In the *D. melanogaster* genome, the alpha-beta repeats are localized not only to the 87C puff, but also to 30 chromosome regions and to the chromocenter, where up to 30% of the total of the relevant silver grains detected in the 87A puff are observed (Lis *et al.*, 1978). Not more than half of the alpha-beta sequences present in the genome occur in the 87C puff; however, all the heat-shock-inducible RNA is read through precisely from this puff sequence (Lis *et al.*, 1981a). The *alpha-beta* sequence does not encode for proteins (Livak *et al.*, 1978). The Hsp70 proteins are normally transcribed in lines with deletions of the alpha-beta sequences (Ish-Horowicz *et al.*, 1977). The presence of the alpha-beta repeat in the chromocenter may be the reason why it hybridizes *in situ* to heat-shock RNA (Spradling *et al.*, 1975, 1977). This repeat is detected only in the chromocenter in *D. simulans* (Livak *et al.*, 1978); it is missing in *D. mauritiana, D. teissieri, D. yakuba* (Leigh Brown and Ish-Horowicz, 1981), and *D. hydei* (Peters *et al.*, 1980). It was therefore suggested that the alpha-beta sequences in the puff 87C were transposed from the chromocenter by recent evolutionary event and that they are activated by virtue of the strong *hsp70* promoter (Hackett *et al.*, 1981; Karch *et al.*, 1981; Lis *et al.*, 1981c; Torok *et al.*, 1982).

The alpha-beta repeats occur not only in the 87C puff, but also in the puff 87A, with a label ratio of 25 : 1 (Lis *et al.*, 1978). The gamma element shows extensive regions of homology to the *hsp70* gene at the heat-shock locus

87A in the Znc regions at positions extending from −484 to +64 bp (83.2%) of the same nucleotide (Torok and Karch, 1980).

The X-beta element of the 87C puff is flanked by short direct repeats and reminds one of a mobile element (Mason *et al.*, 1982).

Regions of homology as high as 96% were identified among the *hsp70* genes located in two puffs (Ingolia *et al.*, 1980; Torok and Karch, 1980; Karch *et al.*, 1981). The *hsp70* and *hsp68* genes within the 95D puff show extensive homology, with about 85% of the nucleotides matching (Holmgren *et al.*, 1979). A region of homology also exists between the heat-shock-cognate (hsc) genes with about 75–82% of the nucleotides matching (Craig *et al.*, 1982).

The primary DNA structure is conserved in the *hsp70* genes isolated from various organisms, such as *Drosophila*, yeast, and *E. coli* (Craig *et al.*, 1982). The heat-shock genes in *Drosophila* (except *hsp83*) all have the same nucleotides at positions −1, +1, +7, +12, +15, and +20 (Hultmark *et al.*, 1986). With this in mind, it is not surprising that the *hsp70* DNA of *D. melanogaster* hybridizes to polytene chromosomes of other species; for example, *D. virilis*, *D. pseudoobscura*, *D. hydei*, and *D. guanche* (Evgen'ev *et al.*, 1978; Peters *et al.*, 1980; Pierce and Lucchesi, 1980; Molto *et al.*, 1991).

The compaction degree of chromatids changes during the formation of the heat-shock puff. The B52 protein (52 kDa) is associated with boundaries of transcriptionally active chromatin; B52 brackets the RNA polymerase signal symmetrically (Champlin *et al.*, 1991).

Subtler changes occur at the chromatin level (Biessmann *et al.*, 1978a,b; Elgin, 1981; Levy, Noll, 1982; Elgin *et al.*, 1983; Karpov *et al.*, 1984), along with visible puffing of chromosome regions where the *hsp70* genes reside. By use of the protein image hybridization technique, it was demonstrated that chromatin of the actively transcribed *hsp70* genes undergoes structural transitions manifest as weakened contacts of DNA with the globular region of histones. The 5′promoter region of the gene, 1 kb away from its 3′ structural part, is particularly free of histones (Nacheva *et al.*, 1989a,b).

Udvardy *et al.* (1985) suggested that the nuclease-hypersensitive sites flanking the *hsp* genes may define the boundaries of the 87A7 chromomere in which the gene resides. These bordering chromatin structures are named *scs* and *scs′*. Each structure is characterized by two sets of nuclease-hypersensitive sites located within moderately G/C-rich DNA. *scs* and *scs′* insulate a *white* reporter gene (in a transgenic strain) from position effects and prevent enhancer–promoter interactions. These and other properties suggest *scs* and *scs′* function as chromatin domain boundaries (Kellum and Schedl, 1991; Vazques *et al.*, 1993; Vazques and Schedl, 1994; review: Gerasimova and Corces, 1996).

A 32-kDa protein termed BEAF-32 has been purified from nuclei of a *Drosophila* cell line. This protein binds to a palindromic sequence that flanks the two hypersensitive regions in the *scs′* sequence (Zhao *et al.*, 1995). In polytene chromosomes BEAF-32 is present in the interband region that separates the highly reproducible and characteristic bands. As expected BEAF-32 is present at the *scs′*-containing border of the 87A7 band. But according to EM analysis the 87A puff originates from four chromomeres (Semeshin *et al.*, 1982). The BEAF-32 antibodies bind with developmental ecdysterone puffs 75B and 74EF. The former puff appear bracketed on one and the latter on either side by BEAF (Zhao *et al.*, 1995). The DNase I hypersensitive regions identified before and during heat treatment are localized between −100 and −400 bp at the 5′ end of the *hsp70* gene (Wu, 1980, 1982; Costlow and Lis, 1984).

P-element-mediated transformation revealed that heat-shock-induced transcription and puff development are normal in lines transformed with the *hsp70* genes having the 5′ fragment extending from 0 to −194 bp (Bonner *et al.*, 1984; Cohen and Meselson, 1984), according to other data to −97 bp (Dudler and Travers, 1984), to −89 bp (Lis *et al.*, 1983; Simon *et al.*, 1985b), and to −51 bp (Corces *et al.*, 1982; Corces and Pellicer, 1984). If DNA fragments are removed from the transcription start site upstream of positions −25, −44, or −52 bp (Cohen and Meselson, 1984), or −73 bp (Simon *et al.*, 1985b), the puffs and transcription are not induced. The TATA domain and a short conserved element, found in the heat-shock genes, are organized within a fragment covering the region from −10 to −66 bp. In an analysis of cells transfected with the *hsp70* genes containing different deletions of the regulatory zones, Pelham (1982) and Mirault *et al.* (1982a,b) indicated that the sequence CTGGAATTTCTAGA required for the induction of the heat-shock response spans positions from −66 to −47 bp. It was called the HSE (the heat-shock element). It was subsequently found that the consensus sequence is composed of 14 nucleotides: C--GAA--TTC--G (Lis *et al.*, 1990). In this 14-bp fragment, the trinucleotide GAA is present in two copies as an inverted repeat (GAA--TTC), and it is thought to be the core of the regulatory region (Figure 142). The protein inducer of heat shock (see Section VI,H,4) binds precisely to this sequence.

Germ-line transformation with an *hsp* gene that contains various deletions of the 5′ region demonstrated that the presence of a single HSE provides a transcription level of not more than 1% of wild type. Two or more HSEs are necessary to ensure normal transcription (see Figure 144) (Dudler and Travers,

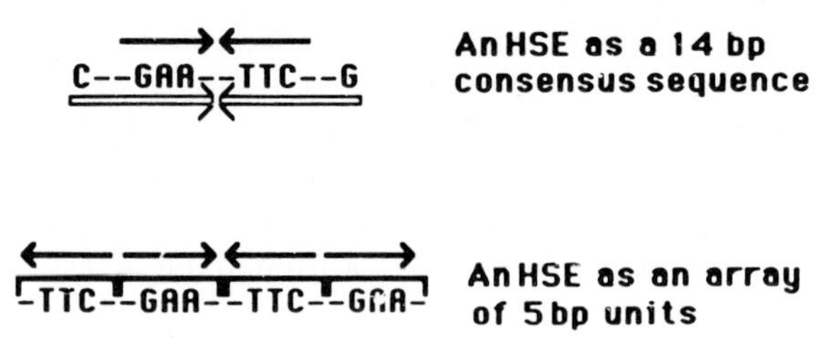

Figure 142. Nucleotide sequences in the HSE (heat-shock control element) (from Lis *et al.*, 1990).

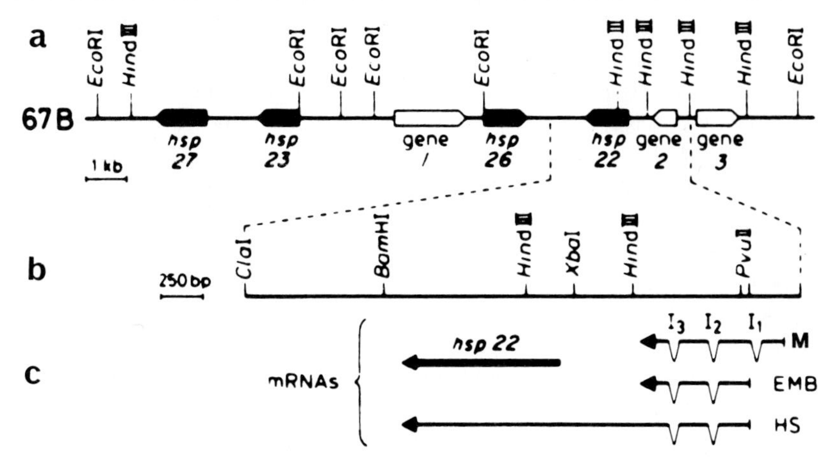

Figure 143. Organization of the DNA fragment included in the 67B puff. (a) A scheme of the disposition of genes and *Eco*RI and *Hind*III restriction sites. Black arrows point to the *hsp27*, *hsp23*, *hsp26*, and *hsp22* genes, white arrows point to genes *1*, *2*, and *3*; (b) a part of the scheme (magnified); (c) localization of transcripts: *hsp22*. EMB, embryonic, M, male-specific; HS, combined transcript under heat shock. I_1–I_3, introns (from Pauli *et al.*, 1988).

1984; Amin *et al.*, 1985, 1987; Simon *et al.*, 1985b; Bienz and Pelham, 1986; Simon and Lis, 1987; Xiao and Lis, 1988, 1990; Lis *et al.*, 1990). Tiny, single-base substitutions in this sequence reduce the transcription level 40-fold (Xiao and Lis, 1988).

Using a UV cross-linking method to examine protein–DNA interactions, it was shown that a molecule of RNA polymerase II is associated with DNA sequences near the transcription start site, even in not heat-shocked cells. The region of this association was defined to lie between nucleotides −12 and +65 bp relative to the transcription start site (Gilmour and Lis, 1986). These RNA polymerase II molecules transcribe a nascent RNA chains of approximately 25 nucleotides; they are arrested at this point, because transcriptionally unengaged in the absence of heat-shock induction (Rougvie and Lis, 1988, 1990; O'Brien and Lis, 1991).

The leader region of the transcript is required to ensure normal translation of *hsp70* in heat-shock cells. When the region from +2 to +205 is deleted, synthesis, but not translation, of these mRNAs proceeds under heat-shock conditions. Translation of the message starts when cells are allowed to recover from heat shock (McGarry and Lindquist, 1985). The first 95 nucleotides from the start site of transcription appear to be important for the efficient translation of *hsp70* (Klemenz *et al.*, 1985). Via purification the heat-shock protein *hsp70* by affinity chromotography on ATP-agarose, it was concluded that the protein has an ATP-binding site and that it can contain ATP (Beaulieu and Tanguay, 1988).

In *D. hydei*, the *hsp70* gene is located in the 2-32A puff (Peters *et al.*, 1980). In *C. thummi*, *hsp70* is encoded by the puff III-A3b. The puff remains transcriptionally active, while the other heat shock-puffs become inactivated after exposure to DRB. A protein of 70 kDa is concurrently synthesized (Barettino *et al.*, 1988b).

In *D. auraria*, a clone isolated from heat-shock cDNA library presented two long, antiparallel coupled open reading frames. One ORF strand is 1929 nucleotides and exhibits 87.5% of nucleotide identity with the *hsp70* gene of *D. melanogaster*. The second strand ORF is 1839 nucleotide long and exhibits 32% identity with putative NAD^+-dependent glutamate dehydrogenase. The overlap of the two ORFs is 1824 nucleotides (Konstantopoulou *et al.*, 1995).

The two genes *hsp83* and *hsp68* encode large heat-shock proteins. Template RNA of 2.4 kb *(hsp 68)* and 3.05 kb *(hsp 83)* hybridizes *in situ* to the 95D *(hsp 68)* and 63BC *(hsp 83)* puffs (McKenzie and Meselson, 1977; Holmgren *et al.*, 1979; O'Connor and Lis, 1981). The puffs originate from the single 63B9 and 95D11 bands (Semeshin *et al.*, 1985b). Homologies were demonstrated between the heat-shock puff at 95D and 63BC in *D. melanogaster* to the *D. hydei* heat-shock loci 2-36A and 4-81B, respectively (Peters *et al.*, 1980). The homology of the heat-shock puff at 63BC in *D. melanogaster* and 23 in *D. pseudoobscura* was also established (Pierce and Lucchesi, 1980). The *hsp83* DNA isolated from the genome of this species hybridizes to the chromosomes of *D. melanogaster* (Sass, 1990). *hsp83* RNA undergoes processing during transcription, because the gene contains a single intron (Holmgren *et al.*, 1981; Hackett and Lis, 1983; Blackman and Meselson, 1986). Three DNase I-hypersensitive sites are localized at a distance of 2.3–3.2 kb upstream and 2 and 1.8 kb downstream of the internal restriction *Sal* site; that is, at the flanks of the coding part of the gene (Wu, 1982). Males and females of *D. pseudoobscura* do not differ in DNase I sensitivity of chromatin of the *hsp83* gene. In this species, the gene is located on the X chromosome; hence, differences due to dosage compensation of the heat-shock genes may be expected (Eissenberg and Lucchesi, 1982). The HSE has a standard structure (Figure 142) (Zimarino and Wu, 1987; Xiao and Lis, 1989; Lis *et al.*, 1990). There is an HSE in the 5′ region of *hsp68* (Pelham and Bienz, 1982). DNase I-sensitive sites were not found in the surroundings of *hsp68* (Wu, 1982).

Seven genes in *D. melanogaster*—including four encoding the small heat-shock proteins Hsp22, Hsp23, Hsp26, and Hsp27 and three other genes (Figure 143), whose functions are not completely understood—are encoded at a DNA fragment 15 kb in length (Corces *et al.*, 1980; Craig and McCarthy, 1980; Ingolia and Craig, 1981; Voellmy *et al.*, 1981; Sirotkin, 1982; Sirotkin and Davidson, 1982; Southgate *et al.*, 1983; Ayme and Tissieres, 1985). The RNA synthesized in response to heat shock, isolated from polysomes, maps to a puff at 67B by *in situ* hybridization (McKenzie and Meselson, 1977; Spradling *et al.*, 1977; Mirault *et al.*, 1978; Moran *et al.*, 1978; Corces *et al.*, 1980; Wadsworth *et al.*, 1980; Lis *et al.*, 1981b; Voellmy *et al.*, 1981). By genetic analysis, the *hsp22–27* loci were also mapped to the region of chromosome 3L containing one 67B heat-shock puff (Peterson *et al.*, 1979a,b). As shown by electron microscopy, the puff at 67B originates from two neighboring diffuse bands 67B1-2 (Semeshin *et al.*, 1984, 1985a).

The genes mapped to the 67B locus share various features:

1. At least six (*hsp22*, *hsp23*, *hsp26*, *hsp27*, and the genes *1* and *3*) are closely homologous to each other. The coding sequence in the four *hsp* genes contains a homologous tract of 108 codons. The sequence of the *hsp22* gene has 35% similarity to the sequences of the *hsp23*, *hsp26*, and *hsp27* genes. The latter three within this fragment are 60% similar. Within the sequence, 80 codons show a high level of similarity to the protein α-cristalline from the lens of the eye of mammals (Ingolia and Craig, 1982b; Southgate *et al.*, 1983; Aime and Tissieres, 1985; Pauli and Tonka, 1987).
2. *hsp22–27* genes are regulated both by heat shock and by the hormone ecdysterone (Ireland and Berger, 1982; Ireland *et al.*, 1982; Larocca, 1986b; Luo *et al.*, 1991). The genes function at different stages of development (Table 19). Northern blot analysis demonstrated that a shorter (950-bp) *hsp23* trancript is translated in response to ecdysterone compared to that induced by shock (two transcripts; one of 2450 bp, the other of 1050 bp). A similar pattern of transcript heterogeneity was found for the *hsp26* and *hsp27* genes. The length of the poly(A)$^+$ tail of mRNA also varies (Berger *et al.*, 1985). Treatment with ecdysterone increased the rate at which the *hsp22*, *hsp23*, *hsp26*, and *hsp27* genes are transcribed 2.3, 2.6, 1.4, and 2.0-fold, respectively; heat shock increases the rates by 12.7, 3.4, 28.4, and 8.3-fold. Exposure to both agents (ecdysterone plus a 37°C heat shock) produces a substantial, although unproportional, elevated levels of transcripts from single genes (Ireland *et al.*, 1982). The genes for the small heat-shock proteins respond differently to ecdysterone. When hormone is added to culture medium, activity of the *hsp27* gene is rapidly induced, even under conditions of inhibited RNA synthesis. High level expression of the *hsp23* gene is detected only after a delay of 6 hr; its maintenance at this level requires the continuous presence of ecdysterone; it is sensitive even to low concentrations of protein synthesis inhibitors. An ecdysterone receptor may be involved in the regulation of both genes (Amin *et al.*, 1991). Taken together, the data indicate that the genes located in two chromomeres (in a single puff) are not transcribed in parallel and coordinately.
3. The DNase I-hypersensitive sites are located at the 5′ end of each gene of the puff 67B (Keene and Elgin, 1982; Thomas and Elgin, 1988; Kelly and Cartwright, 1989).
4. All the small heat-shock genes have HSEs at the 5′ end (Figure 144), and

Table 19. Expression of Genes Located in the 67B Puff Based on Identification of RNA or Protein during Normal Development

Stage of development	*hsp27*	*hsp23*	*Gene 1*	*hsp26*	*hsp22*	*Gene 2*	*Gene 3*
Embryo 8–12 h	+		+	+		++	+
First instar larvae	+ (Brain, gonads)					+	
Second instar larvae	+ (Brain, gonads)					+	
Mid-third instar larvae			–	+	–	–	–
End of third instar–beginning of pupal development	+ (imaginal disks)	+	+++	++++	+++	+	++++
Early pupae	+	+	+	+			
Late pupae		++	–	+	–	–	–
Young adults		+	+				
Adults, male				+ (Spermatocytes)		+ (Testes)	
Adults, females	+ (Ovaries)			+ (Ovarian nurse cells, ovaries)			

Note. The table is based on the data of Sirotkin (1982); Sirotkin and Davidson (1982); Zimmermann *et al.* (1983); Mason *et al.* (1984); Ayme and Tissieres (1985); Glaser *et al.* (1986); Arrigo (1987); Pauli and Tonka (1987); Pauli *et al.* (1988, 1990); Haass *et al.* (1990).

these HSEs contain the nucleotide sequence CT-GAA--TTC-AG standard for the other *hsp* genes. The elements nearest to the start site of transcription lie at a distance of 14–28 bp from the TATA domain (Pelham and Bienz, 1982; Cohen and Meselson, 1985; Riddihough and Pelham, 1986; Hoffman *et al.*, 1987; Lis *et al.*, 1990). All the HSEs lie within the first 400 bp of the transcription start: *hsp22* (Ayme *et al.*, 1985; Klemenz and Gehring, 1986), *hsp23* (Mestril *et al.*, 1985, 1986; Pauli *et al.*, 1986), *hsp26* (Cohen and Meselson, 1985; Pauli *et al.*, 1986; Simon and Lis, 1987; Thomas and Elgin, 1988), and *hsp27* (Hoffman and Corces, 1986; Riddihough and Pelham, 1986).

5. The elements controlling the response to ecdysterone and heat shock lie in the 5′ end of the genes (Figure 145; see also Antoniewski *et al.*, 1993; Dubrovsky *et al.*, 1994, 1996).

Along with common features, the controlling elements have other additional features that distinguish them from other elements. Nor-

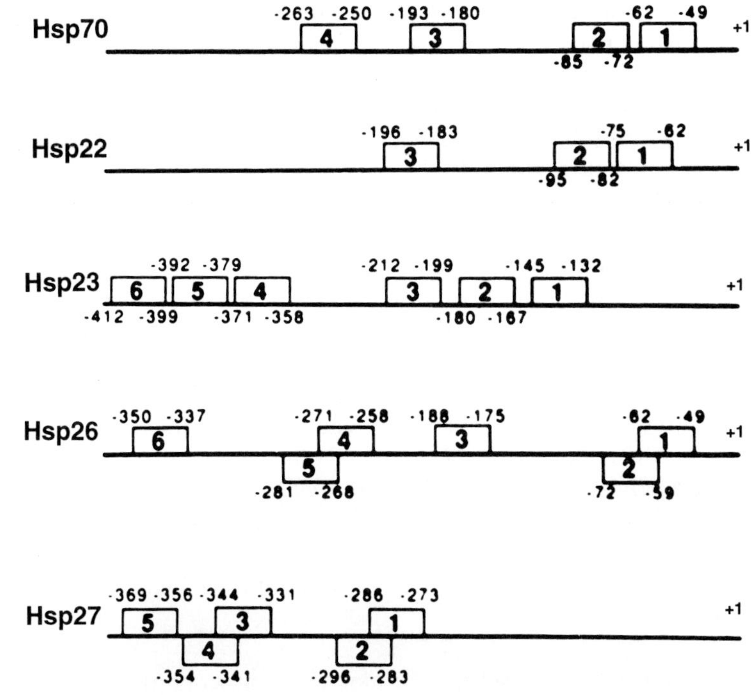

Figure 144. Localization of the HSE elements in the 5′ regions of the *hsp70*, *hsp22*, *hsp23*, *hsp26*, and the *hsp27* heat-shock genes. The HSE elements are depicted as rectangles with numbers. Numbers with minuses designate the position of the beginning and end of the element (as nucleotide pairs). +1, start point of transcription (from Simon, J. A., and Lis, J. T. (1987). A germline transformation analysis reveals flexibility in the organization of heat shock consensus elements. *Nucleic Acids Res.* **15,** 2971–2988 by permission of Oxford University Press).

mal heat-induced expression of the *hsp26* gene requires a sequence containing homopurine–homopyrimidine repeats (dC-dT)·(dG-dA). They are located in the stretch extending from − 85 to − 134 bp and do not contain HSEs. The sequence assumes the structure that is hypersensitive to DNase I (Glaser *et al.*, 1990).

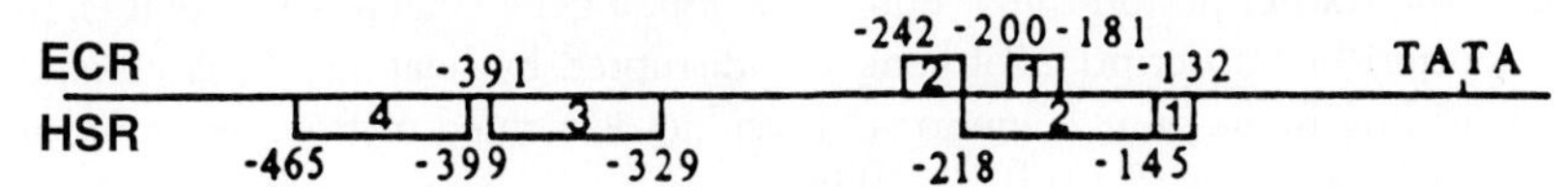

Figure 145. Regulatory sequences in the 5′ nontranscribed region of the *hsp23* gene of *D. melanogaster*. ECR, the binding region of the ecdysterone–receptor complex; HSR, binding region of the protein inductor of heat shock. The figure in the rectangle is the number of the element. Figures with minus mean position of elements in nucleotide pairs upstream of transcription start. TATA, the TATA domain [from Mestril *et al.* (1986). Heat shock and ecdysterone activation of the *Drosophila melanogaster hsp23* gene: A sequence element implied in developmental regulation. *EMBO J.* **5,** 1667–1673, by permission of Oxford University Press].

Besides elements sensitive to ecdysterone and heat-shock factors, the promoter region of *hsps* contain elements controlling organ specificity. Three sites lying between positions − 351 and − 135, − 135 and − 85, and + 11 and + 632 bp are able to induce transcription specific to spermatocyte cells (Glaser *et al.*, 1986; Glaser and Lis, 1990). Transcription specific to ovaries is controlled by a promoter element lying between − 728 and − 341 bp (Cohen and Meselson, 1985). Two regions whose positions coincide with DNase I-hypersensitive sites in embryonic chromatin were identidied in the 5′ region of the *hsp27* gene. Ecdysterone induction is mediated by sequence elements (EcRE) between − 579 and − 455 bp; the consensuses determining sensitivity to heat shock is located between − 370 and − 270 bp. Deletion of these elements abolishes the heat-shock response; the inducibility can be restored by the addition of the HSEs, even by placing them at the 3′ end of *hsp27*. Although more than 2 kb apart, the HSE are still able to interact in a cooperative way and induce *hsp27* (Riddihough and Pelham, 1986).

A protein of 80–90 kDa binds to the ecdysterone response element of the promoter region (Riddihough and Pelham, 1987). The EcRE is inductionally engaged even when inserted into the 3′end of the gene (Dobens *et al.*, 1991).

Performance of organ-specific function (transcription in the ovarian cells) is dependent on the normal state of two elements in the 5′ region of the gene. One element is located in a fragment of 64 bp in the transcribed, but not in the translated, region; the other lies between − 227 and − 958 bp (Hoffman *et al.*, 1987). Gene 2 is transcribed into two polyadenylated RNAs (I_1-I_3 and EMB: see Figure 143); the smaller transcript EMB, 560 nucleotides in size, is synthesized from mid-embryogenesis, then during the first two larval instars; its synthesis resumes at the beginning of pupation. The larger transcript (I_1-I_3), 780 bp in size, has an additional exon. This RNA is identified in pupae and adults. It is specific, occurring only in males, and is synthesized in the testes. Heat shock does not affect the amounts of the two transcripts, but it stimulates the synthesis of a third 2-kb transcript. The latter contains the cap site of the embryonic transcript and the 3′ end of the *hsp22* gene (Figure 143); that is, the termination of transcription of gene 2 is disrupted by heat shock. A common open reading frame of 111 amino acid residues is formed in the new transcript resulting from fusion (Pauli *et al.*, 1988).

The B52 protein specific for chromatin, bracketing in polytene chromosomes the RNA polymerase II fluorescence signal, is located on both edges of the 22–27 *hsp* complex (Champlin *et al.*, 1991).

Besides heat-shock proteins, whose synthesis is increased by heat shock, there are other similar proteins showing high (75–80%) similarity to the nucleotide composition to the *hsp70* gene. However, these genes are not heat inducible; they are expressed during normal development. They are called the heat-shock-cognate genes. They have been mapped to the regions of polytene chromosomes of *Drosophila* by *in situ* hybridization: *hsc1* is present at 70C, *hsc2* at 87D, and *hsc4* at 88E (Craig *et al.*, 1982, 1983; Ingolia and Craig, 1982a; Ingolia *et al.*, 1982; Palter *et al.*, 1986; Perkins *et al.*, 1990).

2. The "unusual" heat-hock puffs

An unusual puff 93D in *D. melanogaster* is distinctive among the puff sites induced by heat shock. For this reason, puffs of this group are called "93D-like." They are also referred to as "unusual" and *hsr*-omega (reviews: Lakhotia, 1987; Bendena *et al.*, 1989b; Gubenko, 1989). The puffs are unusual because, first, the synthesized RNA presumably does not exit from the nucleus and, accordingly, the protein product of the gene is not synthesized; second, the RNA synthesized in these puffs is packaged into "special" granules of giant size (see Table 2). The central globular part is composed of only protein. Smaller granules, made up of RNA and protein, are located on the surface of the globules. In *D. melanogaster*, the globules contain a specific protein (the P11 antigen) detected by indirect immunofluorescence almost exclusively in the 93D heat-shock puff (Dangli *et al.*, 1983; Dangli and Bautz, 1983). Third, the special puffs are inducible by a wide variety of agents independently of the other heat-shock puffs (Table 17).

In other species, the criterion for identification of puffs homologous to 93D is either the presence of giant RNP globules or their inducibility by specific agents; for example, benzamide. The unusual structure is the heat-shock locus 2-48B in *D. hydei*, *Drosophila neohydei*, and *Drosophila eohydei* (Peters *et al.*, 1980); it is located at 20CD in *D. virilis*, *D. texana*, *D. littoralis*, *D. lummei*, *Drosophila montana*, and *Drosophila novamexicana* (Gubenko and Baricheva, 1979); at 2R-48A in *Drosophila nasuta;* at 2L-2C in *Drosophila ananassae;* at E-11B in *Drosophila kikkawai*, at 2-58C in *D. pseudoobscura* (Lakhotia and Singh, 1982), and at 42C in *D. auraria* (Scouras *et al.*, 1986).

The special puff is not induced in *Melanagromyza obtusa* (Agromyzidae). Because this family is thought to be more primitive than Drosophlidae, it was concluded that the function controlled by the 93D puff might have been acquired at the later stages of phylogenesis (Singh, 1988b).

With the use of deletions in *D. melanogaster*, the 93D puff was mapped to the 93D2-93D10 region (Scalenghe and Ritossa, 1977), between 93D4 and 93D9, and presumably to the 93D6–7 region (Mohler and Pardue, 1982a,b). According to electronmicroscopic data, the puff originates from the 93D6-7 band (Semeshin *et al.*, 1985b). The RNA from heat-shock cells hybridizes *in situ* also to the 93D puff. A significant portion of the RNA from 93D remains in the nucleus, while the RNA from the other *hsp* genes is hardly discernible in the nuclei (Lengyel *et al.*, 1980). The organization of the 93D puff was clarified by cloning the relevant DNA. The total length of the transcribed region of the 93D puff is about 10 kb (Figure 146). It contains a region of unique sequences at its 5′ end and a repeat region of 280 bp (Walldorf *et al.*, 1984; Ryseck *et al.*, 1985, 1987; Garbe and Pardue, 1986; Pardue *et al.*, 1987; Bendena *et al.*, 1989b). Expression of this puff is reflected in both poly(A)$^+$ and poly(A)$^-$ RNA (Lengyel *et al.*, 1980). A set of three transcripts is induced from a unique region at a distance of about 2.5 kb from the same transcription start site (Figure 146). Transcripts 1.9 and 1.2 kb in size have termination sites before repeat starts. A large transcript approximately 10 kb in size has a region of 7.5 kb repetitive DNA (Figure 146). The length of the repeated monomer is 280 bp. Differences among the repeats involve single nucleotide substitutions and 1–3 nucleotide deletions or insertions (Hogan *et al.*, 1995).

Transcripts of 10 and 1.9 kb occur only in the nucleus (Figure 147). A 1.2-kb transcript also occurs in the cytoplasm. An intron (710 bp) is spliced from this RNA; it is polyadenylated during processing to become mRNA (Ryseck *et al.*, 1985; Garbe *et al.*, 1986; Hovemann *et al.*, 1986; Pardue *et al.*, 1987, 1990). Nuclear RNA, which is >9 kb, does not undergo processing, but is polyadenylated (Hogan *et al.*, 1994). Analysis of the RNA from many organs that had not been heat shocked demonstrated that the three transcripts are present in each (Bendena *et al.*, 1989a, 1991; Pardue *et al.*, 1990).

The function of the 93D puff is unclear. Protein fractions encoded by the 93D DNA were not found (Table 18). Newly synthesized heat-shock proteins do not accumulate in the cells after specific induction of the puff 93D by benzamide (Lakhotia and Mukherjee, 1982). Individuals in which the 93D puff is deleted [heterozygotes for the overlapping oppositely oriented *Df(3R)e*^{Gp4} and *Df(3R)GC14* deletions] show no changes in the pattern of polypeptide synthesis (Mohler and Pardue, 1982a). The claim that the 93D puff encodes Hsp72 (Scalenghe and Ritossa, 1977) was erroneous because the Hsp70 protein maps to the 87AC puffs (see above). The RNA transcribed from the genes does not have long open reading frames: 27 amino acids are encoded in the omega-3 transcript (Bendena *et al.*, 1989a). The short open reading frames are translated, but this does not lead to accumulation of detectable amounts of protein (Fini *et al.*, 1989).

No complementation groups were detected within the cytological limits of 93D4-93D9, defined by the ends of deletions to which the puff was mapped. Despite this, larvae homozygous for deletions completely removing the 93D puff had 20–30% viability; 5% of the expected adults eclosed, but they died within a few days (Mohler and Pardue, 1982a, 1984). Imaginal disks of larvae homozygous for a 93D deficiency do not differentiate after incubation in an ecdysterone-containing medium. The control disks (from larvae heterozygous for one of the deletions) differentiate normally (S. C. Lakhotia and M. Sandhu,

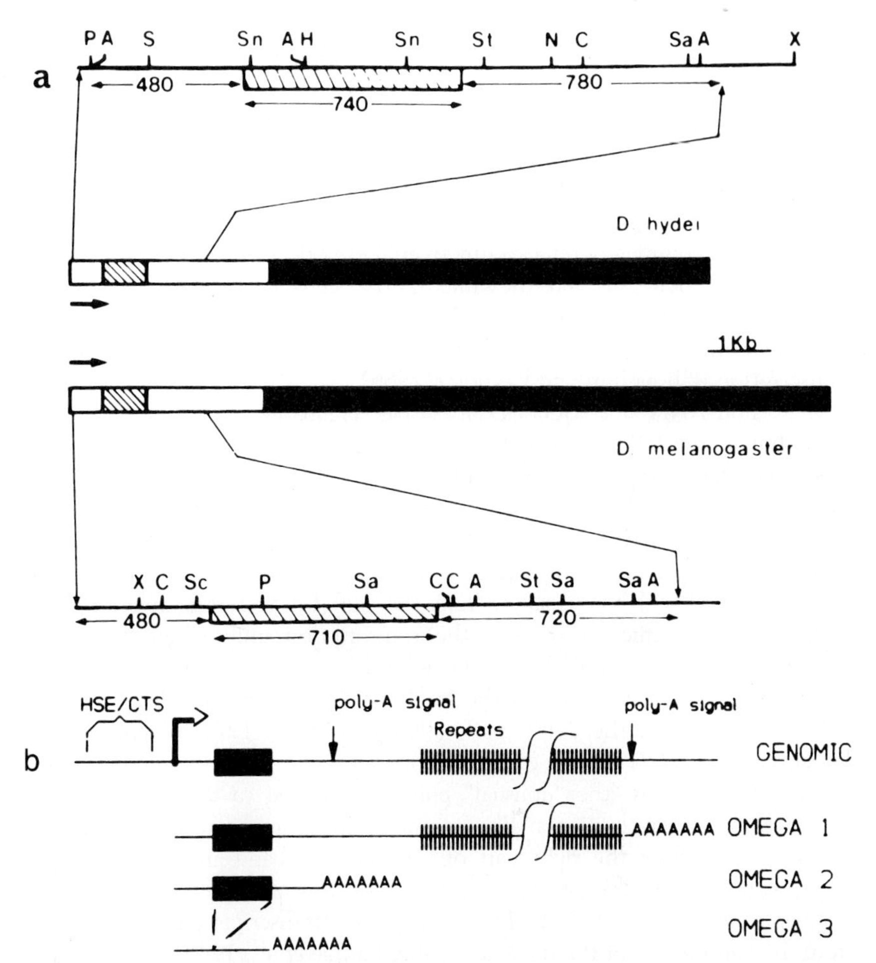

Figure 146. A molecular map of the heat-shock loci 93D in (a) *D. melanogaster* and the 2-48B in *D. hydei* and (b) organization of transcription in the 93D puff. (a) Horizontal lines designate restriction maps of the 5′ ends of the genes, where A, C, H, N, P, S, Sa, Sc, Sn, St, and X are the restriction sites of endonucleases *Asu*II, *Cla*I, *Hind*III, *Nar*I, *Pst*I, *Sal*I, *Sac*I, *Sca*I, *Sna*BI, *Stu*I, and *Xba*I, respectively. Numbers under the maps indicate the extension of DNA fragment (bp). Open rectangles designate fragments of exons; introns are hatched, black rectangles designate tandem repeats; (b, top) genomic DNA, (bottom) three transcripts (omega 1–3). HSE, the heat-shock element; CTS, constitutive signals for transcription organization (reprinted by permission from Pardue *et al.*, 1987; Bendena *et al.*, 1989b, reproduced from *The Journal of Cell Biology*, 1989, **108,** 2017–2028 by copyright permission of The Rockefeller University Press).

cited in Lakhotia, 1987). The suggestion that the *ebony* (*e*) gene may be located in the 93D puff (Scalenghe and Ritossa, 1977) was not confirmed. The *e* gene was subsequently mapped to the 93D1-2 band; that is, somewhat proximal to the puff by deletions (Mohler and Pardue, 1984).

The 87A, 87C, and 93D puffs interact. If the activity of the 93D puff was modified by an agent during heat treatment, the sizes of 87C and 87A changed. If the 93D puff remained in one dose, it was not induced, and the 87C puff became twice as small as 87A under heat induction. In larvae whose 93D puff was completely removed by overlapping deficiencies 87C was smaller in some cells; their sizes were the same in other cells. It was concluded that the 93D and 87C puffs are somewhat related (Burma and Lakhotia, 1986; Ghosh and Mukherjee, 1990). In *D. simulans*, the 93D puff has no effect on 87C transcription (Choudhuri and Lakhotia, 1986).

Puff DNA at 2-48C of *D. hydei* (see Figure 146) shares many features with that at the 93D locus: (1) the greater part of the locus is occupied by repetitive DNA, with the repeat unit being a sequence 115 bp long; (2) the repeats are conserved, divergence not exceeding 10%; (3) the sequence does not have an open reading frame; (4) three transcripts of 9.4, 2, and 1.35 kb are synthesized in the puff. The largest transcript contains repeats (Peters *et al.*, 1983, 1984). The RNA isolated from the puff by micromanipulation has a sedimentation coefficient of 40S; the transcription unit, when the "Miller spreading technique" is used, is 2 μm long. No polyadenylation of this RNA was found. It is not present in the fraction of polysomes RNA; however, it hybridizes *in situ* to the 2-48B puffs (Berendes *et al.*, 1974; Berendes, 1975; Bisseling *et al.*, 1976; Lubsen *et al.*, 1978).

In *D. virilis*, the "unusual" puff was mapped to the 20C region by chromosomal rearrangements. Electronmicroscopic analysis showed that the puff originates from the right part of the 20C6–7 band and that the swelling spreads to the neighboring 20C and 20D sites after prolonged treatment (Gubenko *et al.*, 1991a,b). In *D. pseudoobscura* transcripts are also produced from the "unusual" puff (Garbe *et al.*, 1989; Pardue *et al.*, 1990).

The available data on the organization of the "unusual" puffs indicates that they are functionally similar, but their nucleotide sequences diverge rapidly; length heterogeneity was observed for repeat units (Pardue *et al.*, 1987; Ryseck *et al.*, 1987). In the three species of *Drosophila* studied, there is only a very short open reading frame: the number of amino acids is 27 in *D. melanogaster*, 23 in *D. hydei*, and 24 in *D. pseudoobscura* (Bendena *et al.*, 1989a). The rapid sequence divergence of the "unusual" puffs is remarkable. The heat-shock RNAs from *D. melanogaster* or *D. virilis* do not hybridize to the the *D. hydei* puff 2-48B. *D. hydei* heat-shock RNA hybridized to the chromosomes of only the more closely related species *D. neohydei* and *D. eohydei*; it hybridized neither to the *D. repleta* nor to the *D. virilis* puffs (Peters *et al.*, 1980). There are wide variations in the number of repeat units at a locus, even for inbred lines of *D.*

hydei (Peters *et al.*, 1984). The most conserved sequence in the DNA of the unusual puffs consists of nine nucleotides: ATAGGTAGG. It should be noted that the sequence is present at one copy in a region of 115 bp and two copies in a 280-bp region, so that its concentration in mRNA remains unaltered (Pardue *et al.*, 1987).

The intron sequence was isolated from the puff 93D hybridized to the 2–48B puff DNA (Garbe and Pardue, 1986; Pardue *et al.*, 1987), to the chromosomes of *Drosophila* of the *melanogaster* group (*simulans*, *mauritiana*), as well as to *D. miranda* and *D. pseudoobscura* (Garbe *et al.*, 1989).

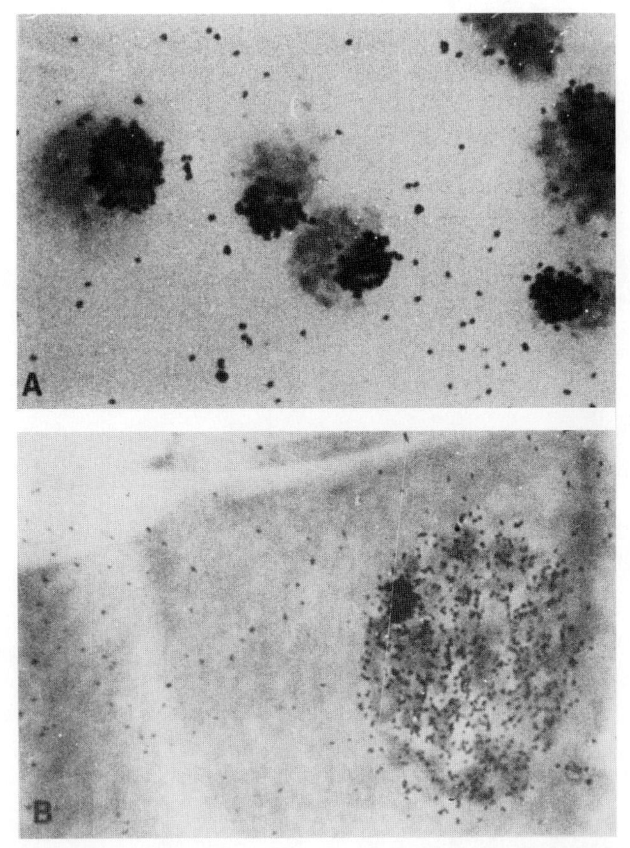

Figure 147. *In situ* hybridization of repetitive DNA complementary to the omega-1 transcript to (a) cell culture and (b) salivary glands. Cells were treated with high temperature (36°C) for (a) 1 hr or (b) not exposed to heat shock. Heat shock increases the amount of omega-1 in the cell but does not affect the nuclear location (reprinted by permission from Pardue *et al.*, 1990).

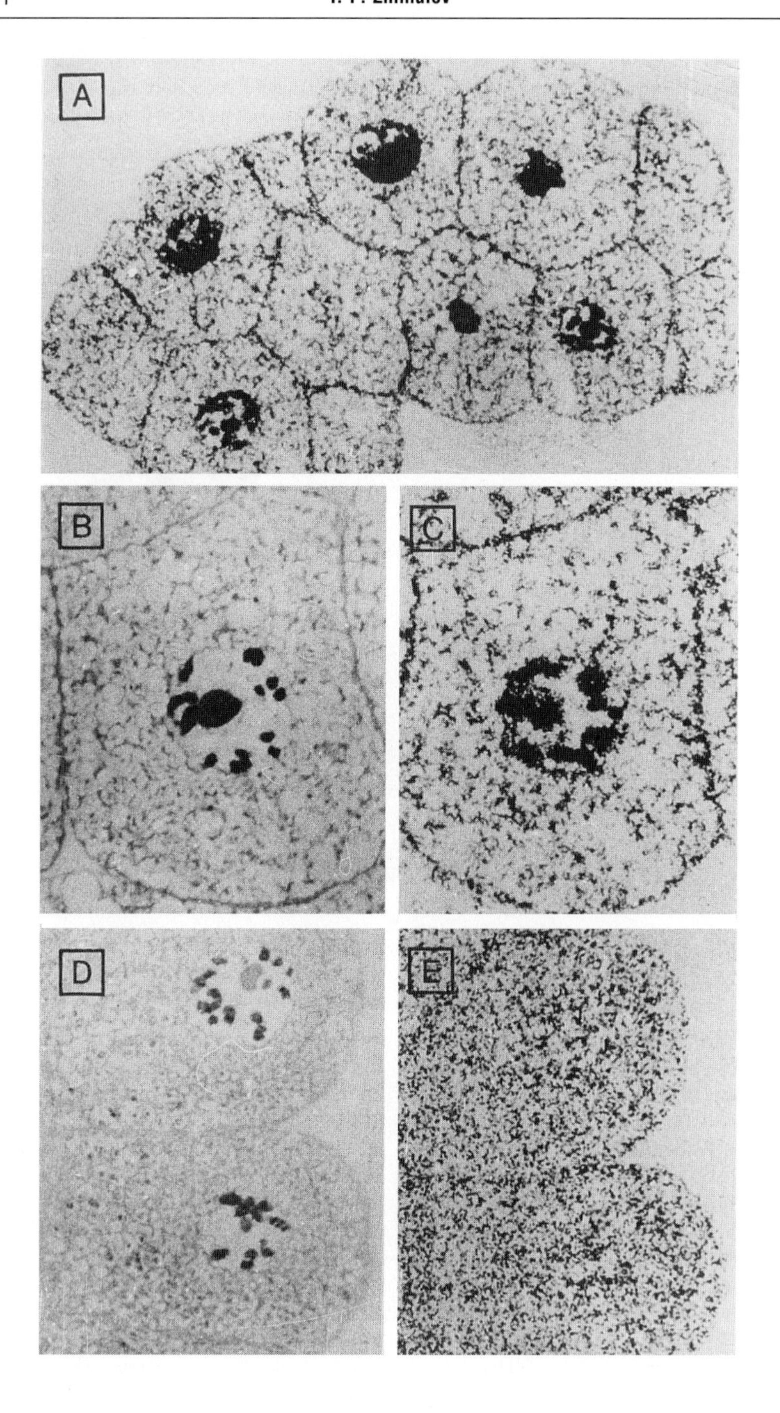
A
B
C
D
E

3. The Hsp proteins during normal development and heat shock

The *hsp* genes are expressed not only when cells are subjected to heat shock, but also during normal development, as introduced above. With the use of antibodies, small amounts of the Hsp83 protein were found in the cells of larval and prepupal salivary glands, larval brain, pupal thoracic epithelium, and ovaries of normally developing adults (Chomyn and Mitchell, 1982; Zimmermann *et al.*, 1983; Mason *et al.*, 1984; Xiao and Lis, 1989). The protein is detected before the time of blastoderm formation in embryos (Zimmermann *et al.*, 1983). Hsp is distributed in the cytoplasm and in small amounts in the nuclear membrane (Carbajal *et al.*, 1986).

There are data indicating that the *hsp70* gene is not expressed under normal conditions (Velazquez *et al.*, 1983), according to other data, this gene constitutively functions in eukaryotes (Pelham, 1985).

The genes encoding the heat-shock proteins contain different regulatory elements in the 5′-regulatory site, responding to tissue-specific inducers as well as to ecdysterone and heat shock. The Hsp28 and Hsp26 proteins are detected in substantial amounts in embryos to the time of blastoderm formation (Zimmermann *et al.*, 1983).

In non-heat-shocked cell cultures, the *hsp23* and *hsp27* genes are expressed only in 2–5% of cells. The Hsp28 protein is then detected in the cytoplasm (Duband *et al.*, 1986; Arrigo, 1987), and Hsp27 in the nuclear region (Beaulieu *et al.*, 1989). In salivary glands, genes located in the 67B puff are expressed differentially. While *hsp23* and *hsp27* transcripts accumulate at a high level without heat shock, those of *hsp22* and *hsp26* are present at low and intermediate levels, respectively (Dubrovsky *et al.*, 1994).

Some of the "cognate" proteins, for example, the Hsc70 protein (the product of the Hsc4 gene) occur in embryos. In cultured embryonic cells, the protein is detected in cytoplasmic fibrils concentrated around the nucleus (Palter *et al.*, 1986).

Experiments involving incorporation of radioactive amino acids demonstrated that the proteins synthesized upon heat shock rapidly move into the nucleus (Figure 148) (Mitchell and Lipps, 1975; Arrigo *et al.*, 1980; Velazquez *et al.*, 1980; DiDomenico *et al.*, 1982). More detailed analysis revealed that the various Hsp proteins of *Drosophila* are nonuniformly distributed in cell organelles. Hsp83 is localized in the cytoplasm, in the periphery near the cell mem-

Figure 148. Localization of labeled proteins in salivary gland cells of *D. virilis*. Glands were incubated 30 min with [^{3}H]leucine and then 30 min without label. (A–C) Incorporation after heat shock or (D and E) without. The B–C and D–E pairs represent consecutive sections of a single cell, stained (B and D) or after autoradiography (C and E) (from Velazquez *et al.*, 1980, © Cell Press).

brane; Hsp70 is almost uniformly localized in the cytoplasm and the nucleus. Several hours after a brief heat shock, antibodies against this protein detect it in the nucleoli. The small heat-shock proteins are concentrated in the nucleus, the chromosomes, and the nucleoli (Tanguay and Vincent, 1979, 1982; Vincent and Tanguay, 1979, 1982; Arrigo, 1980, 1987, 1990; Arrigo *et al.*, 1980, 1982; Pelham, 1984, 1985; Carbajal *et al.*, 1986; Duband *et al.*, 1986; Beaulieu *et al.*, 1989). After recovery from heat shock, the Hsp83, Hsp70, and Hsp23 proteins are excluded from the nucleus and are distributed in the cytoplasm (Velazquez and Lindquist, 1984; Carbajal *et al.*, 1986; Duband *et al.*, 1986). The Hsp22-27 proteins accumulate as particles (with a sedimentation coefficient of 20S) called "prosomes" (Duband *et al.*, 1986).

The function of the Hsp proteins at normal temperatures is not clear. It is generally accepted that Hsp facilitates decrease in the extent of cell damage incurred by stress (Lindquist, 1986; Bienz and Pelham, 1987; Lindquist and Craig, 1988; Morimoto *et al.*, 1990; Georgopoulos and Welch, 1993; Morimoto, 1993; Craig *et al.*, 1994; Glick, 1995; Hartl, 1996).

4. The protein inducing the *hsp* genes

It was deduced that there may exist cytoplasmic factors acting as transmitters linking stressing agents and the cell nucleus. The isolated nuclei of salivary glands do not respond to stressing agents. However, when nuclei are incubated with extracts from heat-shocked (35°C, 30 min) cells, puffing is induced at the heat-shock loci at chromosomal sites 63BC, 67B, 87A, 87C, and 93D; the induction of puffing was not supported when extracts from non-heat-shocked cells were used in an otherwise similar procedure (Compton and Bonner, 1978; Compton and McCarthy, 1978; McCarthy *et al.*, 1978; Bonner, 1981, 1985). The relevant factor presumably preexists in the cytoplasm. Pretreatment of salivary glands for 6 hr with cycloheximide does not prevent puff formation when the temperature is raised from 21° to 35°C (Ashburner, 1974a). The induction of the heat-shock puff occurs in the isolated nuclei of *D. melanogaster* if cytosol (not whole cells) are treated with high temperature. This suggests that the inducer molecules, which mask the temperature-labile elements, may be present in the cells (Bonner, 1985).

The inducing factor is temperature-resistant, because cytoplasmic extract retains its puff-inducing ability in the isolated nuclei even after boiling for 2 min (Bonner, 1982). According to other data, the inducive activity in *D. melanogaster* is retained even after boiling for 10 min; however, it is thermolabile in *D. hydei* (Singh and Lakhotia, 1983).

Puff-inducing activity was associated with various proteins appearing in nuclei in response to heat shock: polypeptides 23, (Helmsing and Berendes, 1971), 39, and 46 kDa (Falkner and Biessmann, 1980; Belew and Brady, 1981b).

However, studies on the nature of the inducive factor became feasible only with the advent of methods of molecular biology. The first crucial finding was that of a 110-kDa polypeptide binding to DNA in the 5′ region of the *hsp* gene. It was called the heat-shock gene-specific transcription factor (abbreviated HSTF, or HSF) (Jack *et al.*, 1981a,b; Jack and Gehring, 1982; Parker and Topol, 1984; Wu, 1984; Bienz, 1985; Pelham, 1985; Topol *et al.*, 1985; Klemenz and Gehring, 1986; Wu *et al.*, 1987; Sorger, 1991). At least two repeat units nGAAn (see Section VI,H,1) are needed for high-affinity binding of heat-shock factor, and the units may be arranged either tail-to-tail or head-to-head (see reviews by Wu *et al.*, 1987; Perisic *et al.*, 1989; Lis *et al.*, 1990; Sorger, 1991). The HSF subunits associate to form homotrimers, and each subunit contacts with the pentanucleotide DNA fragments. HSF binding is cooperative both between multimers and between the subunits of the HSF multimer (Xiao *et al.*, 1991).

The *HSF* gene of the representatives of some species was cloned. Its presence in the genome is essential for viability at all temperatures (reviewed by Sorger, 1991). In *Drosophila,* the gene is localized in the dense 55A band (Clos *et al.*, 1990) in which a puff is not formed (Zhimulev, 1974a).

HSF antigen has been localized by Westwood *et al.* (1991) not only to heat-shock puffs but also to early ecdysone-inducible and 47 additional loci in prepupal salivary gland chromosomes of *D. melanogaster*. However, an immunofluorescent signal at these non-heat-shocked sites was detected only after heat shock; that is, when preexisting puffs at these sites were supposed to have regressed. This suggests that HSF exerts not only a positive, but also a negative control of gene activities (Westwood *et al.*, 1991). Yet in *C. tentans,* HSF may be located at some puffs also at normal temperature. After heat shock, HSF may stay or leave these sites, while appearing in a large number of new loci, mostly heat-shock puffs (Lezzi, 1996).

In most if not all organisms, the heat-shock genes are not induced during early embryogenesis (Bienz, 1985), presumably because heat-shock factor is absent.

The molecular mechanisms involved in the interaction between the HSEs at the 5′ ends of the *hsp* genes and the HSF proteins appear to be highly conserved. When injected into *Xenopus* oocytes, the *Drosophila hsp* genes are transcribed in response to heat shock (Voellmy and Rungger, 1982).

The mechanism of activating the *hsp* genes by stressing factors can be described as follows (Bonner, 1985; Rougvie and Lis, 1988): Although the HSF proteins occur in uninduced cells (Figure 149), they are inactive and not bound to the HSE promoter. RNA polymerase II occupies a part of the gene in the transcription start region. The HSF protein becomes modified by heat shock and binds to the promoter. This activates RNA-polymerase II through either an influence exerted on the TATA domain or release of a hypothetical protein

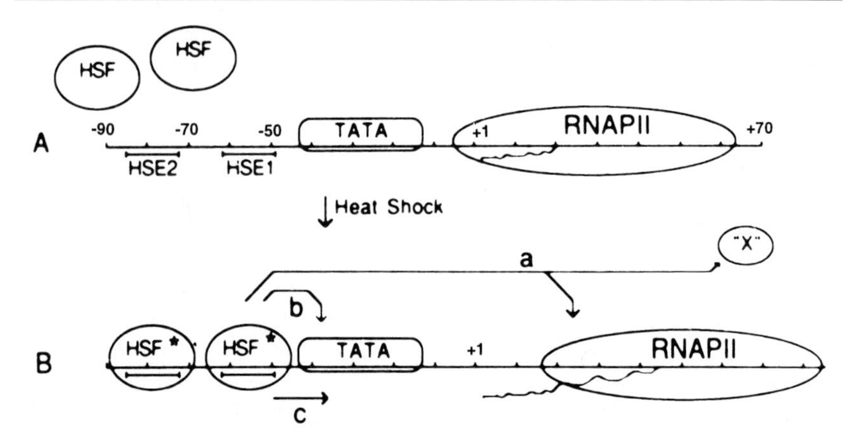

Figure 149. Components of the promoter region of the *hsp70* gene (A) before and (B) after heat-shock induction. The 5′ region of the gene from −90 to +70 bp is shown. HSF and HSE, the heat shock factor and heat shock element, respectively; TATA, the TATA domain; RNAPII, RNA polymerase II. See text for details (from Rougvie and Lis, 1988, © Cell Press).

"x," which would be involved in blocking RNA polymerase II (see Figures 149a and 149c). It is known that local chromatin structure of inducible genes appears to fall into two categories: "preset" and "remodeling." Figure 150 shows examples of preset type genes *hsp26* and *hsp70*. The binding sites for trans-acting factors are accesible (i.e., in a nonnucleosomal, DNAseI-hypersensitive configuration) prior to activation. In response to the activation signal, the factors bind to cis-acting regulatory elements and trigger transcription with no major alteration in the chromatin structure of the promoter region (see the review of Wallrath *et al.*, 1994).

In the case of the *hsp26* gene prior to heat shock, there are two DNAseI-hypersensitive (DH) sites 5′ to the transcription start site. These two DH sites map to the locations of the two functional HSEs (Figure 150). Two stretches of alternating C and T residues immediately adjacent to the HSEs bind GAGA factor *in vitro* (Bhat *et al.*, 1996). A nucleosome is specifically positioned between two DH sites. RNA polymerase II is transcriptionally engaged but paused at position +25 prior to heat shock (Figure 149). Upon heat shock, HSF binds to the HSEs and RNA polymerase II is allowed to continue transcription past the pause site. No major changes in the chromatin structure upstream of the transcription start site are observed upon activation, except the binding of HSF to the HSEs (Wallrath *et al.*, 1994). Similar organization of regulatory region was found for the *hsp70* gene (Figure 150).

The mechanisms by which a heat-shock gene is inactivated upon release from stressing factors are not understood. It was shown that the heat-shock puffs do not regress when protein synthesis is inhibited by cycloheximide

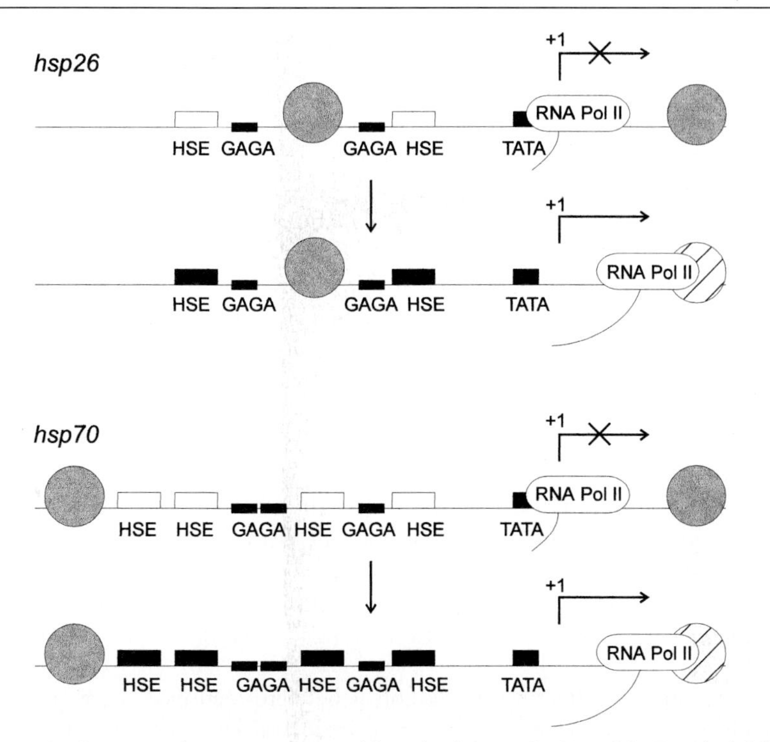

Figure 150. Structure of preset promoters of heat-shock genes *hsp26* and *hsp70*. Shaded large circles represent stable nucleosomes, cross-hatched large circles represent nucleosomes downstream of the transcription start of active genes. HSE, GAGA, and TATA: heat-shock element, GAGA domain, and TATA domain, respectively (from Architectural variations of inducible eukaryotic promoters: Present and remodeling chromatin structures, Wallrath *et al.*, *BioEssays*, Copyright © 1994 John Wiley & Sons, Inc. Reprinted by permission of Wiley-Liss, Inc., a subsidiary of John Wiley & Sons, Inc.).

(Ashburner and Bonner, 1979). This suggested that one of the synthesized Hsp proteins is involved in inhibition of the other *hsp* genes (Bonner, 1982, 1985; DiDomenico *et al.*, 1982). The presence of proteins that inhibit the heat-shock puffs may explain why cells subsequently become insensitive to stressing factor (Berendes, 1969; Myohara and Okada, 1988c). The heat-shock puffs can be reinduced 40 min after release from the effect of the stressing factor (Berendes, 1969; Laran *et al.*, 1990). Support for this idea came from localization of antibodies against Hsp70 protein in salivary gland cells of *D. hydei* exposed to heat shock (37°C, 30 min). Antibodies bind to one of the induced puffs (2-48BC: subdivisions 2-48B3-6 and 2-48C1-2); they do not bind to the other puffs, except subdivision 3-58D although the latter is not a heat-shock puff. The protein is no longer detected in subdivision 2-48C within 10 min after recovery from heat shock (Laran *et al.*, 1990).

5. Genetic control of the synthesis of the HSP proteins

The temperature-sensitive *l(1)ts403* lethal (than allele of the *sbr* gene), located on the X chromosome (Zhimulev *et al.*, 1980a), produces drastic disturbances in the response of *D. melanogaster* cells to heat shock: after a shift to 37°C, the synthesis of proteins Hsp83, Hsp68, Hsp35, Hsp27, Hsp26, and Hsp21 virtually does not occur, but the synthesis of a 72-kDa protein is induced in mutant homozygotes (Evgen'ev *et al.*, 1979, 1985; Evgen'ev and Levin, 1980; Evgen'ev and Denisenko, 1990). High-molecular-weight mRNA is synthesized in the cells of the mutant under heat-shock conditions and RNA transport from the nucleus to the cytoplasm is disrupted (Levin *et al.*, 1984). The heat-shock puffs are normally induced in mutant larvae, and they incorporate [^{3}H]uridine (Evgen'ev *et al.*, 1979). It is believed that the *l(1)ts403* lethal regulates the synthesis of the heat-shock proteins at the translation level (Evgen'ev *et al.*, 1979; Evgen'ev and Denisenko, 1990).

Hochstrasser (1987) has noted that the 87A and 87C heat-shock loci are induced unequally in the midgut polytene chromosomes as result of high temperature administration of early larval development. Similar unequality was found in males aneuploid for the 9A-11A region of the X chromosome. Incorporation of [^{3}H]uridine in 87C was twofold higher than in the 87A (Chatterjee, 1996).

Mutations affecting the *act88F* gene were identified. In homozygous individuals, the *hsp* genes are expressed in a tissue-specific manner (Parker-Thornburg and Bonner, 1987). In the other mutants of *act88F*, the *hsp* genes are not induced (Karlik *et al.*, 1987).

I. Relation between gene activation and puff formation

The formation of the puff from the band is associated with transition of the DNP composing the band into a decompacted state characteristic of the puff. From the 1960s the following stages of activation of the chromosome regions were distinguished (Swift, 1962; Beermann, 1964, 1966; Kroeger and Lezzi, 1966; Berendes, 1967a, 1968a, 1973; Sorsa, 1969; Korge, 1987): (1) Decompaction of DNP included in the bands by modification of histones or changes in molecular interaction of histones with DNA. (2) Accumulation of proteins in the region of the nascent puff. (3) Beginning of transcription.

These stages will be considered in more detail below.

The fraction composition of histones does not change at any time during puff development. However, histones are modified (see Wolffe and Pruss, 1996, for review). Injection of trypsin (which first digests proteins with modified tertiary or quaternary structure) into salivary gland nuclei of *C. thummi* resulted in a threefold increase in [^{3}H]uridine incorporation. New puffs are then not formed. It may be inferred that the structure of histones localized to puffs changes; as a consequence, histones inactivate transcription less efficiently (Robert and Kroeger, 1965; Kroeger and Lezzi, 1966). Despite the numerous attempts at histone acetylation (Ellgaard, 1967; Allfrey *et al.*, 1968; Clever and Ellgaard, 1970), phosphorylation (Benjamin and Goodman, 1969; Ziedler and Emmerich, 1973), methylation or ethylation during puff formation (Holt, 1970) was not demonstrated convincingly on polytene chromosomes.

The supposition that nonhistone proteins may be modified (methylated, phosphorylated) in the region of the activated puff turned out to be incorrect is so (Goodman and Benjamin, 1973; Ziegler and Emmerich, 1973; Schmidt and Goodman, 1976; Goodman *et al.*, 1977).

When transcription is activated in a chromosome region, large quantities of nonhistone proteins are detected by histochemical techniques in the region (Clever, 1962a, 1963b; Berendes, 1867a, 1968a; Berendes and Beermannn, 1969; Berendes and Boyd, 1969; Holt, 1970–1972). Using microspectrophotometry it was shown that during the formation of the 2-48BC puff in *D. hydei*, the DNA-to-protein ratio changes from 1/6, when the puff is in inactive state, to 1/16, when it is in an active state (Pages and Alonso, 1978). According to other data, protein content increases twofold in the activated locus during puff formation in *R. angelae* and *D. hydei* (Rudkin, 1964; Holt, 1970–1972). This is associated with an increase in the staining intensity for nonhistone protein within 3 to 5 min of the induction of a puff.

Despite the considerable increase in protein quality, the induced puffs did not preferentially accumulate label when larvae were injected with a mixture of [^{3}H]amino acids 8–12 hr before induction (Berendes, 1967b)—even 40 hr (Holt, 1970) before it and also when larvae were fed labeled amino acids for 4 days (Holt, 1970). These data suggest a low metabolic rate of proteins involved in puff induction—although there are other, not sufficiently well-grounded data indicating that puffs start to label within 3 hr after the onset of incubation of salivary glands of *S. coprophila* in saline containing [^{35}S]methionine (Goodman *et al.*, 1976).

Direct electrophoretic studies on the nonhistone proteins of chromatin from salivary gland and midgut nuclei of *D. hydei* after puff induction by ecdysone or heat shock revealed the appearance of new proteins synthesized long before in the cytoplasm (Helmsing and Berendes, 1971). The molecular weight of the fraction that appeared in response to heat shock was about 23 kDa; that appearing upon ecdysone treatment was 42 kDa (Helmsing, 1972; Berendes and Helmsing, 1974).

Nonhistone proteins represent a large and diverse group of proteins, such as RNA polymerases, topoisomerases, different proteins involved in packaging of newly synthesized RNA, and various transcription factors (see the reviews of Halle and Meisterernst, 1996; Hanna-Rose and Hansen, 1996; Kingston *et al.*, 1996; Struhl, 1996). Labeled antibodies against cyclic GMP (but not AMP) are detected in all the decondensed regions of polytene chromosomes (the interbands and the puffs). When a fixative removing chromosomal protein was used, cyclic GMP fluorescence was not observed on polytene chromosomes. This suggested that cyclic GMP may be involved in transcriptive activities (Spruill *et al.*, 1978).

The question of whether puff formation may be a consequence of storage of entered proteins and newly synthesized RNA, or whether it is a necessary prerequisite for gene expression, was resolved by using the ecdysterone-induced puffs and the heat-shock puffs as models. These puffs are advantageous because they can be easily induced under controlled conditions and rapidly reach their maximum sizes. In *D. hydei*, RNA synthesis is inhibited by actinomycin D, while the puffs are induced by ecdysterone or heat shock; the puffs are reduced to 1/3 their sizes despite completely blocking RNA synthesis (Berendes, 1967a, 1968a). Similar results were obtained with α-amanitin, another inhibitor of RNA synthesis: heat shock also caused loosening of the material of the chromosome and some increase in its diameter in the puffed region (Holt and Kuijipers, 1972b). In a third study, inhibitors of RNA synthesis, such as actinomycin D, cordycepin (3′-deoxyadenosine), α-amanitin, and DRB, which block 74.2–97.25% of RNA synthesis, caused some increase in chromosome diameter of 74E and of the 75B puffs. However, this is in contrast to Ashburner's belief (Ashburner, 1972c) that there is no increase in puff sizes. The ion-induced puffs are not formed under conditions of inhibited RNA synthesis (Gopalan, 1973a).

A remarkable exception was observed in experiments in which the heat-shock puffs were induced in *C. thummi*. When transcription was completely arrested by any one of the inhibitors (actinomycin D, α-amanitin, or DRB) on the end of the third chromosome, a specific Balbiani ring (T-BRIII) arose. Nonhistone proteins accumulated at this locus (Morcillo *et al.*, 1981; Barettino *et al.*, 1982). The region of this Balbiani ring did not label in *in situ* hybridization of clone t-BRIII DNA to chromosomal RNA, when heat-shock and actynomicin D treatment are combined (Carmona *et al.*, 1985). Taken together, the data show that protein accumulation in an activated puff does not depend on the start of transcription. The crucial event is the binding of nonhistone proteins; for example, of the hormone–receptor complex or heat-shock factor (HSF) to the promoter regions of the genes and, as a consequence, transcription starts.

Puffing development is related to active transcription. From the extensive experimental studies considered in Section III,A, some generalizations emerge. First, [^{3}H]uridine is incorporated into the puffed regions, and the larger the puff, the more intensely it incorporates label. Second, the heat-shock puffs cannot be induced when highly purified bovine pancreatic RNase is injected into the cell nuclei of isolated salivary glands of *D. hydei* before heat treatment. Control injections of a solution of protamine or acidic RNase were without effect (Berendes, 1971; Holt, 1972). Third, using P-element-mediated transformation, deletion analysis of different 5′ promoter regions (see Sections VI,E and VI,H) demonstrated that the puff was not formed when deletions prevented gene expression.

In contrast, many cases are known when chromatin decompaction is uncoupled from gene expression.

1. In autoradiographs of salivary gland, midgut, and Malpighian tubule chromosomes of *C. tentans* larvae, there were considerable (up to three orders of magnitude) variations in the intensity of [^{3}H]uridine incorporation into the chromosomes of different cells (Figure 151). The sizes of Balbiani rings were the same in salivary gland cells (Pelling, 1962–1964). Similar differences in labeling intensity were found between the cells of Malpighian tubules of *R. angelae* (Pavan, 1965a) and salivary glands of *Acricotopus lucidus* (Panitz *et al.*, 1972a). Variations in incorporation level of [^{3}H]uridine into different salivary gland cells of a single larva were found in *D. melanogaster*. In this case, the sizes of puffs in chromosomes of different cells were the same (Zhimulev and Lychev, 1972a).
2. In *Drosophila* salivary glands incubated with [^{3}H]uridine *in vitro*, the proximal gland cells labeled heavier than the distal cells. Heavily labeled cells alternated with weakly labeled ones in 40% of the 250 examined cross sections of salivary glands. Label intensity varied during a 24-hr period.

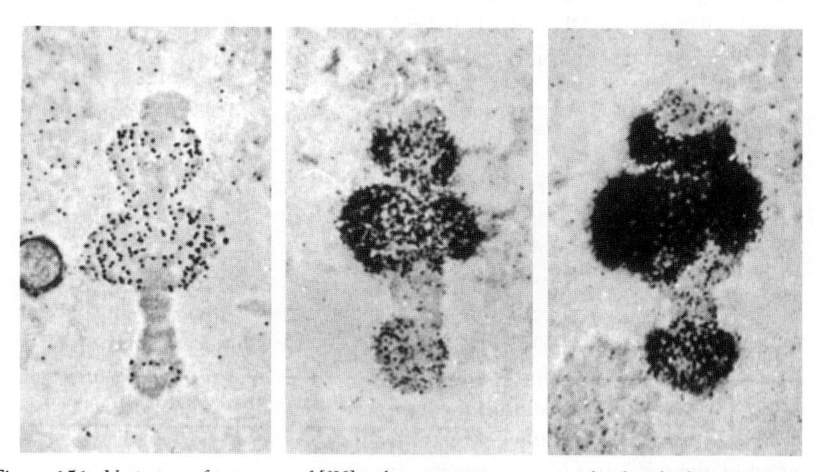

Figure 151. Variation of intensity of [^{3}H]uridine incorporation into the fourth chromosome from various salivary glands of *C. tentans* (reprinted by permission from Pelling, 1964).

Maximal label incorporation into the proximal cells was from 12 to 18 hr; there were two maxima, from 6 to 9 hr and at 21 hr, in the distal cells (Nagel and Rensing, 1974; Probeck and Rensing, 1974). Since these variations occur at unaltered puff sizes, it was concluded that puffing is the necessary, although not the sufficient, condition for RNA synthesis (Pelling, 1962). However, Beermann (1966) offered another explanation: exogenous [^{3}H]uridine possibly may not converted into nucleoside triphosphate, because large amounts of endogenous nucleoside phosphate are pooled. This provides normal RNA synthesis, but RNA is not detected when judgements are based on label incorporation. It may be also suggested that circadian changes in membrane permeability to exogenous nucleoside occur during 24 hr with resulting alterations in label level.

3. Puffing appears in regions of centromeric heterochromatin of *C. thummi* polytene chromosomes when RNA synthesis is inhibited by actinomycin. Decompaction of DNP and formation of the giant puffs occur in regions devoid of genes and when transcription is completely suppressed (see the review of Zhimulev, 1993, 1997). No RNA accumulation was found in these "pseudopuffs" by means of immunofluorescent detection of DNA-RNA hybrids (Valeyeva *et al.*, 1997).
4. Data on the formation of the large Balbiani ring T-BRIII in *C. thummi* in response to heat shock under conditions of inhibited RNA synthesis were described earlier in this section.
5. In polytene chromosomes of *D. melanogaster*, high temperature induced the heat-shock puffs; transcription of the early ecdysterone puffs 74EF and 75B concomitantly ceased; however, the puffs remained large. RNA was detected in these puffs by using the immunofluorescent technique to locate antibodies against DNA/RNA hybrids. Because the RNA is located in the puff region, its morphology presumably remains unaltered (Evgen'ev *et al.*, 1985).
6. In some larvae homozygous for the mutation *l(3)tl* the heat-shock puffs are formed, but RNA is not detected in some of them (Figure 152), possibly because of the higher transport rate of synthesized RNA.
7. If larvae homozygous for the temperature-sensitive $l(1)su(f)^{ts67g}$ mutation incubated from early third instar at 30°C, ecdysterone is not released; and mRNA of the *Sgs3*, *Sgs4*, *Sgs7*, and *Sgs8* glue genes is not synthesized; however, the 3C and 68C puffs are active (Hansson *et al.*, 1981; Hansson and Lambertsson, 1983). In homozygotes for the *l(1)npr-1* mutation at the BR-C locus, the 68C puff is active; however, accumulation of the *Sgs3*, *Sgs7*, and *Sgs8* mRNAs did not occur (Crowley *et al.*, 1984; Meyerowitz *et al.*, 1985a, 1987). Antibodies against DNA/RNA hybrids were not identified in the 68C puff in mutants for this gene (Umbetova, 1991), however, [^{3}H]uridine is incorporated (E. S. Belyaeva and I. F. Zhimulev, 1989, unpublished observations).

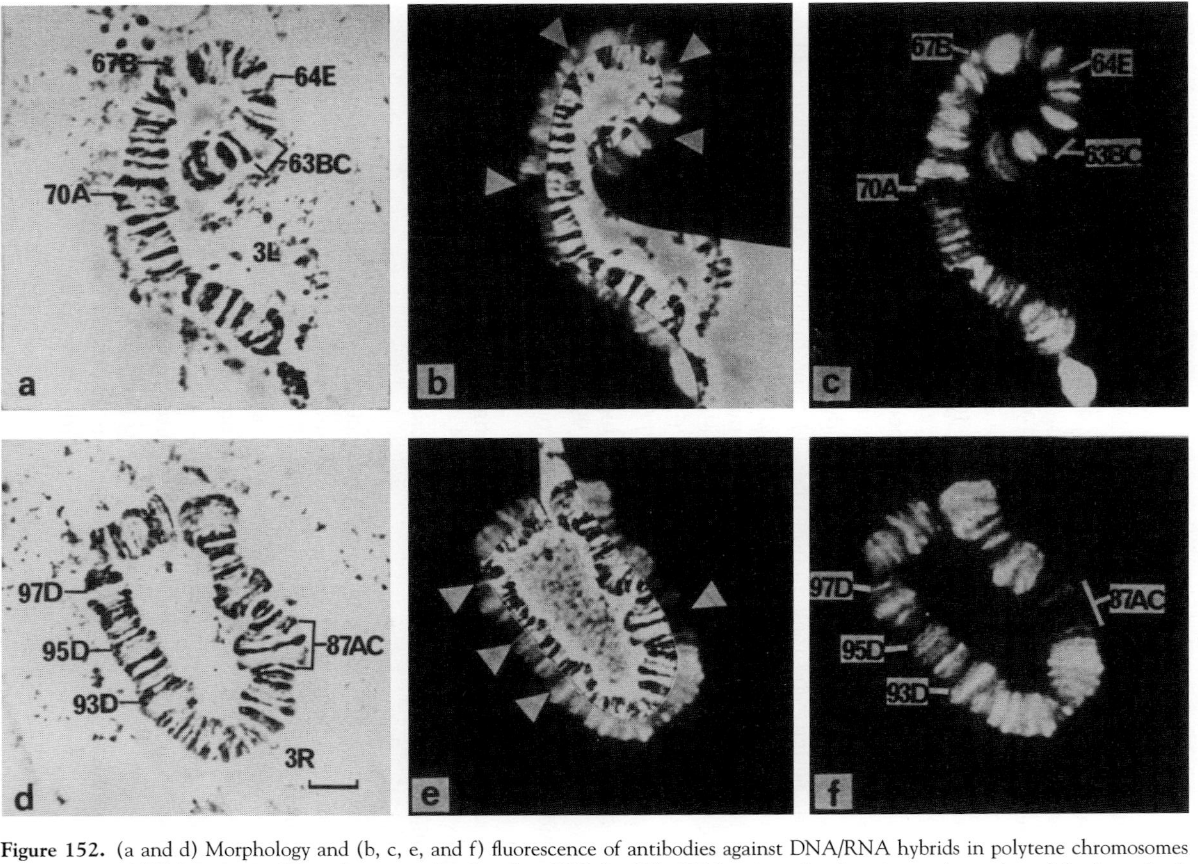

Figure 152. (a and d) Morphology and (b, c, e, and f) fluorescence of antibodies against DNA/RNA hybrids in polytene chromosomes of *l(3)tl* mutants of *D. melanogaster* developed at 18°C for 555–576 hr. Arrowheads point to the region of the heat-shock puffs (from G. H. Umbetova, I. E. Vlassova, and I. F. Zhimulev, unpublished observations).

The speculations offered explain uncoupling of puffing development and gene expression were as follows: (1) the 68C puff may not be correlated with transcription in the region; (2) the 68C puff RNA may differ from normal in the mutant, and for this reason was not detected; (3) transcription at the locus is normal; however, RNA is rapidly degraded; (4) although transcription has started, it is rapidly stopped, the region becomes diffuse and unpuffed, and there is no mature RNA (Crowley *et al.*, 1984).

There is also a correlated change in the amount of synthesized RNA in strains that differ in the sizes of the 3C puff (Korge, 1977; Korge *et al.*, 1990). This correlation implies a causal relation between these processes and simultaneity of their independent enfoldment (Korge *et al.*, 1990). Experiments with lines transformed by transposons containing the DNA of the *Sgs4* gene with different sites of its regulatory region demonstrated three relations between puffing activity and transcription rate: (1) transcription-correlated puffing; (2) transcription without chromosomal puffing; and (3) chromosomal puffing without transcription. In most transpositions, transcription and puffing are correlated and, when a copy of highly expressed gene is integrated into the already puffed locus (3E or 82F), the extent of puffing of the transformed DNA is considerably increased (Korge *et al.*, 1990). In lines *Tf(4)G9–26* and *Tf(3R)A24–1*, the transformed *Sgs4* genes are expressed at high levels without formation of the puffs at the insertion site (Korge *et al.*, 1990). As electronmicroscopic studies demonstrated, inserts of short constructs containing the *Sgs3* gene into the 60B and 67D regions during activation are not converted into puffs. However, some decondensation of the regions, lengthening of the interband region (60B), or activation of micropuff (67D) from DNA within the insert is observed. The diameter of the chromosome does not increase in both cases (Semeshin *et al.*, 1986). For this reason, when no puffs can be observed during gene activation (Korge *et al.*, 1990), decompaction undetectable at the level of light microscopy possibly occurs. In the N-lines, chimeric fragments contained transformed upstream sequences (derived from Oregon-R wild type of *D. melanogaster*), the DNA of the viral oncogene *v-mil*, and a fragment of *v-myc* instead of its structural part. Poly(A)$^+$ or total RNA homologous to these viral fragments could not be detected by Northern hybridization, although puffing was observed at the integration site. This puff was activated and inactivated at the same stage as the endogenous puff. Transcription was presumably disrupted because there was significantly less RNA polymerase II in this region (Korge *et al.*, 1990).

There are data indicating that during puff formation the decompacted region of chromatin is possibly limited at its endpoints; thus, a decompacted domain forms. The evidence for this involves the identification of antibodies against cGMP at the endpoints of the large 63E, 74EF, and 75B puffs (but not of the 62E one) (Spruill *et al.*, 1978); or antibodies against B52, a protein associated with the boundaries of transcriptionally active chromatin (Champlin *et al.*, 1991); or BEAF-32 insulator protein (Zhao *et al.*, 1995).

VII. BALBIANI RINGS

A. Morphology and structure of Balbiani rings

The distinction between the puffs and the larger swellings and the bulbs (see Section VI,F) in *Sciara* was first made by Poulson and Metz (1938). The bulbs seem to be bursting chromosome regions, whereas the usual puffs are readily recognized by an even diffuse increase in diameter. The two similar types of changes in the chromosomes also occur in *Chironomus* species (Beermann, 1956, 1962). A local chromosome region becomes decompacted when certain regions are activated, it converts into a small swelling rapidly increasing in diameter (Figures 153a–153f). On either side of the region, the chromosome unravels into individual filamentous elements or threads: chromatin forms loops and chromatid fibers in the region of maximal decompaction. The thickness of the chromatid fibers increases with distance from the activation site. A large number of loops form a structure which appears to be threaded on the chromosome (see Figures 153h, 153i, and 154). Drawings of such thickenings were presented in the first descriptions of polytene chromosomes (Balbiani, 1881; Alverdes, 1912a,b). These were called Balbiani rings (Erhard, 1910; Alverdes, 1912a).

Several aspects of the organization of Balbiani rings have been amply reviewed (Beermann, 1956, 1962, 1963a,b, 1964, 1967; Mechelke, 1962; Panitz, 1963, 1972a; Ashburner, 1970a; Kiknadze, 1972, 1984, 1985a,b; Pelling, 1972a; Swift, 1973; Daneholt, 1973, 1974, 1975, 1982, 1992; Edstrom, 1974; Lambert, 1974; Case and Daneholt, 1977; Daneholt *et al.*, 1979; Galler *et al.*, 1984a; Grond *et al.*, 1987).

The structure of Balbiani rings was elucidated by many data obtained from light and electron microscopy (Figure 153) (Beermann, 1952a, 1956, 1962, 1963b; Wolstenholme, 1965; Stevens and Swift, 1966; Gaudecker, 1967; Daneholt, 1975; Perov *et al.*, 1976; Stockert, 1979; Kiknadze, 1981a,b; Olins *et al.*, 1982; Istomina *et al.*, 1983).

Balbiani rings occur in the salivary gland cells of the Chironomidae larvae, except in parasitic species (Balbiani, 1881; Erhard, 1910; Alverdes, 1912a; Poulson and Metz, 1938; Beermann, 1952a, 1962; Zacharias, 1979; Kiknadze *et al.*, 1990; Belyanina, 1993). They occur in certain species of *Drosophila* (particularly in species of the *montium* subgroup: *D. auraria* and the closely related species *D. biauraria*, *D. bicornuta*, *D. serrata*, *D. jambulina*, *D. triauraria*, *D. quadraria*, and *D. vulcana;* Figure 154) (Kastritsis and Grossfield, 1971; Scouras and Kastritsis, 1984; Kastritsis *et al.*, 1986; Mavragani-Tsipidou and Kastritsis, 1986; Manousis and Kastritsis, 1987; Mavragani-Tsipidou *et al.*, 1990a, 1992a,b; Scouras and Mavragani-Tsipidou, 1992; Pardali *et al.*, 1996); in the larvae of Culicidae (Redfern, 1981a); in the larvae of Simuliidae (Chubareva and Tsapygina, 1965; Dunbar, 1967; Chubareva, 1984), although these structures are frequently called pseudo- (or para-) Balbiani rings in the midges

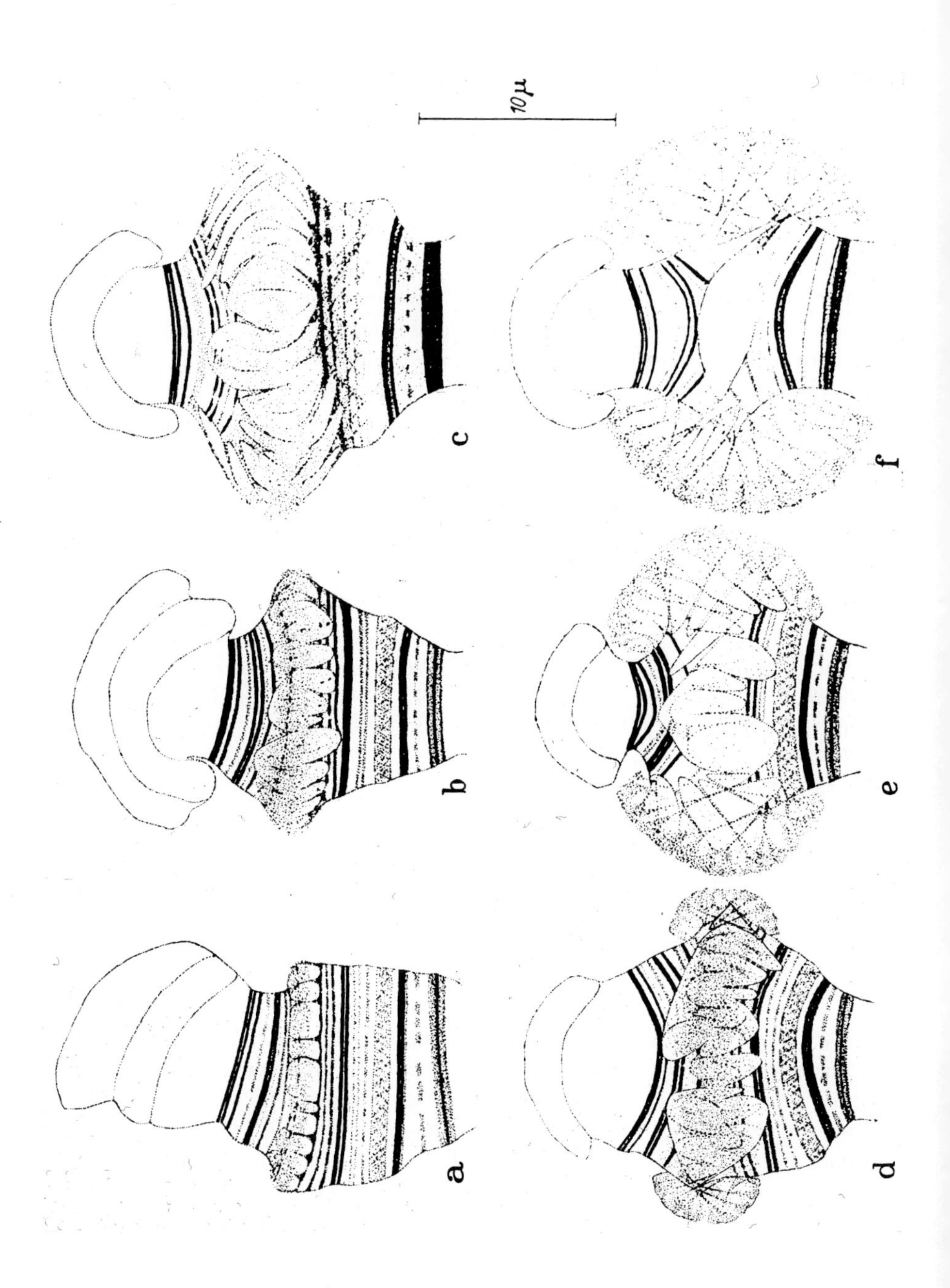
10 μ
a
b
c
d
e
f

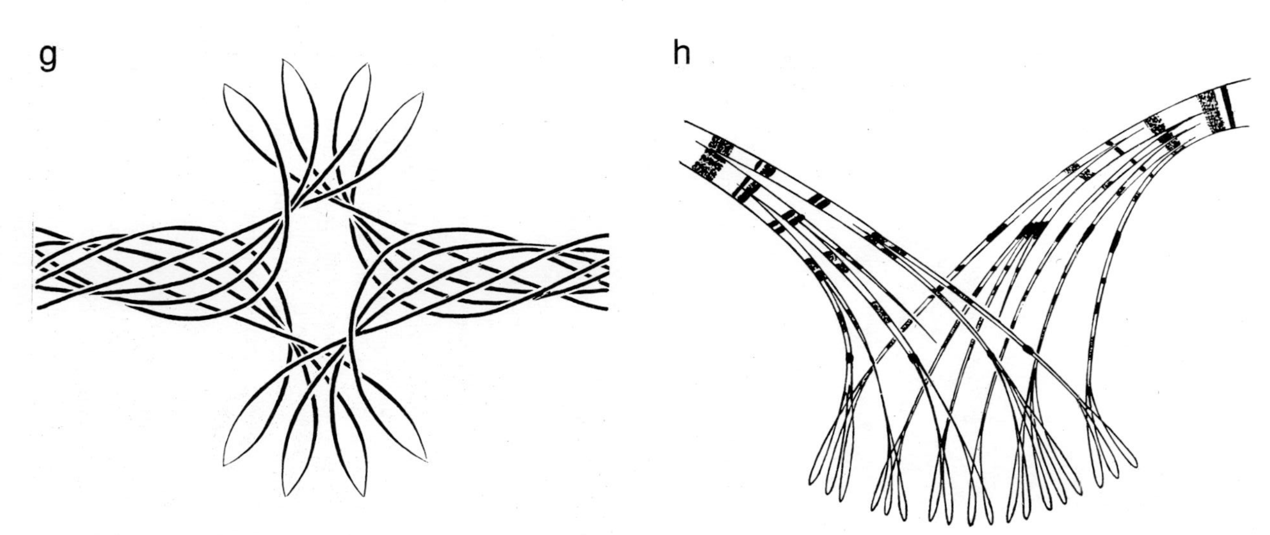

Figure 153. Development of Balbiani rings in C. *tentans*. (a–f) Decompaction of the chromosome region and formation of Balbiani rings (reprinted by permission from Beermann, 1952a); (g) a scheme of the disposition of chromatids in an active Balbiani ring (reprinted by permission from Beermann, 1952a); (h) splitting of chromatid bundles in the vicinity of a developed Balbiani ring (from Beermann, 1956).

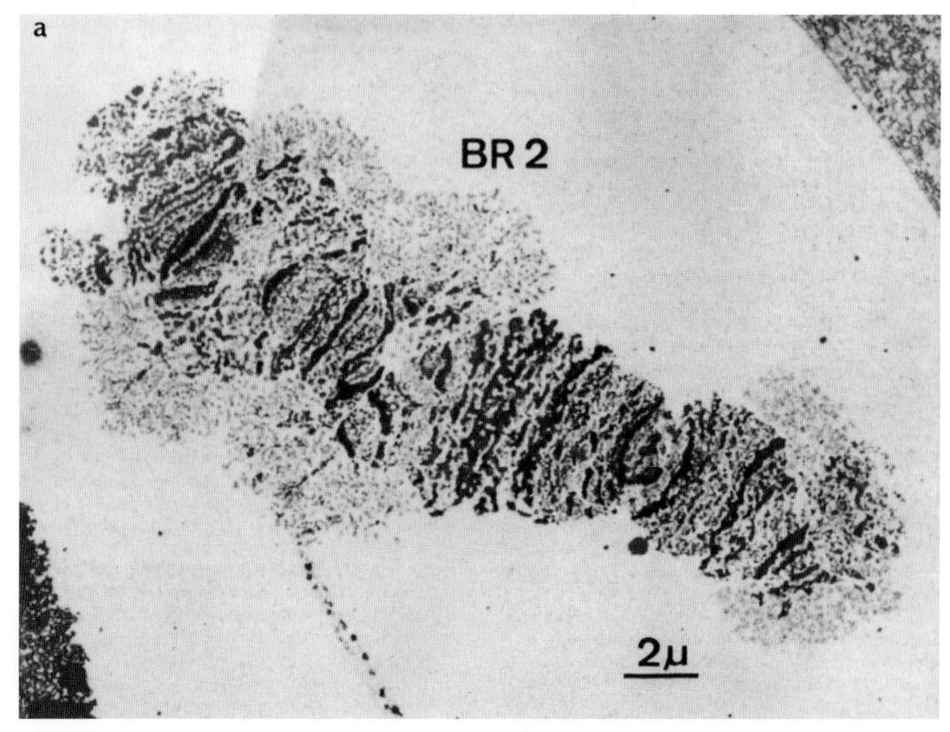

Figure 154. (a, c, d, and f–j) Morphology of Balbiani ring in *C. tentans*, (b) *Bilobella grassei*, (e) *Calliphora erythrocephala*, (k) *Drosophila auraria*, and (l) *Chironomus thummi*. (a) An EM section of the fourth chromosome (from Daneholt, 1975, © Cell Press); (b) a drawing of the second chromosome, numbers designate the regions. ABII, Balbiani ring; EHII and EHII′, heterochromatic regions; n_1, the nucleolus (reprinted by permission from Cassagnau, 1971). (c and d) Morphology of Balbiani ring and incorporation of [^{3}H]uridine (reprinted by permission from Pelling, 1972a); (e) Balbiani ring in ovarian nurse cells (reprinted by permission from Ribbert, 1979). Bar = 10 μm. (f–j) Variation in size of Balbiani rings (BR1-BR3) in (f, h) control and (g, i, j) galactose-treated salivary gland cells (reprinted by permission from Beermann, 1973); (k) (from Z. Scouras, personal communication); (l) telomeric Balbiani ring induced by heat shock (reprinted by permission from Morcillo *et al.*, 1981). Bar = 20 μm.

(Dunbar, 1967; Hunter and Connolly, 1986; Hunter, 1987); and in apterigotan insects of the Collembola order (Figure 154b) (Berendes and Beermann, 1969; Cassagnau, 1971, 1976, Cassagnau *et al.*, 1979; Deharveng and Lee, 1984).

One Balbiani ring was identified in the ovarian nurse cells of adults (Figure 154d) (Ribbert, 1979) and in the trichogen cells of the pupae of *Calliphora erythrocephala* (Ribbert, 1967, 1972). Four Balbiani rings were described in the footpad cells of the pupae of *Sarcophaga bullata* (Whitten, 1969),

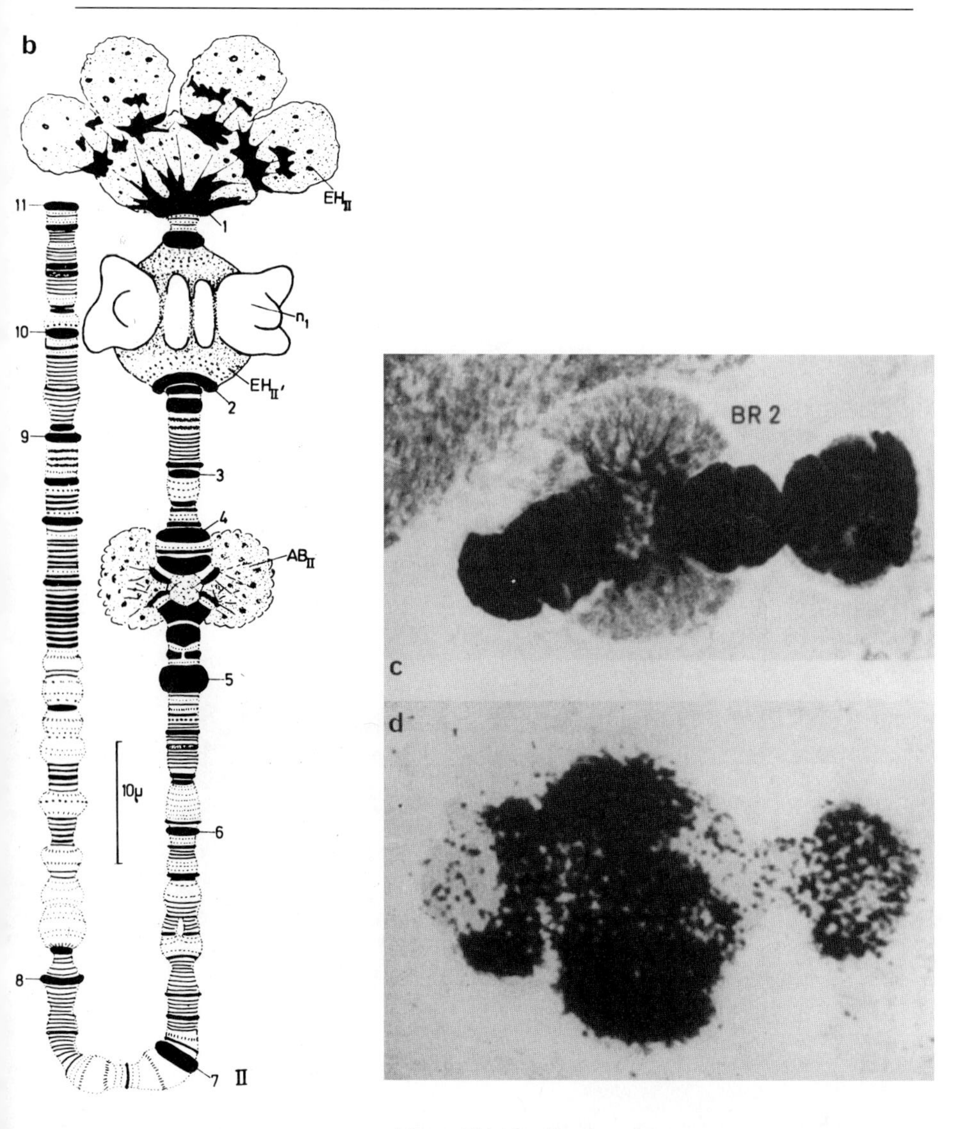

Figure 154. *Continued*

although their occurrence was not mentioned in Bultman and Clever's paper (1969).

The number of Balbiani rings per haploid set varies from one in *Cryptochironomus vulneratus* or *Chironomus pilicornis* to six in *Acricotopus lucidus*;

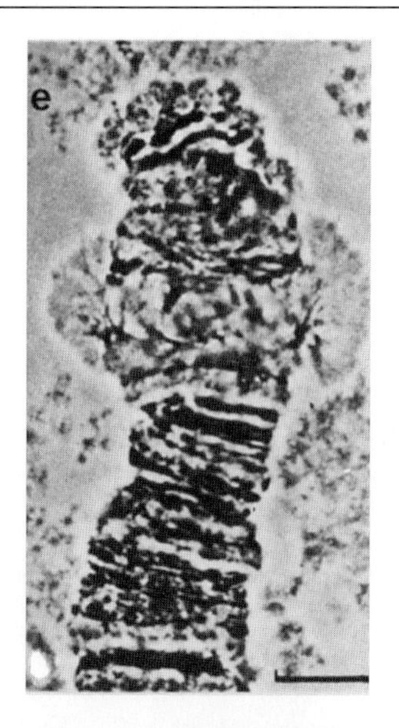

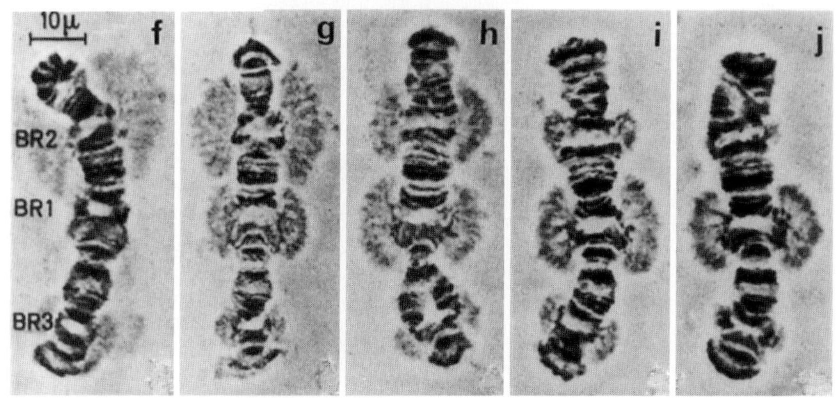

Figure 154. *Continued*

the number is most frequently three (Beermann, 1952a, 1962; Mechelke, 1953, 1962; Konstantinov and Nesterova, 1971; Kiknadze *et al.*, 1989). Balbiani rings are most scarce in the species of the *Cryptochironomus* genus, predators whose larvae do not use glycoprotein secretion to spin a housing tube (Kiknadze *et al.*, 1989).

Balbiani rings in addition to those regularly functioning in the course of normal development can be induced. For example, an X-ray-induced inversion breaks a block of centromeric heterochromatin and transposes a part of it

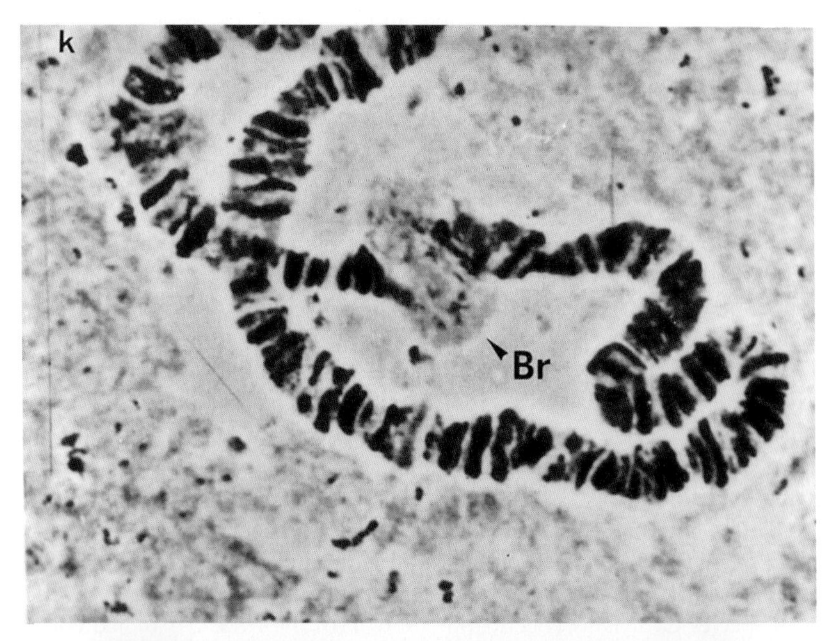

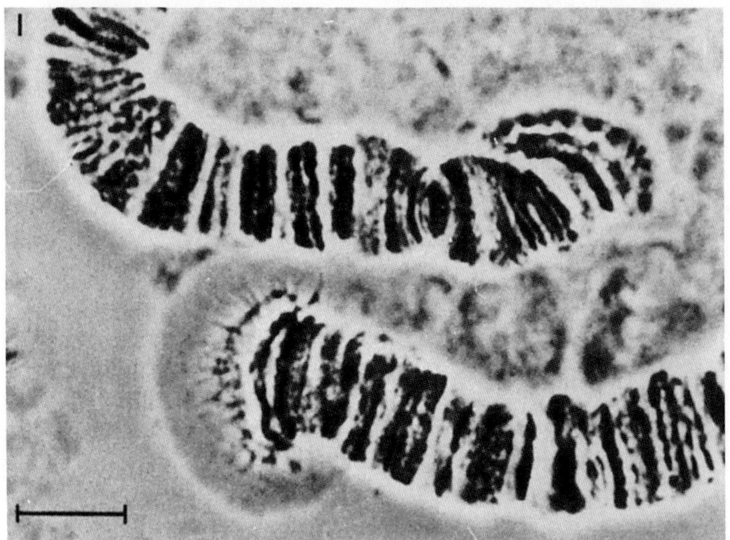

Figure 154. *Continued*

to euchromatin; a Balbiani ring (Staiber, 1982) forms from the transposed part of heterochromatin at its new position. In C. *thummi*, the telomeric Balbiani ring is induced by heat-shock treatment (Figure 154l) (Morcillo *et al.*, 1981; Santa-Cruz *et al.*, 1981; Barettino *et al.*, 1982; Carmona *et al.*, 1985). In C. *pallidivittatus*, Balbiani ring 6 arises after incubation of larvae in a solution of galactose and other sugars (Beermann, 1973; Botella and Edstrom, 1991). An additional Balbiani ring 5 is found in other strains of this species (Beermann, 1973).

Balbiani rings occur only in salivary gland cells (Beermann, 1952a, 1962, 1963b). Based on the presence of a certain set of Balbiani rings, lobes are distinguished in the salivary gland. The highest level of such differentiation was described in *Acricotopus lucidus*, whose salivary gland has three lobes: the main, anterior, and posterior lobes (Figure 81 above). A set of Balbiani rings is specific to each lobe (Figure 155). Differentiation of the salivary gland is less expressed in many other species; it is most frequently reduced to a small group of cells lying near the salivary gland duct; these are the "special cells" according to Beermann (1961).

In *Chironomus pallidivittatus*, Balbiani ring 4 occurs only in these particular cells (Figure 156). However, additional Balbiani rings do not always occur in the special cells. For example, this Balbiani ring is missing in the closely related species C. *tentans* (Beermann, 1956, 1961). Of the 24 studied chironomid species, an additional Balbiani ring was present in only 11 (Kiknadze *et al.*, 1990).

Although many bands lose their integrity in the region of a Balbiani ring (Figure 153i), few become decompacted and puff. Thus, about 10 adjacent bands and interbands are involved in the formation of Balbiani ring 2 in C. *tentans*. However, chromatid fibers can always be found in the region of a Balbini ring (see Figure 153i) that retain the banding pattern observed in the unpuffed region (Beermann, 1952a, 1956; Pelling, 1972a).

Balbiani rings are active in salivary gland cells during most of the larval period when polytene chromosomes can be observed. For this reason, the chromosomes from the cells of other organs—for example, Malphighian tubules—are used to see, in an inactive state, the region where Balbiani rings are formed (Beermann, 1952a). Treatment of larvae with galactose completely inactivates at least Balbiani ring 2 in C. *tentans*, allowing analysis of how salivary gland chromosomes recover their banding patterns (Beermann, 1973). From comparisons of Malpighian tubule and salivary gland chromosomes, it was concluded that Balbiani ring 2 is always formed from a small number of bands (2–5); these are presumably the 3B8 and 3B9 bands (Beermann, 1952a, 1962, 1963b). [^{3}H]RNA from Balbiani ring 2 in rectal chromosomes hybridizes *in situ* to three to five bands (Lambert, 1975). However, 3B10 band is the only site of origin of Balbiani ring 2, as demonstrated in experiments in which Balbiani ring 2 was

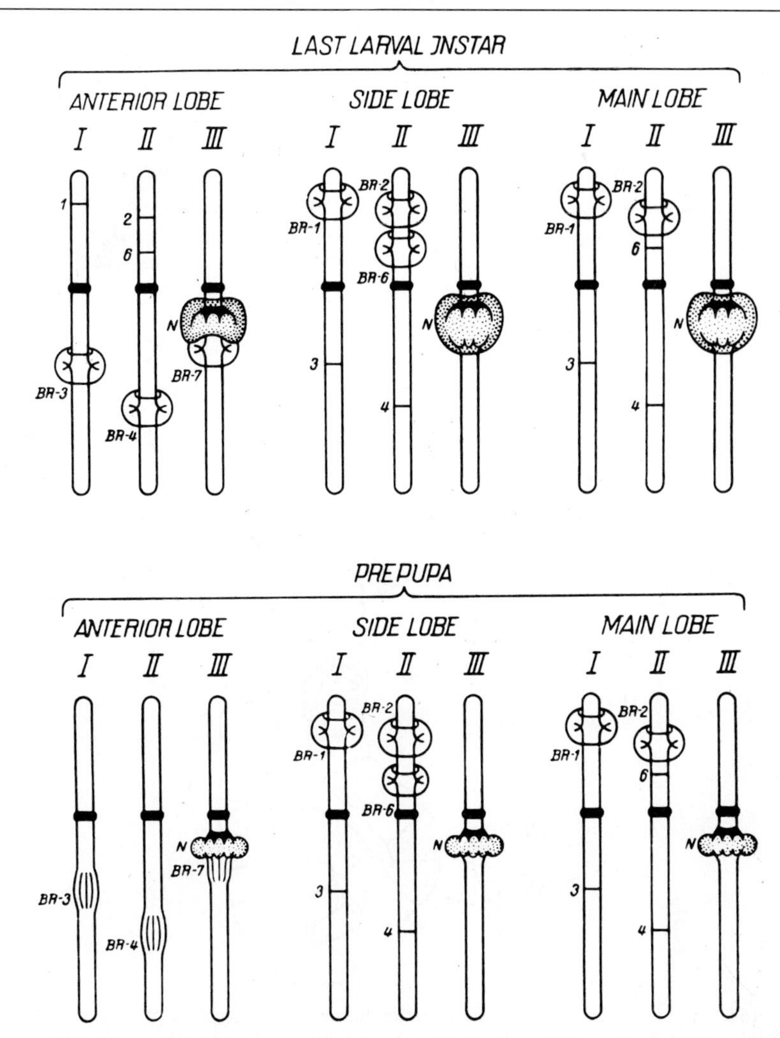

Figure 155. Changes of the activity of Balbiani ring (BR1–6) and the nucleolus (N) in the three lobes of the salivary glands of larvae and prepupae of *Acricotopus lucidus* (from Mechelke, 1963; reprinted by permission from Panitz, 1972a).

repressed by treatment of larvae with galactose solution (Beermann, 1973). This conclusion was substantiated: the DNA from Balbiani ring 2 hybridizes *in situ* to the polytene chromosomes of precisely this band (Edstrom *et al.*, 1978a; Derksen *et al.*, 1980).

Treatment of cells with the transcription inhibitor DRB, or with galactose, produces considerable compaction of DNP from Balbiani ring 2. Hybridiza-

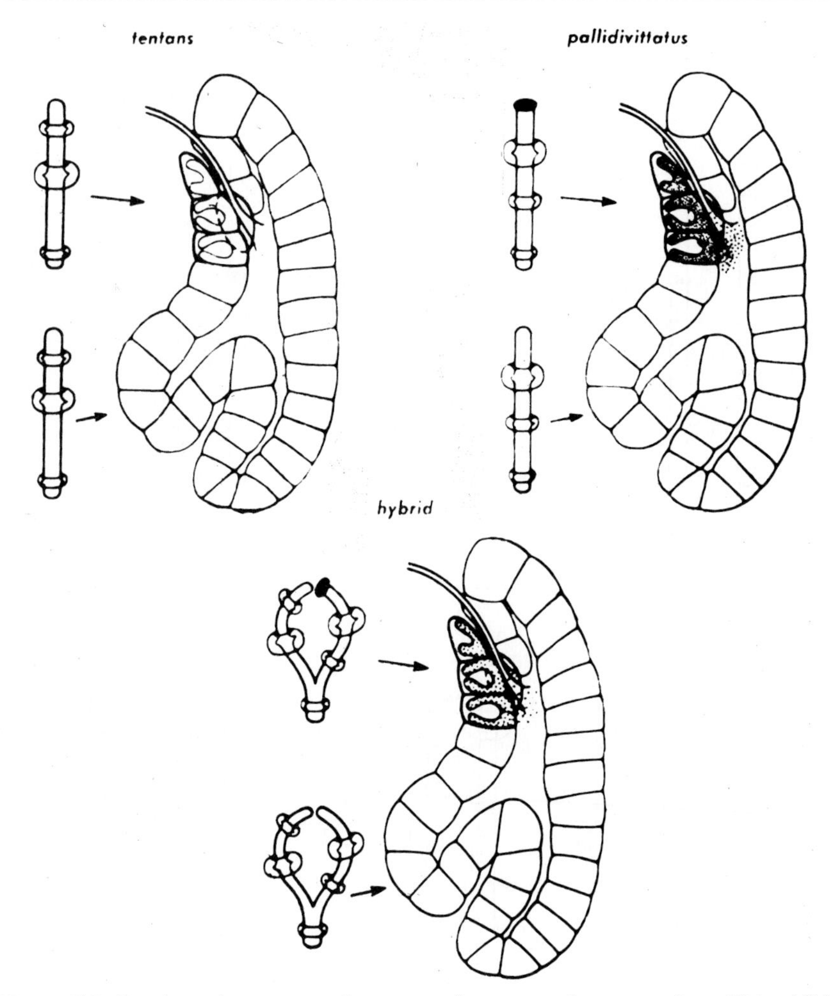

Figure 156. Correlation between granular secretion formation and presence of an additional Balbiani ring in the "special" cells of the larval salivary glands of *Chironomus pallidivittatus* (reprinted by permission from Beermann, 1961, 1963b).

tion *in situ* of a small BR2 DNA clone to such chromosomes revealed most intense labeling of the 3B9/10 interband and much weaker labeling of the 3B10 band (Sass, 1984). The clone contains a fragment of 152 bp, which is repeated 100–200 times in the Balbiani ring 2 gene (Jackle *et al.*, 1982). When one takes into account that the interbands contain not more than 1–2 kb (see: Beermann, 1972; Zhimulev, 1992, 1996), it is hardly comprehensible how the interband can accomodate 15–30 kb DNA. It was shown using similar approaches that BRb and BRc in C. *thummi* originate from one to two bands (Kiknadze *et al.*, 1983b, 1985a,b; Kiknadze, 1984b, 1985a). In C. *pallidivittatus*, Balbiani ring 6 is inactive during normal development; when activated under the effect of galactose, it also develops from one band (Beermann, 1973; Galler *et al.*, 1984a; Botella and Edstrom, 1991).

Cases are known in which Balbiani rings are informationally complex. In chironomids of the *Smittia* genus, a Balbiani ring is broken into two independently functioning parts by a chromosomal rearrangement, suggesting that it contains at least two genes (H. Bauer, 1957, cit. after: Panitz, 1964). Balbiani ring 1 has a composite structure in C. *tentans* and C. *pallidivittatus*. It is composed of adjacent Balbiani rings 1a and 1b originating from two different bands (Beermann, 1973) and containing different genes each (Dreesen *et al.*, 1985a,b).

B. Molecular organization of Balbiani rings

The morphology of Balbiani rings is characteristic, because the transcription of the genes they harbor is exceptionally intense (Figure 154d). According to autoradiographic data, up to 15% of [^{3}H]uridine incorporated into the nucleus of C. *tentans* is accounted for by one Balbiani ring (Pelling, 1972a). The first chromosome in this species contains as much RNA as one Balbiani ring, approximately 20 pg (Edstrom and Beermann, 1962). Balbiani ring 1 contains 7.8 pg RNA and that RNA content is 14.9 pg in Balbiani ring 2 (Edstrom *et al.*, 1978a,b).

The density of RNA polymerase II molecules is 120 per gene, whereas in the usual puff it is about 2 per gene (Pelling, 1972a). Other values for the numbers of such molecules per gene are 84–155 (123 on average), determined during normal development, and up to 370 molecules of RNA polymerase per gene after treatment with pilocarpine (Widmer *et al.*, 1984). These estimates can be interpreted as follows: (1) The number of RNA polymerase molecules per 1 μm of gene length is 16 (Lamb and Daneholt, 1979), 32.8–45.6 (Andersson *et al.*, 1980; Olins *et al.*, 1982), 38 (Widmer *et al.*, 1984), and 82 (Masich, 1992). (2) The average distance between two traversing RNA polymerase molecules is 300 bp during normal development (Lamb and Daneholt, 1979) and 100 bp after treatment with pilocarpine (Widmer *et al.*, 1984). (3) The number of polymerases starting to traverse the whole gene is 6 per minute (Lamb and Daneholt, 1979) or 10–15 (Trepte, 1993). (4) The rate of RNA chain elongation is 21–41 (31, on average) nucleotides per sec (Lamb and Daneholt, 1979; Daneholt *et al.*, 1979). Complete transcription of the Balbiani ring 2 gene takes 20–30 min at 18°C (Edstrom *et al.*, 1978a).

Micromethods provided a unique opportunity to study the molecular organization of Balbiani rings. These methods are based on the isolation of chromosome fragments containing Balbiani rings and cellular and nuclear components with glass needles or a micromanipulator; their accumulation in needed amounts (about 50 Balbiani rings or five nucleoli); followed by their electrophoretic separation and analysis (Edstrom, 1960, 1964; Edstrom and Beermann, 1962; Daneholt *et al.*, 1968; Egyhazi *et al.*, 1968; Daneholt, 1970; Lambert *et al.*, 1973a,b; Lambert and Daneholt, 1975).

In all dipteran insects, nuclear RNA in a cell with polytene chromosomes is represented by three large fractions (Figure 157): (1) preribosomal RNA synthesized in the nucleolus; (2) heterogeneous nuclear RNA synthesized in most chromosome regions; and (3) low-molecular-weight RNA, mainly 4S and 5S RNAs (Edstrom and Daneholt, 1967; Armelin *et al.*, 1969a,b, 1970; Daneholt *et al.*, 1969a–c, 1970; Egyhazi *et al.*, 1969; Greenberg, 1969; Pelling,

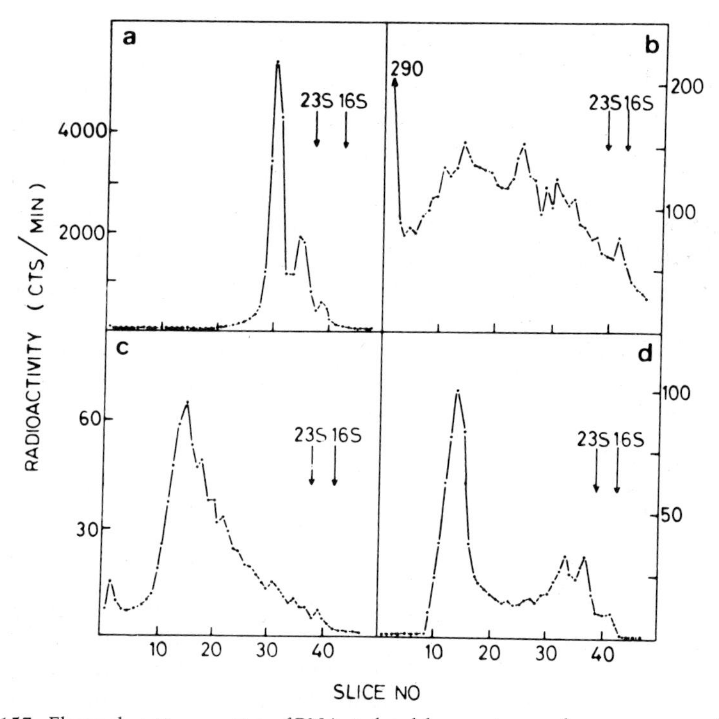

Figure 157. Electrophoretic separation of RNA, isolated from various nuclear components of larval salivary gland cells of *C. tentans:* (a) nucleoli, (b) chromosomes I–III, (c) Balbiani rings 2, and (d) nucleoplasm (from Daneholt, 1973). Salivary glands were incubated with [^{3}H]uridine for 90 min, fixed, and fragments of the nucleus were isolated with a micromanipulator. Labeled RNA was separated in 1% agarose gel. Abscissa, number of the gel fragments; ordinate, radioactivity of the fragment (cpm).

1970; Daneholt and Svedhem, 1971; Elgaard and Clever, 1971; Armelin and Marquez, 1972a,b; Daneholt, 1972, 1975; Rubinstein and Clever, 1972; Serfling *et al.*, 1972, 1974; Balsamo *et al.*, 1973; Edstrom and Tanguay, 1974a,b; Egyhazi, 1976b, 1978; Panitz, 1978a,b; Ribbert and Buddendick, 1984).

A considerable part of heterogeneous RNA synthesized in the polytene chromosomes of chironomides is represented by high-molecular-weight RNA with a sedimentation coefficient of 75S (see Figure 157) (Daneholt, 1972, 1973) and a molecular mass of about 35 $\times$ 10^6 Da, according to electrophoretic data, or 15 $\times$ 10^6 Da, according to centrifugation data (Daneholt, 1972, 1973; Daneholt and Hosick, 1974). Additional results indicated that the molecular weight of 75S RNA from Balbiani ring 2 is (12–14) $\times$ 10^6 Da, corresponding to 37 kb of DNA (Case and Daneholt, 1978). The 75S RNA isolated from Balbiani ring 2, the nuclear sap, the cytoplasm, or polysomes hybridize *in situ* to Balbiani ring 2 (Lambert *et al.*, 1972; Daneholt, 1973; Lambert, 1973, 1974; Wieslander, 1975; Daneholt *et al.*, 1976; Lonn, 1977; Serfling and Huth, 1978; Edstrom *et al.*, 1983). [^{3}H]BR2 RNA appeared in the nuclear sap 90 min after labeling, and it appeared in the cytoplasm during the ensuing 90 min. After a week of continuous labeling, 75S RNA in the cytoplasm constituted approximately 1.5% of the total salivary gland RNA (Daneholt and Hosick, 1973, 1974; Lambert, 1973). The content of 75S RNA in Balbiani ring, the nuclear sap, and the cytoplasm varies in the 1:20:200 range, respectively (Edstrom *et al.*, 1978b).

After completion of transcription, 75S RNA is polyadenylated (Egyhazi *et al.*, 1979), packaged into specific RNP particles, and transported from the Balbiani ring to the cytoplasm. The templates do not undergo considerable processing, because the molecular weights of the newly synthesized and cytoplasmic 75S RNAs are nearly the same. 75S RNA enters the polysomes in the cytoplasm (Kloetzel and Laufer, 1969; Clever and Storbeck, 1970; Daneholt *et al.*, 1976, 1977; Wieslander and Daneholt, 1977; Rydlander and Edstrom, 1980). In *C. tentans*, the polysomes may be giant, containing 30 to 40 and occasionally 100 to 200 ribosomes (Daneholt *et al.*, 1976, 1977); or according to other data, 64–94 ribosomes (79 on average: Franke *et al.*, 1982). Polysomes contain 79–146 ribosomes in *C. thummi* (Kiseleva and Masich, 1991; Masich, 1992).

The excess of 75S RNA in the cytoplasm is due to its stability, whereas total cytoplasmic RNA (from the entirety of chromosomes I–III) is metabolized (Daneholt *et al.*, 1970; Daneholt and Svedhem, 1971).

Metabolic stability of 75S RNA was confirmed by the observation that, after inhibition of transcription by actinomycin D, protein synthesis proceeded uninterrupted without appreciably decreasing in intensity for at least another 24 hr (Clever, 1969; Clever *et al.*, 1969b; Doyle and Laufer, 1969b; Clever and

Storbeck, 1970; Rubinstein and Clever, 1972). The half-life of 75S RNA exceeds 20–35 hr (Daneholt and Hosick, 1973; Edstrom *et al.*, 1975a). Balbiani ring 1 in *C. tentans* also synthesizes 75S RNA (Daneholt *et al.*, 1976).

Northern blotting demonstrated that all the Balbiani-ring genes are expressed only at the larval stage, from the first instar, as soon as larvae form tubes, and exclusively in salivary glands. Passage from the state, when the prepupa is confined to a "tube," to a stage on which it lives free, coincides with the time of transcription inactivation in Balbiani ring (Lendahl and Wieslander, 1987).

Analysis of nucleotide sequences in clones of DNA from different Balbiani rings revealed that a substantial portion (15–25 kb) of each gene consists of repetitive sequences, the so-called core block (Wobus *et al.*, 1980; Degelman and Hollenberg, 1981; Baumlein *et al.*, 1982a,b; Case, 1982; Dereling *et al.*, 1982; Jackle *et al.*, 1982; Sumegi *et al.*, 1982; Wieslander *et al.*, 1982, 1986; Case *et al.*, 1983; Case and Byers, 1983; Wieslander and Lendahl, 1983; Galler *et al.*, 1984a,b; Hamodrakas and Kafatos, 1984; Hoog and Wieslander, 1984; Kolesnikov *et al.*, 1984, 1986; Lendahl and Wieslander, 1984; Panitz *et al.*, 1984; Zainiev *et al.*, 1985; Grond *et al.*, 1987; Lendahl *et al.*, 1987; Saiga *et al.*, 1987, 1988; Botella and Edstrom, 1991).

A tandem repeat of 200–270 bp composed of unique and repetitive parts is common to all Balbiani rings; the repetitive portion is composed of subrepeats of smaller size (Figure 158). It is represented 75–100 times in the

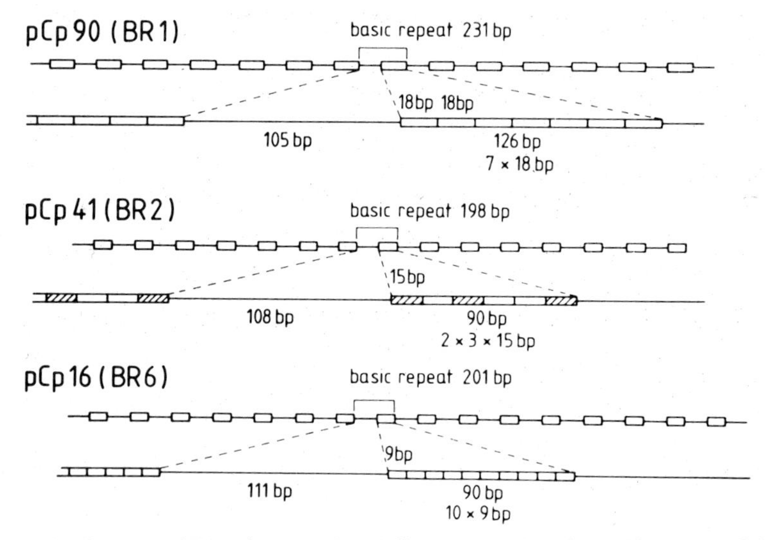

Figure 158. Structure of DNA fragments from Balbiani rings 1, 2, and 6 in *Chironomus pallidivittatus*. pCp90-pCp16, plasmids (from Galler *et al.*, 1984a,b).

genome. The unique portion remains relatively unaltered; it shows 94% homology to nucleotides of the counterpart regions in the other Balbiani rings (Figure 159). The "subrepeat" region evolves rapidly; the differences in nucleotide sequences, length, and subrepeat number are very large. In contrast, the sequences at the 3′ terminal exon and the 3′ nontranslated region are highly conserved (Figure 159). It was suggested that the giant Balbiani ring gene has evolved through duplication and divergence from a short 110- to 120-bp primordial sequence (Sumegi *et al.*, 1982; Pustell *et al.*, 1984; Wieslander *et al.*, 1984).

The BR2.2 gene ends with a 3′-end exon. It is present in RNA, but is not represented in the secretory protein (Botella *et al.*, 1988; Hoog *et al.*, 1988). This protein fragment possesses DNA-binding properties and the cleaved-off C-terminal fragment in the isolated nuclei is detected in BR1 and BR2. It is plausible that this protein fragment may function as a feedback control of Balbiani ring (BR) gene activity (Botella *et al.*, 1988; Botella and Edstrom, 1991).

Similarity in their molecular organization—and, presumably, their common origin—allows one to detect homologous Balbiani rings in various species; for instance, the RNA from BR2 in C. *tentans* hybridizes to Balbiani ring 2 of C. *pallidivittatus* (Lambert and Beermann, 1975). A DNA fragment of 240 bp from the C. *pallidivittatus* genome hybridizes predominantly to Balbiani ring 1 and to a lesser extent to Balbiani rings 2 and 6 (Degelmann and Hollenberg, 1981).

Balbiani ring "c" DNA of C. *thummi* hybridizes to Balbiani rings of C. *piger*, C. *melanescens*, C. *pseudothummi*, C. *plumosus*, and C. *nuditarsis* (Zakhar-

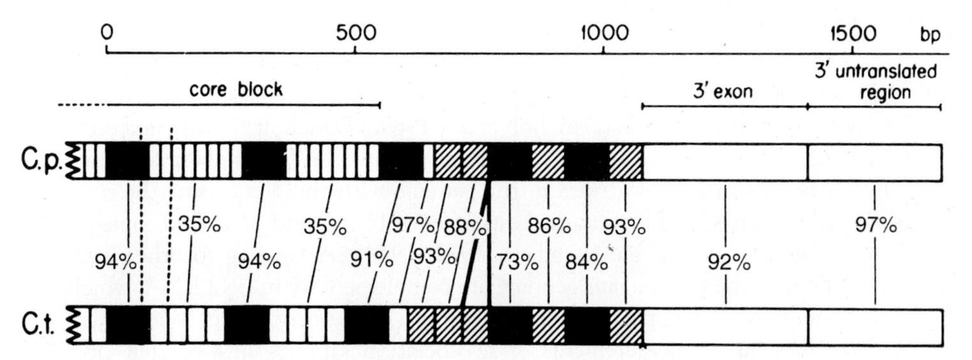

Figure 159. A comparison of the organization of the 3′ region of the BR1 gene limited by the core block and the polyadenylation site in C. *pallidivittatus* (C.p.) and C. *tentans* (C.t.). Solid rectangles designate the constant regions; the short open rectangles refer to the subrepeat regions; the Cys-1-elements are hatched; the long rectangles designate the 3′ exon and the 3′ nontranslated region (from Lendahl *et al.*, 1987).

enko *et al.*, 1988; Kiknadze *et al.*, 1989; Filippova *et al.*, 1993). DNA from Balbiani ring "b" of C. *thummi* shares homology with C. *plumosus*, C. *borokensis*, C. *balatonicus*, C. *entis*, C. *agilis*, C. *nudiventris*, C. *nuditarsis*, C. *annularius*, C. *cingulatus*, C. *sororius*, C. *dorsalis*, and C. *tentans* (Filippova *et al.*, 1992, 1993).

The DNA from the additional Balbiani ring BRa—which is characteristic of the special cells of salivary glands of C. *thummi*—hybridizes *in situ* to the chromosomes of C. *tentans*, C. *t. piger*, and to those of the *Chironomus* species of the *pseudothummi* (C. *dorsalis*) complex—whose additional Balbiani ring is also developed—as well as to the chromosomes of C. *plumosus* (which lack the special lobe) and to an additional, active BR (Kiknadze *et al.*, 1989; Filippova *et al.*, 1993).

Several RNAs are transcribed from Balbiani rings in chironomids. This reveals another feature common to the genetic organization of Balbiani rings in the salivary glands of chironomids. Two transcripts, with molecular weights of 40×10^6 Da (BR1A) and 5×10^6 Da (BR1C), were identified in Balbiani ring 1 of *Acricotopus lucidus* (Panitz, 1979). A second gene encoding poly(A)$^+$ RNA of 6.5 kb, besides the 75S RNA, was detected in Balbiani ring 1 of C. *tentans* (Dreesen *et al.*, 1985a,b; Hoog *et al.*, 1986). Two genes in a Balbiani ring, BR2.1 and BR2.2, were found in C. *tentans* and C. *pallidivittatus* (Galler *et al.*, 1985; Silva *et al.*, 1990); these are not transcribed in a coordinate manner (Case, 1986); two nonallelic genes were found in each BRc (Baumlein *et al.*, 1986) and BRa (Bogachev *et al.*, 1986, 1987, 1989, 1990; Bogachev, 1989; Kolesnikov *et al.*, 1986, 1990).

One of the Balbiani rings, BR3 in C. *tentans*, is structurally different from most Balbiani rings, which are nearly lacking in introns. The Balbiani ring 3 gene encodes a 10.9-kb primary transcript; it contains 38 introns. The BR gene is spliced into 5.5- to 6-kb mRNA (Dignam and Case, 1990; Paulsson *et al.*, 1990).

The question of the number of gene copies in a Balbiani ring has been discussed for a long time. It was suggested at first that the genes are represented many times in the genome (Daneholt *et al.*, 1969c; Daneholt, 1970; Lambert *et al.*, 1972; Sachs and Clever, 1972; Wieslander *et al.*, 1975). Lambert compared DNA amounts in Balbiani rings in the genome with those encoding the ribosomal RNA genes, and he obtained estimates of 15–20 and up to 200 copies of the 75S genes per genome (Lambert, 1972, 1974). According to other data, only 0.04% of the *Chironomus* genome are complementary to BR2 RNA, which corresponds to 1–4 copies of the BR2 genes (Hollenberg, 1976; Wieslander, 1979). Subsequent measurements of reassociation kinetics showed that BR2 DNA (BRc) of C. *thummi* behaves as a single component, although it contains internal repeats (Wobus and Serfling, 1977).

In Balbiani rings 1 and 2 of C. *tentans*, spread on a water surface by the Miller technique, the actively transcribing BR genes are visible (Figure 160). This made feasible the following calculations: In chromosome prepara-

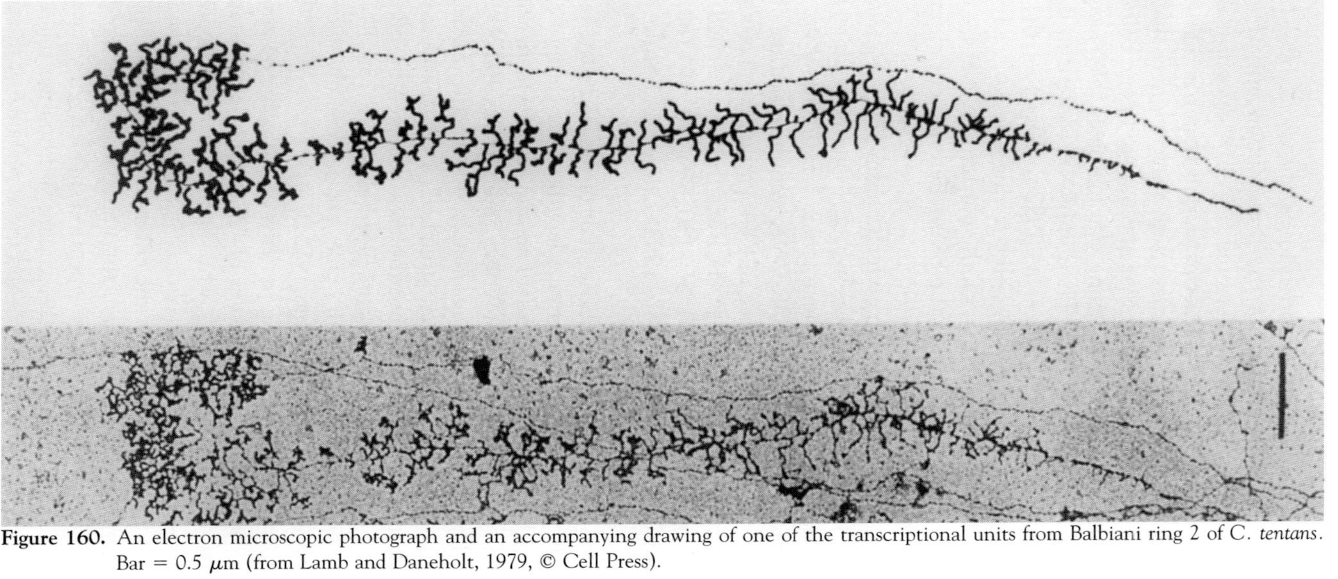

Figure 160. An electron microscopic photograph and an accompanying drawing of one of the transcriptional units from Balbiani ring 2 of C. *tentans*. Bar = 0.5 μm (from Lamb and Daneholt, 1979, © Cell Press).

tions with very well-spread transcription loops, the course of a chromatin fiber could be followed on either side of the transcription unit (Figure 160); no tandemly repetitive units could be detected at a distance of at least 14–16 μm, which is twice the length of the transcription unit (Lamb and Daneholt, 1979). Perhaps the transcription units alternate with spacers longer than 14–16 μm. However, calculations demonstrated that not more than one Balbiani ring gene can occur in the chromatid of polytene chromosomes. DNA content in the transcription unit is 2.28×10^3 kb. The total RNA content in Balbiani ring 2 is 10.8 pg or 1.96×10^7 kb (Edstrom *et al.*, 1978b). Therefore, the total number of transcription units represented in Balbiani ring 2 is 8,600 (1.96×10^7 kb/ 2.28×10^3 kb), in close agreement with the polyteny level of the chromosomes (8,200–16,400). This indicates that the BR2 gene is present in one copy in the chromatid (Lamb and Daneholt, 1979).

The stretching method of polytene chromosomes yielded unexpected results. Labeled RNA or BR DNA was hybridized *in situ* to C. *thummi* polytene chromosomes stretched with a micromanipulator. Estimates of the length of the stretched chromosome occupied by BR DNA were based either on intensity of label incorporation or on the presence of RNP particles, the products of Balbiani rings (Zainiev and Shilova, 1984). The length of a stretched Balbiani ring was variously estimated as 200–300, 56–130, 70–150, 97–180, 70–90, and 200–250 μm (Kiknadze *et al.*, 1983b, 1985; Zainiev and Shilova, 1984; Kiknadze, 1985).

Accepting that the transcription unit of the BR2 gene is 7.7 μm long (Lamb and Daneholt, 1979), the extreme length of the chromosome reaches a 40-fold value, 13 on average. These data are not in agreement with the observational values for both C. *thummi* (Wobus and Serfling, 1977) and C. *tentans* (see above). This discrepancy is presumably due to the inadequacy of the method. It is conceivable that, when stretched, a chromatid may slide along another chromatid, thereby increasing the area occupied by BR DNA in the stretched chromosome. It may be concluded that the 75S RNA genes are present in not more than one to four copies per genome (Wobus and Serfling, 1977; Daneholt *et al.*, 1978; Daneholt, 1982).

Comparisons of the banding patterns of the polytene chromosomes of salivary glands and Malpighian tubules of C. *tentans* demonstrated that BR2 originates from the large band 3B10, which contains approximately 470 kb of DNA according to cytophotometric data (Figure 161). This indicates that the gene occupies not more than 1/10 of the chromomere (Derksen *et al.*, 1980).

The genes located in Balbiani rings encode giant polypeptides. When released from salivary glands, the polypeptides harden in water and serve as fixative during the construction of the feeding and housing tube. Correlations between the location of Balbiani rings and the appearance of the corresponding polypeptide in salivary gland secretion were observed long ago. Balbiani rings are active exclusively in salivary gland cells where secretory proteins (SP) are

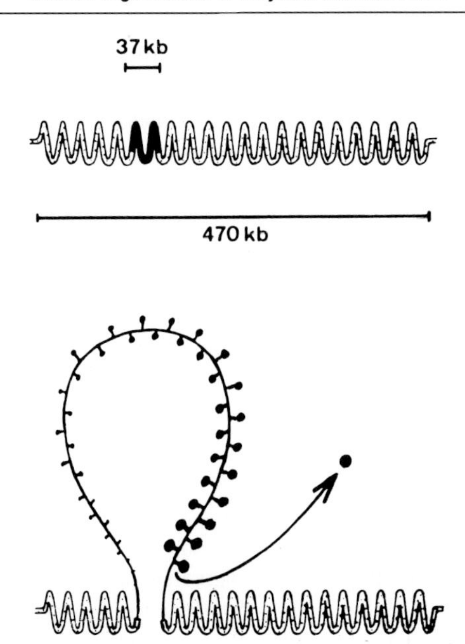

Figure 161. Localization of the 75S RNA gene in the chromomere of the polytene chromosome of C. *tentans* (reprinted by permission from Derksen *et al.*, 1980).

synthesized (Beermann, 1963b). Additional Balbiani rings in the special cells of salivary glands were detected in larvae of 11 of 24 studied *Chironomid* species. Specific secretion was found in 10 species, mainly in those posessing an additional Balbiani ring (Kiknadze *et al.*, 1990). These correlations do not always hold true; the number of Balbiani rings can vary in different chironomids, and there may be no corresponding variations in the number of secretory fractions presumably because Balbiani rings encode more than one gene (Wobus *et al.*, 1970). Clear-cut correlations were established when agents such as pilocarpine or galactose (which selectively alter genomic activity) were used. Pilocarpine results in complete release of secretion by salivary glands. The release is followed by a sharp activation of Balbiani rings 1 and 2 and an enhancement of the synthesis of secretory fractions.

If larvae are kept in medium containing sugars, particularly galactose, ethanol, or glycerol, Balbiani ring 2 of C. *pallidivittatus* regresses, and the sp-Ib protein fraction disappears. While this is occurring, a new Balbiani ring 6 and a new protein sp-Ic are induced (Edstrom *et al.*, 1980, 1982; Galler *et al.*, 1984a; Botella and Edstrom, 1991).

There is a correspondence between the Balbiani ring genes and protein production. In *in vitro* translation experiments, there was a correspondence between the fingerprints of the *in vivo* product and the secretory protein components (Rydlander *et al.*, 1980), these protein products have the same amino acid composition and tandemly arranged repeats (Hertner *et al.*, 1980; Weber *et al.*, 1983). Four giant secretion polypeptides are encoded by the Balbiani ring genes in *C. tentans* and *C. pallidivittatus:* sp-Ia, BR1; sp-Ib, BR2.2; sp-Ic, BR6; sp-Id, BR2.1; 185 kDa 18.5 BD protein, BR3 (Grossbach, 1969; Rydlander and Edstrom, 1980; Hertner *et al.*, 1980; Wieslander *et al.*, 1982; Sumegi *et al.*, 1982; Case and Byers, 1983; Case *et al.*, 1983; Wieslander and Lendahl, 1983; Galler *et al.*, 1984a,b, 1985; Lendahl and Wieslander, 1984; Rydlander, 1984; Galler and Edstrom, 1984; Kao and Case, 1985, 1986; Dreesen *et al.*, 1985a,b, 1988; Case, 1986; Grond *et al.*, 1987; Botella *et al.*, 1988; Hoog *et al.*, 1988; Dignam and Case, 1990; Paulsson *et al.*, 1990; Botella and Edstrom, 1991).

Not all the genes encoding secretory proteins are located in Balbiani rings. The structural gene for SP140 protein is mapped in the usual puff in the *I*-17B region. The nucleotide sequence of the corresponding cDNA contains tandem repeats of 42 bp encoding repetitive domains of 14 amino acids. The *C. tentans* genome contains approximately 70 copies of these repeats organized in a single block 3 kb in size. The *sp140* gene is developmentally regulated; its maximal expression occurs at the prepupal stage (Dignam *et al.*, 1989; Galli *et al.*, 1990).

The location of Balbiani rings and secretory polypeptides is correlated in *A. lucidus* (Baudisch and Panitz, 1968; Wobus *et al.*, 1971a) and in *C. thummi:* spIA is correlated with BR-c (BR1 according to another nomenclature), spIB with BRb (BR2) (Pankow *et al.*, 1976; Serfling, 1976; Serfling and Huth, 1978; Serfling *et al.*, 1983; Kao and Case, 1986; Cortes *et al.*, 1989, 1990).

The chromatid, which is tightly packaged in an inactive Balbiani ring 2 of *C. tentans*, has a nucleosome packing ratio (re : 189 bp per nucleosome) of about 5 (Andersson *et al.*, 1980; Widmer *et al.*, 1987); it is organized into a fiber 10 nm in diameter. The fiber coils once again, forming a fiber with a larger diameter (25 nm) and a packing ratio of about 40 (reviewed by Daneholt *et al.*, 1982). This fiber has one more or several compaction levels with the result that the total compaction ratio reaches 380 (Derksen *et al.*, 1980).

It is believed that the activation of the Balbiani ring gene is a two-step process: (1) puffing is first initiated; as a consequence, the chromatid is decompacted to the state of a 25-nm fiber so that the packing ratio decreases from 380 to 40; (2) this is followed by activation of transcription with a further decrease in DNA packing from 40 to 3.6 (Daneholt *et al.*, 1982).

The active regions of Balbiani rings are represented by loops (Beermann and Bahr, 1954; Stevens and Swift, 1966; Lamb and Daneholt, 1979). When the four chromosomes of *C. tentans* are spread by the Miller technique, the loops are clearly seen (Figure 160). The length of the transcribed DNA region is 7.7 to 12.6 μm (Lamb and Daneholt, 1979; Olins *et al.*, 1980). The 5′ and 3′ ends of the loops are morphologically different (Figure 162). The 5′ nontranscribed region (Figure 162a) is composed of a thin extended nucleofilament, about 5 nm in diameter; the nucleofilament is nucleosome-free and 0.5 kb in length (Ericsson *et al.*, 1989). In the transcribed part of the loop (Figure

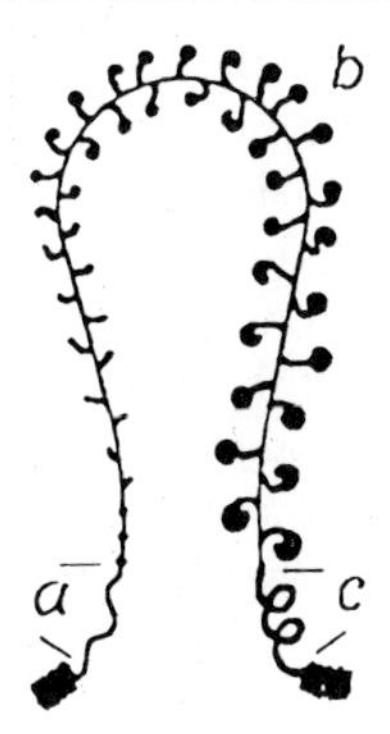

Figure 162. A scheme of the organization of the transcription loop of Balbiani ring 2 of C. *tentans* (a) The 5′ end of the gene; (b) the transcribed template; (c) the 3′ end of the gene (from Daneholt, 1992).

162b), which is also devoid of nucleosomes, the diameter of the filament does not vary during intense transcription; it is 5 nm when the density of RNA polymerase molecules is high; the filament condenses to 10 nm between the sparcely passing RNA polymerase molecules. If transcription initiation is inhibited by DRB, the 10-nm filament is rapidly packaged into a 25-nm fiber upon the passage of the RNA polymerase molecule along the template; that is, the active chromosome fiber is rapidly compacted into a higher-order structure upon cessation of intense transcription (Andersson *et al.*, 1982, 1984; Daneholt *et al.*, 1982; Widmer *et al.*, 1984; Bjorkroth *et al.*, 1988; Daneholt, 1992; Ericsson *et al.*, 1990). It is of interest that histone H1 is present in the loops both at the most active phase of transcriptional activity and in the inactive fibril (Ericsson *et al.*, 1990).

The 3′ end of the loop is 200 nm in length; it is loosely packaged into a short thick chromatin fiber, 30 nm in diameter. Its coiled structure is shown in Figure 162. The loops are open, such that anchoring sites are separated from each other (Ericsson *et al.*, 1989; Masich, 1992).

VIII. NUCLEOLI

A. Morphology and functioning of the nucleolus

Nucleoli in salivary gland cells of the larvae of Diptera were initially described contemporaneously with the discovery of polytene chromosomes (Balbiani, 1881; Alverdes, 1912b; Heitz and Bauer, 1933; King and Beams, 1934; Heitz, 1934; Bauer, 1935; Poulson and Metz, 1938). Early studies of chromosome behavior during the cell cycle demonstrated that the nucleoli are formed in association with a specific chromosome region (Heitz, 1931; Kaufmann, 1934, 1937).

McClintock (1934) found that if a part or all the nucleolar organizer region of the chromosome in maize was removed by irradiation, the nucleolus did not develop. Kaufmann (1938) showed that in *Drosophila* stocks with chromosomal rearrangements affecting the nucleolar organizer region, the nucleolus is transposed together with the chromosome fragment at a new position. Ritossa and Spiegelman (1965) established that the content of ribosomal RNA able to hybridize to the cDNA from *Drosophila* stocks containing various doses of nucleolar organizer is directly proportional to the dose (Figure 163).

It was inferred that the 18S and 28S rRNA genes are encoded in the

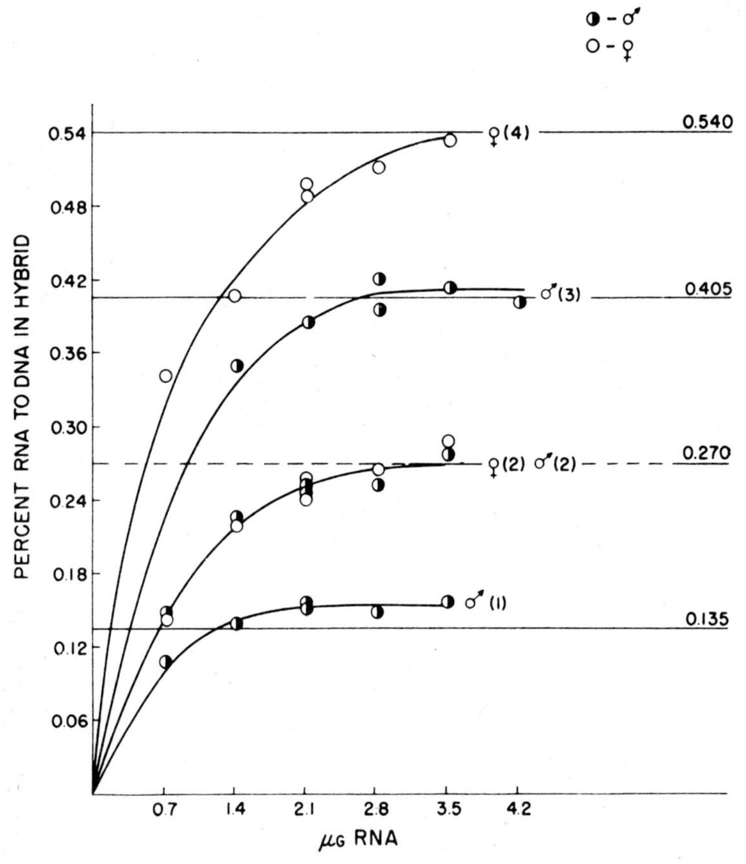

Figure 163. Saturation level of DNA containing different doses of the nucleolus-forming regions in hybridization to ribosomal RNA in *D. melanogaster*. Abscissa, RNA amount taken for hybridization (μg); ordinate, RNA proportion (percentage) in hybrids with DNA. Numbers in parentheses next to sex symbol designate dose of the nucleolar organizer (from Ritossa and Spiegelman, 1965).

nucleolar organizers (Ritossa *et al.*, 1966a,b). The DNA from the mutants of the X-linked *bobbed* gene was similarly studied, and it seemed rather likely that the *bobbed* locus is the nucleolar organizer (Ritossa *et al.*, 1966c; Atwood, 1969; Ritossa, 1976).

Homozygotes for the deletion of nucleolar organizer die as embryos without reaching gastrulation, as expected (Beermann, 1960). In cells with polytene chromosomes, the nucleoli appear as large spherical structures differing in texture from the rest of the nuclear sap (Figure 164). The nucleoli contain large amounts of RNA, for example, 250 pg in *C. thummi* (Serfling *et al.*, 1974). Two types of nucleoli are distinguished in their morphology: *Chironomus* and *Drosophila*.

In chironomids and some other species, particularly sciarids and simulids, the large nucleolus appears to be threaded on the chromosome (Figure 165). The nucleolus is actually a puff resulting from intense synthesis of ribosomal RNA. The nucleolar organizer of this type is quite precisely localized within a definite region, especially when the activity of nucleolus decreases. The nucleolar organizer has a banded appearance when in the fully inactive state (Bauer, 1935; Poulson and Metz, 1938; Melland, 1942; Bauer and Caspersson, 1948; Beermann, 1952a, 1960, 1962; Mechelke, 1953; Wolstenholme, 1965; Kiknadze, 1967; Kiknadze and Belyaeva, 1967; Stockert and Colman, 1975; Brady *et al.*, 1977; Stockert, 1977).

In *C. tentans*, the nucleolar organizer is observed to be a single large doublet (Beermann, 1960; Pelling and Beermann, 1966). In *A. lucidus*, the association between the nucleolus and the massive band is well seen in the case of heterozygosity (Figure 166) for the activity of the nucleolar organizer (Panitz, 1960a). In *C. thummi*, the nucleolus originates from two, up to 6 bands (Kiknadze, 1967; Kiknadze and Belyaeva, 1967; Razmahnin *et al.*, 1982). The morphology of the nucleolar organizer is tissue specific in this species. When the nucleolus possesses the appearance characteristic of chironomuses in salivary glands cells, a large heavily staining body hybridizing to the DNA containing the rRNA genes occurs in the proximity to the nucleolus in Malpighian tubule cells. The body is presumably the compacted part of the nucleolar organizer (Razmakhnin *et al.*, 1982).

In certain Australian species of chironomuses, the nucleolus is formed in a small region, where up to nine bands are counted (Lentzios and Stocker, 1979). Nucleoli have a similar structure in simulids, the nucleolar organizer is associated with several bands in *Odagmia ornata* (Shcherbakov, 1966a; Dunbar, 1967).

In sciarids species, the nucleolar organizer occupies a large block of heterochromatin in the proximal part of the X chromosomes, as shown by *in situ* hybridization experiments. The nucleolus has the appearance of an enormous puff (Jacob and Sirlin, 1963; Gabrusewycz-Garcia and Kleinfeld, 1966;

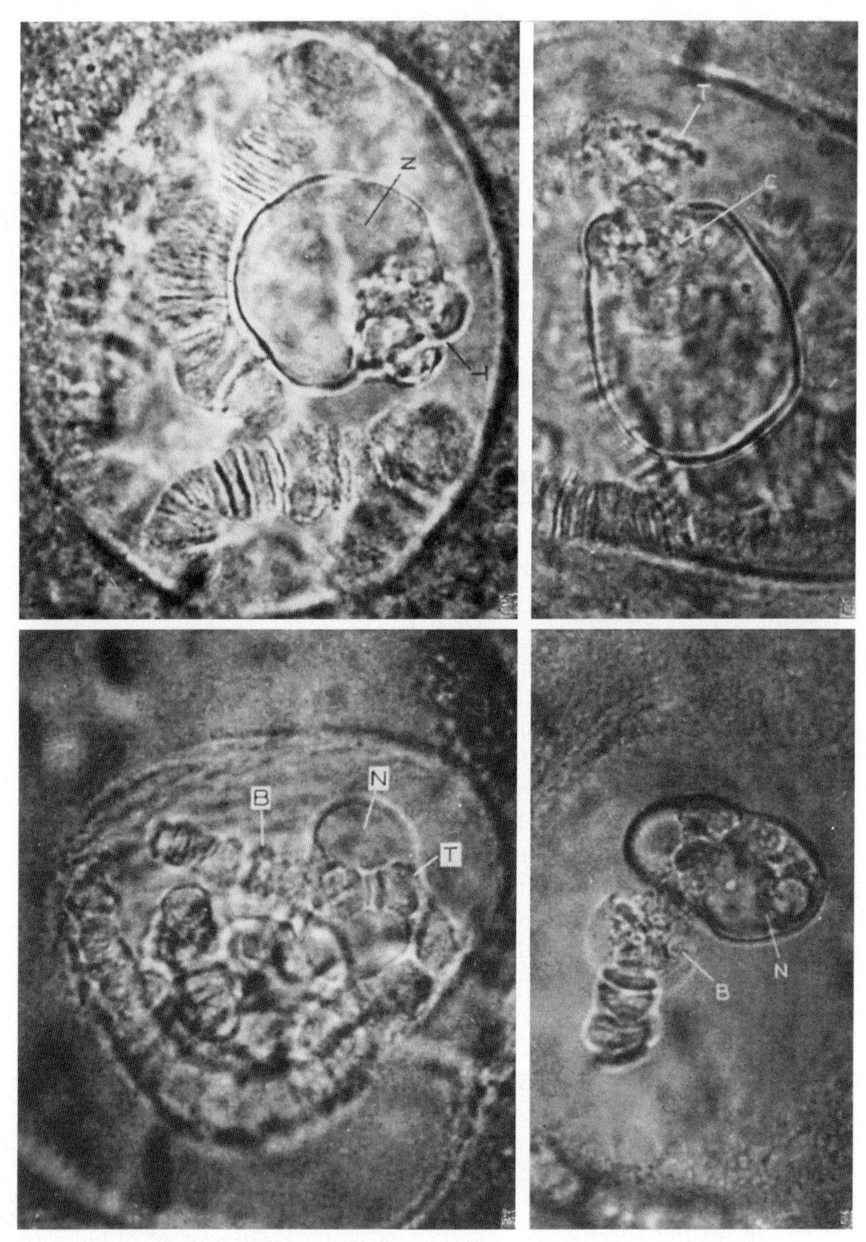

Figure 164. A photograph of the salivary gland nucleolus of a living larva of *Chironomus* sp. N, nucleolus; B, Balbiani ring; T, telomere (from Studies on the structure of the nucleolus-forming regions in the giant salivary gland chromosomes of Diptera, Poulson, D. F., and Metz, C. W., *J. Morphol.*, copyright © 1938 Wiley-Liss, Inc. Reprinted by permission of Wiley-Liss, Inc., a subsidiary of John Wiley & Sons, Inc.).

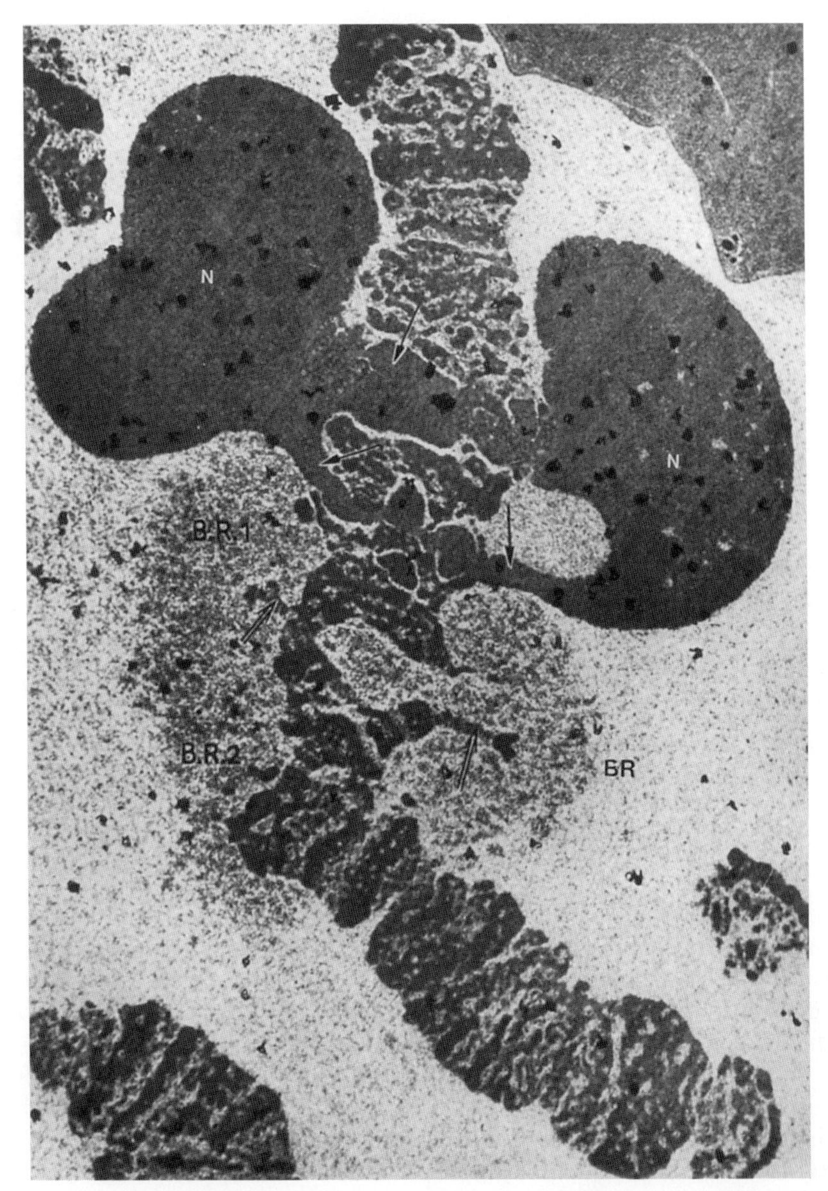

Figure 165. Morphology of the nucleolus (N) and Balbiani rings (BR) of the fourth chromosome of C. *thummi* according to data of electron microscopy (reprinted by permission from Gaudecker, 1967).

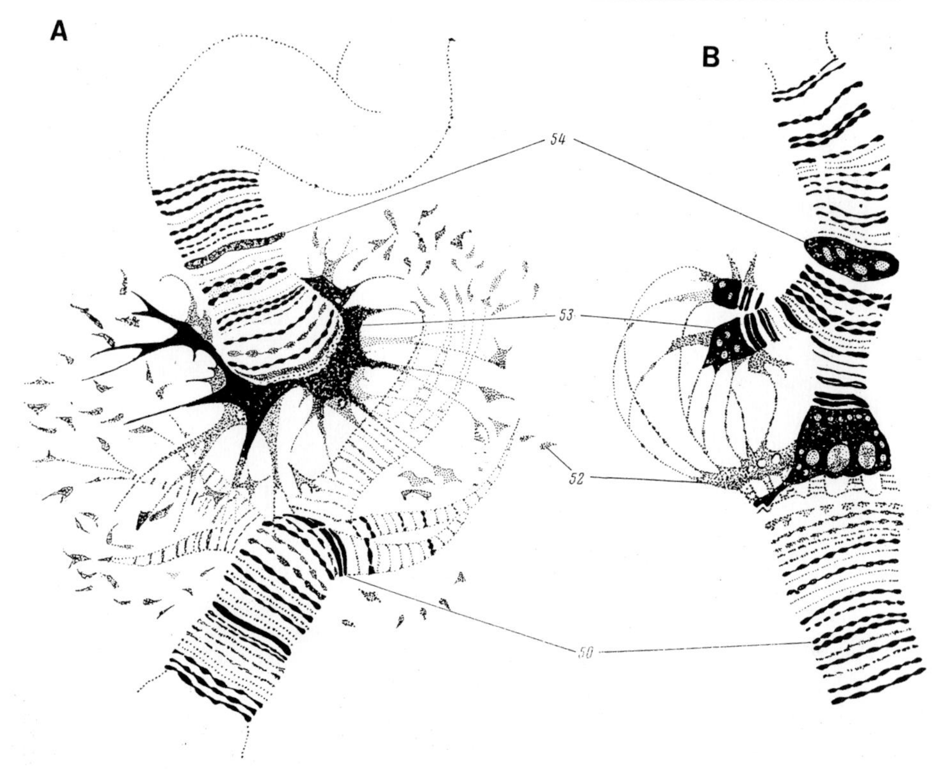

Figure 166. Formation of a heterozygous nucleolus from a part of a band in *Acricotopus lucidus*. (A) A homozygous nucleolus in cells of the anterior gland lobe of a larva, whose nucleolus in the main and lateral lobes is heterozygous; (B) a heterozygous nucleolus in cells of the main lobe of the gland of a young prepupa (reprinted by permission from Panitz, 1960a).

Pardue *et al.*, 1970; Gerbi, 1971; Gabrusewycz-Garcia, 1972; Simoes *et al.*, 1975; Busen *et al.*, 1982; Dessen and Perondini, 1991). The nucleolus in the representatives of the Agromyzidae family appears as a very large puff at the end of the chromosome (Block, 1969).

In *Drosophila* the organization of the nucleolus is of another type. The nucleolar organizers are located in the heterochromatic regions of the X and Y chromosomes lacking distinctive banding patterns. The nucleoli are seen as ball-like structures hanging on threads attached to a chromosome region (Figure 167); it should be noted that no attachment of a fiber to a particular band was, as a rule, revealed (Heitz, 1934; Frolova, 1936; Emmens, 1937; Kaufmann, 1937, 1938; Eloff, 1939; Beermann, 1962; Swift, 1962). In *D. melanogaster*, the nucleolus is most frequently associated with the bands of the X-chromosomal 20CD

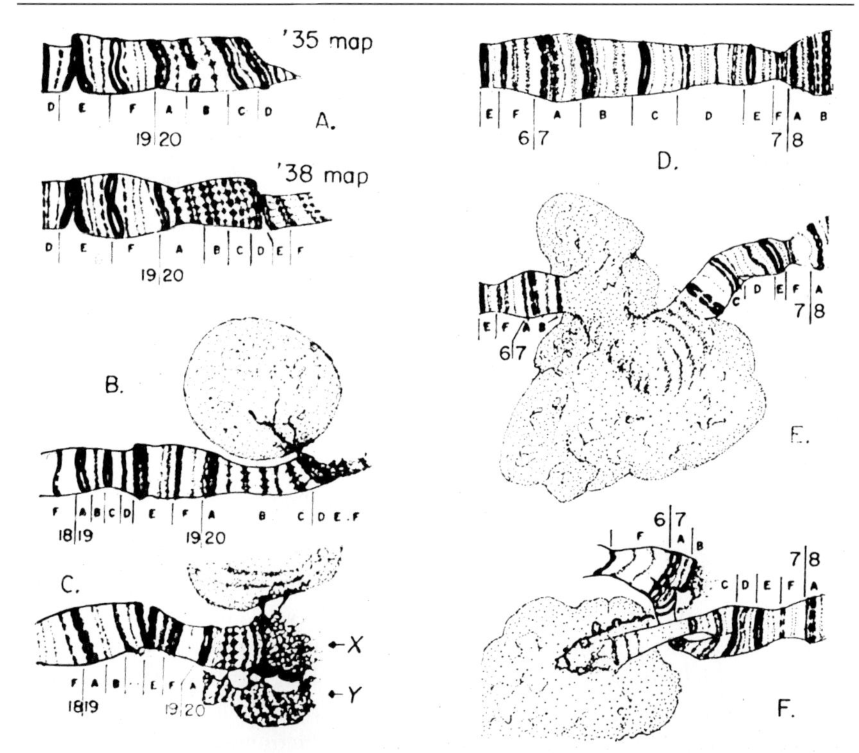

Figure 167. Morphology of nucleoli in the polytene chromosomes of *Drosophila* under (C and D) normal conditions and (F and G) in a chromosome with the ct^{6a1} inversion (reprinted by permission from Hannah-Alava, 1971). (A) Bridges' maps; (B and C) attachment of the fibril from the nucleolus to the 20D region; (E and F) the nucleolus in the chromosome with the ct^{6a1} inversion in (E) a male and (F) a heterozygous female.

region and occasionally with those of the other regions; for example, 20AB or 19E. Association of these regions is possibly secondary; that is, resulting from transposition of the nucleolus during squashing. The term "lateral extrusion" describes well the association mode of the nucleolus with the chromosome. The term emphasizes that chromatin fibers extruded from the band, showing signs of puffing, which provide the attachment of the nucleolus to the chromosome.

In preparations with extrachromosomal RNP, removed by special pretreatment, it is clearly seen that the thin chromatin thread extruded from the band forms a diffuse loop in the nucleolus (Viinikka *et al.*, 1971).

In all the studied species, DNA within the nucleoli was identified by using the Feulgen stain or orcein in *in situ* hybridization of ribosomal DNA (Figure 168), or [^{3}H]thymidine labeling (Heitz, 1934; Kaufmann, 1937; Krivshenko, 1959; Berendes, 1963; Nash and Plaut, 1965; Barr and Plaut, 1966;

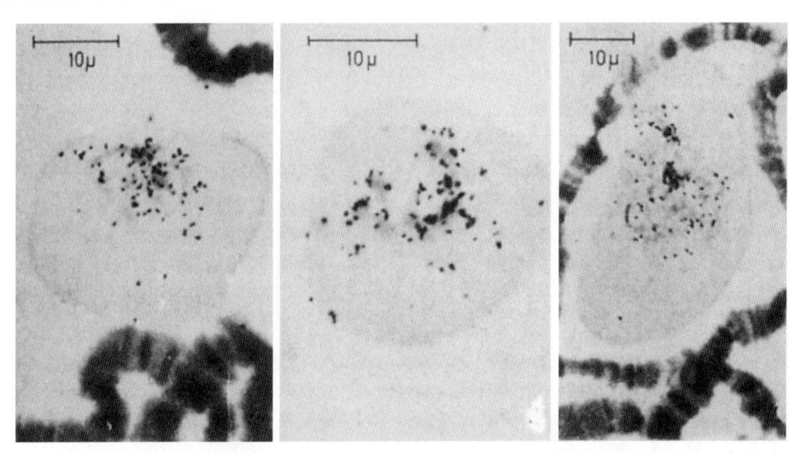

Figure 168. *In situ* hybridization of [³H]ribosomal RNA with the nucleolar organizer (indicated by arrow) of *Drosophila* (reprinted by permission from Pardue *et al.*, 1970).

Painter and Biesele, 1966; Tulchin *et al.*, 1967; Rodman, 1968a, 1969; Olvera, 1969; Ellison and Granholm, 1970; Pardue *et al.*, 1970; Hannah-Alava, 1971; Viinikka *et al.*, 1971; Alonso, 1972b, 1973b; Wimber and Steffensen, 1973; Bicudo and Richardson, 1977; Burkholder, 1977; Nash and Vyse, 1977; Weideli *et al.*, 1978; Bicudo, 1981; Sinibaldi and Cummings, 1981; Chatterjee, 1982; Ghosh and Mukherjee, 1982a,b, 1984a, 1986a,b; Hennig *et al.*, 1982; Ghosh, 1984, 1986; Gupta and Kumar, 1987; Bedo and Webb, 1989).

From the results of *in situ* DNA/RNA hybridization, it was concluded that the nucleolar organizer may be restricted to intranucleolar chromatin. In fact, the labeled material was concentrated in the nucleolus, and it is absent from the chromosome regions from which the attachment thread protruded (Alonso, 1973b).

The nucleolar structure of polytene tissue of the representatives of calliphorids and tephritids is similar (Beckingham and Rubacha, 1984; Bedo and Webb, 1989; Bedo, 1992).

However, nucleolar morphology may be different in Drosophilidae and Chironomidae, for example, when the nucleolar organizer is transposed by an inversion to the euchromatic part of the chromosome (Figure 167). The nucleolar organizers are located in a single band on polytene autosome *3* in two Hawaiian species of *Drosophila*, *D. heteroneura* and *D. silvestris* (Stuart *et al.*, 1981).

Cytological localization of rRNA *in situ* showed that the rRNA gene cluster occupies an extensive region, about half of chromosome length, including the satellite and the nucleolar constriction in the embryo suspensor cells of *Phaseolus coccineus* (Avanzi *et al.*, 1972); after treatment of the cells with actinomycin D, the nucleolus detached from the chromosome and becomes associated with it by a thread, such as in *Drosophila* (Nagl, 1969). The nucleolar organizers are variable in number in various species, their number is as high as six in certain Australian chironomids (Wen *et al.*, 1974; Schafer and Kunz, 1975; Lentzios and Stocker, 1979). Despite these differences, the nucleoli share morphological and structural features. Stereoscopic analysis of the fine structure in diploid nuclei of the larval imaginal disks in *Drosophila* revealed numerous fibrils wound into a loop and extending from the nucleolar organizer (Ashton and Schultz, 1964).

The data concerning the uniform distribution of rDNA in the nucleolar volume are controversial. Large masses of chromatin, most frequently in the center of the nucleolus, are found in *Drosophila* and *C. thummi* (see above). Incorporation of [^{3}H]uridine into salivary gland cells of *C. tentans* demonstrated that only the central part of the nucleolus labels (C. Pelling and W. Beermann, cited in Mechelke, 1963). In contrast, the nucleolus labels uniformly after *in situ* hybridization of ribosomal RNA in this species (Degelmann *et al.*, 1979).

In *Drosophila*, after pulse incorporation of [^{3}H]uridine for 1.5 to 10 min, label was detected mainly at the boundary separating intranucleolar chromatin from the rest of the nucleolus. Using the method of *in vitro* transcription by *E. coli* RNA, it was shown that intranucleolar chromatin is also engaged in transcription (Chatterjee, 1982), DNA clones containing the rRNA genes uniformly label the entire nucleolus (Weideli *et al.*, 1978).

An abundant non-DNA component is also represented in the nucleolar organizer. It differs considerably in total content and composition from ribonucleoproteins of the other chromosome regions. For example, the electron density of Balbiani rings is not high (Figure 165); the rings are optically transparent—although organized in loops, as in the nucleoli—and involved in intense transcription. In contrast, the nucleoli are always composed of electron-dense material (Stevens, 1964; Jacob and Sirlin, 1964; Stevens and Swift, 1966; Gaudecker, 1967; Olins *et al.*, 1980).

Distribution of material in the nucleoli of two types have been described in *Chironomus* (Figure 169). Typically a central fibrillar core alternating with a granular area are distinguished in the nucleolus (the standard type).

The organization of the nucleoli of the segregational type is dual; the internum is compactly packed and the periphery is fibrillar and reticular (Sirlin and Jacob, 1962; Jacob and Sirlin, 1964; Kalnins *et al.*, 1964a,b; Stevens, 1964; Stevens and Swift, 1966; Gaudecker, 1967; Belyaeva, 1971; Burgauer and Stockert, 1975; Stockert and Colman, 1975, 1977; Olins *et al.*, 1980, 1982). The nucleoli stain heavily (Figure 170) with ammoniacal silver (Lentzios and Stockert, 1979; Ananiev *et al.*, 1981).

The size of the nucleoli in the salivary gland cells decreases at the end of the last larval stage or in prepupae; the banding pattern is concomitantly

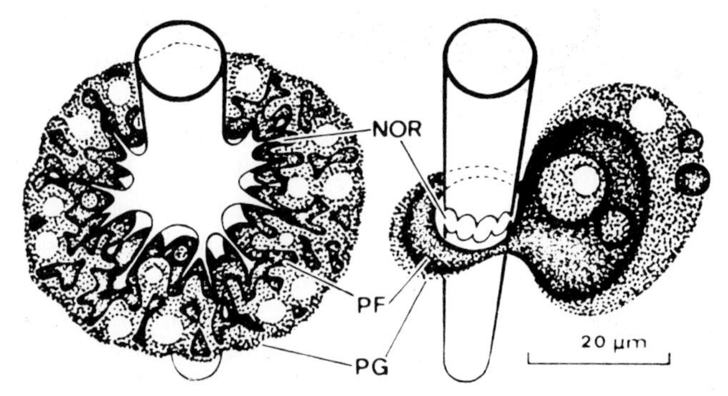

Figure 169. A semischematic depiction of the (left) standard and the (right) segregated nucleolus of *Chironomus*. NOR, nucleolar organizer; PF, fibrillar part; PG, granular part (reprinted by permission from Stockert and Colman, 1975).

recovered in the nucleolar organizer region (Breuer and Pavan, 1955; Kiknadze and Filatova, 1960; Lara and Hollander, 1967; Mattingly and Parker, 1968b). In *D. virilis*, incorporation of [^{3}H]uridine into nuclear RNA is substantially decreased at the end of larval development, and RNA synthesis in the nucleoli of prepupae ceases almost completely (Zakharenko *et al.*, 1976). In *Sciara ocellaris*, the nucleolar organizer puffs during the fourth larval instar; it is rapidly inactivated at metamorphosis onset. This inactivation can be stopped by extirpation of the ring glands and induced by treatment with exogenous ecdysterone

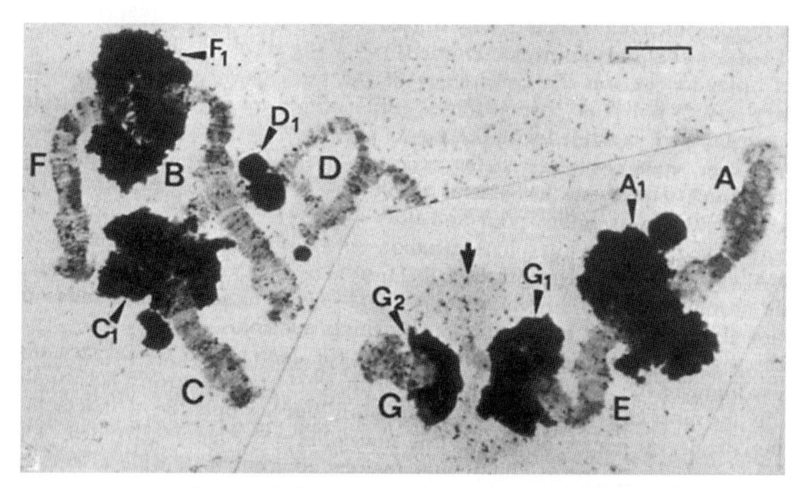

Figure 170. Staining of six nucleolar organizers with ammoniacal silver in the chromosomes of *Chironomus duplex*. Arrowheads designate nucleoli, arrow indicates Balbiani rings. Bar = 10 μm (reprinted by permission from Lentzios and Stocker, 1979).

(Dessen and Perondini, 1976, 1985c). Treatment with ecdysterone leads to a reduction in rRNA synthesis in *D. melanogaster* and *A. lucidus* (Hughes and Kambysellis, 1970; Panitz, 1978b). In *C. thummi*, incorporation of [^{3}H]uridine into the nucleoli decreases during molt periods (Kiknadze *et al.*, 1981). In *Calliphora vicina*, the rate of rRNA synthesis is highest in late larvae. It progressively decreases to the lowest level by the time of adult emergence (Protzel and Levenbook, 1976). Changes in nucleolar volume conform to a pattern during the development of the giant foot pad cells of the pupae of *Sarcophaga bullata*. Two maximum were identified, both occurring on the day of highest puffing activity (Bultmann and Clever, 1969).

The micronucleoli occur in the cells with polytene chromosomes in many dipteran and plants species. Their number can reach several dozens (Bauer, 1935; Beermann, 1952a; Kiknadze and Filatova, 1960; Pavan, 1965a; Gabrusewicz-Garcia and Kleinfeld, 1966; Bier *et al.*, 1969; Da Cunha *et al.*, 1969a, 1973a; Pavan and Da Cunha, 1969a; Ribbert and Bier, 1969; Simoes and Cestari, 1969, 1981; Smith, 1969; Kastritsis, 1971; Pavan *et al.*, 1971a; Santelli *et al.*, 1976; Hill and Watt, 1978; Stocker *et al.*, 1978; Ribbert, 1979; Simoes *et al.*, 1981; Busen *et al.*, 1982; Michailova *et al.*, 1982; Genova and Semionov, 1985; Dessen and Perondini, 1991).

On the basis of the observation that micronucleoli incorporate [^{3}H]uridine it is believed that ribosomal RNA is transcribed in the micronucleoli (Bier *et al.*, 1969; Ribbert and Bier, 1969; Ribbert, 1979; Redfern, 1981a; Dessen and Perondini, 1991); the presence of the rRNA genes was demonstrated by *in situ* hybridization (Pardue *et al.*, 1970; Avanzi *et al.*, 1970, 1972; Gerbi, 1971; Ananiev *et al.*, 1981; Semionov and Kirov, 1986; Mecheva and Semionov, 1989; Bedo, 1992). During development the main nucleolus may become fragmented, giving rise to additional nucleoli (Gabrusewicz-Garcia, 1972; Genova and Semionov, 1985; Dessen and Perondini, 1991; Bedo, 1992).

The formation of micronucleoli is specific: in *C. thummi*, the micronucleoli most frequently occur in two large nuclei at the base of the lateral lobe of the gland (Kiknadze and Filatova, 1960). A decrease in temperature was without significant effect on the formation of additional nucleoli. Individuals grown at three different temperatures (15°, 17°, and 25°C) did not differ in the distribution frequencies of cells according to nucleolus number (Genova and Konova, 1987). The frequencies of additional nucleoli were the same in X0 males and females of wild lines (Semionov, 1988).

The formation of additional nucleoli or micronucleoli is occasionally related to a real increase in RNA mass. A single nucleolus is most frequently encountered in 7- to 9-day-old unpupated larvae homozygous for the *l(3)tl* mutation and up to 24 nucleoli in larvae aged 11–20 days. The total nucleolar area in a nucleus increases six- to seven-fold (Zhimulev and Lychev, 1972c).

Determination of the mass of nuclear RNA by micromethods (Edstrom, 1960) revealed that in zero-hour prepupae of the control *Drosophila* line, RNA content in the nucleus varies from 332 to 348 pg/nucleus; its variation range is from 500 to 515 pg/nucleus in 6-day-old *l(3)tl* larvae, and from 621 to 644 pg/nucleus in 11-day-old larvae. The average RNA content is 755 pg/nucleus in 17-day-old larvae. In contrast, RNA content in the whole cell changes little: it is 3.200 pg/cell in zero-hour prepupae of wild-type strain, 3.317 pg/cell in 6-day-old larvae, and 3.352 to 3.608 in 11-day-old *l(3)tl* larvae (Kopantseva, 1973; Kopantseva *et al.*, 1994).

A special type of nucleolar apparatus was found in polytene nurse cells of imago. In *Calliphora erythrocephala*, besides the nucleolus at chromosome *VI*, large numbers of free nucleoli form during development of oocyte and nurse cells. These nucleoli actively incorporate [^{3}H]uridine. They contain DNA particles, suggesting the possibility of autonomous RNA synthesis (Ribbert and Bier, 1969).

In *D. melanogaster*, mophology of the nurse cell nucleolus changes during the development, from stages 7 to 10B. In stage 7 the nucleolus is composed of a series of interconnected finger-like projections. Later the nucleolar fragments increase in size. These fragments are not randomly distributed throughout the nucleus, but are closer to the surface than to the center of the nucleus. A three-dimensional model of the nucleolus from stage 9 demonstrated that the vast majority of nucleolar segments are interconnected. They form a thick network—a spherical shell whose outer boundary lies close to the inner surface of the nuclear envelope (Dappels and King, 1970; Bishop and King, 1984). A large nucleolus was found when 28S + 18S ribosomal DNA was hybridized *in situ* with nurse cell nuclei of wild type *D. melanogaster* (Hammond and Laird, 1985).

Some interdependence of the nucleolus and B chromosomes was found in black flies. In *Cnephia dacotensis*, a high incidence of B chromosomes is associated with nonexpression of the nucleolar organizer (Procunier, 1982). In *Cnetha djafarovi* an additional nucleolus occurs, alongside the main nucleolus in certain larvae, but only in those animals which had B chromosomes (Kachvoryan *et al.*, 1993).

B. Molecular organization and expression of the ribosomal RNA genes

In 1969, a technique whereby the nucleolar material could be spread on a water surface was developed. With this "Miller spreading method" (introduced in earlier sections), the nucleolar material, the DNA fibrils, and the DNA molecules synthesized on them are straightened out and the nucleolar organizer appears as a group of thin axial fibrils in the electron microscope. Some fibrils are matrix-free; others are matrix-covered (Figure 171). Each matrix unit consists of about 100 thin fibrils connected to the axial fibril at one end. The thin

Figure 171. A part of the spread nucleolar organizer of *Triturus viridescens* (reprinted with permission from Miller, O., and Beatty, B. R. (1969). Visualization of nucleolar gene. *Science* **164**, 955–957. Copyright 1969 American Association for the Advancement of Science).

fibrils show a thick-to-thin graduation along the axial fiber, and the matrix resembles a Christmas tree in its general appearance. Each thin fibril of the matrix represents a separate RNA molecule which is already protein-coated during transcription. Thus, the matrix unit is a separate cistron for the 18S and

28S rRNAs. Visualization of these cistrons is possible because about 100 RNA molecules are simultaneously synthesized on each gene (Miller and Beatty, 1969a,b).

Later studies have demonstrated that the functional unit of many nucleolar organizers have structural features in common (Figure 44). The unit contains genes of 18S and 28S RNA and a the transcribed spacer. Each cistron is separated from the other cistron by an interspersed nontranscribed spacer (Degelmann *et al.*, 1979; Bashkirov, 1980; Bush and Rothblum, 1982; Cross and Dover, 1987). The size of the transcription unit varies from 8.4 kb in *C. tentans* or *S. coprophila* to 17.4 kb in *Sarcophaga bullata* (reviewed by Beckingham, 1982). Such a unit is repeated from dozens to hundreds of times in the genome of various Diptera. For example, the nucleolar organizer of *D. melanogaster* contains about 0.27% of total DNA in the haploid genome; this corresponds to about 200–250 cistrons (Ritossa *et al.*, 1966a–c; Pardue *et al.*, 1970; Tartof, 1971, 1973, 1975; Ritossa, 1976, Gandhi *et al.*, 1982). It seems that the rDNAs of normal X and Y chromosomes have opposite orientations with respect to the centromere (Palumbo *et al.*, 1973). Also, *D. melanogaster* has different ribosomal sequences on X and Y chromosomes (Yagura *et al.*, 1979). Loss of more than 70% of the rRNA genes is lethal in the aforementioned *bobbed* mutants (Ritossa *et al.*, 1966a).

The number of 18S and 28S rRNA genes per haploid set was estimated as 120–320 (Meyer and Henning, 1974) or 460–560 (Glatzer, 1979) in *D. hydei*, 510 in *D. virilis* (S. Endow, 1974, in: Beckingham, 1982), 144 in *S. bullata* (French *et al.*, 1981), 770 for *Calliphora erythrocephala* (Beckingham and Thompson, 1982), and 45 for *S. coprophila* (Gerbi and Crouse, 1976). In *C. tentans*, the number of ribosomal cistrons is ~43 (Hollenberg, 1976) to 72 (Lambert *et al.*, 1973b); in *C. thummi*, their number is from 30 to 50 (Wobus, Serfling, 1977). When stretched with a manipulator, the region containing the material of the nucleolus in the polytene chromosomes of this species becomes about 500 μm long (Shilova *et al.*, 1983). The number of ribosomal cistrons should be not less than 150–200 at this length. The discrepancies are possibly due to sliding of the chromatid along another during stretching and also to artificial lengthening of the gene cluster.

Certain data suggest that the rRNA genes are unintegrated into the nucleolar organizer, but are located in some regions of the genome. In *D. melanogaster*, as a result of gentle lysis of cells and analysis of sucrose gradient sedimentation, the major ribosomal DNA was separated into fractions (5–10) $\times 10^9$ Da. In female adults with one nucleolar organizer, 32% of the detected DNA fragments were of small size (approximately 3×10^8 Da); it was suggested that these fragments are unintegrated into the nucleolar organizer (Zuchowski and Harford, 1976a,b).

Each of the tandemly repeated units of the nucleolar organizer is transcribed into the rRNA precursor, which occurs as 38S RNA (Greenberg, 1967, 1969). Transcription starts from 920 bp upstream of the 5′ end of the sequence encoding the 18S rRNA gene (Long *et al.*, 1981b). The transcript contains the "readthrough" sequence of the 18S gene, the transcribed spacer, and the 28S sequence.

Microdissected nucleoli of the salivary glands of C. *tentans* were incubated in [^{3}H]uridine solution for 45 min; the extracted RNA was then submitted to electrophoresis. The 37–38S, 30S, and 23S RNA fractions incorporated label, the mature 18S and 28S molecules did not (Ringborg *et al.*, 1970a,b). It was inferred that rRNA precursors, the common 38S ones as well as the 23S and 30S species, are synthesized in the nucleolus, whereas the molecules of 18S and 28S rRNAs mature at another extranuclear site. Study of chromosomal RNA demonstrated that both 30S and 23S RNAs migrated to other places: chromosomes properly or nuclear membrane (Edstrom and Daneholt, 1967; Ringborg and Rydlander, 1971). This is in agreement with the results obtained with C. *thummi* in similar experiments (Serfling *et al.*, 1974). Single cistrons contained by the nucleolar organizer can function independently of one another. When the rRNA gene cluster is broken by chromosomal rearangements, some of its fragments can form nucleoli (Kaufmann, 1938; Krivshenko, 1959; Scherbakov, 1965; Pelling and Beermann, 1966; Israelevski, 1975; Nash and Vyse, 1977).

Insertion of the DNA from just one cistron of ribosomal RNA into chromosomal sites other than the nucleolar organizer is associated with the formation of small nucleoli (mininucleoli) (Karpen *et al.*, 1988). In contrast, a case is known in which the whole nucleolus functions in a tissue-specific manner. Thus, in the salivary gland cells of A. *lucidus*, the nucleolus is active on both homologs of the chromosomes in the anterior lobe cells; however, it is heterozygous in the central and lateral lobes (Figure 166). Heterozygosity was also observed for certain cells of Malpighian tubules (Panitz, 1960a). This indicated that the material of the nucleolar organizer is present in all the cells, but turned off as a whole block in some cells.

In *Drosophila* large proportions of the 28S rRNA genes contain insertions called type I and II (R1 and R2 insertions). In the nucleolar organizer of the X chromosome, 60–65% of the 28S rRNA genes have the type II insertions at the 3′ ends (Tartof and Dawid, 1976; Dawid and Botchan, 1977; Glover, 1977, 1981; Glover and Hogness, 1977; Pellegrini *et al.*, 1977; Wellauer and Dawid, 1977; White and Hogness, 1977; Dawid *et al.*, 1978; Dawid and Wellauer, 1978; Wellauer *et al.*, 1978; Endow and Glower, 1979; Roiha and Glover, 1981; Beckingham and Thompson, 1982; Franz *et al.*, 1983; Sharp *et al.*, 1983; Leibovich, 1984). The sizes of the insert are 0.5–5.4 kb, most frequently 5 kb (Wellauer and Dawid, 1977; Pellegrini *et al.*, 1977; Dawid and Wellauer, 1978; Serfling, 1978; Peacock *et al.*, 1981). In the nucleolar organizer of the Y chromosomes, the type I insertions are found in not more than 16% of the 28S rRNA genes or in the transcribed spacer (Tartof and Dawid, 1976; Glover, 1977). In *D. melanogaster,* type I insertions also occur outside the nucleolar organizer, on the centromeric heterochromatin of the X chromosome, in the long autosomes, and in the 100C8-12 region of the third chromosome (Kidd and Glover, 1980; Peacock *et al.*, 1981; Appels and Hilliker, 1982). Type II insertions are about 3.5 kb long; they show no homology to the type I insertions and are integrated into the region of the 28S rRNA gene 5 bp apart from the site of the type I insertions. These insertions occur less frequently, in about 15% of the genes (Glover, 1977; Wellauer and Dawid, 1977; Wellauer *et al.*, 1978; Long and Dawid, 1979; Roiha *et al.*, 1981; George *et al.*, 1996). Type II insertions are located only in the nucleolar organizer in *D. melanogaster* (Roiha *et al.*, 1981).

The closely related species *D. simulans* and *D. mauritiana* contain both types of isertion (Roiha *et al.*, 1983). The type I insertions do not hybridize *in situ* to the 102 region of the fourth chromosome of either species (Mecheva, 1990). The type II insertions are located not only in the nucleolar organizer, but also in the 75B and 80BC regions in *D. simulans*, they are located in the centromeric region of the third chromosome in *D. mauritiana* (Mecheva, 1990). In *D. virilis,* insertions were also found in the 28S rRNA genes at the same position as in *D. melanogaster* (Barnett and Rae, 1979; Rae *et al.*, 1980). In *D. hydei,* an insertion of about 5 kb was detected in the 28S rRNA only in one of the nucleolar organizers (Kunz and Glatzer, 1979; Renkawitz-Pohl *et al.*, 1980, 1981a). In *D. busckii,* insertions of these types were not found (Slavicek and Krider, 1987).

Insertions into the 28S rRNA genes were detected in *S. coprophila* (Renkawitz-Pohl *et al.*, 1981b), *Rhynchosciara americana* (Zaha *et al.*, 1982), *Calliphora erythrocephala* (Beckingham and White, 1980; Beckingham and Rubacha, 1984), *S. bullata* (French *et al.*, 1981), and *D. hydei, D. neohydei, D. eohydei,* and *D. repleta* (Hennig *et al.*, 1982).

The 28S rRNA genes of *Drosophila* having the types I and II insertions are transcribed at a very low rate, if at all (Long *et al.*, 1981a). Analysis of cistrons lengths in Miller spreads of the nucleolar organizer demonstrated few if any larger transcripts whose presence would provide evidence for transcription of genes containing insertions (Glatzer, 1979; Renkawitz *et al.*, 1979; M. Jamrich in: Beckingham, 1982). Chooi (1979) and Endow (1979) found that 13–20% of transcript units are longer than the bulk of ribosomal RNA cistrons with the length of 2.65 μm. This finding led to the supposition that ribosomal genes carrying insertions are capable of being transcribed (Chooi, 1979). However, Beckingham (1982) did not accept this evidence. The insertions in 28S rRNA genes are not transcribed in *D. hydei* (Kunz and Glatzer, 1979), in *Calliphora erythrocephala* (Beckingham and Rubacha, 1984), *D. neohydei, D. eohydei,* and *D. repleta* (Hennig *et al.*, 1982).

The length of the nontranscribed spacer between cistrons in *Drosophila* usually about 5 kb; it is occasionally much longer (Meyer and Hennig, 1974; Pellegrini *et al.*, 1977; Indik and Tartof, 1980; Williams *et al.*, 1987; Tautz *et al.*, 1987, 1988). In contrast, it is shorter in *Chironomus* species, being about 1.8 kb (Derksen *et al.*, 1973; Degelmann *et al.*, 1979).

While the coding sequences of the ribosomal cluster are highly conserved, the sequences of the nontranscribed spacers have been free to diverge; for instance, there is virtually no homology between *D. melanogaster* rDNA spacers and *D. virilis* rDNA spacers and *D. melanogaster* has no spacers that are homologous to those in *D. hydei* as well (Rae *et al.*, 1980, 1981). In 47 lines of *D. melanogaster*, collected around the world, the nontranscribed spacers of the Y chromosome share little, if any, similarity both for total length and their organization (Williams *et al.*, 1987).

A feature common to the organization of the nontranscribed spacers (NTS) in the nucleolar organizers is the presence of short repeats in many species (see review by Beckingham, 1982). In *D. busckii*, two repeat regions are located within the NTS; the region II contains 11–16 *Hinc*II generated repeats, 160 bp each. NTS region III contains a repeat heterogeneous in length (Slavicek and Krider, 1987).

The NTSs of closely related species of *Chironomus* differ greatly in length. The NTS is 9 kb long in *C. t. piger*; its length is 14.5 kb in *C. t. thummi*. The additional 5 kb in the latter species contains an AT-rich *Cla*-fragment 120 bp long repeated about 50 times. A consequence of variation in copy number in the repeat is spacer length heterogeneity (Schmidt *et al.*, 1982; Schmidt and Godwin, 1983). The *Cla* and *Dde* repeats are not confind to the nucleolar organizer in these species. In *C. t. thummi*, the elements map to about 200 different sites in the genome. The *Dde* elements are restricted exclusively to the nucleolar organizer in *C. melanotus* and to one site in the chromosome, not associated with the nucleolar organizer in *C. t. thummi*, and to the nucleolar organizer in several other regions in *C. annularius* (Schmidt and Godwin, 1987).

The number of the ribosomal RNA is subject to variations, owing to magnification, dosage compensation, nucleolar dominance, and amplification.

1. Magnification of the rRNA genes. All the mutant alleles of the *bobbed* gene revert (in a varying manner) to normal, as a consequence of a process referred to as magnification of ribosomal cistrons. The process is restricted to males in which one cluster of the rRNA genes has a deletion in a part of the genes, while the other cluster has a complete deletion (Ritossa, 1968, 1972, 1976; Ritossa and Scala, 1969). Hypotheses to explain magnification were proposed (review: Bashkirov, 1980). First, unequal crossing-over preceded by homolog pairing with some shift of homologous regions relative to one another (Ritossa and Scala, 1969; Ritossa *et al.*, 1974). Second, un-

equal crossing-over may not occur in meioisis, rather in mitotically active germ cells (Tartof, 1973, 1974a,b). Third, the formation of extra copies of rDNA which can be both linear and circulized. The extra copies can integrate anew into the chromosomes (Ritossa, 1972, 1976). Bashkirov (1980) suggested three processes that might account for the formation of extrachromosomal molecules: (a) intrachromatid crossing-over, leading to excision of a certain number of rRNA genes copies, followed by replication outside the chromosome; (b) an extrachromosomal additional copy can result from local overreplication of the rDNA-containing region; and (c) the activity of reverse transcriptases.

2. Dosage compensation. When one of the nucleolar organizers is lost, the gene number increases in the other nucleolar organizer with the result that the same number of rRNA molecules is synthesized per time, both in cells bearing a single and two nucleolar organizers (Kiefer, 1968; Ritossa and Scala, 1969; Mohan and Ritossa, 1970; Tartof, 1971, 1973; Krider and Plaut, 1972a,b; Ritossa, 1976; Strausbaugh and Kiefer, 1979; Dutton and Krider, 1984). Loss of genes located in the X, not the Y chromosomes, is compensated, when the loss is from the nucleolar organizer (Williamson *et al.*, 1973). This phenomenon was called dosage compensation of the rRNA genes. Dosage compensation differs in two features from magnification (Bashkirov, 1980): (a) when compensated, rDNA accumulates already during development in individuals with only one nucleolus and not in subsequent generations, whereas magnification is heritable; (b) the rRNA gene excess is not heritable in the case of dosage compensation (Tartof, 1971, 1973, 1975).

   The formation of extra rDNA is controlled by a specific *compensatory response* (*cr*) locus. If one of the *cr* regions is deleted, an increase in rDNA in this chromosome occurs (Procunier and Tartof, 1978). There is a parallel increase in the number of nucleoli (Toshev and Semionov, 1987). In *D. hydei*, rDNA increases in diploid cells of genotypes containing only one nucleolar organizer (Grimm and Kunz, 1980). When a certain fragment of heterochromatin is removed by deletions (presumably, the nucleolar organizer is gone), the rRNA gene number can increase threefold in diploid cells (Grimm *et al.*, 1984).

3. Nucleolar dominance. Interspecific *D. melanogaster* × *D. simulans* hybrids show a nucleolar constriction only at the *D. melanogaster* nucleolar organizer (Durica and Krieder, 1977, 1978). The rRNA genes of predominantly one of the nucleolar organizers are polytenized (Endow and Glover, 1979; Endow, 1980, 1982, 1983). This phenomenon is termed nucleolar dominance. Deletion of the cr^+ locus has no effect on the dominance, a feature that distinguishes nucleolar dominance from dosage compensation (Endow, 1983).

4. Amplification. The nucleolar genes are subject to amplification; that is, to multiplication of rDNA copies located outside the chromosomes. After several replication rounds, the DNAs increase in number and—presumubly by recombining with chromosomal DNA—integrate into it (Ritossa, 1972, 1976).
5. Incomplete polytenization of nucleolar organizer DNA was shown for both the polytene and the polyploid nuclei of *Drosophila* (Hennig and Meer, 1971; Spear and Gall, 1973; Renkawitz and Kunz, 1975; Kalumuck *et al.*, 1990). The number of ribosomal RNA genes in the polytene nuclei is approximately four times lower than in the diploid nuclei (Spear, 1974; Grimm and Kunz, 1980). The number of the ribosomal RNA genes was estimated as 100 for salivary gland cells of larvae and as 220 for ovaries of adults (Gambarini and Meneghini, 1972). These data may be indicative of either incomplete polytenization in salivary glands or amplification in ovaries. The DNA-containing insertions are much more underrepresented in *D. melanogaster* (Kalumuck *et al.*, 1990); the same is true for *Calliphora erythrocephala* (Beckingham and Thompson, 1982).

In females of *D. melanogaster* having three nucleolar organizers, the rRNA gene number tends to become that observed in individuals having two organizers. This phenomenon was named as retrocompensation (Bashkirov *et al.*, 1984, 1986) and occurs as a stepwise process. In the F1 females with one extra NO, the rRNA content closely corresponds to the expected value calculated upon the assumption that each of the three NO loci has an additional effect. However, considerable loss of rDNA occurred during several generations. Retrocompensation induces a heritable change in rDNA content. In the course of retrocompensation the rDNA content becomes unstable, which is manifested in marked rRNA gene number variations in successive generations (Bashkirov *et al.*, 1984, 1986).

IX. DNA REPLICATION IN POLYTENE CHROMOSOMES

A. Morphological characteristics of replication in polytene chromosomes

The precursor of DNA synthesis is incorporated into only a part of the nuclei after incubating larval salivary glands of Diptera, in a medium containing [^{3}H]thymidine or after injecting [^{3}H]thymidine into larvae (Table 20). The frequencies of labeled nuclei change during development (see below). Darrow and Clever (1970) believed that replicating cells are randomly distributed throughout the gland. However, it was found that cells located in the homolo-

Table 20. Frequencies of Labeling with [^{3}H]Thymidine of Polytene Nuclei of Diptera[a]

Species	Experimental conditions	Percentage of labeled nuclei (%)	References
Anopheles atroparvus		From 2–3 to 91.5	Tiepolo and Laudani (1972)
Chironomus plumosus	Wintering larvae	Not more than 33	Ilyinskaya and Martynova (1978)
C. tentans		Not more than 16	Darrow and Clever (1970)
C. thummi		12–20	Vlassova and Kiknadze (1975)
		22–30	Gunderina *et al.* (1984a)
D. athabasca	Males	40.8–93.2	Meer (1976)
D. hydei		3.53–8.56	Danieli and Rodino (1968)
D. kikkawai		5–90	Roy and Lakhotia (1977)
D. melanogaster		60	Nash and Bell (1968)
		20.8–72	Rodman (1968b)
	10°	16.6–49	Mishra and Lakhotia (1982a,b)
	24°C	25.8–68.9	
	gt, 8 days, 24°C	11.2–23.6	
	gt, 12 days, 24°C	5.9–7.6	
D. virilis		15–62	Gubenko (1974a)
		38.9–43.9	Gubenko (1974b)
		63.7–98	Poluektova *et al.* (1981b)
Sarcophaga bullata	Foot pad cells of pupae	0–90	Roberts *et al.* (1976)
Simulium ornatipes		14.77	Bedo (1982)

[a]Salivary glands, end of third larvae instar, if not stipulated otherwise.

gous regions of the sister salivary gland of *D. kikkawai* begin to replicate synchronously (Roy and Lakhotia, 1977). This is in agreement with the data reported for *C. thummi:* five groups of cells, similar with respect to ploidy level were detectable. The cells of a nuclear group enter the S-period at about the same time (Gunderina *et al.*, 1984a).

The polytene chromosomes of single salivary glands show different types of labeling patterns after exposure to [^{3}H]thymidine: (1) no labeling of the chromosomes; they are beyond the S-period; (2) the chromosomes are uniformly labeled (the phase of continuous labeling) (Figure 172); (3) the chromosomes show discrete spots of label (the phase of discontinuous labeling). The number of such spots varies widely from 1 to 2 to several dozens (Rudkin and Woods, 1959; Sirlin, 1960; Ficq and Pavan, 1961; Sengun, 1961; Woods *et al.*, 1961; Keyl and Pelling, 1963; Plaut, 1963, 1968, 1969; Gabrusewycz-Garcia, 1964; Gay, 1964; Pavan, 1964, 1965c; Plaut and Nash, 1964; Ritossa, 1964b; Swift, 1964; Fujita, 1965; Berendes, 1966b; Pelling, 1966, 1969; Plaut *et al.*,

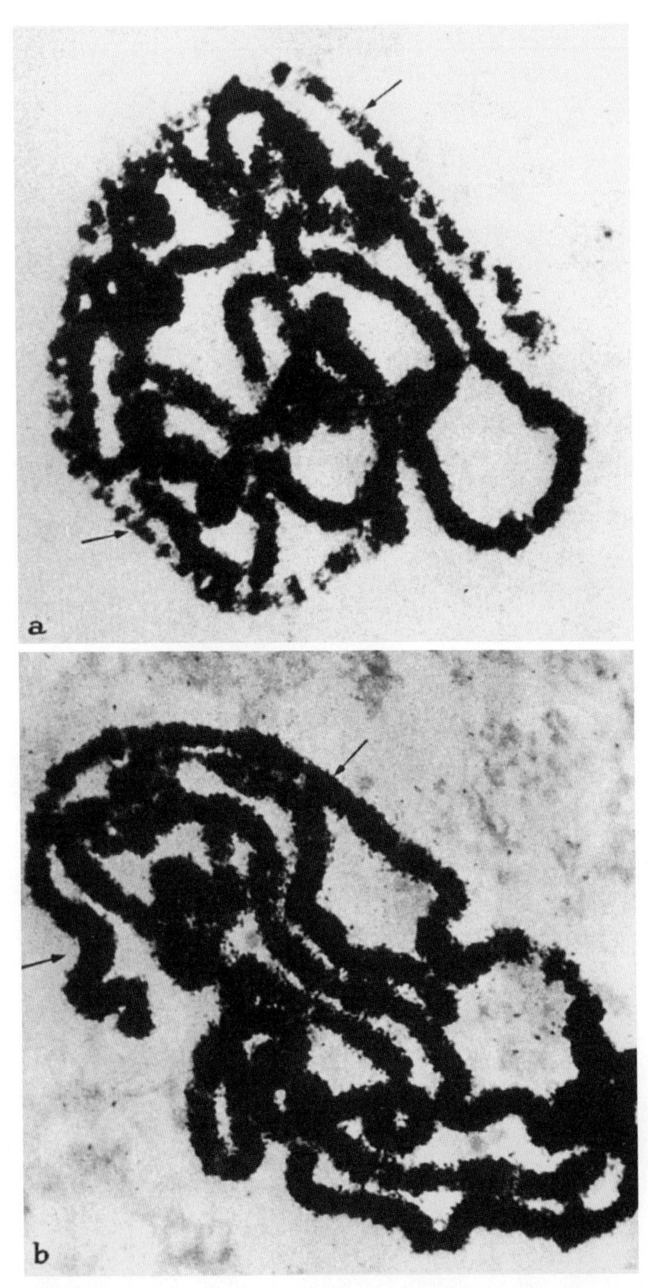

Figure 172. Samples of labeled polytene chromosome of *D. hydei* after incorporation of [^{3}H]thymidine. (a and b) Stage of continuous labeling of (a) male and (b) female chromosomes; (c and d) stage of discrete labelling of (c) male and (d) female chromosomes. Arrows point to the X chromosomes (reprinted by permission from Berendes, 1966b).

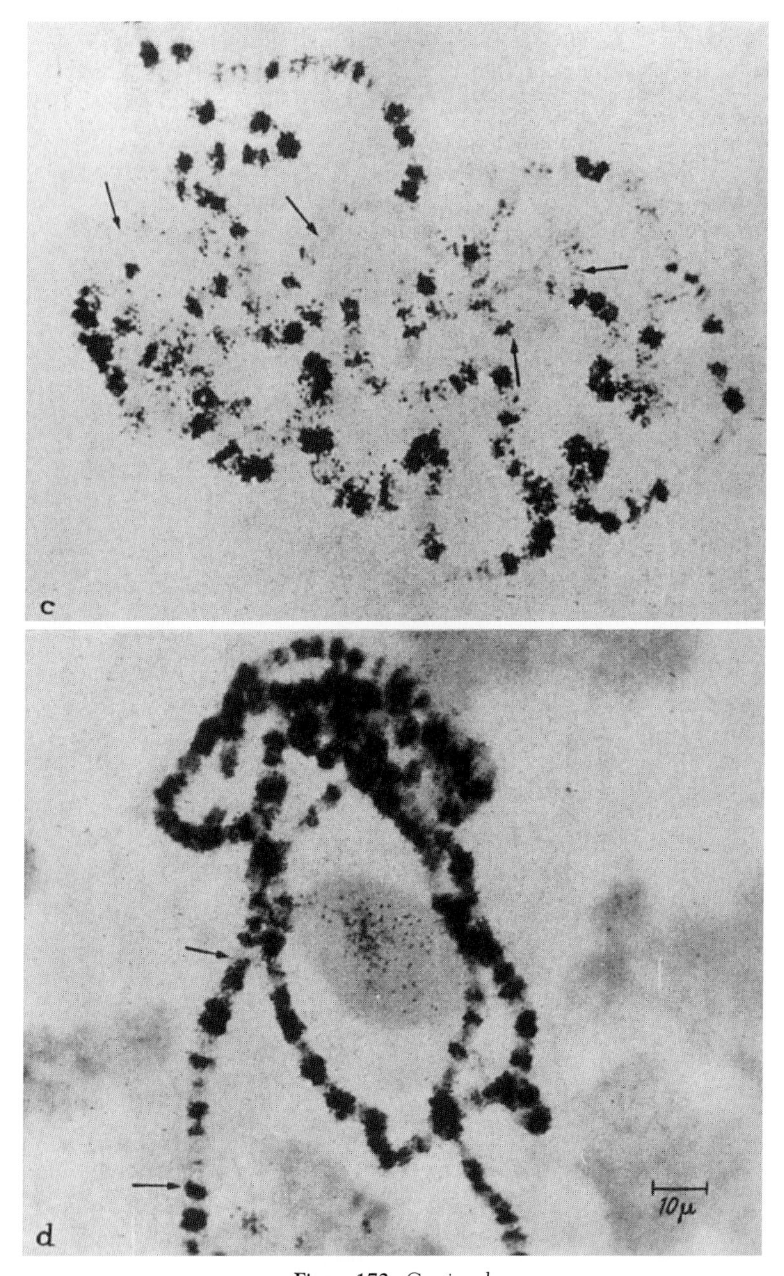

Figure 172. *Continued*

1966; Kiknadze, 1967, 1971b; Tulchin *et al.*, 1967; Howard and Plaut, 1968; Nash and Bell, 1968; Rodman, 1968b; Berendes and Beermann, 1969; Darrow and Clever, 1970; Chatterjee and Mukherjee, 1975; Roberts *et al.*, 1976). Such replication patterns are not observed in polytene chromosomes of bean plants, whose structure becomes more diffuse than that of Diptera during replication and consequently lack distinct banding patterns (Brady and Clutter, 1974).

There has been controversy as to whether polytene chromosome replication begins simultaneously at numerous sites (Keyl and Pelling, 1963; Plaut *et al.*, 1966; Mulder *et al.*, 1968; Nash and Bell, 1968; Rodman, 1968b; Lakhotia and Mukherjee, 1969; Mitra and Mukherjee, 1972; Mukherjee and Mitra, 1973). Hence, does continuous labeling occur during the initial phase or do the different regions begin synthesizing DNA asynchronously? This controversy was settled in favor of the latter assumption. It was found that there occurred complementary labeling types among the discontinuously labeling chromosomes (Figure 173): puffs, not the dense bands, were only labeled in some chromosomes; while the labeling pattern was reversed in other chromosomes (Hagele, Kalisch, 1974). The centromeric regions of the chromosomes were at the phase of continuous labeling; that is, they were not labeled during the replication cycle in C. *thummi* (Hagele, 1970). The heterochromatic regions of the chromosomes are late replicating (Keyl and Pelling, 1963), and this phase could be regarded as the initial in DNA synthesis. It was demonstrated by Plaut *et al.*, (1966) and Nash and Bell (1968) that the patterns with a large number of [^{3}H]thymidine-labeled sites included patterns with a smaller number of such sites, and that all the patterns conformed to a highly ordered sequence.

Based on the experimental data, the sequential [^{3}H]labeling of polytene

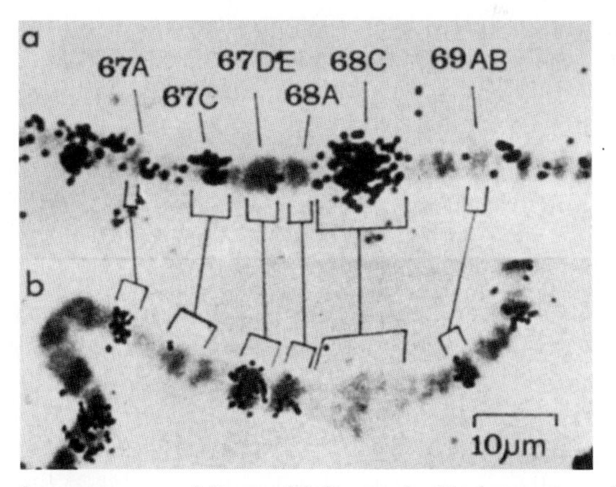

Figure 173. Complementary types of discrete labeling in the 3L chromosome of *D. melanogaster*. (a) [^{3}H]uridine incorporation; (b) [^{3}H]thymidine incorporation (reprinted by permission from Hagele and Kalisch, 1974).

chromosomes is as follows: (1) The phase of discontinuous labeling: (a) the thinnest bands, the interbands, and certain puffs are only labeled (Figure 174); and (b) all the puffs and the interbands are labeled. (2) The phase of continuous labeling: (a) [^{3}H]thymidine is incorporated into all the chromosome regions, with the exception of the chromocenters; and (b) continuous labeling of all the chromosome regions. (3) The phase of discontinuous labeling: (a) certain puffs and interbands are not labeled; (b) many puffs and interbands are not labeled; and (c) toward the end of the S-period only certain regions, the chromocenter in *Drosophila,* or heterochromatin in *Chironomus*, still incorporate [^{3}H]thymidine.

One of the above labeling types is usually observed after incorporation of [^{3}H]thymidine into salivary gland cells (Pavan, 1959b, 1965c; Keyl and Pelling, 1963; Plaut, 1963; Gabrusewycz-Garcia, 1964; Plaut *et al.*, 1966; Tulchin *et al.*, 1967; Arcos-Teran and Beermann, 1968; Nash and Bell, 1968; Rodman, 1968b; Hagele, 1970, 1972, 1973, 1976; Lakhotia, 1970; Simoes, 1970; Chatterjee and Mukherjee, 1971b, 1973, 1975; Arcos-Teran, 1972; Tiepolo and Laudani, 1972; Mukherjee and Mitra, 1973; Gubenko, 1974a,b, 1976a–d; Hess and Hauschteck-Jungen, 1974; Tiepolo *et al.*, 1974; Hagele and Kalich, 1974, 1978, 1980; Chatterjee *et al.*, 1976; Kalisch and Hagele, 1976, 1977; Meer, 1976; Mukherjee *et al.*, 1977; Sokoloff and Zacharias, 1977; Zhimulev and Kulichkov, 1977; Ilyinskaya and Martynova, 1978; Roy and Lakhotia, 1979; Majumdar and Mukherjee, 1980; Mukherjee *et al.*, 1980; Sinha and Lakhotia, 1980; Poluektova *et al.*, 1981b; Simoes *et al.*, 1981, 1982b; Duttagupta *et al.*, 1984b; Mutsuddi *et al.*, 1984; Sinha *et al.*, 1987).

The behavior of the X chromosome in males of *Drosophila* is unusual: it begins replicating some time earlier (Hagele, 1973) and terminates much later than the autosomes. This has been demonstrated for *D. hydei* (Berendes, 1966b; Mulder *et al.*, 1968), *D. melanogaster* (Rodman, 1968b; Lakhotia and Mukherjee, 1969, 1970b, 1972; Hagele and Kalisch, 1974; Ananiev and Gvozdev, 1975; Mukherjee *et al.*, 1977; Duttagupta *et al.*, 1984c), *D. ananassae* (Duttagupta *et al.*, 1973), *D. virilis* (Gubenko, 1976b), *D. pseudoobscura* (Mukherjee and Chatterjee, 1975), *D. persimilis* (Duttagupta *et al.*, 1984a), and *D. azteca* × *D. athabasca* hybrids (Meer, 1976). Thus, in salivary glands of *D. hydei,* the male X chromosome already shows the discontinuous labeling pattern never observed in females, whose autosomes are at the stage of continuous labeling (Berendes, 1966b). In the discontinuously labeled nuclei, the number of replicating regions in the male X chromosome is always smaller than in the female X chromosome (Berendes, 1966b; Mukherjee *et al.*, 1980)—twofold smaller, for instance, in *D. virilis* (Gubenko, 1976b). The same regions of the X chromosome were referred as late replicating in males and females (Berendes, 1966b). The earlier completion of replication in the male X chromosome was associated with its looser packing providing the occurrence of dosage compensation (Duttagupta *et al.*, 1984d; Mutsuddi *et al.*, 1985) (see also Section III,C,2).

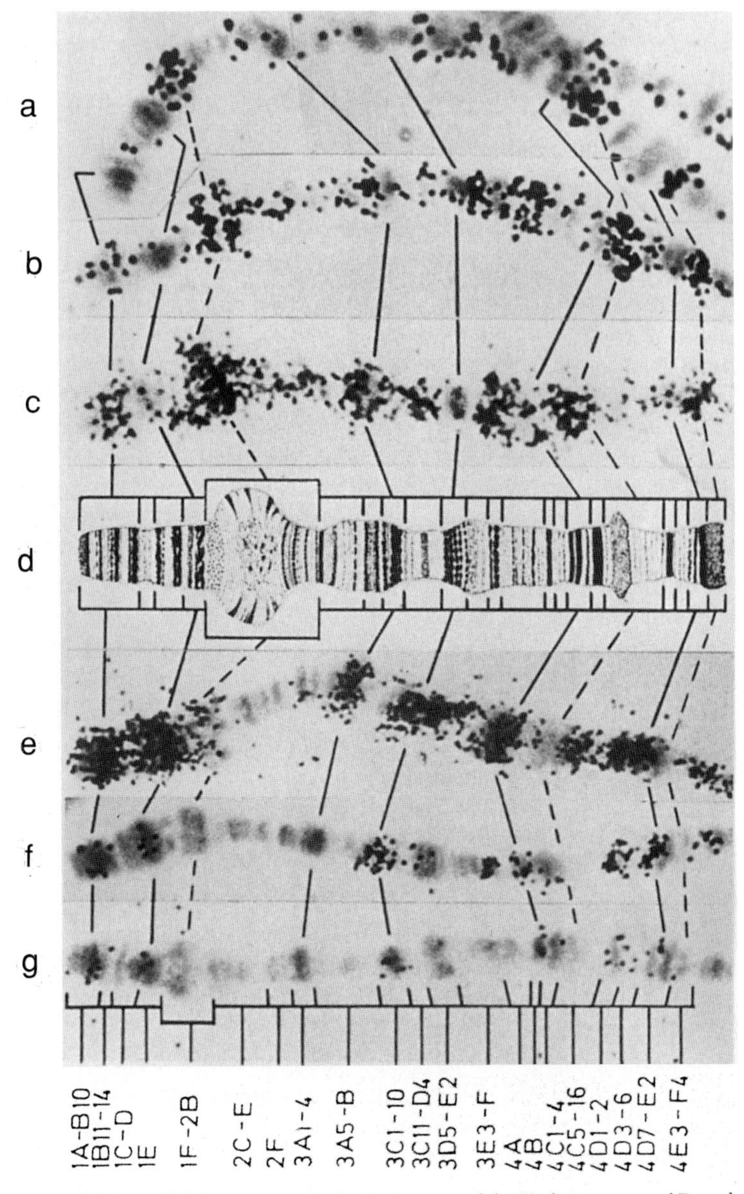

Figure 174. Changes of labeling pattern in the 1–4 region of the X chromosome of *D. melanogaster*. (a–c) Early stage of discontinuous labeling; (d) the map of the X chromosome is given instead of the stage of continuous labeling; (e–g) final stages of discrete labeling (reprinted by permission from Kalisch and Hagele, 1976).

In *D. miranda*, fragments not showing the looser packing specific to the male X are not early replicating in the compound chromosomes that resulted from the fusion of the X and the D elements (X_1) and of the X and the C elements (X_2) (Das *et al.*, 1982). However, the degree of diffusion is not always correlated with replication rate. The 63E1-5 region (on chromosome 3) in *D. melanogaster* is late replicating in larvae when represented by a group of bands in a compact state. A giant puff, which is the last to replicate, is formed in this region in zero-hour prepupae (Hagele and Kalisch, 1980).

There were no appreciable differences in DNA replication patterns in the X chromosomes of *Drosophila* in triploid males (3X3A) and triploid intersexes (2X3A) when compared to diploid males (1X2A) and females (2X2A) (Ananiev and Gvozdev, 1974, 1975). The same (nondifference) was observed for superfemales (3X2A) (Mutsuddi *et al.*, 1987). There were no differences in the frequencies of labeled regions between trisomics for arms 2L and 3L and disomics for arms 2R and 3R (Ghosh and Mukherjee, 1986). In segmental aneuploids of *Drosophila*, the nuclei with 1.05 X, 1.95 X, 2.05 X chromosomes showed no deviations from the normal replication patterns in the X chromosome were found (Duttagupta *et al.*, 1985).

In *Phryne cincta*, DNA replication in the male X chromosome and the autosome terminates at the same time (Sokoloff and Zacharias, 1977). The time when single chromosome regions cease replicating their DNA is mainly invariable (Keyl and Pelling, 1963; Plaut *et al.*, 1966; Arcos-Teran and Beermann, 1968; Rudkin, 1972, 1973; Chatterjee *et al.*, 1976; Zhimulev and Kulichkov, 1977; Lakhotia and Roy, 1979; Zhimulev *et al.*, 1982b); this was also observed for closely related species (Gubenko, 1976b).

Many factors affect the parameters of replication in polytene chromosomes. An attempt was made to determine the degree of autonomy of replication in single regions of the chromosomes by using chromosomal rearrangements. The pertinent results were contraversial. In females of *Drosophila* carrying *T(1;3)ras*^{V}, a duplication in which region 9E to 13C of the X is inserted at 81F of chromosome 3, the labeling patterns were identical in homologous regions of the X and the duplication at the phase of discontinuous labeling (Barr *et al.*, 1968).

The homologous regions of the X chromosome and the *Dp(1;f)1337*, *Dp(1;f)856*, and *Dp(1;4)w*^{m5} duplications replicate synchronously. This is in agreement with the results obtained with the Y chromosome material translocated to euchromatin *T(Y;3)99A* (Valencia and Plaut, 1968). In contrast, there are many cases in which changes occur in replication pattern in the translocated material of the polytene chromosomes. Thus, label is present in material of *Dp(1;f)AM* (translocating the 1A-1C3, 8D1-9C1, and possibly the 16E2-3 regions to the centromere), when the X chromosome regions have terminated replicating their DNA (Bender *et al.*, 1971). The duplicated material in *T(1;3)w*^{vco} terminates replicating later (Majumdar and Mukherjee, 1979). This is presumably a case of position effect. Data concerning the profound changes in replication patterns caused by position effect have been previously considered (Zhimulev, 1993, 1997). Differences were found both in the labeling frequencies and region sets, as well as in incorporation rate, between the material of the *Dp(1;1)Gr* tandem duplication and the X chromosome (Kalisch and Hagele, 1973). Considerable changes in the terminal phases of discontinuous labeling were observed in strains with the *In(1)J1* and *In(1)dl-49* inversions (Chatterjee, 1985b). Study of the replication pattern in the 56E-60F region of chromosome 2R in three *Drosophila* strains with deletions demonstrated no changes in regions remote from the deletions. In contrast, there were significant changes characterized by transition from the late to the early type of replication in regions adjacent to the breakpoints (Chatterjee *et al.*, 1983).

The genotype has an effect on the course of replication. In *D. virilis* × *D. littoralis* hybrids, the replication chronology of the region of the sixth chromosome of *D. littoralis* was defined (Evgeniev *et al.*, 1971, 1972). Experiments on lines with synthetic karyotypes containing various combinations of *D. littoralis* homozygous chromosomes and hybrid chromosome showed that a dominant factor altering the chronology of the replication of the sixth chromosome of *D. littoralis* is located in the sixth chromosome (Evgen'ev and Gubenko, 1977a,b.

In the sex transforming mutants—*sx*, *dsx*, dsx^D, *ix*, and *tra-2*—the X-chromosomal replication pattern is not depended on the mutation, but rather on the degree of chromatin condensation. The X chromosomes of the *sx*/Y mutants and the male intersexes (XY;*dsx*/*dsx*) are early replicating, while the X chromosomes of pseudomales (XX;*tra-2*/*tra-2*), and of the three types of female intersexes, replicate synchronously with the autosomes. This is concurrent with the reproducible deviations in the replication patterns of distinct regions of the X chromosome and the autosomes (Duttagupta *et al.*, 1984c; Mutsuddi *et al.*, 1985).

In an $In(1)BM^2$ strain of *D. melanogaster* the labeling frequencies in females at almost all the relevant X-chromosomal sites (with the except of 6DF and 7AC) were close to normal; the labeling frequencies in all the regions, except the late replicating ones (1A, 3C, 11A, and 12EF), were lower in males than in homozygous females (Ghosh and Mukherjee, 1985).

1. Individual variability in replication termination. As indicated above, although all the chromosomes terminate replicating in a highly ordered temporal sequence, each chromosome region may be an exception to the rule. For example, 3C is the last to replicate in the 1A-3C region of the X chromosome of *Drosophila;* however, 1AB is the last to replicate in one chromosome (Plaut *et al.*, 1966). The 11A region is the last to replicate on the X chromosome, although certain nuclei were detected in which 3C was the last replicating region (Arcos-Teran and Beermann, 1968; Arcos-Teran, 1972). In the asynaptic regions of the third chromosome of *Drosophila* (Ha-

gele and Kalisch, 1978), or *Simulium ornatipes* (Bedo, 1982), replication in individual regions may be occasionally asymmetric; that is, in only one of the homologs. Individual variability was described in the salivary gland chromosomes of *D. melanogaster* (Mishra and Lakhotia, 1982b; Zhimulev *et al.*, 1982b), *D. nasuta* (Lakhotia and Tiwari, 1984), *S. ornatipes* (Bedo, 1982), and in the chromosomes of ovarian nurse cells of the mosquito *Anopheles stephensi* (Redfern, 1981b).

2. Tissue variability. The data on tissue differences in termination rate are to a certain extent controversial. In *A. stephensi*, the [^{3}H]thymidine labeling patterns were essentially similar in the polytene chromosomes of salivary gland and ovarian nurse cells at the end of the S-period (Redfern, 1981b). In the *otu*^{3} mutant of *D. melanogaster*, late replicating sites were identified at 21DE, 25A, 25EF, 32F, 33AC, 34EF, 35CE, 36CD, and 39–40 in chromosome 2L of ovarian nurse cells (Sinha *et al.*, 1987); that is, in the same regions as in salivary gland polytene chromosomes (Zhimulev *et al.*, 1982b).

   Comparisons of this feature in the fragments of the X and second chromosomes of salivary glands and gastric caeca in *D. hydei* revealed differences in the temporal sequence of replication completion in 8 of the 62 regions of the X chromosome, and in 11 of the 55 of the second chromosome (Lakhotia and Tiwari, 1984; Tiwari and Lakhotia, 1984). Similar comparisons of the second and the third chromosomes of *D. nasuta* disclosed considerable differences in the temporal sequence of replication termination in between chromosomes.
3. Interspecific differences. The amount of the late replicating material has been compared among closely related species. The differences are particularly conspicuous in hybrids when one chromosome has the late replicating material and the other, derived from another species, does not. The relevant hybrids are *Anopheles atroparvus* × *A. labranhia* (Tiepolo *et al.*, 1974), *D. virilis* × *D. littoralis* (Evgen'ev and Gubenko, 1977a,b), and *C. t. thummi* × *C. t. piger* (Keyl, 1965, 1966; Hagele, 1970, 1976). In the latter, variations in the late replicating material and DNA amount coincide.
4. Ecological factors. Larval life was prolonged in two strains of *D. melanogaster* (*gt* and Oregon-R wild-type) reared at the low temperature of 10°C. The resulting longer duration of the S-period is due to a considerable asynchrony in the initiation and termination of replication in different sites of the chromosomes. However, there were no differences in the replication sites and the degree of delay in replication completion between larvae raised at the low temperatures and more standard one (24°C: Mishra and Lakhotia, 1982a,b).
5. Effect of chemical compounds. Certain such compounds affect the course of replication in replicating salivary gland cells. In larvae of *Drosophila* fed mitomycin C, the number of [^{3}H]thymidine-incorporating nuclei decreases by 2–3 times, mostly of those at the continuous and the initial phase of

discontinuous labeling (Mukherjee and Mitra, 1973; Banerjee and Chinya, 1982). Novobiocin, a sugar-containing antibiotic, is a strong inhibitor of DNA and RNA syntheses in bacteria. In larvae of *Drosophila* fed novobiocin, the number of labeled nuclei at the phase of terminal discontinuous labeling increased to 77.47% (their number was 27.62% in the controls). The frequencies of the nuclei at the initial phase of continuous labeling sharply drops; that is, the antibiotic synchronizes the cells at the late stage of the S-period (Achary and Duttagupta, 1988).

In larvae treated with 5-fluorodeoxyuridine, polytene chromosomes accumulate at the phase of continuous labeling; that is, replication is blocked in the middle part of the S-period (Achary *et al.*, 1981, 1984). In hydroxyurea-fed larvae, nuclei accumulate at the phase of continuous labeling, so that replication is also delayed in the middle stage of the S-period (Achary and Das, 1986); hydroxyurea simulates the action of 5-fluorodeoxyuridine (Achary *et al.*, 1981), aphidocolin and ricin (Duttagupta and Banerjee, 1984). When salivary glands of *D. virilis* larvae are treated with cycloheximide, the number of discontinuously labeled nuclei increases and that of the continuously labeled nuclei decreases (Gubenko, 1974b). This is in agreement with the results obtained with the chromosomes of *C. thummi* (Vlassova and Kiknadze, 1975). Puromycin, an inhibitor of protein synthesis, has a similar effect (Mukherjee *et al.*, 1977). The inhibitors of RNA synthesis actinomycin D and α-amanitin block the initiation of replication (Mukherjee *et al.*, 1977; Chaterjee, 1985c). α-Methyl-DL-methionine sharply decreases the number of labeled nuclei at the initial phase of discontinuous labeling (Chaudhuri and Mukherjee, 1984).

6. Hormonal control. In the early studies on replication in polytene chromosomes it was noted that, throughout the third larval instar, a progressive decrease in the frequencies of labeled nuclei occurred (Danieli and Rodino, 1967; Rodman, 1968b). Thus, the percentage of [^{3}H]thymidine-labeled nuclei was 72% at 96–110 hr postdeposition; 43.4% at 110–116 hr postdeposition (approximately at the PS1); 20.8% at 116–122 hr postdeposition (the PS3-5); 11.1% in 0- to 0.5-hr prepupae; 3.5% in 1- to 2-hr prepupae; and 2.6% in 2- to 7-hr prepupae (Rodman, 1968b). In *D. virilis*, the labeling index was 48.3% in third-instar larvae, 10.4% in zero-hour prepupae, and 4.4% in orange prepupae (Gubenko, 1974a).

In *Rhynchosciara angelae* and *R.* sp., just prior synthesis of the cocoon, the number of cells synthesizing DNA decreases to almost zero in all tissues of salivary glands, intestine, gastric caeca, Malpighian tubules, and fat body; thereafter the number increases again (Mattingly and Parker, 1968b; Simoes, 1970).

A relation was established between the decrease in the labeling index and the larval molting cycle. In *D. hydei*, the labeling index was 67.74% before the second molt, it dropped to 22–27% at the beginning of the molt, it rose again to 73.96% after the molt, and declined to minimal values to the late third instar (3.53–8.56%) (Danieli and Rodino, 1968). A sharp drop in the labeling index was observed in *C. tentans* (Figure 175) (Darrow and Clever, 1970), *Anopheles atroparvus* (Tiepolo and Laudani, 1972), and *C. thummi* (Kroeger *et al.*, 1973a; Vlassova and Kiknadze, 1975; Gunderina and Sherudilo, 1983).

The question was raised as to whether the decrease in [^{3}H]thymidine incorporation follows the increase in ecdysterone titer (Danieli and Rodino, 1967; Rodman, 1968b; Rudkin, 1972, 1973). The results of direct experiments did not support the suggestion that the decrease in this indicator of replication is under hormonal control. Darrow and Clever (1970) observed no decrease in the frequency of labeled nuclei after injection of ecdysterone, whereas Crouse (1968) observed even an increase in the number of nuclei showing [^{3}H]-incorporation in salivary glands. The frequency of labeled nuclei neither increased nor decreased within 18 hr of incubation of salivary glands of *D. hydei* in an ecdysterone-containing medium (Berendes and de Boer, in: Rudkin, 1973).

Rudkin (Rudkin, 1973; Rudkin and Dunkan, 1984) has expressed the view that nonhormonal factors of the larval internal milieu are involved in the halt of replication cycles at the end of larval development. For example, removal of salivary glands from their normal milieu by transplantation or their prolonged incubation (Staub, 1969; Berendes and Holt, 1965) permits the cells to undergo additional replication rounds. The larval period is extended in homozygotes for the *gt* and *tu-h [l(3)tl]* mutations, and new polytenization cycles are initiated at that time. The chromosomes are thinner in the *l(2)gl* and *translucida* mutants. In Rudkin's opinion, development and polytenization come to a halt under the effect of the same factors (Rudkin, 1973).

The action of juvenile hormone is in general opposite to that of ecdysterone (see Section VI,A,1). After incubation of the salivary glands of *D. nasuta* in a solution containing the juvenile hormone analog (ZR515), the labeling index of the nuclei increased from 5–30% in the control to 50–80% in the treated salivary glands (Sinha and Lakhotia, 1983). Patterns of changes in [^{3}H]thymidine labeling frequencies were observed during the formation of polytene chromosomes in the foot pad and pupal cells of *S. bullata*. The number of labeled nuclei was maximal 4 days after puparium formation; then it declined (Roberts *et al.*, 1976).

B. Correlation between chromomeres and replication units

DNA is replicated during the S-period of the cell cycle. Certain factors accumulate in the cell during the preceding G1-period, transfer it to the S-phase, and cause the initiation of replication. The duration of the S-phase is only 3.6 min during the cleavage divisions of the embryonic cells of *Drosophila;* it is as

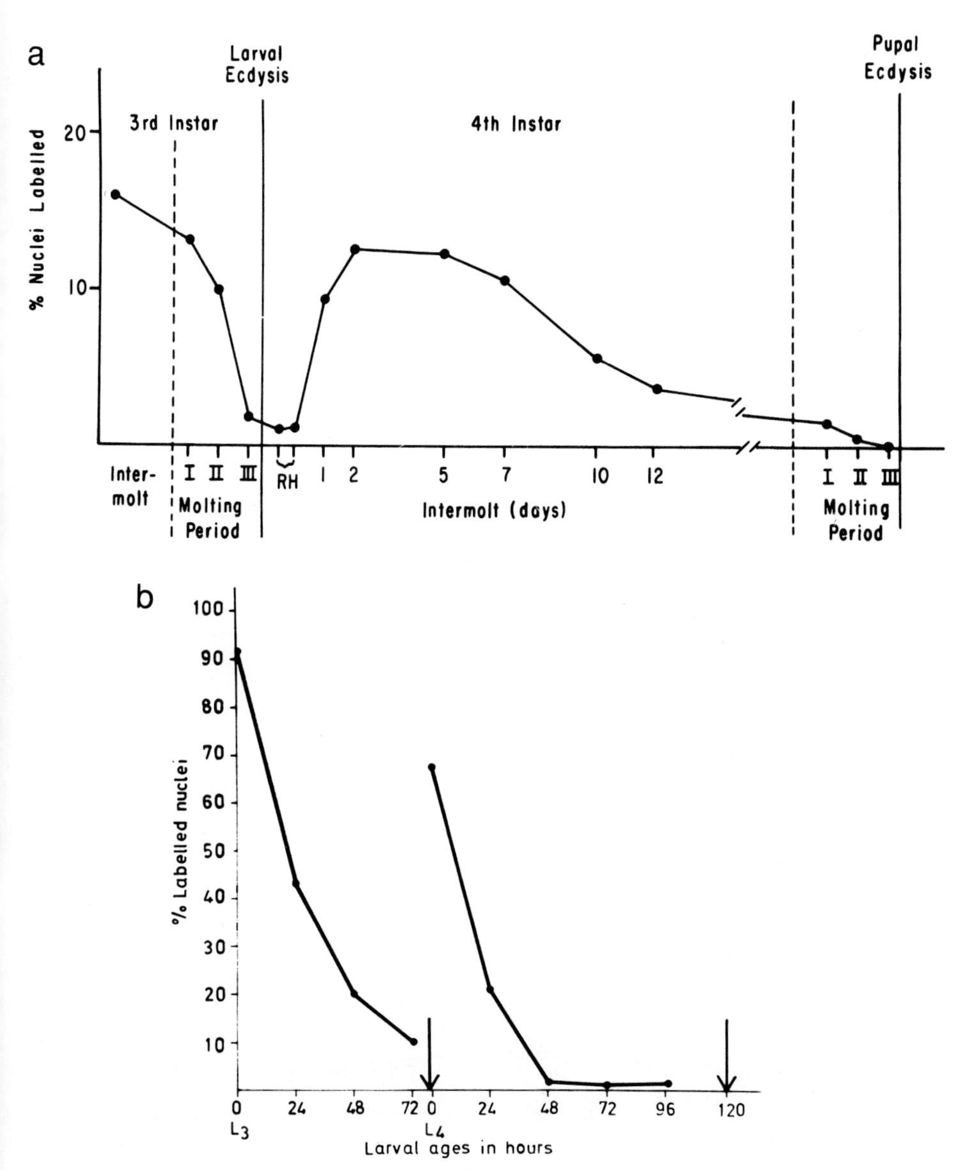

Figure 175. Changes in the occurrence frequencies of [^{3}H] thymidine nuclei in salivary gland cells of (a) *C. tentans* and (b) *Anopheles atroparvus* during development (a, from Darrow and Clever, 1970; b, reprinted by permission from Tiepolo and Laudani, 1972). Abscissa, age (in hours); L3, third instar; L4, fourth instar; arrow points to time of molt. Ordinate, proportion of labeled nuclei (percentage).

Table 21. Duration of the S-period in the Cells of *Dipteran* Insects

Species	Cell type	Duration of the S-period (hr)	References
Aedes aegypti	Diploid cells	8	Marchi and Rai (1978)
A. dorsalis	—	7	Mukherjee and Rees (1969)
Anopheles atroparvus	—	12	Fraccaro *et al.* (1976)
Chironomus thummi	Salivary glands, ploidy degree 2^8C	10	Gunderina and Seraya (1992)
	Ploidy degree, 2^9C	13	Gunderina and Seraya (1992)
	Ploidy degree, 2^{10}C	17	Gunderina *et al.* (1984b)
	Ploidy degree, 2^{11}C	22	Gunderina *et al.* (1984b)
	Imaginal disks	10	Gunderina (1992)
Drosophila melanogaster	Cleaving nuclei of embryos	Interphase takes 3.6 min	Rabinowitz (1941)
	Diploid cells	10.0	Dolfini *et al.* (1970)
	Diploid cells of culture	10	Ananiev *et al.* (1977)
	Diploid cells	8	Kaji and Ishioda (1981)
D. virilis	Diploid cells of brain	11.9 ± 4.3	Steinemann (1980)

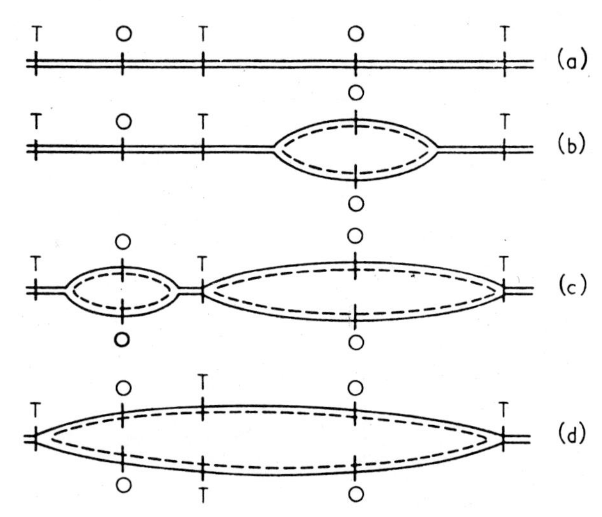

Figure 176. A scheme of the organization of DNA replication. Each pair of horizontal lines represents a part of a DNA molecule. Continuous line depicts the parental DNA strand, the discontinuous represents the newly synthesized strand; (O and T) the regions of initiation (0) or termination (T) of replication. (a) The stage before the start of replication; (b) replication in the right replicon; (c) start of replication in the left replicon, continuation of the process in the right; and (d) termination of replication in both replicons. Sister strands of DNA are connected in the termination site (from Huberman and Riggs, 1968).

long as 22-hr in salivary gland cells of *Chironomus* (Table 21). Within the S-period in *D. virilis*, DNA replication is exclusively restricted to euchromatin during the first hour; this is followed by 8 hr of replication in both euchromatin and heterochromatin; DNA is replicated only in α-heterochromatin during the last 3 hr. Thus, cuchromatin requires 9 hr and α-heterochromatin 11 hr to replicate (Steinemann, 1980). In *D. melanogaster*, euchromatin replicates only during the first 4 hr of the S-period (Ananiev *et al.*, 1977).

DNA replication proceeds at the same time in many chromosome regions. Each region has an initiation point (Figure 176) and a replicon, which provides the convergence and fusion of the two replicating forks (see Lewin, 1994, for details). The rate at which the replication fork moves along the molecule varies widely from one organism to another (Table 22). The rate is very high in *E. coli*, *Triturus*, and *Xenopus* and much lower in mammals and Diptera.

An approach to analysis of replicon size is based on the combined techniques of pulse-labeling and DNA autoradiography (Huberman and Riggs, 1968). The cells are labeled with [^{3}H]thymidine; the labeled DNA is released; a part of it is made to adhere to Millipore filters; the material then stretches out to form fibers. Tracks of labeled regions are seen in DNA fiber autoradiograms

Table 22. Movement rate of DNA replication fork

Species	Cell type	Movement rate of fork	References
Chinese hamster	Cell culture	2.5 μm/min	Huberman and Riggs (1968)
E. coli		21 μm/min	Review: Kraevsky and Kukhanova (1986)
Saccharomyces cerevisiae		0.7 μm/min	Kraevsky and Kukhanova (1986)
Triturus		20 μm/min	Callan (1972)
Xenopus		9 μm/min	Callan (1972)
D. melanogaster	Cleaving nuclei of embryos	0.9 μm/min (2.6 kb/min)	Blumenthal *et al.* (1974)
	Diploid cells of culture	0.5 μm/min (0.63 kb/min)	Ananiev *et al.* (1977)
	Salivary glands of larvae	0.1–0.8, on the average 0.4 μm/min	Yurov (1982)
D. nasuta	Salivary glands of larvae	Replicon type I: 0.95 μm/min (from 0.89 to 1.77) Replicon type II: 0.07 μm/min	Lakhotia and Sinha (1983)
	Ganglia and imaginal disks	I: 0.44–0.52 μm/min II: 0.07–0.08 μm/min	Lakhotia and Tiwari (1985)
D. virilis	Salivary glands of larvae	0.1 μm/min	Steinemann (1981b)
	Brain cells	0.35 μm/min	Steinemann (1981a)
Rhynchosciara angelae	Salivary gland cells	0.025 μm/min	Cordeiro and Meneghini (1973)

obtained after labeling (Figure 177). Because DNA replication proceeds in opposite directions at adjacent growing points, the distance between the centers of the labeled bands (between the origin replication) can be accepted as the size of the replicating unit. The data on replicon size (Table 23) vary greatly.

Table 23. Distance between neighboring regions of DNA replication initiation

Species	Cell type	Distance between initiation regions	References
Chinese hamster	Cell culture	50 μm (from 15 to 120)	Huberman and Riggs (1968)
Xenopus	—	57.5 μm (from 18 to 128)	Callan (1972)
Chironomus tentans	Salivary gland cells of larvae	2–3 μm of 14 kb or $4.75 \cdot 10^6$ Da	Lonn (1980a)
	Salivary gland cells at the "red head" stage and at the end of the fourth instar	Replicons of the same size	Lonn (1980b)
		14 kb	Lonn (1981a)
		Two populations of replicons: $5.5 \cdot 10^6$ Da and $3.3 \cdot 10^6$ Da	Lonn (1981b)
Cochliomyia hominivorax	Early embroys	2.3 μm (from 0.3 to 8.2 μm)	Lee and Pavan (1974)
D. melanogaster		3.87 μm (from 1.2 to 9.7 μm)	Wolstenholme (1973)
	Nuclei at the cleavage stage	9.7 μm	Blumenthal *et al.* (1974); Kriegstein and Hogness (1974)
	Culture cells	70 μm (210 kb) (from 40 to 120 μm)	Ananiev *et al.* (1977)
	Blastoderm	2.32 μm (10.6 kb) (from 0.5 to 5.0 μm)	McKnight and Miller (1977)
	Salivary gland cells of larvae	30 μm (from 5 to 100 μm)	Yurov (1982)
D. nasuta	Salivary gland cells of larvae	Replicons of type I: 64 μm (from 61.51 to 196.22 μm) Replicons of type II: 20.31 μm	Lakhotia and Sinha (1983)
	Ganglia cells	Type I: 22.8 μm Type II: 18.3 μm	Lakhotia and Tiwari (1985)
	Cells of imaginal disks	Type I: 19.5 μm Type II: 18.5 μm	
D. virilis	Brain cells	35.6 μm (from 2 to 238 μm)	Steinemann (1981a)
	Salivary gland cells	46.7 μm (from 5 to 203 μm)	Steinemann (1981b)

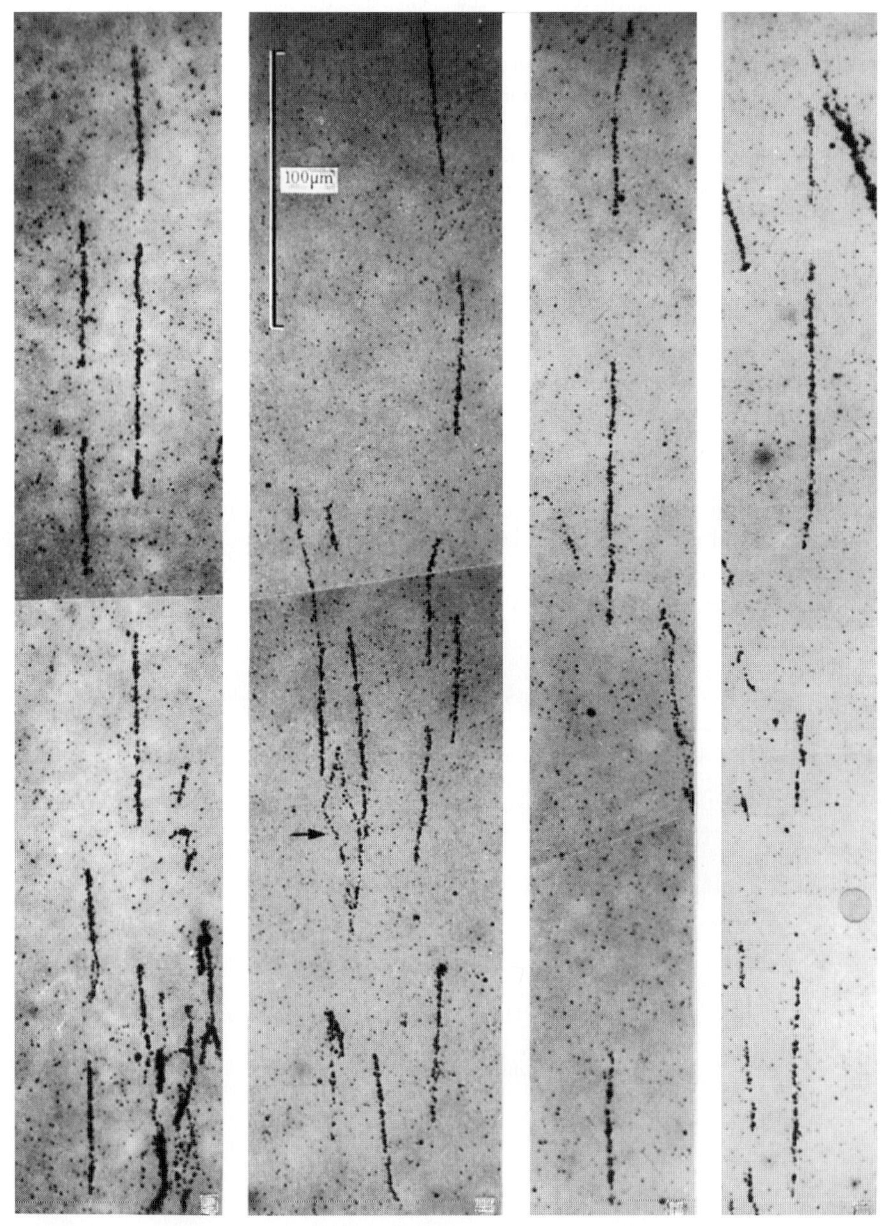

Figure 177. A photograph of replicating DNA from cell culture of *Xenopus laevis* (from Callan, 1972).

This may be because there are at least two distinct types of replicons in the structure of the genome (Lakhotia and Sinha, 1983; Duttagupta and Roy, 1984; Lakhotia and Tiwari, 1985) and also because the number of active initiation points for replication can vary during differentiation (Callan, 1972, 1974).

The idea that the bands of polytene chromosomes are replication units (see Section II,B) was not supported by the experimental evidence (Table 24). The data on polytene chromosomes of *Chironomus* rather favor the idea that there may be a correspondence between the chromomere and the replicon. The duration of DNA synthesis is longest in the bands with high DNA content in *C. t. thummi* in comparison with *C. t. piger*. This suggested that the longest replicons are located in these bands (Hagele, 1971). It is known that DNA content in certain bands of the polytene chromosomes of *C. t. thummi* is 2, 4, 8, or 16 times higher than in the (interspecifically) "homeologous" bands of *C. t. piger*. The bands with doubled DNA content show a twofold higher frequency of [^{3}H]thymidine labeling in hybrids. Although there is no increase in the labeling index in the bands of *C. t. thummi* with 4-, 8-, and 16-fold higher DNA amounts (Hagele, 1976), there may be a correlation between the sizes of the bands and the replicons. However, evidence was presented indicating that the extent of delay in replication completion is not correlated with band size in *D. melanogaster* (Ananiev and Barsky 1975).

Table 24. Ratio of the Size of Replicons to Those of Chromomeres in Polytene Chromosomes

Species	Mean chromomer size	Mean replicon size	Ratio of the number of chromomeres to that of replicons	References
Chironomus tentans			1 band per 5–7.5 replicons	Lonn (1981a)
	30 μm	2–3 μm	10–15 replicons in a chromomere	Lonn (1980a,b)
D. melanogaster			1 band per 3–8 replicons	Blumenthal *et al.* (1974)
	30 kb	210 kb	7 bands per 1 replicon	Ananiev *et al.* (1977)
	1.5–33 μm, in average 10 μm	30 μm	2–3 bands per 1 replicon	Yurov (1982)
D. nasuta			Sizes of replicon I and II corresponds to sizes of 6 and 2 bands, respectively	Lakhotia and Sinha (1983)
D. virilis	13.6 μm	46.7 μm	1 band per 0.29 replicon	Steinemann (1981b)

Acknowledgments

I express deep gratitude to friends and colleagues for assisting in the preparation of this review: particularly I thank E. S. Belyaeva and V. F. Semeshin for discussing the various sections.

I thank very much M. L. Balasov, V. E. Grafodatskaya, V. K. Vasil'ieva, E. A. Dolbak, E. B. Kokoza, N. Yu. Kuznetsova, N. I. Mal'ceva, V. A. Prasolov, and I. P. Selivanova for their invaluable assistance in preparation of the manuscript. I am also grateful to J.-A. Lepesant (Paris, France), M. Ashburner (Cambridge, England), J. Modolell (Madrid, Spain), E. B. Kokoza, I. V. Makunin, O. M. Mazina, T. Yu. Kozlova, A. Gordadze, S. Zakharkin, and V. F. Semeshin for providing me with many reprints and copies of publications; to V. F. Semeshin for supplying me with unpublished photographs, C. S. Thummel (Salt Lake City, USA), H. Sass (Mainz, FRG), and J. Richards (Strasburg, France) for sending me original photographs.

This work was supported by a grant from the Priority Directions in Genetics Program (Russia), the Russian Fund of Basic Research (Grants 95-04-12695 and 96-04-50142), and the INTAS Program.

References

Abbott, M. K., and Kaufman, T. C. (1986). The relationship between the functional complexity and the molecular organization of the *Antennapedia* locus of *Drosophila melanogaster*. *Genetics* **114,** 919–942.

Abdelhay, E., and Miranda, M. (1975). Isolation of active genes from *Rhynchosciara americana*. *An. Acad. Brasil. Cienc.* **47,** 563–564.

Abraham, I., and Lucchesi, J. C. (1974). Dosage compensation of genes of the left and right arms of the X chromosome of *Drosophila pseudoobscura* and *Drosophila willistoni*. *Genetics* **78,** 1119–1126.

Achary, P. M., and Das, C. C. (1986). Evaluation of hydroxyurea as an *in vivo* synchronizing agent of DNA replication on the polytene chromosomes of *Drosophila*. *Dros. Inform. Serv.* **63,** 11.

Achary, P. M., and Duttagupta, A. K. (1988). Effect of novobiocin on the DNA replication of polytene chromosomes of *Drosophila melanogaster in vivo*. *Dros. Inform. Serv.* **67,** 3.

Achary, P. M. R., Majumdar, K., Duttagupta, A., and Mukherjee, A. S. (1981). Replication of DNA on larval salivary glands of *Drosophila* after *in vivo* synchronization. *Chromosoma* **82,** 505–514.

Achary, P. M. R., Dutta, P. K., and Duttagupta, A. K. (1984). Replicon properties of *Drosophila* genome. I. DNA fibre autoradiography of salivary gland cells held at the mid-part of S-phase. *Dros. Inform. Serv.* **60,** 43–44.

Acharyya, M., and Chatterjee, R. N. (1997). *In situ* transcription analysis of chromatin template activity of the X chromosomes of sex-transformed flies of *Drosophila melanogaster*. *Dros. Inform. Serv.* **80,** 33–34.

Adams, C., and Rinne, R. W. (1982). Stress protein formation: Gene expression and environmental interaction with evolutionary significance. *Int. Rev. Cytol.* **79,** 305–315.

Adamson, A. L., and Shearn, A. (1996). Molecular genetic analysis of *Drosophila ash2,* a member of the *Trithorax* group required for imaginal disc pattern formation. *Genetics* **144,** 621–633.

Agapova, O. A., and Kiknadze, I. I. (1978). Puffing and specific function of salivary gland cells of *Chironomus*. II. Comparative study on the ultrastructure of the cells of the special and side lobes of the salivary gland. *Tsitologia* **20**(10), 1107–1111. [in Russian]

Agapova, O. A., and Kiknadze, I. I. (1979). Puffing and specific function of salivary gland cells of *Chironomus thummi*. III. Cell ultrastructure during larval molt. *Tsitologia* **21**(5), 508–513. [in Russian]

Agapova, O. A., and Kiknadze, I. I. (1985a). Structure of the salivary gland in *Chironomidae* and its differentiation in ontogenesis. *In* "Organizatsia i ekspressia genov tkanespetsificheskoy funktsii u *Diptera*," pp. 96–100. Nauka, Novosibirsk. [in Russian]

Agapova, O. A., and Kiknadze, I. I. (1985b). Organization of the secretiry process in salivary gland cells of *Chironomids*. *In* "Organizatsia i ekspressia genov tkanespetsificheskoy funktsii u Diptera," pp. 126–131. Nauka, Novosibirsk. [in Russian]

Aggarwal, S. K., and King, R. C. (1969). A comparative study of the ring glands from wild type and *l(2)gl* mutant *Drosophila melanogaster*. *J. Morphol.* **129,** 171–199.

Agol, L. J. (1931). Step allelomorphism in *Drosophila melanogaster*. *Genetics* **16,** 254–266.

Aimanova, K. G., Pereligina, L. M., and Kolesnikov, N. N. (1987a). A biochemical analysis of the salivary gland secretory glycoproteins of *Drosophila virilis*. *Dros. Inform. Serv.* **66,** 1–3.

Aimanova, K. G., Pereligina, L. M., and Kolesnikov, N. N. (1987b). A genetic analysis of the tissue-specific salivary gland secretory proteins of *Drosophila virilis*. *Dros. Inform. Serv.* **66,** 3–5.

Ait-Ahmed, O., Thomas-Cavallin, M., and Rosset, R. (1987). Isolation and characterization of a region of the *Drosophila* genome which contains a cluster of differentially expressed maternal genes (*yema* gene region). *Devel. Biol.* **122,** 153–162.

Ait-Ahmed, O., Bellon, B., Capri, M., Joblet, C., and Thomas-Delaage, M. (1992). The *yema-nuclein-a:* a new *Drosophila* DNA binding protein specific for the oocyte nucleus. *Mechanisms of Development* **37,** 69–80.

Aizenzon, M. G. (1979). "Cytogenetic Analysis of the Genomic Regions of *Drosophila melanogaster;* Activated by Ecdysone (2B1-2–2B9-10 and 99DE)," Ph.D. thesis. Novosibirsk.

Aizenzon, M. G., and Belyaeva, E. S. (1982). Genetic loci in the X-chromosome region 2A-B. *Dros. Inform. Serv.* **58,** 3–7.

Aizenzon, M. G., and Zhimulev, I. F. (1975). Hormonal control of lysis of the salivary glands of *Drosophila melanogaster* larvae. *Doklady AN SSSR* **221**(6), 1441–1443. [in Russian]

Aizenzon, M. G., Korochkin, L. I., and Zhimulev, I. F. (1975). Hormonal induction of the activity of acid phosphatase in the salivary glands of *Drosophila melanogaster* larvae during metamorphosis. *Doklady Akad. Nauk SSSR* **225**(4), 951–954. [in Russian]

Aizenzon, M. G., Belyaeva, E. S., Kiss, I., Kotska, K., and Zhimulev, I. F. (1980). Cytogenetic analysis of the 2B1-2–2B9-10 regions of the X chromosome of *Drosophila melanogaster*. II. Complementation groups. *Genetika* **16**(2), 251–269. [in Russian]

Aizenzon, M. G., Belyaeva, E. S., and Zhimulev, I. F. (1982). Cytogenetic analysis of the 2B1-2–2B9-10 regions of the X chromosome of *Drosophila melanogaster*. IV. Phenocritical and efficient lethal phase of mutants. *Genetika* **18**(1), 73–83. [in Russian]

Akam, M. E., Roberts, D. B., Richards, G. P., and Ashburner, M. (1978). *Drosophila:* The genetics of two major larval proteins. *Cell* **13,** 215–255.

Akifjev, A. P. (1974). "Silent" DNA and its role in evolution. *Priroda*(9) 49–54. [in Russian]

Akifjev, A. P., and Makarov, V. B. (1972). Genetic and functional organization of chromosomes of eukaryotes. *Uspekhi sovrem. biol.* **74**(3(6)), 401–419. [in Russian]

Alahiotis, S. N. (1983). Heat shock proteins. A new view on the temperature compensation. *Comp. Biochem. Physiol. B* **75,** 379–387.

Alanen, M. (1985). Identification of a heat shock locus by whole mount electron microscopy of polytene chromosomes. *Hereditas* **103,** 39–46.

Alatortsev, V. E. (1997). Genetic loci in the *Pgd-K10* region of the *Drosophila* X chromosome. *Dros. Inform. Serv.* **80,** 34.

Alcover, A., Izquierdo, M., Stollar, B. D., Miranda, M., and Alonso, C. (1981). Analytical studies of chromosomal transcription detected by an endogenous hybridization technique (EHT) using autoradiography and indirect immunoflourescence in *Drosophila hydei*. *Acta Embryol. Morphol. Exp.* **2,** 131–140.

Alcover, A., Izquierdo, M., Stollar, B. D., Kitagawa, Y., Miranda, M., and Alonso, C. (1982). *In situ* immunofluorescent visualization of chromosomal transcripts in polytene chromosomes. *Chromosoma* **87,** 263–277.

Alikhanyan, S. I. (1937). Study on lethal mutations in the left end of the sex chromosome of Drosophila melanogaster. *Zool. zhurn.* **16**(2), 247–279. [in Russian]

Aller P., Santa-Cruz, M. S., Morcillo G., and Diez J. L. (1982). Uptake and incorporation of uridine in salivary gland cells from heat-shock treated *Chironomus* larvae. *Cytobios* **35,** 139–180.

Allfrey, V. G., Pogo, B. G. T., Littau, V. C., Gershey, E. C., and Mirsky, A. E. (1968). Histone acetylation in insect chromosomes. *Science* **159,** 314–316.

Alliaga, B. P., Pueyo, M. T., Noriega-Ponce, P., and Lara, F. J. S. (1975). Isolation and characterization of polyribosomes from the salivary glands of *Rhynchosciara americana*. *Cell Differ.* **4,** 243–255.

Almeida, J. C., and Edstrom, J.-E. (1983). Inhibition of Balbiani ring transcription following differential arrest of Balbiani ring-coded translation. *Chromosoma* **88,** 343–348.

Alonso, C. (1971). The effect of gibberellic acid upon developmental processes in *Drosophila hydei*. *Entomol. Exp. Appl.* **14,** 73–82.

Alonso, C. (1972a). The influence of molting hormone on RNA synthesis in isolated polytene nuclei of *Drosophila*. *Dev. Biol.* **28,** 372–381.

Alonso, C. (1972b). *In situ* hybridization of RNA synthesized in larval salivary glands of *Drosophila hydei* under *in vitro* conditions. *Dros. Inform. Serv.* **48,** 143.

Alonso, C. (1973a). RNA metabolism during metamorphosis of *Drosophila hydei*. *J. Insect Physiol.* **19,** 2135–2142.

Alonso, C. (1973b). Improved conditions for *in situ* RNA/DNA hybridization. *FEBS Lett.* **31,** 85–88.

Alonso, C., and Berendes, H. D. (1975). The location of 5S (ribosomal) RNA genes in *Drosophila hydei*. *Chromosoma* **51,** 347–356.

Alonso, M. C., and Cabrera, C. V. (1988). The *achaete-scute* gene complex of *Drosophila melanogaster* comprises four homologous genes. *EMBO J.* **7,** 2585–2591.

Alonso, C., and Grootegoed, J. A. (1973). RNA synthesis in polytene salivary glands of *Drosophila hydei* maintained under *in vitro* conditions. *Cell Differ.* **2,** 107–118.

Alonso, C., Pages, M., and Alcover A. (1984). Chromosomal transcription: Functional implication of chromatin structure. *Pont. Acad. Sci. Scr. Var.* **51,** 55–73.

Altmann J. (1966). Die Variabilitat der Kernzahlen in den Larvalen Speicheldrusen von *Drosophila melanogaster*. *Z. Zellforsch.* **70,** 35–53.

Alvarenga, C. A. S., Pueyo, M. T., and Lara, F. J. S. (1979). *In vitro* DNA puff induction by beta-ecdysone in salivary glands of *Rhynchosciara amaricana*. *Ann. Brazil. Acad. Sci.* **51,** 182–183.

Alvarenga, C. A. S., Winter, C. E., Stocker, A. J., Pueyo, M. T., and Lara, F. J. S. (1991). *In vivo* effects of ecdysterone on puff formation and RNA and protein synthesis in the salivary glands of *Rhynchosciara americana*. *Brazil. J. Med. Biol. Res.* **24,** 985–1002.

Alverdes, F. (1912a). Die Entwicklung des Kernfadens in der Speicheldruse der Chironomuslarve. *Zool. Anz.* **39,** 1–6.

Alverdes, F. (1912b). Die Kerne in den Speicheldrusen der Chironomuslarve. *Arch. Exp. Zellforch.* **9,** 168–204.

Amabis, D. C., and Amabis, J. M. (1984a). Effects of ecdysterone in polytene chromosomes of *Trichosia pubescens*. *Dev. Biol.* **102,** 1–9.

Amabis, D. C., and Amabis, J. M. (1984b). Hormonal control of gene amplification and transcription in the salivary gland chromosomes of *Trichosia pubescens*. *Dev. Biol.* **102,** 10–20.

Amabis, D. C., Amabis, J. M., and Simoes, L. C. G. (1977). Puffing activity in the salivary gland chromosomes of *Rhynchosciara* under experimental conditions. *Chromosoma* **62,** 139–154.

Amabis, J. M. (1974). Induction of DNA synthesis in *Rhynchosciara angelae* salivary gland. *Cell Differ.* **3,** 199–207.

Amabis, J. M. (1983a). DNA-puffing patterns in the salivary glands of *Trichosia pubescens* (Diptera: Sciaridae). *Genetica* **62,** 3–13.

Amabis, J. M. (1983b). The polytene chromosomes of the salivary gland of *Trichosia pubescens* (Diptera: Sciaridae). *Rev. Bras. Genet.* **6,** 413–424.

Amabis, J. M., and Cabral, D. (1970). RNA and DNA puffs in polytene chromosomes of *Rhynchosciara:* Inhibition by extirpation of prothorax. *Science* **169,** 692–694.

Amabis, J. M., and Janczur, C. (1978). Experimental induction of gene activity in the salivary gland chromosomes of *Trichosia pubescens* (Diptera: Sciaridae). *J. Cell Biol.* **78,** 1–7.

Amabis, J. M., and Simoes, L. C. G. (1971). Puff induction and regression in *Rhynchosciara angelae* by the method of salivary gland implantation. *Genetica* **42,** 404–413.

Amabis, J. M., and Simoes, L. C. G. (1972). Chromosomes studies in a species of *Telmatoscopus* sp. *Caryologia* **25,** 199–210.

Amabis, J. M., Amabis, D. C., Kaburaki, J., and Stollar, B. D. (1990). The presence of an antigen reactive with a human autoantibody in *Trichosia pubescens* (Diptera: Sciaridae) and its association with certain transcriptionally active regions of the genome. *Chromosoma* **99,** 102–110.

Amero, S. A., Elgin, S. C. R., and Beyer, A. L. (1991). A unique zinc finger protein is associated preferentially with active ecdysone-responsive loci in *Drosophila*. *Genes Dev.* **5,** 188–200.

Amin, J., Mestril, R., Lawson, R., Klapper, H., and Voellmy, R. (1985). The heat shock consensus sequence is not sufficcient for *hsp70* gene expression in *Drosophila melanogaster*. *Mol. Cell. Biol.* **5,** 197–203.

Amin, J., Mestril, R., Schiller, P., Dreano, M., and Voellmy, R. (1987). Organization of the *Drosophila melanogaster hsp70* heat shock regulation unit. *Mol. Cell. Biol.* **7,** 1055–1062.

Amin, J., Mestril, R., and Voellmy, R. (1991). Genes for *Drosophila* small heat shock proteins are regulated differently by ecdysterone. *Mol. Cell. Biol.* **11,** 5937–5944.

Amrein, K., Lutz, B., Hertner, T., and Dorsch-Haesler, K. (1986). Complete DNA sequence of the ecdysterone-controlled gene I-18C of *Chironomus tentans*. *Nucleic Acids. Res.* **14,** 7807–7808.

Amrein, K., Dorsch-Hasler, K., Lutz B., and Lezzi, M. (1988). Two transcripts of the same ecdysterone-controlled gene are differentially associated with ribosomes. *Gene* **65,** 277–283.

Ananiev, E. V., and Barsky, V. E. (1975). DNA content, replication rate and transcription intensity in different sections of the 1A1-3C region of the X-chromosome of *Drosophila melanogaster*. *Molekulyarnaya biologia* **9**(5), 752–760. [in Russian]

Ananiev, E. V., and Barsky, V. E. (1978). Localization of RNA synthesis sites in the 1B-3C region of the *Drosophila melanogaster* X chromosome. *Chromosoma* **65,** 359–371.

Ananiev, E. V., and Gvozdev, V. A. (1974). Differences in the replication pattern of DNA in the X chromosome of males, females and triploid intersexes of *Drosophila melanogaster*. *Genetika* **10**(8), 120–127. [in Russian]

Ananiev, E. V., and Gvozdev, V. A. (1975). Differences in DNA replication patterns in the X-chromosome of males, females and intersexes of *Drosophila melanogaster*. *Chromosoma* **49,** 233–241.

Ananiev, E. V., Faizullin, L. Z., and Gvozdev, V. A. (1974a). The role of genetic balance in control of transcription rate in the X-chromosome of *Drosophila melanogaster*. *Chromosoma* **45,** 193–201.

Ananiev, E. V., Faizullin, L. Z., and Gvozdev, V. A. (1974b). Dependence of the transcriptional level of the X chromosome of *Drosophila melanogaster* on the ratio of the X chromosomes to autosome set number. *Genetika* **10**(6), 62–69. [in Russian]

Ananiev, E. V., Polukarova, L. G., and Yurov, Y. B. (1977). Replication of chromosomal DNA in diploid *Drosophila melanogaster* cells cultured *in vitro*. *Chromosoma* **59,** 259–272.

Ananiev, E. V., Barsky, V. E., Ilyin, Y. V., and Churikov, N. A. (1981a). Localization of nucleoli in *Drosophila melanogaster* polytene chromosomes. *Chromosoma* **81,** 619–628.

Ananiev, E. V., Barsky, V. E., and Gvozdev, V. A. (1981b). Transcription and replication in polytene chromosomes. *In* "Biochemical genetics of *Drosophila*" (M. D. Golubovsky, and L. I. Korochkin, eds.), pp. 36–68. Nauka, Novosibirsk. [in Russian]

Anastasia-Sawicki, J. (1974). Ultrastructural histochemical localization of acid phosphatase in salivary glands of *Drosophila melanogaster*. *Dros. Inform. Serv.* **51,** 40–41.

Anderson, K. V., and Lengyel, J. A. (1979). Rates of synthesis of major classes of RNA in *Drosophila* embryos. *Dev. Biol.* **70,** 217–231.

Anderson, K. V., and Lengyel, J. A. (1984). Histone gene expression in *Drosophila* development: Multiple levels of gene regulation. *In* "Histone Genes: Structure, Organization, and Regulation" (G. S. Stein, J. L. Stein, and W. F. Marzluff, eds.), pp. 135–161. Wiley, Chicester.

Andersson, K., Bjorkroth, B., and Daneholt, B. (1980). The *in situ* structure of the active 75S RNA genes in Balbiani Rings of *Chironomus tentans*. *Exp. Cell Res.* **130,** 313–326.

Andersson, K., Mahr, R., Bjorkroth, B., and Daneholt, B. (1982). Rapid information of the thick chromosome fiber upon completion of RNA synthesis at the Balbiani Rings genes in *Chironomus tentans*. *Chromosoma* **87,** 33–48.

Andersson, K., Bjorkroth, B., and Daneholt, B. (1984). Packing of a specific gene into higher order structures following repression of RNA synthesis. *J. Cell Biol.* **98,** 1296–1303.

Andres, A. J., and Cherbas, P. (1994). Tissue-specific regulation by ecdysone: Distinct patterns of *Eip28/29* expression are controlled by different ecdysone response elements. *Dev. Genet.* **15,** 320–331.

Andres, A. J., and Thummel, C. S. (1992). Hormones, puffs and flies: The molecular control of metamorphosis by ecdysone. *Trends Genet.* **8**(4), 132–138.

Andres, A. J., and Thummel, C. S. (1995). The *Drosophila* 63F early puff contains *E63-1*, an ecdysone-inducible gene that encodes a novel Ca^{2+} binding protein. *Development* **121,** 2667–2679.

Andres, A. J., Fletcher, J. C., Karim, F. D., and Thummel, C. S. (1993). Molecular analysis of the initiation of insect metamorphosis: A comparative study of *Drosophila* ecdysteroid-regulated transcription. *Dev. Biol.* **160,** 388–404.

Antoniewski, C., Laval, M., and Lepesant, J.-A. (1993). Structural features critical to the activity of an ecdysone receptor binding site. *Insect Biochem. Mol. Biol.* **23,** 105–114.

Antoniewski, C., Laval, M., Dahan, A., and Lepesant J.-A., (1994). The ecdysone response enhancer of the *Fbp1* gene of *Drosophila melanogaster* is a direct target for the EsR/USP nuclear receptor. *Mol. Cell. Biol.* **14,** 4465–4474.

Antoniewski, C., Mugat, B., Delbac, F., and Lepesant J.-A. (1996). Direct repeats bind the EcR/USP receptor and mediate ecdysteroid responses in *Drosophila melanogaster*. *Mol. Cell. Biol.* **16,** 2977–2986.

Appels, R., and Hilliker, A. J. (1982). The cytogenetic boundaries of the rDNA region within heterochromatin of the X chromosome of *Drosophila melanogaster* and their relation to male meiotic pairing sites. *Genet. Res. Cambr.* **39,** 149–156.

Aquadro, C. F., Weaver, A. L., Schaeffer, S. W., and Anderson, W. W. (1991). Molecular evolution of inversions in *Drosophila pseudoobscura:* The amylase gene region. *Proc. Natl. Acad. Sci. USA* **88,** 305–309.

Arbona, M., and de Frutos, R. (1987a). Induction of new puffs during recovery from anoxia in *Drosophila subobscura*. *Dros. Inform. Serv.* **66,** 15.

Arbona, M., and de Frutos, R. (1987b). Stress response in *Drosophila subobscura*. II. Puff activity during anoxia and recovery from anoxia. *Biol. Cell* **60,** 173–182.

Arbona, M., de Frutos, R. and Diez, J. L. (1990). Localization of DNA-RNA hybrids in polytene chromosomes of *Drosophila subobscura* by indirect immunofluorescence. *Dros. Inform. Serv.* **70,** 24–25.

Arcos-Teran, L. (1972). DNS-replication und die Natur der spat replizierende Orte im X-Chromosom von *Drosophila melanogaster*. *Chromosoma* **37,** 233–296.

Arcos-Teran, L., and Beermann, W. (1968). Changes of DNA replcation behaviour associated with intragenic changes of the *white* region in *Drosophila*. *Chromosoma* **25,** 377–391.

Arking, R., Putnam, R. L., and Schubiger, M. (1975). Phenogenetics of *eyeless-dominant* mutant of *Drosophila melanogaster*. II. Involvement of the nervous systems. *J. Exp. Zool.* **193,** 301–312.

Armelin, H. A., and Marques, N. (1972a). Transcription and processing of ribonucleic acid in *Rhynchosciara* salivary glands. I. Rapidly labelled ribonucleic acid. *Biochemistry* **11,** 3663–3672.

Armelin, H. A., and Marques, N. (1972b). Transcription and processing of ribonucleic acid in *Rhynchosciara* salivary glands. II. Hybridization of nuclear and cytoplasmic ribonucleic acid with nuclear deoxyribonucleic acids. Indication of deoxyrobonucleic acid amplification. *Biochemistry* **11,** 3672–3679.

Armelin, H. A., Meneghini, R., and Lara, F. J. S. (1969a). Patterns of ribonucleic acid synthesis in salivary glands of *Rhynchosciara angelae* larvae during development. *Genetics* **61** (Suppl. 1), 351–360.

Armelin, H. A., Meneghini, R., and Lara, F. J. S. (1969b). Extraction and characterization of newly synthesized RNA from whole cells and cellular fractions of *Rhynchosciara angelae* salivary glands. *Biochem. Biophys. Acta* **190,** 358–367.

Armelin, H. A., Meneghini, R., Marques, N., and Lara, F. J. S. (1970). Extraction and characterization of nuclear rapidly labelled RNA fraction from *Rhynchosciara* salivary glands. *Biochem. Biophys. Acta* **217,** 426–433.

Arnold, G. (1965). An autoradiographic study of RNA synthesis in isolated salivary glands of *Drosophila hydei*. I. Autoradiographic studies. *J. Morphol.* **116,** 65–88.

Arribas, C., and Izquierdo, M. (1984). Sequences isolated from a *Drosophila* early-ecdysone puff are expressed in rat liver. *Biosci. Rep.* **4,** 387–396.

Arribas, C., Sampredo, J., and Izquierdo, M. (1986). The ubiquitin genes in *Drosophila melanogaster:* Transcription and polymorphism. *Biochem. Biophys. Acta* **868,** 119–127.

Arrigo, A.-P. (1980). Investigation of the function of the heat shock proteins in *Drosophila melanogaster* tissue culture cells. *Mol. Gen. Genet.* **178,** 517–524.

Arrigo, A.-P. (1987). Cellular localization of *hsp23* during *Drosophila* development and following subsequent heat shock. *Dev. Biol.* **122,** 39–48

Arrigo, A.-P. (1990). The monovalent ionophore monensin maintains the nuclear localization of the human stress protein *hsp28* during heat shock recovery. *J. Cell Sci.* **96,** 419–427.

Arrigo, A.-P., Fakan S., and Tissieres, A. (1980). Localization of the heat shock-induced proteins in *Drosophila melanogaster* tissue culture cells. *Dev. Biol.* **78,** 86–103.

Artavanis-Tsakonas, S., Schedl, P., Tschudi, C., Pirrotta, V., Steward, R., and Gehring, W. (1977). The 5S genes of *Drosophila melanogaster*. *Cell* **12,** 1057–1067.

Artavanis-Tsakonas, S., Schedl, P., Mirault, M.-E., Moran, L., and Lis, J. (1979a). Genes for the 70,000 Dalton heat shock protein in two cloned *Drosophila melanogaster* DNA segments. *Cell* **17,** 9–18.

Artavanis-Tsakonas, S., Schedl, P., Steward, R., Gehring, W. J., Mirault, M.-E., Moran, L., Goldschmidt-Clermont, M., Arrigo, P., Tissieres, A., and Lis, J. (1979b). Heat activated genes of *Drosophila melanogaster*. "Specific Eukaryotic Genes, Alfred Benzon Symposium," Vol. 13, pp. 72–84. Munksgaard, Copenhagen.

Artavanis-Tsakonas, S., Muskavitch, M. A. T., and Yedvobnick, B. (1983). Molecular cloning of *Notch*, a locus affecting neurogenesis in *Drosophila melanogaster*. *Proc. Natl. Acad. Sci. USA* **80,** 1977–1981.

Ashburner, M. (1967a). Gene activity dependent on chromosome synapsis in the polytene chromosomes of *Drosophila melanogaster*. *Nature* **214,** 1159–1160.

Ashburner, M. (1967b). Patterns of puffing activity in the salivary gland chromosomes of *Drosophila*. I. Autosomal puffing patterns in a laboratory stock of *Drosophila melanogaster*. *Chromosoma* **21,** 398–428.

Ashburner, M. (1969a). Genetic control of puffing in polytene chromosomes. *Chrom. Today* **2,** 99–106.

Ashburner, M. (1969b). Patterns of puffing activity in the salivary gland chromosomes of *Drosophila*. II. The X-chromosome puffing patterns of *Drosophila melanogaster* and *Drosophila simulans*. *Chromosoma* **27,** 47–63.
Ashburner, M. (1969c). Patterns of puffing activity in the salivary gland chromosomes of *Drosophila*. III. A comparison of the autosomal puffing patterns of the sibling species of *Drosophila melanogaster* and *Drosophila simulans*. *Chromosoma* **27,** 64–85.
Ashburner, M. (1969d). Patterns of puffing activity in the salivary gland chromosomes of *Drosophila*. IV. Variability of puffing patterns. *Chromosoma* **27,** 156–177.
Ashburner, M. (1969e). On the problem of genetic similarity between sibling species puffing patterns in *Drosophila melanogaster* and *Drosophila simulans*. *Am. Natur.* **103,** 189–191.
Ashburner, M. (1969f). Puff loci in the salivary gland chromosomes of *Drosophila melanogaster*. *Dros. Inform. Serv.* **44,** 55.
Ashburner, M. (1970a). Function and structure of polytene chromosomes during insect development. *In* "Advances in insect physiology" (J. W. L. Beament, J. E. Treherne, V. B. Wigglesworth, eds.), Vol. 7, pp. 1–95. London, New York.
Ashburner, M. (1970b). The genetic analysis of puffing in polytene chromosomes of *Drosophila*. *Proc. Roy. Soc. Lond. B* **176,** 319–327.
Ashburner, M. (1970c). Patterns of puffing activity in the salivary gland chromosomes of *Drosophila*. V. Responses to environmental treatments. *Chromosoma* **31,** 356–376.
Ashburner, M. (1970d). A prodromus to the genetic analysis of puffing in *Drosophila melanogaster*. *Cold Spring Harbor Symp. Quant. Biol.* **35,** 533–538.
Ashburner, M. (1971). Induction of puffs in polytene chromosomes of *in vitro* cultured salivary glands of *Drosophila melanogaster* by ecdysone and ecdysone analogues. *Nature New Biol.* **230,** 222–224.
Ashburner, M. (1972a). Puffing patterns in *Drosophila melanogaster* and related species. *In* "Results and Problems in Cell Differentiation" (W. Beermann, ed.), Vol. 4, pp. 101–151. Springer-Verlag, Berlin/Heidelberg/New York.
Ashburner, M. (1972b). Patterns of puffing activity in the salivary gland chromosomes of *Drosophila*. VI. Induction by ecdysone in salivary glands of *Drosophila melanogaster* cultured *in vitro*. *Chromosoma* **38,** 255–281.
Ashburner, M. (1972c). Ecdysone induction of puffing in polytene chromosomes of *Drosophila melanogaster*. Effects of inhibitors of RNA synthesis. *Exp. Cell Res.* **71,** 433–440.
Ashburner, M. (1972d). *N*-ethylmaleimide inhibition of the induction of gene activity by the hormone ecdysone. *FEBS Lett.* **22,** 265–269.
Ashburner, M. (1972e). The hormonal control of gene activity in polytene chromosomes of *Drosophila melanogaster*. *Int. Congr. Ser. Endocrinol.* **N 273,** 315–318.
Ashburner, M. (1973). Sequential gene activation by ecdysone in polytene chromosomes of *Drosophila melanogaster*. I. Dependence upon ecdysone concentration. *Dev. Biol.* **35,** 47–61.
Ashburner, M. (1974a). Sequential gene activation by ecdysone in polytene chromosomes of *Drosophila melanogaster*. II. The effect of inhibitors of protein synthesis. *Dev. Biol.* **39,** 141–157.
Ashburner, M. (1974b). Genetic and hormonal control of puffing in the polytene chromosomes of *Drosophila melanogaster*. *Ontogenez* **5**(2), 107–121. [in Russian]
Ashburner, M. (1975). The puffing activities of salivary gland chromosomes. *In* "Handbook of Genetics" (R. C. King, ed.), Vol. 3, pp. 793–811. Plenum, New York/London.
Ashburner, M. (1976). Aspects of polytene chromosomes structure and function. *In* "Life Sciences Research Report. 4. Organization and expression of chromosomes" (V. G. Allfrey, E. K. F. Bautz, B. J. McCarthy, R. T. Schimke, and A. Tissieres, eds.), pp. 81–95. Dahlem Konferenzen, Berlin.
Ashburner, M. (1977). Happy birthday-puffs! *In* "Chromosomes today" (A. de la Chapelle and M. Sorsa, eds.), Vol. 6, pp. 213–222. Elsevier, Amsterdam.
Ashburner, M. (1978). Patterns of puffing activity in the salivary gland chromosomes of *Drosophila*.

VIII. A revision of the puffing patterns of the proximal region of chromosome arm 2L of *D. melanogaster*. *Chromosoma* **68,** 195–203.

Ashburner, M. (1979). Genetic variation in insect endocrine systems. *In* "Genetic variation in hormone systems" (J. G. M. Shire, ed.), Vol. 2, pp. 91–122. CRC Press, Boca Raton, Florida.

Ashburner, M. (1980). Chromosomal action of ecdysone. *Nature* **285,** 435–436.

Ashburner, M. (1989). "*Drosophila:* A Laboratory Handbook." Cold Spring Harbor Laboratory Press, New York.

Ashburner, M. (1990). Puffs, genes and hormones revisited. *Cell* **61,** 1–3.

Ashburner, M., and Berendes, H. D. (1978). Puffing of polytene chromosomes. *In* "The genetics and biology of *Drosophila*" (M. Ashburner and T. R. F. Wright, eds.), Vol. 26, pp. 316–395. Academic Press, London/New York/San Francisco.

Ashburner, M., and Bonner J. (1979). The induction of gene activity in *Drosophila* by heat shock. *Cell* **17,** 241–254.

Ashburner, M., and Cherbas P. (1976). The control of puffing by ions—the Kroeger hypothesis: A critical review. *Mol. Cell Endocrinol.* **5,** 89–107.

Ashburner, M., and Garcia-Bellido, A. (1973). Ecdysone induction of puffing activity in salivary glands of *Drosophila melanogaster* grown in adult abdomens. *Wilhelm Roux' Arch.* **172,** 166–170.

Ashburner, M., and Lemeunier, F. (1972). Patterns of puffing activity in the salivary gland chromosomes of *Drosophila*. VII. Homology of puffing patterns on chromosome arm 3L in *D. melanogaster* and *D. yakuba,* with notes on puffing in *D. teissieri*. *Chromosoma* **38,** 283–295.

Ashburner, M., and Richards G. (1976a). Sequential gene activation by ecdysone in polytene chromosomes of *Drosophila melanogaster*. III. Cosequences of ecdysone withdrawal. *Dev. Biol.* **54,** 241–255.

Ashburner, M., and Richards, G. (1976b). The role of ecdysone in the control of gene activity in the polytene chromosomes of *Drosophila*. *In* "Insect Development" (A. Lawrence, ed.), pp. 203–225. Blackwell, Edinburgh/Melbourn/Oxford/London,

Ashburner, M., Chihara C., Meltzer, P., and Richards, G. (1974). On the temporal control of puffing activity in polytene chromosomes. *Cold Spring Harbor Symp. Quant. Biol.* **38,** 655–662.

Ashburner, M., Aaron, C. S., and Tsubota, S. (1982). The genetics of a small autosomal region of *Drosophila melanogaster,* including the structural gene for alcohol dehydrogenase. V. Characterization of X-ray-induced *Adh* null mutations. *Genetics* **102,** 421–435.

Ashton, F. T., and Schultz, J. (1964). Stereoscopic analysis of the fine structure of chromosomes in diploid *Drosophila* nuclei. *J. Cell Biol.* **23,** 7A.

Ashton, F. T., and Schultz, J. (1971). The three-dimensional fine structure of chromosomes in a prophase *Drosophila* nucleus. *Chromosoma* **35,** 383–392.

Asoaka, M., Myohara, M., and Okada, M. (1994). Digitonin activates different sets of puff loci depending on developmental stages in *Drosophila melanogaster* salivary glands. *Dev. Growth Differ.* **36,** 605–614.

Atkinson, B. G., and Walden, D., B. (eds.) (1985). "Changes in Eukaryotic Gene Expression in Response to Environmental Stimuli." Academic Press, New York.

Atwood, K. C. (1969). Some aspects of the *bobbed* problem in *Drosophila*. *Genetics* **61**(Suppl.1), 319–327.

Avanzi, S., Cionini, P. G., and D'Amato, F. (1970). Cytochemical and autoradiographic analysis on the embryo suspensor cells of *Phaseolus coccineus*. *Caryologia* **23,** 605–638.

Avanzi, S., Durante, M., Cionini, P. G., and D'Amato, F. (1972). Cytological localization of ribosomal cistrons in polytene chromosomes of *Phaseolus coccineus*. *Chromosoma* **39,** 191–203.

Ayme, A., and Tissieres, A. (1985). Locus 67B of *Drosophila melanogaster* contains seven, not four, closely related heat-shock genes. *EMBO J.* **4,** 2949–2954.

Ayme, A., Southgate, R., and Tissieres, A. (1985). Nucleotide sequences responsible for the thermal inducibility of the *Drosophila* small heat-shock protein genes in monkey COS cells. *J. Mol. Biol.* **182,** 469–475.

Baehrecke, E. H., and Thummel, C. S. (1995). The *Drosophila E93* gene from the 93F early puff displays stage- and tissue-specific regulation by 20-hydroxyecdysone. *Dev. Biol.* **171,** 85–97.

Baillie, D. L., and Chovnick, A. (1971). Studies on the genetic control of tryptophan pyrrolase in *Drosophila melanogaster. Mol. Gen. Genet.* **112,** 341–353.

Bainbridge S.P., and Bownes, M. (1988). Ecdysteroid titers during *Drosophila* metamorphosis. *Insect Biochem.* **18,** 185–197.

Baird, D. H., Schalet, A. P., and Wyman, R. J. (1990). The *Passover* locus in *Drosophila melanogaster:* Complex complementation and different effects on the giant fiber neural pathway. *Genetics* **126,** 1045–1059.

Baker, B. S., and Belote, J. M. (1983). Sex determination and dosage compensation in *Drosophila melanogaster. Annu. Rev. Genet.* **17,** 345–393.

Baker, B. S., Hoff, G., Kaufman, T. C., Wolfner, M. F., and Hazelrigg, T. (1991). The *doblesex* locus of *Drosophila melanogaster* and its flanking regions: A cytogenetic analysis. *Genetics* **127,** 125–138.

Baker, B. S., Gorman, M., and Marin, I. (1994). Dosage compensation in *Drosophila. Annu. Rev. Genet.* **28,** 491–521.

Balakirev, E. S., and Ayala, F. J. (1996). Is Esterase-P encoded by a cryptic pseudogene in *Drosophila melanogaster? Genetics* **144,** 1511–1518.

Balasov, M. L., and Bgatov, A. V. (1992). Mapping of the action focus of the lethal allele of the *ecs*$\sim^{t76}$ gene by the method of genetic mosaics. *Genetika* **28**(11), 40–47. [in Russian]

Balasov, M. L., and Zhimulev, I. F. (1997). Resque *in vivo* of puffing and histolysis of salivary glands in *rbp* mutants by product of ecdysterone dependent locus Broad-Complex. *Tsitologia* **39**(1), 38–39. [in Russian]

Balbiani, E. G. (1881). Sur la structure du noyau des cellules salivares chez les larves de *Chironomus. Zool. Anz.* **4,** 637–641 and 662–666.

Ballard, J. W. O., and Bedo, D. G. (1991). Population cytogenetic of *Austrosimulium bancrofti* (Diptera, Simuliidae) in eastern Australia. *Genome* **34,** 338–353.

Ballinger, D., and Pardue, M. L. (1985). Mechanism of translational control in heat-shocked *Drosophila* cells. *In* "Changes in eukaryotic gene expression in response to environmental stress" (B. G. Atkinson and D. B. Walden, eds.), pp. 53–70. Academic Press, New York.

Bally-Cuif, L., Payant, V., Abukashawa, S., Benkel, B. F., and Hickey, D. A. (1990). Molecular cloning and partial sequence characterization of the duplicated amylase genes from *Drosophila erecta. Genet. Sel. E* **22,** 57–64.

Balsamo, J., Hierro, J. M., Birnstiel, M. L., and Lara, F. J. S. (1973a). *Rhynchosciara angelae* salivary gland DNA: kinetic compexity and transcription of repetitive sequences. *In* "Gene Expression and Its Regulation" (F. T. Kenney, B. A. Hamkalo, and L. Favelukas., eds.), 101–122. Plenum, New York/London.

Balsamo, J., Hierro, J. M., and Lara, F. J. S. (1973b). Transcription of repetitive DNA sequences in *Rhynchosciara* salivary glands. *Cell Differ.* **2,** 119–130.

Balsamo, J., Hierro, J. M., and Lara, F. G. S. (1973c). Further studies on the characterization of repetitive *Rhynchosciara* DNA. *Cell Differ.* **2,** 131–141.

Banerjee, S., and Chinya, P. K. (1982). Effect of mitomycin C on polytene chromosome replication of *Drosophila. Dros. Inform. Serv.* **58,** 19–20.

Banerjee, U., Renfranz, P. J., Pollock, J. A., and Benzer, S. (1987). Molecular characterization and expression of *sevenless,* a gene in ved in neuronal pattern formation in the *Drosophila eye. Cell* **49,** 281–291.

Banga, S. S., Bloomquist, B. T., Brodberg, R. K., Pye, Q. N., Larrivee, D. C., and Mason, J. M. (1986). Cytogenetic characterization of the 4BC region on the X chromosome of *Drosophila melanogaster:* Localization of the *mei-9, norpA* and *omb* genes. *Chromosoma* **93,** 341–346.

Banga, S. S., Yamamoto, A. H., Mason, J. M., and Boyd, J. B. (1995). Molecular cloning of *mei-41,* a gene that influence both somatic and germline chromosome metabolism of *Drosophila melanogaster*. *Mol. Gen. Genet.* **246,** 148–155.

Barbas, J. A., Galceran, J., Krah-Jentgens, I., de la Pompa, J. L., Canal, I., Pongs, O., and Ferrus, A. (1991). Troponin I is encoded in the haplolethal region of the *Shaker* gene complex of *Drosophila*. *Genes Dev.* **5,** 132–140.

Barettino, D., Morcillo, G., and Diez, J. L. (1982). Induction of heat-shock Balbiani Rings after RNA synthesis inhibition in polytene chromosomes of *Chironomus thummi*. *Chromosoma* **87,** 507–517.

Barettino, D., Morcillo, G., and Diez, J. L. (1988a). Induction of the heat-shock response by carbon dioxide in *Chironomus thummi*. *Cell Differ.* **23,** 27–36.

Barettino, D., Morcillo, G., Diez, J. L., Carretero, M. T., and Carmona, M. J. (1988b). Correlation between the activity of a 5,6-dichloro-1-b-D-ribofuranosylbenzimidazole insensitive puff and the synthesis of major heat-shock, polypeptide, *hsp70*, in *Chironomus thummi*. *Biochem. Cell Biol.* **66,** 1177–1185.

Baricheva, E. M., and Semeshin, V. F. (1985). Electron microscopic study on the formation of some puffs in *Drosophila melanogaster* 3L chromosome. *Tsitologiya* **27**(1), 50–53. [in Russian]

Baricheva, E. M., Semeshin, V. F., Zhimulev, I. F., and Belyaeva, E. S. (1987). Electron microscopical mapping of some puffs of *Drosophila melanogaster* polytene chromosomes. *Dros. Inform. Serv.* **66,** 18–22.

Barker, R. J., and Alexander, B. H. (1958). Acid and alkaline phosphatases in house flies of different ages. *Ann. Entomol. Soc. Am.* **51,** 255–257.

Barnett, S. W., Flynn, K., Webster, M. K., and Beckendorf, S. K. (1990). Noncoordinate expression of *Drosophila* glue genes: *Sgs-4* is expressed at many stages and in two different tissues. *Dev. Biol.* **140,** 362–373.

Barnett, T., and Rae, P. M. M. (1979). A 9,6 kb intervening sequence in *D. virilis* rDNA, and sequence homology in rDNA interruptions of diverse species of *Drosophila* and other Diptera. *Cell* **16,** 763–775.

Barnett, T., Pachel, C., Gergen, J. P., and Wensink, P. C. (1980). The isolation and characterization of *Drosophila* yolk protein genes. *Cell* **21,** 729–738.

Barr, H. J., and Plaut, W. (1966). Comparative morphology of nucleolar DNA in *Drosophila*. *J. Cell Biol.* **31**(3), 17–22.

Barr, H. J., Valencia, J. I., and Plaut, W. (1968). On temporal autonomy of DNA replications in a chromosome translocation. *J. Cell Biol.* **39,** 8A.

Barret, J. A. (1980). The estimation of the number of mutationally silent loci in saturation-mapping experiments. *Genet. Res. Cambr.* **35,** 33–44.

Bashkirov, V. N. (1980). Regilation of the number of ribosomal RNA genes in *Drosophila*. *Genetika* **16**(1), 7–29. [in Russian]

Bashkirov, V. N., Kubaneishvili, M. S., Ialakas, M. E., Karpov, I. A., and Shuppe, N. G. (1984). Regulation of rDNA quantity in *Drosophila melanogaster*. Retrocompensation of rDNA in trinucleolar females. *Doklady Acad. Nauk USSR* **275**(3), 754–758. [in Russian]

Bashkirov, V. N., Kubaneishvili, M. S., Ialakas, M. E., and Shuppe, N. G. (1986). Instability of the rRNA gene number in trinucleolar *Drosophila* females. *Genetika* **22**(9), 2276–2285. [in Russian]

Basile, R., Casartelli, C., and Guevara, M. (1970). Nucleic acids synthesis during the 3 rd molt period of *Rhynchosciara americana*. *Rev. Brasil. Biol.* **30,** 291–296.

Baudisch, W. (1960). Spezifisches Vorkmmen von Carotinoiden und Oxyprolin in den Speicheldrusen von *Acricotopus lucidus*. *Naturwissenschaften* **47,** 498–499.

Baudisch, W. (1961). Synthese von Oxyprolin in den Speicheldrusen von *Acricotopus lucidus*. *Naturwissenschaften* **48,** 2.

Baudisch, W. (1963a). Untersuchungen zur physiologischen Charakterisierung der einzelnen Speicheldrussenlappen von *Acricotopus lucidus*. *Struct. Funkt. genetisch. Mater.* **Bd. III,** 231–234.

Baudisch, W. (1963b). Aminosaurezammensetzung der Speicheldrusen von *Acricotopus lucidus*. *Biol. Zbl.* **82,** 351–361.

Baudisch, W. (1963c). Chemisch-physiologische Untersuchungen an den Speicheldrusen von *Acricotopus lucidus*. 100 Jahre Landwirtsohaftl. *Inst. der Universitat Halle*, 152–159.

Baudisch, W. (1967). Spezifische Hydroxyprolinsynthese in den Speicheldrusen von *Acricotopus lucidus*. *Biol. Zbl.* **86**(Suppl.), 157–162.

Baudisch, W. (1972). A study on the relationships between the gene loci and the products of their synthesis in the salivary glands of *Acricotopus lucidus*. *In* "Kletochnoe yadro: Morfologia, fiziologiya, biochimiya. (Sbornik pod redakts. Zbarskogo, I. B., i Georgieva, G. P.), pp. 202–208. Nauka, Moscow. [in Russian]

Baudisch, W. (1977). Balbiani ring pattern and biochemical activities in the salivary gland of *Acricotopus lucidus* (Chironomidae). *Results Prob. Cell Differ.* **8,** 197–212.

Baudisch, W., and Herrmann, F. (1972). Uber Proteine mit ungewehnlichem Loslichkeitsverhalten im Speicheldrusensekret von *Acricotopus lucidus*. *Acta Biol. Med. Germ.* **29,** 521–525.

Baudisch, W., and Panitz R. (1968). Kontrolle eines Biochemischen Merkmals in den Speicheldrusen von *Acricotopus lucidus* durch einen Balbiani-Ring. *Exp. Cell Res.* **49,** 470–476.

Bauer H. (1935). Der Aufbau der Chromosomen aus den Speicheldrusen von *Chironomus thummi* Kiefer (Untersuchungen an den Riesenchromosomen der Dipteren I). *Z. Zellforsch.* **23,** 280–313.

Bauer, H. (1936). Beitrage zur vergleichenden Morphologie der Speicheldrusenchromosomen (Untersuchungen an den Riesenchromosomen der Dipteren II). *Zool. Jahrb. Physiol.* **56,** 239–276.

Bauer, H. (1953). Die Chromosomen im Soma der Matazoen. *Zool. Anz.* **17**(Suppl.), 252–268.

Bauer, H., and Cassperson, T. (1948). Cytochemical observations on nucleolus formation in *Chironomus*. *Proc. 8th Intern. Congr. Genetics (Hereditas)*, Suppl., 533–534.

Baumann, A., Krah-Jentgens, I., Muller, R., Muller-Holtkamp, F., Seidel, R., Kecskemethy, N., Casal, J., Ferrus, A., and Pongs, O. (1987). Molecular organization of the maternal effect region of the *Shaker* complex of *Drosophila:* Characterization of an IA channel transcript with homology to vertebrate Na^+ channel. *EMBO J.* **6,** 3419–3429.

Baumlein, H., and Wobus, U. (1976). Chromosomal localization of ribosomal 5S RNA genes in *Chironomus thummi* by *in situ* hybridization of iodinated 5S RNA. *Chromosoma* **57,** 199–204.

Baumlein, H., Wobus, U., Gerbi, S., and Kafatos, F. C. (1982a). Characterization of a 249-bp tandemly repetitive, satellite-like repeat in the translated portion of Balbiani Ring c of *Chironomus thummi*. *EMBO J.* **1,** 641–647.

Baumlein, H., Wobus, U., Gerbi, S. A., and Kafatos, F. C. (1982b). The basic repeat unit of a *Chironomus* Balbiani ring gene. *Nucleic. Acids Res.* **10,** 3893–3904.

Baumlein, H., Pustell, J., Wobus, U., Case, S. T., and Kafatos, F. C. (1986). The 3′ ends of two genes in the Balbiani Ring c locus of *Chironomus thummi*. *J. Mol. Evol.* **24,** 72–82.

Bauren, G., and Wieslander, L. (1994). Splicing of Balbiani ring 1 gene pre-mRNA occurs simultaneously with transcription. *Cell* **76,** 183–192.

Bautz, E. K. F. and Kabisch, R. (1983). Polytene chromosomes. *In* "Eukaryotic Genes: Structure, Activity and Regulation" (S. P. Maclean and R. A. Gregory, eds), pp. 101–113. Butterworths, London.

Bayer, C., von Kalm, L., and Fristrom, J. W. (1995). Gene regulation in imaginal disc and salivary gland development during *Drosophila* metamorphosis. *In* "Metamorphosis: Postembryonic Reprogramming of Gene Expression in Amphibian and Insect Cells" (L. Gilbert, J. Tata, and B. Atkison, B., eds.), pp. 321–361, Academic Press, San Diego.

Bayer, C. A., Holley, B., and Fristrom, J. W. (1996). A switch in *Broad-Complex* Zinc-finger isoform expression is regulated posttranscriptionally during the metamorphosis of *Drosophila* imaginal discs. *Dev. Biol.* **177,** 1–14.

Bayer, C. A., von Kalm, L., and Fristrom, J. W. (1997). Relationships between protein isoforms and genetic functions demonstrate functional redundancy at the *Broad-Complex* during *Drosophila* metamorphosis. *Devel. Biol.* **187,** 267–282.

Beams, H. W., and Goldsmith, J. B. (1930). Golgi bodies, vacuole, and mitochondria in the salivary glands of the *Chironomus* larva. *J. Morphol.* **50,** 497–511.

Beaton, A. H., Kiss, I., Fristrom, D., and Fristrom, J. W. (1988). Interaction of the *Stubble-stubbloid* locus and the *Broad-Complex of Drosophila melanogaster*. *Genetics* **120,** 453–464.

Beaulieu, J.-F., and Tanguay, R. M. (1988). Members of the *Drosophila HSP70* family share ATP-binding properties. *Eur. J. Biochem.* **172,** 341–347.

Beaulieu J.-F., Arrigo, A. P., and Tanguay R. M. (1989). Interaction of *Drosophila* 27000Mr heat-shock protein with the nucleus of heat-shocked and ecdysone-stimulated culture cells. *J. Cell Science* **92,** 29–36.

Beckendorf, S. K., and Kafatos, F. C. (1976). Differentiation in the salivary glands of *Drosophila melanogaster:* Characterization of the glue proteins and their developmental appearance. *Cell* **9,** 365–373.

Becker, H. J. (1959). Die Puffs der Speicheldrusenchromosomen von *Drosophila.* I Mitt. Beobachtungen zum Verhalten des Puffmusters im Normalstamm und bei zwei Mutanten, *giant* und *lethal-giant-larva*. *Chromosoma* **10,** 654–678.

Becker, H. J. (1962a). Die Puffs der Speicheldrusenchromosomen von *Drosophila melanogaster* II. Die Auslosung der Puffbildung, ihre Spezifitat und ihre Beziehung zur Funktion der Ringdruse. *Chromosoma* **13,** 341–384.

Becker, H. J. (1962b). Stadienspezifische Genaktivierungen in Speicheldrusen nach Transplantation bei *Drosophila melanogaster*. *Zool. Anz. Suppl.* **25,** 92–101.

Becker, H. J. (1965). Chromosome function in specific regions. *Nat. Cancer Inst. Monogr.* **18,** 293–307.

Beckingham, K. (1982). Insect rDNA. *In* "The Cell Nucleus, X, rDNA, Part A" (H. Busch and L. Rothblum, eds.), pp. 205–269. Academic Press, New York.

Beckingham, K., and Rubacha, A. (1984). Different chromatin states of the intron$^-$ and type 1 intron$^+$ rRNA genes of *Calliphora erythrocephala*. *Chromosoma* **90,** 311–316.

Beckingham, K., and Thompson, N. (1982). Under-replication of intron$^+$ rDNA cistrons in polyploid nurse cell nuclei of *Calliphora erythrocephala*. *Chromosoma* **87,** 177–196.

Beckingham, K., and White, R. (1980). The ribosomal DNA of *Calliphora erythrocephala:* An analysis of hybrid plasmids containing ribosomal DNA. *J. Mol. Biol.* **137,** 349–373.

Bedo, D. G. (1976). Polytene chromosomes in pupal and adult blackflies (Diptera: Simuliidae). *Chromosoma* **57,** 387–396.

Bedo, D. G. (1978). Band and nucleolar polymorphisms in polytene chromosomes of *Simulium ornatipes* (Diptera, Simuliidae). *Cytobios* **21,** 113–133.

Bedo, D. G. (1982). Patterns of polytene-chromosome replication in *Simulium ornatipes* (Diptera: Simuliidae). *Genetica* **59,** 9–21.

Bedo, D. G. (1986). Polytene and mitotic chromosome analysis in *Ceratitis capitata* (Diptera; Tephritidae). *Can. J. Genet. Cytol.* **28,** 180–188.

Bedo, D. G. (1987). Polytene chromosome mapping in *Ceratitis capitata* (Diptera: Tephritidae). *Genome* **29,** 598–611.

Bedo, D. G. (1992). Nucleolar fragmentation in polytene trichogen cells of *Lucilia cuprina* and *Chrysomya bezziana* (Diptera: Calliphoridae). *Genome* **35,** 283–293.

Bedo, D. G., and Webb, G. C. (1989). Observation of nucleolar structure in polytene tissues of *Ceratitis capitata* (Diptera: Tephritidae). *Chromosoma* **98,** 443–449.

Bedo, D. G., and Webb, G. C. (1990). Localization of the 5S RNA genes to arm 2R in polytene chromosomes of *Lucilia cuprina* (Diptera: Calliphoridae). *Genome* **33,** 941–943.

Bedo, D. G., and Zacharopoulou, A. (1988). Inter-tissue variability of polytene chromosome banding patterns. *Trends Genet.* **4**(4), 90–91.

Been, A. C., and Rasch, E. M. (1972). Cellular and secretory proteins of the salivary glands of *Sciara coprophila* during the larval–pupal transformation. *J. Cell Biol.* **55,** 420–432.

Beermann, W. (1950). Chromomerenkonstanz bei *Chironomus*. *Naturwissenschaften* **37,** 543–544.

Beermann, W. (1952a). Chromomerenkonstanz und spezifische Modifikationen der Chromosomenstruktur in der Entwicklung und Organdifferenzierung von *Chironomus tentans*. *Chromosoma* **5,** 139–198.

Beermann, W. (1952b). Chromosomenstruktur und Zelldifferenzierung in der Speicheldruse von *Trichocladius vitripennis*. *Z. Naturforsch.* **7B,** 237–242.

Beermann, W. (1956). Nuclear differentiation and functional morphology of chromosomes. *Cold Spring Harbor Symp. Quant. Biol.* **21,** 217–232.

Beermann, W. (1959). Chromosomal differentiation in insects. *In* "Developmental Cytology" (D. Rudnick, ed.), pp. 83–103. Ronald, New York.

Beermann, W. (1960). Der Nucleolus als lebenswichtiger Bestandteil des Zellkernes. *Chromosoma* **11,** 263–296.

Beermann, W. (1961). Ein Balbiani—Ring als Locus einer Speicheldrusenmutation. *Chromosoma* **12,** 1–25.

Beermann, W. (1962). Riesenchromosomen. *In* "Protoplasmatologia," Vol. 6. Springer, Wien.

Beermann, W. (1963a). Kontrollierte RNS-Synthese in Riesenchromosomen. *In* "Struktur und Funktion des genetischen Material (III)", pp. 211–223. Erwin Bauer, Berlin.

Beermann, W. (1963b). Cytological aspects of information transfer in cellular differentiation. *Am. Zool.* **3,** 23–32.

Beermann, W. (1964). Control of differentiation at the chromosomal level. *J. Exp. Zool.* **157,** 49–62.

Beermann, W. (1965). Structure and function of interphase chromosomes. *In* "Genetics Today: XI International Congress of Genetics" (S. J. Gerts, ed.), Vol. 2, pp. 375–384. Pergamon Press, Oxford/London/Edinburgh/New York/Paris/Frankfurt.

Beermann, W. (1966). Differentiation at the level of the chromosomes. *In* "Cell Differentiation and Morphogenesis," pp. 24–54. North Holland, Amsterdam.

Beermann, W. (1967). Gene action at the level of the chromosome. *In* "Heritage from Mendel" (R. A. Brink, and E. D. Styles, eds.), pp. 179–201. Univ. of Wisconsin Press, Madison.

Beermann, W. (1971). Effect of a-amanitin on puffing and intranuclear RNA synthesis in *Chironomus* salivary glands. *Chromosoma* **34,** 152–167.

Beermann, W. (1972a). Chromomeres and genes. *In* "Results and Problems in Cell Differentiation" (W. Beermann, ed.), Vol. 4, pp. 1–33. Springer, Berlin/Heidelberg/New York.

Beermann, W. (1972b). Functional aspects of the chromomere organization of eukaryotic chromosomes. *In* "Kletochnoe yadro. Morfologia, fiziologia, biochimia", (Sbornik pod redakts. Zbarskogo, I. B. i Georgieva, G. P.), pp. 190–202. Nauka, Moscow. [in Russian]

Beermann, W. (1973). Directed changes in the pattern of Balbiani Ring puffing in Chironomus: Effects of a sugar treatment. *Chromosoma* **41,** 297–326.

Beermann, W., and Bahr, G. F. (1954). The submicrscopic structure of the Balbiani Ring. *Exp. Cell Res.* **6,** 195–201.

Beermann, W., and Clever, U. (1964). Chromosome puffs. *Sci. Am.* **210,** 50–58.

Beermann, W., and Clever, U. (1966). Chromosome puffs. *In* "Molekuly i kletki," pp. 145–158. Mir, Moscow. [in Russian]

Behnel, H. J. (1978). Changes of enzyme activity in larval salivary glands following the induction of respiration dependent puffs in giant chromosomes of *Drosophila melanogaster*. *Cell Differ.* **7,** 215–222.

Behnel, H. J. (1982). Comparative study of protein synthesis and heat-shock puffing activity in *Drosophila* salivary glands treated with chloramphenicol. *Exp. Cell Res.* **142,** 223–228.

Behnel, H. J., and Rensing, L. (1975). Respiratory functions involved in the induction of puffs in Drosophila salivary glands. *Exp. Cell Res.* **91,** 119–124.

Behnel, H. J., and Seydewitz, H. H. (1980). Changes of the membrane potential during formation of heat shock puffs induced by ion carriers in *Drosophila* salivary glands. *Exp. Cell Res.* **127,** 133–141.

Belew, K., and Brady, T. (1981a). Induction of tyrosine aminotransferase by pyridoxine in *Drosophila hydei*. *Chromosoma* **82,** 99–106.

Belew, K., and Brady, T. (1981b). Changes in phenol-soluble nuclear proteins correlated with puff induction in *Drosophila hydei*. *Cell Differ.* **10,** 229–235.

Belikov, S. V., Belgovsky, A. I., Preobrazhenskaya, O. V., Karpov, V. L., and Mirzabekov, A. D. (1993). Two non-histone proteins are associated with the promoter region and histone H1 with the transcribed region of active *hsp-70* genes as revealed by UV- induced DNA-proyein crosslinking *in vivo*. *Nucleic Acids Res.* **21,** 1031–1034.

Belling, J. (1928). The ultimate chromomeres of *Lilium* and *Aloe* with regard to the numbers of genes. *Univ. Calif. Publ. Bot.* **14,** 307–318.

Bello, B., and Couble, P. (1990). Specific expression of a silk-encoding gene of *Bombyx* in the anterior salivary gland of *Drosophila*. *Nature* **346,** 480–482.

Belote, J. M. (1992). Sex determination in *Drosphila melanogaster:* from the X:A ratio to *doublesex*. *Seminars Devel. Biol.* **3,** 319–330.

Belote, J. M., and Lucchesi, J. C. (1980a). Control of X chromosome transcription by the *maleless* gene in *Drosophila*. *Nature* **285,** 573–575.

Belote, J. M., and Lucchesi, J. C. (1980b). Male-specific lethal mutations of *Drosophila melanogaster*. *Genetics* **96,** 165–186.

Belyaeva, E. S. (1971). Structure and functioning of the nucleolus. *Tsitologiya* **13**(6), 733–744. [in Russian]

Belyaeva, E. S. (1982). "Cytogenetic study on the organization and functioning of the transcriptionally active regions of the chromosomes," SciD. thesis. Novosibirsk. [in Russian]

Belyaeva, E. S., and Zhimulev, I. F. (1973). Puff size variability in *Drosophila melanogaster*. *Dros. Inform. Serv.* **50,** 41.

Belyaeva, E. S., and Zhimulev, I. F. (1974). On variability of puff sizes in *Drosophila melanogaster*. *Genetika* **10**(5), 74–80. [in Russian]

Belyaeva, E. S., and Zhimulev, I. F. (1976). RNA synthesis in the *Drosophila melanogaster* puffs. *Cell Differ.* **4,** 415–427.

Belyaeva, E. S., and Zhimulev, I. F. (1982). Cytogenetic analysis of the X-chromosome region 2B3-4–2B11 of *Drosophila melanogaster*. IV. Mutation at the *swi (singed wings)* locus interfering with the late 20-OH ecdysone puff system. *Chromosoma* **86,**251–263.

Belyaeva, E. S., Korochkina, L. S., Zhimulev, I. F., and Nazarova, N. K. (1974). A characteristics of puffs of the X chromosome in *Drosophila melanogaster* females. *Tsitologia* **16**(4), 440–446. [in Russian]

Belyaeva, E. S., Pokholkova, G. V., and Zhimulev, I. F. (1976). A cytogenetic analysis of complementation groups in the 10A1-2 band of the X chromosome of *Drosophila melanogaster*. *Doklady Akad. Nauk SSSR* **228**(5), 1208–1211. [in Russian]

Belyaeva, E. S., Aizenzon, M. G. Ilyina, O. V., and Zhimulev, I. F. (1978). A cytogenetic analysis of puffs in the 2B1-10 region of the X chromosome of *Drosophila melanogaster*. *Doklady Akad. Nauk. SSSR* **240**(5), 1219–1222.

Belyaeva, E. S., Aizenzon, M. G., Semeshin, V. F., Kiss, I., Koczka, K., Baricheva, E. M., Gorelova, T. D., and Zhimulev, I. F. (1980a). Cytogenetic analysis of the 2B3-4–2B11 region of the X-chromosome of *Drosophila melanogaster*. I. Cytology of the region and mutants complementation groups. *Chromosoma* **81,** 281–306.

Belyaeva, E. S., Zhimulev, I. F., and Semeshin, V. F. (1980b). A cytogenetic analysis of the 2B1-2–2B9-10 region of the X chromosome of *Drosophila melanogaster*. I. Cytological localization of rearrangements and puffing patterns. *Genetika* **16**(1), 103–114. [in Russian]

Belyaeva, E. S., Vlassova, I. E., Biyasheva, Z. M., Kakpakov, V. T., Richards, G., and Zhimulev, I. F. (1981). Cytogenetic analysis of the 2B3-4–2B11 region of the X-chromosome of *Drosophila melanogaster*. II. Changes in 20-OH ecdysone puffing caused by genetic defects of puff 2B5. *Chromosoma* **84,** 207–219.

Belyaeva, E. S., Aizenzon, M. G., Kiss, I., Gorelova, T. D., Pak, S. D., Umbetova, G. H., Kramers, P. G. N., and Zhimulev, I. F. (1982). Mutations and chromosome rearrangements in the 1E3-4–2B18 region of *Drosophila melanogaster* X chromosome. *Dros. Inform. Serv.* **58,** 184–190.

Belyaeva, E. S., Protopopov, M. O., Baricheva, E. M., Semeshin, V. F., Izquierdo, M. L., and Zhimulev, I. F. (1987). Cytogenetic analysis of region 2B3-4–2B11 of the X-chromosome of *Drosophila melanogaster*. VI. Molecular and cytological mapping the *ecs* locus and the 2B puff. *Chromosoma* **95,** 295–310.

Belyaeva, E. S., Protopopov, M. O., Dubrovsky, E. B., and Zhimulev, I. F. (1989). Cytogenetic analysis of ecdysteroid action. *In* "Ecdysone" (J. Koolman, ed.), pp. 368–376. Georg Thieme-Verlag, Stuttgart/New York.

Belyanina, S. I. (1977). Chromosomal polymorphism of *Chironomus plumosus* from different parts of distribution area. II. Karyotypic structure of three geographical distinct populations. *Tsitologiya* **19**(5), 565–570. [in Russian]

Belyanina, S. I. (1993). The karyotype of *Demeijerea rufipes* L. - endoparasite of sponges from the Volga river. *In* "Kariosistematika bespozvonochnych zhivotnykh II," pp. 55–58. Trudy Inst. Zool. Ross. Acad. Nauk. [in Russian]

Bendena, W. G., Fini, M. E., Garbe, J. C., Kidder, G. M., Lakhotia, S. C., and Pardue, M. L. (1989a). *hsr:* Different sort of heat shock locus. Stress-induced proteins. *UCLA Symp. Mol. Cell New Ser.* **96,** 3–14.

Bendena, W. G., Garbe, J. C., Traverse, K. L., Lakhotia, S. C., and Pardue, M. L. (1989b). Multiple inducers of the *Drosophila* heat shock locus 93D *(hsr):* Inducer-specific patterns of the three transcripts. *J. Cell Biol.* **108,** 2017–2028.

Bendena, W. G., Ayme-Southgate, A., Garbe, J. C., and Pardue, M. L. (1991). Expression of heat shock locus *hsr*-omega in nonstressed cells during development in *Drosophila melanogaster*. *Dev. Biol.* **144,** 65–77.

Bender, H. A., Barr, H. J., and Ostrowski, R. S. (1971). Asynchronous DNA synthesis in a duplicated chromosomal region of *Drosophila melanogaster*. *Nature New Biol.* **231,** 217–219.

Bender, M. (1996). New lethal loci in the 42AB region of the *Drosophila melanogaster* 2nd chromosome. *Dros. Inform. Serv.* **77,** 143.

Bender, M., Imam, F. B., Talbot, W. S., Ganetzky, B., and Hogness, D. S. (1997). *Drosophila* ecdysone receptor mutations reveal functional differences among receptor isoforms. *Cell* **91,** 777–788.

Bender, W., Akam, M., Karch, F., Beachy, P. A., Peifer, M., Spierer, P., Lewis, E. B., and Hogness, D. S. (1983a). Molecular genetics of the *bithorax* complex in *Drosophila melanogaste*. *Science* **221,** 23–29.

Bender, W., Spierer, P., and Hogness, D. S. (1983b). Chromosomal walking and jumping to isolate DNA from the *Ace* and *rosy* loci and the *Bithorax* complex in *Drosophila melanogaster*. *J. Mol. Biol.* **168,** 17–33.

Bender, W., Weiffenbach, B., Karch, F., and Peifer, M. (1985). Domains of cis-interaction in the *Bithorax* complex. *Cold Spring Harbor Symp. Quant. Biol.* **50,** 173–180.

Benjamin, W. B., and Goodman, R. M. (1969). Phosphorylation of Dipteran chromosomes and rat liver nuclei. *Science* **166,** 629- 631.

Benkel, B. F., Abukashawa, S., Boer, P. H., and Hickey, D. A. (1987). Molecular cloning of DNA complementary to *Drosophila melanogaster* a-amylase mRNA. *Genome* **29,** 510–515.

Berendes, H. D. (1963). The salivary gland chromosomes of *Drosophila hydei* Sturtevant. *Chromosoma* **14,** 195–206.

Berendes, H. D. (1965a). Salivary gland function and chromosomal puffing patterns in *Drosophila hydei*. *Chromosoma* **17,** 35–77.

Berendes, H. D. (1965b). The induction of changes in chromosomal activity in different polytene types of cell in *Drosophila hydei*. *Devel. Biol.* **11,** 371–384.

Berendes, H. D. (1965c). Functional structures in the salivary glands of *Drosophila hydei*. *Arch. Neerland. Zool.* **16**(3), 404–405.

Berendes, H. D. (1965d). Gene homologies in different species on the *repleta* group of the genus *Drosophila*. I. Gene activity after temperature shocks. *Gen. Phaenen.* **10,** 32–41.

Berendes, H. D. (1966a). Gene activities in the Malpighian tubules of *Drosophila hydei* at different development stages. *J. Exp. Zool.* **162,** 209–217.

Berendes, H. D. (1966b). Differential replication of male and female X-chromosome in *Drosophila*. *Chromosoma* **20,** 32–43.

Berendes, H. D. (1966c). The effect of temperature shocks in some related species of the genus *Drosophila*. *Dros. Inform. Serv.* **41,** 168–169.

Berendes, H. D. (1967a). The hormone ecdysone as effector of specific changes in the pattern of gene activities of *Drosophila hydei*. *Chromosoma* **22,** 274–293.

Berendes, H. D. (1967b). Amino acid incorporation into giant chromosomes of *Drosophila hydei*. *Dros. Inform. Serv.* **42,** 102.

Berendes, H. D. (1968a). Factors involved in the expression of gene activity in polytene chromosomes. *Chromosoma* **24,** 418–437.

Berendes, H. D. (1968b). The effect of ecdysone analogues on the puffing pattern of *Drosophila hydei*. *Dros. Inform. Serv.* **43,** 145.

Berendes, H. D. (1969). Induction and control of puffing. *Ann. Embryol. Morphogen.* **1**(Suppl.), 153–164.

Berendes, H. D. (1970). Polytene chromosome structure at the submicroscopic level. I. A map of region X, 1-4E of *Drosophila melanogaster*. *Chromosoma* **29,** 118–130.

Berendes, H. D. (1971). Gene activation in dipteran polytene chromosomes. *In* "Control Mechanisms of Growth and Differentiation," pp. 145–161. Cambridge Univ. Press.

Berendes, H. D. (1972a). The control of puffing in *Drosophila hydei*. *In* "Developmental Studies on Giant Chromosomes" (W. Beermann, ed.), pp. 181–207. Springer-Verlag, Berlin.

Berendes, H. D. (1972b). Changes in specialized cell functions induced by ecdysone. *Proc. 4th. Int. Congr. Endocrinol.*, Series N273, 311–314.

Berendes, H. D. (1973). Synthetic activity of polytene chromosomes. *Int. Rev. Cytol.* **35,** 61–116.

Berendes, H. D. (1975). Experimental puff induction and its consequences in *Drosophila*. *In* "Regulationsmechanismen der Genaktivitat und Replikation bei Riesenchromosomen" (L. Rensing, H. H. Trepte, G. Birukow, eds.), Vol. 11, pp. 185–186. Nachrich. Akad. Wiss. Gottingen. II. Math.-Phys. Klasse.

Berendes, H. D., and Ashburner, M. (1978). The salivary glands. *In* "The Genetics and Biology of *Drosophila*" (M. Ashburner and T. R. F. Wright, eds.), Vol. 2, pp. 453–492. Academic Press, London/New York/San Francisco.

Berendes, H. D., and Beermann, W. (1969). Biochemical activity of interphase chromosomes (polytene chromosomes). *In* "Handbook of Molecular Cytology" (A. Lima-de-Faria, ed.), Vol. 15, pp. 500–519. North Holland, Amsterdam.

Berendes, H. D., and Boyd, J. B. (1969). Structural and functional properties of polytene nuclei isolated from salivary glands of *Drosophila hydei*. *J. Cell Biol.* **41,** 591–599.

Berendes, H. D., and de Bruyn, W. C. (1963). Submicroscopic structure of *Drosophila hydei* salivary gland cells. *Z. Zellforsch.* **59,** 142–152.

Berendes, H. D., and Helmsing, P. J. (1974). Nonhistone proteins of dipteran polytene nuclei. *In* "Acidic Proteins of the Nucleus" (I. L. Cameron and J. R. Jeter, Jr., eds.), pp. 191–212. Academic Press, New York.

Berendes, H. D., and Holt, T. K. H. (1964). The induction of chromosomal activities by temperature shock. *Gen. Phaenen.* **9,** 1–7.

Berendes, H. D., and Holt, T. K. H. (1965). Differentiation of transplanted larval salivary glands of *Drosophila hydei* in adults of the same species. *J. Exp. Zool.* **160,** 299–318.

Berendes, H. D., and Lara, F. J. S. (1975). RNA synthesis: A requirement for hormone-induced DNA amplification in *Rhynchosciara americana*. *Chromosoma* **50,** 259–274.

Berendes, H. D., and Thijssen, W. T. M. (1971). Developmental changes in genome activity in *Drosophila lebanonensis casteeli* Pipkin. *Chromosoma* **33,** 345–360.

Berendes, H. D., van Breugel, F. M. A., and Holt, T. K. H. (1965). Experimental puffs in salivary gland chromosomes of *Drosophila hydei*. *Chromosoma* **16,** 35–46.

Berendes, H. D., Alonso, C., Helmsing, H. J., Leenders, H. J., and Derksen, J. (1974). Structure and function in the genome of *Drosophila hydei*. *Cold Spring Harbor Symp. Quant. Biol.* **38,** 645–654.

Berger, C. A. (1940). The uniformity of the gene complex. *J. Hered.* **31,** 3–4.

Berger, E. M., Vitek, M. P., and Morganelli, C. M. (1985). Transcript length heterogeneity at the small heat shock protein genes of *Drosophila*. *J. Mol. Biol.* **186,** 137–148.

Bernhard, W. (1969). A new staining procedure for electron microscopical cytology. *J. Ultrastruct. Res.* **27,** 250–265.

Berreur, P., Porcheron, P., Berreur-Bonnenfant, J., and Simpson, P. (1979). Ecdysteroid levels and pupariation in *Drosophila melanogaster*. *J. Exptl. Zool.* **210,** 347–352.

Berreur, P., Porcheron, P., Moriniere, M., Berreur-Bonnenfant, J., Belinski-Peuch, S., Busson, D., and Lamour-Audit, C. (1984). Ecdysteroids during the third larval instar in l(3)ecd-1^{ts} a temperature sensitive mutant of *Drosophila melanogaster*. *Gen. Comp. Endocrinol.* **54,** 76–84.

Bgatov, A. V. (1985). "Phenotypic manifestation and action focus of mutations of closely linked genes in the 9F12–10A7 interval of the *Drosophila melanogaster* chromosome," *Ph.D. thesis*. Inst. Cytology and Genetics, Novosibirsk. [in Russian]

Bgatov, A. V., and Zhimulev, I. F. (1982). Cytogenetic study on the 9E-10-A region of the X chromosome of *Drosophila melanogaster*. V. Phenotypic manifestation of the mutations located in the 9F12–10A7 interval. *Genetika* **18**(4), 613–624. [in Russian]

Bgatov, A. V., and Zhimulev, I. F. (1985). Action foci of genes localized in the region of the 9F12 band of the X *Drosophila melanogaster* chromosome. *Doklady Akad. Nauk SSSR* **283**(2), 465–469. [in Russian]

Bgatov, A. V., Zharkikh, A. A., and Zhimulev, I. F. (1983). A cytogenetic study on the 9E–10A region of the X chromosome of *Drosophila melanogaster*. VI. Action focus of the *l(1)BP4*. *Genetika* **19**(11), 1790–1800.

Bgatov, A. V., Zharkikh, A. A., and Zhimulev, I. F. (1984). Fine cytogenetical analysis of the band 10A1-2 and the adjoining regions in the *Drosophila melanogaster* X chromosome. III. The fate mapping of the lethal focus of the *l(1)BP4*. *Mol. Gen. Genet.* **196,** 110–116.

Bgatov, A. V., Pokholkova, G. V., and Zhimulev, I. F. (1986). Fine cytogenetic analysis of the band 10A1-2 and adjoining regions in the *Drosophila melanogaster* X-chromosome. IV. Phenotypic expression of mutations located in region 9F12–10A7. *Biol. Zentralbl.* **105,** 389–405.

Bhadra, M., Bhadra, U., and Chatterjee, R. N. (1992). $MgCl_2$ induced activity pattern of 93D and other puff sites in *Drosophila melanogaster*. *Dros. Inform. Serv.* **71,** 242.

Bhadra, U., and Chatterjee, R. N. (1986). Dosage compensation and template organization in *Drosophila: In situ* transcriptional analysis of the chromatin template activity of the X and autosomes of *D. melanogaster* strains trisomic for the left arm of the second and the third chromosomes. *Chromosoma* **94,** 285–292.

Bhakta, R. K., and Mukherjee, A. S. (1984). *Genetics* and epigenetics of some sex-linked lethals in *Drosophila melanogaster*. *Dros. Inform. Serv.* **60,** 60–61.

Bhakta, R. K., and Mukherjee, A. S. (1991). Replication cytology of polytene chromosomes in a sex linked recessive lethal *(l^4)* in *Drosophila melanogaster*. *Dros. Inform. Serv.* **70,** 32–33.

Bhat, K. M., Farkas, G., Karch, F., Guyrkovics, H., Gausz, J., and Schedl, P. (1996). The GAGA factor is required in the early *Drosophila* embryo not only for transcriptional regulation but also for nuclear division. *Development* **122,** 1113–1124.

de Bianchi, A. G., and Terra, W. R. (1975). Chemical composition and rate of synthesis of the larval secretion of the fly *Rhynchosciara americana*. *J. Insect Physiol.* **21,** 643–657.

de Bianchi, A. G., Terra, W. R., and Lara, F. J. S. (1973). Formation of salivary secretion in *Rhynchosciara americana*. I. Kinetics of labeled amino acid incorporation. *J. Cell Biol.* **58,** 470–476.

Bicudo, H. E. M. C. (1981a). Nucleolar organiser activity and its regulatory mechanisms in *Drosophila* species of the "*Mulleri*" complex and their hybrids. *Caryologia* **34,** 231–253.

de Bicudo, H. E. M. C., (1981b). Is there dosage compensation in *Drosophila* nucleolar organizing activity? *Rev. Brasil. Genet.* **4,** 347–355.

de Bicudo, H. E. M. (1983). Evidence for gene amplification operating in dosage compensation in *Drosophila arizonensis*. *Biol. Zbl.* **102,** 685–695.

de Bicudo, H. E. M. C., and Richardson R. H. (1977). Gene regulaion in *Drosophila mulleri, D. arizonensis*, and their hybrids: The nucleolar organizer. *Proc. Natl. Acad. Sci USA* **74,** 3498–3502.

Bienz, M. (1985). Transient and developmental activation of heat-shock genes. *Trend Biochem. Sci.* **10,** 157–161.

Bienz, M., and Pelham, H. R. B. (1986). Heat shock regulatory elements function as an inducible enhancer in the *Xenopus hsp 70* gene and when linked to a heterologous promoter. *Cell* **45,** 753–760.

Bienz, M., and Pelman, H. R. B. (1987). Mechanisms of heat shock gene activation in higher eukaryotes. *Adv. Genet.* **24,** 31–72.

Bienz-Tadmor, B., and Gerbi, S. A. (1990). The promoter of DNA puff gene II"9-1 of *Sciara coprophila* is inducible by ecdysone in late prepupal salivary glands of *Drosophila melanogaster*. *J. Cell Biol.* **111**(5), 123.

Bienz-Tadmor, B., Smith, H. S., Gerbi, S. A. (1991). The promoter of DNA puff gene II/9-1 of *Sciara coprophila* is inducible by ecdysone in late prepupal salivary glands of *Drosophila melanogaster*. *Cell Regul.* **2,** 875–888.

Bier, K. (1960). Der Karyotyp von *Calliphora erythrocephala* Meigen unter besonderer berucksichtigung der Nahrzellenkernchromosomen im debundelten und gepaarten zustand. *Chromosoma* **11,** 335–364.

Bier, K., Kunz, S., and Ribbert, D. (1969). Insect oogenesis with and without lampbrush chromosomes. *In* "Chromosomes Today" (C. D. Darlington and K. R. Lewis, eds.), Vol. 2, 107–115. Oliver and Boyd, Edinburgh.

Biessmann, H., Levy, B. W., and McCarthy, B. J. (1978a). *In vitro* transcription of heat-shock specific RNA from chromatin of *Drosophila melanogaster* cells. *Proc. Natl. Acad. Sci. USA* **75,** 759–763.

Biessmann, H., Wadsworth, S., Levy, W., and McCarthy, B. J. (1978b). Correlation of structural changes in chromatin with transcription in the *Drosophila* heat-shock response. *Cold Spring Harbor Symp. Quant. Biol.* **42,** 829–834.

Birchler, J. A., Hiebert, J. C., and Krietzman, M. (1989). Gene expression in adult metafemales of *Drosophila melanogaster*. *Genetics* **122,** 869–879.

Birnstiel, M. L. (ed.) (1988). "Small Nuclear Ribonucleoprotein Particles." Springer-Verlag, Berlin.

Birnstiel, M. L., Weinberg, E. S., and Pardue, M. L. (1973). Evolution of 9S mRNA sequences. *In* "Molecular Cytogenetics" (B. A. Hamkalo, and J. Papaconstantinou, eds.), pp. 75–93. Plenum, New York/London.

Bishoff, W. L., and Lucchesi, J. C. (1971). Genetic organization in *Drosophila melanogaster:* Complementation and fine structure analysis of the *deep orange locus*. *Genetics* **69,** 453.

Bishop, D. L., and King, R. C. (1984). An ultrastructural study of ovarian development in the otu^7 mutant of *Drosophila melanogaster*. *J. Cell Sci.* **67,** 87–119.

Bishop, J. O. (1974). The gene numbers game. *Cell* **2,** 81–86.

Bisseling, T., Berendes, H. D., and Lubsen, N. H. (1976). RNA synthesis in puff 2-48BC after experimental induction in *Drosophila hydei*. *Cell* **8,** 299–304.

Biyasheva, Z. M. (1987). "A gene cluster in the region in the 2B region of the X chromosome of *Drosophila melanogaster* involved in the control of ecdysone induction," *Ph. D. thesis*. Inst. of Cytology and Genetics, Novosibirsk. [in Russian]

Biyasheva, Z. M., Zhimulev, I. F., Kakpakov, V. T., Richards, G., Aizenzon, M. G., and Belyaeva, E. S. (1981). Two closely linked loci controlling ecdysone-dependent puffing in *Drosophila melanogaster*. *Doklady Akad. Nauk SSSR* **256**(4), 983–986. [in Russian]

Biyasheva, Z. M., Belyaeva, E. S., and Zhimulev, I. F. (1985). Cytogenetic analysis of the X-chromosome region 2B1-2–2B9-10 of *Drosophila melanogaster*. V. Puffing disturbances carried by lethal mutants of the gene *dor* are due to ecdysterone deficiency. *Chromosoma* **92,** 351–356.

Biyasheva, Z. M., Belyaeva, E. S., and Zhimulev, I. F. (1986). Cytogenetic analysis of the X chromosome region 2B1-2–2B9-10 of *Drosophila melanogaster*. VI. Puffing disturbances in the polytene chromosomes of the lethal mutant of the *dor* gene are due to ecdysterone deficiency. *Genetika* **22**(6), 975–982. [in Russian]

Biyasheva, Z. M., Protopopov, M. O., and Belyaeva, E. S. (1991). Some characteristics of mutations in *sta* and *HM40* loci in the 2AB region of *Drosophila melanogaster* X-chromosome. *Dros. Inform. Serv.* **70,** 37–38.

Bjorkroth, B., Ericsson, C., Lamb, M. M., and Daneholt, B. (1988). Structure of the chromatin axis during transcription. *Chromosoma* **96,** 333–340.

Black, B. C., Pentz, E. S., and Wright, T. R. F. (1987). The alpha methyl dopa hypersensitive gene, *l(2)amd,* and two adjacent genes in *Drosophila melanogaster:* Physical location and direct effects of *amd* on catecholamine metabolism. *Mol. Gen. Genet.* **209,** 306–312.

Blackman, R. K., and Meselson, M. (1986). Interspecific nucleotide sequence comparisons used to identify regulatory and structural features of the *Drosophila hsp82 gene*. *J. Mol. Biol.* **188,** 499–515.

Block, K. (1969). Chromosomal variation in Agromyzidae. I. *Phytomyza abdominalis* Zett.—Two incipient species and their natural hybrid. *Hereditas* **62,** 131–152.

Blumel, J., and Kirby, H. (1948). Amino acid constituents of tissues and isolated chromosomes of *Drosophila*. *Proc. Natl. Acad. Sci. USA* **34,** 561–566.

Blumenthal, A. B., Kriegstein, H. J., and Hognes, D. S. (1974). The units of DNA replication in *Drosophila melanogaster* chromosomes. *Cold Spring Harbor Symp. Quant. Biol.* **38,** 205–223.

Bode, J., Schlake, T., Rios-Ramires, M., Mielke, C., Stengert, M., Kay, V., and Klehr-Wirth, D. (1995). Scaffold/Matrix-attached regions: Structural properties creating transcriptionally active loci. *Int. Rev. Cytol.* **162A,** 389–454.

Bodenstein, D. (1943). Factors influencing growth and metamorphosis of the salivary gland in *Drosophila*. *Biol. Bull.* **84,** 13–33.

Bodenstein, D. (1950). The postembryonic development of *Drosophila*. *In* "Biology of *Drosophila*" (M. Demerec, ed.), pp. 275–367. Wiley, New York.

Boer, P. H., and Hickey, D. A. (1986). The a-amylase gene in *Drosophila melanogaster:* Nucleotide sequence, gene structure and expression motifs. *Nucleic Acids Res.* **14,** 8399–8411.

Bogachev, S. S. (1989). "Analysis of the organization and expression of coding sequences ring BRa of *Chironomus thummi*," *Ph.D. thesis*. Novosibirsk. [in Russian]

Bogachev, S. S., Blinov, A. G., Blinov, V. M., Gaidamakova, E. K., Kiknadze, I. I., and Shakhmuradov, I. A. (1986). Some structural elements of DNA sequences from Balbiani Ring (BRa) region IV of *Chironomus thummi*. *Doklady Akad. Nauk SSSR* **288**(4), 230–233. [in Russian]

Bogachev, S. S., Blinov, A. G., Kolesnikov, N. N., Blinov, V. M., Fedorov, S. P., Gaidamakova, E. K., Panova, T. M., and Kiknadze, I. I. (1987). Analysis of DNA sequences from the tissue-specific puffs BRa of *Chironomus thummi*. *Doklady Akad. Nauk SSSR* **296**(6), 1473–1476. [in Russian]

Bogachev, S. S., Shcherbik, S. V., Taranin, A. V., and Sebeleva, T. E. (1989). Analysis of the expression of the gene from the tissue- specific puff BRa of *Chironomus thummi* encoding low molecular weight secretory polypeptide. *Genetika* **25**(9), 1541–1550. [in Russian]

Bogachev, S. S., Blinov, A. G., Kolesnikov, N. N., Scherbik, S. V., Taranin, A. V., Sebeleva, T. E., Baiborodin, S. I., and Kiknadze, I. I. (1990). A tissue-specific puff (Balbiani Ring a) in *Chironomus thummi* may contain a gene encoding a 67 kDa protein which exhibits non-tissue-specific expression. *Gene* **96,** 241–247.

Bonaldo, M. F., Santelli, R. V., and Lara, F. J. S. (1979). The transcript from a DNA puff of *Rhyn chosciara* and its migratio to the cytoplasm. *Cell* **17,** 827–833.

Bone, J. R., and Kuroda, M. I. (1996). Dosage compensation regulatory proteins and the evolution of sex chromosomes in *Drosophila*. *Genetics* **144,** 705–713.

Bone, J. R., Lavender, J., Richman, R., Palmer, M. J., Turner, B. M., and Kuroda, M. I. (1994). Acetylated histone H4 on the male X chromosome in *Drosophila*. *Genes Dev.* **8,** 96–104.

Bonner, J. J. (1981). Induction of *Drosophila* heat shock puffs in isolated polytene nuclei. *Dev. Biol.* **86,** 409–418.

Bonner, J. J. (1982). Regulation of the *Drosophila* heat-shock response. *In* "Heat Shock from Bacteria to Man" (M. L. Schlesinger, M. Ashburner, and A. Tissieres, eds.), pp. 147–153. Cold Spring Harbor Laboratory Press, Cold Spring Harbor, NY.

Bonner, J. J. (1985). Mechanism of transcriptional control during heat shock. *In* "Changes in Eukaryitic Gene Expression in Response to Environmental Stress" (B. G. Atkinson and D. B. Walden, eds.), pp. 31–51. Academic Press, New York.

Bonner, J. J., and Kerby, R. L. (1982). RNA polymerase II transcribes all of the heat shock induced genes of *Drosophila melanogaster*. *Chronosoma* **85,** 93–108.

Bonner, J. J., and Pardue, M. L. (1976a). Ecdysone stimulated RNA synthesis in imaginal discs of *Drosophila melanogaster*. *Chromosoma* **58,** 87–99.

Bonner, J. J., and Pardue, M. L. (1976b). The effect of heat shock on RNA synthesis in *Drosophila* tissues. *Cell* **8,** 43–50.

Bonner, J. J., and Pardue, M. L. (1977a). Ecdysone-stimulated RNA synthesis in salivary glands of *Drosophila melanogaster:* assay by *in situ* hybridization. *Cell* **12,** 219–225.

Bonner, J. J., and Pardue, M. L. (1977b). Polytene chromosome puffing and *in situ* hybridization measure different aspects of RNA metabolism. *Cell* **12,** 227–234.

Bonner, J., and Wu, J.-R. (1973). A proposal for the structure of the *Drosophila* genome. *Proc. Natl. Acad. Sci. USA* **70,** 535–537.

Bonner, J. J., Berninger, M., and Pardue, M. L. (1978). Transcription of polytene chromosomes and of the mitochondrial genome in *Drosophila melanogaster*. *Cold Spring Harbor Symp. Quant. Biol.* **42,** 803–814.

Bonner, J. J., Parks, C., Parker-Thornburg, J., Mortin, M. A., and Pelham, H. R. B. (1984). The use of promoter fusions in *Drosophila* genetics: Isolation of mutations affecting the heat shock response. *Cell* **37,** 979–991.

Bonorino, C. B. C., Pereira, M., Alonso, C. E. V., Valente, V. L. S., and Abdelhay, E. (1993a). *In situ* mapping of the *hsp70* locus in seven species of the *willistoni* group of *Drosophila*. *Rev. Brasil. Genet.* **16,** 561–571.

Bonorino, C. B. C., Silva, T. C., Abdelhay, E., and Valente, V. L. S. (1993b). Heat shock genes in the *willistoni* group of *Drosophila:* Induced puffs and proteins. *Cytobios* **73,** 49–64.

Borst, D. W., Bollenbacher, W. E., O'Connor, J. D., King, D. S., and Fristrom, J. W. (1974). Ecdysone levels during metamorphosis of *Drosophila melanogaster*. *Dev. Biol.* **39,** 308–316.

Bossy, B., Hall, L. M. C., and Spierer, P. (1984). Genetic activity along 315 kb of the *Drosophila* chromosome. *EMBO J.* **3,** 2537–2541.

Botella, L. M., and Edstrom, J.-E. (1991). The Balbiani Ring 6 induction in *Chironomus*. *Biol. Cell* **71,** 11–16.

Botella, L. M., Grond, C., Saiga, H., and Edstrom, J.-E. (1988). Nuclear localization of a DNA-binding C-terminal domain from Balbiani Ring coded secretory protein. *EMBO J.* **7,** 3881–3888.

Botella, L. M., Donoro, C., Sanchez, L., Segarra, C., and Granadino, B. (1996). Cloning and characterization of the *scute (sc)* gene of *Drosophila subobscura. Genetics* **144,** 1043–1051.

Bourouis, M., and Richards, G. (1985a). Remote regulatory sequences of the *Drosophila* glue gene *Sgs3* as revealed by P-element transformation. *Cell* **40,** 349–357.

Bourouis, M., and Richards, G. (1985b). Hybrid genes in the study of glue gene regulation in *Drosophila. Cold Spring Harbor Symp. Quant. Biol.* **50,** 355–360.

Bowman, J. T., and Simmons, J. R. (1973). Gene modulation in *Drosophila:* dosage compensation of Pgd^+ and Zw^+ genes. *Biochem. Genet.* **10,** 319–331.

Bownes, M. (1989). Vitellogenesis. *In* "Ecdysone: From Chemistry to Mode of Action" (J. Koolman, ed.), pp. 414–420. Georg Thieme-Verlag, Stuttgart/New York.

Bownes, M. (1994). The regulation of the yolk protein genes in *Drosophila melanogaster. BioEssays* **16,** 745–752.

Bowtell, D. D. L., Simon, M. A., and Rubin, G. M. (1988). Nucleotide sequence and structure of the *sevenless* gene of *Drosophila melanogaster. Gen. Dev.* **2,** 620–634.

Boyd, J. B., and Logan, W. R. (1972). Developmental variations of a deoxyribonuclease in the salivary glands of *Drosophila hydei* and *D. melanogaster. Cell Differ.* **1,** 107–118.

Boyd, L., and Thummel, C. S. (1993). Selection of CUG and AUG initiator codons for *Drosophila* E74A translation depends on downstream sequences. *Proc. Natl. Acad. Sci. USA* **90,** 9164–9167.

Boyd, L., O'Toole, E., and Thummel, C. S. (1991). Patterns of E74A RNA and protein expression at the onset of metamorphosis in *Drosophila. Development* **112,** 981–995.

Boyd, M., and Ashburner, M. (1977). The hormonal control of salivary gland secretion in *Drosophila melanogaster:* Studies *in vitro. J. Insect Physiol.* **23,** 517–523.

Bozhkova, V. P., Boitsova, L. Y., Kovalev, S. A., Mittelman, L. A., Chailakhyan, L. M., Sharovskaya, Y. Y., and Shilyanskaya, E. N. (1970a). High penetrance of contact membranes, a possible mechanism of cell interaction. *In* "Mezhkletochnoe vzaimodeistvie v differentsirovke i roste," pp. 183–193. Nauka, Moscow. [in Russian]

Bozhkova, V. P., Kovalev, S. A., Mittelman, L. A., and Shilyanskaya, E. N. (1970b). Changes in conductance of intercellular contacts in the process of differentiation of the salivary gland cells of *Drosophila virilis* larvae at the third larval instar. *Tsitologia* **12**(9), 1108–1115. [in Russian]

Bradshaw, G. J., and Baker, B. S. (1995). The msl-2 dosage compensation gene of *Drosophila* encodes a putative DNA-binding protein whose expression is sex specifically regulated by Sex-lethal. *Development* **121,** 3245–3258.

Bradshaw, G. J., and Baker, B. S. (1996). Dosage compensation and chromatin structure in *Drosophila. Curr. Opin. Genet. Dev.* **6,** 496–501.

Bradshaw, W. S., and Papaconstantinou, J. (1970). Differential incorporation of 5-bromodeoxyuridine into DNA puffs of larval salivary gland chromosomes in *Rhynchosciara. Biochem. Biophys. Res. Commun.* **41,** 306–312.

Brady, T., and Belew, K. (1981). Pyridoxine induced puffing (II-48C) and synthesis of a 40kD protein in *Drosophila hydei. Chromosoma* **82,** 89–98.

Brady, T., and Clutter, M. E. (1974). Structure and replication of Phaseolus polytene chromosomes. *Chromosoma* **45,** 63–79.

Brady, T., Berendes, H. D., and Kuijpers, A. M. C. (1974). Gene activation following microinjection of b-ecdysone into nuclei and cytoplasm of larval salivary gland cells of *Drosophila. Mol. Cell Endocrinol.* **1,** 249–257.

Brady, T., Bailey, J. F., and Payne, M. B. (1977). Scanning electron microscopy of isolated *Chironomus* chromosomes. *Chromosoma* **60,** 179–186.

Breen, T. R., and Lucchesi, J. C. (1986). Analysis of the dosage compensation of a specific transcript in *Drosophila melanogaster. Genetics* **112,** 483–491.

Breuer, M. E. (1967). Cromossomos politenicos das glandulas salivares de "*Rhynchosciara angelae*" Nonato et Pavan, 1951 (Diptera, Sciaridae). *Rev. Brasil. Biol.* **1,** 105–108.

Breuer, M. E., and Pavan, C. (1952). Gens na diferenciacao. *Cienc. Cult.* **4,** 115.

Breuer, M. E., and Pavan, C. (1954). Salivary chromosome and differentia tion. Proc. IX Intern. Congr. Genet. *Caryologia* **6,** (Suppl.), 778.

Breuer, M. E., and Pavan, C. (1955). Behaviour of polytene chromosomes of *Rhynchosciara angelae* at different stages of larval development. *Chromosoma* **7,** 371–386

van Breugel, F. M. A. (1965). Experimental puffs in *D. hydei* salivary gland chromosomes observed after treatment with CO, CO2 and N2. *Dros. Inform. Serv.* **40,** 62.

van Breugel, F. M. A. (1966). Puff induction in larval salivary gland chromosomes of *Drosophila hydei* Sturtevant. *Genetica* **37,** 17–28.

van Breugel, F. M. A., and Bos, H. J. (1969). Some notes on differential chromosome activity in *Drosophila*. *Genetica* **40,** 359–378.

Bridges, C. B. (1922). The origin of variations in sexual and sex- limited characters. *Am. Natur.* **56,** 51.

Bridges, C. B. (1935). Salivary chromosome maps with a key to the banding of the chromosomes of *Drosophila melanogaster*. *J. Hered.* **26,** 60–64.

Bridges, C. B. (1937). Correspondences between linkage maps and salivary chromosome structure, as illustrated in the tip of chromosome 2R. *Cytologia* 745–755.

Bridges, C. B. (1938). A revised map of the salivary gland X-chromosome of *Drosophila melanogaster*. *J. Hered.* **29,** 11–13.

Britten, R. J., and Davidson, E. H. (1973). Organization, transcription and regulation in the animal genome. *Quart. Rev. Biol.* **48,** 565–613.

Brock, H. W., and Roberts, D. B. (1983). Location of the *LSP-1* genes in *Drosophila* species by *in situ* hybridizaion. *Genetics* **103,** 75–92.

Brown, D. D., and Gurdon, J. B. (1978). Cloned single repeatig units of 5S DNA direct accurate transcription of 5S RNA when injected into *Xenopus* oocytes. *Proc. Natl. Acad. Sci. USA* **75,** 2849–2853.

Brumley, L. L., Bogachev, S., Kolesnikov, N. N., Waite, J. H., and Case, S. T. (1993). Divergence and conservation of epitopes in intermediate size secretory proteins from three species of *Chironomus*. *Comp. Biochem. Physiol.* B**104**(4), 731–738.

Brun, C., Qi, D., and Miassod, R. (1990). Studies of an 800-kilobase DNA stretch of the *Drosophila* X chromosome: Comapping of a subclass of scaffold-attached regions with sequences able to replicate autonomously in *Saccharomyces cerevisiae*. *Mol. Cell. Biol.* **10,** 5455–5463.

Buetti, E. (1968). "Einfluss von Elektrolyten auf das Puff-Muster der Riesenchromosomen in den Malpighigefassen von *Chironomus thummi*," *Diploma-Thesis Eidgenossischen Technischen Hochschule. Zurich*, Switzerland.

Bukhteeva, N. M. (1974). Karyotype and inversional polymorphism characteristics of *Chironomus plumosus* L. from Eastern Siberia. *Tsitologiya* **16**(3), 358–361. [in Russian]

Bultmann, H. (1986a). Heat shock responses in polytene foot pad cells of *Sarcophaga bullata*. *Chromosoma* **93,** 347–357.

Bultmann, H. (1986b). Induction of a heat shock puff by hypoxia in polytene foot pad chromosomes of *Sarcophaga bullata*. *Chromosoma* **93,** 358–366.

Bultmann, H., and Clever, U. (1969). Chromosomal control of foot pad development in *Sarcophaga bullata*. I. The puffing pattern. *Chromosoma* **28,** 120–135.

Burdette, W. J. (1969a). Causality, casuistry and clinical carcinogenesis. *Progr. Exp. Tumor Res.* **11,** 395–430.

Burdette, W. J. (1969b). Tumors, hormones and viruses in *Drosophila*. *Natl. Cancer Inst. Monogr.* **31,** 303–321.

Burdette, W. T. (1964). The significance of invertebrate hormones in relation to differentiation. *Cancer Res.* **24,** 521–536.

Burdette, W. T., and Anderson, R. (1965a). Puffing of salivary gland chromosomes after treatment with carbon dioxide. *Nature* **208,** 409–410.

Burdette, W. T., and Anderson, R. (1965b). Conditioned response of salivary gland chromosomes of *Drosophila melanogaster* to ecdysone. *Genetics* **51,** 628–633.

Burdette, W. T., and Bullock, M. W. (1963). Ecdysone: five biologically active fractions from *Bombyx*. *Science* **140,** 1311.

Burdette, W. T., and Carver, J. E. (1971). A procedure for quantitative analysis of RNA synthesis at puffing sites in salivary gland chromosomes of *Drosophila melanogaster*. *Dros. Inform. Serv.* **46,** 159–160.

Burdette, W. T., and Kobayashi, M. (1969). Response of chromosomal puffs to crystalline hormones *in vivo*. *Proc. Soc. Biol. Med.* **131,** 209–213.

Burdette, W. T., and Yoon, J. S. (1967). Mutations, chromosomal aberrations, and tumors in insects treated with oncogenic virus. *Science* **155,** 340–341.

Burdon, R. H. (1986). Heat shock and the heat shock proteins. *Biochem. J.* **240,** 313–324.

Burdon, R. H. (1988). The heat shock proteins. *Endeavour* **2,** 133–138.

Burgauer, S. A., and Stockert, J. C. (1975). Observations on the selective demnostration of nucleolar material by protein staining techniques in epon thick sections. *Histochemistry* **41,** 241–247.

Burkhart, B. D., Montgomery, E., Langley, C. H., and Voelker, R. A. (1984). Characterization of allozyme null and low activity alleles from two natural populations of *Drosophila melanogaster*. *Genetics* **107,** 295–306.

Burkholder, G. D. (1977). Whole mount electron microscopy of the nucleolus in salivary gland cells of *Drosophila melanogaster*. *Can. J. Genet. Cytol.* **19,** 21–29.

Burma, P. K., and Lakhotia, S. C. (1984). Cytological identity of 93D-like and 87C-like heat shock loci in *Drosophila pseudoobscura*. *Ind. J. Exp. Biol.* **22,** 577–580.

Burma, P. K., and Lakhotia, S. C. (1986). Expression of 93D heat shock puff of *Drosophila melanogaster* in deficiency genotypes and its influence on activity of the 87C puff. *Chromosoma* **94,** 273–278.

Burnet, B., and Hartmann-Goldstein, I. (1971). Environmental influences on puffing in the salivary gland chromosomes of *Drosophila melanogaster*. *Genet. Res. Cambr.* **17,** 113–124.

Burtis, K. C., Thummel, C. S., and Hogness, D. (1984). Genetic and molecular analysis of an ecdysone inducible gene from *Drosophila melanogaster*. *Genetics* **107,** S15.

Burtis, K. C., Thummel, C. S., Jones, C. W., Karim, F. D., and Hogness, D. S. (1990). The *Drosophila* 74EF early puff contains E74, a complex ecdysone-inducible gene that encodes two ets-related proteins. *Cell* **61,** 85–99.

Busch, H., and Rothblum, L. (eds.) (1982). "The cell nucleus X., rDNA: Part A." Academic Press, New York.

Busen, W., Amabis, J. M., Leoncini, O., Stollar, B. D., and Lara, F. J. S. (1982). Immunofluorescent characterization of DNA-RNA hybrids on polytene chromosomes of *Trichosia pubescens* (Diptera, Sciaridae). *Chromosoma* **87,** 247–262.

Bush, G. L., and Taylor, S. C. (1969). The cytogenetics of Procecidochares. I. The mitotic and polytene chromosomes of the pamakani fly, *P. utilis* Stone (Tephritidae-Diptera). *Caryologia* **22,** 311–321.

Butenandt, A., and Karlson, P. (1954). Uber die Isolierung eines Metamorphose-Hormons der Insekten in Kristallisierten Form. Zeitschrift. *Naturforsch* **96,** 389–391.

Butler, B., Pirrotta, V., Irminger-Finger, I., and Nothiger, R. (1986). The sex-determining gene *tra* of *Drosophila:* Molecular cloning and transformation studies. *EMBO J.* **5,** 3607–3613.

Butterworth, F. M., and Berendes, H. D. (1974). Ecdysone-binding proteins in haemolymph and tissues of *Drosophila hydei* larvae. *J. Insect Physiol.* **20,** 2195–2205.

Buzin, C. H., and Petersen, N. S. (1982). A comparison of the multiple *Drosophila* heat shock proteins in cell lines and larval salivary glands by two-dimensional gel electrophoresis. *J. Mol. Biol.* **158,** 181–201.

Caggese, C., Caizzi, R., Morea, M., Scalenghe, F., and Ritossa, F. (1979). Mutation generating a fragment of the major heat shock-inducible polypeptide in *Drosophila melanogaster*. *Proc. Natl. Acad. Sci. USA* **76,** 2385–2389.

Caggese, C. M., Bozzetti, M., Palumbo, G., and Barsanti, P. (1983). Induction by hydrocortisone-21-sodium succinate of 70kD heat-shock polypeptide in isolated salivary glands of *Drosophila melanogaster* larvae. *Experientia* **39,** 1143–1144.

Caizzi, R., Ritossa, F., Ryseck, R.-P., Richter, S., and Hovemann, B. (1987). Characterization of the *ebony* locus in *Drosophila melanogaster*. *Mol. Gen. Genet.* **206,** 66–70.

Callan, H. G. (1963). The nature of lampbrush chromosomes. *Int. Rev. Cytol.* **15,** 1–34.

Callan, H. G. (1972). Replication of DNA in the chromosomes of eukaryotes. *Proc. Roy. Soc. Lond. Ser. B* **181,** 19–41.

Callan, H. G. (1974). DNA replication in the chromosomes of eukaryotes. *Cold Spring Harbor Symp. Quant. Biol.* **38,** 195–203.

Callan, H. G. (1978). Functional units. Introductionary remarks. *Phil. Trans. Roy. Soc. Lond. B* **283,** 381–382.

Callan, H. G. (1986). "Lampbrush Chromosomes." Springer-Verlag, Berlin.

Camargo, E. P., and Plaut, W. (1967). The radioautographic detection of DNA with tritiated actinomycin D. *J. Cell Biol.* **35,** 713–716.

Campbell, S. D., Hilliker, A. J., and Philips, J. P. (1984). Genetic analysis of the 68A-C region in *Drosophila melanogaster* using hybrid dysgenesis and EMS. *Genetics* **107**(3), 16–17.

Campbell, S. D., Hilliker, A. J., and Philips, J. P. (1985). Genetic analysis of the cSOD microregion in *Drosophila melanogster*. *Genetics* **112,** 205–215.

Campos-Ortega, J. A., and Hartenstein, V. (1985). "The Embryonic Development of *Drosophila melanogaster*." Springer-Verlag, Berlin.

Campuzano, S., and Modolell, J. (1992). Patterning of the *Drosophila* nervous system: The *achaete-scute* gene complex. *Trends Genet.* **8**(6), 202–208.

Campuzano, S., Carramolino, L., Cabrera, C. V., Ruiz-Gomez, M., Villares, R., Boronat, A., and Modolell, J. (1985). Molecular genetics of the *achaete-scute* gene complex of *Drosophila melanogaster*. *Cell* **40,** 327–338.

Cannon, G. B. (1965). Puff development and RNA synthesis in *Sciara* salivary gland chromosomes in tissue culture. *J. Cell. Comp. Physiol.* **65,** 163–182.

Carbajal, M. E., Duband, J.-L., Lettre, F., Valet, J.-P., and Tanguay, R. M. (1986). Cellular localization of *Drosophila* 83- kilodalton heat shock protein in normal heat-shocked, and recovering cultured cells with a specific antibody. *Biochem. Cell Biol.* **64,** 816–825.

Carmona, M. J., Morcillo, G., Galler, R., Martinez-Salas, E., de la Campa, A. G., Diez, J. L., and Edstrom, J.-E. (1985). Cloning and molecular characterization of a telomeric sequence from a temperature induced Balbiani ring. *Chromosoma* **92,** 108–115.

Carretero, M. T., Carmona, M. J., Morcillo, G., Barettino, D., and Diez, J. L. (1986). Asynchronous expression of heat-shock genes in *Chironomus thummi*. *Biol. Cell* **56,** 17–21.

Carthew, R. W., and Rubin, G. M. (1990). *Seven in absentia*, a gene required for specification of R7 cell fate in the *Drosophila* eye. *Cell* **63,** 561–577.

Casartelli, C., Basile, R., and Dos Santos, E. P. (1969). DNA and RNA synthesis in *Rhynchosciara angelae* Nonato and Pavan (1951) and the problem of dosage compensation. *Caryologia* **22,**203–211.

Case, S. (1982). Selective deletion of large segments of Balbiani Ring DNA during molecular cloning. *Gene* **20,** 169–176.

Case, S. (1986). Correlated changes in steady-state levels of Balbiani Ring mRNA and secretory polypeptides in salivary glands of *Chironomus tentans*. *Chromosoma* **94,** 483–491.

Case, S., and Bower, J. R. (1983). Characterization of a cloned, moderately repeated sequence from Balbiani ring 2 in *Chironomus Tentans*. *Gene* **22,** 85–93.

Case, S., and Byers, M. R. (1983). Repeated nucleotide sequence arrays in Balbiani ring 1 of *Chironomus tentans* contain internally nonrepeating and subrepeating elements. *J. Biol. Chem.* **258,** 7793–7799.

Case, S., and Daneholt, B. (1977). Cellular and molecular aspects of genetic expression in *Chironomus* salivary glands. *Int. Rev. Biochem.* **15,** 45–77.

Case, S., and Daneholt, B. (1978). The size of the transcription unit in Balbiani Ring 2 of *Chironomus tentans* as derived from analysis of the primary transcript and 75S RNA. *J. Mol. Biol.* **124,** 223–241.

Case, S., Summers, R. L., and Jones, A. G. (1983). A variant tandemly repeated nucleotide sequence in Balbiani ring 2 of *Chironomus tentans*. *Cell* **36,** 555–562.

Cassagnau, P. (1971). Les chromosomes salivaires polytenes chez *Bilobella grassei* (Denis) (Collemboles: Neanuridae). *Chromosoma* **35,** 57–83.

Cassagnau, P. (1976). La variabilite des chromosomes polytenes chez *Bilobella aurantiaca* Caroli (Collembola Neanuridae) et ses rapports avec la biogeographie et l'ecologie de l'espece. *Arch. Zool. Exp. Genet.* **117,** 511–572.

Cassagnau, P., Dallai, R., and Deharveng, L. (1979). Le polymorphisme des chromosomes polytenes de *Lathriopyga longiseta* Caroli (Collembola Neanuridae). *Caryologia* **32,** 461–483.

Cavener, D., Corbett, G., Cox, D., and Whetten, R. (1986a). Isolation of the eclosion gene cluster and the developmental expression of the *Gld* gene in *Drosophila melanogaster*. *EMBO J.* **5,** 2939–2948.

Cavener, D. R., Otteson, D. C., and Kaufman, T. C. (1986b). A rehabilitation of the genetic map of the 84B-D region in *Drosophila melanogaster*. *Genetics* **114,** 111–123.

Champlin, D., Frasch, M., Saumweber, H., and Lis, J. T. (1991). Characterization of a *Drosophila* protein associated with boundaries of transcriptionally active chromatin. *Genes Dev.* **5,** 1611–1621.

Champlin, D. T., and Lis, J. T. (1994). Distribution of B52 within a chromosomal locus depends on the level of transcription. *Mol. Biol. Cell* **5,** 71–79.

Chao, A. T., and Guild, G. M. (1986). Molecular analysis of the ecdysterone-inducible 2B5 "early" puff in *Drosophila melanogaster*. *EMBO J.* **5,** 143–150.

Charles, J.-P., Chihara, C., Nejad, S., and Riddiford, L. M. (1997). A cluster of cuticle protein genes of *Drosophila melanogaster* at 65A: sequence, structure and evolution. *Genetics* **147,** 1213–1224.

Chatterjee, C., Mukherjee, A. S., Prasad, J., and Duttagupta, A. K. (1983). Genetic dissection of replicaton unit of *Drosophila*. Part I—Autonomy of control of termination. *Ind. J. Exp. Biol.* **21,** 475–484.

Chatterjee, R. N. (1982). Transcription of nucleolus: localization of initial trancription site of the nucleolus of the giant cells of *Drosophila hydei*. *Proc. Zool. Soc. Calcutta* **34,** 37–53.

Chatterjee, R. N. (1985a). X-chromosomal organization and dosage compensation: *In situ* transcription of chromatin template activity of X-chromosome hyperploids of *Drosophila melanogaster*. *Chromosoma* **91,** 259–266.

Chatterjee, R. N. (1985b). Changes of DNA replicaton pattern of the polytene chromosomes of *Drosophila melanogaster* resulting from chromosomal rearrangements. *Dros. Inform. Serv.* **61** pp. 50–53.

Chatterjee, R. N. (1985c). Effect of α-amanitin on the DNA synthesis in the polytene chromosome of *Drosophila melanogaster*. *Dros. Inform. Serv.* **61,** 53–55.

Chatterjee, R. N. (1991). Puffwise analysis of the gene activity in the 4th chromosome of male and female of *Drosophila hydei*. *Dros. Inform. Serv.* **70,** 46–47.

Chatterjee, R. N. (1992). Influence of X chromosomal aneuploidy on X chromosomal organization in *Drosophila melanogaster*. *Dros. Inform. Serv.* **71,** 202.

Chatterjee, R. N. (1993). Investigation of transcriptional activity of the X chromosome of *Drosophila hydei* carrying a male lethal mutation. *Dros. Inform. Serv.* **72,** 107.

Chatterjee, R. N. (1996). The 87A and 87C heat shock loci are induced unequally in a strain of X chromosomal aneuploid of *Drosophila melanogaster. Dros. Inform. Serv.* **77,** 104–105.

Chatterjee, R. N., and Bhadra, U. (1985). *In situ* transcription analysis of chromatin template activity of X-chromosome of *Drosophila melanogaster* carrying *maleless* gene *(mle^{ts}). Ind. J. Exp. Biol.* **23,** 353–355.

Chatterjee, R. N., and Ghosh, S. (1985). Transcription of the X chromosomes in hybrids of *Drosophila melanogaster* and *D. simulans. Ind. J. Exptl. Biol.* **23,** 293–297.

Chatterjee, R. N., and Mukherjee, A. S. (1982). Effect of a-amanitin on the ^{3}H-uridine incorporation of the *Drosophila melanogater* salivary gland chromosomes. *Dros. Inform. Serv.* **58,** 35–37.

Chatterjee, R. N., and Mukherjee, A. S. (1984). Effect of a-amanitin on RNA synthesis in *Drosophila* polytene chromosome: a probe into the regulation of dosage compensation. *Proc. Ind. Natl. Acad. Sci. B* **50,** 464–472.

Chatterjee, R. N., and Mukherjee, R. (1992). RNA synthesis in *Drosophila melanogaster* polytene chromosomes after *in vitro* treatment of cAMP. *Dros. Inform. Serv.* **71,** 202–204.

Chatterjee, R. N., Mukherjee, A. S., Derksen, J., and van der Ploeg, M. (1980). Role of nonhistone chromosomal protein in attainment of hyperactivity of X chromosome of male *Drosophila:* a quantitative cytochemical study. *Ind. J. Exp. Biol.* **18,** 574–575.

Chatterjee, R. N., Dube, D. K., and Mukherjee, A. S. (1981). *In situ* transcription analysis of chromatin template activity of the X-chromosome of *Drosophila* following high molar NaCl treatment. *Chromosoma* **82,** 515–523.

Chatterjee, S. N., and Mukherjee, A. S. (1971a). Chromosomal basis of dosage compensation in *Drosophila.* V. Puffwise analysis of gene activity in the X-chromosome of male and female *Drosophila hydei. Chromosoma* **36,** 46–59.

Chatterjee, S. N., and Mukherjee, A. S. (1971b). DNA replication pattern of the puffing sites in the X-chromosome of *Drosophila hydei. Dros. Inform. Serv.* **46,** 55.

Chatterjee, S. N., and Mukherjee, A. S. (1973). Chromosomal basis of dosage compensation in *Drosophila.* VII. DNA replication patterns of the puffs in the male and female larval polytene chromosomes. *Cell. Differ.* **2,** 1–19.

Chatterjee, S. N., and Mukherjee, A. S. (1975). DNA replication in polytene chromosomes of *Drosophila pseudoobscura:* New facts and their implications. *Ind. J. Exp. Biol.* **13,** 452–459.

Chatterjee, S. N., Dutta Gupta, A. K., and Mukherjee, A. S. (1974). Lack of differential transcriptional activity in the left and right arms of the X chromosome of *Drosophila pseudoobscura. Dros. Inform. Serv.* **51,** 84.

Chatterjee, S. N., Mondal, S. N., and Mukherjee, A. S. (1976). Interchromosomal asynchrony of DNA replication in polytene chromosmes of *Drosophila pseudoobscura. Chromosoma* **54,** 117–125.

Chaudhuri, A. R. (1969). Lipo-protein nature of so-called vacuole in salivary gland cells of the mutant "*fat*" in *Drosophila melanogaster. Dros. Inform. Serv.* **44,** 118.

Chaudhuri, A. R., and Mukherjee, A. S. (1969). Certain aspects of genetic physiology of the mutant "*fat*" in *Drosophila melanogaster.* A preliminary report. *The Nucleus* **12**(1), 75–80.

Chaudhuri, A. R., and Mukherjee, A. S. (1970). Developmental changes in puffing pattern in the mutant "*fat*" in *Drosophila melanogaster. Dros. Inform. Serv.* **45,** 76.

Chaudhuri, G. K., and Mukherjee, A. S. (1984). Affect of a-methyl-DL-methionine on the replication of polytene chromosomes in *Drosopohila melanogaster. Dros. Inform. Serv.* **60,** 74–75.

Chelysheva, L. A., Solovei, I. V., Rodionov, A. V., Yakovlev, A. F., and Gaginskaya, E. R. (1990). The lampbrush chromosomes of the chicken. The cytological map of macrobivalents. *Tsitologia* **32**(4), 303–316. [in Russian]

Chen, C.-N., Denome, S., and Davies, R. L. (1986). Molecular analysis of cDNA clones and the corresponding genomic coding sequences of the *Drosophila dunce*$^{+}$ gene, the structural gene for cAMP phosphodiesterase. *Proc. Natl. Acad. Sci. USA* **83,** 9313–9317.

Chen, C.-N., Malone, T., Beckendorf, S. K., and Davies, R. L. (1987). At least two genes reside within a large intron of the *dunce* gene of *Drosophila*. *Nature* **329,** 721–724.

Chen, T., Bunting, M., Karim, F. D., and Thummel, C. S. (1992). Isolation and characterization of five *Drosophila* genes that encode an ets-related DNA binding domain. *Dev Biol.* **151,** 176–191.

Cherbas, L., Benes, H., Bourouis, M., Burtis, K., Chao, A., Cherbas, P., Crosby, M., Garfinkel, M., Guild, G., Hogness, D., Jami, J., Jones C. W., Koehler, M., Lepesant, J.-A., Martin, C., Maschat, F., Mathers, P., Meyerowitz, E., Moss, R., Pictet, R., Rebers, J., Richards, G., Roux, J., Schulz, R., Segraves, W., Thummel, C., and Vijayraghavan, K. (1986a). Structural and functional analysis of some moulting hormone-responsive genes from *Drosophila*. *Insect Biochem.* **16,** 241–248.

Cherbas, L., Schulz, R. A., Koehler, M. M. D., Savakis, C., and Cherbas, P. (1986b). Structure of the *Eip28''29* gene, an ecdysone inducible gene from *Drosophila*. *J. Mol. Biol.* **189,** 617–631.

Cherbas, L., Lee, K., and Cherbas, P. (1991). Identification of ecdysone response elements by analysis of the *Drosophila Eip28''29* gene. *Genes Dev.* **5,** 120–131.

Cherbas, P. (1993). The IV-th Karlson Lecture: Ecdysone-responsive genes. *Insect Biochem. Mol. Biol.* **23,** 3–11.

Cherbas, P., Cherbas, L., and Williams, C. M. (1977). Induction of acetylcholine esterase activity by β-ecdysone in a *Drosophila* cell line. *Science* **197,** 275–277.

Chomyn, A., and Mitchell, H. K. (1982). Synthesis of the 84,000 Dalton protein in normal and heat shocked *Drosophila melanogaster* cells as detected by specific antibody. *Insect Biochem.* **12,** 105–114.

Chomyn, A., Moller, G., and Mitchell, H. K. (1979). Patterns of protein synthesis following heat shock in pupae of *Drosophila melanogaster*. *Dev. Genet.* **1,** 77–95.

Chooi, W. Y. (1979). The occurrence of long transcription units among the X and Y ribosomal genes of *Drosophila melanogaster:* transcription of insertion sequences. *Chromosoma* **74,** 57–74.

Chovnick, A., McCarron, M., Hilliker, A., O'Donnell, J., Gelbart, W., and Clark, S. (1978). Gene organization in *Drosophila*. *Cold Spring Harbor Symp. Quant. Biol.* **42,** 1011–1021.

Chowdhuri, D. K., and Lakhotia, S. C. (1986). Different effects of 93D on 87C heat shock puff activity in *Drosophila melanogaster* and *D. simulans*. *Chromosoma* **94,** 279–284.

Christensen, M. E., LeStourgeon, W., Jamrich, M., Howard, G. C., Serunian, L. A., Silver, L. M., and Elgin, S. C. R. (1981). Distribution studies on polytene chromosomes using antibodies directed against hn RNP. *J. Cell. Biol.* **90,** 18–24.

Chubareva, L. A. (1984). Comparative analysis of the karyotypes of black flies and Chironomidae. *Zool. zhurn.* **63**(11), 1652–1660. [in Russian]

Chubareva, L. A., and Tsapygina, R. I. (1965). Some data on triploids in natural populations of the *Odagmia ornata* Mg. (Simuliidae, Dip. *Genetika* **1**(3), 15–18. [in Russian]

Cionini, P. G., Cavallini, A., Corsi, R., and Fogli, M. (1982). Comparison of homologous polytene chromosomes in *Phaseolus coccineus* embryo suspensor cells: Morphological, autoradiographic and cytophotometric analyses. *Chromosoma* **86,** 383–396.

Clark, A. G., and Wang, L. (1997). Molecular population genetics of *Drosophila* immune system genes. *Genetics* **147,** 713–724.

Clark, A. G., Szumski, F. M., and Lyckergaard, E. M. S. (1990). Population genetics of the Y chromosome of *Drosophila melanogaster:* rRNA variation and phenotypic correlates. *Genet. Res. Cambr.* **58,** 7–13.

Claycomb, W. C., LaFond, R. E., and Villee, C. A. (1971). Autoradiographic localization of ^{3}H-b-ecdysone in salivary gland cells of *Drosophila virilis*. *Nature* **234,** 302–304.

Clegg, N. J., Whitenhead, I. P., Brock, J. K., Sinclair, D. A., Mottus, R., Stromotich, G., Harrington, M. J., and Grigliatti, T. A. (1993). A cytogenetic analysis of chromosomal region 31 of *Drosophila melanogaster*. *Genetics* **134,** 221–230.

Clever, U. (1961). Genaktivitaten in den Riesenchromosomen von *Chironomus tentans* und ihre Bezichungen zur Entwicklung. I. Mitt. Genaktivierung durch Ecdyson. *Chromosoma* **12,** 607–675.

Clever, U. (1962a). Genaktivitaten in den Riesenchromosomen von *Chironomus tentans* und ihre Bezichungen zur Entwicklung. II. Mitt. Das Verhalten der Puffs wahrend des letzten Larven stadium und der Puppenhautung. *Chromosoma* **13,** 385–436.

Clever, U. (1962b). Genaktivitaten in den Riesenchromosomen von *Chironomus tentans* und ihre Bezichungen zur Entwicklung. III. Das Aktivitatsmuster in Phasen der Entwicklungsruhe. *J. Insect. Physiol.* **8,** 357–376.

Clever, U. (1963a). Genaktivitaten in den Riesenchromosomen von *Chironomus tentans* und ihre Bezichungen zur Entwicklung. IV. Das Verhalten der Puffs in der Larvenhautung. *Chromosoma* **14,** 651–675.

Clever, U. (1963b). Von der Ecdysonkonzentration abhangige Genaktivitatsmuster in den Speicheldrusenchromosomen von *Chironomus tentans*. *Dev. Biol.* **6,** 73–98.

Clever, U. (1963c). Studies of nucleo-cytoplasmic interrelations in giant chromosomes of Diptera. *Proc. XVI. Intern. Congr. Zool.* **3,** 210–215.

Clever, U. (1964a). Actinomycin and puromycin: Effect on sequential gene activation by ecdysone. *Science* **146,** 794–795.

Clever, U. (1964b). Genaktivitaten und ihre Kontrolle in der tierischen Entwicklung. *Naturwissenschaften* **19,** 449–459.

Clever, U. (1964c). Genaktivitaten und ihre Kontrolle in der tierischen Entwicklung. *Naturwissenschaften* **51,** 449–459.

Clever, U. (1964d). Puffing in giant chromosomes of Diptera and mechanism of its control. *In* "The Nucleohistones" (J. Bonner and P. Ts'o, eds.), pp. 317–334. Holden Day, San Francisco/London/Amsterdam.

Clever, U. (1965a) Chromosomal changes associated with differetiation. *Brookhav. Sympd. Biol.* **18,** 242–253.

Clever, U. (1965b) Puffing changes in incubated and ecdysone treated *Chironomus tentans* salivary glands. *Chromosoma* **17,** 309–322.

Clever, U. (1966a). Gene activity patterns and cellular differentiation. *Am. Zool.* **6,** 33–41.

Clever, U. (1966b). Induction and repression of a puff in *Chironomus*. *Dev. Biol.* **14,** 421–438.

Clever, U. (1967) Control of chromosome puffing. *In* "The Control of Nuclear Activity" (L. Goldstein, ed.), pp. 161–186. Prentice-Hall, Englewood Cliffs, NJ.

Clever, U. (1968). Regulation of chromosome function. *Ann. Rev. Genet.* **2,** 11–30.

Clever, U. (1969). Chromosome activity and cell function in polytenic cells. II. The formation of secretion in the salivary glands of *Chironomus*. *Exp. Cell. Res.* **55,** 317–322.

Clever, U., and Ellgaard, E. G. (1970). Puffing and histone acetylation in polytene chromosomes. *Science* **169,** 373–374.

Clever, U., and Karlson, P. (1960). Induktion von Puff-Veranderungen in den Speicheldrusenchromosomen von *Chironomus tentans* durch Ecdyson. *Exp. Cell Res.* **20,** 623–626.

Clever, U., and Romball, C. G. (1966). RNA and protein synthesis in the cellular response to a hormone ecdysone. *Proc. Natl. Acad. Sci. USA* **56,** 1470–1476.

Clever, U., and Storbeck, I. (1970). Chromosome activity and cell function in polytenic cells. IV. Polyribosomes and their sensitivity to actinomycin. *Biochim. Biophys. Acta* **217,** 108–119.

Clever, U., Bultmann, H., and Darrow, J. M. (1969a). The immediacy of genomic control in polytenic cells. *In* "Problems in Biology: RNA in Development" (E. W. Hanley, ed.), pp. 403–423. Univ. of Utah Press, Salt Lake City, UT.

Clever, U., Storbeck, I., and Romball, C. G. (1969b). Chromosome activity and cell function in polytenic cells. I. Protein synthesis at various stages of larval development. *Exp. Cell Res.* **55,** 306–316.

Clever, U., Clever, I., Storbeck, I., and Young, N. L. (1973). The apparent requirement of two hormones a- and b-ecdysone for moulting induction in insects. *Dev. Biol.* **31,** 47–60.

Cline, T. (1988). Evidence that *sisterless-a* and *sisterless-b* are two of several discrete "numerator elements" of the X/A determination signal in *Drosophila* that switch *Sxl* between two alternative stable expression stat. *Genetics* **119,** 829–868.

Cline, T. W., and Meyer, B. J. (1996). Vive la difference: males vs females in flies vs worms. *Annu. Rev. Genet.* **30,** 637–702.

Clos, J., Westwood, J. T., Becker, P. B., Wilson, S., Lambert, K., and Wu C. (1990). Molecular cloning and expression of a hexameric *Drosophila* heat shock factor subject to negative regulation. *Cell* **63,** 1085–1097.

Clos, J., Rabindran, S., Wisniewski, J., and Wu, C. (1993). Induction temperature of human heat shock factor is reprogrammed in a *Drosophila* cell environment. *Nature* **364,** 252–255.

Cohen, C. J. (1977). Characterization of the resting potential in *Chironomus* salivary gland cells. *Exp. Cell. Res.* **106,** 15–30.

Cohen, M., Jr. (1976). Evolution of 5S ribosomal RNA genes in the chromosomes of the *virilis* group of *Drosophila*. *Chromosoma* **55,** 359–371.

Cohen, R. S., and Meselson, M. (1984). Inducible transcription and puffing in *Drosophila melanogaster* transformed with *hsp70*-phage 1 hybrid heat shock genes. *Proc. Natl. Acad. Sci. USA* **81,** 5509–5513.

Cohen, R. S., and Meselson, M. (1985). Separate regulatory elements for the heat-inducible and ovarian expression of the *Drosophila hsp26* gene. *Cell* **43,** 737–746.

Collier, G. E., and James, J. M. (1987). Arginine kinase and pyruvate kinase are coordinately regulated by 75B "early" gene in *Drosophila* imaginal discs. *Genetics* **116**(1), 12.

Colman, O. D., and Stockert, J. C. (1975). Puffing patterns during fourth larval instar in *Chironomus pallidivittatus* salivary glands. *Chromosoma* **53,** 381–392.

Coluzzi, M., and Kitzmiller, J. B. (1975). Anopheline mosquitos. *In* "Handbook of Genetics" (R. C. King, ed.), Vol. 3, pp. 285–309. Plenum, New York/London.

Coluzzi, M., Cancrini, G., and DiDeco, M. (1970). The polytene chromosomes of *Anopheles superpictus* and relationship with *Anopheles stephensi*. *Parassitologia* **12,** 101–111.

Comings, D. E., and Okada, T. A. (1974). Some aspects of chromosome structure in eukaryotes. *Cold Spring Harbor Symp. Quant. Biol.* **38,** 145–153.

Compton, J. L., and Bonner, J. J. (1978). An *in vitro* assay for the specific induction and regression of puffs in isolated polytene nuclei of *Drosophila melanogaster*. *Cold Spring Harbor Symp. Quant. Biol.* **42,** 835–838.

Compton, J. L., and McCarthy, B. J. (1978). Induction of the *Drosophila* heat shock response on isolated polytene nuclei. *Cell* **14,** 191–201.

Cooper, M. K., Hamblen-Coyle, M. J., Liu, X., Rutila, J. E., and Hall, J. C. (1994). Dosage compensation of the period gene in *Drosophila melanogaster*. *Genetics* **138,** 721–732.

Corces, V., and Pellicer, A. (1984). Identification of sequences involved in the transcriptional control of a *Drosophila* heat-shock gene. *J. Biol. Chem.* **259,** 14812–14817.

Corces, V., Holmgren, R., Freund, R., Morimoto, R., and Meselson, M. (1980). Four heat shock proteins of *Drosophila melanogaster* coded within a 12-kilobase region in chromosome subdivision 67B. *Proc. Natl. Acad. Sci. USA* **77,** 5390–5393.

Corces, V., Pellicer, A., Axel, R., Mei, S. Y., and Meselson, M. (1982). Approximate localization of sequences controlling transcription of a *Drosophila* heat shock gene. *In* "Heat Shock from Bacteria to Man" (M. L. Schlesinger, M. Ashburner, and A. Tissieres, eds.), pp. 27–34. Cold Spring Harbor Laboratory Press, Cold Spring Harbor, NY.

Cordeiro, M., and Meneghini, R. (1973). The rate of DNA replication in the polytene chromosomes of *Rhynchosciara angelae*. *J. Mol. Biol.* **78,** 261–274.

Cortes, E., Botella, L. M., Barettino, D., and Diez, J. L. (1989). Identification of the spI products of Balbiani ring genes in *Chironomus thummi*. *Chromosoma* **98,** 428–432.

Cortes, E., Serrano, M., Yanez, R. J., and Diez, J. L. (1990). Induction of Balbiani Ring puffing changes by sugars and alcohols in *Chironomus thummi*. *Insect Biochem.* **20,** 523–529.

Costa, D., Ritossa, F., and Scalenghe, F. (1977). Production of deletions of the puff forming regions 87A and 87B in *Drosophila melanogaster*. *Dros. Inform. Serv.* **52,** 140.

Costlow, N., and Lis, J. T. (1984). High-resolution mapping of DNAase I-hypersensitive sites of *Drosophila* heat shock genes in *Drosophila melanogaster* and *Saccharomyces cerevisiae*. *Mol. Cell. Biol.* **4,** 1853–1863.

Courgeon, A.-M., Maisonhaute, C., and Best-Belpomme, M. (1984). Heat shock proteins are induced by cadmium in *Drosophila* cells. *Exp. Cell. Res.* **153,** 515–521.

Courgeon, A.-M., Rollet, E., Becker, J., Maisonhaute, C., and Best-Belpomme, M. (1988). Hydrogen peroxide (H_2O_2) induces actin and some heat-shock proteins in *Drosophila* cells. *Eur. J. Biochem.* **171,** 163–170.

Craig, E. A. (1984). The heat shock response. *CRC Crit. Rev. Biochem.* **18,** 239–280.

Craig, E. A., and McCarthy, B. J. (1980). Four *Drosophila* heat shock genes at 67B: characterization of recombinant plasmids. *Nucleic Acids Res.* **8,** 4441–4450.

Craig, E. A., McCarthy, B. J., and Wadsworth, S. C. (1979). Sequence organization of two recombinant plasmids containing genes for the major heat shock-induced protein of *Drosophila melanogaster*. *Cell* **16,** 575–588.

Craig, E. A., Ingolia, T. D., Slater, M., Manseau, L. J., and Bardwell, J. (1982). *Drosophila,* yeast, and *E. coli* genes related to the *Drosophila* heat-shock genes. *In* "Heat Shock from Bacteria to Man" (M. L. Schlesinger, M. Ashburner, and A. Tissieres, eds.), pp. 11–18. Cold Spring Harbor Laboratory Press, Cold Spring Harbor, NY.

Craig,, E. A., Ingolia, T. D., and Manseau, L. J. (1983). Expression of *Drosophila* heat-shock cognate genes during heat shock and development. *Dev. Biol.* **99,** 418–426.

Craig, E. A., Weismann, J. S., and Horwich, A. L. (1994). Heat shock proteins and molecular chaperones: Mediators of protein conformation and turnover in the cell. *Cell* **78,** 365–372.

Cribbs, D. L., Pultz, M. A., Johnson, D., Mazzulla, M., and Kaufman, T. C. (1992). Structural complexity and evolutionary conservation of the *Drosophila* homeotic gene *proboscipedia*. *EMBO J.* **11,** 1437–1449.

Crick, F. (1971). General model for the chromosomes of higher organisms. *Nature* **234,** 25–27.

Crick, F. (1972). A general model of the chromosomes of higher organisms. *Priroda* **4,** 65–68. [in Russian]

Crosby, M. A., and Meyerowitz, E. M. (1986a). Lethal mutations flanking the 68C glue gene cluster on chromosome 3 of *Drosophila melanogaster*. *Genetics* **112,** 785–802.

Crosby, M. A., and Meyerowitz, E. M. (1986b). *Drosophila* glue gene *Sgs-3:* sequences required for puffing and transcriptional regulation. *Dev. Biol.* **118,** 593–607.

Cross, N. C. P., and Dover, G. A. (1987). Tsetse fly rDNA: an analysis of strucuture and sequence. *Nucleic Acids Res.* **15,** 15–30.

Crossgrove, K., Bayer, C. A., Fristrom, J. W., and Guild, M. (1996). The *Drosophila Broad-Complex* early gene directly regulated late gene transcription during the ecdysone-induced puffing cascade. *Dev. Biol.* **180,** 745–758.

Crouse, H. V. (1968). The role of ecdysone in "DNA-puff" formation of *Sciara coprophila*. *Proc. Natl. Acad. Sci. USA* **61,** 971–978.

Crouse, H. V., and Keyl, H. G. (1968). Extra replications in the "DNA-puffs" of *Sciara coprophila*. *Proc. Natl. Acad. Sci. USA* **61,** 971–978.

Crowley, T. E., and Meyerowitz, E. M. (1984). Steroid regulation of RNAs transcribed from the *Drosophila* 68C polytene chromosome puff. *Dev. Biol.* **102,** 110–121.

Crowley, T. E., Bond, M. W., and Meyerowitz, E. M. (1983). The structural genes for three *Drosophila* glue proteins reside at a single polytene chromosome puff locus. *Mol. Cell Biol.* **3,** 623–634.

Crowley, T., Mathers, P. H., and Meyerowitz, E. M. (1984). A transacting regulatory product necessary for expression on the *Drosophila melanogaster* 68C glue gene cluster. *Cell* **39,** 149–156.

Crowley, T. E., Hoey, T., Liu, J.-K., Jan, Y. N., Jan, L. Y., and Tjian, R. (1993). A new factor related to TATA-binding protein has highly restricted expression patterns in *Drosophila*. *Nature* **361,** 557–561.

Dabbs, C. K., and King, R. C. (1980). The differentiaion of pseudonurse cells in the ovaries of *fs231* females of *Drosophila melanogaster* Meigen (Diptera: Drosophilidae). *Int. J. Insect Morphol. Embryol.* **9,** 215–229.

Da Cunha, A. B., and Pavan, C. (1969). Alguns problemas da diferenciacao celular em Sciarideos. *Cienc. cult.* **21**(1), 18–19.

Da Cunha, A. B., Morgante, J. S., Pavan, C., and Garrido, M. C. (1969a). Studies on cytology and differentiation in Sciaridae. III. Nuclear and cytoplasmatic differentiation in the salivary glands of *Bradysia* sp.. *Univ. Texas Publ.* **6918,** 1–11.

Da Cunha, A. B., Morgante, J. S., Pavan, C., Garrido, M. C., and Marques, J. (1973a). Studies on cytology and differentiation in Sciaridae. IV. Nuclear and cytoplasmatic differentiation in the salivary glands of *Bradysia elegans* (Diptera, Sciaridae). *Caryologia* **26,** 83–100.

Da Cunha, A. B., Pavan, C., Morgante, J. S., and Garrido, M. C. (1969b). Studies on cytology and differentiation in Sciaridae. II. DNA redundancy in salivary glands cells of *Hybosciara fragilis* (Diptera, Sciaridae). *Genetics* **61** (Suppl.) 335–349.

Da Cunha, A. B., Riess, R. W., Biesele, J. J., Morgante, J. S., Pavan, C., and Garrido, V. C. (1973b). Studies on cytology and differentiation in Sciaridae. VI. Nuclear cytoplasmic transfers in *Hybosciara fragilis* Morgante (Diptera, Sciaridae). *Caryologia* **26,** 549–561.

Da Cunha, A. B., Riess, R. W., Pavan, C., Biesele, J. J., Morgante, J. S., and Garrido, M. C. (1973c). Studies on cytology and differentiation in Sciaridae. 5. Electron microscopic studies on the salivary glands of Bradysia elegans Morgante (Diptera, Sciaridae). *Differentiation* **1,** 145–166.

Da Cunha, P. R., Granadino, B., Perondini, A. L. P., and Sanchez, L. (1994). Dosage compensation in Sciarids is achieved by hypertranscription of the single X chromosome in males. *Genetics* **138,** 787–790.

Dainou, O., Cariou, M. L., David, J. R., and Hickey, D. (1987). Amylase gene duplication: An ancestral trait in the *Drosophila melanogaster* species subgroup. *Heredity* **59,** 245–251.

DaLage, J.-L., Lemmeunier, F., Cariou, M.-L., and David, J. R. (1992). Multiple amylase genes in *Drosophila ananassae* and related species. *Genet. Res. Gamb.* **59,** 85–92.

Daneholt, B. (1970). Base ratios in RNA molecules of different sizes from a Balbiani Ring. *J. Mol. Biol.* **49,** 381–391.

Daneholt, B. (1972). Giant RNA transcript in a Balbiani Ring. *Nature New Biol.* **240,** 229–232.

Daneholt, B. (1973). The giant RNA transcript in a Balbiani ring of *Chironomus tentans*. *In* "Molecular Cytogenetics" (B. A. Hamkalo and J. Papaconstantinou, eds.), pp. 155–165. Plenum, New York/London.

Daneholt, B. (1974). Transfer of genetic information in polytene cells. *Int. Rev. Cytol.* **4**(Suppl.), 417–462.

Daneholt, B. (1975). Transcription in polytene chromosomes. *Cell* **4,** 1–9.

Daneholt, B. (1982). Structural and functional analysis of Balbiani ring genes in the salivary glands of *Chironomus tentans*. *In* "Insect Ultrastructure" (R. C. King and H. Akai, eds.), Vol. 1, pp. 382–401. Plenum, New York.

Daneholt B. (1992). The transcribed template and the transcription loop in Balbiani Rings. *Cell Biol. Int. Reports* **16,** 709–715.

Daneholt, B., and Edstrom, J.-E. (1967). The content of desoxyribonucleic acid in individual polytene chromosomes of *Chironomus tentans*. *Cytogenetics* **6,** 350–356.

Daneholt, B., and Hosick, H. (1973). Evidence for transport of 75S RNA from a discrete chromosome region via nuclear sap to cytoplasm in *Chironomus tentans*. *Proc. Natl. Acad. Sci. USA* **70,** 442–446.

Daneholt, B., and Hosick, H. (1974). The transcription unit in Balbiani Ring 2 of *Chironomus tentans*. *Cold Spring Harbor Symp. Quant. Biol.* **38,** 629–635.

Daneholt, B., and Svedhem, L. (1971). Differential representation of chromasomal RNA in nuclear sap. *Exp. Cell Res.* **67,** 263–272.

Daneholt, B., Ringborg, U., Egyhazy, E., and Lambert, B. (1968). Microelectrophoretic technique for fractionation of RNA. *Nature* **218,** 292–293.

Daneholt, B., Edstrom, J.-E., Egyhazi, E., Lambert, B., and Ringborg, U. (1969a). Physico-chemical properties of chromosomal RNA in *Chironomus tentans* polytene chromosomes. *Chromosoma* **28,** 379–398.

Daneholt, B., Edstrom, J.-E., Egyhazi, E., Lambert, B., and Ringborg, U. (1969b). Chromosomes RNA synthesis in polytene chromosomes of *Chironomus tentans*. *Chromosoma* **28,** 399–417.

Daneholt, B., Edstrom, J.-E., Egyhazi, E., Lambert, B., and Ringborg, U. (1969c). RNA synthesis in a Balbiani Ring in *Chironomus tentans* salivary gland cells. *Chromosoma* **28,** 418–429.

Daneholt, B., Edstrom, J.-E., Egyhazi, E., Lambert, B., and Ringborg, U. (1970). RNA synthesis in a Balbiani Ring in *Chironomus tentans*. *Cold Spring Harbor Symp. Quant. Biol.* **35,** 513–519.

Daneholt, B., Case, S. T., and Wieslander, L. (1976). Gene expression in the salivary glands of *Chironomus tentans*. *In* "Progress in Differentiation Research" (N. Muller-Berat *et al.*, eds.), pp. 125–133. North Holland, Amsterdam.

Daneholt, B., Andersson, K., and Fagerlind, M. (1977). Large-sized polysomes in *Chironomus tentans* salivary glands and their relation to Balbiani Ring 75S RNA. *J. Cell. Biol.* **73,** 149–160.

Daneholt, B., Case, S. T., Derksen, J., Lamb, M. M., Nelson, L., and Wieslander, L. (1978). The size and chromosomal location of the 75S RNA transcription unit in Balbiani Ring 2. *Cold Spring Harbor Symp. Quant. Biol.* **42,** 867–876.

Daneholt, B., Case, S. T., Derksen, J., Lamb, M. M., Nelson, L., and Wieslander, L. (1979). The transcription unit in Balbiani Ring 2 and its relation to the chromomeric subdivision of the polytene chromosome. *In* "Specific Eukaryotic Genes, Structural Organization and Function" (J. Engberg, H. Klenow, and V. Leick, eds.), Vol. 13, pp. 39–54. Alfred Benzon Symp., Copenhagen.

Daneholt, B., Andersson, K., Bjorkroth, B., and Lamb, M. M. (1982). Visualization of active 75S RNA genes in the Balbiani Rings of *Chironomus tentans*. *Eur. J. Cell. Biol.* **26,** 325–332.

Dangli, A., and Bautz, E. K. F. (1983). Differential distribution of nonhistone proteins from polytene chromosomes of *Drosophila melanogaster* after heat shock. *Chromosoma* **88,** 201–207.

Dangli, A., Grond, C., Kloetzel, P., and Bautz, E. K. F. (1983). Heat-shock puff 93D from *Drosophila melanogaster:* Accumulation of a RNP-specific antigen associated with giant particles of possible storage function. *EMBO J.* **2,** 1747–1751.

Danieli, G. A., and Rodino, E. (1967). Larval moulting cycle and DNA synthesis in *Drosophila hydei* salivary glands. *Nature* **213,** 424–425.

Danieli, G. A., and Rodino, E. (1968). Incubazione *in vitro* con timidina-^{3}H di ghiandole salivari di *Drosophila hydei* St (Diptera), isolate a vari stadi di sviluppo. *Atti. Acad. Naz. Lincei, Cl. Sci. fis. mat e natur.* **44,** 123–126.

Dapples C. C., and King., R. C. (1970). The development of the nucleolus of the ovarian nurse cell of *Drosophila melanogaster*. *Z. Zellforsch.* **103,** 34–47.

Darrow, I. M., and Clever, U. (1970). Chromosome activity and cell function in polytenic cells. *Dev. Biol.* **21,** 331–348.

Das, M., Mutsuddi, D., Duttagupta, A. K., and Mukherjee, A. S. (1982). Segmental heterogeneity in replication and transcription of the X_2 chromosome of *Drosophila miranda* and conservativeness in the evolution of dosage compensation. *Chromosoma* **87,** 373–388.

D'Avino, P. P., Crispi, S., and Furia, M. (1995a). Hormonal regulation of the *Drosophila melanogaster ng*-genes. *Eur. J. Enthomol.* **92,** 259–261.

D'Avino, P. P., Crispi, S., Polito, L. C., and Furia, M. (1995b). The role of the *BR-C* locus on the expression of genes located at the ecdysone regulated 3C puff of *Drosophila melanogaster*. *Mech. Dev.* **49,** 161–171.

Davis, P. S., and Judd, B. H. (1995). Molecular characterization of the 3C region between *white* and *roughest* loci of *Drosophila melanogaster*. *Dros. Inform. Serv.* **76,** 130–134.

Davis, R. L., and Dauwalder, B. (1991). The *Drosophila dunce* locus: learning and memory genes in the fly. *Trends Genet.* **7**(7), 224–229.

Davis, R. L., and Davidson, N. (1984). Isolation of the *Drosophila melanogaster dunce* chromosomal region and recombinational mapping of *dunce* sequences with restriction site polymorphisms as genetic markers. *Mol. Cell. Biol.* **4,** 358–367.

Dawid, I. B., and Botchan, P. (1977). Sequences homologous to ribosomal insertions occur in the *Drosophila* genome outside the nucleolus organizer. *Proc. Natl. Acad. Sci. USA* **74,** 4233–4237.

Dawid, I. B., and Wellauer, P. K. (1978). Ribosomal DNA and related sequences in *Drosophila melanogaster*. *Cold Spring Harbor Symp. Quant. Biol.* **42,** 1185–1194.

Dawid, I. B., Wellauer, P. K., and Long, E. O. (1978). Ribosomal DNA in *Drosophila melanogaster*. I. Isolation and characterization of cloned fragments. *J. Mol. Biol.* **126,** 749.

De Lotto, R., Louis, C., Wormington, M., and Schedl, P. (1982). Intimate association of 5SRNA and tRNA genes in *Drosophila melanogaster*. *Mol. Gen. Genet.* **188,** 299–304.

De Roda, E., Barbera, E., and Alonso, C. (1980). Molecular dimorphism of the X chromsome of male and female larvae in *Drosophila hydei*. *Acta Embryol. Morphol. Exp.* **1,** 119–127.

Deak, P., and Laufer, H. (1995). Ecdysteroid receptor in *Chironomus thummi* (Diptera: Chironomidae). *Eur. J. Entomol.* **92,** 251–257.

Degelmann, A., and Hollenberg, C. P. (1981). A structural analysis of BR DNA sequences in *Chironomus tentans*. *Chromosoma* **83,** 295–313.

Degelmann, A., Royer, H. D., and Hollenberg, C. P. (1979). The organization of the ribosomal RNA genes of *Chironomus tentans* and some closely related species. *Chromosoma* **71,** 263–281.

Deharveng, L., and Lee, B. H. (1984). Polytene chromosomal variability of *Bilobella aurantiaca* (Collembola) from Sainte Baume population (France). *Caryologia* **37,** 51–67.

Demakov, S. A., Semeshin, V. F., and Zhimulev, I. F. (1991). Cloning and molecular analysis of interband DNA in *Drosophila* polytene chromosomes. *Doklady AN SSSR* **317**(4), 989–992. [in Russian]

Demakov, S. A., Semeshin, V. F., and Zhimulev, I. F. (1993). Cloning and moleculargenetic analysis of *Drosophila melanogaster* interband DNA. *Mol. Gen. Genet.* **238,** 437–443.

Demakov, S. A., Nebrat, L. T., Schwartz, Yu. B. and Zhimulev, I. F. (1997). Organization of DNA sequences in interbands of *Drosophila melanogaster* polytene chromosomes. *Tsitologia* **39**(1), 53–54. [in Russian]

Demerec, M. (1934). The gene and its role in ontogeny. *Cold Spring Harbor Symp. Quant. Biol.* **2,** 110–115.

Demerec, M., and Hoover, M. E. (1936). Three related X-chromosome deficiencies in *Drosophila*. *J. Hered.* **27,** 207–212.

Denell, R. E., Hummels, K. R., Wakimoto, B. T., and Kaufman, T. C. (1981). Developmental studies of lethality associated with the *Antennapedia* gene complex in *Drosophila melanogaster*. *Dev. Biol.* **81,** 43–50.

Dennhofer, L. (1975). Die Speicheldrusenchromosomen der Stechmucke *Culex pipiens* L. IV. Der Chromosomal Geschlechtsdimorphismus. *Genetica* **45,** 163–175.

Dennhofer, L. (1978). Wirkungen von L-Glutaminsaure auf Grosse und Puff-Muster der larvalen Speicheldrusenchromosomen von *Drosophila melanogaster in vivo*. *Genetica* **48,** 107–116.

Dennhofer, L. (1979). Die Kontrolle prapupaler Puffs in Speicheldrusenchromosomen von *Drosophila melanogaster*. I. Die Auslosung prapupaler Puffs *in vitro*. *Genetica* **51,** 93–101.

Dennhofer, L., and Muller, P. (1978). Die Puff-Musterfolgen der Speicheldrusenchromosomen aus Wanderlarven und Vorpuppen von *Drosophila melanogaster in vitro*. *Genetica* **48,** 117–128.

DeReggi, M. L., Hirn, M. H. H., and Delaage, M. A. (1975). Radioimmunoassay of ecdysone—An application to *Drosophila* larvae and pupae. *Biochem. Biophys. Res. Commun.* **66,** 1307–1314.

Dereling, E., Meyer, L., Baumlein, H., and Wobus, U. (1982). Structure and expression of Balbiani ring genes from *Chironomus*. *Biol. Cell.* **45,** 174.

Derksen, J. (1975a). The submicroscopic structure of synthetically active units in a puff of *Drosophila hydei* giant chromosomes. *Chromosoma* **50,** 45–52.

Derksen, J. (1975b). Induced RNP production in different cell types of *Drosophila*. *Cell. Differ.* **4,** 1–10.

Derksen, J., and Willart, E. (1976). Cytochemical studies on RNP complexes produced by puff 2–48BC in *Drosophila hydei:* Uranyl acetate and phosphotungstic acid stainig. *Chromosoma* **55,** 57–68.

Derksen, J., Berendes, H. D., and Willart E. (1973a). Production and release of a locus-specific ribonucleoprotein product in polytene nuclei of *Drosophila hydei*. *J. Cell. Biol.* **59,** 661–668.

Derksen, J., Trendelenburg, M. F., Scheer, U., and Franke, W. W. (1973b). Spread chromosomal nucleoli of *Chironomus* salivary glands. *Exp. Cell. Res.* **80,** 476–479.

Derksen, J., Wieslander, L., Ploeg, M. van der, and Daneholt, B. (1980). Identification of the Balbiane Ring 2 chromomere and determination of the content and compaction of its DNA. *Chromosoma* **81,** 65–84.

Desai, L., and Tencer, R. (1968). Effects of histones and polylyzine on the synthetic activity of the giant chromosomes of the salivary glands of Dipteran larvae. *Exp. Cell. Res.* **52,** 185–197.

Desrosiers, R., and Tanguay, R. M. (1985). The modification in the methylation patterns of H2B and H3 after heat shock can be correlated with the inactivation of normal gene expression. *Biochem. Biophys. Res. Commun.* **133,** 823–829.

Dessen, E. M. B., and Perondini, A. L. P. (1973). Hormonal control of nucleolar and puff behavior in *Sciara ocellaris*. *Genetics* **74**(2), 61–62.

Dessen, E. M. B., and Perondini, A. L. P. (1976). Evidence of hormonal control of the nucleolar activity in *Sciara ocellaris*. *Cell. Differ.* **5,** 275–282.

Dessen, E. M. B., and Perondini, A. L. P. (1985a). Polytene chromosomes and puffing patterns in the salivary glands of *Sciara ocellaris*. *Rev. Brasil. Genet.* **8,** 465–478.

Dessen, E. M. B., and Perondini, A. L. P. (1985b). Hormonal control of RNA and DNA puffs in polytene chromosomes of *Sciara ocellaris*. *Rev. Brasil. Genet.* **8,** 669–679.

Dessen, E. M. B., and Perondini, A. L. P. (1985c). Effects of ecdysterone on nucleolar activity in salivary glands of *Sciara ocellaris*. *Caryologia* **38,** 309–317.

Dessen, E. M. B., and Perondini, A. L. P. (1991). Nucleolar subunits in the salivary gland nuclei of *Sciara ocellaris* (Diptera, Sciaridae). *Rev. Brasil. Genet.* **14,** 673–683.

Devlin, R. H., Holm, D. G., and Grigliatti, T. A. (1982). Autosomal dosage compensation in *Drosophila melanogaster* strains trisomic for the left arm of chromosome 2. *Proc. Natl. Acad. Sci. USA* **79,** 1200–1204.

Devlin, R. H., Grigliatti, T. A., and Holm, D. G. (1984). Dosage compensation is transcriptionally regulated in autosomal trisomies of *Drosophila*. *Chromosoma* **91,** 65–73.

Devlin, R. H., Holm, D. G., and Grigliatti, T. A. (1985). Regulation of dosage compensation in X-chromosomal trisomies of *Drosophila melanogaster*. *Mol. Gen. Genet.* **198,** 422–426.

Diaz, M., and Pavan, C. (1965). Changes in chromosomes induced by microorganism infection. *Proc. Natl. Acad. Sci. Wash.* **54,** 1321–1326.

DiBartolomeis, S. M., and Gerbi, S. A. (1989). Molecular characterization of DNA puff II''9A genes in *Sciara coprophila*. *J. Mol. Biol.* **210,** 531–540.

DiBello, P. R., Withers, D. A., Bayer, C. A., Fristrom, J. W., and Guild, G. M. (1991). The *Drosophila* Broad-Complex encodes a family of related proteins containing zinc fingers. *Genetics* **129,** 385–397.

Dickinson, W. J., and Sullivan, D. T. (1975). "Gene–enzyme Systems in *Drosophila*." Springer-Verlag, Berlin/Heidelberg/New York.

DiDomenico, B., Bugaisky, G. E., and Lindquist, S. (1982). The heat shock response is self-regulated at both the transcriptional and posttranscriptional levels. *Cell* **31,** 593–603.

Diez, J. L. (1973). Effect of cordycepin (3′-desoxyadenosine) on polytene chromosomes of *Chironomus pallidivittatus* salivary glands. *Chromosoma* **42,** 345–358.

Diez, J. L., and Barettino, D. (1984). DNA-RNA hybrids and transcriptional activity in *Chironomus* polytene chromosomes. *Chromosoma* **90,** 103–110.

Diez, J. L., Santa-Cruz, M. C., and Aller, P. (1977). Effect of cycloheximide on RNA synthesis in *Chironomus* polytene chromosomes. *Chromosoma* **61,** 369–379.

Diez, J. L., Santa-Cruz, M. C., Villanueva, A., and Aller, P. (1980). Dependence of Balbiani Ring puffing on protein synthesis. *Chromosoma* **81,** 263–269.

Diez, J. L., Cortes, E., Merino, Y., and Santa-Cruz, M. C. (1990). Galactose-induced puffing changes in *Chironomus thummi* Balbiani Rings and their dependence on protein synthesis. *Chromosoma* **99,** 61–70.

Digilio, A., Giordano, E., Artiaco, D., D'Avino, P. P., Crispi, S., and Furia, M. (1990). A putative new member of the *Drosophila melanogaster Sgs* gene family. *Atti Assoc. Genet. Ital.* **36,** 197–198.

Dignam, S. S., and Case, S. T. (1990). Balbiani ring 3 in *Chironomus tentans* encodes a 185-kDa secretory protein which is synthesized throughout the fourth larval instar. *Gene* **88,** 133–140.

Dignam, S. S., Yang, L., Lezzi, M., and Case, S. T. (1989). Identification of a developmentally regulated gene for a 140 kDa secretory protein in salivary glands of *Chironomus tentans* larvae. *J. Biol. Chem.* **264,** 9444–9452.

Dobel, P. (1968). Uber die plasmatischen Zellstrukturen der Speicheldruse von *Acricotopus lucidus*. *Die Kulturpflanze* **16,** 203–214.

Dobens, L., Rudolph, K., and Berger, E. M. (1991). Ecdysterone regulatory elements function as both transcriptional activators and repressors. *Mol. Cell. Biol.* **11,** 1846–1853.

Dobzhansky, T. (1944). Distribution of heterochromatin in the chromosomes of *Drosophila pallidipennis*. *Am. Nat.* **776,** 192–213.

Dolfini, S., Courgeon, A. M., and Tiepolo, L. (1970). The cell cycle of an established line of *Drosophila melanogaster* cells *in vitro*. *Experientia* **26,** 1020–1021.

Domier, L. L., Rivard, J. J., Sabatini, L. M., and Blumenfeld, M. (1986). *Drosophila virilis* histone gene clusters lacking H1 coding segments. *J. Mol. Evol.* **23,** 149–158.

Donahue, P. R., Palmer, D. K., Condie, J. M., Sabatini, L. M., and Blumenfeld, M. (1986). *Drosophila* histone H2A. 2 is associated with the interbands of polytene chromosomes. *Proc. Natl. Acad. Sci. USA* **83,** 6878–6882.

Dorsch-Hasler, K., Lutz, B., Spindler, K.-D., and Lezzi, M. (1990). Structural and developmental analysis of a gene cloned from the early ecdysterone-inducible puff site, I-18C, in *Chironomus tentans*. *Gene* **96,** 233–239.

Doyle, D., and Laufer, H. (1968). Analysis of secretory processes in Dipteran salivary glands *in vitro*. *In* "Differentiation and Defense Mechanisms in Lower Organisms: Symposium of the Tissue Culture Asociation" (M. M. Siegel, ed.), Vol. 3, pp. 93–103.

Doyle, D., and Laufer, H. (1969a). Sources of larval salivary gland secretion in the Dipteran *Chironomus tentans*. *J. Cell. Biol.* **40,** 61–78.

Doyle, D., and Laufer, H. (1969b). Requirements of ribonucleic acid synthesis for the formation of salivary gland specific proteins in the larvae *Chironomus tentans*. *Exp. Cell. Res.* **57,** 205–210.

Doyle, W. L., and Metz, C. W. (1935). Sructure of the chromosomes in the salivary gland cells in *Sciara* (Diptera). *Biol. Bull.* **69,** 126–135.

Dreesen, T. D., Bower, F. R., and Case, S. T. (1985a). A second gene in a Balbiani ring. *J. Biol. Chem.* **260,** 11824–11832.

Dreesen, T. D., Bower, F. R., and Case, S. T. (1985b). A second gene in a Balbiani ring: *Chironomus* salivary glands contain a 6,5 kb poly(A)$^+$ RNA that is transcribed from a hierarchy of tandemly repeated sequences in Balbiani ring 1. *J. Cell. Biol.* **101**(5), 311.

Dreesen, T. D., Lezzi, M., and Case, S. T. (1988). Developmentally regulated expression of a Balbiani Ring 1 gene for a 180-kDa secretory polypeptide in *Chironomus tentans* salivary glands before larval/pupal ecdysis. *J. Cell. Biol.* **106,** 21–27.

Drosdovskaya, L. N. (1971). Effect of boron on the nucleolus, chromocenter and chromosomes of the salivary glands of *Drosophila melanogast*. *Genetika* **7**(11), 84–90. [in Russian]

Drosdovskaya, L. N. (1974). Effect of boron on the development of puffs in the polytene chromosomes of the salivary glands of *Drosophila melanogaster* at the early prepupal stage. *Genetika* **10**(5), 57–64. [in Russian]

Drosdovskaya, L. N. (1975). Effect of long-term heat exposure on the activity of salivary gland polytene chromosomes. *Ontogenez* **6**(1), 95–99. [in Russian]

Drosdovskaya, L. N. (1988a). Change in the frequency of puffs in the polytene chromosomes of *Drosophila* under the effect of colchicine. *In* "Novye sorta, sozdannye metodom khimicheskogo mutageneza," pp. 226–230. Moscow. [in Russian]

Drosdovskaya, L. N. (1988b). Effect of colchicine on the spectrum of puffs of the polytene chromosomes of salivary glands of male and female *Drosophila*. *In* "Novye sorta, sozdannye metodom khimicheskogo mutageneza," pp. 231–236. Moscow.

Drosdovskaya, L. N. (1988c). Effect of colchicine on the spectrum of puffs in the polytene chromosomes of salivary glands of *Drosophila*. *Izvestiya AN SSSR Ser. Biol. Nauk.* **4,** 618–622.

Drosdovskaya, L. N. (1989). Effect of low doses of paraaminobenzoic acid on the spectrum of puffs of the polytene chromosomes in the salivary glands of *Drosophila*. *In* "Khimicheskiy mutagenez i para-amino-benzoynaya kislota v povyshenii urozhaynosti sel'skokhozyaistvennyx rasteniy," pp. 179–188. Nauka, Moscow. [in Russian]

Drosdovskaya, L. N. and Rapoport, I. A. (1974). Effect of sodium arsenate on the polytene chromosomes of the salivary glands of *Drosophila melanogaster* at the stage of early prepupa. *Genetika* **10**(5), 65–73. [in Russian]

Drosdovskaya, L. N., and Rapoport, I. A. (1991). Spectrum of puffs of the autosomes of the salivary glands of early prepupa under the effect of paraaminobenzoic acid at different periods of larval age of *Drosophila*. *In* "Khimicheskiy mutagenez i problemy selektsii," pp. 267–270. Nauka, Moscow. [in Russian]

Drosdovskaya, L. N., Beknazar'yants, M. M., and Rapoport, I. A. (1977a). Study on the effect of uretane on the puffs of the polytene chromosomes of the salivary glands of *Drosophila*. *In* "Khimicheskiy mutagenez i sozdanie sortov intensivnogo tipa," pp. 212–219. Nauka, Moscow. [in Russian]

Drosdovskaya, L. N., Vishnyakova, N. M., and Rapoport, I. A. (1977b). Effect of urea on the chromosomes of the salivary glands of *Drosophila melanogaster*. *In* "Khimicheskiy mutagenez i sozdanie sortov intensivnogo tipa," pp. 220–226. Nauka, Moscow. [in Russian]

Drosdovskaya, L. N., Rapoport, I. A., and Beknazar'yants, M. M. (1977c). Study on triphenylstibine as a polymodifier agent. *Genetika* **13**(12), 2159–2164. [in Russian]

Drosopoulou, E., and Scouras, Z. G. (1995). The beta-tubulin gene family evolution in the *Drosophila montium* subgroup of the *melanogaster species group*. *J. Mol. Evol.* **41,** 293–298.

Drosopoulou, E., Konstantopoulou, I., and Scouras, Z. G. (1996). The heat shock genes in the *Drosophila montium* subgroup: Chromosomal localization and evolutionary implications. *Chromosoma* **105,** 104–110.

Drosopoulou, E., Tsiafouli, M., Mavragani-Tsipidou, P., and Scouras, Z. G. (1997). The glutamate dehydrogenase, E74 and putative actin gene loci in the *Drosophila montium* subgroup. *Chromosoma* **106,** 20–28.

Duband, J. L., Lettre, F., Arrigo, A. P., and Tanguay, R. M. (1986). Expression and localization of *hsp23* in unstressed and heat-shocked *Drosophila* cultured cells. *Can. J. Genet. Cytol.* **28,** 1088–1092.

Dubinin, N. P. (1932). Step-allelomorphism and the theory of centres of the gene *achaete-scute*. *J. Genet.* **26,** 37–58.

Dubinin, N. P. (1933). Step-allelomorphism in *Drosophila melanogaster*. *J. Genet.* **27,** 443–464.

Dubrovsky, E. B., and Zhimulev, I. F. (1985). Cytogenetic analysis of the 2B1-2–2B9-10 region of the X chromosome of *Drosophila melanogaster*. V. Changes in the synthesis pattern of polypeptides in the salivary glands under the effect of mutations located in the puff 2B5. *Genetika* **21**(8), 1282–1289. [in Russian]

Dubrovsky, E. B., and Zhimulev, I. F. (1988). Trans-regulation of ecdysterone-induced protein synthesis in *Drosophila melanogaster* salivary glands. *Dev. Biol.* **127,** 33–44.
Dubrovsky, E. B., Zhimulev, I. F., and Belyaeva, E. S. (1985). Parallel changes in puffing and protein synthesis in the salivary glands of larvae and pupa of *Drosophila melanogaster. Ontogenez* **16**(1), 26–33. [in Russian]
Dubrovsky, E. B., Kozlova, T. Yu. and Rzhetskaya, M. A. (1990). Regulation of the developmental expression of the heat shock genes from the puff 67B in the salivary glands of *Drosophila melanogaster. Ontogenez* **21**(5), 508–516. [in Russian]
Dubrovsky, E. B., Dretzen, G., and Bellard, M. (1994). The *Drosophila Broad-Complex* regulates developmental changes in transcription and chromatin structure of the 67B heat shock gene cluster. *J. Mol. Biol.* **241,** 353–362.
Dubrovsky, E., Dretzen, G., and Berger, E. M. (1996). The *Broad-Complex* gene is a tissue-specific modulator of the ecdysone response of the *Drosophila hsp23* gene. *Mol. Cell. Biol.* **16,** 6542–6552.
Dudick, M. E., Wright, T. R. F., and Brothers, L. L. (1974). The developmental genetics of the temperature-sensitive lethal allele of the *suppressor of forked, l(1)su(f)ts67g,* in *Drosophila melanogaster. Genetics* **76,** 487–510.
Dudler, R., and Travers, A. A. (1984). Upstream elements necessary for optimal function of the *hsp70* promoter in transformed flies. *Cell* **38,** 391–398.
Dudley, I. (1983). "Characterization of larval serum protein genes in *Drosophila pseudoobscura.*" Abstracts of 8th European Drosophila Research Conference, Cambridge.
Duffy, J. B., Wells, J., and Gergen, J. P. (1996). Dosage-sensitive maternal modifiers of the *Drosophila* segmentation gene *runt. Genetics* **142,** 839–852.
Dunbar, R. W. (1967). The salivary gland chromosomes of six closely related black flies near *Eusimulium congareenarum* (Diptera: Simuliidae). *Can. J. Zool.* **45,** 377–396.
Duncan, I. W., and Kaufman, T. C. (1975). Cytogenetic analysis of chromosome 3 in *Drosophila melanogaster*: Mapping of the proximal portion of the right arm. *Genetics* **80,** 733–752.
Durica, D. S., and Krider, H. M. (1977). Studies on the ribosomal RNA cistroms in interspecific *Drosophila* hybrids. I. Nucleolar dominance. *Dev. Biol.* **59,** 62–74.
Durica, D. S., and Krider, H. M. (1978). Studies on the ribosomal RNA cistroms in interspecific *Drosophila* hybrids. II. Heterochromatic regions mediating nucleolar dominance. *Genetics* **89,** 37–64.
Duttagupta, A. K., and De, S. (1986). Transcriptive acrivity of X-chrosomome in *Drosophila virilis, D. americana* and in their hybrid. *Dros. Inform. Serv.* **63,** 48–49.
Duttagupta, A. K., and De, S. (1987). Studies in the heat shock loci in some *Drosophila* species. *Dros. Inform. Serv.* **66,** 50–51.
Duttagupta, A. K., and Roy, I. (1984). Isolation of nascent DNA from polytene chromosomes of *Drosophila melanogaster. Dros. Inform. Serv.* **60,** 98–99.
Duttagupta, A. K., Mitra, N., and Mukherjee, A. S. (1973). Salivary gland chromosomes of *Drosophila ananassae.* I. Banding pattern and intercalary heterocharomatin. *The Nucleus* **16,** 130–146.
Duttagupta, A. K., Mutsuddi, D., and Mutsuddi, M. (1984a). Conservativeness in the regulation of replication in three related species of *obscura* group. *Dros. Inform. Serv.* **60,** 94–95.
Duttagupta, A. K., Mutsuddi, D., and Mutsuddi, M. (1984b). Unequal diameter of the homologous chromosomal elements in the hybrids of *Drosophila mulleri* and *Drosophila arizonensis. Dros. Inform. Serv.* **60,** 95–96.
Duttagupta, A. K., Mutsuddi, M. (Das), and Mutsuddi, D. (1984c). Effect of transforming mutants on the X chromosomal replication pattern in *Drosophila melanogaster. Dros. Inform. Serv.* **60,** 96–97.
Duttagupta, A. K., Mutsuddi, M. (Das), and Mutsuddi, D. (1984d). X-chromosome replication in *Drosophila. Dros. Inform. Serv.* **60,** 97–98.

Duttagupta, A. K., Mutsuddi, D., and Mutsuddi, M. (Das). (1985). Replication in X chromosomal segmental aneuploids in *Drosophila*. *Dros. Inform. Serv.* **61,** 69–70.
Duttagupta, A. L., and Banerjee, S. (1984). *In vivo* synchronization by aphidicolin and ricin in *Drosophila*. *Dros. Inform. Serv.* **60,** 89–90.
Dutton, F. L., and Krider, H. M. (1984). Factors influencing disproportionate replication of the ribosomal RNA cistrons in *Drosophila melanogaster*. *Genetics* **107,** 395–404.
Dworniczak, B., Kobus, S., Eiteljirge, K., and Pongs, O. (1982). Regulation of development of salivary glands in 3rd instar *Drosophila melanogaster* larvae by ecdysone. *In* "Biochemistry of Differentiation and Morphogenesis" (L. Jaenicke, ed.), pp. 163–176. Springer-Verlag, Berlin/Heidelberg.
Dworniczak, B., Seidel, R., and Pongs, O. (1983). Puffing activities and binding of ecdysteroid to polytene chromosomes of *Drosophila melanogaster*. *EMBO J.* **2**(8), 1323–1330.
Dybas, L. K., Harden, K. K., Machnicki, J. L., and Geer, B. W. (1983). Male fertility in *Drosophila melanogaster:* Lesions of spermatogenesis associated with male sterile mutations of the *vermilion* region. *J. Exp. Zool.* **226,** 293–302.
Eastman, E. M., Goodman, R. M., Erlanger, B. F., and Miller, O. J. (1980). 5-Methylcytosine in the DNA of the polytene chromosomes of the Diptera *Sciara coprophila*, *Drosophila melanogaster* and *D. persimilis*. *Chrosomoma* **79,** 225–239.
Eberl, D. F., and Hilliker, A. J. (1988). Characterization of X-linked recessive lethal mutations affecting embryonic morphogenesis in *Drosophila melanogaster*. *Genetics* **118,** 109–120.
Eberl, D. F., Perkins, L. A., Engelstein, M., and Hilliker, A. J. (1992). Perrimon N. Genetic and developmental analysis of polytene section 17 of the X chromosome of *Drosophila melanogaster*. *Genetics* **130,** 569–583.
Eberl, D. F., Lorenz, L. J., Melnick, M. B., Sood, V., Lasko, P., and Perrimon, N. (1997). A new enhancer of position-effect variegation in *Drosophila melanogaster* encodes a putative RNA helicase that binds chromosomes and is regulated by the cell cycle. *Genetics* **146,** 951–963.
Ebstein, B. E. (1967). Tritiated actinomycin D as a cytochemical label for small amounts of DNA. *J. Cell. Biol.* **35,** 709–713.
Edstrom, J.-E. (1960). Extraction, hydrolysis, and electrophoretic analysis of ribonucleic acid from microscopic tissue units (microphoresis). *J. Biophys. Biochem. Cytol.* **8,** 39–46.
Edstrom, J.-E. (1964). Microextraction and microelectrophoresis for determination and analysis of nucleic acids in isolated cellular units. *In* "Methods in Cell Physiology" (D. M. Prescott, ed.), Vol. 1, pp. 417–447. Academic Press, New York.
Edstrom, J.-E. (1974). Polytene chromosomes in studies of gene expression. *In* "The Cell Nucleus" (H. Bush, ed.), Vol. 2, pp. 293–332. Academic Press, New York.
Edstrom, J.-E. (1975). Eukaryotic evolution based on information in chromosomes on allele frequencies. *J. Theor. Biol.* **52,** 163–174.
Edstrom, J.-E., and Beermann, W. (1962). The base composition of nucleic acids in chromosomes, puffs, nucleoli and cytoplasm of *Chironomus* salivary gland cells. *J. Cell. Biol.* **14,** 371–380.
Edstrom, J.-E., and Daneholt, B. (1967). Sedimentation properties of newly synthesized RNA from isolated nuclear components of *Chironomus tentans* salivary gland cells. *J. Mol. Biol.* **28,** 331–343.
Edstrom, J.-E., and Tanguay, R. (1974a). Chromosome products in *Chironomus tentans* salivary gland cells. *Cold Spring Harbor Symp. Quant. Biol.* **38,** 693–699.
Edstrom, J.-E., and Tanguay, R. (1974b). Cytoplasmic ribonucleic acids with messenger characteristics in salivary gland cells of *Chironomus tentans*. *J. Mol. Biol.* **84,** 569–583.
Edstrom, J.-E., Daneholt, B., Egyhazy, E., Lambert, B., and Ringborg, U. (1969). Formation and processing of ribonucleic acid in subnuclear components of *Chironomus tentans*. *Biochem. J.* **114,** 51–52.
Edstrom, J.-E., Egyhazi, E., Daneholt, B., Lambert, B., and Ringborg, U. (1971). Localization of

newly synthesized low molecular weight RNA in polytene chromosomes from *Chironomus tentans*. *Chromosoma* **35,** 431–442.

Edstrom, J.-E., Ericson, E., Lindgren, S., Lonn, U., and Rydlander, L. (1978a). Fate of Balbiani-Ring RNA *in vivo*. *Cold Spring Harbor Symp. Quant. Biol.* **42,** 877–884.

Edstrom, J.-E., Lindgren, S., Lonn, U., and Rydlander, L. (1978b). Balbiani Ring RNA content and half-life in nucleus and cytoplasm of *Chironomus tentans* salivary gland cells. *Chromosoma* **66,** 33–44.

Edstrom, J.-E., Rydlander, L., and Francke, C. (1980). Concomitant induction of a Balbiani Ring and a giant secretory protein in *Chironomus* salivary glands. *Chromosoma* **81,** 115–124.

Edtrsom, J.-E., Sierakowska, H., and Burvall, K. (1982). Dependence of Balbiani Ring induction in *Chironomus* salivary glands on inorganic phosphate. *Dev. Biol.* **91,** 131–137.

Edstrom, J.-E., Rydlander, L., and Thyberg, J. (1983). Spread of Balbiani ring derived messenger RNA in *Chironomus* salivary gland cell. *Eur. J. Cell. Biol.* **29,** 281–287.

Eeken, J. C. J. (1974). Circadian control of the cellular response to b-ecdysone in *Drosophila lebanonensis*. I. Experimental puff induction and its relation to puparium formation. *Chromosoma* **49,** 205–217.

Eeken, J. C. J. (1977). Ultrastructure of salivary glands of *Drosophila lebanonensis* during normal development and *in vivo* ecdysterone administration. *J. Insect. Physiol.* **23,** 1043–1056.

Egyhazi, E. (1974). A tentative inhibition of chromosomal heterogeneous RNA synthesis. *J. Mol. Biol.* **84,** 173–183.

Egyhazi, E. (1975). Inhibition of Balbiani Ring RNA synthesis at the initiation level. *Proc. Natl. Acad. Sci. USA* **72,** 947–950.

Egyhazi, E. (1976a). Initiation inhibition and reinitiation of the synthesis of heterogeneous nuclear RNA in living cells. *Nature* **262,** 319–321.

Egyhazi, E. (1976b). Quantitation of turnover and export to the cytoplasm of hnRNA transcribed in the Balbiani Rings. *Cell* **7,** 507–515.

Egyhazi, E. (1978). Kinetic evidence that a discrete messenger-like RNA is formed by post-transcriptional size reduction of heterogeneous nuclear RNA. *Chromosoma* **65,** 137–152.

Egyhazi, E., and Holst, M. (1981). Adenosine selectively inhibits labeling of chromosomal RNA, especially *hn*RNA probably by acting at the near the site of chain initiation. *J. Cell. Biol.* **89,** 1–8.

Egyhazi, E., and Pigon, A. (1983). Rapidly phosphorylated histone-like proteins are modulated in the course of transcription block by 5,6-dichloro-1-b-D-ribofuranosylbenzimidazole. *Eur. J. Cell. Biol.* **31,** 354–359.

Egyhazi, E., and Pigon, A. (1986). Selective repression of RNA polymerase II by microinjected phosvitin. *Chromosoma* **94,** 329–336.

Egyhazi, E., Ringborg, U., Daneholt, B., and Lambert, B. (1968). Extraction and fractionation of low molecular weight RNA on the microscale. *Nature* **220,** 1036–1037.

Egyhazi, E., Daneholt, B., Edstrom, J.-E., Lambert, B., and Ringborg, U. (1969). Low molecular weight RNA in cell components of *Chironomus tentans* salivary glands. *J. Mol. Biol.* **44,** 517–532.

Egyhazi, E., Daneholt, B., Edstrom, J.-E., Lambert, B., and Ringborg, U. (1970). Differential inhibitory effect of a substituted benzimidazole on RNA labelling in polytene chromosomes. *J. Cell. Biol.* **47,** 516–520.

Egyhazi, E., D'Monte, B., and Edstrom, J.-E. (1972). Effects of a-amanitin on *in vitro* labelling of RNA from defined nuclear components in salivary gland cells from *Chironomus tentans*. *J. Cell. Biol.* **53,** 523–531.

Egyhazi, E., Holst, M., and Ossoinak, A. (1979). The size distribution of poly(A) in newly synthesized and old Balbiani ring RNA. *Mol. Biol. Reports* **5,** 105–114.

Egyhazi, E., Pigon, A., and Rydlander, L. (1982). 5,6-dichlororibofuranosylbenzimidazole inhibits the rate of transcription initiation in intact *Chironomus* cells. *Eur. J. Biochem.* **122,** 445–451.

Egyhazi, E., Pigon, A., Ossoinak, A., Holst, M., and Tayip, U. (1984). Phosphorylation of some chromosomal nonhistone proteins in active genes in blocked by the transcription inhibitor 5,6-dichloro-1-b-D-ribofuranosylbenzimidazole (DRB). *J. Cell. Biol.* **98,** 954–962.

Eickbush, D.G., Eickbush, T.H. (1995). Vertical transmission of the retrotransposable elements R1 and R2 during the evolution of the *Drosophila melanogaster* species subgroup. *Genetics* **139,** 671–684.

Eickbush, D. G., Lathe, W. C. III., Francino, M. P., and Eickbush, T. H. (1995). R1 and R2 retrotransposable elements of *Drosophila* evolve at rates similar to those of nuclear genes. *Genetics* **139,** 685–695.

Eigenbrod, J. (1978). Differences in the number of nucleolus organizers in *Chironomus tepperi* shown by in situ hybridization. *Chromosoma* **67,** 63–66.

Eissenberg, J. C., and Lucchesi, J. C. (1982). Chromatin structure and dosage-compensated heat-shock genes in *Drosophila pseudoobscura*. *In* "Heat Shock from Bacteria to Man" (M. J. Schlesinger, M. Ashburner, and A. Tissieres, eds.), pp. 109–114. Cold Spring Harbor Laboratory Press, Cold Spring Harbor, NY.

Elgin, S. C. R. (1981). DNAase I-hypersensitive sites of chromatin. *Cell* **27,** 413–415.

Elgin, S. C. R., Serunian, L. A., and Silver, L. M. (1978). Distribution patterns of *Drosophila* nonhistone chromosomal proteins. *Cold Spring Harbor Symp. Quant. Biol.* **42,** 839–850.

Elgin, S. C. R., Cartwright, I. L, Fleischmann, G., Lowenhaupt, K., and Keene, M. A. (1983). Cleavage reagents as probes of DNA sequence organization and chromatin structure: *Drosophila melanogaster* locus 67B1. *Cold Spring Harbor Symp. Quant. Biol.* **47,** 529–538.

Elgin, S. C. R., Amero, S. A., Eissenberg, J. C., Fleischmann, G., Gilmour, D. S., and James, T. C. (1988). Distribution patterns of nonhistone chromosomal proteins on polytene chromosomes: functional correlations. *In* "Chromosome Structure and Function: Impact of New Concerts" (J. P. Gustafson and R. Appels, eds.), pp. 145–156. Plenum, New York/London.

Ellengorn, Ya. E. (1937). Chromomeres as indices of the morphological features of mitotic chromosomes. *Biol. Zhurn.* **6**(3), 633–644. [in Russian]

Ellgaard, E. G. (1967). Gene activation without histone acetylation in *Drosophila melanogaster*. *Science* **157,** 1070–1072.

Ellgaard, E. G. (1972). Similarities in chromosomal puffing induced by temperature shocks and dinitrophenol in *Drosophila*. *Chromosoma* **37,** 417–422.

Ellgaard, E. G., and Brosseau, G. E., Jr. (1969). Puff forming ability as a function of chromosomal position in *Drosophila melanogaster*. *Genetics* **62,** 337–341.

Ellgaard, E. G., and Clever, U. (1971). RNA metabolism during puff induction in *Drosophila melanogaster*. *Chromosoma* **36,** 60–78.

Ellgaard, E. G., and Kessel, R. G. (1966). Effects of high salt concentration on salivary gland cells of *Drosophila virilis*. *Exp. Cell. Res.* **42,** 302–307.

Ellgaard, E. G., and Maxwell, B. L. (1975). Nucleotide metabolism in *Drosophila melanogaster* salivary glands during temperature- and dinitrophenol-induced puffig. *Cell. Differ.* **3,** 379–387.

Ellison, J. R., and Granholm, N. A. (1970). Multi-stranded nuclolar DNA in polytene salivary gland cells of *Samoaia leonensis* Wheeler (Drosophilidae). *Dros. Inform. Serv.* **45,** 172.

Eloff, G. (1939). Mode of attachment of nucleolus to chromocenter in salivary gland nuclei of some Transvaal Drosophilids. *Dros. Inform Serv.* **10,** 64.

Emery, I. F., Bedian, V., and Guild, G. M. (1994). Differential expression of *Broad-Complex* transcription factors may forecast tissue-specific developmental fates during *Drosophila* metamorphosis. *Development* **120,** 3275–3287.

Emmens, C. W. (1937). The morphology of the nucleus in the salivary glands of four species of *Drosophila (D. melanogaster, D. immigrans, D. funebris and D. subobscura)*. *Z. Zellforsch.* **26,** 1–20.

Emmerich, H. (1969a). Distribution of tritiated ecdysone in salivary gland cells of *Drosophila*. *Nature* **221,** 954–955.

Emmerich, H. (1969b). Anricherung von tritiummarkierten Ecdyson in den Zellkernen der Speicheldrusen von *Drosophila hydei*. *Exp. Cell. Res.* **58,** 261–270.

Emmerich, H. (1970). Ecdysonbindende Proteinfraktionen in den Speicheldrusen von *Drosophila hydei*. *Z. Vergl. Physiol.* **68,** 385–402.

Emmerich, H. (1972). Ecdysone binding proteins in nuclei and chromatin from *Drosophila* salivary glands. *Gen. Comp. Endocrinol.* **19,** 543–551.

Emmerich, H. (1975). Puffinduktion durch Ecdyson bei *Drosophila hydei*. Regulationsmechanismen der Genaktivitat und Replikation bei Riesenchromosomen. *Nachr. Akad. Wiss. Gottingen. II. Matem.-Physik. kl.* **11,** 150–151.

Emmerich, H., and Schmialek, P. (1966). Die lokalisation cancerogener Kohlenwasserstoffe in der Speicheldruse von *Drosophila hydei*. *Exp. Cell. Res.* **43,** 228–231.

Endow, S. A. (1980). On ribosomal gene compensation in *Drosophila*. *Cell 22,* 149–155.

Endow, S. A. (1982). Polytenization of the ribosomal genes on the X and Y chromosomes of *Drosophila melanogaster*. *Genetics* **100,** 375–385.

Endow, S. A. (1983). Nucleolar dominance in polytene cells of *Drosophila*. *Proc. Natl. Acad. Sci. USA* **80,** 4427–4431.

Endow, S. A., and Glover, D. M. (1979). Differential replication of ribosomal gene repeats in polytene nuclei of *Drosophila*. *Cell* **17,** 597–605.

Enghofer, E., and Kress, H. (1980). Glucosamine metabolism in *Drosophila virilis* salivary glands: ontogenetic changes of enzyme activities and metabolite synthesis. *Dev. Biol.* **78,** 63–75.

Enghofer, E., Kress, H., and Linzen, B. (1978). Glucosamine metabolism in *Drosophila* salivary glands. Separation of metabolites and some characteristics of three enzymes involved. *Biochim. Biophys. Acta* **544,** 245.

Erhard, H. (1910). Uber den Aufbau der Speicheldrusenkerne der Chironomuslarve. *Arch. Mikr. Natl.* **76,** 114–124.

Ericsson, C., Mehlin, H., Bjorkroth, B., Lamb, M. M., and Daneholt, B. (1989). The ultrastructure of upstream and downstream regions of an active Balbiani Ring gene. *Cell* **56,** 631–639.

Ericsson, C., Grossbach, U., Bjorkroth, B., and Daneholt, B. (1990). Presence of histone H1 on an active Balbiani Ring gene. *Cell* **60,** 73–83.

Evans, R. M. (1988). The steroid and thyroid hormone receptor superfamily. *Science* **240,** 889–895.

Eveleth, D. D., and Marsh, J. L. (1986a). Sequence and expression of the Cc gene, a member of the dopa decarboxylase gene cluster of *Drosophila:* Possible translational regulation. *Nucleic Acids Res.* **14,** 6169–6183.

Eveleth, D. D., and Marsh, J. L. (1986b). Evidence for evolutionary duplication of genes in the dopa decarboxylase region of *Drosophila*. *Genetics* **114,** 469–483.

Evgen'ev, M. B., and Denisenko, O. N. (1990). Effect of the ts-mutation on the expression of the heat shock genes induced by heat shock in *Drosophila melanogaster*. III. Synthesis of proteins, related to the *hsp70*. *Genetika* **26**(2), 266–271. [in Russian]

Evgen'ev, M. B., Gubenko, I. S. (1997a). Genetic regulation of the replication pattern of polytene chromosomes in interspecific hybrids of *Drosophila*. *Chromosoma* **63,** 89–100.

Evgen'ev, M. B., and Gubenko, I. S. (1977b). Genetic analysis of the system controlling chronology of replication of the polytene chromosomes. *Genetika* **13**(12), 2149–2158. [in Russian]

Evgen'ev, M. B., and Levin, A. V. (1980). Effect of ts-mutation on the expression of genes induced by heat shock in *Drosophila melanogaster*. I. Analysis of protein synthesis. *Genetika* **16**(6), 1026–1029. [in Russian]

Evgen'ev, M. B., Lubennikova, E. I., and Shapiro, I. M. (1971a). Autoradiographic study on the polytene chromosomes. Chronology of DNA replication in the presence of inversion and in interspecific hybrids. *Doklady AN SSSR* **200**(5), 1215–1217. [in Russian]

Evgen'ev, M. B., Lubennikova, E. I., and Shapiro, I. M. (1971b). Autoradiographic study on reduplication of the polytene chromosomes of the salivary glands of *Drosophila*. Chronology of DNA reduplication in the norm and at translocation. *Doklady Akad. Nauk SSSR* **204**(1), 215–217. [in Russian]

Evgen'ev, M. B., Kolchinski, A., Levin, A., Preobrazhenskaya, O., and Sarkisova, E. (1978). Heat-shock DNA homology in distantly related species of *Drosophila*. *Chromosoma* **68,** 357–365.

Evgen'ev, M. B., Levin, A., and Lozovskaya, E. (1979). The analysis of a temperature-sensitive (ts) mutation influencing the expression of heat shock-inducible genes in *Drosophila melanogaster*. *Mol. Gen. Genet.* **176,** 275–280.

Evgen'ev, M. B., Zatsepina, O. L., and Titarenko, H. (1985a). Autoregulation of heat-shock system in *Drosophila melanogaster*. Analysis of heat-shock response in a temperature-sensitive cell-lethal mutant. *FEBS Lett.* **188,** 286–290.

Evgen'ev, M. B., Zatsepina, O. G., Kakpakov, V. T., and Vlassova, I. E. (1985b). Combined effect of heat shock and ecdysone on the transcription of polytene chromosomes of *Drosophila melanogaster*. *Mol. Biol.* **19**(2), 483–488. [in Russian]

Faizullin, L. Z., and Gvozdev, V. A. (1973a). Dosage compensation of sex-linked genes in *Drosophila melanogaster*. *Mol. Gen. Genet.* **126,** 233–245.

Faizullin, L. Z., and Gvozdev, V. A. (1973b). Effect of quantitative ratio of the number of X chromosomes to autosomes on the activity of the X chromosomes genes in *Drosophila melanogaster*. *Genetika* **9**(9), 107–117. [in Russian]

Falk, R. (1970). Evidence against the one-to-one correspondence between bands of the salivary gland chromosomes and genes. *Dros. Inform. Serv.* **45,** 112.

Falkner, F.-G., and Biessmann, H. (1980). Nuclear proteins in *Drosophila melanogaster* cells after heat shock and their binding to homologous DNA. *Nucleic Acids Res.* **8,** 943–955.

Faussek, W. (1913). Zur Frage uber den Bau des Zellkerns in den Speicheldrusen der Larvae von *Chironomus*. *Arch. Mikr. Anat.* **82,** 39–60.

Fedoroff, N., and Milkman, R. (1964). Specific puff induction by tryptophan in *Drosophila* salivary chromosomes. *Biol. Bull.* **127,** 369.

Fedoroff, N., and Milkman, R. D. (1965). Induction of puffs in *Drosophila* salivary chromosomes by amino acids. *Dros. Inform. Serv.* **40,** 48.

Feigl, G., Gram, M., and Pongs, O. (1989). A member of the steroid hormone receptor family is expressed in the 20-OH-ecdysone inducible puff 75B in *Drosophila melanogaster*. *Nucleic Acids Res.* **17,** 7167–7178.

Felten, T. L. (1978). Observations of the influence of cAMP in puff induction along the salivary gland chromosomes of *Drosophila melanogaster* (Canton S). *Dros. Inform. Serv.* **53,** 187.

Ferreira, J. F., and Amabis, J. M. (1983). Correlation between a DNA puff and a polypeptide fraction in the salivary gland of *Trichosia pubescens* larvae. *Arquivos Biol. Technol.* **27,** 119.

Ferrus, A., Lamazares, S., de la Pompa, J. L., Tanouye, M.A., Pongs, O. (1990). Genetic analysis of the *Shaker* gene complex of *Drosophila melanogaster*. *Genetics* **125,** 383–398.

Ficq, A. (1959). Autoradiographic study of the relation between nucleic acids and protein synthesis. *Lab. Invest.* **8,** 237–244.

Ficq, A., and Pavan, C. (1957). Autoradiography of polytene chromosomes of *Rhynchosciara angelae* at different stages of larval development. *Nature* **180,** 983–984.

Ficq, A., and Pavan, C. (1961). Metabolisme des acides nucleiques et des proteines dans les chromosomes geants. *Pathol. Biol. Paris* **9,** 756–757.

Ficq, A., Pavan, C., and Brachet, J. (1958). Metabolic processes in chromosomes. *Exptl. Cell. Res.* **6**(Suppl.), 105–114.

Filippova, M. A. (1985). Activity of acid and alkaline phosphatases at the stages of larvae and prepupae of *Chironomus thummi*. *Ontogenez* **16**(2), 127–134. [in Russian]

Filippova, M. A., Kiknadze, I. I., Aimanova, K. G., Fischer, J., and Blinov, A. G. (1992). Homology of Balbiani rings among *Chironomid species and localization of a new mobile element on the polytene chromosomes*. *Netherlands J. Aquat. Ecol.* **26,** 123–128.

Filippova, M. A., Kiknadze, I. I., and Aimanova, K. G. (1993). Homology of Balbiani rings among the *Chironomus* species (Chironomidae, Diptera). *In* "Kariosistematika bespozvonochnykh zhivotnykh II. Trudy Inst. Zool. Ross. Acad. Nauk, pp. 39–42. [in Russian]

Fini, M. E., Bendena, W. G., and Pardue, M. L. (1989). Unusual behavior of the cytoplasmic transcript of hsr: An abundant, stress- inducible RNA that is translated but yields no detectable protein product. *J. Cell. Biol.* **108,** 2045–2057.

Firling, C. E., and Kobilka, B. K. (1979). A medium for the maintenance of *Chironomus* salivary gland *in vitro.*. *J. Insect Physiol.* **25,** 93–103.

Firtel, R.A. (1972). Changes in the expression of single copy DNA during development of cellular slime mold *Dictyostelium discoideum*. *J. Mol. Biol.* **66,** 363–377.

Firtel, R. A., and Bonner, J. (1972). Characterization of the genome of the cellular slime mold *Dictyostelium discoideum*. *J. Mol. Biol.* **66,** 339–361.

Firtel, R. A., and Lodish, H. F. (1973). A small nuclear precursor of messenger RNA in the cellular slime mold *Dictyostelium discoideum*. *J. Mol. Biol.* **79,** 295–314.

Fisk, G. J., and Thummel, C. S. (1995). Isolation, regulation, and DNA-binding properties of three novel *Drosophila* nuclear hormone receptor superfamily members. *Proc. Natl. Acad. Sci. USA* **92,** 10604–10608.

Fitch, D. H. A., Strausbaugh, L. D., and Barrett, V. (1990). On the origins of tandemly repeated genes: Does histone gene copy number in *Drosophila* reflect chromosomal location. *Chromosoma* **99,** 118–124.

Fleischmann, G., Pflugfelder, G., Steiner, E. K., Javaherian, K., Howard, G. C., Wang, J. C., and Elgin, S. C. R. (1984). *Drosophyila* DNA topoisomerase I is associated with transcriptionally active regions of the genome. *Proc. Natl. Acad. Sci. USA* **81,** 6958–6962.

Flemming, W. (1882). "Zellsubstanz, Kern und Zellteilung." Leipzig.

Fletcher, J. C., and Thummel, C. S. (1995a). The *Drosophila E74* gene is required for the proper stage- and tissue-specific transcription of ecdysone-regulated genes at the onset of metamorphosis. *Development* **121,** 1411–1421.

Fletcher, J. C., and Thummel, C. S. (1995b). The ecdysone-inducible *Broad-Complex* and *E74* early genes interact to regulate target gene transcription and *Drosophila* metamorphosis. *Genetics* **141,** 1025–1035.

Fletcher, J. C., Burtis, K. C., Hogness, D. S., and Thummel, C. S. (1995). The *Drosophila E74* gene in reguired for metamorphosis and play a role in the polytene chromosome puffing response to ecdysone. *Development* **141,** 1455–1465.

Fletcher, J. C., D'Avino, P. P., and Thummel, C. S. (1997). A steroid-triggered switch in E74 transcription factor isoforms regulates the timing of secondary-response gene expression. *Proc. Natl. Acad. Sci. USA* **94,** 4582–4586.

Fontes, A. M., Monesi, N., and Paco-Larson, M. L. (1990). Isolation of a cDNA clone related to DNA puff B10 of *Bradysia hygida*. *Proc. VI Congr. Pan-Amer. Assoc. Biochem. Soc.*, 187.

Fortebraccio, M., Scalenghe, F., and Ritossa, F. (1977). Cytological localization of the "ebony" in *Drosophila melanogaster*. *Dros. Inform. Serv.* **52,** 102.

Fraccaro, M., Laudani, U., Marchi, A., and Tiepolo, L. (1976). Karyotype, DNA replication and origin of sex chromosomes in *Anopheles atroparvus*. *Chromosoma* **55,** 27–36.

Fraenkel, G. (1935). A hormone causing pupation in the blowfly *Calliphora erythrocephalla*. *Proc. Roy. Soc. Lond. B* **118,** 1–12.

Fraenkel, G. (1952). A function of the salivary glands of the larvae of *Drosophila* and other flies. *Biol. Bull.* **103,** 285–286.

Fraenkel, G., and Brookes, V. J. (1953). The process by which the puparia of many species of flies become attached to a substrate. *Biol. Bull.* **105,** 442–449.

Franke, C., Edstrom, J.-E., McDowall, A. W., and Miller, O. L., Jr. (1982). Electron microscopic vizualization of a discrete class of giant translation units in salivary gland cells of *Chironomus tentans*. *EMBO J.* **1,** 59–62.

Franz, F., Kunz, W., and Grimm, C. (1983). Determination of the region of rDNA involved in polytenization in salivary glands of *Drosophila hydei*. *Mol. Gen. Genet.* **191,** 74–80.

Franz, G., and Kunz, W. (1981). Intervening sequences in ribosomal RNA genes and *bobbed* phenotype in *Drosophila hydei*. *Nature* **292,** 683–640.

Freitas, F. de, Floeter-Winter, L. M., and Leoncini, O. (1985). Cytological and cytochemical characterization of the polytene chromosomes of *Chironomus sancticaroli* (Diptera: Chironomidae). *Rev. Brasil Genet.* **8,** 47–60.

French, C. K., Fouts, D. L., and Manning, J. E. (1981). Sequence arrangement of the rRNA genes of the dipteran *Sarcophaga bullata*. *Nucleic Acids Res.* **9,** 2563–2576.

Fresquez, C. L. (1979). Nucleic acid synthesis in *Rhynchosciara hollaenderi* polytene chromosomes: I. Dose response and temporal sequence after injection of 20-hydroxyecdysone. *Insect Biochem.* **9,** 517–523.

Freund, J. N., and Jarry, B. P. (1987). The rudimentary gene of *Drosophila melanogaster* encodes four enzymic functions. *J. Mol. Biol.* **193,** 1–13.

Fridell, R. A., Pret, A.-M., and Searles, L. L. (1990). A retrotransposon 412 insertion within an exon of the *Drosophila melanogaster vermilion* gene is spliced from the precursor RNA. *Genes Dev.* **4,** 559–566.

Fridell, Y.-W. C., and Searles, L. L. (1992). *In vivo* transcriptional analysis of the TATA-less promoter of the *Drosophila melanogaster vermilion* gene. *Mol. Cell. Biol.* **12,** 4571–4577.

Friedman, T. B., Owens, K. N., Burnett, J. B., Saura, A. O., and Wallrath, L. L. (1991). The faint band interband region 28C2 to 28C4-5 of the *Drosophila melanogaster* salivary gland polytene chromosomes is rich in transcripts. *Mol. Gen. Genet.* **226,** 81–87.

Fristrom, D., and Fristrom, J. W. (1993). The metamorphic development of the adult epidermis. *In* "The Development of *Drosophila melanogaster*" (M. Bate and A. M. Arias, eds.), pp. 843–897. Cold Spring Harbor Laboratory Press, Cold Spring Harbor, NY.

Fristrom, J. W., Logan, W. R., and Murphy, C. (1973). The synthetic and minimal culture requirements for evagination of imaginal discs of *Drosophila melanogaster in vitro*. *Dev. Biol.* **33,** 441–456.

Fristrom, D. K., Fekete, E., and Fristrom, J. W. (1981). Imaginal disc development in a non-pupariating lethal mutant in *Drosophila melanogaster*. *Wilhelm Roux's Arch.* **190,** 11–21.

Frolov, M. V., and Alatortsev, V. E. (1993). Region of the *Drosophila melanogaster* gene prune is transcriptionally saturated. *Genetika* **29,** 1460–1476. [in Russian]

Frolov, M. V., and Alatortsev, V. E. (1994). Cluster of cytochrome P450 genes on the X chromosome of *Drosophila melanogaster*. *DNA Cell Biol.* **13,** 663–668.

Frolova, S. L. (1936). Structure of the nuclei in the salivary gland cells of *Drosophila*. *Nature* **137,** 319.

de Frutos, R., and Latorre, A. (1982a). Patterns of puffing activity in the polytene chromosomes of *Drosophila subobscura*. *In* "Advances in Genetics, Development and Evolution of *Drosophila*" (S. Lakavaara, ed.), pp. 33–46. Plenum, New York.

de Frutos, R., and Latorre, A. (1982b). Patterns of puffing activity and chromosomal polymorphism in *Drosophila subobscura*. I. J and U chromosomes. *Genetica* **58,** 177–188.

de Frutos, R., Latorre, A., and Pascual, L. (1984a). Patterns of puffing activity and chromosomal polymorphism in *Drosophila subobscura*. III. Puffing activity depression by inbreeding. *Theor. Appl. Genet.* **69,** 101–110.

de Frutos, R., Latorre, A., and Pascual, L. (1984b). Differential puffing activity in two E chromosomal arrangements of *Drosophila subobscura*. *Dros. Inform. Serv.* **66,** 82–83.

de Frutos, R., Latorre, A., and Pascual, L. (1987). Patterns of puffing activity and chromosomal polymorphism in *Drosophila subobscura*. IV. Position effect at the boundaries of the E_{12} inversion. *Genetica* **75,** 11–22.

Frydman, H. M., Cadavid, E. O., Yokosawa, J., Silva, F. H., Navarro-Cattapan, L. D., Santelli, R. V., Jacobs-Lorena, M., Graessmann, M., Graessmann, A., Stocker, A. J., and Lara, F. J. S. (1993). Molecular characterization of the DNA puff C-8 gene of *Rhynchosciara americana*. *J. Mol. Biol.* **233,** 799–803.

Fujita, S. (1965). Chromosomal organization as a genetic basis of cytodifferentiation in multicellular organisms. *Nature* **206,** 742–744.

Fujita, S., and Takamoto, K. (1963). Synthesis of messenger RNA on the polytene chromosomes of Dipteran salivary gland. *Nature* **200,** 494–495.

Furia, M., Kaiser, K., and Glover, D. (1984). Molecular cloning and characterization of two *D. melanogaster Sgs4* alleles defective in dosage compensation. *Atti Assoc. Genet. Ital.* **30,** 101–102.

Furia, M., Del Pozzo, G., Aurilia, V., and Polito, L. C. (1985). Organization of flanking sequences in two *sgs4* variant strains of *Drosophila melanogaster* defective in dosage compensation. *Atti Assoc. Genet. Ital.* **31,** 87–88.

Furia, M., Digilio, F. A., Artiaco, D., Giordano, E., and Polito, L. C. (1990). A new gene nested within the *dunce* genetic unit of *Drosophila melanogaster*. *Nucleic Acids. Res.* **18,** 5837–5841.

Furia, M., Digilio, F. A., Artiaco, D., D'Avino, P. P., Cavaliere, D., and Polito, L. C. (1991). Molecular organization of the *Drosophila melanogaster Pig-1* gene. *Chromosoma* **101,** 49–54.

Furia, M., D'Avino, P. P., Digilio, F. A., Crispi, S., Giordano, E., and Polito, L. C. (1992). Effect of ecd^1 mutation on the expression of genes mapped at the *Drosophila melanogaster* 3C11-12 intermoult puff. *Genet. Res. Cambr.* **59,** 19–26.

Gabrusewycz-Garcia, N. (1964). Cytological and autoradiographic studies in *Sciara coprophila* salivary gland chromosomes. *Chromosoma* **15,** 312–344.

Gabrusewycz-Garcia, N. (1968). RNA metabolism of polytene chromosomes of *Sciara coprophila*. *J. Cell. Biol.* **39**(2), 49a.

Gabrusewycz-Garcia, N. (1971). Studies in polytene chromosomes of Sciarids. I. The salivary chromosomes of *Sciara* (*Lycoriella*) *pauciseta* (II), Felt. *Chromosoma* **33,** 421–435.

Gabrusewycz-Garcia, N. (1972). Further studies of the nucleolar material in salivary gland nuclei of *Sciara coprophila*. *Chromosoma* **38,** 237–254.

Gabrusewycz-Garcia, N. (1975). *Sciara coprophila*. *In* "Handbook of Genetics" (R. C. King, ed.), Vol. 3, pp. 257–267. Plenum, New York/London.

Gabrusewycz-Garcia, N., and Garcia, A. M. (1974). Studies on the fine structure of puffs in *Sciara coprophila*. *Chromosoma* **47,** 385–401.

Gabrusewycz-Garcia, N., and Kleinfeld, R. G. (1966). A study of the nucleolar material in *Sciara coprophila*. *J. Cell. Biol.* **29,** 347–359.

Gabrusewycz-Garcia, N., and Margles, S. (1969). Induction of DNA puffs by ecdysterone. *J. Cell. Biol.* **43**(2), 41a.

Galceran, J., Gimenes, C., Edstrom, J.-E., and Izquierdo, M. (1986). Microcloning and characterization of the early ecdysone puff region 2B of the X chromosome of *Drosophila melanogaster*. *Insect Biochem.* **16,** 249–254.

Galceran, J., Llanos, J., Sampedro, J., Pongs, O., and Izquierdo, M. (1990). Transcription at the ecdysone-inducible locus 2B5 in *Drosophila*. *Nucleic Acids Res.* **18,** 539–545.

Gall, J. G. (1954). Lampbrush chromosomes from oocyte nuclei of the newt. *J. Morphol.* **94,** 283–352.

Gall, J. G. (1963). Chromosomes and cytodifferentiation. *In* "Cytodifferential and Macromolecular Synthesis," pp. 119–143. Academic Press, New York.

Gall, J. G. (1991). Spliceosomes and snurposomes. *Science* **252,** 1499–1500.

Galler, R., and Edstrom, J.-E. (1984). Phosphate oncorporation into secretory protein of *Chironomus* salivary glands occurs during translation. *EMBO J.* **3,** 2851–2855.

Galler, R., Riedel, N., Rydlander, L., and Edstrom, J.-E. (1984a). Balbiani ring genes and theur induction. *Chromo. Today* **8,** 153–160.

Galler, R., Rydlander, L., Riedel, N., Kluding, H., and Edstrom, J.-E. (1984b). Balbiani Ring induction in phospate metabolism. *Proc. Natl. Acad. Sci. USA* **81,** 1448–1452.

Galler, R., Saiga, H., Widmer, R. M., Lazzi, M., and Edstrom, J.-E. (1985). Two genes in Balbiani Ring 2 with metabolically different 75S transcript. *EMBO J.* **4,** 2977–2982.

Galli, J., Lendahl, U., Paulsson, G., Ericsson, C., Bergman, T., Carlquist, M., and Wieslander, L. (1990). A new member of a secretory protein gene family in the Dipteran *Chironomus tentans* has a variant repeat structure. *J. Mol. Evol.* **31,** 40–50.

Gambarini, A.G., and Maneghini, R. (1972). Ribosomal RNA genes in salivary gland and ovary of *Rhynchosciara angelae*. *J. Cell. Biol.* **54,** 421–426.

Gandhi, V., Sharp, Z., and Procunier, J. (1982). Analysis of Y chromosome nucleolar organizer mutants in *Drosophila melanogaster*. *Biochem. Biophys. Res. Commun.* **104,** 778–784.

Ganguly, R., Ganguly, N., and Manning, J. E. (1985). Isolation and characterization of glucose 6-phosphate dehydrogenase gene of *Drosophila melanogaster*. *Gene* **35,** 91–101.

Garbe, J. C., and Pardue, M. L. (1986). Heat shock locus 93D of *Drosophila melanogaster:* A spliced RNA most strongly conserved in the intron sequence. *Proc. Natl. Acad. Sci. USA* **83,** 1812–1816.

Garbe, J. C., Bendena, W. G., Alfano, M., and Pardue, M. L. (1986). A *Drosophila* heat shock locus with a rapidly diverging sequence but a conserved structure. *J. Biol. Chem.* **261,** 16889–16894.

Garbe, J. C., Bendena, W. G., and Pardue, M. L. (1989). Sequence evolution of the *Drosophila* heat shock locus *hsr*. I. The nonrepeated portion of the gene. *Genetics* **122,** 403–415.

Garber, R. L., Kuroiwa, A., and Gehring, W. J. (1983). Genomic and cDNA clones of the homeotic locus *Antennapedia* in *Drosophila*. *EMBO J.* **2,** 2027–2036.

Garcia-Bellido, A. (1979). Genetic analysis of the achaete-scute system of *Drosophila melanogaster*. *Genetics* **91,** 491–520.

Garcia-Bellido, A., and Ripoll, P. (1978). The number of genes in *Drosophila melanogaster*. *Nature* **273,** 399–400.

Garen, A., and Gehring, W. (1972). Repair of the lethal developmental defect in deep orange embryos of *Drosophila* by injection of normal egg cytoplasm. *Proc. Natl. Acad. Sci. USA* **69,** 2982–2985.

Garen, A., Kauver, L., and Lepesant, J.-A. (1977). Roles of ecdysone in *Drosophila* development. *Proc. Natl. Acad. Sci. USA* **74,** 5099–5103.

Garfinkel, M. D., Pruitt, R. E., and Meyerowitz, E. M. (1983). DNA sequences, gene regulation and modular protein evolution in the *Drosophila* 68C glue gene cluster. *J. Mol. Biol.* **168,** 765–789.

Garzino, V., Pereira, A., Laurenti, P. Graba, Y., Levis, R. W., Le Parco, Y., and Pradel, J. (1992). Cell lineage-specific expression of *modulo*, a dose dependent modifier of variegation in *Drosophila*. *EMBO J.* **11,** 4471–4479.

Gasser, S. M., and Laemmli, U. K. (1987). A glimpse at chromosomal order. *Trends Genet.* **3,** 16–22.

Gatenby, J. B. (1932). Neutral-red staining in *Chironomus* salivary-gland cells, and the so-called "vacuome". *Am. J. Anat.* **51,** 253–262.

Gatti, M., and Pimpinelli, S. (1992). Functional elements in *Drosophila melanogaster* heterochromatin. *Ann. Rev. Genet.* **26,** 239–275.

von Gaudecker, B. (1967). RNA synthesis in the nucleolus of *Chironomus thummi* as studied by high resolution autoradiography. *Z. Zellforsch.* **82,** 536–557.

von Gaudecker, B. (1972). Der Strukturwandel der larvalen Speicheldruse von *Drosophila melanogaster*. Ein Beitrag zur Frage nach der steuernden Wirkung aktiver Gene auf das Cytoplasma. *Z. Zellforsch* **127,** 50–86.

von Gaudecker B., and Schmale, E.-M. (1974). Substrate-histochemical investigations and ultrahistochemical demonstrations of acid posphatase in larval and prepupal salivary glands of *Drosophila melanogaster*. *Cell. Tiss. Res.* **155,** 75–89.

Gausz, J., Bencze, G., Gyurkovics, H., Ashburner, M., Ish-Horowicz, D., and Holden, J. J. (1979). Genetic characterization of the 87C region of the third chromosome of *Drosophila melanogaster*. *Genetics* **93,** 917–934.

Gausz, J., Gyurkovics, H., Bencze, G., Awad, A. A. M., Holden, J. J., and Ish-Horowicz, D. (1981). Genetic characterization of the region between 86F1-2 and 87B15 on chromosome 3 of *Drosophila melanogaster*. *Genetics* **98,** 775–789.

Gausz, J., Hall, L. M. C., Spierer, A., and Spierer, P. (1986). Molecular genetics of the rosy-Ace region of *Drosophila melanogaster*. *Genetics* **112,** 65–78.

Gautam, N. (1983). Identification of two closely located larval salivary protein genes in *Drosophila melanogaster*. *Mol. Gen. Genet.* **189**(3), 495–500.

Gautam, N. (1984). The allelic forms of three larval salivary proteins from *Drosophila melanogaster*. *Mol. Gen. Genet.* **195,** 336–340.

Gay, H. (1956). Nucleocytoplasmic relations in *Drosophila*. *Cold Spring Harbor Symp. Quant. Biol.* **21,** 257–269.

Gay, H. (1964). Chomosome structure and function. *Ybk. Carnegie Inst. Wash. Publ.* **63,** 605–614.

Geer, B. W., Lischwe, T. D., and Murphy, K. G. (1983). Male fertility in *Drosophila melanogaster:* Genetics of the *vermilion* region. *J. Exp. Zool.* **225,** 107–118.

Gehring, W. J. (1985). Homeotic genes, the homeo box, and the genetic control of development. *Cold Spring Harbor Symp. Quant. Biol.* **50,** 243–251.

Gehring, W. J., Klemenz, R., Weber, U., and Kloter, U. (1984). Functional analysis of the white$^+$ gene of *Drosophila* by P-factor-mediated transformation. *EMBO J.* **3,** 2077–2085.

Gelbart, W. M., Irish, V. F., Johnston, R. D. St., Hoffmann, F. M., Lackman, R. K., Segal, D., Posakony, L. M., and Grimaila, R. (1985). The Decapentaplegic gene Complex in *Drosophila melanogater*. *Cold Spring Harbor Symp. Quant. Biol.* **50,** 119–125.

Gelbart, W., Rindone, W., Chillemi, J., Russo, S., Ashburner, M., Drysdale, R., Grey, A.de, Whitfield, E., Welbourne, E., Kaufman, T., Matthews, K., Gilbert, D., Merriam, J., Matthews, B., McAlpine, G., and Tolstoshev, C. (1994). Fly Base: The *Drosophila* Genetic Database; Genetic Loci: Part A. *Dros. Inform. Serv.* **73,** 9–515.

Gemmill, R. M., Levy, J. N., and Doane, W. W. (1983). Molecular structure of the *Drosophila* amylase locus. *J. Cell. Biol.* **97**(5), 132.

Gemmill, R. M., Schwartz, P. E., and Doane, W. W. (1986). Structural organisation of the Amy locus in seven strains of *Drosophila melanogaster*. *Nucleic Acids. Res.* **14,** 5337–5352.

Genova, G. K., and Konova, O. M. (1987). Additional nucleoli in the salivary gland cells of *Drosophila melanogaster* at decreased temperature. *Doklady Bolgar. Akad. Nauk* **40**(1), 137–138. [in Russian]

Genova, G. K., and Semionov, E. P. (1985a). Study on additional nucleoli in the cells with polytene chromosomes in *Drosophila melanogaster*. *Tsitologia* **27**(1), 70–75. [in Russian]

Genova, G. K., and Semionov, E. P. (1985b). Unstably localized nucleoli in *Drosophila melanogaster* salivary gland cells in various stocks. *Dros. Inform. Serv.* **61,** 76.

George, H., and Terracol, R. (1997). The *vrille* gene of *Drosophila* is a maternal enhancer of *decapentaplegic* and encodes a new member of the bZIP family of transcription factors. *Genetics* **146,** 1345–1363.

George, J. A., Burke, W. D., and Eickbush, T. H. (1996). Analysis of the 5′ junctions of R2 insertions with the 28S gene: Implications for non-LTR retrotransposition. *Genetics* **142,** 853–863.

Georgel, P., Ramain, P., Giangrande, A., Dretzen, G., Richards, G., and Bellard, M. (1991). *Sgs-3* chromatin structure and trans-activators: Developmental and ecdysone induction of a glue enhancer-binding factor, GEBF-I, in *Drosophila* larvae. *Mol. Cell. Biol.* **11,** 523–532.

Georgel, P., Bellard, F., Dretzen, G., Jagla, K., Richards, G., and Bellard, M. (1992). GEBF-I in *Drosophila* species and hybrids: the co-evolution of an enhancer and its cognate factor. *Mol. Gen. Genet.* **235,** 104–112.

Georgel, P., Dretzen, G., Jagla, K., Bellard, F., Dubrovsky, E., Calco, V., and Bellard, M. (1993). GEBF-1 activates the *Drosophila Sgs3* gene enhancer by altering a positioned nucleosomal core particle. *J. Mol. Biol.* **234,** 319–330.

Georgopoulos, C., and Welch, W. J. (1993). Role of the major heat shock proteins as molecular chaperones. *Annu. Rev. Cell Biol.* **9,** 601–634.

Gerasimova, T. I., and Corces, V. G. (1996). Boundary and insulator elements in chromosomes. *Curr. Opin. Genet. Dev.* **6,** 185–192.

Gerbi, S. A. (1971). Localization and characterization of the ribosomal RNA cistrons in *Sciara coprophila*. *J. Mol. Biol.* **58,** 499–511.

Gerbi, S. A., and Crouse, H. V. (1976). Further studies on the ribosomal RNA cistrons of *Sciara coprophila* (Diptera). *Genetics* **83,** 81–90.

Gerbi, S. A., Liang, C., Wu, N., DiBartolomeis, S. M., Bienz-Tadmore, B., Smith, H. S., and Urnov, E. D. (1993). DNA amplification in DNA puff II/9A of *Sciara coprophila*. *Cold Spring Harbor Symp. Quant. Biol.* **58,** 487–494.

Gersh, E. S. (1965). A new locus in the white-Notch region of the *Drosophila melanogaster* X chromosome. *Genetics* **51,** 477–480.

Gersh, E. S. (1967). Genetic effects associated with band 3C1 of the salivary gland X chromosome of *Drosophila melanogaster*. *Genetics* **56,** 309–319.

Gersh, E. S. (1972a). Studies of the sal locus in *Drosophila pseudoobscura*. I. The banding pattern of the tip of the salivary chromosome 3 in larvae homozygous and heterozygous for sal. *Genetics* **71,** 369–371.

Gersh, E. S. (1972b). Studies on the sal locus in *Drosophila pseudoobscura*. II. Modes of regulation of the sal banding pattern. *Genetics* **72,** 531–536.

Gersh, E. S. (1973). Studies on the sal locus in *Drosophila pseudoobscura*. III. The molecular pattern of DNA in active and inactive chromosome regions. *Can. J. Genet. Cytol.* **15,** 497–507.

Gersh, E. S. (1975). Sites of gene activity and of inactive genes in polytene chromosomes of Diptera. *J. Theor. Biol.* **50,** 413–428.

Gersh, E. S., and Gersh, I. (1973a). Some observations on the distribution and molecular arrangement of nucleic acids in salivary gland nuclei of *Drosophila melanogaster*. *Can. J. Genet. Cytol.* **15,** 509–522.

Gersh, E. S., and Gersh, I. (1973b). Cytochemical study of nucleic acids in salivary gland chromosomes of *Drosophila* larvae. *In* "Submicroscopic Cytochemistry, Proteins and Nucleic Acids" (I. Gersh, ed.), pp. 98–144. Academic Press, New York/London.

Ghosh, A. K. (1985). Transcriptional activity of an autosomal arm (2L) in trisomic condition in *Drosophila melanogaster*. *Dros. Inform. Serv.* **61,** 79.

Ghosh, A. K., and Mukherjee, A. S. (1986). Replication behaviour of trisomy for 2L and 3L arm in *Drosophila melanogaster*. *Dros. Inform. Serv.* **63,** 59–60.

Ghosh, A. K., and Mukherjee, A. S. (1987). Lack of dosage compensation phenomenon in haplo-4 chromosome in *Drosophila melanogaster*. *Dros. Inform. Serv.* **66,** 63.

Ghosh, A. K., and Mukherjee, A. S. (1988). Transcriptive activity of right arm of 4th chromosome in *Drosophila melanogaster* in different doses and problem of autosomal dosage compensation. *Ind. J. Exp. Biol.* **26,** 329–332.

Ghosh, A. K., and Mukherjee, A. S. (1990). Expression of 87A, 87C and 93D heat shock puffs in trisomic (2L and 3L) strains of *Drosophila melanogaster:* Evidence for a functional relation between 93D and 87C. *Chromosoma* **99,** 71–75.

Ghosh, P., and Mukherjee, A. S. (1996). Induction of a new round of DNA replication in some major heat shock puffs in the 3R arm of polytene chromosomes of *Drosophila melanogaster* at 37° C. *Dros. Inform. Serv.* **77,** 139–141.

Ghosh, A. K., Ghosh, M., and Mukherjee, A. S. (1989). Autosomal dosage compensation in interspecific hybrids of *Drosophila:* Transcriptional activity in diploid and triploid hybrids. *Current Sci.* **58,** 907–911.

Ghosh, M. (1984). Nucleolar chromatin thread in different species in *Drosophila. Dros. Inform. Serv.* **60,** 112–113.

Ghosh, M. (1986). Characterization of nucleolar chromatin thread (NCT) in salivary gland cells of *Drosophila melanogaster. Dros. Inform. Serv.* **63,** 60–61.

Ghosh, M., and Mukherjee, A. S. (1982a). Organization of nucleolar chromatin thread (NCT) of *Drosophila. Cell. Chromos. Res.* **5,** 7–22.

Ghosh, M., and Mukherjee, A. S. (1982b). Nucleolar chromatin structures in *Drosophila hydei. Dros. Inform. Serv.* **58,** 66–67.

Ghosh, M., and Mukherjee, A. S. (1984a). Evolutionarily related species and their NCT structures. *Dros. Inform. Serv.* **60,** 113–115.

Ghosh, M., and Mukherjee, A. S. (1984b). Nucleolar chromatin thread (NCT) in hybrids of *Drosophila. Dros. Inform. Serv.* **60,** 116–117.

Ghosh, M., and Mukherjee, A. S. (1985). DNA replication in the X chromosome of *In(1)BM*[2] (rv, mosaic) of *Drosophila melanogaster. Dros. Inform. Serv.* **61,** 82–83.

Ghosh, M., and Mukherjee, A. S. (1986a). Nucleolar chromatin thread in different mutant stocks of *Drosophila hydei. Dros. Inform. Serv.* **63,** 62–63.

Ghosh, M., and Mukherjee, A. S. (1986b). Nucleolar chromatin thread in different mutant strains of *Drosophila melanogaster. Dros. Inform. Serv.* **63,** 63–64.

Ghosh, S., and Mukherjee, A. S. (1986c). Transcriptive and replicative activity of the X chromosome in an autosomal segmental hyperploid in *Drosophila* and its significance. *J. Cell. Sci.* **81,** 267–281.

Ghosh, S., Chatterjee, R. N., Bunick, D., Manning, J. E., and Lucchesi, J. C. (1989). The LSP1-a gene of *Drosophila melanogaster* exhibits dosage compensation when it is located to a different site on the X-chromosome. *EMBO J.* **8,** 1191–1196.

Ghysen, A., and Dambly-Chaudiere, C. (1988). From DNA to form: the achaete-scute complex. *Genes Dev.* **2,** 495–501.

Giangrande, A., Mettling, C., and Richards, G. (1987). *Sgs-3* transcript levels are determined by multiple remonte sequence elements. *EMBO J.* **6,** 3079–3984.

Giangrande, A., Mettling, C., Martin, M., Ruiz, C., and Richards, G. (1989). *Drosophila Sgs3* TATA: Effects of point mutations on expression in vivo and protein bindin in vitro with staged nuclear extracts. *EMBO J.* **8,** 3459–3466.

Giardina, C., Perez-Riba, M., and Lis, J. T. (1992). Promoter melting and TFIID complexes on *Drosophila* genes *in vivo*. *Genes Dev.* **6,** 2190–2200.

Gilbert, D., Hirsh, J., and Wright, T. R. F. (1984). Molecular mapping of a gene cluster flanking the *Drosophila Dopa* decarboxylase gene. *Genetics* **106,** 679–694.

Gilbert, E. F., and Pistei, W. R. (1966). Chromosomal puffs in *Drosophila* induced by hydrocortisone phosphate. *Proc. Soc. Exp. Biol. Med.* **121,** 831–832.

Gilmour, D. S., and Lis, J. T. (1986). RNA polymerase II interacts with the promoter region of the noninduced hsp70 gene in *Drosophila melanogaster* cells. *Mol. Cell. Biol.* **6,** 3984–3989.

Glaser, R. L., and Lis, J. T. (1990). Multiple, compensatory regulatory elements specify spermatocyte-specific expression of the *Drosophila melanogaster hsp26* gene. *Mol. Cell. Biol.* **10,** 131–137.

Glaser, R., Wolfner, M., and Lis, J. (1986). Spatial and temporal pattern of *hsp26* expression during normal development. *EMBO J.* **5,** 747–754.

Glaser, R. L., Thomas, G. H., Siegfried, E., Elgin, S. C. R., and Lis, J. T. (1990). Optimal heat-induced expression of the *Drosophila hsp26* gene requires a promoter sequence containing (CT)n. (GA)n repeats. *J. Mol. Biol.* **211,** 751–761.

Glatzer, K. H. (1979). Lengths of transcribed rDNA repeating units in sparmatocytes of *Drosophila hydei:* Only genes without an intervening sequences are expressed. *Chromosoma* **75,** 161–175.

Glick, B. S. (1995). Can Hsp70 proteins act as force-generating motors? *Cell* **80,** 11–14.

Glover, D. M. (1977). Cloned segment of *Drosophila* rDNA containing new types of sequence insertion. *Proc. Natl. Acad. Sci. USA* **74,** 4932–4936.

Glover, D. M. (1981). The rDNA of *Drosophila melanogaster*. *Cell* **26,** 297–298.
Glover, D. M., and Hogness, D. S. (1977). A novel arrangement of the 18S and 28S sequences in a repeating unit of *Drosophila melanogaster* rDNA. *Cell* **10,** 167–176.
Glover, D. M., Zaha, A., Stocker, A. J., Santelli, R. V., Pueyo, M. T., Toledo, S. M. de, and Lara, F. J. S. (1982). Gene amplification in *Rhynchosciara* salivary gland chromosomes. *Proc. Natl. Acad. Sci. USA* **79,** 2947–2951.
Goldberg, D. A., Posakony, J. W., and Maniatis, T. (1983). Correct developmental expression of a cloned alcohol dehydrogenase gene tranduced into the *Drosophila* germ line. *Cell* **34,** 59–73.
Goldschmidt, E. (1942). The pattern of salivary gland chromosomes in a hybrid in the genus *Chironomus*. *J. Hered.* **33,** 265–272.
Goldschmidt-Clermont, M. (1980). Two genes for the major heat-shock protein of *Drosophila melanogaster* arranged as an inverted repeat. *Nucleic Acids. Res.* **8,** 235–252.
Gonzales, F., Romani, S., Cubas, P., Modolell, J., and Campuzano, S. (1989). Molecular analysis of the asense gene, a member of the achaete-scute complex of *Drosophila melanogaster*, and its novel role in optic lobe development. *EMBO J.* **8,** 3553–3562.
Gonzy-Treboul, G., Lepesant, J.-A., and Deutsch, J. (1995). Enhancer-trap targeting at the *Broad-Complex* locus of *Drosophila melanogaster*. *Genes Dev.* **9,** 1137–1148.
Goodman, R. M. (1969). Ecdysone and DNA puffs. *Nature* **222,** 605.
Goodman, R. M., and Benjamin, W. B. (1973). Nucleoprotein methylation in salivary gland chromosomes of *Sciara coprophila:* Correlation with DNA synthesis. *Exp. Cell. Res.* **77,** 63–72.
Goodman, R. M., and Henderson, A. (1986). Sine waves induce cellular transcription. *Bioelectromagnetic* **7,** 23–30.
Goodman, R. M., Goidl, J., and Richart, R. M. (1967). Larval development in *Sciara coprophila* without the formation of chromosomal puffs. *Proc. Natl. Acad. Sci. USA* **58,** 553–559.
Goodman, R. M., Schmidt, E. C., and Benjamin, W. B. (1976). The effect of ecdysterone on nonhistone proteins in salivary gland chromosomes of *Sciara coprophila*. *Cell. Differ.* **5,** 233–246.
Goodman, R. M., Schmidt, E. C., and Benjamin, W. B. (1977). Posttranscriptional modifications of nonhistone proteins of salivary gland cells of *Sciara coprophila*. *Methods Cell. Biol.* **14,** 343–359.
Goodman, R. M., Bassett, C. A. L., and Henderson, A. S. (1983). Pulsing electromagnetic fields induce cellular transcription. *Science* **220,** 1283–1285.
Goodman, R., Abbott, J., and Henderson, A. S. (1987a). Transcriptional patterns in the X chromosome of *Sciara coprophila* following exposure to magnetic fields. *Bioelectromagnetic* **8,** 1–7.
Goodman, R., Krim, A., Henderson, A., and Weisbrot, D. (1987b). Gene activation in *Drosophila* salivary gland cells exposed to low frequency non-ionizing radiation. *Genetics* **116**(1), S50.
Goodwin, S. W. (1955). Effect of ring-gland transplantation on nonpupating tumor-bearing larvae. *Dros. Inform. Serv.* **29,** 121.
Goodwin, T. W., Horn, D. H. S., Karlson, P., Koolman, J., Nakanishi, K., Robbins, W. E., Siddal, J. B., and Takemoto, T. (1978). Ecdysteroids: a new generic term. *Nature* **272,** 122.
Gopalan, H. N. B. (1973a). Cordycepin inhibits induction of puffs by ions in *Chironomus* salivary gland chromosomes. *Experientia* **29,** 724–726.
Gopalan, H. N. B. (1973b). Amino acid induced changes of puffing patterns in *Chironomus* salivary gland chromosomes. *Chromosoma* **44,** 25–32.
Gopalan, H. N. B., and Dass, C. M. S. (1972a). Polyteny and salivary gland secretion in the melonfly, *Dacus cucurbitae*. *Experientia* **28,** 684–686.
Gopalan, H. N. B., and Dass, C. M. S. (1972b). Puffing patterns in the salivary gland chromosomes of the melonfly, *Dacus cucurbitae*. *Genetica* **43,** 61–75.
Gorman, M., Kuroda, M. I., and Baker, B. S. (1993). Regulation of the sex-specific binding of the maleless dosage compensation protein to the male X chromosome in *Drosophila*. *Cell* **72,** 39–49.
Gorman, M., Franke, A., and Baker, B. S. (1995). Molecular characterization of the *male-specific*

lethal-3 gene and investigations of the regulation of dosage compensation in *Drosophila*. *Development* **121,** 463–475.

Gorman, M. J., and Kaufman, T. C. (1995). Genetic analysis of embryonic cis-acting regulatory elements of the *Drosophila* homeotic gene *Sex combs reduced*. *Genetics* **140,** 557–572.

Gotwals, P. J., and Fristrom, J. W. (1991). Three neighboring genes interact with the Broad-Complex and the Stubble-stubbloid locus to affect imaginal disc morphogenesis in *Drosophila*. *Genetics* **127,** 747–759.

Gowen, J. W., and Gay, E. H. (1933). Gene number, kind and size in *Drosophila*. *Genetics* **18,** 1–31.

Graessmann, A., Graessmann, M., and Lara, F. J. S. (1973). Involvement of RNA in the process of puff induction in polytene chromosomes. *In* "Molecular Cytogenetics" (B. A. Hamkalo and J. Papaconstantinou, eds.), pp. 209–215. Plenum, New York/London.

Granok, H., Leibovitch, B. A., Shaffer, C. D., and Elgin, S. C. R. (1995). Ga-ga over GAGA factor. *Current Biol.* **5,** 238–241.

Greenberg, J. R. (1967). Sedimentation studies on *Drosophila virilis* salivary gland RNA. *J. Cell. Biol.* **35**(2), 49A–50A.

Greenberg, J. R. (1969). Synthesis and properties of ribosomal RNA in *Drosophila*. *J. Mol. Biol.* **46,** 85–98.

Greenleaf, A. L., and Bautz, E. K. F. (1975). RNA polymerase B from *Drosophila melanogaster* larvae. Purification and partial characterization. *Eur. J. Biochem.* **60,** 169–179.

Greenleaf, A. L., Plagens, U., Jamrich, M., and Bautz, E. K. F. (1978). RNA polymerase B (or II) in heat induced puffs of *Drosophila* polytene chromosomes. *Chromosoma* **65,** 127–136.

Greenleaf, A. L., Borsett, L. M., Jiamachello, P. F., and Coulter, D. E. (1979). a-Amanitin-resistant *D. melanogaster* with an altered RNA polymerase II. *Cell* **18,** 613–622.

Gregg, T. G., McCrate, A., Reveal, G., Hall, S., and Rypstra, A. L. (1990). Insectivory and social digestion in *Drosophila*. *Biochem. Genet.* **28,** 197–207.

Grigliatti, T. A., White, B. N., Tener, G. M., Kaufman, T. C., Holden, J. J., and Suzuki, D. T. (1974). Studies on the transfer RNA genes of *Drosophila*. *Cold Spring Harbor Symp. Quant. Biol.* **38,** 461–474.

Grimm, C., and Kunz, W. (1980). Disproportionate rDNA replication does occur in diploid tissue in *Drosophila hydei*. *Mol. Gen. Genet.* **180,** 23–26.

Grimm, C., Kunz, W., and Franz, G. (1984). The organ-specific rRNA gene number in *Drosophila hydei* is controlled by sex heterochromatin. *Chromosoma* **89,** 48–54.

Grimwade, B. G., Muskavitch, M. A. T., Welshons, W. J., Yedvobnick, B., and Artavanis-Tsakonas, S. (1985). The molecular genetic of the Notch locus in *Drosophila melanogaster*. *Dev. Biol.* **107,** 503–519.

Grinchuk, T. M. (1967). A study of polymorphism in the polytene chromosomes of the *Prosimulium hirtipes* (Simulidae, Diptera) speci. *Genetika* **3**(1), 165–172.

Grinchuk, T. M. (1968). Karyological study on a natural population of the midge *Wilhelmia equina*. *Tsitologiya* **10**(8), 1002–1007. [in Russian]

Grinchuk, T. M. (1969). Karyological analysis of *Wilhelmia equina* with reference to its seasonal variation. *Tsitologiya* **11**(3), 320–327. [in Russian]

Grinchuk, T. M. (1977). Morphology of the polytene chromosomes of the gall midge *Wyattella ussuriensis*. I. Karyotype structure. *Tsitologiya* **19**(2), 210–214. [in Russian]

Grond, C. J., and Derksen, J. (1983). The banding pattern of the salivary gland chromosomes of *Drosophila hydei*. *Eur. J. Cell. Biol.* **30,** 144–148.

Grond, C., Saiga, H., and Edstrom, J.-E. (1987). The sp-I genes in the Balbiani Rings of *Chironomus* salivary glands. *In* "Results and Problems in Cell Differentiation" Vol. 14, pp. 69–80.

Gronemeyer, H., and Pongs, O. (1980). Localization of ecdysterone on polytene chromosomes of *Drosophila melanogaster*. *Proc. Natl. Acad. Sci. USA* **77,** 2108–2112.

Gronemeyer, H., Hameister, H., and Pongs, O. (1981). Photoinduced binding of endogeneous ecdysterone to salivary gland chromosomes of *Chironomus tentans*. *Chromosoma* **82,** 543–559.

Gross, J. D. (1957). Incorporation of phosphorus-32 into salivary type chromosomes which exhibit "puffs." *Nature* **180,** 440.

Grossbach, U. (1968). Cell differentiation in the salivary glands of *Camptochironomus tentans* and *C. pallidivittatus*. *Ann. Zool. Fennici* **5,** 37–40.

Grossbach, U. (1969). Chromosomen-Aktivitat und biochemische Zelldifferenzierung in den Speicheldrusen von *Camptochironomus*. *Chromosoma* **28,** 136–187.

Grossbach, U. (1974). Chromosome puff and gene expression in polytene cells. *Cold Spring Harbor Symp. Quant. Biol.* **38,** 619–627.

Grossbach, U. (1975). Korrelation zwischen Chromosomen-Aktivitat und Genexpression in Differezierten Zellen. *Nachr. Akad. Wissensch. Gottingen*. II. Math.-phys. Kl, 176–179.

Grossbach, U. (1977). The salivary gland of Chironosmus (Diptera) a model system for the study of cell differentiation. *In* "Results and problems in cell differentiation. Biochemical differentiation in insect glands" (W. Beermann, ed.), Vol. 8, pp. 147–196. Springer-Verlag, Berlin/Heidelberg/New York.

Grossman, G. L., and James, A. A. (1993). The salivary glands of the vector mosquito, *Aedes aegypti*, express a novel member of the amylase gene family. *Insect Mol. Biol.* **1,** 223–232.

Grossniklaus, U., Pearson, R. K., and Gehring, W. J. (1992). The *Drosophila* sloppy paired locus encodes two proteins involved in segmentation that show homology to mammalian transcription factors. *Genes Dev.* **6,** 1030–1051.

Guaraciaba, H. L. B., and de Toledo, L. F. A. (1967). Age determination of *Rhynchosciara angelae* larvae. *Rev. Brasil. Biol.* **27,** 321–332.

Guaraciaba, H. L. B., and de Toledo, L. F. A. (1968). Morphological changes in the salivary chromosomes of *Rhyncosciara angelae*. *Arq. Inst. Biol. (San Paulo)* **35,** 89–98.

Guay, P. S., and Guild, G. M. (1991). The ecdysone-induced puffing cascade in *Drosophila* salivary glands: A Broad-Complex early gene regulates intermolt and late gene transcription. *Genetics* **129,** 169–175.

Gubenko, I. S. (1976a). Autoradiographic identificaton of the late replicating regions of polytene chromosomes of the salivary gland cells of *Drosophila virilis*. *Tsitologia* **18**(8), 964–968. [in Russian]

Gubenko, I. S. (1976b). Distribution of late replicating in polytene chromosomes of salivary gland cells of *Drosophila virilis* males and females. *Tsitologia* **18**(8), 969–974. [in Russian]

Gubenko, I. S. (1976c). Late replicating regions of the polytene chromosomes of the salivary glands of different species and interspecific hybrids of *Drosophila virilis* group. *Genetika* **12**(5), 72–80. [in Russian]

Gubenko, I. S. (1989). Unusual loci of the *Drosophila* genome activated under heat shock and other stress conditions. *Biopolimery i kletka* **5**(3), 5–22. [in Russian]

Gubenko, I. S., and Baricheva, E. M. (1979). Puffs of *Drosophila virilis* induced by temperature and other endogenous actions. *Genetika* **15**(8), 1399–1414. [in Russian]

Gubenko, I. S., Subbota, R. P., and Semeshin, V. F. (1991a). Unusual *Drosophila virilis* stress-puff at 20CD: Cytological localization of a heat sensitive locus and some peculiarities of the heat shock response. *Hereditas* **115,** 283–290.

Gubenko, I. S., Subbota, R. P., and Semeshin, V. F. (1991b). The "unusual" activated under stress conditions puff 20CD of *Drosophila virilis:* Cytogenetical localization of the heat-sensitive locus and features of the response to heat shock. *Genetika* **27**(1), 61–69. [in Russian]

Guevara, M., and Basile, R. (1973). DNA and RNA puffs in *Rhynchosciara*. *Caryologia* **26,** 273–295.

Guild, G. M. (1984). Molecular analysis of a developmentally regulated gene which is expressed in the larval salivary gland of *Drosophila*. *Dev. Biol.* **102,** 462–470.

Guild, G., and Richards, G. (1992). Ecdysone and the onion. *Nature* **357,** 539.

Guild, G. M., and Shore, E. M. (1984). Larval salivary gland secretion proteins in *Drosophila:* Identification and characterization of *Sgs-5* structural gene. *J. Mol. Biol.* **179,** 289–314.

Gundacker, D., Phannavong, B., Vef, O., Gateff, E., and Kurzik- Dumke, U. (1993). Genetic and molecular characterization of breakpoints of five deficiencies in the genomic region 59F-60A. *Dros. Inform Serv.* **72,** 129–131.

Gunderina, L. I. (1992). The DNA cycle during chromosome polytenization in the salivary gland cells of *Chironomus thummi.* IV. The duration of S-phase of diploid cells. *Tsitologia* **34,** 46–53. [in Russian]

Gunderina, L. I., and Aimanova, K. G. (1998). Genetic effects of gamma-irradiation on *Chironomus thummi.* III. Aberrations in polytene chromosomes. *Genetika* **34**(3), 355–363. [in Russian]

Gunderina, L. I., and Seraya, E. I. (1992). The DNA replication cycle during chromosome polytenization in the salivary gland cells of *Chironomus thummi.* V. The duration of S-phase of 2^8C and 2^9C cells. *Tsitologia* **34,** 54–64. [in Russian]

Gunderina, L. I., Sherudilo, A. I., and Mitina, P. I. (1983). DNA reduplication cycle during polytenization of chromosomes of the salivary gland cells of the larvae of *Chironomus thummi.* I. Pattern of ^{3}H-thymidine incorporation into the polytene chromosomes at different phases of larval development during fourth instar. *Tsitologia* **25**(5), 527–533. [in Russian]

Gunderina, L. I., Sherudilo, A. I., and Mitina, R. L. (1984a). DNA reduplication cycle during polytenization of chromosomes of the salivary gland cells of *Chironomus thummi.* II. Cytophotometric and autoradiographic analysis of the dynamics of DNA synthesis in individual salivary gland cells. *Tsitologiya* **26**(7), 794–801. [in Russian]

Gunderina, L. I., and Sherudilo, A. I. (1984b). DNA reduplication cycle during polytenization of chromosomes of the salivary gland cells of *Chironomus thummi.* III. Determination of the duration of the period of DNA synthesis. *Tsitologia* **26**(8), 927–935. [in Russian]

Gupta, J. P., and Kumar, A. (1987). Cytogenetics if *Zaprionus indianus*(Diptera; Drosophilidae): Nucleolar organizer regions, mitotic and polytene chromosomes and inversion polymorphism. *Genetica* **74,** 19–25.

Gupta, J. P., and Singh, O. P. (1983). Puffing patterns in polytene chromosomes of different tissues of *Melanagromyza obtusa* during development. *J. Heredity* **74,** 365–370.

Gvozdev, V. A. (1968). Regulation of gene effects in *Drosophila melanogaster* (Prospects of biochemical studies). *Uspekhi Sovr. Biol.* **65**(3), 398–423. [in Russian]

Gvozdev, V. A. (1970). Relation between gene dosage and synthesis of macromolecules in animals. *Izvestiya AN SSSR. Ser. Biol. Nauk.* **4,** 499–507. [in Russian]

Gvozdev, V. A., Gostimsky, S. A., Gerasimova, T. I., and Gavrina, E. S. (1973). Complementation and fine structure analysis at the 2D3–2F5 region of the X-chromosome of *Drosophila melanogaster. Dros. Inform. Serv.* **50,** 34.

Gvozdev, V. A., Gostimsky, S. A., Gerasimova, T. I., Dubrovskaya, E. S., and Braslavskaya, O. Yu. (1975a). Fine genetic structure of the 2D3–2F5 region of the X-chromosome of *Drosophila melanogaster.* Mol. Gen. Genet. **141,** 269–275.

Gvozdev, V. A., Gostimsky, S. A., Gerasimova, T. I., Dubrovskaya, E. S., and Braslavskaya, O. Yu. (1975b). Fine genetic structure of the 2D3–2F5 region of the X chromosome of *Drosophila melanogaster. Genetika* **11**(1), 73–79. [in Russian]

Gvozdev, V. A., Gerasimova, T. I., Kovalev, Yu. M., and Ananiev, E. V. (1977). Cytogenetic structure of the 2D3–2F5 region of the X-chomosome of *Drosophila melanogaster. Dros. Inform. Serv.* **52,** 67–68.

Gvozdev, V. A., Ananiev, E. V., Kotelyanskaya, A. E., and Zhimulev, I. F. (1980). Transcription of intercalary heterochromatin regions in *Drosophila melanogaster* cell culture. *Chromosoma* **80,** 177–190.

Gvozdev, V. A., Leibovitch, B. A., and Ananiev, E. V. (1983). Gene dosage compensation in the X chromosome of *Drosophila melanogaster:* Transcription levels in metafemales and metamales and the amount of 6-phosphogluconate dehydrogenase in metafemales. *Dros. Inform. Serv.* **59,** 48–49.

Haass, C., Klein, U., and Kloetzel, P.-M. (1990). Developmental expression of *Drosophila melanogaster* small heat-shock proteins. *J. Cell. Sci.* **96,** 413–418.

Hackett, R. W., and Lis, J. T. (1981). DNA sequence analysis reveals extensive homologies of regions preceding *hsp70* and *ab* heat shock genes in *Drosophila melanogaster. Proc. Natl. Acad. Sci. USA* **78,** 6196–6200.

Hackett, R. W., and Lis, J. T. (1983). Localization of the *hsp83* transcript within a 3292 nucleotide-sequence from the *63B* heat shock locus of *Drosophila melanogaster. Nucleic Acids Res.* **11,** 7011–7030.

Hadorn, E. (1937). An accelerating effect of normal "ring-glands" on puparium-formation in lethal larvae of *Drosophila melanogaster. Proc. Natl. Acad. Sci. USA* **23,** 478–484.

Hadorn, E., and Faulhaber, I. (1962). Range of variability in cell number of larval salivaries. *Dros. Inform. Serv.* **36,** 71.

Hadorn, E., Gehring, W., and Staub, M. (1963). Extensives Wachstum der larvalen Speicheldrusen-chromosomen von *Drosophila melanogaster* in Adultmilieu. *Experientia* **19,** 530–531.

Hafen, E., Basler, K., Edstroem, J.-E., and Rubin, G. M. (1987). *Sevenless,* a cell-specific homeotic gene of *Drosophila,* encodes a putative transmembrane receptor with a tyrosine kinase domain. *Science* **236,** 55–63.

Hagele, K. (1970). DNS-Replikationsmuster der Speicheldrusen- chromosomen von Chironomiden. *Chromosoma* **31,** 91–138.

Hagele, K. (1971). Srucutrverandernde Wirkung von FUdR auf polytane Chromosomen und Beziehungen zwischen Replikationsdauer und Bruchhaufigkeit von Querscheiben. *Chromosoma* **33,** 297–318.

Hagele, K. (1972). Querscheibenspezifische Unterscheiede in der ^{3}H-Thymidin- markierung wahrend der Larvenen-twicklung von *Chironomus thummi* piger. *Chromosoma* **39,** 63–82.

Hagele, K. (1973). Komplementare DNA Replikationsmuster bei *Drosophila melanogaster. Chromosoma* **41,** 231–236.

Hagele, K. (1976). Prolongation of replication time after doubling of the DNA content of polytene chromosome bands of *Chironomus. Chromosoma* **55,** 253–258.

Hagele, K., and Kalisch, W.-E. (1974). Initial phases of DNA synthesis in *Dosophila melanogaster.* I. Differential participation in replication of the X chromosomes in males and females. *Chromosoma* **47,** 403–413.

Hagele, K., and Kalisch, W.-E. (1978). ^{3}H-thymidine labelling over homologous asynapsed loci polytene chromosomes. *Dros. Inform. Serv.* **53,** 188–189.

Hagele, K., and Kalisch, W.-E. (1980). DNA replication of a polytene chromosome section in *Drosophila* prior to and during puffing. *Chromosoma* **79,** 75–83.

Hager, E. J., Miller, O. L., Jr. (1991). Ultrastructural analysis of polytene chromatin of *Drosophila melanogaster* reveals clusters of tightly linked co-expressed genes. *Chromosoma* **100,** 173–186.

Hall, L. M. C., Mason, P. J., and Spierer, P. (1983). Transcripts, genes and bands in 315,000 base-pairs of *Drosophila* DNA. *J. Mol. Biol.* **169,** 83–96.

Halle, J.-P., and Meisterernst, M. (1996). Gene expression: Increasing evidence for a transcriptosome. *Trends Genet.* **12,** 161–163.

Hameister, H. (1977). RNA synthesis inisolated polytene nuclei from *Chironomus tentans. Chromosoma* **62,** 217–242.

Hammond, M. P., and Laird, C. D. (1985). Chromosome structure and DNA replication in nurse and follicle cells of *Drosophila melanogaster. Chromosoma* **91,** 267–278.

Hamodrakas, S. J., and Kafatos, F. C. (1984). Structural implications of primary sequences from a family of Balbiani Ring-encoded proteins in *Chironomus. J. Mol. Evol.* **20,** 296–303.

Han, S., Udvardy, A., and Schedl, P. (1985). Novobiocin blocks the *Drosophila* heat shock response. *J. Mol. Biol.* **183,** 13–29.

Handler, A. M. (1982). Ecdysteroid titers during pupal and adult development in *Drosophila melanogaster. Devel. Biol.* **93,** 73–82.

Hankeln, T., and Schmidt, E. R. (1990). New foldback transposable element TFB1 found in histone genes of the midge *Chironomus thummi*. *J. Mol. Biol.* **215**, 477–482.

Hankeln, T., and Schmidt, E. R. (1991). The organization, localization and nucleotide sequence of the histone genes of the midge *Chironomus thummi*. *Chromosoma* **101**, 25–31.

Hanna-Rose, W., and Hansen, U. (1996). Active repression mechanisms of eukaryotes transcription repressors. *Trends Genet.* **12**, 229–234.

Hannah-Alava, A. (1971). Cytogenetics of nucleolus transpositions in *Drosophila melanogaster*. *Mol. Gen. Genet.* **113**, 191–203.

Hansson, L., and Lambertsson, A. (1983). The role of *su(f)* gene function and ecdysterone in transcription of glue polypeptide mRNAs in *Drosophila melanogaster*. *Mol. Gen. Genet.* **192**, 395–401.

Hansson, L., and Lambertsson, A. (1984). Ecdysterone responsive functions in the mutant *l(1)su(f)*^{ts67g} of *Drosophila melanogaster*. *Wilhelm Roux's Arch. Dev. Biol.* **193**, 48–51.

Hansson, L., and Lambertsson, A. (1989). Steroid regulation of glue protein genes in *Drosophila melanogaster*. *Hereditas* **110**, 61–67.

Hansson, L., Lineruth, K., and Lambertsson, A. (1981). Effect of the *l(1)su(f)*^{ts67g} mutation of *Drosophila melanogaster* on glue protein synthesis. *Wilhelm Roux's Arch. Dev. Biol.* **190**(6), 308–312.

Hardy, P. A., and Pelling, C. (1980). Cell-free synthesis and immunological characterization of salivary proteins from *Chironomus tentans*. *Chromosoma* **81**, 403–417.

Harris, H. (1968). "Nucleus and Cytoplasm," pp. 78–79 and 100–103. Clarendom Press, Oxford.

Harris, W. A., Stark, W. S., and Walker, J. A. (1976). Genetic dissection of the photoreceptor system in the compound eye of *Drosophila melanogaster*. *J. Physiol.* **256**, 415–439.

Harrison, S. D., Solomon, N., and Rubin, G. M. (1995). A genetic analysis of the 63E-64A genomic region of *Drosophila melanogaster:* Identification of mutations in a replication factor C subunit. *Genetics* **139**, 1701–1709.

Harrod, M. J. E., and Kastritsis, C. D. (1972a). Developmental studies in *Drosophila*. II. Ultrastructural analysis of the salivary glands of the *Drosophila pseudoobscura* during some stages of development. *J. Ultrastruct. Res.* **38**, 482–499.

Harrod, M. J. E., and Kastritsis, C. D. (1972b). Developmental studies in *Drosophila*. VI. Ultrastructural analysis of the salivary glands of the *Drosophila pseudoobscura* during the late larval period. *J. Ultrastruct. Res.* **40**, 292–312.

Hart, M. C., Wang, L., and Coulter, D. E. (1996). Comparison of the structure and expression of *odd-skipped* and two related genes that encode a new family of zinc finger proteins in *Drosophila*. *Genetics* **144**, 171–182.

Hartl, F. U. (1996). Molecular chaperons in cellular protein folding. *Nature* **381**, 571–580.

Hatzopulos, P., and Kambysellis, M. P. (1987). Isolation and structural analysis of *Drosophila grimshawi* vitellogenin genes. *Mol. Gen. Genet.* **206**, 475–484.

Hawley, R. S., and Marcus, C. H. (1989). Recombinatinal controls of rDNA redundancy in *Drosophila*. *Ann. Rev. Genet.* **23**, 87–120.

Hazelrigg, T., and Kaufman, T. C. (1983). Revertants of dominant mutations associated with the *Antennapedia* gene complex of *Drosophila melanogaster*. *In* "Molecular aspects of early development" (G. Malacinski and W. Klein, eds.), pp. 189–218. Plenum Press, New York.

Hazelrigg, T., Levis, R., and Rubin, G. M. (1984). Transformation of white locus DNA in *Drosophila:* Dosage compensation, zeste interaction, and position effects. *Cell* **36**, 469–481.

Hegdekar, B. M., and Smallman, B. N. (1967). Lysosomal acid phosphatase during metemorphosis of *Musca domestica* (Linn). *Can. J. Biochem.* **45**(7), 1202–1205.

Heino, T. I. (1989). Polytene chromosomes from ovarian pseudonurse cells of the *Drosophila melanogaster otu* mutant. I. Photographic map of chromosome 3. *Chromosoma* **97**, 363–373.

Heino, T. (1994). Polytene chromosomes from ovarian pseudonurse cells of the *Drosophila melanogaster otu* mutant. II. Photographic map of the X chromosome. *Chromosoma* **103**, 4–15.

Heino, T. I., Saura, A. O., and Sorsa, V. (1994). Maps of the salivary gland chromosomes of *Drosophila melanogaster*. *Dros. Inform. Serv.* **73,** 619–738.

Heino, T. I., Lahti, V. P., Tirronen, M., and Roos, C. (1995). Polytene chromosomes show normal gene activity but some mRNA are abnormally accumulated in the pseudonurse cell nuclei of *Drosophila melanogaster otu* mutants. *Chromosoma* **104,** 44–55.

Heitz, E. (1931). Die Ursache der gesetzmassigen Zahl. Lage, Form, und Grose pflanzlicher Nukleolen. *Planta Arch. Wiss. Bot.* **12,** 775–844.

Heitz, E. (1934). Uber a- and b-Heterochromatin sowie Konstanz und Bau der Chromomeren bei *Drosophila*. *Biol. Zbl.* **54,** 588–609.

Heitz, E., and Bauer, H. (1933). Beweise fur die Chromosomennatur der Kernscleifen in den Knauelkernene von *Bibio hortulanus*. *Z. Zellforsch.* **17,** 67–82.

Heitz, J. G., and Plaut, W. (1975). Alteration of Ringer pH and its effect on precursor incorporation into *Drosophila* salivary gland nuclei. *Chromosoma* **51,** 147–155.

Heling, W., and Ruqi, L. (1985). An observation of the polytene chromosomes in salivary gland cells *Chironomus* sp. II. The reversible changes of the chromosome I and II from larval to adult stage. *Acta Genetica Sinica.* **12,** 132–136. [in Chinese].

Helmsing, P. J. (1972). Induced acumulation of non-histone proteins in polytene nuclei of *Drosophila hydei*. II. Accumulation of proteins in polytene nuclei and chromatin of different larval tissues. *Cell. Differ.* **1,** 19–24.

Helmsing, P. J., and Berendes, H. D. (1971). Induced accumulation of nonhistone proteins on polytene nuclei of *Drosophila hydei*. *J. Cell. Biol.* **50,** 893–896.

Henderson, S. A. (1967). The salivary gland chromosomes of *Dasyneura crataegy* (Diptera, Cecidomiidae). *Chromosoma* **23,** 38–58.

Hendrickson, J. E., and Sakonju, S. (1995). Cis and Trans interaction between the *iab* regulatory regions and *abdominal-A* and *Abdominal-B* in *Drosophila melanogaster*. *Genetics* **139,** 835–849.

Henikoff, S., and Meneely, P. M. (1993). Unwinding dosage compensation. *Cell* **72,** 1–2.

Henikoff, S., and Meselson, M. (1977). Transcription at two heat shock loci in *Drosophila*. *Cell* **12,** 441–451.

Henikoff, S., Keene, M. A., Fechtel, K., and Fristrom, J. W. (1986). Gene within a gene: Nested *Drosophila* genes encode unrelated proteins on opposite DNA strands. *Cell* **44,** 33–42.

Henkel-Tigges, J., and Davis, R. L. (1990). Rat homologs of the *Drosophila dunce* gene code for cyclic AMP phosphodiesterases sensitive to rolipram and *ro 20-1724*. *Mol. Pharmacol.* **37,** 7–10.

Hennig, W. (1980). Functional units of chromosomes. In "Well-Being of Mankind and Genetics (Proc XIV Intern. Congr. Genet.)," Vol. 1, pp. 117–128. Mir Publishers, Moscow.

Hennig, W., and Meer, B. (1971). Reduced polyteny of ribosomal RNA cistrons in giant chromosomes of *Drosophila hydei*. *Nature New Biol.* **233,** 70–72.

Hennig, W., Vogt, P., Jacob, G., and Siegmund, I. (1982). Nucleolus organizer regions in *Drosophila* species of the repleta group. *Chromosoma* **87,** 279–292.

Henrich, V. C., Tucker, R. L., Maroni, G., and Gilbert, L. I. (1987). The ecdysoneless (*ecd1*^{ts}) mutation disrupts ecdysteroid synthesis autonomously in the ring gland of *Drosophila melanogaster*. *Dev. Biol.* **120,** 50–55.

Henrich, V. C., Sliter, T. J., Lubahn, D. B., MacIntyre, A., and Gilbert, L. I. (1990). A steroid/thyroid hormone receptor superfamily member in *Drosophila melanogaster* that shares extensive sequence similarity with a mammalian homologue. *Nucleic Acids Res.* **18,** 4143–4148.

Henrich, V. C., Szekely, A. A., Kim, S. J., Brown, N. E., Antoniewski, C., Hayden, M. A., Lepesant, J.-A., and Gilbert, L. I. (1994). Expression and function of the *ultraspiracle (usp)* gene during development of *Drosophila melanogaster*. *Dev. Biol.* **165,** 38–52.

Henrikson, P. A., and Clever, U. (1972). Protease activity and cell death during metamorphosis in the salivary gland of *Chironomus tentans*. *J. Insect Physiol.* **18,** 1981–2004.

Hershey, N. D., Conrad, S. E., Sodja, A., Yen, P. H., Cohen, M., Jr., and Davidson, N. (1977). The sequence arrangement of *Drosophila melanogaster* 5S DNA cloned in recombinant plasmids. *Cell* **11,** 585–598.

Herskowitz, I. H. (1950). An estimate of the number of loci in the X chromosome of *Drosophila melanogaster*. *Am. Natur.* **84,** 255–260.

Hertner, T., and Mahr, R. (1979). Electrophoretic analysis of the salivary proteins of *Chironomus larvae*. *Experienta* **35,** 966.

Hertner, T., Meyer, B., Eppenberger, H. M., and Mahr, R. (1980). The secretion proteins in chironomus tentans salivary glands: Electrophoretic characterization and molecular weight estimation. *Wilhelm Roux's Arch.* **189,** 69–72.

Hertner, T., Eppenberger, H. M., and Lezzi, M. (1983). The giant secretory proteins of *Chironomus tentans* salivary glands: the organization of their primary structure, their amino acid and carbohydrate composition. *Chromosoma* **88,** 194–200.

Hertner, T., Hertner-Meyer, B., Sogo, J. M., Schlapfer, G., Jackle, H., Edstrom, J.-E., and Lezzi, M. (1986). Microcloning of a long segment of a gene from the primary ecdysterone-controlled puff region I-18C of *Chironomus tentans:* Molecular characterization and effect of hormone on its transcription. *Dev. Biol.* **113,** 29–39.

Hess, C., and Hauschteck-Jungen, E. (1974). Replication paterns of the distal end of the A-chromosome from the salivary gland of *Drosophila subobscura*. *Dros. Inform. Serv.* **51,** 118.

Heuer, J. G., Li, K., and Kaufman, T. C. (1995). The *Drosophila* homeotic target gene *centrosomin (cnn)* encodes a novel centrosomal protein with leuzine zipper and maps to a genomic region required for midgut morphogenesis. *Development* **121,** 3861–3876.

Hickey, D. A., Bally-Guif, L., Abukashawa, S., Payant, V., and Benkel, B. F. (1991). Concerted evolution of duplicated protein-coding genes in *Drosophila*. *Proc. Natl. Acad. Sci. USA* **88,** 1611–1615.

Hiebert, J. C., and Birchler, J. A. (1992). Dosage compensation of the copia retrotransposon in *Drosophila melanogaster*. *Genetics* **130,** 539–545.

Hill, R. J., and Rudkin, G. T. (1987). Polytene chromosomes: The status of the band–interband question. *BioEssays* **7,** 35–40.

Hill, R. J., and Watt, F. (1978). "Native" salivary chromosomes of *Drosophila melanogaster*. *Cold Spring Harbor Symp. Quant. Biol.* **42,** 859–865.

Hill, R. J., and Whytock, S. (1993). Cytological structure of the native polytene salivary gland nucleus of *Drosophila melanogaster:* A microsurgical analysis. *Chromosoma* **102,** 446–456.

Hill, R. J., Watt, F., Yund, M. A., and Fristrom, J. W. (1982). The effect of 20-hydroxyecdysone on synthesis of chromosomal and cytosol proteins in imaginal discs. *Dev. Biol.* **90,** 340–351.

Hilliker, A. J., Clark, S. H., Chovnick, A., and Gelbart, W. M. (1980). Cytogenetic analysis of the chromosomal region immediately adjacent to the rosy locus in *Drosophila melanogaster*. *Genetics* **95,** 95–110.

Hilliker, A. J., Chovnick, A., and Clark, S. H. (1981a). The relative mutabilities of vital genes in *Drosophila melanogaster*. *Dros. Inform. Serv.* **56,** 64–65.

Hilliker, A. J., Clark, S. H., Gelbart, W. M., and Chovnick, A. (1981b). Cytogenetic analysis of the rosy micro-region polytene chromosome interval 87D2-4; 87E12-F1 of *Drosophila melanogaster*. *Dros. Inform. Serv.* **56,** 65–72.

Hiraoka, L., Newfeld, S., Akui, A., and Stuart, W. D. (1983). A monoclonal ELISA technique for identifying RNA-DNA hybrids. *Genetics* **104**(1), 35.

Hirsh, J., and Davidson, N. (1981). Isolation and characterization of the Dopa Decarboxylase gene of *Drosophila melanogaster*. *Mol. Cell. Biol.* **1,** 475–485.

Hochman, B. (1971). Analysis of chromosome 4 in *Drosophila melanogaster*. II. Ethyl methanolsulfonate induced lethals. *Genetics* **67,** 235–252.

Hochman, B. (1972). The detection of four more vital loci on chromosome 4 in *Drosophila melanogaster*. *Genetics* **71,** 71.

Hochman, B. (1974). Analysis of a whole chromosome in *Drosophila*. *Cold Spring Harbor Symp. Quant. Biol.* **38,** 581–589.

Hochman, B. (1976). The fourth chromosome of *Drosophila melanogaster*. *In* "The Genetica and

Biology of *Drosophila*" (M. Ashbirner and E. Novitsky. eds.), Vol. 1b, pp. 903–928. Academic Press, London/New York/San Francisco.

Hochman, B., Gloor, H., and Green, M. M. (1964). Analysis of chromosome 4 in *Drosophila melanogaster*. I. Spontaneous and X-ray-induced lethals. *Genetica* **35,** 109–126.

Hochstrasser, M. (1987). Chromosome structure in four wild-type polytene tissues of *Drosophila melanogaster*. The 87A and 87C heat shock loci are induced unequally in the midgut in a manner dependent on growth temperature. *Chromosoma* **95,** 197–208.

Hodgetts, R. B., Sage, B., and O'Connor, J.D. (1977). Ecdysone titres during postembryonic development of *Drosophila melanogaster*. *Dev. Biol.* **60,** 310–317.

Hodgetts, R. B., Clark, W. C., O'Keefe, S. L., Schouls, M., Crossgrove, K., Guild, G. M., and Kalm, L. von. (1995). Hormonal induction of Dopa decarboxylase in the epidermis of *Drosophila* is mediated by the *Broad-Complex*. *Development* **121,** 3913–3922.

Hoffman, E. P., and Korge, G. (1987). Upstream sequences of dosage-compensated and noncompensated alleles of the larval secretion protein gene *Sgs-4* in *Drosophila*. *Chromosoma* **96,** 1–7.

Hoffman, E. P., Gerrring, S. L., and Corces, V. G. (1987a). The ovarian, ecdysterone, and heat-shock-responsive promrters of the *Drosophila melanogaster hsp27* gene react very differently to perturbations of DNA sequence. *Mol. Cell. Biol.* **7,** 973–981.

Hoffman, E. P., Keimhorst, A., Krumm, A., and Korge, G. (1987b). Regulatory sequences of the *Sgs-4* gene of *Drosophila melanogaster* analysed by P-element-mediated transformation. *Chromosoma* **96,** 8–17.

Hofmann, A., Garfinkel, M. D., Meyerowitz, E. M. (1991). cis-Acting sequences reguired for expression of the divergently transcribed *Drosophila melanogaster Sgs-7* and *Sgs-8* glue protein genes. *Mol. Cell. Biol.* **11,** 2971–2979.

Hogan, N. C., Traverse, K. L., Sullivan, D. E., and Pardue, M. L., (1994). The nucleus-limited *Hsr-omega-n* transcript is a polyadenilated RNA with a regulated intranuclear turnover. *J. Cell. Biol.* **125,** 21–30.

Hogan, N. C., Slot, F., Traverse, K. L., Garbe, J. C., Bendena, W. G., and Pardue, M.L. (1995). Stability of tandem repeats in the *Drosophila melanogaster Hsr-omega* nuclear RNA. *Genetics* **139,** 1611–1621.

Hogness, D. S., Lipshitz, H. D., Beachy, P. A., Peattie, D. A., Saint, R. B., Goldschmidt-Clermont, M., Harte, P. J., Gavis, E. R., and Helfand, S. L. (1985). Regulation and products of the *Ubx* domain of the bithorax complex. *Cold Spring Harbor Symp. Quant. Biol.* **50,** 181–194.

Holden, J. J., and Ashburner, M. (1978). Patterns of puffing activity in the saivary gland chromosomes of *Drosophila*. IX. The salivary and prothoracic gland chromosomes of a dominant temperature sensitive lethal of *D. melanogaster*. *Chromosoma* **68,** 205–227.

Holden, J. J. A., Walker, V. K., Maroy, P., Watson, K. L., White, B. N., and Gausz, J. (1986). Analysis of moulting and methamorphosis in the ecdysteroid-deficient mutant, *L(3)3*^{DTS} of *Drosophila melanogaster*. *Dev. Genet.* **6,** 153–162.

Holderegger, Ch. (1973). Scanning electron microscopy of isolated *Chironomus* salivary gland chromosomes. *Experienta* **29,** 773–774.

Holderegger, Ch., and Lezzi, M. (1972). Juvenile hormone-induced puff formation in chromosomes of Malpighian tubules of *Chironomus tentans* pharate adults. *J. Insect Physiol.* **18,** 2237–2250.

Hollenberg, C. P. (1976). Proportionate representation of rDNA and Balbiani Ring DNA in polytene chromosomes of *Chironomus tentans*. *Chromosoma* **57,** 185–197.

Holmgren, R., Livak, K., Morimoto, R., Freund, R., and Meselson, M. (1979). Studies of cloned sequences from four *Drosophila* heat shock loci. *Cell* **18,** 1359–1370.

Holmgren, R., Corces, V., Morimoto, R., and Meselson, M. (1981). Sequence homologies in the 5′ regions of four *Drosophila* heat shock genes. *Proc. Natl. Acad. Sci. USA* **78,** 3775–3778.

Holmquist, G. (1971). Removal of RNA from polytene chromosomes by lacto-aceto orcein. *Dros. Inform. Serv.* **46,** 151.

Holmquist, G. (1972). Transcriptional rates of individual polytene chromosome bands: Effects of gene dose and sex in *Drosophila*. *Chromosoma* **36,** 413–452.

Holst, M., and Egyhazi, E. (1985). Posttranslational phospharylation of spetic chromosomal proteins and transcription of *hn* RNA genes in isolated nuclei: Retention of *in vivo* sensitivity to 5,6-Dichloro-1-b-D-ribofuranosylbenzimidazole (DRB). *J. Cell. Biochem.* **29,** 115–126.

Holt, T. K. H. (1970). Local protein accumulation during gene activation. I. Quantotative measurements on dye binding capacity at subsequent stages of puff formation in *Drosophila hydei*. *Chromosoma* **32,** 64–78.

Holt, T. K. H. (1971). Local protein accumulation during gene activation. II. Interferometric measurements of the amount of solid material in temperature induced puffs of *Drosophila hydei*. *Chromosoma* **32,** 428–435.

Holt, T. K. H. (1972). Acidic protein and RNA in the initiation of puff formation. *Chromos. Today* **3,** 56–62.

Holt, T. K. H., and Berendes, H. D. (1965). Experimental puffs in *Drosophila hydei* polytene chromosomes, induced by temperature shocks. *Dros. Inform. Serv.* **40,** 59.

Holt, T. K. H., and Kuijpers, A. M. C. (1972a). Effects of a-aminitin on nucleolar structure and metabolism in *Drosophila hydei*. *Experienta* **28,** 899–900.

Holt, T. K. H., and Kuijpers, A. M. C. (1972b). Induction of chromosome puffs in *Drosophila hydei* salivary glands after inhibition RNA synthesis by a-amanitin. *Chromosoma* **37,** 423–432.

Homyk, T., Jr., and Sheppard, D.E. (1977). Behavioral mutants of *Drosophila melanogaster*. I. Isolation and mapping of mutants which decrease flight ability. *Genetics* **87,** 95–104.

Homyk, T., Jr., Szidonya, J. J., and Suzuki, D. T. (1977). Behavioral mutants of *Drosophila melanogaster*. III. Isolation and mapping of mutations by direct visual observations of behavioral phenotypes. *Mol. Gen. Genet.* **177,** 553–565.

Hong, C.-S., and Ganetzky, B. (1996). Molecular characterization of neurally expressing genes in the *para* sodium channel gene cluster of *Drosophila*. *Genetics* **142,** 879–892.

Hoog, C., and Wieslander, L. (1984). Different evolutionary behaviour of structurally related, repetitive sequences occuring in the same Balbiani ring gene in *Chironomus tentans*. *Proc. Natl. Acad. Sci. USA* **81,** 5165–5169.

Hoog, C., Engberg, C., and Wieslander, L. (1986). A BR1 gene in *Chironomus tentans* has a composite structure: a large repetitive core block is separated from a short unrelated 3′-terminal domain by a small intron. *Nucleic Acids Res.* **14,** 703–719.

Hoog, C., Daneholt, B., and Wieslander, L. (1988). Tandem repeats in long repeat arrays are likely to reflect the early evolution of Balbiani ring genes. *J. Mol. Biol.* **200,** 655–664.

Hoogwerf, A. M., Akam, M., and Roberts, D. (1988). A genetic analysis of the *rose-gespleten* region (68C8-69B5) of *Drosophila melanogaster*. *Genetics* **118,** 665–670.

Hooper, J. E., Perez-Alonso, M., Bermingham, J. R., Prout, M., Rocklein, B. A., Wagenbach, M., Edstrom, J.-E., de Frutos, R., and Scott, M. P. (1992). Comparative studies of *Drosophila Antennapedia* genes. *Genetics* **132,** 453–469.

Horner, M. A., Chen, T., and Thummel, C. S. (1995). Ecdysteroid regulation and DNA binding properties of *Drosophila* nuclar hormone receptor superfamily members. *Dev. Biol.* **168,** 490–502.

Hoshizaki, D. K., Dlott, B. M., Joslyn, G. L., and Beckendorf, S. K. (1987). Genetic localization of a regulatory site necessary for the production of the glue protein P5 in *Drosophila melanogaster*. *Genet. Res. Cambr.* **49,** 111–119.

Hovemann, B., Walldorf, U., and Ryseck, R.-P. (1986). Heat shock locus 93D of *Drosophila melanogaster:* An RNA with limited coding capacity accumulates precursor transcropts after heat shock. *Mol. Gen. Genet.* **204,** 334–340.

Hovemann, B. T., Dessen, E., Mechler, H., and Mack, E. (1991). *Drosophila snRNP* associated protein P11 which specifically binds to heat shock puff 93d reveals strong homology with *hnRNP* core protein A1. *Nucleic Acids Res.* **19,** 4909–4914.

Howard, E. F., and Plaut, W. (1968). Chromsomal DNA synthesis in *Drosophila melanogaster*. *J. Cell. Biol.* **39,** 415–429.

Hsu, T. C., and Liu, T. T. (1948). Microgeographic analysis of chromosomal variation in a chinese species of *Chironomus* (Diptera). *Evolution* **2,** 49–57.

Hsu, W. S. (1948). The Golgi material and mitochondria in the salivary glands of the larva of *Drosophila melanogaster*. *Q. J. Microsc. Sci.* **89,** 401–414.

Huang, R.-Y., and Orr, W. C. (1992). Broad - Complex function during oogenesis in *Drosophila melanogaster*. *Dev. Genet.* **13,** 277–288.

Huberman, J. A., and Riggs, A. D. (1968). On the mechanism of DNA replication in mammalian chromosomes. *J. Mol. Biol.* **32,** 327–341.

Huet, F., Ruiz, C., and Richards, G. (1993). Puffs and PCR: The *in vivo* dynamics of early gene expression during ecdysone responses in *Drosophila*. *Development* **118,** 613–627.

Huet, F., Ruiz, C., and Richards, G. (1995). Sequential gene activation by ecdysone in *Drosophila melanogaster:* The hierarchical equivalence of early and early late genes. *Development* **121,** 1195–1204.

Huet, F., DaLage, J.-L., Ruiz, C., and Richards, G. (1996). The role of ecdysone in the induction and maintenance of *hsp27* transcripts during larval and prepupae development of *Drosophila*. *Dev. Genet. Evol.* **206,** 326–332.

Hughes, M., and Kambysellis, M. P. (1970). Effects of ecdusone on RNA synthesis. *Dros. Inform. Serv.* **45,** 102.

Hugle, B., Guldher, H., Bautz, F. A., and Alonso, A. (1982). Cross reaction of *hn* RNP proteins of Hela cells with nuclear proteins of *D. melanogaster* demonstrated by a monoclonal antibody. *Exp. Cell. Res.* **142,** 119–126.

Hultmurk, D., Klemenz, R., and Gehring, W. (1986). Translational and transcroptional control elements in the untranslated leader of the heat shock gene *hsp22*. *Cell* **44,** 429–438.

Hunter, F. F. (1987). Cytotaxonomy of four European species in the *Eusimulium vernum* group (Diptera: Simuliidae). *Can. J. Zool.* **65,** 3102–3115.

Hunter, F. F., and Connoly, V. (1986). A cytotaxonomic investigation of seven species in the *Eusimulium vernum* group (Diptera: Simuliidae). *Can. J. Zool.* **64,** 296–311.

Hurban, P., and Thummel, C. S. (1993). Isolation and characterization of fifteen ecdysone-inducible *Drosophila* genes reveal unexpected complexities in ecdysone regulation. *Mol. Cell. Biol.* **13,** 7101–7111.

Hutter, P., and Karch, F. (1994). Molecular analysis of a candidate gene for the reproductive isolation between sibling species of *Drosophila*. *Experientia* **50,** 749–762.

Iarovaia, O., Hancock, R., Lagarcova, M., Miassod, R., and Razin, S. V. (1996). Mapping of genomic DNA loop organization in a 500- kilobase region of the *Drosophila* X chromosome by the topoisomerase II-mediated DNA loop excision protocol. *Cell. Biol.* **16,** 302–308.

Ilyina, O. V., Sorokin, A. V., Belyaeva, E. S., and Zhimulev, I. F. (1980). Report on *Drosophila melanogaster* new mutants. *Dros. Inform. Serv.* **55,** 205.

Ilyinskaya, N. B. (1980). Functional organization of polytene chromosomes and problems of karyosystematics. In: Novye dannye po kariosistematike dvukrylykh nasekomykh. *Trudy Zool. Inst. AN SSSR,* **95,** 14–22. [in Russian]

Ilyinskaya, N. B. (1981). Effect of maintenance temperature on wintering larvae of *Chironomus plumosus* on the morphology of polytene chromosome. *Tsitologiya* **23**(3), 264–269. [in Russian]

Ilyinskaya, N. B. (1994). Seasonal changes in the polytene chromosomes of Chironomidae. *Tsitologia* **36**(7), 605–622. [in Russian]

Ilyinskaya, N. B., and Maksimova, F. L. (1976). Changes in polytene chromosomes of the salivary glands of *Chironomus plumosus* larvae in different months of the year. *Tsitologiya* **18**(7), 847–851. [in Russian]

Ilyinskaya, N. B., and Maksimova, F. L. (1978). Changes in banding and number of bands in the polytene chromosomes of Chironomus plumosus larvae during different seasons of the year. *Tsitologiya* **20**(3), 291–297. [in Russian]

Ilyinskaya, N. B., and Martynova, M. G. (1978). Incorporation of ^{3}H-thymidine into the polytene chromosomes of salivary glands of wintering *Chironomus* larvae. *Tsitologiya* **20**(5), 522–526. [in Russian]

Ilyinskaya, N. B., and Nilova, V. K. (1989). Electron microscopic study on myocytes containing polytene chromosomes in *Chironomus* larvae. *Tsitologiya* **31**(10), 1193–1199. [in Russian]

Ilyinskaya, N. B., Demin, S.Yu., and Martynova, M. G. (1991). Effect of increase in temperature on the compactness level and transcriptional activity pattern in the polytene chromosomes in wintering larvae of *Chironomus plumosus*. *Tsitologiya* **33**(3), 48–56. [in Russian]

Imhof, M. O., Wegmann, I. S., Rusconi, S., Dorsch-Hasler, K., and Lezzi, M. (1992). Cloning of a *Chironomus tentans* protein belonging to the steroid and thyroid hormone receptor superfamily. Absracts Xth Ecdysone Workshop, Liverpool, 49 pp.

Imhof, M. O., Rusconi, S., and Lezzi, M. (1993). Cloning of a *Chironomus tentans* cDNA encoding a protein (cEcRH) homologous to the *Drosophila melanogaster* ecdysteroid receptor (dEcR). *Insect Biochem. Mol. Biol.* **23,** 115–124.

Indik, Z., and Tartof, K. (1980). Long spacers among ribosomal genes of *Drosophila melanogaster*. Nature **234,** 477–479.

Ineichen, H. (1978). Photoperiodische Kontrolle der Entwicklung von Chironomus tentans und Entwicklungesspecifische Puff Veranderungen (*in vivo* und *in vitro*). *Revue suisse Zool.* **85,** 807–809.

Ineichen, H., Riesen-Wollo, U., and Fisher, J. (1979). Experimental contribution to the ecology of *Chironomus* (Diptera). II. The influence of the photoperiod on the development of *Chironomus plumosus* in the 4th larval instar. *Oecologia* **39,** 161–183.

Ingolia, T. D., and Craig, E. A. (1981). Primary sequence of the 5′ flanking region of the *Drosophila* heat shock genes in chromosome subdivision 67B. *Nucleic Acids Res.* **9,** 1627–1642.

Ingolia, T. D., and Craig, E. A. (1982a). *Drosophila* gene related to the major heat shock-induced gene is transcribed at normal temperatures and not induced by heat shock. *Proc. Natl. Acad. Sci. USA* **79,** 525–529.

Ingolia, T. D., and Craig, E. A. (1982b). Four small *Drosophila* heat shock proteins are related to each other and to mammalian a- crystallin. *Proc. Natl. Acad. Sci. USA* **79,** 2360–2364.

Ingolia, T. D., Craig, E. A., and McCarthy, B. J. (1980). Sequence of three copies of the gene for the major *Drosophila* heat shock induced protein and their flanking regions. *Cell* **21,** 669–679.

Ingolia, T. D., Slater, M. J., and Craig, E. A. (1982). *Saccharomyces cerevisiae* contains a complex multigene family related to the major heat shock inducible gene of *Drosophila*. *Mol. Cell. Biol.* **2,** 1388–1398.

Ireland, R. C., and Berger, E. M. (1982). Synthesis of low molecular weight heat shock peptides stimulated by ecdysterone in cultured *Drosophila* cell line. *Proc. Natl. Acad. Sci. USA* **79,** 855–859.

Ireland, R. C., Berger, E. M., Sirotkin, K., Yund, M. A., Osterbur, D., and Fristrom, J. (1982). Ecdysterone induces the transcription of four heat-shock genes in *Drosophila* S3 cells and imaginal discs. *Dev. Biol.* **93,** 498–507.

Irick, H. A. (1983). Estimation ofthe number of genes in a region. *Dros. Inform. Serv.* **59,** 59–61.

Ish-Horowicz, D., and Pinchin, S. M. (1980). Genomic organization of the 87A7 and 87C1 heat-induced loci of *Drosophila melanogaster*. *J. Mol. Biol.* **142,** 231–245.

Ish-Horowicz, D., Holden, J. J., and Gehring, W. J. (1977). Deletion of two heat activated loci in *Drosophila melanogaster* and their effects on heat-induced protein synthesis. *Cell* **12,** 643–652.

Ish-Horowicz, D., Gausz, J., Gyurkovicz, H., Bencze, G., Goldschmidt-Clermont, M., and Holden J. J. (1979a). Deletion mapping of two *D. melanogaster* loci that code for the 70,000 Dalton heat-induced protein. *Cell* **17,** 565–571.

Ish-Horowicz, D., Pinchin, S. M., Schedl, P., Artavanis-Tsakonas, S., and Mirault, M.-E. (1979b). Genetic and molecular analysis of the 87A7 and 87C1 heat-inducible loci of *Drosophila melanogaster*. *Cell* **18,** 1351–1358.

Israelewski, N. (1975). X-ray induced visible alterations in the giant chromosomes of *Phryne cincta* (Nematocera, Diptera). Relation of radiation sensitivity to ptonuclear chromosome structure. *Chromosoma* **53,** 243–263.

Istomina. A., and Kiknadze, I. (1989). Electron microscopy of Balbiani rings and nucleoli of *Chironomus* species of the *plumosus* group. *Acta Biol. Oecol. Hung.* **2,** 93–101.

Istomina, A. G., and Kiknadze, I. I. (1993). Karyotype of *Stictochironomus* sp. ex gr. *histrio* from Mountain Altai. *In* "Kariosistematika bespozvonochnykh zhivotnykh II." *Trudy Inst. Zool. Ross. Acad. Nauk*, 37–39. [in Russian]

Istomina, A. G., Kiknadze, I. I., Khristolyubova, N. B. (1983). Ultrastructural organization of tissue-specific Balbiani ring 3 of *Chironomus thummi*. *Tsitologia* **25**(9), 1037–1041. [in Russian]

Istomina, A. G., Kiknadze, I. I., and Kerkis, I. E. (1992). The karyotype of three *Cryptochironomus* species of West Siberia, USSR. *Netherlands J. Aquat. Ecol.* **26**(2–4), 139–144.

Izquierdo, M., and Bishop, J. O. (1979). An analysis of cytoplasmic RNA population in *Drosophila melanogaster*, Oregon-R. *Biochem. Genet.* **17,** 473–497.

Izquierdo, M., Arribas, C., and Alonso, C. (1981). Isolation of a structural gene mapping to subregion 63F of *Drosophila melanogaster* and 90B of *D. hydei* polytene chromosomes. *Chromosoma* **83,** 353–366.

Izquierdo, M., Arribas, C., Galceran, J., Burke, J., and Cabrera, V. M. (1984). Characterization of a *Drosophila* repeat mapping at the early-ecdysone puff 63F and present in many eucaryotic genomes. *Biochim. Biophys. Acta.* **783,** 114–121.

Izquierdo, M., Galceran, J., Sampedro, J., and Llanos, J. (1988). The role of 2B5 in the control of gene activity by ecdysone in *Drosophila*. *J. Insect Physiol.* **34,** 685–689.

Jack, R. S., and Gehring, W. J. (1982). A *Drosophila* DNA-binding protein showing specificity for sequences close to heat-shock genes. *In* "Heat Shock from Bacteria to Man" (M. L. Schlesinger, M. Ashburner, and A. Tissieres, eds.), pp. 155–159. Cold Spring Harbor Laboratory Press, Cold Spring Harbor, NY.

Jack, R. S., Brack, C., and Gehring, W. J. (1981a). A sequence specific DNA binding protein from *Drosophila melanogaster* which recognises sequences associated with two structural genes for the major heat shock protein. *Fortschr. Zool.* **26,** 271–285.

Jack, R. S., Gehring, W. J., and Brack, C. (1981b). Protein component from *Drosophila* larval nuclei showing sequence specificity for a short region near a major heat shock protein gene. *Cell* **24,** 321–331.

Jackle, H., de Almeida, J. C., Galler, R., Kluding, H., Lehrach, H., and Edstrom, J.-E. (1982). Constant and variable parts in the Balbiani Ring 2 repeat unit and the translation termination region. *EMBO J.* **1,** 883–888.

Jackson, D. A. (1991). Structure-function relationship in eukaryotic nuclei. *BioEssays* **13**(1), 1–10.

Jacob, J., and Jurand, A. (1963). Electron microscope studies on salivary gland cells of *Bradysia mycorum* Frey (Sciaridae). III. The structure of cytoplasm. *J. Insect. Physiol.* **9,** 849–857.

Jacob, J., and Jurand, A. (1965). Electron microscope studies on salivary gland cells. V. The cytoplasm of *Smittia parthenogenetica* (Chironomidae). *J. Insect. Physiol.* **11,** 1337–1343.

Jacob, J., and Sirlin, J. L. (1963). Electron microscope studies on salivary gland cells. I. The nucleus of *Bradysia mycorum* Frey (Sciaridae), with special reference to the nucleolus. *J. Cell. Biol.* **17,** 153–165.

Jacob, J., and Sirlin, J. L. (1964). Electron microscope studies on salivary gland cells. IV. The nucleus of *Smittia parthenogenetica* (Chironomidae) with special reference to the nucleolus and the effects of actynomycin thereon. *J. Ultrastruct. Res.* **11,** 315–328.

Jakubczak, J. L., Xiong, Y., Eickbush, T. H. (1990). Type I (R1) and Type II (R2) ribosomal DNA insertions of *Drosophila melanogaster* are retrotransposable elements closely related to those of *Bombyx mori*. *J. Mol. Biol.* **212,** 37–52.

James, A. A., Grossman, G. L., Marinotti, O., and Stark, K. R. (1992). Gene expression in the salivary glands of mosquitoes. *J. Cell. Biochem.* **16A** (Suppl.), 114.

Jamrich, M., Greenleaf, A. L., and Bautz, E. K. F. (1977a). Localization of RNA-polymerase in *Drosophila melanogaster* polytene chromosomes. *Proc. Natl. Acad. Sci. USA* **74,** 2079–2083.

Jamrich, M., Haars, R., Wulf, E., and Bautz, F. A. (1977b). Correlation of RNA polymerase B and transcriptional activity in the chromosomes of *Drosophila melanogaster*. *Chromosoma* **64,** 319–326.

Jamrich, M., Greenleaf, A. L., Bautz, F. A., and Bautz, E. K. F. (1978). Functional organization of polytene chromosomes. *Cold Spring Harb. Symp. Quant. Biol.* **42,** 389–396.

Janca, F. C., Woloshyn, E. P., and Nash, D. (1986). Heterogeneity of lethals in a "simple" lethal complementation group. *Genetics* **112,** 43–64.

Janike, J., Grimaldi, D. A., Sluder, A. E., and Greenleaf, A. L. (1983). A-amanitin tolerance in mycophagous *Drosophila*. *Science* **221,** 165–167.

Janknecht, R., Taube, W., Ludecke, H.-J., and Pongs, O. (1989). Characterization of putative transcription factor gene expressed in the 20-OH-ecdysone inducible puff 74EF in *Drosophila melanogaster*. *Nucleic Acid Res.* **17,** 4455–4464.

Jarry, B. P. (1979). Genetical and cytological location of the structural parts coding for the first three steps of pyrimidine biosythesis in *Drosophila melanogaster*. *Mol. Gen. Genet.* **172,** 199.

Jeffery, D. E. (1976). Comparative salivary-gland chromosome puffing patterns in the *planitiba* subgroup of the Hawaiian picture-winged *Drosophila*. *Genetics* **83**(Suppl.), 35.

Jiang, C., Baehrecke, E. H., and Thummel, C. S. (1997). Steroid regulated programmed cell death during *Drosophila* metamorphosis. *Development* **124,** 4673–4683.

Johnston, R. D. St., Hoffmann, R. M., Blackman, R. K., Segal, D., Grimaila, R., Pedgett, R. W., Irick, H. A., and Gelbart, W. M. (1990). Molecular organization of the *decapentapledgic* gene in *Drosophila melanogaster*. *Genes Dev.* **4,** 1114–1127.

Jones, C. W., Dalton, M. W., and Townley, L. H. (1991). Interspecific comparisons of the structure and regulation of the *Drosophila* ecdysone-inducible gene E74. *Genetics* **127,** 535–543.

Jongens, T. A., Fowler, T., Shermoen, A. W., and Beckendorf, S. K. (1988). Functional redundancy in the tissue-specific enhancer of the *Drosophila Sgs-4* gene. *EMBO J.* **7,** 2559–2567.

Joplin, K. H., and Denlinger, D. L. (1990). Developmental and tissue specific control of the heat shock induced 70kDa related proteins in the flesh fly, *Sacrophaga crassipalpis*. *J. Insect Physiol.* **36,** 239–249.

Judd, B. H. (1962). An analysis of mutations contined to a small region of the X-chromosome of *Drosophila melanogaster*. *Science* **138,** 990–991.

Judd, B. H. (1975). Genes and chromomeres of *Drosophila*. *In* "The Eukaryote Chromosome" (W. Y. Peacock and R. D. Brock, eds.), pp. 169–184. Australian National University Press, Canberra.

Judd, B. H. (1976). Genetic units of *Drosophila-complex* loci. *In* "The Genetics and Biology of *Drosophila*" (M. Ashburner and E. Novitsky, eds.), Vol. 1b, pp. 767–799. Academic Press, London/New York/San Francisco.

Judd, B. (1979). Mapping the functional organization of eukaryotic chromosomes. *In* "Cell Biology. A Comprehensive Treatise" (D. Prescott and L. Goldstein, eds.), Vol. 2, 223–265. Academic Press, New York.

Judd, B. H. (1995). Report on *Drosophila melanogaster* new mutants. *Dros. Inform. Serv.* **76,** 79.

Judd, B. H., and Young, M. W. (1974). An examination of the one cistron: One chromomere concept. *Cold Spring Harbor Symp. Quant. Biol.* **38,** 573–578.

Judd, B. H., Shen, M. W., and Kaufman, T. C. (1972). The anatomy and function of a segment of the X-chromosome of *Drosophila melanogaster*. *Genetics* **71,** 139–156.

Jurand, A., and Pavan, C. (1975). Ultrastructural aspects of histolytic processes in the salivary gland cells during metamorphic stages in *Rhynchosciara hollaenderi* (Diptera, Sciaridae). *Cell. Differ.* **4,** 219–236.

Kabisch, R., and Bautz, E. K. F. (1983). Differential distribution of RNA polymerase B and nonhistone chromosomal proteins in polytene chromosomes of *Drosophila melanogaster. EMBO J.* **2,** 395–402.

Kabisch, R., Krause, J., and Bautz, E. K. F. (1982). Evolutionary changes in non-histone chromosomal proteins within the *Drosophila melanogaster* group revealed by monoclonal antibodies. *Chromosoma* **85,** 531–538.

Kachvoryan, E. A. (1989) Nucleolus polymorphism in natural populations of *Tetisimulium condici. Parasitologia* **23,** 134–138. [in Russian]

Kachvoryan, E. A. (1990). Nucleolar polymorphysm and puff hetezygosity in natural populations of *Cnetha djafarovi* Rubz. (Diptera, Simuliidae). *Biol. Zhurn. Armenii* **5**(43), 378–382. [in Russian]

Kachvoryan, E. A., Usova, Z. V., and Oganesyan, V. S. (1993). The genetic relations of the two close species of the blackfly genus *Cnetha* end. (Diptera, Simuliidae). *In* "Karyosistematika bespozvonochnykh zhivotnykh II" (L. A. Chubareva and V. G. Kuznetsova, eds.), pp. 85–88, Zool. Inst. Press, Sankt-Petersburg. [in Russian]

Kaiser, K., Furia, M., and Glover, D. M. (1986). Dosage compensation at the *Sgs4* locus of *Drosophila melanogaster. J. Mol. Biol.* **187,** 529–536.

Kaji, S., and Ushioda, Y. (1981). The cell cycle during the development of the imaginal eye dick cell in *Drosophila melanogaster. Ann. Zool. Jpn.* **54,** 164–170.

Kalisch, W.-E., and Hagele, K. (1973). Different DNA replication bahaviour of a tandem duplication in *Drosophila melanogaster. Chromosoma* **44,** 265–283.

Kalisch, W.-E., and Hagele, K. (1976). Correspondence of banding patterns to ^{3}H-thymidine labelling patterns in polytene chromosomes. *Chromosoma* **57,** 19–23.

Kalisch, W.-E., and Hagele, K. (1977). Determination of ^{3}H-thymidine background labelling over polytene chromosomes in *Drosophila melanogaster. Dros. Inform. Serv.* **52,** 127–128.

Kalisch, W.-E., and Ramesh, S. R. (1997a). X chromosomal linkage of larval secretion fractions in the *Drosophila nasuta* subgroup. *Dros. Inform. Serv.* **80,** 47–49.

Kalisch, W.-E., and Ramesh, S. R. (1997b). Salivary gland secretion in *Drosophila hydei. Dros. Inform. Serv.* **80,** 49–50.

Kalisch, W.-E., and Ramesh, S. R. (1997c). Stain-specific characterization of larval secretion fraction. *Dros. Inform. Serv.* **80,** 51–53.

Kalisch, W.-E., and Ramesh, S.R. (1988). Individual empty pupal cases used for genotypic characterization of larval glue protein fractions in *Drosophila nasuta nasuta. Dros. Inform. Serv.* **67,** 51–52.

Kalm, L., von, Crossgrove, K., Seggern, D., von, Guild, G. M., and Beckendorf, S. K. (1994). The *Broad-Complex* directly controls a tissue-specific response to the steroid hormone ecdysone at the onset of *Drosophila* metamorphosis. *EMBO J.* **13,** 3505–3516.

Kalnins, V. I., Stich, H. F., and Bencosme, S. A. (1964a). Fine structure of nucleolar organizer of salivary gland chromosomes of Chironomids. *J. Ultrastruct. Res.* **11,** 282–291.

Kalnins, V. I., Stich, H. F., and Bencosme, S. A. (1964b). Fine structure of nucleoli and RNA containing chromosome regions of salivary gland chromosomes of Chironomids and their interrelationship. *Can. J. Zool.* **42,** 1147–1155.

Kalumuck, K. E., Wetzel, F. L., and Procunier, J. D. (1990). Relative abundance of various rDNA repeat types in polytene nuclei of *Drosophila melanogaster. Genome* **33,** 240–246.

Kamb, A., Iverson, L. E., and Tanouye, M. A. (1987). Molecular characterization of *Shaker*, a *Drosophila* gene that encodes a potassium channel. *Cell* **50,** 405–413.

Kamyshev, N. G., Smirnova, G. P., and Ponomarenko, V. V. (1988). Effect of mutations blocking the sequential steps of the metabolic pathway of tryptophan-xantommatin on the motor activity of *Drosophila melanogaster. Zhurn. obshch. biol.* **49**(4), 501–511. [in Russian]

Kao, W.-Y., and Case, S. T. (1985). A novel giant secretion polypeptide in *Chironomus* salivary glands: Implications for another *Balbiani Ring* gene. *J. Cell. Biol.* **101,** 1044–1051.

Kao, W.-Y., and Case, S. T. (1986). Individual variations in the content of giant secretory polypeptides in salivary glands of *Chironomus*. *Chromosoma* **94,** 475–482.

Kaplan, R. A., and Plaut, W. (1968). A radioautographic study of dosage compensation in *Drosophila melanogaster*. *J. Cell. Biol.* **39,** 71a.

Kapoun, A. M., and Kaufman, T. C. (1995). Regulatory regions of the homeotic gene proboscipedia are sensitive to chromosomal pairing. *Genetics* **140,** 643–658.

Karakin, E. I. (1985). Change in activity of the genes for secretory proteins of the salivary glands of *Drosophila melanogaster* in ontogenesis and their regulation. *In* "Organizatsia i ekspressia genov tkanespetsificheskoy funktsii u *Diptera*," pp. 71–80. Nauka, Novosibirsk. [in Russian]

Karakin, E. I., and Sviridov, S. M. (1981). Phenogenetics of watersoluble antigenes of *Drosophila*. *In* "Biochemical Genetics of *Drosophila*" (M. D. Golubovsky and L. I. Korochkin, eds.), pp. 156–209. Nauka, Novosibirsk. [in Russian]

Karakin, E. I., Lerner, T. Ya., Kiknadze, I. I., Korochkin, L. I., and Sviridov, S. M. (1977a). Antigenes of the larval organs of *Drosophila*. *Tsitologia* **19**(1), 111–119. [in Russian]

Karakin, E. I., Lerner, T. Ya., Kokoza, V. A., and Sviridov, S. M. (1977b). Antigenes of the secretion of the salivary glands of *Drosophila melanogaster* larvae. *Doklady Akad. Nauk* **233**(4), 698–701. [in Russian]

Karakin, E. I., Lerner, T. Ya., Kokoza, V. A., Kiknadze, I. I., Korochkin, L. I., and Sviridov, S. M. (1983). Antigenes of the larval organs and the salivary gland secretion of *Drosophila melanogaster*. *Biol. Zbl.* **102,** 187–199.

Karch, F., Torok, I., and Tissieres, A. (1981). Extensive regions of homology in front of the two *hsp* heat shock genes in *Drosophila melanogaster*. *J. Mol. Biol.* **148,** 218–230.

Karess, R. E., and Rubin, G. M. (1984). Analysis of P transposable element functions in *Drosophila*. *Cell* **38,** 135–146.

Karim, F. D., and Thummel, C. S. (1991). Ecdysone coordinates the timing and amounts of E74A and E74B transcription in *Drosophila*. *Genes Dev.* **5,** 1067–1079.

Karim, F. D., and Thummel, C. (1992). Temporal coordination of regulatory gene expression by the steroid hormone ecdysone. *EMBO J.* **11,** 4083–4093.

Karim, F. D., Urness, L. D., Thummel, C. S., Klemsz, M. J., McKercher, S. R., Celada, A., van Beveren, C., Maki, R. A., Gunther, C. V., Nye, J. A., and Graves, B. J. (1990). The ETS-domain: a new DNA-binding motif that recognizes a purine-rich core DNA sequence. *Genes Dev.* **4,** 1451–1453.

Karim, F. D., Guild, G. M., and Thummel, C. S. (1993). The *Drosophila Broad-Complex* plays a key role in controlling acdysone- regulated gene expression at the onset of metamorphosis. *Development* **118,** 977–988.

Karlik, C. C., Saville, D. L., and Fyrberg, E. A. (1987). Two missense alleles of the *Drosophila melanogaster act88F* actin gene are strongly antimorphic but only weakly induce synthesis of heat shock proteins. *Mol. Cell. Biol.* **7,** 3084–3091.

Karlson, P. (1963). New concepts on the mode of action of hormones. *Perspect. Biol. Med.* **6,** 203–214.

Karlson, P. (1965). Biochemical studies of ecdysone control of chromosomal activity. *J. Cell. Comp. Physiol.* **66,** 69–76.

Karlson, P. (1966). Ecdyson, das Hautungshormon der Insekten. *Naturwissenschaften* **53,** 445–453.

Karlson, P. (1967). The effects of ecdysone on giant chromosomes, RNA metabolism and enzyme induction. *Mem. Soc. Endocrin.* **15,** 67–76.

Karlson, P. (1975). Genaktivierung durch Hormone. *Kachrichten Akad. Wissen. Gottingen. II. Matem. Physik.* Kl. **11,** 9–17.

Karlson, P., and Bode, C. (1969). Die Inaktiviering des Ecdysons bei der Schmeissfliege *Calliphora erythrocephala* Meigen. *J. Insect Physiol.* **15,** 111–118.

Karlson, P., and Hanser, G. (1952). Uber die Wirkung des Puparisierungshormons bei der Wildform und Mutante *lgl* von *Drosophila melanogaster*. *Z. Naturforsch.* **7B,** 80–83.

Karlson, P., and Sekeris, C. E. (1966). Ecdysone, an insect steroid hormone, and its mode of action. *Rec. Progr. Horm. Res.* **22,** 473–493.

Karlson, P., Sekeris, C. E., and Maurer, R. (1964). Verteilung von Tritium-markiertem Ecdyson in Larven von *Calliphora erythocephala. Hoppe-Seylers Z. Physiol. Chem.* **336,** 100–106.

Karpen, G. H., and Spradling, A. C. (1990). Reduced DNA polytenization of a minichromosome region undergoing position-effect variegation in *Drosophila. Cell* **63,** 97–107.

Karpen, G. H., and Spradling, A. C. (1992). Analysis of subtelomeric heterochromatin in the *Drosophila* minichromosome Dp1187 by single P element insertional mutagenesis. *Genetics* **132,** 737–753.

Karpen, G. H., Schaefer, J. E., and Laird, C. D. (1988). A *Drosophila* rDNA gene located in euchomatin is active in transcription and nucleolus formation. *Genes Dev.* **2,** 1745–1763.

Karpov, V. L., Preobrazhenskaya, O. V., and Mirazbekov, A. D. (1984). Chromatin structure of *hsp70* genes activated by heat shock: selective removal of histones from coding region and their absence from the 5′region. *Cell* **36,** 423–431.

Kas, E., and Laemmli, U. K. (1992). *In vivo* topoisomerase II cleavage of the *Drosophila* histone and satellite III repeats: DNA sequence and structural characteristics. *EMBO J.* **11,** 705–716.

Kastritsis, C. D. (1971). "Micronucleoli" in *Drosophila paulistorum? Dros. Inform. Serv.* **46,** 137–138.

Kastritsis, C. D., and Grossfield, J. (1971). Balbiani-Rings in *Drosophila auraria. Dros. Inf. Serv.* **47,** 123.

Kastritsis, C. D., Scouras, Z. G., and Ashburner, M. (1986). Duplications in the polytene chromosomes of *Drosophila auraria. Chromosoma* **93,** 381–385.

Kato, K.-I., and Sirlin, J. L. (1963). Aspects of mucopolysaccharide production in larval insect salivary cells. *J. Histochem. Cytochem.* **11,** 163–168.

Kato, K.-I., Perkowska, E., and Sirlin, J. L. (1963). Electro- and immunoelectrophoretic patterns in the larval salivary secretion of *Chironomus thummi. J. Histochem. Cytochem.* **11,** 485–488.

Kaufman, T. C. (1972). Characterization of 3 new alleles of the giant locus of *Drosophila melanogaster. Genetics* **71**(Suppl.), 28–29.

Kaufman, T. C. (1978). Cytogenetic analysis of chromosome 3 in *Drosophila melanogaster:* Isolation and characterization of four new alleles of the *proboscipedia (pb)* locus. *Genetics* **90,** 579–596.

Kaufman, T. C., Shen, M. W., and Judd, B. H. (1969). The complementation map of mutations in a small region of the X- chromosome of *Drosophila melanogaster. Genetics* **61**(Suppl.), 30–31.

Kaufman, T. C., Lewis, R., and Wakimoto, B. (1980). Cytogenetic analysis of chromosome 3 in *Drosophila melanogaster:* the homoetic gene complex in polytene chromosome interval 84A-B. *Genetics* **94,** 115–133.

Kaufman, T. C., Seeger, M. A., and Olsen, G. (1990). Molecular and genetic organization of the *Antennapedia* gene complex of *Drosophila melanogaster. Adv. Genet.* **27,** 309–362.

Kaufmann, B. P. (1934). Somatic mitoses of *Drosophila melanogaster. J. Morphol.* **56,** 125–155.

Kaufmann, B. P. (1937). Morphology of the chromosomes of *Drosophila ananassae. Cytologia, Fujii Jub.*, 1043–1055.

Kaufmann, B. P. (1938). Nucleolus-organizing regions in salivary gland chromosmes of *Drosophila melanogaster. Z. Zellforsch.* **28,** 1–11.

Kaufmann, B. P., and Gay, H. (1958). The nuclear membrane as an intermediary in gene-controlled reactions. *The Nucleus* **1,** 57–74.

Kaul, D., Tewari, R. R., and Gaur, P. (1981). The chromosomes of sarcophagid flies. *La Kromosomo* **II-24,** 697–706.

Kaul, D., Agrawal, U. R., and Tewari, R. R. (1983). Puffing activity in the foot pad chromosomes of *Parasarcophaga ruficornis* (Fab.) (Sarcophagidae: Diptera). *La Kromosomo* **II-32,** 961–970.

Kavenoff, R., and Zimm, B. H. (1973). Chromosome-sized DNA molecules from *Drosophila. Chromosoma* **41,** 1–27.

Kavenoff, R., Klotz, L. C., and Zimm, B. H. (1974). On the nature of chromosome-sized DNA molecules. *Cold Spring Harbor Symp. Quant. Biol.* **38,** 1–8.

Kawata, Y., Fujiwara, H., and Ishikawa, H. (1988a). Low molecular weight RNA of *Drosophila* cells which is induced by heat shock. I. Synthesis and its effect on protein synthesis. *Comp. Biochem. Physiol. B* **91,** 149–153.

Kawata, Y., Fujiwara, H., Shiba, T., Miyake, T., and Ishikawa, H. (1988b). Low molecular weight RNA of *Drosophila* cells which is induced by heat shock. II. Structural properties. *Comp. Biochem. Physiol. B* **91,** 155–157.

Kazazian, H. H., Young, W. J., and Childs, B. (1965). X-linked 6-phosphogluconate dehydrogenase in *Drosophila:* Subunit associations. *Science* **150,** 1601–1602.

Keene, M. A., and Elgin, S. C. R. (1982). Perturbations of chromatin structure associated with gene expression. *In* "Heat Shock from Bacteria to Man" (M. J. Schlesinger, M. Ashburner, and A. Tissieres, eds.), pp. 83–89. Cold Spring Harbor Laboratory Press, Cold Spring Harbor, NY.

Kelley, R. L., and Kuroda, M. I. (1995). Equality for X chromosomes. *Science* **270,** 1607–1610.

Kelley, R. L., Solovyeva, I., Lyman, L. M., Richman, R., Solovyev, V., and Kuroda, M. I. (1995). Expression of Msl-1 causes assembly of dosage compensation regulators on the X chromosomes and female lethality in *Drosophila. Cell* **81,** 867–877.

Kellum, R., and Schedl, P. (1991). A position-effect assay for boundaries of higher order chromosomal domains. *Cell* **64,** 941–950.

Kelly, S. E., and Cartwright, I. L. (1989). Perturbation of chromatin architecture on ecdysterone induction of *Drosophila melanogaster* small heat shock protein genes. *Mol. Cell. Biol.* **9,** 332–335.

Kenney, J. A., and Hunter, A. S. (1977). Effect of elevated temperature on salivary chromosomes of *Drosophila virilis. Dros. Inform. Serv.* **52,** 138.

Keppy, D. O., and Welshons, W. J. (1977). The cytogenetics of a recessive visible mutant associated with a deficiency adjacent to the *Notch* locus in *Drosophila melanogaster. Genetics* **85,** 497–500.

Keppy, D. O., and Welshons, W. J. (1980). The synthesis of compound bands in *Drosophila melanogaster* salivary gland chromosomes. *Chromosoma* **76,** 191–200.

Kerkis, A. Yu., Zhimulev, I. F., and Belyaeva, E. S. (1975). A method of electron microscopic autoradiography for studying the structural and functional organization of definite regions of polytene chromosomes. *Tsitologiya* **17**(11), 1330–1331. [in Russian]

Kerkis, A. Yu., Zhimulev, I. F., and Belyaeva, E. S. (1977). EM autoradiographic study of ^{3}H-uridine incorporated into *Drosophila melanogaster* salivary gland chromosomes. *Dros. Inform. Serv.* **52,** 14–17.

Kerkis, I. E., Kiknadze, I. I., and Istomina, A. G. (1989). A comparative analysis of the karyotypes of three sibling species of Chironomids of the *plumosus* group. *Tsitologiya* **31**(6), 713–721. [in Russian]

Keyl, H.-G. (1961). Chromosomen evolution bei *Chironomus.* I. Strukturabwandlungen an Speicheldrusen-chromosomen. *Chromosoma* **12,** 26–47.

Keyl, H.-G. (1965). Duplikationen von untereinheiten der chromosomalen DNS wahrend der Evolution von *Chironomus thummi. Chromosoma* **17,** 139–180.

Keyl, H.-G. (1966). Lokale DNS-replikatonen in Riesenchromosomen. *In* "Problem der Biologischen Reduplikation," pp. 55–69. Springer-Verlag, Berlin.

Keyl, H.-G. (1975). Lampbrush chromosomes in spermatocytes of *Chironomus. Chromosoma* **51,** 75–91.

Keyl, H.-G., and Pelling, C. (1963). Differentielle DNS-Replikation in den Speicheldrusenchromosopmen von *Chironomus thummi. Chromosoma* **14,** 347–359.

Khabibullaev, P. K., Shcherbakov, D. Yu., Koldatova, Z. T., Karakin, E. I., Kokoza, V. A., and Kiknadze, I. I. (1986). Phosphorylation of proteins in larvae salivary glands of *Drosophila melanogaster. Doklady Akad. Nauk SSSR* **287**(5), 1259–1261. [in Russian]

Khesin, R. B. (1980). Transcription regulation. *In* "Well-being of mankind and genetics" (Proc. XIV Intern. Congr. Genet.), Vol. 1, pp. 128–146. Mir Publishers, Moscow.

Khesin, R. B., and Leibovitch, B. A. (1974). Synthesis of RNA by *Escherichia coli* RNA polymerase on the chromosomes of *Drosophila melanogaster*. *Chromosoma* **44,** 161–172.

Kidd, S. J., and Glover, D. M. (1980). A DNA segment from *Drosophila melanogaster* which contains five tandemly repeating units homologous to the major rDNA insertion. *Cell* **19,** 103–119.

Kidd, S., Lockett, T. J., and Young, M. W. (1983). The *Notch* locus of *Drosophila melnogaster*. *Cell* **34,** 421–433.

Kidd, S., Kelley, M. R., and Young, M. W. (1986). Sequence of the *Notch* locus of *Drosophila melanogaster:* Relationship of the encoded protein to mammalian clotting and drowth factors. *Mol. Cell. Biol.* **6,** 3094–3108.

Kiefer, B. I. (1968). Dosage regulation of ribosomal DNA in *Drosophila melanogaster*. *Proc. Natl. Acad. Sci. USA* **61,** 85–89.

Kiger, J. A., and Golanty, E. (1979). A genetically distinct form of cyclic AMP phospho-diesterase associated with chromomere *3D4* in *Drosophila melanogaster*. *Genetics* **91,** 521–535.

Kiknadze, I. I. (1965a). The structural and cytochemical characteristics of chromosome puffs. *In* "Genetic Variations in Somatic Cells" (Proc. of Symp. Mutat. Proc Praha, August 9–11, Praha), pp. 177–181. Academia-Publishing House of the Czechoslovak Academy of Sciences, Prague.

Kiknadze, I. I. (1965b). Analysis of puff function in the course of larval development of Diptera on the background of experimental effects. *In* "Kletochnaya differentsirovka i induktsionnye mekhanizmy," pp. 78–99. Nauka, Moscow. [in Russian]

Kiknadze, I. I. (1965c). Functional changes in giant chromosomes under conditions of inhibition of RNA synthesis. *Tsitologiya* **7**(3), 311–318. [in Russian]

Kiknadze, I. I. (1966). Functioning of the chromosomes. *In* "Rukovodstvo po tsitologii," Vol. 2, pp. 329–347. Nauka, Moscow/Leningrad. [in Russian]

Kiknadze, I. I. (1967). Structure of the nuceolar organizar in *Chironomus dorsalis*. *Tsitologia* **9**(8), 902–911. [in Russian]

Kiknadze, I. I. (1970). Cytogenetic aspects of ontogenesis: Patterns of transcriptional activity of chromosomes during differentiation. *Ontogenez* **1**(1), 17–27. [in Russian]

Kiknadze, I. I. (1971a). Polytene chromosomes as a model of the interphase chromosome. *Tsitologiya* **13**(6), 716–733. [in Russian]

Kiknadze, I. I. (1971b). Functional organization of the chromosomes. *In* "Uspekhi sovremen. genetiki," Vol. 3, pp. 175–205. Nauka, Moscow. [in Russian]

Kiknadze, I. I. (1972). Functional organization of the chromosomes. *In* "Uspekhi sovremen. genetiki." Nauka, Leningrag. otdelen. [in Russian]

Kiknadze, I. I. (1976). A comparative characterization puffing in the salivary gland chromosomes of *Chironomus thummi* during larval development and metamorphosis. I. Puffing in chromosome IV. *Tsitologia* **18**(11), 1322–1329. [in Russian]

Kiknadze, I. I. (1978). A comparative characterization puffing in the salivary gland chromosomes of *Chironomus thummi* during larval development and metamorphosis. II. Puffing in chromosomes I, II and III. *Tsitologia* **20**(5), 514–521. [in Russian]

Kiknadze, I. I. (1981a). Puffing and transcriptional activity of polytene chromosomes. *In* "Molekulyarnye osnovy geneticheskikh protsessov," pp. 370–389. Trudy 14 Mezhdunar. genet. kongr. Nauka, Moscow. [in Russian]

Kiknadze, I. I. (1981b). The puffing pattern and transcriptional acrivity of polutene chromosomes. *In* "Molecular Bases of Genetic Processes" (Proc. 14 Intern. Genet. Congr.), Vol. 3(2), pp. 223–256. Mir, Moscow.

Kiknadze, I. I. (1984). Organization and expression of multigene families encoding tissue-specific proteins of *Diptera*. *Trudy 3-iey nats. konf. tsitogenetikov. Plovdiv* **1,** 39–48. [in Russian]

Kiknadze, I. I. (1985). Molecular cytogenetic organization of Balbiani rings and genes located in them. *In* "Organizatsia i ekspressia genov tkanespetsificheskoy funktsii u *Diptera*," pp. 115–126. Nauka, Sib. otd-e., Novosibirsk.

Kiknadze, I. I. (1989). Ontogenetic changes in the activity of the Balbiani rings and genes located in them. *In* "Organizatsia i ekspressia genov tkanespetsificheskoy funktsii u *Diptera*," pp. 131–138. Nauka, Sib. otd-e., Novosibirsk. [in Russian]

Kiknadze, I. I., and Belyaeva, E. S. (1967). The nucleolus, patterns of its formation and genetic role. *Genetika* **7**(8), 149–161. [in Russian]

Kiknadze, I. I., and Filatova, I. T. (1960). Functinal changes in RNA content in the nuclei of the salivary glands of *Chironomus dorsalis* during metamorphosis. *Izvestia AN SSSR* **12,** 131–134. [in Russian]

Kiknadze, I. I., and Filatova, I. T. (1963). Changes in RNA in the giant chromosomes of the midge during metamorphosis and under experimental treatment. *Doklady Akad. Nauk SSSR* **152**(2), 450–453. [in Russian]

Kiknadze, I. I., and Panova, T. M. (1972). On heteromorphism of puffs in *Chironomus thummi. Tsitologia* **14**(9), 1084–1091. [in Russian]

Kiknadze, I. I., and Valeyeva, F. S. (1980). Changes in the structure of *Chironomus thummi* polytene chromosomes under the effect of hight concentrations of actinimycin D. *Eur. J. Cell. Biol.* **22,** 36.

Kiknadze, I. I., and Valeyeva, F. S. (1983). Conditions of the formation of pseudopuffs during inhibition of RNA synthesis in the polytene chromosomes of *Chironomus thummi. Tsitologia* **25**(9), 1042–1048. [in Russian]

Kiknadze, I. I., Lychev, V. A., and Tonayan, O. P. (1965). The influence of genetic factors on the effect of puffs in *Drosophila melanogaster. In* "Mechanism of Mutation and Inducing Factors" (Symp. Mutat. Proc. Sess 1/a), pp. 70–80. Academia-Publishing House of the Czechoslovak Academy of Sciences, Prague.

Kiknadze, I. I., Korochkina, L. S., and Muradov, S. V. (1972). Puffing pattern in the salivary gland polytene chromosomes of *Chironomus thummi* in ontogenesis. *In* "Kletochnoye yadro. Morfologia, fiziologia, biochimia" (Pod red. I. B. Zbarskogo, i G. P. Georgieva), pp. 209–210. Nauka, Moscow. [in Russian]

Kiknadze, I. I., Kolesnikov, N. N., and Lopatin, O. E. (1975). Chironomus *Chironomus thummi* Kieff (a laboratory culture). *In* "Ob'ekty biologii razvitia," pp. 95–127. Nauka, Moscow. [in Russian]

Kiknadze, I. I., Perov, N. A., and Chentsov, Yu. S. (1977). Electron microscopic study on puff formation in *Chironomus. Tsitologiya* **19**(3), 259–262. [in Russian]

Kiknadze, I. I., Valeyeva, F. S., Vlassova, I. E., Panova, T. M., Sebeleva, T. E., and Kolesnikov, N. N. (1979). Puffing and specific function of the salivary gland cells of *Chironomus thummi.* I. Quantitative changes in protein and glycoproteins in salivary glands at different developmental stages of larvae. *Ontogenez* **10**(2), 161–172. [in Russian]

Kiknadze, I. I., Panova, T. M., and Zakharenko, L. P. (1981). Comparative characterization of puffing in the salivary gland chromosomes of *Chironomus thummi* during larval development and metamorphosis. III. Transcriptional activity of the nucleolus and Balbiani rings. *Tsitologia* **26**(5), 531–538. [in Russian]

Kiknadze, I. I., Razmakhnin, E. P., Zakharenko, L. P., Shilova, I. E., Panova, T. M., Zainiev, G. A., and Mertvetsov, N. P. (1983a). Molecular cytological organization of Balbiani ring I (BRc) in *Chironomus thummi. Doklady AN SSSR* **269**(4), 951–954. [in Russian]

Kiknadze, I. I., Gunderina, L. I., and Valeyeva, F. S. (1983b). Morphology of heterochromatic centromeric regions of polytene chromosomes of *Chironomus thummi* and its change during inhibition of RNA synthesis. *Tsitologia* **25**(10), 1159–1165. [in Russian]

Kiknadze, I. I., Panova, T. M., and Agapova, O. A. (1984). Effect of ecdysone on the activity of Balbiani rings and ultrascructure of the salivary gland cells of *Chironomus. Tsitologia* **26**(3), 285–291. [in Russian]

Kiknadze, I. I., Zainiev, G. A., Panova, T. M., Istomina, A. G., Zacharenko, L. P., and Potapov, W. A. (1985a). Identification of Balbiani ring chromomeres in *Chironomus thummi* polytene chromosomes. *Biol. Zbl.* **104,** 113–123.

Kiknadze, I. I., Zainiev, G. A., Panova, T. M., Zakharenko, L. P., Istomina, A. G., and Potapov, V. A. (1985b). Identification of the chromomeres of Balbiani rings BRa, BRb, BRc and their structural-functional organization in *Chironomus thummi*. *Tsitologiya* **17**(4), 376–382. [in Russian]

Kiknadze, I. I., Gunderina, L. I., Filippova, M. A., and Seraya, E. I. (1988). Chromosomal polymorphism in natural and laboratory populations of *Chironomus thummi thummi* Kieff. *Genetika* **24**(10), 1795–1805. [in Russian]

Kiknadze, I. I., Blinov, A. G., and Kolesnikov, N. N. (1989). Molecular-genetic organization of the chironomid genome. *In* "Strukturno-funktsionalnaya organizatsia genoma," pp. 3–58. Nauka, Sib. otd-e., Novosibirsk. [in Russian]

Kiknadze, I. I., Istomina, A. G., Spirin, M. T., and Sebeleva, T. E. (1990). Karyotype and system of Balbiani rings of the chironomid *Fleuria lacustris*. *Tsitologia* **32**(4), 371–377. [in Russian]

King, D. S., and Siddall, J. B. (1969). Conversion of a-ecdysone to b-ecdysone by Crustaceans and Insects. *Nature* **221,** 955–956.

King, R. C. (1975). *Drosophila melanogaster:* An introduction. *In* "Handbook of Genetics" (R. C. King, ed.), Vol. 3, pp. 625–652. Plenum, New-York/London.

King, R. C., and Riley, S. F. (1982). Ovarian pathologies generated by various alleles of the *otu* locus in *Drosophila* mutant. *Dev. Genet.* **3,** 69–89.

King, R. C., Riley, S. F., Cassidy, J. D., White, P. E., and Paik, Y. K. (1981). Giant polytene chromosmomes from the ovaries of a *Drosophila melnogaster*. *Science* **212,** 441–443.

King, R. C., Mohler, D., Riley, S. F., Storte, P. D., and Nicolazzo, P. S. (1986). Complementation between alleles at the ovarian tumor locus of *Drosophila melnogaster*. *Dev. Genet.* **7,** 1–20.

King, R. L., and Beams, H. W. (1934). Somatic synapsis in *Chironomus* with special reference to the individuality of the chromosomes. *J. Morphol.* **56,** 577–588.

King, R. S. (1970). "Ovarian Development in *Drosophila melanogaster*." Academic Press, New York/London/San Francisco.

Kingston, R. E., Bunker, C. A., and Imbalzano, A. N. (1996). Repression and activation by multiprotein complexes that alter chromatin structure. *Genes Dev.* **10,** 905–920.

Kirchhoff, C. (1981). Electrophoretic analysis of newly synthesized and stored maternal RNA during oogenesis of *Calliphora erythrocephala* (Dipt.). *Wilhelm Roux's Arch.* **190,** 331–338.

Kiseleva, E. V. (1989). Secretory protein synthesis in *Chironomus* salivary gland cells is not coupled with protein translocation across endoplasmic reticulum membranes. Electrone micriscopic evidence. *FEBS Lett.* **257**(2), 251–253.

Kiseleva, E. V., and Masich, S. V. (1991). Electron microscopic analysis of the structure of giant polyribosomes in the salivary gland cells of *Chironomus thummi*. *Molekulyar. biol.* **25**(5), 1258–1265. [in Russian]

Kiseleva, E., Wurtz, T., Visa, N., and Daneholt, B. (1994). Assembly and disassembly of spliceosomes along a specific pre-messenger RNP fiber. *EMBO J.* **13,** 6054–6061.

Kiseleva, E., Goldberg, M. W., Daneholt, B., and Allen, T. D. (1996). RNP export is mediated by structural reorganization of the nuclear pore basket. *J. Mol. Biol.* **260,** 304–311.

Kislov, A. N., and Veprintsev, B. N. (1970). On the question of the existence of a dormancy potential in the nuclei of the salivary gland cells of *Drosophila funebris* larvae. *Biofisika* **15**(1), 99–103. [in Russian]

Kiss, I., and Molnar, I. (1980). Metamorphic changes of wild type and mutant *Drosophila* tissues induced by 20-hydroxy ecdysone in vitro. *J. Insect Physiol.* **26,** 391–401.

Kiss, I., Bencze, G., Fekete, E., Fodor, A., Gausz, J., Maroy, P., Szabad, J., and Szidonya, J. (1976a). Isolation and characterization of X-linked lethal mutants affecting differentiation of the imaginal discs in *Drosophila melanogaster*. *Theoret. Appl. Genet.* **48,** 217–226.

Kiss, I., Bencze, G., Fodor, A., Szabad, J., and Fristrom, J. (1976b). Prepupal-larval mosaics in *Drosophila melanogaster*. *Nature* **262,** 136–138.

Kiss, I., Major, J., and Szabad, J. (1978). Genetic and developmental analysis of puparium formation in *Drosophila melonagaster*. *Mol. Gen. Genet.* **164,** 77–83.

Kiss, I., Szabad, J., Belyaeva, E. S., Zhimulev, I. F., and Major, J. (1980). Genetic and developmental analysis of mutants in an early ecdysone-inducible puffing region in *Drosophila melanogaster*. *In* "Development and Neurobiology of Drosophila" (O. Siddiqi, P. Babu, L. M., Hall, and J. C., Hall, eds.), pp. 163–181. Plenum, New York/London.

Kiss, I., Beaton, A. H., Tardiff, J., Fristrom, D., and Fristrom, J. W. (1988). Interactions and developmental effects of mutations in the *Broad-Complex* of *Drosophila melanogaster*. *Genetics* **118,** 247–259.

Kitagawa, Y., and Stollar, B. D. (1982). Comparison of poly(A) x poly(dT) and poly(I) x poly(dC) as immunogens for the induction of antibodies to RNA-DNA hybrids. *Mol. Immunol.* **19,** 413–420.

Klemenz, R., and Gehring, W. J. (1986). Sequence requirement for expression of the *Drosophila melanogaster* heat shock protein hsp22 gene during heat shock and normal development. *Mol. Cell. Biol.* **6,** 2011–2019.

Klemenz, R., Hultmark, D., and Gehring, W. J. (1985). Selective translation of heat shock mRNA in *Drosophila melanogaster* depends on sequence information in the leader. *EMBO J.* **4,** 2053–2060.

Kloetzel, J. A., and Laufer, H. (1968). Fine structure analysis of larval salivary gland function in *Chironomus thummi*. *J. Cell. Biol.* **39,** 74a.

Kloetzel, J. A., and Laufer, H. (1969). A fine structural analysis of larval salivary gland function in *Chironomus thummi* (Diptera). *J. Ultrastruct. Res.* **29,** 15–36.

Kloetzel, J. A., and Laufer, H. (1970). Developmental changes in fine structure associated with secretion in larval salivary glands of *Chironomus*. *Exp. Cell. Res.* **60,** 327–337.

Klose, W., Gateff, E., Emmerich, H., and Beikirch, H. (1980). Developmental studies on two ecdysone deficient mutants of *Drosophila melanogaster*. *Wilhelm Roux's Arch.* **189,** 57–67.

Knibb, W. R., Tearle, R. G., Elizur, A., and Saint, R. (1993). Genetic analysis of chromosomal region 97D2-9 of *Drosophila melanogaster*. *Mol. Gen. Genet.* **239,** 109–114.

Knipple, D. C., Fuerst, T. R., and McIntyre, R. J. (1991). Molecular cloning and analysis of the chromosomal region 26A of *Drosophila melanogaster*. *Mol. Gen. Genet.* **226,** 241–249.

Knust, E., Schrons, H., Grawe, F., and Campos-Ortega, J. A. (1992). Seven genes of the *Enhancer of split* complex of *Drosophila melanogaster* encode helix-loop-helix protein. *Genetics* **132,** 505–518.

Koana, T., and Hotta, Y. (1978). Isolation and characterization of flightless mutants in *Drosophila melanogaster*. *J. Embryol. Exp. Morphol.* **45,** 123–143.

Kobel, H. R., and Breugel, F. M. A., van. (1967). Observations on *ltl* (*lethal tumorous larvae*) of *Drosophila melanogaster*. *Genetica* **38,** 305–327.

Kodani, M. (1948). The protein of the salivary gland secretion in *Drosophila*. *Proc. Natl. Acad. Sci. USA* **34,** 131–135.

Koelle, M. R., Talbot, W. S., Segraves, W. A., Bender, M. T., Cherbas, P., and Hogness, D. S. (1991). The *Drosophila* EcR gene encodes an ecdysone receptor, a new member of the steroid receptor superfamily. *Cell* **67,** 59–77.

Koelle, M. R., Segraves, W. A., and Hogness, D. S. (1992). DHR3: A new *Drosophila* steroid receptor homolog. *Proc. Natl. Acad. Sci USA* **89,** 6167–6171.

Kogan, M. E., Moshkov, D. A., Sakson, M. E., Mazul, V. M., Yanchevskaya, T. G., and Konev, S. V. (1982). Ultrastructure of intercellular contacts of the salivary glands of *Drosophila* larvae. *Tsitologia* **24**(7), 802–804. [in Russian]

Kokoza, E. B., Kozlova, T. Yu., Umbetova, G. H., Dubrovsky, E. B., Pirrotta, V., and Zhimulev, I. F. (1990). Molecular and genetic analysis of the 10A1-2 band in *Drosophila melanogaster* X chromosome. *Genetika* **26,** 1361–1369. [in Russian]

Kokoza, E. B., Belyaeva, E. S. and Zhimulev, I. F. (1991). Localization of the *ecs*, *dor*, and *swi* genes in eight *Drosophila* species. *Genetika* (Moscow) **27**(12), 2082–2090. [in Russian]

Kokoza, E. B., Belyaeva, E. S., and Zhimulev, I. F. (1992). Localization of genes *ecs*, *dor* and *swi* in eight *Drosophila* species. *Genetica* **87**, 79–85.

Kokoza, E. B., Sedelnikova, A. V., and Zhimulev, I. F. (1997). An analysis of repeated DNA sequences in the 10A1-2 band of polytene chromosome of *Drosophila melanogaster*. *Tsitologia* **39**(1), 69. [in Russian]

Kokoza, V. A. (1985). Organization of genes incoding secretion proteins of the salivary gland of *Drosophila melanogaster*. *In* "Organizatsia i ekspressia genov tkanespetsificheskoy funktsii u *Diptera*," pp. 46–59. Nauka, Novosibirsk. [in Russian]

Kokoza, V. A., and Karakin, E. I. (1981). Electrophoretic analysis of the secretion proteins of the salivary glands of some strains isolated from natural populations of *Drosophila melanogaster*. *Genetika* **17**(7), 935–939. [in Russian]

Kokoza, V. A., and Karakin, E. I. (1982). The investigation of some biochemical characteristics of the salivary gland secretion proteins of *Drosophila melanogaster*. *Dros. Inform. Serv.* **58**, 93.

Kokoza, V. A., and Karakin, E. I. (1985). Physicotechnical characteristics of the secretion proteins of the salivary gland of *Drosophila melanogaster*. *In* "Organizatsia i ekspressia genov tkanespetsificheskoy funktsii u *Diptera*," pp. 36–46. Nauka, Novosibirsk. [in Russian]

Kokoza, V. A., Karakin, E. I., and Sviridov, S. M. (1980). Genetic bichemical analysis of secretory proteins of *Drosophila melanogaster* larvae. *Doklady Akad. Nauk SSSR* **252**(3), 735–738. [in Russian]

Kokoza, V. A., Kazakova, S. G., and Karakin, E. I. (1982). The genetic localization of genes responsible for two salivary gland secretion proteins of *Drosophila melanogaster*. *Dros. Inform. Serv.* **58**, 94–95.

Kolesnikov, N. N. (1977). "Functional activity of the salivary glands of *Drosophila melanogaster* in the course of metamorphosis." *Ph. D. thesis*, Novosibirsk. [in Russian]

Kolesnikov, N. N. (1990). "Organization and expression of the genes for the tissue-specific function of the salivary glands in dipteran (*Drosophila*, *Chironomus*)." *Doctor of Science thesis*, Novosibirsk. [in Russian]

Kolesnikov, N. N., and Aimanova, K. Y. (1985). A characterization of the secretion of glycoproteins of the salivary glands and their genetic control in *Drosophila virilis*. *In* "Organizatsia i ekspressia genov tkanespetsificheskoy funktsii u *Diptera*," pp. 80–96. Nauka, Novosibirsk. [in Russian]

Kolesnikov, N. N., and Kiknadze, I. I. (1976). Secretion proteins in the salivary glands of *Drosophila*. *Doklady Akad. Nauk SSSR* **229**(4), 979–982. [in Russian]

Kolesnikov, N. N., and Kokoza, V. A. (1985). Structure of the salivary gland of *Drosophila melanogaster* and its defferentiation in ontogenesis. *In* "Organizatsia i ekspressia genov tkanespetsificheskoy funktsii u *Diptera*," pp. 30–36. Nauka, Novosibirsk. [in Russian]

Kolesnikov, N. N., and Zhimulev, I. F. (1975). Synthesis of mucoprotein secretion in the salivary glands of *Drosophila melanogaster* during the third larval instar. *Ontogenez* **6**(2), 177–182. [in Russian]

Kolesnikov, N. N., Slobodyanyuk, S. Y., Berdnikov, V. A., and Kiknadze, I. I. (1975). Electrophoretic analysis of the proteins of the salivary glands and fat body of *Drosophila melanogaster* during development. *Doklady Akad. Nauk SSSR* **225**(1), 201–204. [in Russian]

Kolesnikov, N. N., Slobodyanyuk, S. Y., Berdnikov, V. A., and Kiknadze, I. I. (1976a). Comparative electrophoretic analysis of proteins of defferent tissues of *Drosophila melanogaster* during development. *Tsitologia* **18**(7), 862–870. [in Russian]

Kolesnikov, N. N., Slobodyanyuk, S. Y., and Kiknadze, I. I. (1976b). Functional activity of the salivary glands of *Drosophila melanogaster* during metamorphosis: A comparative electrophoretic analysis of the proteins of the salivary glands fat body and hemolimph. *Ontogenez* **7**(6), 547–557. [in Russian]

Kolesnikov, N. N., Karakin, E. I., Sebeleva, T. E., Meyer, L., and Serfling, E. (1981). Cell-specific synthesis and glycosylation of secretory proteins in larval salivary glands of *Chironomus thummi. Chromosoma* **83,** 661–677.

Kolesnikov, N. N., Kiknadze, I. I., Panova, T. M., Zakharenko, I. N., Blinov, A. G., Baiborodin, S. I., and Khristolyubova, N. B. (1984). Microcloning of DNA specific regions of the polytene chromosomes of *Chironomus thummi. Tr. 3-ei natsion. Konf. tsitogenet. Plovdiv.* **1,** 211–214. [in Russian]

Kolesnikov, N. N., Kiknadze, I. I., Bogachev, S. S., Blinov, A. G., Blinov V. M., Panova, T. M., Gaidamakova, E. K., Fedorova, S. P., Sobanov, Yu. V., Shakhmuradov, I. A., Chekaev, N. A., and Nazarenko, N. A. (1986). Molecular cytogenetic approaches to study on specific loci of polytene chromosomes. *In* "Evolutsia, vidoobrazovanie i sistematika khironomid," pp. 146–156. Inst. tsitologii i genetiki, Novosibirsk. [in Russian]

Kolesnikov, N. N., Bogachev, S. S., Scherbic, S. V., Taranin, A. V., Baiborodin, S. I., Donchenko, A. P., Sebeleva, T. E., and Kiknadze, I. I. (1993). Structural elements of Balbiani ring BRa of *Chironomus tummi. In* "Nuclear Structure and Function" (J. R. Harris, and I. B. Zbarsky, eds.), pp. 53–56. Plenum, New York.

Kolmer, W., and Fleischmann, W. (1928). Beobachtungen an den Speicheldrussen von Chironomusarten. *Protoplasma* **4,** 358–366.

Koltzoff, N. K. (1934). The structure of the chromosomes in the salivary glands of *Drosophila. Science* **80,** 312–313.

Komma, D. J. (1966). Effect of sex transformation genes on glucose-6-phosphate dehudrogenase activity in *Drosophila melanogaster. Genetics* **54,** 497–503.

Konev, A. Yu., Varentsova, E. R., Sarantseva, S. V., and Khromykh, Yu. M. (1991a). Cytogenetic analysis of the chromosome region containing the radiosensitive gene of *Drosophila:* A study of radiation mutagenesis in the 44-45 region of chromosome 2. *Genetika* **27**(1), 77–87. [in Russian]

Konev, A. Yu., Varentsova, E. R., and Khromykh, Yu. M. (1991b). Cytogenetic analysis of the chromosome region containing the radiosensitive gene of *Drosophila:* Effect of centromeric heterochromatin on mutagenesis in the 44-45 region of chromosome 2. *Genetika* **27**(4), 667–675. [in Russian]

Konev, A. Yu., Varentsova, E. R., Levina, V. V., Sarantseva, S. V., and Khromykh, Yu. M. (1994a). Cytogenetic analysis of the chromosome region containing radiosensitivity gene in *Drosophila.* I. A cytogenetic mapping of radiosensitivity gene. *Genetika* **30**(2), 192–200. [in Russian]

Konev, A. Yu., Varentsova, E. R., and Khromykh, Yu. M. (1994b). Cytogenetic analysis of the chromosome region containing radiosensitivity gene in *Drosophila.* II. A study of vital loci of 44F-45C region of chromosome 2. *Genetika* **30**(2), 201–211. [in Russian]

Koninkx, J. F. J. G. (1975). Induced transcription dependent synthesis of mitochondrial reduced nicotinamide-adenine dinucleotide dehydrogenase in *Drosophila. Biochem. J.* **152,** 17–22.

Koninkx, J. F. J. G. (1976). Protein synthesis in salivary glands of *Drosophila hydei* after experimental gene induction. *Biochem. J.* **158,** 623–628.

Koninkx, J. F. J. G., Leenders, H. J., and Birt, L. M. (1975). A correlation between newly induced gene activity and an enhahcement of mitochondrial enzyme activity in the glands of *Drosophila. Exp. Cell. Res.* **92,** 275–282.

Konopka, R. J., and Benzer, S. (1971). Clock mutants of *Drosophila melanogaster. Proc. Natl. Acad. Sci. USA* **68,** 2112–2116.

Konstantinov, A. S., and Nesterova, S. I. (1971). Identification by anatomical and karyotypical parameters in the systematics of Chironomids. *Limnologica* **8,** 19–25.

Konstantopoulou, I., Ouzounis, C. A., Drosopoulou, E., Yiangou, M., Sideras, P., Sander, C., and Scouras, Z. G. (1995). A *Drosophila* hsp70 gene contains long, antiparallel, coupled open reading frames (LAC ORFs) conserved in homologous loci. *J. Mol. Evol.* **41,** 414–420.

Konstantopoulou, I., Drosopoulou, E., and Scouras, Z. G. (1997). Variations in the heat-induced protein pattern of several *Drosophila montium* subgroup species (Diptera: Drosophilidae). *Genome* **40,** 132–137.

Kontermann, R., Sitzler, S., Seifarth, W., Petersen, G., and Bautz, E. K. F. (1989). Primary structure and functional aspects of the gene coding for the second—Largest subunit of RNA polymerase III of *Drosophila*. *Mol. Gen. Genet.* **219,** 373–380.

Koolman, J. (ed.). (1989). Ecdysone, from chemistry to mode of action. Georg Theime-Verlag, Stuttgart/New York.

Kopantseva, T. I. (1973). "Study on RNA weight in the cell nuclei of the salivary glands of the *l(3)tl* mutant (*lethal tumorous larvae*) *Drosophila melanogaster*." *Diploma thesis*, Novosibirsk. [in Russian]

Kopantseva, T. I., Maximovsky, L. F., and Zhimulev, I. F. (1994). Increase of RNA quantity in salivary gland nuclei of nonpupariating larvae homozygous for *l(3)tl* mutation in *Drosophila melanogaster*. *Dros. Inform. Serv.* **75,** 171–172.

Kopantzev, E. P., Karakin, E. I., Panova, T. M., and Kiknadze, I. I. (1979). Antigenes of the secretion of the salivary glands of *Chironomus thummi* larvae. *Doklady Akad. Nauk SSSR* **249,** 1477–1480. [in Russian]

Kopantzev, E. P., Karakin, E. I., and Kiknadze, I. I. (1985a). Tissue-specific secretory proteins of the salivary glands of *Chironomus thummi:* An electrophoretic and immunochemical analysis. *Chromosoma* **92,** 283–289.

Kopantzev, E. P., Karakin, E. I., and Kiknadze, I. I. (1985b). Phisicochemical characteristics of the tissue-specific secretion proteins of salivary glands of *Chironomus thummi*. *In* "Organizatsia i ekspressia genov tkanespetsificheskoy funktsii u *Diptera*," pp. 100–109. Nauka, Novosibirsk. [in Russian]

Korge, G. (1970a). Dosage compensation and effect for RNA synthesis in chromosome puffs of *Drosophila melanogaster*. *Nature* **225,** 386–388.

Korge, G. (1970b). Dosiskompensation und Dosiseffekt fur RNS-synthese in Chromosomen-Puffs von *Drosophila melanogaster*. *Chromosoma* **30,** 430–464.

Korge, G. (1975). Chromosome puff activity and protein synthesis in larval gland of *Drosophila melanogaster*. *Proc. Natl. Sci Acad. USA* **72,** 4550–4554.

Korge, G. (1977a). Direct correlation between a chromosome puff and the synthesis of a larval saliva protein in *Drosophila melanogaster*. *Chromosoma* **62,** 155–174.

Korge, G. (1977b). Larval saliva in *Drosophila melanogaster:* Production, composition and relationship to chromosome puffs. *Dev. Biol.* **58,** 339–355.

Korge, G. (1980). Gene activities in larval salivary glands of insects. *Verh. Dtsh. Zool. Ges.* **1980,** 94–110.

Korge, G. (1981). Genetic analysis of the larval secretion gene *Sgs-4* and its regulatory chromosome sites in *Drosophila melanogaster*. *Chromosoma* **84,** 373–390.

Korge, G. (1986). Analyse der expression eines Gens bei *Drosophila*. *Naturwissenschaften* **73,** 719–727.

Korge, G. (1987). Polytene chromosomes. *In* "Results and Problems in Cell Differentiation. Structure and Function of Eukaryotic Chromosomes" (W. Hennig, ed.), Vol. 14, pp. 27–58. Springer-Verlag, Berlin/Heidelberg.

Korge, G., Heide, I., Sehnert, M., and Hofmann, A. (1990). Promotor is an important determinant of developmentally regulated puffing at the *Sgs-4* locus of *Drosophila melanogaster*. *Dev. Biol.* **138,** 324–337.

Kornher, S. J., and Brutlag, D. (1986). Proximaty-dependent enhancement of *Sgs-4* gene expression in *Drosophila melanogaster*. *Cell* **44,** 879–883.

Korochkin, L. I. (1976). Role of temporal factors in the regulation of development. *Zhurnal obshch. biologii* **37**(2), 184–191. [in Russian]

Korochkin, L. I. (1979). A hypothesis of the temporal principle of the organization of gene systems regulating individual development. *Issledovania po genetike* **8,** 150–160. [in Russian]

Korochkin, L. I., and Belyaeva, E. S. (1972). Differential activity of homologous chromosomes during development. *Ontogenez* **3**(1), 11–20. [in Russian]

Korochkin, L. I., and Evgen'ev, M. B. (1982). Cytogenetic localization of the gene encoding organ-specific esterase in *Drosophila*. *Doklady Akad. Nauk SSSR* **263**(2), 471–474. [in Russian]

Korochkin, L. I., and Poluektova, E. V. (1983). Interrelation of puffing and esterase isozyme spectrum in the salivary gland of *Drosophila* of the *virilis* group. *Doklady Akad. Nauk SSSR* **273**(5), 1243–1246. [in Russian]

Korochkina, L. S. (1972). Change of the ring gland in post-natal development of *Drosophila melanogaster Canton S* and in *l(2)gl* mutant. *Dros. Inform. Serv.* **48,** 74–75.

Korochkina, L. S. (1977). Morphology and some functional characteristics of the chromosomes of the *Drosophila* genus. *In* "Problemy genetiki v issledovaniyakh na drosofile," pp. 112–151. Nauka, Sibirsk. otdel., Novosibirsk. [in Russian]

Korochkina, L. S., and Nazarova, N. K. (1977). Comparative characteristics of the puffing pattern and the endocrine system in an *Oregon* laboratory stock and in *l(2)gl Drosophila melanogaster* mutants differing in the time of their death. *Chromosoma* **62,** 175–190.

Korochkina, L. S. and Nazarova, N. K. (1978). Comparative characteristics of the puffing pattern and the endocrine system in an *Oregon* laboratory stock and in *l(2)gl Drosophila melanogaster* mutants differing in the time of their death. *Genetika* **14**(1), 93–102. [in Russian]

Korochkina, L. S., Kiknadze, I. I. and Muradov, S. V. (1972). Effect of hormons on puffing in *Chironomus thummi thummi* Kieffer. *Ontogenez* **3**(2), 177–186. [in Russian]

Kosswig, C., and Sengun, A. (1947a). Intraindividual variability of chromosome IV of *Chironomus*. *J. Hered.* **38,** 235–239.

Kosswig, C., and Sengun, A. (1947b). Neuere Untersuchungen uber den Bau der Riesenchromosomen der Dipteren. *Rev. de la Fac. Sci. Univ. Istanbul. Ser. B.* **12,** 107–121.

Kosswig, C., and Sengun, A. (1947c). Verglekhende Untersuchungen uber die Riesenchromosomen der verschiedenen gewebearten verschienderner Dipteren. C.r. ann. et arch. *Soc. Turque Sci. Phys. et Maturelles.* **12,** 94.

Kostoff, D. (1930). Discoid structure of the spireme and irregular cell division in *Drosophila melanogaster. J. Hered.* **21,** 323–324.

Kotarski, M. A., Pickert, S., and McIntyre, R. J. (1983). A cytogenetic analysis of the chromosomal region surrounding the *alpha-glycerophosphate dehydrogenase* locus of *Drosophila melanogaster*. *Genetics* **105,** 371–386.

Kozlova, T. Yu., Umbetova, G. H., Dubrovsky, E. B., Pirrota, V. and Zhimulev, I. F. (1990). Molecular cytogenetic analysis of the 10A1-2 band of the X chromosome of *Drosophila melanogaster*. *Genetika* **26**(8), 1361–1369. [in Russian]

Kozlova, T. Yu., Semeshin, V. F., Tretyakova, I. V., Kokoza, E. B., Pirrotta, V., Grafodatskaya, V. E., Belyaeva, E. S., and Zhimulev, I. F. (1994). Molecular and cytogenetical characterization of the 10A1-2 band and adjoining region in the *Drosophila melanogaster* polytene X chromosome. *Genetics* **136,** 1063–1073.

Kozlova, T., Zhimulev, I. F., and Kafatos, F. C. (1997). Molecular organization of an individual *Drosophila* polytene chromosome chromomere: Transcribed sequences in the 10A1-2 band. *Mol. Gen. Genet.* **257,** 55–61.

Kraevsky, A. A., and Kukhanova, M. K. (1986). DNA replication in eukaryotes. *In* "Itogi nauki i tekhniki. Molekulyarnaya biologiya," Vol. 22, pp. 5–164. VINITI, Moscow. [in Russian]

Kramer, A., Haars, R., Kabisch, R., Wilkl, H., Bautz, F. A. A., and Bautz, E. K. F. (1980). Monoclonal antibody directed against RNA polymerase II of *Drosophila melanogaster*. *Mol. Gen. Genet.* **180,** 193–199.

Kramer, H., and Phistry, M. (1996). Mutations in the *Drosophila hook* gene inhibit endocytosis of the boss transmembrane ligand into multivesicular bodies. *J. Cell. Biol.* **133,** 1205–1215.

Kramerov, A. A., and Gvozdev, V. A. (1984). Molecular features of ecdysone effect on cells. *In:* "Itogi nauki i tekhniki. Molekulyarnaya biologiya," Vol. 20, pp. 106–141. VINITI, Moscow. [in Russian]

Kraminsky, G. P., Clark, W. C., Estelle, M. A., Gietz, R. D., Sage, B. A., O'Connor, J. D., and Hodgetts, R. B. (1980). Induction of translatable mRNA for dopa decarboxylase in *Drosophila:* an early response to ecdysterone. *Proc. Natl. Acad. Sci. USA* **77,** 4175–4179.

Kremer, H., and Hennig, W. (1990). Isolation and characterization of a *Drosophila hydei* histone DNA repeat unit. *Nucleic Acids Res.* **18,** 1573–1580.

Kress, H. (1972). "Das Puffmaster der Riesenchromosomen in den larvalen Speicheldrusen von *Drosophila virilis:* Seine Veranderungen in der Normalentwicklung und nach Injection von Ecdyson." Bayerische Academie der Wissenchaften. Math.-Natur. Klasse, pp. 129–149.

Kress, H. (1973). Specific repression of a puff in the salivary gland chromosomes of *Drosophila virilis* after injection of glucosamine. *Chromosoma* **40,** 379–386.

Kress, H. (1974). Temporal relationships between leaving food, ecdysone release, mucoprotein extrusion, and puparium formation in *Drosophila virilis*. *J. Insect Physiol.* **20,** 1041–1055.

Kress, H. (1977). Transcriptional control of developmental processes by ecdysone in *Drosophila virilis* salivary glands. *Wilhelm Roux's Arch.* **182,** 107–116.

Kress, H. (1979). Ecdysone-induced changes in glycoprotein synthesis and puff activities in *Drosophila virilis* salivary glands. *Chromosoma* **72,** 53–66.

Kress, H. (1981). Ecdysone-induced puffing in *Drosophila:* A model. *Naturwisenschaften* **68,** 28–33.

Kress, H. (1982). Biochemical and ontogenetic aspects of glycoprotein synthesis in *Drosophila virilis* salivary glands. *Dev. Biol.* **93,** 231–239.

Kress, H. (1986). Stoffwechselaktivaten und Transkripionsmuster in den larvalen Speicheldrusen von *Drosophila virilis*. *Naturwissenschaften* **73,** 180–187.

Kress, H. (1993). The salivary gland chromosomes of *Drosophila virilis:* A cytological map, pattern of transcription and aspects of chromosome evolution. *Chromosoma* **102,** 734–742.

Kress, H. (1996). Polytene chromosomes: a general model for eucaryotic interphase state. *Int. J. Insect Morphol. Embryol.* **25,** 63–91.

Kress, H., and Enghofer, E. (1975). Kinetic characteristics and ontogeny of glutamine-fructose-6-phosphate-aminotransferase in *Drosophila* salivary glands. *Insect Biochem.* **5,** 171–181.

Kress, H., and Swida, U. (1990). *Drosophila* glue protein gene expression. A proposal for its ecdysone-dependent developmental control. *Naturwissenschaften* **77,** 317–324.

Kress, H., Meyerowitz, E. M., and Davidson, N. (1985). High resolution mapping of *in situ* hybridized biotinylated DNA to surface-spread *Drosophila* polytene chromosomes. *Chromosoma* **93,** 113–122.

Kress, H., Lucka, L., Swida, U., Thuroff, E., and Klemm, U. (1990). Genes from two intermoult puffs in *Drosophila virilis* polytene chromosomes are differentially transcribed during larval development. *Development* **108,** 261–267.

Krider, H. M., and Plaut, W. (1972a). Studies on nucleolar RNA synthesis in *Drosophila melanogaster*. I. The relationship between number of nucleolar organizers and rate of synthesis. *J. Cell. Sci.* **11,** 675–687.

Krider, H. M., and Plaut, W. (1972b). Studies on nucleolar RNA synthesis in *Drosophila melanogaster*. II. The influence of conditions result in a *bobbed* phenotype on rate of synthesis and secondary constriction formation. *J. Cell. Sci.* **11,** 689–697.

Kriegstein, H. J., and Hogness, D. S. (1974). Mechanism of DNA replication in *Drosophila* chromosomes: Structure of replication forks and evidence for bidirectionality. *Proc. Natl. Acad. Sci. USA* **71,** 135–139.

Krishnan, R., Swanson, K. D., and Ganguly, R. (1991). Dosage compensation of a retina-specific gene in *Drosophila miranda*. *Chromosoma* **100,** 125–133.

Krivshenko, J. (1959). The divisibility of the nucleolus organizer and the strucuture of the nucleolus in *Drosophila busckii*. *Genetics* **44**(4), 520–521.

Kroeger, H. (1960). The induction of new puffing patterns by transplantation of salivary gland nuclei into egg cytoplasm of *Drosophila*. *Chromosoma* **11,** 129–145.

Kroeger, H. (1963). Chemical nature of the system controllong gene activities in insect cells. *Nature* **200,** 1234–1235.

Kroeger, H. (1964). Zellphysiologische mechanismen beu der Regulation von Genaktivitaten in den Riesenchromosomen von *Chironomus thummi*. *Chromosoma* **15,** 36–70.

Kroeger, H. (1966). Potentialdifferenz und Puffmuster Electrophysiologische und cytologische Untersuchungen an der Speicheldrusen von *Chironomus thummi*. *Exp. Cell. Res.* **41,** 64–80.

Kroeger, H. (1967). Hormones, ion balances and gene activity in Dipteran chromosomes. *Mem. Soc. Endocrinol.* **15,** 55–66.

Kroeger, H. (1968), Gene activities during insect metamorphosis and their control by hormones. *In* "Metamorphosis: A problem of Developmental Biology" (W. Etkin, and L. I. Gilbert, eds.), pp. 185–219. North-Holland, Amsterdam.

Kroeger, H. (1977). The control of puffing by ions: A reply. *Mol. Cell. Endocrinol.* **7,** 105–110.

Kroeger, H., and Lezzi, M. (1966). Regulation of gene action in insect development. *Ann. Rev. Entomol.* **11,** 1–22.

Kroeger, H., and Muller, G. (1973). Control of puffing activity in three chromosomal segments of explanted salivary gland cells of *Chironomus thummi* by variations in extracellular Na^+, K^+, and Mg^{2+}. *Exp. Cell. Res.* **82,** 89–94.

Kroeger, H., and Trosch, W. (1974). Influence of the explantation milieu on intranuclear [Na], [K], and [Mg] of *Chironomus thummi* salivary gland cells. *J. Cell. Physiol.* **83,** 19–25.

Kroeger, H., Gettmann, W., and Kalter, Ch. (1973a). Zur Kontrolle der DNA-Replikation in Riesenchromosomen explantierter Speicheldrusen. I. Zeitablauf und Einfluss verschiedener Medien. *Cytobiology* **7,** 117–126.

Kroeger, H., Trosch, W., and Muller, G. (1973b). Changes in nuclear salivary gland cells during development. *Exp. Cell. Res.* **80,** 329–339.

Krum, A., Roth, G. E., and Korge, G. (1985). Transformation of salivary gland secretion protein gene *Sgs-4* in *Drosophila:* stage- and tissue-specific regulation, dosage compensation, and position effect. *Proc. Natl. Acad. Sci USA* **85,** 5055–5059.

Kryukova, Z. I. (1929). Cytological observations of the salivary glands of *Chironomus* larvae. *Arkh. anatomii, gistologii i embriologii* **8**(1), 25–43. [in Russian]

Kubli, E. (1982). The genetics of transfer RNA in *Drosophila. Adv. Genet.* **21,** 123–172.

Kumar, A., and Rai, K. S. (1990). Chromosomal localization and copy number of 18S+28S ribosomal RNA genes in evolutionary diverse mosquitoes (Diptera, Culicidae). *Hereditas* **113,** 277–289.

Kunz, W., and Glatzer, K. H. (1979). Similarities and differences in the gene structure of the three nucleoli of *Drosophila hydei*. *Hoppe-Seyler's Z. Physiol. Chem.* **360,** 313.

Kunz, W., Petersen, G., Renkawitz-Pohl, R., Glatser, K.H., and Schafer, M. (1981). Distribution of spacer length classes and the intervaning sequence among different nucleolus organizers in *Drosophila hydei*. *Chromosoma* **83,** 145–158.

Kurazhkovskaya, T. P. (1969). Structure of the salivary glands of the larvae of *Chironomids*. *In* "Fiziologiya vodnykh organizmov i ikh rol v krugovorote organicheskogo veshchestva," pp. 185–195. Nauka, Leningrad. [in Russian]

Kuroda, M., Kernan, M. J., Kreber, R., Ganetzky, B., and Baker, B. S. (1991). The maleless ptotein associates with the X chromosome to regulate dosage compensation in *Drosophila*. *Cell* **66,** 935–947.

Kuroda, M. I., Palmer, M. J., and Lucchesi, J. C. (1993). X chromosome dasage compensation in *Drosophila*. *Semin. Dev. Biol.* **4,** 107–116.

Kurzik-Dumke, U., Gundacker, D., Rentrop, M., and Gateff, E. (1995). Tumor suppression in *Drosophila* is causally related to the function of the *lethal (2) tumorous imaginal discs* gene, a DNA J homolog. *Dev. Genet.* **16,** 64–76.

Kylsten, P., Samakovlis, C., and Hultmark, D. (1990). The cecropin locus in *Drosophila:* A compact gene cluster involved in the response to infection. *EMBO J.* **9,** 217–224.

Laemmli, U. K., Cheng, S. M., Adolph, K. W., Paulson, J. R., Brown, J. A., and Baumbach, W. R. (1978). Metaphase chromosome structure: the role of nonhistone proteins. *Cold Spring Harbor Symp. Quant. Biol.* **42,** 351–360.

Laicine, E. M., Alves, M. A. R., de Almeida, J. C., Albernaz, W. C., and Sauaia, H. (1982). Expressao genica no desenvolvimento da glandula salivar de *Bradysia hygida:* Significado biologico dos pufes de DNA. *Cienc. Cult.* **34,** 488–492.

Laicine, E. M., Alves, M. A. R., de Almeida, J. C., Rizzo, E., Albernaz, W. C., and Sauaia, H. (1984). Development of DNA puffs and patterns of polypeptide synthesis in the salivary glands of *Bradysia hygida. Chromosoma* **89,** 280–284.

Laird, C. D. (1971). Chromatid structure: Relationship between DNA content and nucleotide sequence diversity. *Chromosoma* **32,** 378–406.

Laird, C. D. (1980). Structural paradox of polytene chomosomes. *Cell* **22,** 869–874.

Lakhotia, S. C. (1970). Chromosomal basis of dosage compensation in *Drosohpila.* II. The DNA replication patterns of the male X-chromosome in an autosome X insertion in *D. melanogaster. Genet. Res. Cambr.* **15,** 301–307.

Lakhotia, S. C. (1974). Ultrastructure of the chromocentre region in salivary gland polytene nuclei of *Drosophila hydei. The Nucleus* **17,** 100–104.

Lakhotia, S. C. (1979). Bands and condensed chromatin as sites of transcription in polytene chromosomes of *Drosophila. Ind. J. Exp. Biol.* **17,** 239–248.

Lakhotia, S. C. (1987). The 93D heat shock locus in *Drosophila:* A review. *J. Genet.* **66,** 139–157.

Lakhotia, S. C., and Jacob, J. (1974). EM autoradiographic studies on polytene nuclei of *Drosophila melanogaster.* II. Organization and transcriptive activity of the chromocentre. *Exp. Cell. Res.* **86,** 253–263.

Lakhotia, S. C., and Mukherjee, A. S. (1969). Chromosomal basis of dosage compensation in *Drosophila.* I. Cellular autonomy of hyperactivity of the male X-chromosome in salivary glands and sex differentiation. *Genet. Res. Cambr.* **14,** 137–150.

Lakhotia, S. C., and Mukherjee, A. S. (1970a). Activation of a specific puff by benzamide in *Drosophila melanogaster. Dros. Inform. Serv.* **45,** 108.

Lakhotia, S. C., and Mukherjee, A. S. (1970b). Chromosomal basis of dosage compensation in *Drosophila.* III. Early completion of replication by the polytene X-chromosome in male: Further evidence and its implications. *J. Cell. Biol.* **47,** 18–33.

Lakhotia, S. C., and Mukherjee, A. S. (1971). Hyperactivity of the polytene X-chromosome in male *D. kikkawai* and *D. bipectinata. Dros. Inform. Serv.* **46,** 65.

Lakhotia, S. C., and Mukherjee, A. S. (1972). Chromosomal basis of dosage compensation in *Drosophila.* VI. Transcription and replication in male X-chromosome of *D.melanogaster* after X-irradiation. *The Nucleus* **15,** 189–200.

Lakhotia, S. C., and Mukherjee, T. (1980). Specific activation of puff 93D of *Drosophila melanogaster* by benzamide and the effect of benzamide treatment on the heat shock induced puffing activity. *Chromosoma* **81,** 125–136.

Lakhotia, S. C., and Mukherjee, T. (1982). Absence of novel translation products in relation to induced activity of the 93D puff in *Drosophila melanogaster. Chromosoma* **85,** 369–374.

Lakhotia, S. C., and Mukherjee, T. (1984). Specific induction of the 93D puff in polytene nuclei of *Drosophila melanogaster* by colchicine. *Ind. J. Exp. Biol.* **22,** 67–70.

Lakhotia, S. C., and Roy, S. (1979). Replication in *Drosophila* chromosomes. I. Replication of intranucleolar DNA in polytene cells of *D. nasuta. J. Cell. Sci.* **36,** 185–197.

Lakhotia, S. C., and Sharma, A. (1995). RNA metabolism *in situ* at the 93D heat shock locus in polytene nuclei of *Drosophila melanogaster* after various treatments. *Chromos. Res.* **3,** 151–161.

Lakhotia, S. C., and Singh, A. K. (1982). Conservation of the 93D puff of *Drosophila melanogaster* in different species of *Drosophila. Chromosoma* **86,** 265–278.

Lakhotia, S. C., and Singh, A. K. (1985). Non-inducibility of the 93D heat-shock puff in cold-reared larvae of *Drosophila melanogaster. Chromosoma* **92,** 48–54.

Lakhotia, S. C., and Sinha, P. (1983). Replication in *Drosophila* chromosomes. X. Two kinds of active replicons in salivary gland polytene nuclei and their relation to chromosomal replication patterns. *Chromosoma* **88,** 265–276.

Lakhotia, S. C., and Tiwari, P. K. (1984). Replication in *Drosophila* chromosomes. XI. Differences in temporal order of replication in two polytene cell types in *Drosophila nasuta*. *Chromosoma* **89,** 212–217.

Lakhotia, S. C., and Tiwari, P. K. (1985). Replication in *Drosophila* chromosomes: part XV - organization of active replicons in brain ganglia and wing imaginal disks of *Drosophila nasuta* larvae. *Ind. J. Exp. Biol.* **23,** 413–420.

Lakhotia, S. C., Mishra, A., and Sinha, P. (1981). Dosage compensation of X-chromosome activity in interspecific hybrids of *Drosophila melanogaster* and *D. simulans*. *Chromosoma* **82,** 229–236.

Lakhotia, S. C., Chowdhuri, D. K., and Burma, P. K. (1990). Mutations affecting b-alanine metabolism influence inducibility of the 93D puff by heat shock in *Drosophila melanogaster*. *Chromosoma* **99,** 296–305.

Lam, G. T., Jiang, C., and Thummel, C. S. (1997). Coordination of larval and prepupal gene expression by the DHR3 orphan receptor during *Drosophila* metamorphosis. *Development* **124,** 1757–1769.

Lamb, M. M., and Daneholt, B. (1979). Characterization of active transcription units in Balbiani ring in *Chironomus tentans*. *Cell* **17,** 835–848.

Lambert, B. (1972). Repeated DNA sequences in a Balbiani ring. *J. Mol. Biol.* **72,** 65–85.

Lambert, B. (1973). Tracing of RNA from a puff in the polytene chromosomes to the cytoplasm in *Chironomus tentans* salivary gland cells. *Nature* **242,** 51–53.

Lambert, B. (1974). Repeated nucleotide sequences in a single puff of *Chironomus tentans* polytene chromosomes. *Cold Spring Harbor Symp. Quant. Biol.* **38,** 637–644.

Lambert, B. (1975). The chromosomal distribution of Balbiani ring DNA in *Chironomus tentans*. *Chromosoma* **50,** 193–200.

Lambert, B., and Beermann, W. (1975). Homology of Balbiani ring DNA in two closely related *Chironomus species*. *Chromosoma* **51,** 41–47.

Lambert, B., and Daneholt, B. (1975). Microanalysis of RNA from defined cellular components. *Methods in Cell Biology* **10,** 17–47.

Lambert, B., Wieslander, L., Daneholt, B., Egyhazi, E., and Ringborg, U. (1972). *In situ* demonstration of DNA hybridizing with chromosomal and nuclear RNA in *Chironomus tentans*. *J. Cell. Biol.* **53,** 407–418.

Lambert, B., Daneholt, B., Edstrom, J.-E., Egyhazi, E., and Ringborg, U. (1973a). Comparison between chromosomal and nuclear sap RNA from *Chironomus tentans salivary* gland cells by RNA"DNA hybridization. *Exp. Cell. Res.* **76,** 381–389.

Lambert, B., Egyhazi, E., Daneholt, B., and Ringborg, U. (1973b). Quantitative micro-assay for RNA"DNA hybrids in the study of nucleolar RNA from *Chironomus tentans* salivary gland cells. *Exp. Cell. Res.* **76,** 369–380.

Lane, N. J., Carter, Y. R., and Ashburner, M. (1972). Puffs and salivary gland function: the fine structure of the larval and prepupal salivary glands of *Drosophila melanogaster*. *Wilhelm Roux' Arch.* **169,** 216–238.

Lapta, G. E., and Shakhbazov, V. G. (1984). Induction of puffs during heat shock in inbred strains and hybrids of *Drosophila melanogaster*. *Vestnik Khar'kov. Univer.* **262,** 54–55. [in Russian]

Lara, F. J. S. (1987). Gene amplification in *Rhynchosciara* (1955-1987). *Mem. Inst. Oswaldo Cruz, Rio de Janeiro,* **82**(Suppl. III), 125–128.

Lara, F. J. S. (1990). A importancia de descoberta dos pufes de DNA para a ciencia Brasiliera. *Rev. Brasil. Genet.* **13,** 1–4.

Lara, F. J. S., and Hollander, F. M. (1967). Changes in RNA metabolism during the development of *Rhynchosciara angelae*. *Natl. Cancer. Inst. Monogr.* **27,** 235–242.

Lara F. J. S. , Millar S., Hauward D. C., Read C. A., Browne M. J., Santelli, R. V., Vallejo, F. G., Pueyo, M. T., Zaha, A., and Glover, D. M. (1985). Gene amplification in the DNA puffs of *Rhynchosciara americana* salivary gland chromosomes. *In* "Cellular Regulation and Malignant Growth" (S. Ebashi, ed.), pp. 339–349. Japan Sci. Soc. Press, Berlin: Springer-Verlag/Tokyo.

Lara, F. J. S., Stocker, A. J., and Amabis, J. M. (1991). DNA sequence amplification in Sciarid flies: Results and perspectives. Brazil. *J. Med. Biol. Res.* **24,** 233–248.

Laran, E., Requena, J. M., Jimenez-Ruiz, A., Lopez, M. C., and Alonso, C. (1990). The heat shock protein hsp70 binds *in vivo* to subregions 2–48BC and 3–58D of the polytene chromosomes of *Drosophila hydei*. *Chromosoma* **99,** 315–320.

Larocca, D. (1986). Ecdysterone and heat shock induction of transfecting and endogenous heat shock genes in cultured *Drosophila* cells. *J. Mol. Biol.* **191,** 563–567.

Laski, F. A., Rio, D. C., and Rubin, G. M. (1986). Tissue specificity of *Drosophila* P element transposition is regulated at the level of mRNA splicing. *Cell* **44,** 7–19.

Lasko, P. F., and Pardue, M. L. (1988). Studies of the genetic organization of the *vestigial* microregion of *Drosophila melanogaster*. *Genetics* **120,** 495–502.

Latorre, A., Pascual, L., and de Frutos, R. (1983). Loci active in two strains of *Drosophila subobscura*. *Dros. Inform. Serv.* **59,** 73–74.

Latorre, A., de Frutos, R., and Pascual, L. (1984). Loci activity in three A chromosomal arrangements of *Drosophila subobscura*. *Dros. Inform. Serv.* **60,** 136–138.

Latorre, A., Moya, A., and de Frutos, R. (1988). Patterns of puffing activity and chromosomal polymorphism in *Drosophila subobscura*. IV. Effect of inversions on gene expression. *Evolution* **42,** 1298–1308.

Laufer, H. (1963). Hormones and the development of insects. *Proc. of XVIth Int. Congr. Zool.* **4,** 215–220.

Laufer, H. (1965). Developmental studies of the Dipteran salivary gland. III. Relationship between chromosomal puffing and cellular function during development. *In* "Developmental and Metabolic Control Mechanisms and Neoplasia. A Collection of Papers Presented at the 9th Annual Symposium on Fundamental Cancer Research," pp. 237–250. Williams & Wilkins, Baltimore, Maryland.

Laufer, H. (1968). Developmental interactions on the Dipteran salivary gland. *Am. Zool.* **8,** 257–271.

Laufer, H., and Calvet, J. P. (1972). Hormonal effects on chromosomal puffs and insect development. *Gen. Compar. Endocrinol.* **3**(Suppl.), 137–148.

Laufer, H., and Goldsmith, M. (1965). Ultrastructural evidence for a protein transport system in *Chironomus* salivary glands and its implication for chromosomal puffing. *J. Cell. Biol.* **27,** 57A.

Laufer, H., and Greenwood, H. (1969). The effects of juvenile hormone on larvae of the Dipteran, *Chironomus thummi*. *Am. Zool.* **9,** 603.

Laufer, H., and Holt, T. K. H. (1970). Juvenile hormone effects on chromosomal puffing and development in *Chironomus thummi*. *J. Exp. Zool.* **173,** 341–352.

Laufer, H., and Nakase, Y. (1964). Chromosomal puffing as an expression of protein transport by the Dipteran salivary gland. *J. Cell. Biol.* **23,** 52A.

Laufer, H., and Nakase, Y. (1965a). Developmental studies of the Dipteran salivary gland. II. DNAse activity in *Chironomus thummi*. *J. Cell. Biol.* **25,** 97–102.

Laufer, H., and Nakase, Y. (1965b). Salivary gland secretion and its relation to chromosomal puffing in the Dipteran, *Chironomus thummi*. *Proc. Natl. Acad. Sci. USA* **53,** 511–516.

Laufer, H., and Schin, K.S. (1971). Quantitative studies of hydrolytic enzyme activity in the salivary gland of *Chironomus tentans* (Diptera: Chironomidae) during metamorphosis. *Can. Entomol.* **103,** 454–457.

Laufer, H., Nakase, Y., and Vanderberg, J. (1963). Enzyme activities in the Dipteran salivary gland during matamorphosis. *Am. Zool.* **3,** 486.

Laufer, H., Nakase, Y., and Vanderberg, J. (1964). Developmental studies of the Dipteran salivary glands. I. The effects of actynomicin D on larval development, enzyme activity, and chromosomal differentiation in *Chironomus thummi*. *Dev. Biol.* **9,** 367–384.

Laufer, H., Rao, B., and Nakase, Y. (1965). Dipteran salivary gland metamorphosis, a self regulating system. *Am. Zool.* **5,**642.

Laufer, H., Rao, B., and Nakase, Y. (1968). Developmental studies of the Dipteran salivary glands. IV. Changes in DNA content. *J. Exp. Zool.* **166,** 71–76.

Laughon, A., Carroll, S. B., Storfer, F. A., Riley, P. D., and Scott, M.P. (1985). Common properties of proteins encoded by the *Antennapedia* complex genes of *Drosophila melanogaster*. *Cold Spring Harbor Symp. Quant. Biol.* **50,** 253–262.

Laughon, A., Boulet, A. M., Bermingham, J. R., Laymon, R. A., and Scott, M. P. (1986). Structure of transcript from homeotic *Antennapedia* gene of *Drosophila melanogaster:* Two promoters control the major protein-coding region. *Mol. Cell. Biol.* **6,** 4676–4689.

Laurie-Ahlberg, C. C., and Stam, L. F. (1987). Use of P-mediated transformation to identify the molecular basis of naturally occuring variants affecting *Adh* expression in *Drosophila melanogaster*. *Genetics* **115,** 129–140.

Lavorgna, G., Ueda, H., Clos, J., and Wu, C. (1991). FTZ-F1, a steroid hormone receptor-like protein implicated in the activation of *fushi tarazu*. *Science* **252,** 848–851.

Lavorgna, G., Karim, F. D., Thummel, C. S. and Wu, C. (1993). Potential role for a FTZ-F1 steroid receptor superfamily member in the control of *Drosophila* metamorphosis. *Proc. Natl. Acad. Sci. USA* **90,** 3004–3008.

Law, R. E., Sinibaldi, R. M., Cummings, M. R., and Ferro, A. J. (1976). Inhibition of RNA synthesis in salivary glands of *Drosophila melanogaster* by 5'-methylthioadenosine. *Biochem. Biophys. Res. Commun.* **73,** 600–606.

Law, R. E., Sinibaldi, R. M., Ferro, A. J., and Cummings, M. R. (1979). Effect of 5′-methylthioadenosine on gene action during heat shock in *Drosophila melanogaster*. *FEBS Lett.* **99,** 247–250.

Lawrence, P. (1992). The making of a fly. In "The Genetics of Animal Design." Blackwell Scientific Publications, Oxford/London.

Lea, D. E. (1947). "Actions of Radiation on Living Cells." MacMillan, New York.

Lee, C.-G., and Hurwitz, J. (1993). Human RNA helicase A is homologous to the maleless protein of *Drosophila*. *J. Biol. Chem.* **268,** 16822–16830.

Lee, C. S., and Pavan, C. (1974). Replicating DNA molecules from fertilized eggs of *Cochliomyia hominivorax* (Diptera). *Chromosoma* **47,** 429–437.

Lee, H.-S., Kraus, K. W., Wolfner, M. F., and Lis, J. T. (1992). DNA sequence requirements for generating paused polymerase at the start of *hsp70*. *Gene Dev.* **6,** 284–295.

Leenders, H. J. (1971). Temperature induced puffs in *Drosophila:* their possible physiological origin. *Dros. Inform. Serv.* **46,** 64.

Leenders, H. J. (1975). Gene activation as a consequence of deficiencies in the mitochondrial respiratory metabolism. *In* "Regulationsmechanismen der Genaktivitat und Replikation bei Riesenchromosomen" (L. Rensong, H.-H. Trepte, and G. Birukow, eds.), Vol. II, (11), pp. 55–56. Nachrichten Akad. Wissensch., Gottingen.

Leenders, H. J., and Beckers, P. J. A. (1972). The effect of changes in the respiratory metabolism upon genome activity: A correlation between induced gene activity and an increase in activity of a respiratory enzyme. *J. Cell. Biol.* **55,** 257–265.

Leenders, H. J., and Berendes, H. D. (1972). The effect of changes in the respiratory metabolism upon genome activity in *Drosophila*. *Chromosoma* **37,** 433–444.

Leenders, H. J., and Knoppien, W. G. (1973). Respiration of larval salivary glands of *Drosophila* in relation to the activity of specific genome loci. *J. Insect Physiol.* **19,** 1793–1800.

Leenders, H. J., Wullems, G. J., and Berendes, H. D. (1970). Competitive interaction of adenosine 3′, 5′-monophosphate on gene activation by ecdysterone. *Exp. Cell. Res.* **63,** 159–164.

Leenders, H. J., Derksen, J., Maas, P. M. J. M., and Berendes, H. D. (1973). Selective induction of a giant puff in *Drosophila hydei* by vitamin B6 and derivatives. *Chromosoma* **41,** 447–460.

Leenders, H. J., Kemp, A., Koninx, J. F. J. G., and Rosing, J. (1974). Changes in cellular ATP, ADP, and AMP levels following treatments affecting cellular respiration and the activity of certain nuclear genes in *Drosophila* salivary glands. *Exp. Cell. Res.* **86,** 25–30.

Lefevre, G. Jr. (1969). The eccentricity of *vermilion* deficiencies in *Drosophila melanogaster*. *Genetics* **63,** 589–600.

Lefevre, G., Jr. (1971). Salivary chromosome bands and the frequency of crossing over in *Drosophila melanogaster*. *Genetics* **67,** 497–513.

Lefevre, G., Jr. (1974a). Exceptions of the one band–one gene hypothesis. *Genetics* **77**(Suppl.), 39.

Lefevre, G., Jr. (1974b). The one band–one gene hypothesis: Evidence from a cytogenetic analysis of mutant and nonmutant rearrangement breakpoints in *Drosophila melanogaster*. *Cold Spring Harbor Symp. Quant. Biol.* **38,** 591–599.

Lefevre, G., Jr. (1974c). The relationship between genes and polytene chomosome bands. *Ann. Rev. Genet* **8,** 51–62.

Lefevre, G., Jr. (1976a). A photagraphic representation and interpretation of the polytene chromosomes of *Drosophila melanogaster* salivary glands. *In* "The Genetics and Biology of *Drosophila*" (M. Ashburner and E. Novitski, eds.), Vol. 1a, pp. 31–66. Academic Press, London.

Lefevre, G., Jr. (1976b). Reflections on the one gene–one band hypothesis. *In* "Chromosomes: From Simple to Complex; Proceedings of the 35th Annual Biology Colloqium," pp. 43–54. Oregon State Univ. Press.

Lefevre, G., Jr. (1981). The distribution of randomly recovered X-ray-induced sex-linked genetic effects in *Drosophila melanogaster*. *Genetics* **99,** 461–480.

Lefevre, G., Jr., and Green, M. M. (1972). Genetic duplication in the white-split interval of the X chromosome in *Drosophila melanogaster*. *Chromosoma* **36,** 391–412.

Lefevre, G., Jr., and Watkins, W. (1986). The question of the total gene number in *Drosophila melanogaster*. *Genetics* **113,** 869–895.

Lefevre, G., Jr., and Wiedenheft, K. B. (1974). Two genes in one band? *Dros. Inform. Serv.* **51,** 83.

Lehmann, M., and Koolman, J. (1988). Ecdysteroid receptors of the blowfly *Calliphora vicina:* Partial purification and characterization of ecdysteroid binding. *Mol. Cell. Endocrinol.* **57,** 239–249.

Lehmann, M., and Korge, G. (1995). Ecdysone regulation of the *Drosophila Sgs-4 gene* is mediated by the synergistic action of ecdysone receptor and SEBP3. *EMBO J.* **14,** 716–726.

Lehmann, M., and Korge, G. (1996). The *fork head* product directly specifies the tissue-specific hormone responsiveness of the *Drosophila Sgs-4* gene. *EMBO J.* **15,** 4825–4834.

Lehmann, M., Wattler, F., and Kroge, G. (1997). Two new regulatory elements controlling the *Drosophila Sgs-3* gene are potential ecdysone receptor and fork head binding sites. *Mechanisms of Development* **62,** 15–27.

Leibovitch, B. A. (1984). Differential replication in *Drosophila*. Itogi nauki i tekhniki. *Moleculyar biol.* **20,** 186–223. [in Russian]

Leibovitch, B. A., and Bogdanova, E. S. (1983). Interaction of histones H1 of mouse and *Drosophila* with chromosomes and DNA. *Molekulyar. biol.* **17**(1), 162–171. [in Russian]

Leibovitch, B. A., and Khesin, R. B. (1974). Template activity in male and female X chromosomes of *Drosophila* during RNA synthesis in bacterial RNA polymerase. *Molekulyar. biol.* **8**(3), 467–474. [in Russian]

Leibovitch, B. A., Belyaeva, E. S., Zhimulev, I. F., and Khesin, R. B. (1974). Comparison of RNA synthesis in bacterial RNA polymerase on *Drosophila* polytene chromosomes with transcription in living cells. *Ontogenez* **5**(6), 544–556. [in Russian]

Leibovitch, B. A., Belyaeva, E. S., Zhimulev, I. F., and Khesin, R. B. (1976). Comparison of *in vivo* and *in vitro* RNA synthesis on polytene chromosomes of *Drosophila*. *Chromosoma* **54,** 349–362.

Leicht, B. G., and Bonner, J. J. (1986). Genetic analysis of the 67 region of *Drosophila melanogaster*. *Genetics* **113**(1), 61.

Leicht, B. G., and Bonner, J. J. (1988). Genetic analysis of chromosomal region 67A-D of *Drosophila melanogaster*. *Genetics* **119,** 579–593.

Leigh Brown, A. J. (1981). Genetic variation at the molecular level in a natural population of *Drosophila melanogaster*. *Genetics* **97**(Suppl.), 63.

Leigh Brown, A. J., and Ish-Horowicz, D. (1980). Organization of *hsp70* coding sequence in related species of *Drosophila*. *Genetics* **94**(4), 58.

Leigh Brown, A. J., and Ish-Horowicz, D. (1981). Evolution of the 87A and 87C heat shock in *Drosophila*. *Nature* **290,** 677–682.

LeMaire, M. F., and Thummel, C. S. (1990). Splicing precedes polyadenilation during *Drosophila* E74A transcription. *Mol. Cell. Biol.* **10,** 6059–6063.

Lendahl, U., and Wieslander, L. (1984). Balbiani Ring 6 gene in *Chironomus tentans:* a diverged member of the Balbiani ring gene family. *Cell* **36,** 1027–1034.
Lendahl, U., and Wieslander, L. (1987). Balbiani Ring (BR) genes exhibit different patterns of expression during development. *Devel. Biol.* **121,** 130–138.
Lendahl, U., Saiga, H., Hoog, C., and Edstrom, J.-E. (1987). Rapid and concerted evolution of repeat units in a Balbiani Ring gene. *Genetics* **117,** 43–49.
Lengycl, J., and Penman, S. (1975). HhRNA size and processing as related to different DNA content in two Dipterans: *Drosophila* and *Aedes*. *Cell* **5,** 281–290.
Lengyel, J., Ransom, L. J., Graham, M. L., and Pardue, M. L. (1980). Transcription and metabolism of RNA from the *Drosophila melanogaster* heat shock puff site 93D. *Chromosoma* **80,** 237–252.
Lentzios, G., and Stocker, A. J. (1979). Nucleolar relationships in some Australian *Chironomus* species. *Chromosoma* **75,** 235–258.
Lepesant, J.-A., and Richards, G. (1989). Ecdysteroid-regulated genes. *In* "Ecdysone: From Chemistry to Mode of Action" (J. Koolman. ed.), pp. 355–367. Georg Thieme-Verlag, Stutgart/New York.
Lepesant, J.-A., Kejzlarova-Lepesant, J., and Garen, A. (1978). Ecdysone-inducible functions of larval fat bodies in *Drosophila*. *Proc. Natl. Acad. Sci. USA* **75**(11), 5570–5574.
Lepesant, J.-A., Maschat, F., Kejzlarova-Lepesant, J., Benes, H., and Yanicostas, C. (1986). Developmental and ecdysteroid regulation of gene expression in the larval fat body of *Drosophila melanogaster*. *Arch. Insect Biochem. Physiol.* **9**(Suppl.), 133–141.
Lesher, S. (1951). Studies on the larval salivary gland of *Drosophila*. I. The nucleic acids. *Exp. Cell. Res.* **2,** 577–585.
Lesher, S. (1952). Studies on the larval salivary gland of *Drosophila*. III. The histochemical localization and possible significance of ribonucleic acid, alkaline phosphatase and polysacharide. *Anat. Res.* **114,** 633–651.
Levin, A. V., Lozovskaya, E. R., and Evgen'ev, M. B. (1984). Effect of the *ts*-mutation on the expression of genes induced by heat shock in *Drosophila melanogaster*. II. Analysis of the effect of the *ts*-mutation. *Genetika* **20**(6), 949–953. [in Russian]
Levine, L., and van Vallen, L. (1962). A gene in *Drosophila* that produces a new chromosomal banding pattern. *Science* **137,** 993–994.
Levis, R., Bingham, P.M., and Rubin, G.M. (1982). Physical map of the *white* locus of *Drosophila melanogaster*. *Proc. Natl. Acad. Sci. USA* **79,** 564–568.
Levy, A., and Noll, M. (1982). Chromatin structure of *hsp70* genes of *Drosophila*. In "Heat Shock from Bacteria to Man" (M. L. Schlesinger, M. Ashburner, and A. Tissieres, eds.), pp. 99–107. Cold Spring Harbor Laboratory Press, Cold Spring Harbor, NY.
Levy, J. N., Gemmill, R. M., and Doane, W. W. (1985). Molecular cloning of a-amylase genes from *D. melanogaster*. II. Clone organization and verification. *Genetics* **110,** 313–324.
Levy, W. B., and McCarthy, B. J. (1975). Messenger RNA complexity in *Drosophila melanogaster*. *Biochemistry* **14,** 2440–2446.
Levy, W. B., Johnson, C. B., and McCarthy, B. J. (1976). Diversity of sequences in total and polyadenylated nuclear RNA from *Drosophila* cells. *Nucleic Acids Res.* **3,** 1777–1789.
Lewin, B. (1974). Lethal mutants and gene numbers. *Nature* **251,** 373–375.
Lewin, B. (1975). Units of transcription and translation: sequence components of heterogeneous nuclear RNA and messenger RNA. *Cell* **4,** 77–93.
Lewin, B. (1980). "Gene Expression," Vol. 2. Wiley, New York.
Lewin, B. (1994). "Genes," pp. 1–1272. Oxford Univ. Press, Oxford/New York/Tokyo.
Lewis, E. B. (1978). A gene complex controlling segmentation in *Drosophila*. *Nature* **276,** 565–570.
Lewis, E. B. (1985). Regulation of the genes of the Bithorax complex in *Drosophila*. *Cold Spring Harbor Symp. Quant. Biol.* **50,** 155–164.
Lewis, E. B., Knafels, J. D., Mathog, D. R., and Celniker, S. E. (1995). Sequence analysis of the cis-

regulatory regions of the *bithorax* complex of *Drosophila. Proc. Natl. Acad. Sci. USA* **92,** 8403–8407.

Lewis, M., Helmsing, P. J., and Ashburner, M. (1975). Parallel changes in puffing activity and patterns of protein synthesis in salivary glands of *Drosophila. Proc. Natl. Acad. Sci. USA* **72,** 3604–3608.

Lewis, R. A., Kaufman, T. C., and Denell, R. E. (1980a). Genetic analysis of the *Antennapedia* gene complex (ANT-C): Mutant screen of proximal 3R, section 84B-D. *Dros. Inform. Serv.* **55,** 85–87.

Lewis, R. A., Kaufman, T. C., Denell, R. E., and Tallerico, P. (1980b). Genetic analysis of the *Antennapedia* gene complex (ANT-C) and adjacent chromosomal regions of *Drosophila melanogaster*. I. Polytene chromosome segments 84B-D. *Genetics* **95,** 367–381.

Lewis, R. A., Wakimoto, B. T., Denell, R. E., and Kaufman, T. C. (1980c). Genetic analysis of the *Antennapedia* gene complex (ANT-C) and adjacent chromosomal regions of *Drosophila melanogaster*. II. Polytene chromosome segments 84A-84B1-2. *Genetics* **95,** 383–397.

Lezzi, M. (1966). Induktion eines Ecdyson-aktivierbaren Puff in isolierten Zellkernen von *Chironomus* durch KCl. *Exp. Cell. Res.* **43,** 571–577.

Lezzi, M. (1967a). Cytochemische Untersuchungen an Puff isolierter Speicheldrusen-Chromosomen von *Chironomus*. *Chromosoma* **21,** 89–108.

Lezzi, M. (1967b). Spezifische Aktivitatssteigerung eines Balbianringes dirch Mg^{2+} in isolierten Zellkernen von *Chironomus*. *Chromosoma* **21,** 109–122.

Lezzi, M. (1974). *In vitro* effects of juvenile hormone on puffing in *Chironomus* salivary glands. *Mol. Cell. Endocrinol.* **1,** 179–207.

Lezzi, M. (1975). Einfluss des Juvenilhormones auf die Genaktivierung von Polytanchromosomen. Gottingen, *Nachrichten Akad. Wissensch.*, II. Mat.-Phys. Klasse, N 11, 26–30.

Lezzi, M. (1980). Two steps in the activation of an ecdysone-sensitive chromosome region. *Experientia* **38,** 743–744.

Lezzi, M. (1984). Heat-shock phenomena in *Chironomus tentans*. II. In vitro effects of heat and over heat on puffing and their reversal. *Chromosoma* **90,** 198–203.

Lezzi, M. (1996). Chromosome puffing: supramolecular aspects of ecdysone action. *In* "Metamorphosis. Postembryonic Reprogramming of Gene Expression in Amphibian and Insect Cells" (L. I. Gilbert, J. R. Tata, and B. G. Atkinson, eds.), pp. 145–173. Academic Press, San Diego/New York/Boston/London/Sydney/Tokyo/Toronto.

Lezzi, M., and Frigg, M. (1971). Specific effects of juvenile hormone on chromosome function. *Mitt. Schweiz. Entomol. Ges.* **44,** 163–170.

Lezzi, M., and Gilbert, L. I. (1969). Control of gene activities in the polytene chromosomes of *Chironomus tentans* by ecdysone and juvenile hormone. *Proc. Natl. Acad. Sci. USA* **64,** 498–503.

Lezzi, M., and Gilbert, L. I. (1970). Differential effects of K^+ and Na^+ on specific bands of isolated polytene chromosomes of *Chironomus tentans*. *J. Cell. Sci.* **6,** 615–627.

Lezzi, M., and Gilbert, L. I. (1972). Hormonal control of gene activity in polytene chromosomes. *Gen. Comp. Endocrinol.* **3**(Suppl.) 159–167.

Lezzi, M., and Kroeger, H. (1966). Aufnahme von ^{22}Na in die Zellkerme der Speiheldrusen von *Chironomus thummi*. *Z. Naturforsch.* **21b,** 274–277.

Lezzi, M., and Richards, G. (1989). Salivary glands. *In* "Ecdysone: From Chemistry to Mode of Action" (J. Koolman, ed.), pp. 393–406. Georg Thieme-Verlag, Stuttgart/New York.

Lezzi, M., and Robert, M. (1972). Chromosomes isolated from unfixed salivary glands of *Chironomus*. *In* "Results and Problems in Cell Differentiation: Developmental Studies on Giant Chromosomes" (W. Beermann, ed.), Vol. 4, pp. 35–57. Springer-Verlag, Berlin/Heidelberg/New York.

Lezzi, M., and Wyss, C. (1976). The antagonism between juvenile hormone and ecdysone. *In* "The Juvenile Hormones" (L. I. Gilbert, ed.), pp. 252–269. Plenum, New York.

Lezzi, M., Meyer, B., and Mahr, R. (1981). Heat shock phenomena in *Chironomus tentans*. I. *In vivo* effects of heat, overheat and quenching of salivary chromosome puffing. *Chromosoma* **83,** 327–339.

Lezzi, M., Gatzka, F., and Meyer, B. (1984). Heat-shock phenomena in *Chironomus tentans*. III. Quantitative autoradiographic studies on ^{3}H-uridine incorporation into *Balbiani Ring* 2 and heat-shock puff IV-5C. *Chromosoma* **90,** 204–210.

Lezzi, M., Gatzka, F., and Robert-Nicoud, M. (1989a). Developmental changes in the responsiveness to ecdysterone of chromosome region I-18C *Chironomus thummi*. *Chromosoma* **98,** 23–32.

Lezzi, M., Lutz, B., and Dorsch-Haesler, K. (1989b). The ecdysterone-controlled chromosome region I-18C in *Chironomus tentans:* From cytology to molecular biology. *Acta Biol. Debr. Oecol. Hung.* **2,** 129–139.

Lezzi, M., Gatzka, F., Ineichen, H., and Gruzdev, A. D. (1991). Transcriptional activation of puff site I-18C of *Chironomus tentans:* Hormonal responsiveness changes in parallel with diurnal decondensation cycle. *Chromosoma* **100,** 235–241.

Lezzi, M., Gatzka, F., and Lutz, B. (1992). Dual effect of heat shock on the transcription of the early ecdysterone induced gene locus I-18C of *Chironomus tentans:* Suppression and [re-]induction. *Abstr. Xth ecdysone workshop*. Liverpool, 105.

Lifschytz, E. (1967). Induced X-chromosome lethals covered by Y.w$^+$. *Dros. Inform. Serv.* **42,** 89.

Lifschytz, E. (1971). Fine-structure analysis of the chromosome recombinational patterns at the base of the X-chromosome of *Drosophila melanogaster*. *Mutat. Res.* **13,** 35–47.

Lifschytz, E. (1973). Genetical aspects of gene-band relationship in *Drosophila melanogaster:* Critical review. *Chromos. Today* **4,** 143–148.

Lifschytz, E. (1978). Fine structure analysis and genetic organization at the base of the X-chromosome in *Drosophila melanogaster*. *Genetics* **88,** 457–467.

Lifschytz, E., and Falk, R. (1968a). Fine structure analysis of a chromosome segment in *Drosophila melanogaster*. Analysis of X- ray-induced lethals. *Mutat. Res.* **6,** 235–244.

Lifschytz, E., and Falk, R. (1968b). A system for fine structure analysis of chromosome segments. *Dros. Inform. Serv.* **43,** 168–169.

Lifschytz, E., and Falk, R. (1969). Fine structure analysis of a chromosome segment in *Drosophila melanogaster*. Analysis of ethyl-methanesulphonate-induced lethals. *Mutat. Res.* **8,** 147–155.

Lifschytz, E., and Yakobovitz, N. (1978). The role of X-linked lethal and viable male-sterile mutations in male gametogenesis of *Drosophila melanogaster:* Genetic analysis. *Mol. Gen. Genet.* **161,** 275–284.

Lifton, R. P., Goldberg, M. L., Karp, R. W., and Hogness, D. S. (1978). The organization of the histone genes in *Drosophila melnogaster:* Functional and evolutionary implications. *Cold Spring Harbor Symp. Quant. Biol.* **42,** 1047–1051.

Lim, J. K., and Snyder, L. A. (1974). Cytogenetic and complementation analysis of recessive lethal mutations induced in the X- chromosome of *Drosophila* by three alkylating agents. *Genet. Res. Cambr.* **24,** 1–10.

Lima-de-Faria, A. (1954). Chromosome gradient and chromosome field in *Agapanthus*. *Chromosoma* **6,** 330–370.

Lima-de-Faria, A. (1975). The relation between chromomeres, replicons, operons, transciption units, genes, viruses and palindromes. *Hereditas* **81,** 249–284.

Linares, A. R., Bowen, T., and Dover, G. A. (1994). Aspects of nonrandom turnover involved in the concerted evolution of intergenic spacers within the ribosomal DNA of *Drosophila melanogaster*. *J. Mol. Evol.* **39,** 151–159.

Lindquist, S. (1980). Varying patterns of protein synthesis in *Drosophila* during heat shock: Implications for regulation. *Devel. Biol* **77,** 463–479.

Lindquist, S. (1986). The heat shock response. *Ann. Rev. Biochem.* **55,** 1151–1191.

Lindquist, S., and Craig, E. A. (1988). The heat shock proteins. *Ann. Rev. Genet.* **22,** 631–677.

Lindsley, D. L., and Grell, E. H. (1968). Genetic variations of *Drosophila melanogaster*. *Carnegie Inst. Wash. Publ.* **627.**

Lindsley, D. L., and Zimm, G. G. (1992). "The Genome of *Drosophila Melanogaster*." Academic Press, San Diego/New York/Boston/London/Sydney/Tokyo/Toronto.

Lis, J. T., Prestidge, L., and Hogness, D. S. (1978). A novel arrangement of tandemly repeated genes at a major heat site in *Drosophila melanogaster*. *Cell* **14,** 901–919.

Lis, J. T., Ish-Horowicz, D., and Pinchin, S. M. (1981a). Genomic organization and transciption of the "alfa" and "beta" heat shock DNA in *Drosophila melanogaster*. *Nucleic Acids Res.* **9,** 5297–5310.

Lis, J. T., Neckameyer, W., Dubensky, R., and Costlow, N. (1981b). Cloning and characterization of nine heat-shock-induced mRNAs of *Drosophila melanogaster*. *Gene* **15,** 67–80.

Lis, J. T., Neckameyer, M. E., Mirault, M.-E., Artavanis-Tsakonas, S., Lall, P., Martin, G., and Schedl, P. (1981c). DNA sequences flanking the starts of the *hsp70* and a b heat-shock genes are homologous. *Dev. Biol.* **83,** 291–300.

Lis, J. T., Simon, J. A., and Sutton, C. A. (1983). New heat shock puff and b-galactosidase activity resulting from transformation of *Drosophila* with an *hsp70-lacZ* hybrid gene. *Cell* **35,** 403–410.

Lis, J. T., Xiao, H., and Perisic, O. (1990). Modular units of heat shock regulatory regions: Structure and function. *In* "Stress proteins in biology and medicine," pp. 411–428. Cold Spring Harbor Laboratory Press, Cold Spring Harbor, NY.

Liu, C. P., and Lim, J. K. (1975). Complementation analysis of methyl methanesulfonate-induced recessive lethal mutations in the *zeste-white* region of the X-chromosome of *Drosophila melanogaster*. *Genetics* **79,** 601–611.

Livak, K. J., Freund, R., Schweber, M., Wensink, P. C., and Meselson, M. (1978). Sequence organization and transcription at two heat shock loci in *Drosophila*. *Proc. Natl. Acad. Sci. USA* **75,** 5613–5617.

Lockshin, R. A. (1985). Programmed cell death. *In* "Comprehensive Insect Physiology, Biochemistry and Pharmacology" (G. A. Kerkut and L. I. Gilbert, eds.), Vol. 2, pp. 301–317. Pergamon Press, Oxford/New York/Toronto/Sydney/Paris/Frankfurt.

Loewenstein, W. R., and Kanno, Y. (1962). Some electrical properties of the membrane of a cell nucleus. *Nature* **195,** 462–464.

Loewenstein, W. R., and Kanno, Y. (1964). Studies on an epithelial (gland) cell junction. I. Modifications of surface membrane permeability. *J. Cell. Biol.* **22,** 565–586.

Long, E. O., and Dawid, I. (1979). Restriction analysis of spacers in ribosomal DNA insertions in *Drosophila melanogaster*. *Nucleic Acids Res.* **7,** 205–215.

Long, E. O., and Dawid, I. B. (1980). Repeated genes on eukaryotes. *Ann. Rev. Biochem.* **49,** 727–764.

Long, E. O., Rebbert, M. L., and Dawid, I. B. (1981a). Structure and expression of ribosomal RNA genes of *Drosophila melanogaster* interrupted by type-2 insertions. *Cold Spring Harbor. Symp. Quant. Biol.* **45,** 667–672.

Long, E. O., Rebbert, M. L., and Dawid, I. B. (1981b). Nucleotide sequence of the initiation site for ribosomal RNA transcription in *Drosophila melanogaster:* Comparison of genes with and without insertions. *Proc. Natl. Acad. Sci. USA* **78,** 1513–1517.

Lonn, U. (1977). Direct association of Balbiani ring 75S RNA with membranes of the endoplasmic reticulum. *Nature* **270,** 630–631.

Lonn, U. (1978). Delayed flow-through cytoplasm of newly synthesized Balbiani ring 75S RNA. *Cell* **13,** 727–733.

Lonn, U. (1980a). Isolation and partial characterization of replication intermediates in *Chironomus* polytene chromosomes. *Chromosoma* **77,** 29–40.

Lonn, U. (1980b). Similar size of the replicons in *Chironomus* polytene chromosomes at two developmental stages. *J. Cell. Sci.* **46,** 387–397.

Lonn, U. (1981a). Double-stranded DNA with the size expected of replicons can be released from *Chironomus* polytene chromosomes. *Chromosoma* **81,** 641–653.

Lonn, U. (1981b). The double-stranded DNA replication intermediates formed in *Chironomus* polytene chromosomes consists of two populations. *Chromosoma* **84,** 221–229.

Lonnroth, A., Wurtz, T., Skoglund, U., and Daneholt, B. (1987). Structural and biochemical characterization of specific premessenger RNP particle. *Mol. Biol. Reports* **12,** 178–179.

Lonnroth, A., Alexciev, K., Mehlin, H., Wurtz, T., Skoglund, U., and Daneholt, B. (1992). Demonstration of a 7-nm RNP fiber as the basic structural element in a premessenger RNP particle. *Exp. Cell. Res.* **199,** 292–296.

Loreto, E. L. S., Valente, V. L. S., and Oliveira, A. K. (1988). Differential gene activation in *Drosophila melanogaster* populations selected for developmental rate. *Rev. Brasil. Genet.* **11,** 519–534.

Lotto, R. de, Louis, C., Wormington, M., and Schedl, P. (1982). Intimate association if 5SRNA and tRNA genes in *Drosophila melanogaster*. *Mol. Gen. Genet.* **188,** 299–304.

Lozovskaya, E. R., and Evgen'ev, M. B. (1984). Heat shock in Drosophila and regulation of the activity of the genome. In "Itogi nauki i tekhniki Molekulyarnaya biologiya," Vol. 20, pp. 142–185. VINITI, Moscow. [in Russian]

Lozovskaya, E. R., Levin, A. V., and Evgen'ev, M. B. (1982). Heat shock in *Drosophila* and regulation of the activity of the genome. *Genetika* **11,** 1749–1762. [in Russian].

Lu, Q., Wallrath, L. L., Allan, B. D., Glaser, R. L., Lis, J. T., and Elgin, S. C. R. (1992). A promoter sequence containing (CT)nx(GA)n repeats is critical for the formation of the DNase I hypersensitive sites in the *Drosophila hsp26* gene. *J. Mol. Biol.* **225,** 985–998.

Lu, Q., Wallrath, L. L., Granok, H., and Elgin, S. C. R. (1993). (CT)n.(GA)n repeats and heat shock elements have distinct roles in chromatin structure and transcriptional activation of the *Drosophila hsp26* gene. *Mol. Cell. Biol.* **13,** 2802–2814.

Lubsen, N. H., and Sondermeijer, P. J. A. (1978). The product of the "heat-shock" loci of *Drosophila hydei:* Correlation berween locus 2–36A and the 70000 MW "heat-shock" peptide. *Chromosoma* **66,** 115–125.

Lubsen, N.H., Sondermeijer, P.J.A., Pages, M., and Alonso, C. (1978). *In situ* hybridization of nuclear and cytoplasmic RNA to locus 2–48BC in *Drosophila hydei*. *Chromosoma* **65,** 199–212.

Lucchesi, J. C. (1973). Dosage compensation in *Drosophila*. *Ann. Rev. Genet.* **7,** 225–237.

Lucchesi, J. C. (1977). Dosage compensation: transcription level regulation of X-linked genes in *Drosophila*. *Am. Zool.* **17,** 685–693.

Lucchesi, J. C. (1996). Dosage compensation in *Drosophila* and the "complex" world of transcriptional regulation. *BioEssays* **18**(N7), 541–547.

Lucchesi, J. C., and Manning, J. E. (1987). Gene dosage compensation in *Drosophila melanogaster*. *Adv. Genet.* **24,** 371–429.

Lucchesi, J. C., and Rawls, J. M., Jr. (1973a). Regulation of gene function: A comparison of X-linked enzyme activity levels in normal and intersexual triploids of *Drosophila melanogaster*. *Genetics* **73,** 459–464.

Lucchesi, J. C., and Rawls, J. M., Jr. (1973b). Regulation of gene function: A comparison of enzyme activity levels in relation to gene dosage in diploids and triploids of *Drosophila melanogaster*. *Biochem. Genet.* **9,** 41–51.

Lucchesi, J. C., and Skripsky, T. (1981). The link between dosage compensation and sex differentiation in *Drosophila melanogaster*. *Chromosoma* **82,** 217–227.

Lucchesi, J. C., Belote, J. M., and Maroni, G. (1977). X-linked gene activity in metamales (XY; 3A) of *Drosophila*. *Chromosoma* **65,** 1–7.

Lucchesi, J. C., Skripsky, T., and Tax, F. E. (1982). A new male-specific lethal mutation in *Drosophila melanogaster*. *Genetics* **100,** 42.

Luhrmann, R., Kastner, B., and Bach, M. (1990). Structure of spliceosomal *sn* RNPs and their role in pre-mRNA splicing. *Biochim. Biophys. Acta* **1087,** 265–292.

Luo, Y., Amin, J., and Voellmy, R. (1991). Ecdysterone receptor is a sequence-specific transcription factor involved in the developmental regulation of heat shock genes. *Mol. Cell. Biol.* **11,** 3660–3675.

Lychev, V. A. (1965). Study on the activity of chromosomes under deep inbreeding in *Drosophila*. *Tsitologia* **7**(3), 325–333. [in Russian]

Lychev, V. A. (1968). A study on the size and distribution of puffs of *Drosophila melanogaster* during normal development and under the effect of inbreeding and radiation. Ph.D. Thesis, Obninsk. [in Russian]

Lychev, V. A., and Medvedev, Zh. A. (1967). Some problems of methods of study of puffs of the polytene chromosomes of the salivary glands of *Diptera* in studies of the effect of different factors on chromos. *Genetika* **3** (8), 53–59. [in Russian]

Lyckergaard, E. M. S., and Clark, A. G. (1989). Ribosomal DNA and Stellate gene copy number variation on the Y chromosome of *Drosophila melanogaster*. *Proc. Natl. Acad. Sci. USA* **86,** 1944–1948.

Lyman, L. M., Copps, K., Rastelli, L., Kelley, R. L., and Kuroda, M. I. (1997). *Drosophila* male-specific lethal-2 protein: Structure/function analysis and dependence on MSL-1 for chromosome association. *Genetics* **147,** 1743–1753.

MacGregor, H. C., and Mackie, J. B. (1967). Fine structure of the cytoplasm in salivary glands of Simulium. *J. Cell Sci.* **2,** 137–144.

Machado-Santelli, G. M., and Basile, R. (1973). Studies of DNA puff development in *Rhynchosciara*. *Genetics* **74**(Suppl.), S168.

Machado-Santelli, G. M., and Basile, R. (1975). DNA replication and DNA puffs in salivary chromosomes of *Rhynchosciara*. *Cien. Cult.* **27,** 167–174.

Machado-Santelli, G. M., and Basile, R. (1978). Action of hydroxyurea on chromosome physiology in *Rhynchosciara angelae*. *Genetica* **48,** 131–135.

Macheva, I., and Semionov, E. (1991). Localization of ribosomal DNA in diploid and polytene nuclei of *Drosophila simulans* and *Drosophila mauritiana*. *Dros. Inform Serv.* **70,** 141–143.

Mackensen, O. (1934). A cytological study of short deficiencies in the X chromosome of *Drosophila melanogaster*. *Am. Natur.* **67,** 76.

Mackensen, O. (1935). Locating genes on the salivary chromosomes. Cytogenetic methods demonstrated in determining position of genes on the X-chromosome of *Drosophila melanogaster*. *J. Hered.* **26,** 163–174.

Mahmood, F., and Sakai, R. K. (1985). An ovarian chromosome map of *Anopheles stephensi*. *Cytobios* **43,** 79–86.

Mahr, R., Meyer, B., Daneholt, B., and Eppenberger, H. M. (1980). Activation of Balbiani Ring genes in *Chironomus tentans* after a pilocarpine-induced depletion of the secretory products from the salivary gland lumen. *Dev. Biol.* **80,** 409–418.

Mainx, F. (1976). Some special forms of polytene chromosomes in two species of Diptera (Trypetidae and Rhagionidae). *Caryologia* **29,** 81–97.

Majumdar, D., and Mukherjee, A. S. (1979). Autonomous determination and regulation of termination of replication in polytene chromosomes of *Drosophila*. *Genetics* **91**(4(2)), 73S.

Majumdar, D., and Mukherjee, A. S. (1980). Analysis of early stages of replication in *Drosophila*. *Genetics* **94**(4(2)), 70S.

Majumdar, D., Chatterjee, R., and Mukherjee, A. S. (1975). Effect of a-amanitin and cordycepin on the RNA synthesis in the polytene chromosomes of *Drosophila melanogaster*. *J. Cytol. Genet. Congr.* (Suppl.), 128–132.

Makino, S. (1938). A morphological study of the nucleus in various kinds of somatic cells of *Drosophila virilis*. *Cytologia* **9,** 272–282.

Maksimova, F.L. (1974). On some features of the morphology of puffs of the polytene chromosomes of *Chironomus plumosus* larvae. In "Funktsional'naya morfologia, genetika i biokhimia kletki," pp. 62–64. Institut tsitologii AN SSSR, Leningrad. [in Russian]

Maksimova, F. L. (1983). Morphometric analysis of puffing in the polytene chromosomes of *Chironomus plumosus* L. larvae. *Tsitologiya* **25**(9), 1049–1052. [in Russian]

Maksimova, F. L., and Ilyinskaya, N. B. (1983). A characteristics of puffing in the polytene chromosomes of *Chironomus plumosus* larvae during the summer period of development. *Tsitologiya* **25**(3), 320–327. [in Russian]

Mala, J. (1970). K otazce zdurelin na polytennich chromosomech a genova aktivita v postembryonalnim vyvoji hmyzu. Cs. *Fysiol.* **19,** 317–322.

Mal'ceva, N. I., and Zhimulev, I. F. (1993). Extent of polyteny in the pericentric heterochromatin of polytene chromosomes of pseudonurse cells of otu (ovarian tumor) mutants of *Drosophila melanogaster*. *Mol. Gen. Genet.* **240,** 273–276.

Mal'ceva, N. I., Gyurkovics, H., and Zhimulev, I. F. (1995). General characteristics of the polytene chromosomes from ovarian pseudonurse cells of the *Drosophila melanogaster* otu^{11} and *fs(2)B* mutants. *Chromos. Res.* **3,** 191–200.

Malevanchuk, O. A., Peunova, N. I., Sergeev, P. V., and Yenikolopov, G. N. (1991). The Est^{S} locus of *D. virilis* contains two related esterase-like gene. *Dros. Inform. Serv.* **70,** 138, 140.

Mamayev, B. M. (1968). "Evolution of Gall Forming Insects, Gall Midges." Nauka, Leningrad. [in Russian]

Manglesdorf, D. J., Thummel, C., Beato, M., Herrlich, P., Schultz, G., Umesono, K., Blumberg, B., Kastner, P., Mark, M., Chambon, P., and Evans, R. M. (1995). The nuclear receptor superfamily: the second decade. *Cell* **83,** 835–839.

Manousis, T. H. (1985). Larval saliva in several *Drosophila* species. *Dros. Inform. Serv.* **61,** 113–114.

Manousis, T. H., and Kastritsis, C. D. (1987). Possible correlation of polypeptides and Balbiani rings in the salivary glands of *Drosophila auraria* Peng. *Genetica* **74,** 31–40.

Manseau, L. J., Ganetsky, B., and Craig, E. A. (1988). Molecular and genetic characterization of the *Drosophila melanogaster* 87E actin gene region. *Genetics* **119,** 407–420.

Marchi, A., and Rai, K. (1978). Cell cycle and DNA synthesis in the mosquito *Aedes aegypti*. *Can. J. Genet. Cytol.* **20,** 243–247.

Maroni, G., and Lucchesi, J. C. (1980). X-chromosome transcription in *Drosophila*. *Chromosoma* **77,** 253–261.

Maroni, G., and Plaut, W. (1973a). Dosage compensation in *Drosophila melanogaster* triploids. I. Autoradiographic study. *Chromosoma* **40,** 361–377.

Maroni, G., and Plaut, W. (1973b). Dosage compensation in *Drosophila melanogaster* triploids. II. Glucose-6-phosphate dehydrogenase activity. *Genetics* **74,** 331–342.

Maroni, G., Kaplan, R., and Plaut, W. (1974). RNA synthesis in *Drosophila melanogaster* polytene chromosomes. Indication of simultaneous dosage compensation and dosage effect in X chromosome. *Chromosoma* **47,** 203–212.

Maroy, P., Dennis, R., Beckers, C., Sage, B. A., and O'Connor, J. D. (1978). Demonstration of an ecdysteroid receptor in a cultured cell line of *Drosophila melanogaster*. *Proc. Natl. Acad. Sci. USA* **75,** 6035–6038.

Maroy, P., Koczka, K., Fekete, E., and Vargha, J. (1980). Molting hormone titer of *Drosophila melanogaster* larvae. *Dros. Inform. Serv.* **55,** 98–99.

Marsh, J. L., and Wright, T. R. F. (1979). The genetics and cytology of dopa decarboxylase region of *Drosophila*. *Genetics* **91**(Suppl.), 74.

Marsh, J. L., Erfle, M. P., and Leeds, C. A. (1986). Molecular localization, developmental expression and nucleotide sequence of the alpha-methyldopa hypersensitive gene of *Drosophila*. *Genetics* **114,** 453–467.

Marshak, A. (1936). The stucture of the chromosomes of the salivary gland of *Drosophila melanogaster*. *Am. Natur.* **70,** 181–184.

Martin, C. H., and Meyerowitz, E. M. (1986). Characterization of the boundaries between adjacent rapidly and slowly evolving genomic regions in *Drosophila*. *Proc. Natl. Acad. Sci. USA* **83,** 8654–8658.

Martin, C. H., and Meyerowitz, E. M. (1988). Mosaic evolution in the *Drosophila* genome. *BioEssays* **9**(2/3), 65–69.

Martin, C. H., Monsma, S. A., Romac, J. M.-J., and Leser, G. P. (1987). The intranuclear states of snRNP complexes. *Mol. Biol. Reports* **12,** 180–181.

Martin, C. H., Mayeda, C. A., and Meyerowitz, E. M. (1988). Evolution and expression of the *Sgs-3* glue gene of *Drosophila*. *J. Mol. Biol.* **201,** 273–287.

Martin, C. H., Giangrande, A., Ruiz, C., and Richards, G. (1989a). Induction and repression of the *Drosophila Sgs-3* glue gene are mediated by distinct sequences in the proximal promoter. *EMBO J.* **8,** 561–568.

Martin, C. H., Mettling, C., Giangrande, A., Ruiz, C., and Richards, G. (1989b). Regulatory elements and interactions in the *Drosophila* 68C glue gene cluster. *Dev. Genet.* **10,** 189–197.

Martin, C. H., Mayeda, C. A., Davis, C. A., Ericsson, C. L., Knafels, J. D., Mathog, D. R., Celniker, S. E., Lewis, E. B., and Palazzolo, M. J. (1995). Complete sequence of the bithorax complex of *Drosophila*. *Proc. Natl. Acad. Sci. USA* **92,** 8398–8402.

Martin, J. (1974). A review of the genus *Chironomus* (Diptera: Chironomidae). IX. The cytology of *Chironomus tepperi* Skuse *Chromosoma* **45,** 91–98.

Masich, S. V. (1992). Structural organization of the transcriptional and translational units of the 75S RNA genes of chironomus. Autoreferat Ph.D. Thesis, Novosibirsk. [in Russian]

Masich, S. V., Skoglund, U., Eufferstedt, L.-G., and Daneholt, B. (1997). Three-dimensional organization of RNP, containing mRNA from the Balbiani rings in *Chironomus tentans*. *Tsitologia* **39**(N1), 84. [in Russian]

Mason, P. J., Torok, I., Kiss, I., Karch, F., and Udvardy, A. (1982). Evolutionary implicetions of a complex pattern of DNA sequence homology extending far upstream of the *hsp70* genes at loci 87A7 and 87C1 in *Drosophila melanogaster*. *J. Mol. Biol.* **156,** 21–35.

Mason, P. J., Hall, L. M. C., and Gausz, J. (1984). The expression of heat shock genes during normal development in *Drosophila melanogaster*. *Mol. Gen. Genet.* **194,** 73–78.

Mattingly, E., and Parker, C. (1968a). Sequence of puff formation in *Rhynchosciara* polytene chromosomes. *Chromosoma* **23,** 255–270.

Mattingly, E., and Parker, C. (1968b). Nucleic acid synthesis during larval development of *Rhynchosciara*. *J. Insect Physiol.* **14,** 1077–1083.

Matunis, E. M., Matunis, M. J., and Dreyfuss, G. (1992). Characterization of the major hn RNP proteins from *Drosophila melanogaster*. *J. Cell Biol.* **116,** 257–269.

Matunis, M. J., Matunis, E. M., and Dreufuss, G. (1992). Isolation of hn RNP complexes from *Drosophila melanogaster*. *J. Cell Biol.* **116,** 245–255.

Mavragani-Tsipidou, P., and Kastritsis, C. D. (1986). The role of the hormone ecdysterone in the control of the activity of the Balbiani rings and the other puffs of *Drosophila auraria*. *Chromosoma* **94,** 505–513.

Mavragani-Tsipidou, P., and Scouras, Z. G. (1991). Developmental changes in fat body and midgut chromosomes of *Drosophila auraria*. *Chromosoma* **100,** 443–452.

Mavragani-Tsipidou, P., and Scouras, Z. G. (1992). Puffing activity of the salivary gland chromosomes of *Drosophila bicornuta*. *Cytobios* **69,** 171–177.

Mavragani-Tsipidou, P., Kyrpides, N., and Scouras, Z. G. (1990a). Evolutionary implications of duplications and Balbiani Rings in *Drosophila:* A study of *Drosophila serrata*. *Genome* **33,** 478–485.

Mavragani-Tsipidou, P., Scouras, Z. G., and Kastritsis, C. D. (1990b). Comparison of the polytene chromosomes of the salivary gland, the fat body and the midgut nuclei of *Drosophila auraria*. I. Banding patterns. *Genetica* **81,** 99–108.

Mavragani-Tsipidou, P., Scouras, Z. G., Haralampides, K., Lavrentiadou, S., and Kastritsis, C. D. (1992a). The polytene chromosomes of *Drosophila triauraria* and *D. quadraria*, sibling species of *D. auraria*. *Genome* **35,** 318–326.

Mavragani-Tsipidou, P., Scouras, Z. G., and Natsiou-Voziki, A. (1992b). The Balbiani ring and the polytene chromosomes of *Drosophila bicornuta*. *Genome* **35,** 64–67.

Mavragani-Tsipidou, P., Zambetaki, A., Kleanthous, K., Pangou, E., and Scouras, Z. G. (1994).

Cytotaxonomic differentiation of the Afrotropical *Drosophila* montium subgroup: *D. diplacantha* and *D. seguyi*: The major role of reverse tandem duplications. *Genome* **37,** 935–944.

Mazina, O. M., and Dubrovsky, E. B. (1990). On the relation of female fertility function of the *ecs* locus and the gene of the minor chorion protein S70 of *Drosophila melanogaster*. *Genetika* **26**(12), 2156–2165. [in Russian]

Mazina, O. M., and Korochkina, S. E. (1991). A study on the nature of female sterility in mutants for the *ecs* locus controlling sensitivity to ecdysterone in *Drosophila melanogaster*. *Genetika* **27**(11), 1920–1927. [in Russian]

Mazina, O. M., Belyaeva, E. S., Grafodatskaya, V. E., and Zhimulev, I. F. (1990). Effect of the *ecs* locus of *Drosophila melanogaster* on female fertility. *Genetika* **26**(6), 1038–1045. [in Russian]

Mazina, O. M., Belyaeva, E. S., and Zhimulev, I. F. (1991) Cytogenetical analysis of the 2B3-4–2B11 region of the X chromosome of *Drosophila melanogaster*. VII. Influence of the *ecs* locus on female fertility. *Mol. Gen. Genet.* **225,** 99–105.

McCarthy, B. J., Compton, J. L., Craig, E. A., and Wadsworth, S. C. (1978). Transcription at the heat shock loci of *Drosophila*. Tenth Miami Winter Symposium, pp. 317–333. Academic Press, New York.

McClintock, B. (1934). The relation of a particular chromosomal element to the development of the nucleoli in *Zea mays*. *Zeit. Zellforsch.* **21,** 294–328.

McGarry, T. J., and Lindquist, S. (1985). The preferential translation of *Drosophila hsp70* mRNA requires sequences in the untranslated leader. *Cell* **42,** 903–911.

McGinnis, W., Farrell, J., Jr., and Beckendorf, S.K. (1980). Molecular limits on the size of a genetic locus in *Drosophila melanogaster*. *Proc. Natl. Acad. Sci. USA* **77,** 7367–7371.

McGinnis, W., Shermoen, A. W., Heemskerk, J., and Beckendorf, S. K. (1983). DNA sequence changes in an upstream DNase I hypersensitive region are correlated with reduced expression. *Proc. Natl. Acad. Sci. USA* **80,** 1063–1067.

McKee, B. D., Habera, L., and Vrana, J. A. (1992). Evidence that intergenic spacer repeats of *Drosophila melanogaster* rRNA genes function as X-Y pairing sites in male meiosis, and a general model for achiasmatic pairing. *Genetics* **132,** 529–544.

McKenzie, S. L. (1977). Translation control of protein synthesis in *Drosophila*. *J. Cell Biol.* **75**(2), 336a.

McKenzie, S. L., and Meselson, M. (1977). Translation in vitro of *Drosophila* heat shock messages. *J. Mol. Biol.* **117,** 279–283.

McKenzie, S. L., Henikoff, S., and Meselson, M. (1975). Localization of RNA from heat induced polysomes at puff sites in *Drosophila melanogaster*. *Proc. Natl. Acad. Sci. USA* **72,** 1117–1121.

McKeown, M., Belote, J. M., and Baker, B. S. (1987). A molecular analysis of transformer, a gene in *Drosophila melanogaster* that controls female sexual differentiation. *Cell* **48,** 489–499.

McKim, K. S., Dahmus, J. B., and Hawley, R. S. (1996). Cloning of the *Drosophila melanogaster* meiotic recombination gene *mei-218*, a genetic and molecular analysis of interval 15E. *Genetics* **144,** 215–228.

McKnight, S., and Miller, O. L., Jr. (1977). Electron microscopic analysis of chromatin replication in the cellular blastoderm *Drosophila melanogaster* embryo. *Cell* **12,** 795–804.

McLean, J. R., Boswell, R., and O'Donnel, J. (1990). Cloning and molecular characterization of a metabolic gene with developmental functions in *Drosophila*. I. Analysis of the head function of *Paunch*. *Genetics* **126,** 1007–1019.

McMaster-Kaye, R. (1962). Synthetic processes in the cell nucleus. III. The metabolism of nuclear ribonucleic acid in salivary glands of *Drosophila repleta*. *J. Histochem. Cytochem.* **10,** 154–161.

McMaster-Kaye, R., and Taulor, J. H. (1959). The metabolism of chromosomal ribonucleic acid in *Drosophila* salivary glands and its relation to the synthesis of desoxyribonucleic acid. *J. Biophys. Biochem. Cytol.* **5,** 461–467.

McNabb, S. L., and Beckendorf, S. K. (1986). Cis-acting sequences which regulate expression of the *Sgs-4* glue protein gene of *Drosophila*. *EMBO J.* **5,** 2331–2340.

McNabb, S., Greig, S., and Davis, T. (1996). The alcohol dehydrogenase gene is nested in the outspread locus of *Drosophila melanogaster*. *Genetics* **143,** 897–911.
Mechelke, F. (1953). Reversible Strukturmodificationen der Speicheldrusenchromosomen von *Acricotopus lucidus*. *Chromosoma* **5,** 511–543.
Mechelke, F. (1958). The timetable of physiological activity of several loci in the salivary gland chromosomes of *Acricotopus lucidus*. *Proc. Intern. Congr. Genet.* **10**(2), 185.
Mechelke, F. (1959). Beziehungen zwischen der Menge der DNA und dem Ausmass der potentiellen Oberflachenentfaltung von Riesenchromosomenloci. *Naturwissenschaften* **46,** 609.
Mechelke, F. (1960). Structurmodifikationen in Speicheldrusenchromosomen von Acricotopus als Manifestation eines Positionseffektes. *Naturwissenschaften* **14,** 334–335.
Mechelke, F. (1961). Das Wandern des Aktivitatsmaximums im BR_4-Locus von *Acricotopus lucidus* als Modell fur die Wirkungsweise eines komplexen Locus. *Naturwissenschaften* **48,** 29.
Mechelke, F. (1963). Spezielle Funktionzustande des genetischen Materials. In "Funktionelle und Morphologische Organization der Zelle. Wisenschaftliche Konferenz der Gesellschaft deutscher Naturforscher und Arzte in Rottach-Egern," Berlin-Gottingen-Heidelberg, 1962, pp. 1–29.
Mecheva, I. S. (1990). Multiplicity of nucleoli unrelated to the nucleolar organizer in interspecific hybrids of *Drosophila*. *Doklady Bolg. Akad. Nauk* **43**(12), 87–90 [in Russian]
Mecheva, I. S., and Semionov, E. P. (1989). Comparison of the localization of ribosomal DNA in there closely related *Drosophila* species (the *melanogaster* subgroup). *Doklady Bolgar. Akad. Nauk* **42**(6), 105–108. [in Russian]
Mecheva, I., and Semionov, E. (1991). Localization of ribosomal DNA in diploid and polytene nuclei of *Drosophila simulans* and *D. mauritiana*. *Dros. Inform Serv.* **70,** 141–143.
Meer, B. (1976). Anomalous development and differential DNA replication in the X-chromosome of a *Drosophila* hybrid. *Chromosoma* **57,** 235–260.
Mehlin, H., and Daneholt, B. (1993). The Balbiani ring particle: a model for the assembly and export of RNPs from the nucleus? *Trends in Cell Biol.* **3,** 443–447.
Mehlin, H., Lonnroth, A., Skoglund, U., and Daneholt, B. (1988). Structure and transport of a specific premessenger RNP particle. *Cell Biol. Intern. Reports* **12,** 729–736.
Mehlin, H., Skoglund, U., and Daneholt, B. (1991). Transport of Balbiani Ring granules through nuclear pores in *Chironomus tentans*. *Exp. Cell Res.* **193,** 72–77.
Mehlin, H., Daneholt, B., and Skoglund, U. (1992). Translocation of a specific premessenger ribonucleoprotein particle through the nuclear pore studied with electron microscope tomography. *Cell* **69,** 605–613.
Mehlin, H., Daneholt, B., and Skoglund, U. (1995). Structural interaction between the nuclear pore complex and a specific translocating RNP particle. *J. Cell Biol.* **129,** 1205–1216.
Melland, A. M. (1942). Types of development of polytene chromosomes. *Proc. Roy. Soc. Edinb. B* **61,** 316–327.
Melnick, M. B., Noll, E., and Perrimon, N. (1993). The *Drosophila stubarista* phenotype is associated with a dosage effect of the putative ribosome-associated protein D-p40 on *spineless*. *Genetics* **135,** 553–564.
Meneghini, R., and Cordeiro, M. (1972). DNA replication in polytene chromosomes of *Rhynchosciara angelae*. *Cell Differ.* **1,** 167–177.
Meneghini, R., Armelin, H. A., Balsamo, J., and Lara, F. J. S. (1971). Indication of gene amplification in *Rhynchosciara* by RNA-DNA hybridization. *J. Cell Biol.* **49,** 913–916.
Mestril, R. D., Rungger, D., Schiller, P., and Voellmy, R. (1985). Identification of a sequence element in the promoter of the *Drosophila melanogaster hsp23* gene that is required for its heat activation. *EMBO J.* **4,** 2971–2976.
Mestril, R. D., Schiller, P., Amin, J., Klapper, H., Ananthan, J., and Voellmy, R. (1986). Heat shock and ecdysterone activation of the *Drosophila melanogaster hsp23* gene: A sequence element implied in developmental regulation. *EMBO J.* **5,** 1667–1673.
Mettling, C., Bourouis, M., and Richards, G. (1985). Allelic variation at the nucleotide level in *Drosophila* glue genes. *Mol. Gen. Genet.* **201,** 265–268.

Mettling, C., Giangrande, A., and Richards, G. (1987). The *Drosophila Sgs 3* gene: An *in vivo* test of intron function. *J. Mol. Biol.* **196,** 223–226.

Mettling, C., Giangrande, A., and Richards, G. (1988). The use of oligonucleotide probes in studies of insect gene activity in development. *J. Insect Physiol.* **34,** 679–684.

Metz, C. W. (1935). Structure of the salivary gland chromosomes in *Sciara*. *J. Hered.* **26,** 176–188.

Metz, C. W. (1937). Small deficiencies and the problem of genetic units in the giant chromosomes. *Genetics* **22,** 543–556.

Metz, C. W. (1941). Structure of salivary gland chromosomes. *Cold Spring Harbor Symp. Quant. Biol.* **9,** 23–39.

Meyer, B., Hertner, T., Eppenberger, H. M., Daneholt, B., and Mahr, R. (1980). Activation of Balbiani ring genes in *Chironomus tentans* salivary glands. *Eur. J. Cell Biol.* **22,** 36.

Meyer, B., Mahr, R., Eppenberger, H. M., and Lezzi, M. (1983). The activity of Balbiani rings 1 and 2 in salivary glands of *Chironomus tentans* larvae under different modes of development and after pilocarpine treatment. *Dev. Biol.* **98,** 265–277.

Meyer, G. F., and Hennig, W. (1974). The nucleolus in primary spermatocytes of *Drosophila hydei*. *Chromosoma* **46,** 121–144.

Meyerowitz, E. M., and Hogness, D. S. (1982). Molecular organization of a *Drosophila puff* site that responds to ecdysone. *Cell* **28,** 165–176.

Meyerowitz, E. M., and Martin, C. H. (1984). Adjacent chromosomal regions can evolve at very different rates: Evolution of the *Drosophila* 68C glue gene cluster. *J. Mol. Evol.* **20,** 251–264.

Meyerowitz, E. M., Crosby, M. A., Garfinkel, M. D., Martin, C. H., Mathers, P. H., and Raghavan, K. V. (1985a). The 68C glue puff of *Drosophila*. *Cold Spring Harbor Symp. Quant. Biol.* **50,** 347–353.

Meyerowitz, E. M., Crosby, M. A., Garfinkel, M. D., Martin, C. H., Mathers, P. H., and Raghavan, K. V. (1985b). The 68C glue gene cluster of *Drosophila*. *In* "Eukaryotic Transcription: The Role of *cis*- and *trans*-Acting Elements in Initiation" (Y. Gluzman, ed.), pp. 189–193. Cold Spring Harbor Laboratory Press, Cold Spring Harbor, NY.

Meyerowitz, E. M., Raghavan, K. V., Mathers, P. H., and Roark, M. (1987). How *Drosophila* larvae make glue: Control of *Sgs-3* gene expression. *Trends Genet.* **3,** 288–293.

Mglinets, V. A. (1973). Structure of the genetic locus in multicellular organisms. *Uspekhi sovr. biol.* **76**(2(5)), 189–198. [in Russian]

Michael, W. M., Bowtell, D. D. L., and Rubin, G. M. (1990). Comparison of the *sevenless* gene of *Drosophila virilis* and *Drosophila melanogaster*. *Proc. Natl. Acad. Sci. USA* **87,** 5351–5353.

Michailova, P. (1992). *Endochironomus tendens* (F) (Chironomidae, Diptera) an example of stasipatric speciation. *Netherlands J. Aquat. Ecol.***26**(2–4), 173–180.

Michailova, P. V., Semionov, E. P., Genova, G. K., and Konstantinov, G. Kh. (1982a). Supernumerary nucleoli and their relation to the polytene chromosomes in the salivary gland. *Tsitologia* **24**(3), 248–251. [in Russian]

Michailova, P. V., Semionov, E. P., Genova, G. R., and Konstantinov, G. Kh. (1982b). Additional nucleolar material in the polytene salivary gland chromosomes of *Drosophila melanogaster*. *Tsitologia* **24**(3), 248–251. [in Russian]

Midelfort, M. E., and Rasch, R. W. (1968). The effect of exogenous cortisone acetate on development and "chromosome puffing" in *Drosophila virilis*. *J. Cell Biol.* **39,** 91a.

Miklos, G. L. G., and Rubin, G. M. (1996). The role of the genome project in determining gene function: Insight from model organisms. *Cell* **86,** 521–529.

Miklos, G. L. G., Kelly, L. E., Coombe, P. E., Leeds, C., and Lefevre, G. (1987). Localization of the genes *shaking-B, small optic lobes, sluggish-A, stoned* and *stress sensitive-C* to a well defined region on the X-chromosome of *Drosophila melanogaster*. *J. Neurogenetics* **4,** 1–19.

Millar, S., Hayward, D. C., Read, C. A., Browne, M. J., Santeli, R. V., Vallejo, F. G., Pueyyo, M. T., Zaha, A., Glover, D. M., and Lara, F. J. S. (1985). Segments of chromosomal DNA from *Rhynchosciara americana* that undergo additional rounds of DNA replication in the salivary gland DNA puffs have only weak ARS activity in yeast. *Gene* **34,** 81–86.

Miller, O., and Beatty, B. R. (1969a). Portrait of a gene. *J. Cell Physiol.* **74,** 225–232.

Miller, O., and Beatty, B. R. (1969b). Visualization of nucleolar gene. *Science* **164,** 955–957.

Miranda, M., Garcia, M. L., and Alonso, C. (1981). A novel method for the analysis of active chromosomal loci by an endogenous hybridization technique (EHT). *Brother. Genet.* **2,** 57–64.

Mirault, M.-E., Goldschmidt-Clermont, M., Moran, L., Arrigo, A.-P., and Tissieres, A. (1978). The effect of heat shock on gene expression in *Dosophila melanogaster. Cold Spring Harbor Symp. Quant. Biol.* **42,** 819–827.

Mirault, M.-E., Goldschmidt-Clermont, M., Artavanis-Tsakonas, S., and Schedl, P. (1979). Organization of the multiple genes for the 70.000 dalton heat-shock protein in *Drosophila melanogaster. Proc. Natl. Acad. Sci. USA* **76,** 5254–5258.

Mirault, M.-E., Delwart, E., and Southgate, R. (1982a). A DNA sequence upstream of *Drosophila hsp70* genes is essential for their heat induction in monkey cells. *In* "Heat shock from bacteria to man" (M. L. Schlesinger M. L., M. Ashburner, A. and Tissieres, eds.), pp. 35–42. Cold Spring Harbor Laboratory Press, Cold Spring Harbor, NY.

Mirault, M.-E., Southgate, R., and Delwart, E. (1982b). Regulation of heat-shock genes: a DNA sequence upstream of *Drosophila hsp70* genes is essential for their induction in monkey cells. *EMBO J.* **1,** 1279–1285.

Mirkovitch, J., Spierer, P., and Laemmli, U. K. (1986). Genes and loops in 320,000 base-pairs of the *Drosophila melanogaster* chromosome. *J. Mol. Biol.* **190,** 255–258.

Mirkovitch, J., Gasser, S. M., and Laemmli, U. K. (1987). Relation of chromosome structure and gene expression. *Phil. Trans. R. Soc. Lond. B* **317,** 563–574.

Misch, D. W. (1962). Localization of acid phosphatase in tissues of metamorphosing flesh-fly larvae. *J. Histochem. Cytochem.* **10,** 666.

Miseiko, G. N., and Popova, V. S. (1970a). Karyotypic study on *Cryptochironomus* gr. *defectus*. I. General characteristics of karyotypic diversity. *Tsitologiya* **12**(2), 158–165. [in Russian]

Miseiko, G. N., and Popova, V. S. (1970b). Karyotypic study on *Cryptochironomus* gr. *defectus*. II. Characteristics of the first karyotype. *Tsitologia* **12**(9), 1170–1182. [in Russian]

Mishra, A., and Lakhotia, S. C. (1982a). Replication in *Drosophila* chromosomes. VII. Influence of prolonged larval life on patterns of replication in polytene chromosomes of *Drosophila melanogaster. Chromosoma* **85,** 221–236.

Mishra, A., and Lakhotia, S. C. (1982b). Replication in *Drosophila chromosomes*. VIII. Temporal order of replication of specific sites on polytene chromosomes from 24°C and 10°C reared larvae of *D. melanogaster. Ind. J. Exp. Biol.* **20,** 652–659.

Mitchell, H. K., and Lipps, L. S. (1975). Rapidly labeled proteins on the salivary gland chromosomes of *Drosophila melanogaster. Biochem. Genet.* **13,** 585–602.

Mitchell, H. K., and Lipps, L. S. (1978). Heat shock and phenocopy induction in *Drosophila. Cell* **15,** 907–918.

Mitchell, H. K., Tracy, U. W., and Lipps, L. S. (1977). The prepupal salivary glands of *Drosophila melanogaster. Biochem. Genet.* **15,** 563–573.

Mitchell, H. K., Moller, G., Peterson, N. S., and Lipps-Sarmiento, L. (1979). Specific protection from phenocopy induction by heat shock. *Dev. Genet.* **1,** 181–192.

Mitra, N., and Mukherjee, A. S. (1972). Continuous ^{3}H-TdR labelling pattern as the beginning of replication cycle: further evidence. *Dros. Inform. Serv.* **48,** 114–115.

Mitrofanov, V. G., and Poluektova, E. V. (1977). Age specificity of the effect of 5-bromdeoxyuridine on the differentiation and spectrum of puffs of the salivary gland chromosomes of *Drosophila virilis*. I. Dependence of survival of individuals and induction of new puffs on the time of action of 5-bromdeoxyuridine. *Genetika* **13**(7), 1202–1209. [in Russian]

Mlodzik, M., Fjose, A., and Gehring, W. J. (1988). Molecular structure and spatial expression of a homeobox gene from the labial region of the *Antennapedia-complex. EMBO J.* **7,** 2569–2578.

Mohan, J., and Ritossa, F. M. (1970). Regulation of ribosomal RNA synthesis and its bearing in the bobbed phenotype in *Drosophila melanogaster. Dev. Biol.* **22,** 495–512.

Mohler, D., and Carroll, A. (1984). Sex-linked female-sterile mutations in the Iowa collection. *Dros. Inform. Serv.* **60,** 236–241.

Mohler, J. D. (1977). Developmental genetics of the *Drosophila* egg. I. Identification of 59 sex-linked cistrons with maternal effects on embryonic development. *Genetics* **85,** 259–272.

Mohler, J., and Pardue, M. L. (1982a). Genetic analysis of the region of the 93D heat shock locus. *In* "Heat Shock from Bacteria to Man" (M. L. Schlesinger, M. Ashburner, and A. Tissieres, eds.), pp. 77–82. Cold Spring Harbor Laboratory Press, Cold Spring Harbor, NY.

Mohler, J., and Pardue, M. L. (1982b). Deficiency mapping of the 93D heat-shock locus in *Drosophila melanogaster*. *Chromosoma* **86,** 457–467.

Mohler, J., and Pardue, M. L. (1984). Mutational analysis of the region surrounding the 93D heat shock locus of *Drosophila melanogaster*. *Genetics* **106,** 249–265.

Molto, M. D., de Frutos, R., and Martinez-Sebastian, M. J. (1987a). The banding pattern of polytene chromosomes of *Drosophila guanche* compared with that of *D. subobscura*. *Genetica* **75,** 55–70.

Molto, M. D., de Frutos, R., and Martinez-Sebastian, M. J. (1987b). Characteristic puffing patterns of *Drosophila guanche*. *Dros. Inform. Serv.* **66,** 105.

Molto, M. D., Pascual, L., and de Frutos, R. (1987c). Puff activity after heat shock in two species of the *Drosophila obscura* group. *Experientia* **43,** 1225–1227.

Molto, M. D., de Frutos, R., and Martinez-Sebastian, M. J. (1988). Gene activity of polytene chromosomes in *Drosophila* species of the *obscura* group. *Chromosoma* **96,** 382–390.

Molto, M. D, Martinez-Sebastian, M. J., and de Frutos, R. (1991). Localization of the *hsp 70* gene in Drosophila guanche. *Dros. Inform. Serv.* **70,** 150–151.

Montell, C., Jones, K., Zuker, C., and Rubin, G. (1987). A second opsin gene expressed in the ultraviolet-sensitive R7 photoreceptor cells of *Drosophila melanogaster*. *J. Neurosci.* **7,** 1558–1566.

Moorfield, H. H., and Fraenkel, G. (1954). The character and ultimate fate of the larval salivary secretion of *Phormia regina* Meig. (Diptera, Callophoridae). *Biol. Bull.* **106,** 178–184.

Moran, L., Mirault, M.-E., Arrigo, A.-P., Goldschmidt-Clermont, M., and Tissieres, A. (1978). Heat shock in *Drosophila melanogaster* induced synthesis of new mRNAs and proteins. *Trans. R. Soc.* **283,** 391–406.

Moran, L., Mirault, M.-E., Tissieres, A., Lis, J., Schedl, P., Artavanis-Tsakonas, S., and Gehring, W. J. (1979). Physical map of two *Drosophila melanogaster* DNA segments containing sequences coding for the 70,000 dalton heat shock protein. *Cell* **17,** 1–8.

Morcillo, G., Santa-Cruz, M. C., and Diez, J. L. (1981). Temperature-induced Balbiani rings in *Chironomus thummi*. *Chromosoma* **83,** 341–352.

Morcillo, G., Barettino, D., and Diez, J. L. (1982). Heat shock puffs in isolated salivary glands of *Chironomus thummi*. *Biol. Cell* **44,** 221–227.

Morcillo, G., Barettino, D., Carmona, M. J., Carretero, M. T., and Diez, J. L. (1988). Telomeric DNA sequences differentially activated by heat shock in two *Chironomus* subspecies. *Chromosoma* **96,** 139–144.

Morcillo, G., Diez, J. L., Carbajal, M. E., and Tanguay, R. M. (1993). Hsp90 associates with specific heat shock puff (hsrw). *Chromosoma* **102,** 648–659.

Moreira, M. A., da Silva, L. A. F., Bezerra, J. P., and Leoncini, O. (1992). Analysis of the rDNA of *Chironomus sancticaroli* (Diptera, Chironomidae). *Rev. Brasil Genet.* **15,** 821–829.

Morimoto, R. (1993). Cells in stress: Transcriptional activation of heat shock genes. *Science* **259,** 1409–1410.

Morimoto, R. I., Tissieres, A., and Georgopoulos, C. (1990). "Stress Proteins in Biology and Medicine," Cold Spring Harbor Laboratory Press, Cold Spring Harbor, NY.

Moritz, Th., Edstrom, J.-E., and Pongs, O. (1984). Cloning of a gene localized and expressed at the ecdysteroid regulated puff 74EF in salivary glands of *Drosophila* larvae. *EMBO J.* **3,** 289–295.

Moriwaki, D., and Ito, S. (1969). Studies on puffing in the salivary gland chromosomes of *Drosophila ananassae*. *Jpn. J. Genet.* **44,** 129–139.

Mott, M. R., and Hill, R. J. (1986). The ultrastructural morphology of native salivary gland chromosomes of *Drosophila melanogaster:* The band-interband question. *Chromosoma* **94,** 403–411.

Mott, M. R., Burnett, E. J., and Hill, R. J. (1980). Ultrastructure of polytene chromosomes of *Drosophila* isolated by microdissection. *J. Cell Sci.* **45,** 15–30.

Mougneau, E., von Seggeren, D., Fowler, T., Rosenblatt, J., Jongens, T., Rogers, B. T., Gietzen, D., and Beckendorf, S. K. (1993). A transcriptional switch between the *Pig-1* and *Sgs-4* genes of *Drosophila melanog. Mol. Cell. Biol.* **13,** 184–195.

Mukherjee, A. S. (1966). Dosage compensation in *Drosophila:* an autoradiographic study. *The nucleus* **9,** 83–96.

Mukherjee, A. S. (1968). Effect of dicyandiamide on puffing activity and morphology of salivary gland chromosomes of *Drosophila melanogaster. Ind. J. Exp. Biol.* **6,** 49–51.

Mukherjee, A. S., and Beerman, W. (1965). Synthesis of ribonucleic acid by the X-chromosomes of *Drosophila melanogaster* and the problem of dosage compensation. *Nature* **207,** 785–786.

Mukherjee, A. S., and Chatterjee, R. N. (1977). Application and efficiency of scintillation autoradiography for *Drosophila* polytene chromosomes. *Histochemistry* **52,** 73–84.

Mukherjee, A. S., and Chatterjee, S. N. (1975). Chromosomal basis of dosage compensation in *Drosophila.* VIII. Faster replication and hyperactivity of both arms of the X-chromosome in males of *Drosophila pseudoobscura* and their possible significance. *Chromosoma* **53,** 91–105.

Mukherjee, A. S., and Chatterjee, S. N. (1976). Hypeactivity and faster replicating property of the two arms of the male X of *Drosophila pseudoobscura. J. Microscopy* **106**(2), 199–208.

Mukherjee, A. S., and Lakhotia, A. S. (1979). ^{3}H-uridine incorporation in the puff 93D and in chromocentric heterochromatin of heat shocked salivary glands of *Drosophila melanogaster. Chromosoma* **74,** 75–82.

Mukherjee, A. S., and Lakhotia, A. S. (1981). Specific induction of the 93D puff in *Drosophila melanogaster* by a homogenate of heat shocked larval salivary glands. *Ind. J. Exp. Biol.* **19,** 1–4.

Mukherjee, A. S., and Lakhotia, A. S. (1982). Heat shock puff activity in salivary glands of *Drosophila melanogaster* larvae during recovery from anoxia at two different temperatures. *Ind. J. Exp. Biol.* **20,** 437–439.

Mukherjee, A. S., and Mitra, N. (1973). ^{3}H-thymidine labelling patterns in polytene chromosomes of mitomycin-treated *Drosophila melanogaster. Exp. Cell Res.* **76,** 47–54.

Mukherjee, A., and Rees, D. (1969). Duration of mitotic cycle in brain cells of mosquito *Aedes dorsalis. Can. J. Genet. Cytol.* **11,** 673–676.

Mukherjee, A. S., Lakhotia, A. S., and Chatterjee, S. (1968). On the molecular and chromosomal basis of dosage compensation in *Drosophila. The Nucleus (Symposial Vol.)*, 161–173.

Mukherjee, A. S., Chatterjee, S. N., Chatterjee, R. N., Majumdar, D., and Nag, A. (1977). Replication cytology of *Drosophila* polytene chromosomes. *The Nucleus* **20,** 171–177.

Mukherjee, A. S., Duttagupta, A. K., Chatterjee, C., Ghosh, M., Achary, P. M., Dey, A., and Banerjee, I. (1980). Regulation of DNA replication in *Drosophila. In* "Development and Neurobiology of *Drosophila*" (O. Siddiqi, P. Babu, L. M. Hall, and J. C. Hall, eds.), pp. 57–83. Plenum, New York/London.

Mukherjee, A. S., Singh, A. K., and Lakhotia, S. C. (1982). Effects of microtubule poisons on the activity of heat shock puff loci in *Drosophila. Abstr. Book of VI All India Cell Biology Confer.*, 82–83.

Mulder, M. P., van Duijn, P., and Gloor, H. J. (1968). The replicative organization of DNA in polytene chromosomes of *Drosophila hydei. Genetica* **39,** 385–428.

Muller, H. J. (1929). The gene as the basis of life. *Proc. Int. Congr. Plant. Sci. 4th, 1926* **1,** 897–921.

Muller, H. J. (1932). Further studies on the nature and causes of gene mutations. *Proc. Sixth Intern. Congr. Genet.* **1,** 213–255.

Muller, H. J. (1950). Evidence of the precision of genetic adaptation. *In* "Harvey Lecture Series XLIII, 1947–1948" (C. Thomas Charles, ed.), Vol. 1, pp. 165–229. Springfield, Illinois.

Muller H. J., and Altenberg E. (1919). The rate of change of heredity factors in *Drosophila*. *Proc. Soc. Exp. Biol.* **17,** 10–14.

Muller, H. J., and Kaplan, W. D. (1966). The dosage compensation of *Drosophila* and mammals as showing the accuracy of the normal type. *Genet Res. Cambr.* **8,** 41–59.

Muller, H. J., and Prokofyeva, A. A. (1935). The individual gene in relation to the chromomere and the chromosome. *Proc. Natl. Acad. Sci. USA* **21,** 16–26.

Muller, H. J., League, B. B., and Offermann, C. A. (1931). Effects of dosage changes of sex-linked genes and the compensatory effect of other gene differences between male and female. *Anat. Rec.* **51**(Suppl.), 110.

Muskavitch, M. A. T., and Hogness, D. S. (1980). Molecular analysis of a gene in a developmentally regulated puff of *Drosophila melanogaster*. *Proc. Natl. Acad. Sci USA* **77,** 7362–7366.

Muskavitch, M. A. T., and Hogness, D. S. (1982). An expandable gene that encodes a *Drosophila* glue protein is not expressed in variants lacking remote upstream sequences. *Cell* **29,** 1041–1051.

Mutsuddi, D., Mutsuddi, (nee Das) M., and Duttagupta, A. K. (1985). Role of altered sexual physiology on X chroosomal replication pattern in *Drosophila melanogaster*. *Ind. J. Exp. Biol.* **23,** 8–16.

Mutsuddi, D., Mutsuddi, (nee Das) M., and Duttagupta, A. K. (1987). Replication and transcription of X chromosome in metafemales (3X;2A) of *Drosophila melanogaster*. *Ind. J. Exp. Biol.* **25,** 579–586.

Mutsuddi, M., and Mukhergee, A. S. (1991). Specificity of the interruption of X chromosome in determining male and or female level activity vis-a-vis level of compaction in *Drosophila*. *Dros. Inform Serv.* **70,** 156–157.

Mutsuddi, M. (nee Das), Mutsuddi, D., Mukherjee, A. S., and Duttagupta, A. K. (1984). Conserved autonomy of replication of the X chromosomes in hybrids of *Drosophila miranda* and *Drosophila persimilis*. *Chromosoma* **89,** 55–62.

Mutsuddi, M. (nee Das), Mutsuddi, D., and Duttagupta, A. K. (1985). Conservation in regulation of replication and transcription in three Nearctic species of *obscura* group in *Drosophila*. *Ind. J. Exp. Biol.* **23,** 616–624.

Mutsuddi, M. (Das), Mutsuddi, D., and Duttagupta, A. K. (1988). X-autosomal dosage compensation in *Drosophila*. *Dros. Inform. Serv.* **67,** 61–62.

Myohara, M., and Okada, M. (1987). Induction of ecdysterone-stimulated chromosomal puffs in permeabilized *Drosophila* salivary glands: A new method for assaying the gene-regulating activity of cytoplasm. *Dev. Biol.* **122,** 386–406.

Myohara, M., and Okada, M. (1988a). Puff induction in permeabilized *Drosophila* salivary glands in a chemically defined medium. *Dev. Biol.* **125,** 462–465.

Myohara, M., and Okada, M. (1988b). Digitonin treatment activates specific genes including the heat-shock genes in salivary glands of *Drosophila melanogaster*. *Dev. Biol.* **130,** 348–355.

Myohara, M., and Okada, M. (1988c). Activation of heat-shock genes by digitonin is selectively repressed in preheated *Drosophila* salivary glands. *Dev. Growth. Differ.* **30,** 629–638.

Myohara, M., and Okada, M. (1990). Permeabilization of *Drosophila* salivary glands with mild detergents. *Bull. Natl. Inst. Seric. Entomol. Sci.* **1,** 1–11.

Nacheva, G. A., Guschin, D. Y., Preobrazhenskaya, O. V., Karpov, V. L., Ebralidse, K. K., and Mirzabekov, A. D. (1989a). Change in the pattern of histone binding to DNA upon transcriptional activation. *Cell* **58,** 27–36.

Nacheva, G. A., Preobrazhenskaya, O. V., Mel'nikova, A. F., and Karpov, V. L. (1989b). Structural changes in chromatin of the gene encoding heat shock protein in Drosophila during transcription. *Mol. Biol.* **23**(3), 879–888. [in Russian]

Nagel, G., and Rensing, L. (1971). Puffing pattern and puff size of Drosophila salivary gland chromosome *in vitro*. *Cytobiologie* **3,** 288–292.

Nagel, G., and Rensing, L. (1972). Struktur und Funktion von Riesenchromosomen und Puffs. *Naturw. Rdsch.* **25,** 53–65.

Nagel, G., and Rensing, L. (1974). Circadian rhythm in size and ^{3}H-uridine incorporation of single puffs of *Drosophila* salivary glands *in vitro*. *Exp. Cell Res.* **89,** 436–439.
Nagl, W. (1969). Correlation of structure and RNA synthesis in the nucleolus-organizing polytene chromosomes of *Phaseolus vulgaris*. *Chromosoma* **28,** 85–92.
Nagl, W. (1990). Gene amplification and related events. *In* "Biotechnology in Agriculture and Forestry" (Y. P. S. Bajaj, ed.), Vol. 11, pp. 153–201. Springer-Verlag, Berlin/Heidelberg/New York.
Nagoshi, R. N., and Gelbart, W. M. (1979). Molecular and recombinational mapping of mutations in the *Ace* locus of *Drosophila melanogaster*. *Genetics* **117,** 487–502.
Nash, D., and Bell, J. (1968). Larval age and the pattern of DNA synthesis in polytene chromosomes. *Can. J. Genet. Cytol.* **10,** 82–90.
Nash, D., and Plaut, W. (1965). On the presence of DNA in larval salivary gland nucleoli in *Drosophila melanogaster*. *J. Cell Biol.* **27,** 682–686.
Nash, D., and Vyse, E. R. (1977). On the polytene morphology of a *Drosophila melanogaster* chromosomal inversion including the nucleolus organizer. *Can. J. Genet. Cytol.* **19,** 637–644.
Nath, B. B., and Lakhotia, S. C. (1989). Heat shock response in ovarian nurse cells of *Anopheles stephensi*. *J. Biosci.* **14,** 143–152.
Nath, B. B., and Lakhotia, S. C. (1991). Search for a *Drosophila-93D-like locus in Chironomus* and *Anopheles*. *Cytobios*. **65,** 7–13.
Natzle, J. E. (1993). Temporal regulation of *Drosophila* imaginal disc morphogenesis: A hierarchy of primary and secindary 20-hydroxyecdysone-responsive loci. *Dev. Biol.* **155,** 516–532.
Natzle, J. E., Hammonds, A. S., and Fristrom, J. W. (1986). Isolation of genes active during hormone-induced morphogenesis in *Drosophila melanogaster* imaginal discs. *J. Biol. Chem.* **261,** 5575–5583.
Neidhardt, F. C., VanBogelen, R. A., and Vaughn, V. (1984). The genetics and regulation of heat shock proteins. *Ann. Rev. Genet.* **18,** 295–329.
Nelson, C. R., and Szauter, P. (1992). Cytogenetic analysis of chromosome region 89A of *Drosophila melanogaster:* Isolation of deficiencies and mapping of *Po, Aldox-1* and transposon insertions. *Mol. Gen. Genet.* **235,** 11–21.
Nelson, L. G., and Daneholt, B. (1981). Modulation of 75S RNA synthesis in the Balbiani Rings of *Chironomus tentans* with galactose treatment. *Chromosoma* **83,** 645–659.
Nelson, R. J., Odell, G. M., Chistiansen, A. E., and Laird, C. D. (1991). Hormonal control of gene expression: interaction between two trans-acting regulators in *Drosophila*. *Dev. Biol.* **144,** 152–166.
Nesterova, S. I. (1967). A study on giant chromosomes in some Chironomidae (Diptera) species. *Tsitologiya* **9**(5), 524–529. [in Russian]
Nicklas, J. A., and Cline, T. W. (1983). Vital genes that flank Sex-lethal, an X-linked sex-determining gene of *Drosophila melanogaster*. *Genetics* **103,** 617–631.
Nikiforov, Yu. L., Sakharov, M. N., Rapoport, I. A., and Beknazar'yants, M. M. (1970). Specific effect of chrome on the pattern of puffs formation in *Drosophila melanogaster*. *Doklady Akad. Nauk SSSR* **194**(2), 441–444. [in Russian]
Nilsen, T. W. (1994). RNA-RNA interaction in the spliceosome: unraveling the ties that bind. *Cell* **78,** 1–4.
Nishida, Y., Hata, M., Ayaki, T., Ryo, H., Yamagata, M., Shimizu, K., and Nishizuka, Y. (1988). Proliferation of both somatic and germ cells is affected in the *Drosophila* mutants of *raf* proto-oncogene. *EMBO J.* **7,** 775–781.
Nissani, M. (1975). Cell lineage analysis of kinurenine producing organs in *Drosophila melanogaster*. *Genet. Res. Cambr.* **26,** 63–72.
Nivard, M. J. M., Pastink, A., and Vogel, E. W. (1992). Molecular analysis of mutations induced in the vermilion gene of *Drosophila melanogaster* by methylmethanesulfonate. *Genetics* **131,** 673–682.

Nothiger, R. (1992). Genetic control of sexual development in *Drosophila*. *Verh. Dtsch. Zool. Ges.* **85**(2), 177–183.

Nover, L. (editor). (1984). "Heat Shock Response of Eukaryotic Cells." Springer-Verlag, Berlin/Heidelberg/New York/Tokyo.

Nover, L. (1990). Molekulare Zellbiologie der Hitzestressantwort. *Naturwissenschaften* **77,** 359–365.

Nover, L., Hellmund, D., Neumann, D., Scharf, K.-D., and Serfling, E. (1984). The heat shock response of eukaryotic cells. *Biol. Zbl.* **103,** 357–435.

Novozhilov, Yu. K., and Slepyan, E.I. (1976). Morphofunctional changes in the salivary glands of the larvae of the gall midges *Harmandia loewi* in development. *Tsitologiya* **18**(10), 1194–1197. [in Russian]

Nusslein-Volhard, C. (1994). Of flies and fishes. *Science* **266,** 571–574.

O'Brien, S. J. (1973). On estimating functional gene number in eukaryotes. *Nature New Biol.* **242,** 52–54.

O'Brien, T., and Lis, J. T. (1991). RNA polymerase II pauses at the 5′ end of the transcriptionally induced *Drosophila hsp70* gene. *Mol. Cell. Biol.* **11,** 5285–5290.

O'Connor, D., and Lis, J. T. (1981). Two closely linked transcription units within the 63B heat shock puff locus of *Drosophila melanogaster*. *Nucleic Acids Res.* **9,** 5075–5092.

O'Connor, M. B., Binari, R., Perkins, L. A., and Bender, W. (1988). Alternative RNA products from the *Ultrabithorax* domain of the bithorax complex. *EMBO J.* **7,** 435–445.

O'Donnel, J., Mandel, H. C., Krauss, M., and Sofer, W. (1977). Genetic and cytogenetic analysis of the ADH region in *Drosophila melanogaster*. *Genetics* **86,** 553–566.

O'Donnell, J., Boswell, R., Reynolds, T., and Mackay, W. (1989). A cytogenetic analysis of the *Punch-tudor* region of chromosome 2R in *Drosophila melanogaster*. *Genetics* **121,** 273–280.

O'Hare, K. (1986). Genes within genes. *Trends Genet.* **2,** 33.

O'Hare, K., and Rubin, G. M. (1983). Structures of P transposable elements and their sites of insertion and exscision in the *Drosophila melanogaster* genome. *Cell* **34,** 25–35.

O'Hare, K., Levis, R., and Rubin, G. M. (1983). Transcription of the white locus in *Drosophila melanogaster*. *Proc. Natl. Acad. Sci. USA* **80,** 6917–6921.

O'Hare, K., Murphy, C., Levis, R., and Rubin, G. M. (1984). DNA sequence of the white locus of *Drosophila melanogaster*. *J. Mol. Biol.* **180,** 437–455.

Ohno, C. K., and Petkovitch, M. (1992). *FTZ-F1b*, a novel member of the *Drosophila* nuclear receptor family. *Mech. Dev.* **40,** 13–24.

O'Kane, C. J., and Gehring, W. S. (1987). Detection *in situ* of genomic regulatory elements in *Drosophila*. *Proc. Natl. Acad. Sci. USA* **84,** 9123–9127.

O'Keefe, S., Schouls, M., and Hodgetts, R. (1995). Epidermal cell-specific quantitation of DOPA decarboxylase mRNA in *Drosophila* by competitive RT-PCR: An effect of *Broad-Complex* mutants. *Dev. Genet.* **16,** 77–84.

Okretic, M. C., Penoni, J. S., and Lara, F. J. S. (1977). Messenger-like RNA synthesis and DNA chromosomal puffs in the salivary gland of *Rhynchosciara americana*. *Arch. Biochem. Biophys.* **178,** 158–165.

Okuno, T., Satou, T., and Oishi, K. (1984). Studies on the sex-specific lethal of *Drosophila melanogaster*. VII. Sex-specific lethals that do not affect dosge compensation. *Jpn. J. Genet.* **59,** 237–247.

Olenov, Yu M. (1974). On the organization of the genome in Metazoa. *Tsitologiya* **16**(4), 403–420. [in Russian]

Olins, A. L., Olins, D. E., and Franke, W. W. (1980). Stereo-electron microscopy of nucleoli Balbiani rings and endoplasmic reticulum in *Chironomus* salivary gland cells. *Eur. J. Cell Biol.* **22,** 714–723.

Olins, A. L., Olins, D. E., and Lezzi, M. (1982). Ultrastructural studies of *Chironomus* salivary gland cells in different states of Balbiani ring activity. *Eur. J. Cell Biol.* **27,** 161–169.

Olins, A. L., Olins, D. E., and Bazett-Jones, D. P. (1992). Balbiani ring hn RNP substructure vizualized by selective staining and electron spectroscopic imaging. *J. Cell Biol.* **117,** 483–491.

Olins, A. L., Olins, D. E., Levy, H. A., Shah, M. B., and Bazett-Jones, D. P. (1993). Electron microscope tomography of Balbiani ring hn RNP structure. *Chromosoma* **102,** 137–144.

Olins, A. L., Olins, D. E., Olman, V., Levy, H. A., and Bazett-Jones, D. P. (1994). Modeling the 3-D RNA distribution in the Balbiani ring granule. *Chromosoma* **103,** 302–310.

Olvera, O. R. (1969). The nucleolar DNA of three species of *Drosophila* in the *hydei* complex. *Genetics* **61**(Suppl.), 245–249.

Onischenko, A. M., Mikichur, N. I., and Maksimovsky, L. F. (1976). The nucleolar apparatus and differentiation of the salivary glands of *Drosophila virilis* larvae. I. Change in the content of DNA and RNA in different cell structures at the stage of secretion formation. *Ontogenez* **7**(1), 76–81. [in Russian]

O'Reilly, D. R., and Miller, L. K. (1989). A baculovirus blocks insect molting by producing ecdysteroid UDP-glucosyl transferase. *Science* **245,** 1110–1112.

Oro, A. E., McKeown, M. and Evans, R. M. (1990). Relationship between the product of the *Drosophila ultraspiracle* locus and verterbrate retinoid X receptor. *Nature* **347,** 298–301.

Oro, A. E., McKeown, M., and Evans, R. M. (1992). The *Drosophila* nuclear receptors: new insight into the actions of nuclear receptors in development. *Curr. Opin. Genet. Dev.* **2,** 269–274.

Orr, C. W. M., Hudson, A., and West, A. S. (1961). The salivary glands of *Aedes aegypti*: Histological-histochemical studies. *Can. J. Zool.* **39,** 265–272.

Orr, W. C., Galanopoulos, V. K., Romano, C. P., and Kafatos, F. C. (1989). A female sterile screen of the *Drosophila melanogaster* X chromosome using hybrid dysgenesis: Identification and characterization of egg morphology mutants. *Genetics* **122,** 847–858.

Orr-Weaver, T. L. (1991). *Drosophila* chorion genes: Cracking the eggshell's secrets. *BioEssays* **13,** 97–105.

Osorio, E., and Lara, F. J. S. (1990). Partial characterizarion of the transcription unit of the C-8 DNA puff of *Rhynchosciara americana*. *Proc. VI Congr. Pan-Am. Assoc. Biochem. Soc.*, 187.

Paco-Larson, M. L., de Almeida, J. C., Alves, M. A. R. Laicine, E. M., and Sauaia, H. (1990a). Gene amplification in polytene chromosomes. *Proc. VI Congr. Pan-Am. Assoc. Biochem. Soc.*, 89.

Paco-Larson, M. L., de Almeida, J. C., and Nakanishi, Y. (1990b). cDNA cloning of a developmental amplified gene from DNA puff C4 of *Bradysia hygida*. *Proc. VI Congr. Pan-Am. Assoc. Biochem. Soc.*, 186.

Pages, M., and Alonso, C. (1976). Activity of the endogenous RNA polymerase on fixed polytene chromosomes. *Exp. Cell Res.* **98,** 120–126.

Pages, M., and Alonso, C. (1978). Chemical and conformational changes in chromosome regions being actively transcribed. *Nucleic Acids Res.* **5,** 549–562.

Painter, T. S. (1934). Salivary chromosomes and the attack on the gene. *J. Hered.* **25,** 465–476.

Painter, T. S. (1935). The morphology of the third chromosome in the salivary gland of *Drosophila melanogaster* and a new cytological map of this element. *Genetics* **20,** 301–326.

Painter, T. S. (1941). An experimental study of salivary chromosomes. *Cold Spring Harbor Symp. Quant. Biol.* **9,** 47–54.

Painter, T. S. (1945). Nuclear phenomena associated with secretion in certain gland cells with especial reference to the origin of cytoplasmic nucleic acid. *J. Exp. Zool.* **100,** 523–541.

Painter, T. S., and Biesele, J. J. (1966). Endomitosis and polyribosome formation. *Proc. Natl. Acad. Sci. USA* **56,** 1920–1925.

Palmer, M. J., Mergner, V. A., Richman, R., Manning, J. E., Kuroda, M. I., and Lucchesi, J. C. (1993). The *male-specific lethal-one (msl-1)* gene of *Drosophila melanogaster* encodes a novel protein that associated with the X chromosome in males. *Genetics* **134,** 545–557.

Palmer, M. J., Richman, R., Richter, L., and Kuroda, M. I. (1994). Sex-specific regulation of the *male-specific lethal-one* dosage compensation gene in *Drosophila*. *Genes Dev.* **8,** 698–706.

Palter, K., Watanabe, M., Stinson, L., Mahowald, A. P., and Craig, E. A. (1986). Expresion and localization of *Drosophila melanogaster hsp70* cognate proteins. *Mol. Cell. Biol.* **6,** 1187–1203.

Palumbo, G., Caizzi, R., and Ritossa, F. (1973). Relative orientation with respect to the centromere of ribosomal RNA genes of the X and Y chromosomes of *Drosophila melanogaster*. *Proc. Natl. Acad. Sci. USA* **70,** 1883–1885.

Panitz, R. (1960a). Gewebespezifische Manifestierung einer Heterozygotie des Nucleolus in Speicheldrusenchromosomen von *Acricotopus lucidus*. *Naturwissenschaften* **47,** 359.

Panitz, R. (1960b). Innersekretorische Wirkung auf Strukturmodifikationen der Speicheldrusenchromosomen von *Acricotopus lucidus*. *Naturwissenschaften* **47,** 383.

Panitz, R. (1963). Experimentell induzierte Inaktivierung Balbiani-Ring bildender Gen-Loci in Riesenchromosomen. *In* "Structuren und Funktion des genetischen Material," Vol. III, pp. 225–229. Erwin Bauer, Berlin.

Panitz, R. (1964). Hormonkontrollierte Genaktivitaten in den Riesenchromosomen von *Acricotopus lucidus*. *Biol. Zentrbl.* **83,** 197–230.

Panitz, R. (1965). Heterozygote Funktionsstrukturen in den Riesenchromosomen von *Acricotopus lucidus*. Puffs als Orte unikoler Strukturmutationen. *Chromosoma* **17,** 199–218.

Panitz, R. (1967). Funktionelle Veranderungen an Riesenchromosomen nach Behandlung mit Gibberellinen. *Biol. Zbl.* **86**(Suppl.), 147–156.

Panitz, R. (1972). Balbiani Ring activities in *Acricotopus lucidus*. *Results Prob. Cell Differ.* **4,** 209–227.

Panitz, R. (1978a). Nucleo-cytoplasmic relationship and cell- and developmental-dependent differences in the RNA synthesis of the *Acricotopus* salivary gland. *Biol. Zbl.* **97,** 549–559.

Panitz, R. (1978b). Cell specific effect of ecdysone on RNA synthesis in the differentiated salivary gland of *Acricotopus lucidus*. *Cell Differ.* **7,** 387–398.

Panitz, R. (1979). Two transcripts of an individual Balbiani Ring from the salivary gland cells of *Acricotopus lucidus* (Diptera, Chironomidae). *Chromosoma* **74,** 253–268.

Panitz, R., Serfling, E., and Wobus, U. (1972a). Autoradiographische Untersuchungen zur RNA-Syntheseleistung von Balbiani-Ringen. *Biol. Zbl.* **91,** 359–380.

Panitz, R., Wobus, U., and Serfling, E. (1972b). Effect of ecdysone and ecdysone analogues on two Balbiani Rings of *Acricotopus lucidus*. *Exp. Cell Res.* **70,** 154–160.

Panitz, R., Baumlein, H., Wobus, U., and Serfling, E. (1984). Self-complementary DNA sequences within the BRc gene of *Chironomus thummi*. *Chromosoma* **89,** 254–262.

Pankow, W., Lezzi, M., and Holderegger-Mahling, I. (1976). Correlated changes of Balbiani ring expansion and secretory protein synthesis in larval salivary glands of *Chironomus tentans*. *Chromosoma* **58,** 137–153.

Panzer, S., Weigel, D., and Beckendorf, S.K. (1992). Organogenesis in *Drosophila melanogaster:* Embryonic salivary gland determination is controlled by homeotic and dorsoventral patterning gene. *Development* **114,** 49–57.

Parat, M. (1928). Contribution a l'etude morphologique et physiologique du cytoplasme. *Arch. Anat. Microsc.* **24,** 73–357.

Parat, M., and Painleve, J. (1924a). Constitution du cytoplasme d'une cellule glandulaire: la cellule des glandes salivaires de la larve du Chironome. *C. R. Acad. Sci.* **179,** 543–544

Parat, M., and Painleve, J. (1924b). Observation vitale d'une cellule glandulaire en activite. Nature et role de l'appareil reticulaire interne de Goldi et de l'appareil de Holmgren. *C. R. Acad. Sci.* **179,** 612–614.

Pardali, E., Feggou, E., Drosopoulou, E., Konstantopoulou, I., Scouras, Z. G., and Mavragani-Tsipidou, P. (1996). The Afrotropical *Drosophila montium* subgroup: Balbiani ring 1, polytene chromosomes, and heat shock response of *Drosophila vulcana*. *Genome* **39,** 588–597.

Pardue, M. L., and Gall, J. G. (1972). Molecular cytogenetic. *In* "Molecular Genetics and Developmental Biology" (M. Sussman, ed.), pp. 65–99. Prentice-Hall, Englewood Cliffs, NJ.

Pardue, M. L., Gerbi, S. A., Eckhart, R. A., and Gall, J. G. (1970). Cytological localization of DNA complementation to ribosomal RNA in polytene chromosomes of Diptera. *Chromosoma* **29,** 268–290.

Pardue, M. L., Kedes, L. H., Weinberg, E. S., and Birnstiel, M. L. (1977a). Localization of sequences coding for histone messenger RNA in the chromosomes of *Drosophila melanogaster*. *Chromosoma* **63,** 135–151.

Pardue, M. L., Bonner, I. I., Lengyel, I. A., and Spradling, A. S. (1977b). *Drosophila* salivary gland polytene chromosomes studies *in situ* hybridization. *Mol. Cytogenet. Eukaryot.* 509–519.

Pardue, M. L., Bendena, W. G., and Garbe, J. C. (1987). Heat shock: Puffs and response to environmental stress. *In* "Results and Problems in Cell Differentiation: Structure and Function of Eucaryotic Chromosomes" (Hennig W., ed.), Vol. 14, pp. 121–131. Springer-Verlag, Berlin/Heidelberg.

Pardue, M. L., Bendena, W. G., Fini, M. E., Garbe, J. C., Hogan, N. C., and Traverse, K. L. (1990). *Hsr-omega,* a novel gene encoded by a *Drosophila* heat shock puff. *Biol. Bull.* **179,** 77–86.

Parker, C. S., and Topol, J. (1984). A *Drosophila* polymerase II transcription factor binds to the regulatory site of an *hsp70* gene. *Cell* **37,** 273–283.

Parker-Thornburn, J., and Bonner, J. J. (1987). Mutations that induce the heat shock response of *Drosophila*. *Cell* **51,** 763–772.

Pascual, L., and de Frutos, R. (1984). Heat shock puffs in *Drosophila subobscura* polytene chromosomes. *Dros. Inform. Serv.* **60,** 158–159.

Pascual, L., and de Frutos, R. (1985). *In vitro* puffs after heat shocks at two different temperatures in *Drosophila subobscura*. *Dros. Inform. Serv.* **61,** 134–135.

Pascual, L., and de Frutos, R. (1986). Stress response in *Drosophila*. I. Puff activity after heat shock. *Biol. Cell* **57,** 127–134.

Pascual, L., and de Frutos, R. (1988). Stress response in *Drosophila*. IV. Differential gene activity induced by heat shock. *Chromosoma* **97,** 164–170.

Pascual, L., de Frutos, R., and Latorre, A. (1983). Polytene chromosomes at the end of the prepupal stage. *Dros. Inform. Serv.* **59,** 98–99.

Pascual, L., de Frutos, R., and Latorre, A. (1985). Patterns of puffing activity and chromosomal polymorphism in *Drosophila subobscura*. II. Puffing patterns at the prepupa stage. *Genetica* **66,** 123–138.

Pascual, L., Latorre, A., and de Frutos, R. (1987). Stress response in *Drosophila subobscura*. III. Variablilty of heat shock puffs. *Biol. Cell* **61,** 15–21.

Patrushev, L.I. (1997). Altruistic DNA. About protective functions of the abundant DNA in the eukaryotic genome and its role in stabilizing genetic information. *Biochem. Mol. Biol. Intern.* **41,** 851–860.

Paul, J. (1972). General theory of chromosomes structure and gene activation in eukaryotes. *Nature* **238,** 444–446.

Pauli, D., and Tonka, C.-H. (1987). A *Drosophila* heat shock gene from locus 67B is expressed during embryogenesis and pupation. *J. Mol. Biol.* **198,** 235–240.

Pauli, D., Spierer, A., and Tissieres, A. (1986). Several hundred base pairs upstream of *Drosophila hsp23* and *26* genes are required for their heat induction in transformed flies. *EMBO J.* **5,** 755–761.

Pauli, D., Tonka, C.-H., and Ayme-Southgate, A. (1988). An unusual split *Drosophila* heat shock gene expressed during embryogenesis, pupation and in testis. *J. Mol. Biol.* **200,** 47–53.

Pauli, D., Tonka, C.-H., Tissieres, A., and Arrigo, A.-P. (1990). Tissue-specific expression of the heat shock protein *hsp27* during *Drosophila melanogaster* development. *J. Cell Biol.* **111,** 817–828.

Paulsson, G., Lendahl, U., Galli, J., Ericsson, E., and Wieslander, L. (1990). The Balbiani ring 3 gene in *Chironomus tentans* has a diverged repetitive structure split by many introns. *J. Mol. Biol.* **211,** 331–349.

Pavan, C. (1959a). Organization of the chromosome. *In* "Organization of the Chromosome in Biological Organization: Cellular and Subcellular" (C. H. Waddington, ed.), pp. 72–89. Pergamon Press, London.

Pavan, C. (1959b). Morphological and physiological aspects of chromosomal activities. *Proc. X Intern. Congr. Genet.* **1,** 321–336.

Pavan, C. (1964). Modern concept of chromosome structure and function. *Triangle* **6,** 287–293.

Pavan, C. (1965a). Nucleic acid metabolism in polytene chromosomes and the problem of diferentiation. *Brookhav. Symp. Biol.* **18,** 222–241.

Pavan, C. (1965b). Chromosomal differentiation. *Natl. Cancer Inst. Monogr.* **18,** 309–323.

Pavan, C. (1965c). Synthesis. *In* "Genetics Today: Proc 11th Intern. Congr. Genetics, The Hague, The Netherlands, 1963," Vol. 2, pp. 335–342. Pergamon Press, Oxford/London/Edinburgh/New York/Paris/Frankfurt.

Pavan, C., and Breuer, M.E. (1952). Polytene chromosomes in different tissues of *Rhynchosciara angelae*. *J. Hered.* **43,** 150–157.

Pavan, C., and da Cunha, A. B. (1969a). Chromosomal activities in *Rhynchosciara* and other Sciaridae. *Annu. Rev. Genet.* **3,** 425–450.

Pavan, C., and da Cunha, A. B. (1969b). Gene amplification in ontogeny and phylogeny of animals. *Genetics* **61**(Suppl.), 289–304.

Pavan, C., and Perondini, A. L. P. (1967). Heterozygous puffs and bands in *Sciara ocellaris* Comstock (1882). *Exp. Cell Res.* **48,** 202–205.

Pavan, C., Perondini, A. L. P., and Amabis, J. M. (1968). Heterozygous puffs, bands, hetrochromatin and nucleolus in natural population of Diptera. *Proc. XII Int. Congr. Genet.* **1,** 14.

Pavan, C., da Cunha, A. B., and Morsoletto, C. (1971a). Virus-chromosome relationships in cells of *Rhynchosciara* (Diptera, Sciaridae). *Caryologia* **24,** 371–389.

Pavan, C., Sanders, J. L., and Richmond, R. C. (1971b). Effect of irradiation on neuroendocrine glands and puffs in *Rhynchosciara*. *Abst. 11th Ann. Meet. Amer. Soc. Cell Biol.*, 220.

Pavan, C., da Cunha, A. B., and Sanders, P. (1975). *Rhynchosciara*. *In* "Handbook of Genetics" (R. C. King, ed.), Vol. 3, pp. 207–256. Plenum, New York/London.

Payre, F., Yanicostas, C., and Vincent, A. (1989). Serendipity Delta, a *Drosophila* zinc finger protein present in embryonic nuclei at the onset of zygotic gene transcription. *Dev. Biol.* **136,** 469–480.

Peacock, W. J., Appels, R., Endow, S., and Glover, D. (1981). Chromosomal distribution of the major insert in *Drosophila melanogaster* 28S rRNA genes. *Genet. Res. Cambr.* **37,** 209–214.

Pelc, S. R., and Howard, A. (1956). Metabolic activity of salivary gland chromosomes in Diptera. *Exp. Cell Res.* **10,** 549–552.

Pelham, H. R. B. (1982). A regulatory upstream promoter element in the *Drosophila hsp70* heat shock gene. *Cell* **30,** 517–528.

Pelham, H. R. B. (1984). *Hsp70* accelerates the recovery of nucleolar morphology after heat shock. *EMBO J.* **3,** 3095–3100.

Pelham, H. (1985). Activation of heat-shock genes in eukaryotes. *Trends Genet.* **1,** 31–35.

Pelham, H. R. B. (1986). Speculations on the functions of the major heat shock and glucose-regulated proteins. *Cell* **46,** 959–961.

Pelham, H. (1988). Coming in from the cold. *Nature* **332,** 776–777.

Pelham, H., and Bienz, M. (1982). DNA sequences required for transcriptional regulation of the *Drosophila hsp70* heat shock gene in monkey cells and Xenopus oocytes. *In* "Heat Shock from Bacteria to Man" (M. L. Schlesinger, M. Ashburner, and A. Tissieres, eds.), pp. 43–48. Cold Spring Harbor Laboratory Press, Cold Spring Harbor, NY.

Pellegrini, M., Manning, J., and Davidson, N. (1977). Sequence arrangement of the rDNA of *Drosophila melanogaster*. *Cell* **10,** 213–224.

Pelling, C. (1959). Chromosomal synthesis of ribonucleic acid as shown by incorporated of uridine labelled with tritium. *Nature* **184,** 655–656.

Pelling, C. (1962). Application of tritiated compounds to the midge *Chironomus* and some aspects of the metabolism of salivary gland chromosomes. *In* "Tritium in the Physical Sciences," Vol. 2, 327–334. Intern. Atom. Energy Agency, Vienna.

Pelling, C. (1963). Variability of RNA synthesis in polytene tissues. *Proc. Intern. Congr. Genet. 11th.* (The Hague) **1,** 108.

Pelling, C. (1964). Ribonucleinsaure-Synthese der Riesenchromosomen. Autoradiographische Untersuchungen an *Chironomus tentans*. *Chromosoma* **15,** 71–122.

Pelling, C. (1966). A replicate and synthetic chromosomal unit—The modern concept of the chromomere. *Proc. R. Soc. B.* **164,** 279–289.
Pelling, C. (1969). Synthesis of nucleic acids in giant chromosomes. *In* "Progress in Biophysics" (J. A. V. Butler, D. Noble, eds.), Vol. 19(1), 239–270. Pergamon Press, Oxford/London.
Pelling, C. (1970). Puff RNA in polytene chromosomes. *Cold Spring Harbor Symp. Quant. Biol.* **35,** 521–531.
Pelling, C. (1972a). Transcription in giant chromosomal puffs. *In* "Results and Problems in Cell Differentiation" (W. Beermann, ed.), Vol. 4, pp. 87–99. Springer-Verlag, Berlin/Heidelberg/New York.
Pelling, C. (1972b). RNA synthesis in giant chromosomal puffs and the mode of puffing. *FEBS Symp.* **24,** 77–89.
Pelling, C., and Beermann, W. (1966). Diversity and variation of the nucleolar organizing regions in Chironomids. *Natl. Cancer Inst. Monogr.* **23,** 393–409.
Pennypacker, M. (1950). Large "anal gill" chromosomes of a chironomid larva. *J. Hered.* **41,** 155, 164.
Perisic, O., Xiao, H., and Lis, J. T. (1989). Stable binding of *Drosophila* heat shock factor head-to-head and tail-to-tail repeats of a conserved 5bp recognition unit. *Cell* **59,** 797–806.
Perkins, L. A., Doctor, J. S., Zhang, K., Stinson, L., Perrimon, N., and Craig, E. A. (1990). Molecular and developmental characterization of the heat shock cognate 4 gene of *Drosophila melanogaster*. *Mol. Cell. Biol.* **10,** 3232–3238.
Perkowska, E. (1963a). The salivary gland secretion in *Bradysia*. *Exp. Cell Res.* **30,** 432–436.
Perkowska, E. (1963b). Some characteristics of the salivary gland secretion of *Drosophila virilis*. *Exp. Cell Res.* **32,** 259–271.
Perondini, A. L. P. (1971). Origin of bulb in the polytene chromosomes of *Sciara ocellaris* (Diptera, Sciaridae). *Cien. Cult.* **23,** 84–85.
Perondini, A. L. P., and Desen, E. M. (1969). Heterozygous puffs in *Sciara ocellaris*. *Genetics* **61**(Suppl.), 251–260.
Perondini, A. L. P., and Desen, E. M. (1983). Pufes assimtricos nos chromossomos politenicos de *Sciara ocellaris*. *Cienc. Cult.* **35,** 1929–1935.
Perondini, A. L. P., and Desen, E. M. B. (1985). Polytene chromosomes and the puffing patterns in the salivary glands of *Sciara ocellaris*. *Rev. Brasil. Genet.* **8,** 465–478.
Perondini, A. L. P., and Desen, E. M. (1988). The asymmetric bands of the polytene chromosomes of *Sciara ocellaris* (Diptera; Sciaridae). *Rev. Brasil. Genet.* **11,** 13–26.
Perondini, A. L. P., and Otto, P. A. (1991). Evidence for selective differences in single-band polytene chromosome polymorphism in *Sciara ocellaris*. *Heredity* **82,** 275–281.
Perov, N. A. (1971). "Electron microscopic study of active chromosomes with dipteran polytene chromosomes and chromosomes of lampbrush type as examples." Ph.D. Thesis, Moscow.
Perov, N. A., and Chentsov, Yu. S. (1971). Electron microscopic study on the salivary gland chromosomes of *Chironomus plumosus*. *Doklady AN SSSR* **196**(6), 1452–1455. [in Russian]
Perov, N. A., and Kiknadze, I. I. (1980). Differential reaction of regions of polytene chromosomes to the effect of high doses of actinomycin D. *Tsitologiya* **22**(3), 254–259. [in Russian]
Perov, N. A., Kiknadze, I. I., and Chentsov, Yu. S. (1975a). Specific ultrastructure of puffs in the 4E–2b-e region in *Chironomus thummi*. *Doklady Akad. Nauk SSSR* **223**(6), 1465–1467. [in Russian]
Perov, N. A., Kiknadze, I. I., and Chentsov, Yu S. (1975b). Ultrastructural organization of the polytene chromosomes of salivary glands of *Chironomus thummi*. *Tsitologiya* **17**(4), 390–397. [in Russian]
Perov, N. A., Kiknadze, I. I., and Chentsov, Yu. S. (1976). Ultrastructure of puffs in *Chironomus thummi* polytene chromosomes. *Tsitologiya* **18**(7), 840–846. [in Russian]
Perrimon, N., Engstrom, L., and Mahowald, A. P. (1984). Developmental genetics of the 2E-F region of the *Drosophila* X chromosome: A region rich in "developmentally important" genes. *Genetics* **108,** 559–572.

Perrimon, N., Engstrom, L., and Mahowald, A. P. (1985). Developmental genetics of the 2C-D region of the *Drosophila* X chromosome. *Genetics* **111,** 23–41.

Peters, F. P. A. M. N., Lubsen, N. H., and Sondermejer, P. J. A. (1980). Rapid sequence divergence in a heat shock locus of *Drosophila*. *Chromosoma* **81,** 271–280.

Peters, F. P. A. M. N., Grond, C. J., Sondermejer, P. J. A., and Lubsen, N. H. (1982). Chromosomal arrangement of heat shock locus 2–48B in *Drosophila hydei*. *Chromosoma* **85,** 237–249.

Peters, F. P. A. M. N., Lubsen, N. H., Walldorf, U., Moormann, R. J. M., and Hovemann, B. (1984). The unusual structure of heat shock locus 2–48B of *Drosophila hydei*. *Mol. Gen. Genet.* **197,** 392–398.

Peterson, J. S., and Petri, W. H. (1986). Update of linkage information on the *s70* chorion gene of *Drosophila melanogaster*. *Dros. Inform. Serv.* **63,** 158.

Peterson, N. S., Moller, G., and Mitchell, H. K. (1979a). Genetic mapping of the coding regions for three heat-shock proteins in *Drosophila melanogaster*. *Genetics* **92,** 891–902.

Peterson, N. S., Moller, G., and Mitchell, H. K. (1979b). Three *Drosophila melanogaster* genes which are expressed coordinately after heat shock are also located in the same region of the chromosome. *Fed. Proc. Soc. Am.* **38,** 297.

Petri, W. H., Wyman, A. R., and Henikoff, S. (1977). Synthesis of "heat shock" mRNA by *Drosophila melanogaster* follicle cells under standard organ culture conditions. *Dros. Inform. Serv.* **52,** 80.

Petrukhina, T. E. (1968). Study on chromosomal polymorphism in natural population of the red head midge. *Tsitologiya* **10**(9), 1148–1154. [in Russian]

Phillips, D. M. (1965). An ordered filamentous component in *Sciara* (Diptera) salivary gland nuclei. *J. Cell Biol.* **26,** 677–683.

Phillips, D. M., and Swift, H. (1965). Cytoplasmic fine structure of *Sciara* salivary glands. I. Secretion. *J. Cell Biol.* **27,** 395–409.

Phillips, J. P., Willms, J., and Pitt, A. (1982). a-amanitin resistance in three wild strains of *Drosophila melanogaster*. *Can. J. Genet. Cytol.* **24,** 151–162.

Pierce, D. A., and Lucchesi, J. C. (1980). Dosage compensation of X-linked heat-shock puffs in *Drosophila pseudoobscura*. *Chromosoma* **76,** 245–254.

Piper, P. (1987). How cells respond and adapt to heat stress through alterations in gene expression. *Sci. Progr.* **71,** 531–543.

Pironcheva, G. L. (1985). Heterogeneity of the nuclei of RNP particles. *Uspekhi mol. biol.* **4,** 28–37. (in Bulgarian)

Pirotta, V., Bonaldo, M. F., Amabis, J. M., Santelli, R. V., Frydman, H. M., Cattapan, L. D. N., and Lara, F. J. S. (1988). Cloning of sequences from the B-2 DNA puff of *Rhynchosciara* salivary chromosomes. *Arquives Biol. Technol.* **31,** 41.

Plagens, U., Greenleaf, A.L., and Bautz, E. K. F. (1976). Distribution of RNA polymerase on *Drosophila* polytene chromosomes as studied by indirect immunofluorescence. *Chromosoma* **59,** 157–165.

Plaut, W. (1963a). On the replicative organization of DNA in the polytene chromosome of *Drosophila melanogaster*. *J. Mol. Biol.* **7,** 632–635.

Plaut, W. (1963b). Pulse labelling studies of nucleoprotein synthesis on *Drosophila* polytene chromosomes. *Proc. Intern. Congr. Genet., 11th The Hague* **1,** 108.

Plaut, W. (1968). On the replication in polytene chromosomes. *In* "Replication and Recombination of Genetic Material" (W. J. Peacock and R. D. Brock, eds.), pp. 87–91. Australian Academy Science, Canberra.

Plaut, W. (1969). On ordered DNA replication in polytene chromosomes. *Genetics* **61**(Suppl.), 239–244.

Plaut, W., and Nash, D. (1964). Localized DNA synthesis in polytene chromosomes and its implications. *In* "The Role of Chromosomes in Development" (M. Locke, ed.), pp. 113–135. Academic Press, New York.

Plaut, W., Nash, D., and Fanning, T. (1966). Ordered replication of DNA in polytene chromosomes of *Drosophila melanogaster*. *J. Mol. Biol.* **16,** 85–93.

Poels, C. L. M. (1970). Time sequence in the expression of various developmental characters induced by ecdysterone in *Drosophila hydei*. *Dev. Biol.* **23,** 210–225.

Poels, C. L. M. (1972). Mucopolysaccharide secretion from *Drosophila* salivary gland cells as a consequence of hormone induced gene activity. *Cell Differ.* **1,** 63–78.

Poels, C. L. M., de Loof, A., and Berendes, H. D. (1971). Functional and structural changes in *Drosophila* salivary gland cells triggered by ecdysterone. *J. Insect Physiol.* **17,** 1717–1729.

Poels, C. L. M., Alonso, C., and de Boer, S. B. (1972). Functional capacities of isolated salivary glands in a chemically defined medium. *Dros. Inform. Serv.* **48,** 54–55.

Poeting, A., Koerwer, W., and Pongs, O. (1982). Ecdysterone induced protein synthesis in salivary glands and in fat body of *Drosophila* larvae. *Chromosoma* **87,** 89–102.

Pokholkova, G. V., and Solovyova, I. V. (1989). A characterization of insertional mutations produced in the P-M system of hybrid dysgenesis in the 9F12–10A7 region of the X chromosome of *Drosophila melanogaster*. *Genetika* **25**(10), 1776–1785. [in Russian]

Pokholkova, G. V. and Zhimulev, I. F. (1984). Genetic characterization of the 9F12–10A7 region of the X chromosome of *Drosophila melanogaster*. *Doklady Akad. Nauk SSSR* **274**(4), 934–938. [in Russian]

Pokholkova. G. V., Zhimulev, I. F., Grafodatsakaya, V. E., Fomina, O. V., Baricheva, E. M., Semeshin, V. F., and Belyaeva, E. S. (1982). A cytogenetic study fo the 9E–10A region of the X chromosome of *Drosophila melanogaster*. III. Genetic mapping of loci in the *ras-dsh* interval. *Genetika* **18**(2), 255–262. [in Russian]

Pokholkova, G. V., Solovjeva, I. V., and Belyaeva, E. S. (1991). Lethal mutations of the X chromosome 9F12–10A7 region induced by P-M hybrid dysgenesis. *Dros. Inform. Serv.* **70,** 179–180.

Pokholkova, G. V., Shloma, V. V., Semeshin, V. F., Belyaeva, E. S., and Zhimulev, I. F. (1997). Characteristics of the *DHR38* gene coding for one of the ecdysterone receptor in *Drosophila melanogaster*. *Tsitologia* **39**(1), 94–95. [in Russian]

Poluektova, E. V. (1969). Features of the functional morphology of the third chromosome in nuclei of the salivary glands of two related species *Drosophila virilis* and *Drosophila texana* at the stage of puparium formation. *Genetika* **5**(9), 124–128. [in Russian]

Poluektova, E. V. (1970a). A comparison of puff sizes in the homologous loci of the third chromosome in related species of *Drosophila* of the *virilis* group. *Genetika* **6**(5), 68–75. [in Russian]

Poluektova, E. V. (1970b). Structural modifications of homologous loci in the third chromosome of the salivary gland nuclei of hybrids from crosses between *Drosophila virilis* and *Drosophila texana*. *Genetika* **6**(9), 110–117. [in Russian]

Poluektova, E. V. (1975a). Change in the spectrum of puffs during the developmental process in *Drosophila virilis*. *Ontogenez* **6**(3), 263–268. [in Russian]

Poluektova, E. V. (1975b). Expression of puffs depending on genotype. *Genetika* **11**(7), 87–94. [in Russian]

Poluektova, E. V., and Mitrofanov, V. G. (1977). Age specificity of the action of 5-bromdeoxyuridine on the differentiation and spectrum of puffs in salivary gland chromosomes of *Drosophila virilis* Sturt. II. Changes in puffs spectrum in the salivary gland chromosomes. *Genetika* **13**(9), 1596–1604. [in Russian]

Poluektova, E. V., Mitrofanov, V. G., Burychenko, G. M., Myasnyankina, E. N., and Bakunina, E. D. (1975a). *Drosophila*. *In* "Ob'ekty biolgii razvitia," pp. 128–146. Nauka, Moscow. [in Russian]

Poluektova, E. V., Mitrofanov, V. G., and Gauze, G. G. (1975b). Changes in the spectrum of puffs of the salivary glands of *Drosophila virilis* under the effect of 5-bromdeoxyuridine. *Doklady AN SSSR* **222**(3), 720–722. [in Russian]

Poluektova, E. V., Mitrofanov, V. G. and Gauze, G. G. (1976). Heterogeneity for sensitivity to 5-bromdeoxyuridine in *Drosophila virilis* strains. *Genetika* **12**(4), 73–78. [in Russian]

Poluektova, E. V., Bakulina, E. D., and Mitrofanov, V. G. (1978a). Dependence of the puffing response in the chromosomes of the salivary glands of *Drosophila virilis* Sturt on age and sex of the recipient during transplantation into abdomen of adult. *Genetika* **14**(10), 1730–1739. [in Russian]

Poluektova, E. V., Mitrofanov, V. G., Bakunina, E. D., Polikarpova, S. I., and Kakpakov, V. G. (1978b). Effect of heat shock on the puffing of salivary gland chromosomes of *Drosophila virilis* Sturt under different conditions of culture. *Genetika* **14**(6), 1005–1015. [in Russian]

Poluektova, E. V., Mitrofanov, V. G., and Kakpakov, V. T. (1980a). Effect of insect hormones on puffing of the salivary gland chromosomes of *Drosophila virilis* Sturt cultured *in vitro*. I. Change in puffing during cilturing of glands in medium C-50. *Ontogenez* **11**(2), 175–180. [in Russian]

Poluektova, E. V., Mitrofanov, V. G., and Kakpakov, V. T. (1980b). Effect of insect hormones on puffing of the salivary gland chromosomes of *Drosophila virilis* Sturt cultured *in vitro*. II. Effect of ecdysterone and the juvenile hormone. *Ontogenez* **11**(4), 392–401. [in Russian]

Poluektova, E. V., Mitrofanov, V. G., and Kakpakov, V. T. (1980c). Effect of insect hormones on puffing of the salivary gland chromosomes of *Drosophila virilis* Sturt cultured *in vitro*. III. Primary effect of hormones. *Ontogenez* **11**(6), 600–607. [in Russian]

Poluektova, E. V., Bakulina, E. D., Mitrofanov, V. G. and Kakpakov, V. T. (1981a). Effect of insect hormones on puffing in the chromosomes of the salivary glands of *Drosophila virilis* Sturt cultures *in vitro*. IV. Modification of the hormonal posteffect in the abdomen of females. *Ontogenez* **12**(3), 273–282. [in Russian]

Poluektova, E. V., Polikarpova, S. I., Mitrofanov, V. G., and Svaninskaya, R. A. (1981b). Incotporation of 5-brodeozyuridine and ^{3}H-uridine into the salivary glands of the larvae of *Drosophila virilis* Sturt. *Ontogenez* **12**(6), 572–578. [in Russian]

Poluektova, E. V., Kakpakov, V. T., and Mitrofanov, V. G. (1985). Differentiation of puffing of the salivary gland chromosomes of *Drosophila virilis* in different *in vitro* cultivation media. *Ontogenez* **16**(4), 375–380. [in Russian]

Poluektova, E. V., Mitrofanov, V. G., Mukhovatova, L. M., and Kakpakov, V. T. (1986). Expression of ecdysterone puffs in the salivary gland chromosomes of *Drosophila virilis* Sturt larvae. *Ontogenez* **17**(5), 478–485. [in Russian]

Poluektova, E. V., Kozhanov, N. I., Chudakova, I. V., Mitrofanov, V. G., and Burov, V. N. (1987a). Expression of the "early" ecdysterone puffs in the chromosomes of the salivary glands of *Drosophila virilis* cultured in the abdomen of different insects. *Ontogenez* **18**(1), 19–27. [in Russian]

Poluektova, E. V., Mukhovatova, L. M., Kakpakov, V. T., and Mitrofanov, V. G. (1987b). Kinetics of puffing in regions indicators of ecdysterone in the salivary gland chromosomes of *Drosophila virilis* Sturt cultured *in vitro*. *Ontogenez* **18**(2), 140–146. [in Russian]

Polyakova, E. V., Mamadalieva, M. A., Yankulova, E. D., Poluektova, E. V., and Korochkin, L. I. (1989). Phenogenetic analysis of the absence of p-esterase character in *Drosophila lummei*. *Genetika* **25**(4), 641–649. [in Russian]

Polyakova, E. V., Poluektova, E. V., Yankulova, E. D., and Korochkin, L. I. (1990). Puffing in the regions of the esterase and ecdysterone genes in *Drosophila lummei* strains differing in the expression of p-esterase. *Ontogenez* **21**(1), 41–46. [in Russian]

Pongs, O. (1984a). Ecdysteroid-regulated puffs and genes in *Drosophila*. *In* "Biosynthesis, Metabolism and Mode of Action if Invertebrate Hormones" (J. Hoffmann and M. Porchett, eds.), pp. 285–292. Springer-Verlag, Heidelberg.

Pongs, O. (1984b) Active hormones and puffs. *Chromos. Today* **8,** 161–168.

Pongs, O. (1988). Ecdysteroid-regulated gene expression in *Drosophila melanogaster*. *Eur. J. Biochem.* **175,** 199–204.

Pongs, O. (1989). Biochemical properties of ecdysteroid receptors. *In* "Ecdysone: From Chemistry to Mode of Action" (J. Koolman, ed.), pp. 338–344. Georg Thieme-Verlag, Stuttgart/New York.

Pongs, O., Kecskemethy, N., Muller, R., Krah-Jentgens, I., Baumann, A., Kiltz, H. H., Canal, I., Lamazares, S., and Ferrus, A. (1988). Shaker encodes a family of putative potassium channel proteins in the nervous system of *Drosophila*. *EMBO J.* **7,** 1087–1096.

Posakony, J. W., Ficher, J. A., and Maniatis, T. (1985). Identification of DNA sequences required for the regulation of *Drosophila* Alcohol Dehydroganase gene expression. *Cold Spring Harbor Symp. Quant. Biol.* **50,** 515–520.

Poulson, D. F., and Metz, C. W. (1938). Studies on the structure of the nucleolus-forming regions in the giant salivary gland chromosomes of Diptera. *J. Morphol.* **63,** 363–395.

Prasad, J., Duttagupta, A. K., and Mukherjee, A. S. (1981). Transcription in the X-chromosomal segmental aneuploids of *Drosophila melanogaster* and regulation of dosage compensation. *Genet. Res. Cambr.* **38,** 103–113.

Prensky, W., Steffensen, D. M., and Hughes, W. L. (1973). The use of iodinated RNA for gene localization. *Proc. Natl. Acad. Sci. USA* **70,** 1860–1864.

Preston, R. A., Manolson, M. F., Becherer, K., Weidenhammer, E., Kirkpatrick, D., Wright, R., and Jones, E. W. (1991). Isolation and characterization of PEP3, gene required for vacuolar biogenesis in *Saccharomyces cerevisae*. *Mol. Cell. Biol.* **11,** 5801–5812.

Pret, A.-M., and Searles, L. L. (1991). Splicing of retrotransposon insertions from transcripts of the *Drosophila melanogaster vermilion* gene in a revertant. *Genetics* **129,** 1137–1145.

Price, G. M. (1974). Protein metabolism by the salivary glands and other organs of the larva of the blowfly, *Calliphora erythrocephala*. *J. Insect Physiol.* **20,** 329–347.

Pritchard, J. K., and Schaeffer, S. W. (1997). Plymorphism and divergence at a *Drosophila* pseudogene locus. *Genetics* **147,** 199–208.

Probeck, H.-D., and Rensing, L. (1974). Cellular patterns of differing circadian rhythms and levels of RNA synthesis in *Drosophila salivary* glands. *Cell Differ.* **2,** 337–345.

Procunier, J. D., and Dunn, R. J. (1978). Genetic and molecular organization of the 5S locus and mutants in *Drosophila melanogaster*. *Cell* **15,** 1087.

Procunier, J. D., and Tartof, K. D. (1975). Genetic analysis of the 5S RNA genes in *Drosophila melanogaster*. *Genetics* **81,** 515–523.

Procunier, J. D., and Tartof, K. D. (1976). Restriction map of 5S RNA genes of *Drosophila melanogaster*. *Nature* **263,** 255–257.

Procunier, J. D., and Tartof, K. D. (1978). A genetic locus having trans and contiguous cis functions that control the disproportionate replication of ribosomal RNA genes in *Drosophila melanogaster*. *Genetics* **88,** 67–69.

Procunier, W. S. (1982). The interdependence of B chromosomes, nucleolar organizer expression, and larval development in the blackfly species *Cnephia dacotensis* and *Cnephia ornitophilia* (Diptera: Simuliidae). *Can. J. Zool.* **60,** 2879–2896.

Procunier, W. S., and Post, R. J. (1986). Development of a method for the cytological identification of man-biting sibling species within the *Simulium damnosum* complex. *Trop. Med. Parasit.* **37,** 49–53.

Prokofyeva-Belgovskaya, A. A., and Bogdanov, Yu. F. (1963). Organization of the chromosome. *Zhurn. Vsesoyuzn. Khimich. obva im D. I. Mendeleyeva* **8**(1), 33–46. [in Russian]

Prostakova, T. M. (1968). A study of the effect of ribonuclease on RNA synthesis in the giant chromosomes of *Chironomus*. *Tsitologia* **10**(2), 157–161. [in Russian]

Protopopov, M. O., Belyaeva, E. S., Kokoza, E. B., Sharapova, E. M., Baricheva, E. M., Demakova, O. V., Graphodatskaya, V. E., and Zhimulev, I. F. (1988). Cytogenetic analysis of the 2B1-2–2B9-10 region in the X chromosome of *Drosophila melanogaster*. VII. Cloning the *ecs*, *dor* and *swi* loci. *Genetika* **24**(9), 1614–1623. [in Russian]

Protopopov, M. O., Belyaeva, E. S., Tretyakova, I. V., and Zhimulev, I. F. (1991). Molecular map of the 2B region of *Drosophila melanogaster* X-chromosome. *Dros. Inform. Serv.* **70,** 182–184.

Protzel, A., and Levenbook, L. (1976). Rate of rRNA synthesis during metamorphosis of the blowfly, *Calliphora vicina*. *Insect Biochem.* **6,** 631–635.

Prud'homme, N., Gans, M., Masson, M., Trezian, C., and Bucheton, A. (1995). Flamenco, a gene controlling the gypsy tetrovirus of *Drosophila melanogaster*. *Genetics* **139,** 697–711.

Puchalla, S. (1994). Polytene chromosomes of monogenic and amphogenic *Chrysomya* species (Calliphoridae, Diptera): Analysis of banding patterns and *in situ* hybridization with *Drosophila* sex determining gene sequences. *Chromosoma* **103,** 16–30.

Pulver, U., and Fischer, J. (1980). Uber eine Heterochromatin-Mutation aus einer Wildpopulation von *Chironomus nuditarsis*. II. Zum Replikationsverhalten des Veranderten Genomabschnittes. *Genetica* **54,** 87–90.

Pustell, J., Kafatos, F. C., Wobus, U., and Baumlein, H. (1984). Balbiani Ring DNA: Sequence comparisons and evolutionary history of a family of hierarchically repetitive protein-coding genes. *J. Mol. Evol.* **20,** 281–295.

Qian, S., and Pirrotta, V. (1995). Dosage compensation of the *Drosophila* gene required both the X chromosome environment and multiple intragenic elements. *Genetics* **139,** 733–744.

Qiu, Y., Chen, Ch.-N., Malone, T., Richter, L., Beckendorf, S. K., and Davis, R. L. (1991). Characterization of the memory gene *dunce* of *Drosophila melanogaster*. *J. Mol. Biol.* **222,** 553–565.

Quincey, R. V. (1971). The number and location of genes for 5S ribonucleic acid within the genome of *Drosophila melanogaster*. *Biochem. J.* **123,** 227–233.

Rabinowitz, M. (1941). Studies on the cytology and early embryology of the egg of *Drosophila melanogaster*. *J. Morphol.* **69,** 1–49.

Rae, P. M. M., Kohorn, B. D., and Wade, R. P. (1980). The 10kb *Drosophila virilis* 28S rDNA intervening sequence is flanked by a direct repeat of 14 base pairs of coding sequence. *Nucleic Acids Res.* **8,** 3491–3504.

Rae, P. M. M., Barnett, T., and Murtif, V. L. (1981). Nontranscribed spacers in *Drosophila* ribosomal DNA. *Chromosoma* **82,** 637–655.

Raghavan, K. V., Crosby, M. A., Mathers, P. H., and Meyerowitz, E. M. (1986). Sequences sufficient for correct regulation of Sgs-3 lie close to or within the gene. *EMBO J.* **5,** 3321–3326.

Raikow, R. B. (1973). Puffing in salivary gland chromosomes of picture-winged Hawaiian *Drosophila*. *Chromosoma* **41,** 221–230.

Ralcheva, N. (1977). Puff heterozigosity and nucleolar polymorphism in natural populations of Simuliidae. *Dokl. Bolgar. Acad. Sci.* **30,** 1217–1220. [in Russian]

Ramain, P., Bourouis, M., Dretzen, G., Richards, G., Sobkowiak, A., and Bellard, M. (1986). Changes in the chromatin structure of *Drosophila* glue genes accompany developmental cessation of transcription in wild type and transformed strains. *Cell* **45,** 545–553.

Ramain, P., Giangrande, A., Richards, G., and Bellard, M. (1988). Analysis of a DNA-aes I-hypersensitive site in transgenic *Drosophila* reveals a key regulatory element of Sgs3. *Proc. Natl. Acad. Sci. USA* **85,** 2718–2722.

Rambousek, F. J. (1912). Cytologicke pomery slinnych zlaz u larev *Chironomus plumosus Lin*. *Vestnik Kral. Ces. Spol. Nauk.* **2,** 1–25.

Ramesh, S. R., and Kalisch, W.-E. (1987). Salivary gland proteins in members of the *Drosophila nasuta* subgroup. *Dros. Inform. Serv.* **66,** 117.

Ramesh, S. R., and Kalisch, W.-E. (1988a). SDS-PAGE technique for demonstrating sex linked genes. *Dros. Inform. Serv.* **67,** 107–108.

Ramesh, S. R., and Kalisch, W.-E. (1988b). Glue proreins in *Drosophila nasuta*. *Biochem. Genet.* **26,** 527–541.

Ramesh, S. R., and Kalisch, W.-E. (1989). Taxonomic identification of *Drosophila nasuta* subgroup strains by glue protein analysis. *Genetica* **78,** 63–72.

Ramos, R. G. P., Grimwade, B. G., Wharton, K. A., Scottgale, T. N., and Artavanis-Tsakonas, S. (1989). Physical and functional definition of the *Drosophila Notch* locus by P element transformation. *Genetics* **123,** 337–348.

Ranjan, A., and Kaul, D. (1988). A heat shock induced puff in the foot pad cells of *Parasarcophaga ruficornis* (Fab.) (Sarcophagidae: Diptera). *Chromos. Inform. Serv.* **44,** 5–6.

Rapoport, I. A., Sakharova, M. N., Nikiforov, Yu. Z. and Beknazar'ants, M. M. (1970). Induction of new and specific puffs under the effect of perchlorate. *Doklady Akad. Nauk SSSR* **194**(3), 701–704. [in Russian]

Rapoport, I. A., Drozdovskaya, L. N., and Ivanitskaya, E. A. (1971a). Boron induced new puffs and modification localization of gene. *Genetika* **7**(8), 182–185. [in Russian]

Rapoport, I. A., Ivanitskaya, E. A., Nikiforov, Yu. L., and Beknazar'ants, M. M. (1971b). Cobalt induced new puffs and effect of cobalt on enzymes. *Doklady Akad. Nauk SSSR* **201**(2), 473–476. [in Russian]

Rasch, E. M. (1970). Two-wavelength cytophotometry of *Sciara* salivary gland chromosomes. *In* "Introduction to Quantitative Cytochemistry" (G. L. Wield and G. F. Bahr, eds.), pp. 335–355. Academic Press, New York/London.

Rasch, E. M., and Gawlik, S. (1964). Cytolisosomes in tissues of metamorphosing *Sciarid* larvae. *J. Cell Biol.* **23,** 123A.

Rasch, E. M., and Lewis, A. L. (1968). Effects of cortisone, gibberellic acid and starvation on puffing patterns in polytene chromosomes of *Sciara*. *J. Histochem. Cytochem.* **16,** 508.

Rasch, E. M., and Rasch, R. W. (1967). Extra desoxyribonucleic acid synthesis at specific sites of *Sciara* salivary gland chromosomes. *J. Histochem. Cytochem.* **15,** 793.

Rasmuson, B. (1967). Modulation of the puff in the tip of the X-chromosome in *Drosophila melanogater*. *Dros. Inform. Serv.* **42,** 72.

Rathore, H. S., and Swarup, H. (1980). Studies on the effects of neomycin on puffing in *Chironomus*. *Acta Histochem.* **67,** 86–94.

Rathore, H. S., and Swarup, H. (1982a). Toxicity of lead nitrate to *Chironomus* sp. larvae: A cytogenetic investigation. *Pakistan J. Zool.* **14,** 118–121.

Rathore, H. S., and Swarup, H. (1982b). Preliminary observations on the influence of thyroxine on puffing in *Chironomus*. *Pakistan J. Zool.* **14,** 243–248.

Rawls, J. M., and Fristrom, H. W. (1975). A complex locus that controls the first three steps of pyrimidine biosynthesis in *Drosophila*. *Nature* **255,** 738–740.

Rawls, J. M., Freund, J.-N., Jarry, B. P., Louis, C., Segraves, N. A., and Schedl, P. (1986). Organization of transcription units around the *Drosophila melanogaster rudimentary* locus and temporal pattern of expression. *Mol. Gen. Genet.* **202,** 493–499.

Rayle, R. E. (1967a). A new mutant in *Drosophila melanogaster* causing an ecdysone—Curable interruption of prepupal moult. *Genetics* **56**(Suppl.), 583.

Rayle, R. E. (1967b). "Mutation and chromosomal puffing in the vicinity of the *white* locus of *Drosophila melanogaster*." Ph.D. Thesis, University of Illinois, Urbana, Illinois.

Rayle, R. E. (1972). Genetic analysis of a short X-chromosomal region in *Dosophila melanogaster*. *Genetics* **71**(Suppl.), 50.

Rayle, R. E., and Green, M. M. (1968). A contribution to the genetic fine structure of the region adjacent to *white* in *Drosophila melanogaster*. *Genetica* **39,** 497–507.

Rayle, R. E., and Hoar, D. I. (1969). Gene order and cytological localization of several X-linked mutants of *Drosophila melanogaster*. *Dros. Inf. Serv.* **44,** 94.

Razin, S. V., Gromova, I. I., and Iarovaia, O. V. (1995). Specificity and functional significance of DNA interaction with the nuclear matrix: New approaches to clarify the old questions. *Int. Rev. Cytol. B* **162,** 405–448.

Razmakhnin, E. P., Kiknadze, I. I., Panova, T. M., Mertvetsov, N. P., Ammosov, A. D., and Sidorov, V. N. (1982). Study on the functional state of the nucleolar organizer in the polytene chromosomes of different tissues of *Chironomus* by in situ hybridization of nucleic acids. *Tsitologiya* **24**(8), 863–868. [in Russian]

Redfern, C. P. F. (1981a). Homologous banding patterns in the polytene chromosomes from the larval salivary glands and ovarian nurse cells of *Anopheles stephensi Liston (Culicidae)*. *Chromosoma* **83,** 221–240.

Redfern, C. P. F. (1981b). DNA replication in polytene chromosomes: similarity of termination patterns in somatic and germ-line derived polytene chromosomes of *Anopheles stephensi Liston* (Diptera: Culicidae). *Chromosoma* **84,** 33–47.

Redfern, C. P. F., and Bownes, M. (1983). Pleiotropic effects of the "ecdysoneless-1" mutation of *Drosophila melanogaster*. *Mol. Gen. Genet.* **189,** 432–440.

Renkawitz, R., and Kunz, W. (1975). Independent replication of the ribosomal RNA genes in the

polytrophic-meroistic ovaries of *calluphora erythrocephala*, *Drosophila hydei* and *Sarcophaga barbata*. *Chromosoma* **53,** 131–140.
Renkawitz, R., Gerbi, S. A., and Glatzer, K. H. (1979). Ribosomal DNA of the fly *Sciara coprophila* has a very small and homogenous repeat unit. *Mol. Gen. Genet.* **173,** 1–13.
Renkawitz-Pohl, R. (1978). Number of the repetitive euchromatic 5SRNA genes in polyploid tissues of *Drosophila hydei*. *Chromosoma* **66,** 249–258.
Renkawitz-Pohl, R., Glatzer, K. H., and Kunz, W. (1980). Characterization of cloned ribosomal DNA from *Drosophila hydei*. *Nucleic Acids Res.* **8,** 4593–4611.
Renkawitz-Pohl, R., Glatzer, K. H., and Kunz, W. (1981a). Ribosomal RNA genes with an intervening sequence are clustered within the X chromosomal ribosomal DNA of *Drosophila hydei*. *J. Mol. Biol.* **148,** 95–101.
Renkawitz-Pohl, R., Matsumoto, L., and Gerbi, S. A. (1981b). Two distinct intervening sequences in different ribosomal DNA repeat units of *Sciara coprophila*. *Nucleic Acids Res.* **9,** 3747–3764.
Rensing, L. (1969). Circadiane Rhythmik von *Drosophila—Speicheldrusen in vivo, in vitro* and nach Ecdysonzugabe. *J. Insect Physiol.* **15,** 2285–2303.
Rensing, L. (1971). Hormonal control of circadian rhythms in *Drosophila*. *In* "Biochronometry," pp. 527–540. Natl. Acad. Sci., Washington, D.C.
Rensing, L. (1973). Effects of 2,4-dinitrophenol and dinactin on heat-sensitive and ecdysone-specific puffs of *Drosophila* gland chromosomes *in vitro*. *Cell Differ.* **2,** 221–228.
Rensing, L. (1975). Wirkungen von monovalenten Kationen auf Indyktion und Regression von Puffs. *In* "Regulationsmechanismen der Gnaktivitat und Replikation bei Riesenchromosomen" (L. Rensing, H.-H. Trepte, and G. Birukow, eds.), Vol. II, pp. 38–39. Mat.-Physik kl. Nachr. Akad. Wissen, Gottingen.
Rensing, L., and Fischer, M. (1975). The effects of sodium, potassium and ATP on a developmental puff sequence in *Drosophila* salivary glands *in vitro*. *Cell Differ.* **4,** 209–217.
Rensing, L., and Hardeland, R. (1967). Zur Wirkung der circadian Rhythmik auf die Entwicklung von *Drosophila*. *J. Insect Physiol.* **13,** 1547–1568.
Rensing, L., and Hardeland, R. (1972). Effects of adenosine 3′, 5′-cyclic monophosphate on membrane potential, nuclear volume, and puff size in *Drosophila* salivary gland *in vitro*. *Exp. Cell Res.* **73,** 311–318.
Rensing, L., Trepte, H.-H., and Birukow, G. (1975). (editors) "Regulationsmechanismen der Genaktivitat und Replikation bei Riesenchromosomen," Vol. II. Nachrichten der Akademie der Wissenschaften in Gottingen. Math.-Physik.
Rensing, L., Olomski, R., and Drescher, K. (1982). Genetics and models of the *Drosophila* heat-shock system. *Systems* **15,** 341–356.
Restifo, L., and Guild, G. M. (1986a). An ecdysone-responsive puff site in *Drosophila* contains a cluster of seven differentially regulated genes. *J. Mol. Biol.* **18,** 517–528.
Restifo, L., and Guild, G. M. (1986b). Poly (A) shortening of coregulated transcripts in *Drosophila*. *Dev. Biol.* **115,** 507–510.
Restifo, L., and Merrill, V. K. L. (1994). Two *Drosophila* regulatory genes, *Deformed* and the *Broad-Complex*, share common functions in development of adult CNS, head and salivary glands. *Dev. Biol.* **162,** 465–485.
Restifo, L., and White, K. (1991). Mutations in a steroid hormone-regulated gene disrupt the metamorphosis of the central nervous system in *Drosophila*. *Dev. Biol.* **148,** 1–21.
Restifo, L., and White, K. (1992). Mutations in a steroid hormone-regulated gene disrupt the metamorphosis of internal tissues in *Drosophila:* salivary glands, muscle, and gut. *Roux's Arch. Devel. Biol.* **201,** 221–234.
Restifo, L. L., Estes, P. S., and Dello Russo, C. (1995). Genetisc of ecdysteroid-regulated central nervous system metamorphosis in *Drosophila* (Diptera, Drosophilidae). *Eur. J. Entomol.* **92,** 169–187.
Reznik, N. A., Kerkis, I. E., and Gruzdev, A. D. (1981). Temperature induction in *Chironomus*. *Doklady Akad. Nauk SSSR* **261**(1), 202–203. [in Russian]

Reznik, N. A., Kerkis, I. E., and Gruzdev, A. D. (1983). Effect of short-term temperature shock on puffing of the polytene chromosomes of the salivary glands of *Chironomus thummi*. *Tsitologia* **25**(8), 918–926. [in Russian]

Reznik, N. A., Gruzdev, A. S., and Kerkis, I. E. (1984). Puffing and structural changes following brief temperature shocks in the polytene chromosomes of *Chironomus thummi*. *Biol. Zbl.* **103,** 267–282.

Reznik, N. A., Yampol, G. P., Kiseleva, E. V., and Gruzdev, A. S., Khristolyubova N. B. (1991a). Functional and structural units in the chromomere. *Genetica* **83,** 293–299.

Reznik, N. A., Yampol, Y. P., Kiseleva, E. V., Gruzdev, A. D., and Khristolyubova, N. B. (1991b). The possible functional role of the chromosomes structure. *In* "Molekulyarnye mekhanizmy geneticheskikh processov," pp. 51–58. Nauka, Moscow. [in Russian]

Ribbert, D. (1967). Die Polytanchromosomen der Borstenbildungzellen von *Calliphora erythrocephala* unter besonderer Berucksichtigung der geschlechts gebundenen Structurheterozygotie und des Puffmusters wahrend der Metamorphose. *Chromosoma* **21,** 296–344.

Ribbert, D. (1972). Relation of puffing to bristle and footpad differentiation in *Calliphora* and *Sarcophaga*. *In* "Results and problems in Cell Differentiation" (W. Beermann, ed.), pp. 153–179. Springer-Verlag, Berlin/Heidelberg/New York.

Ribbert, D. (1975). Unterschiedliche Chromomerenmuster von Polytanchromosomen in Keibahn und Soma der Fliege *Calliphora erythrocephala*. *In* "Regulationsmechanismen der Genaktivitat und Replikation bei Riesenchromosomen" (L. Rensing, H.-H. Trepte, and G. Birukow, eds.), pp. 189–192. Vandenhoeck and Ruprecht, Gottingen.

Ribbert, D. (1979). Chromomeres and puffing in experimentally induced polytene chromosomes of *Calliphora erythrocephala*. *Chromosoma* **74,** 269–298.

Ribbert, D., and Brier, K. (1969). Multiple nucleoli and enhanced nucleolar activity in the nurse cells of the insect ovary. *Chromosoma* **27,** 178–197.

Ribbert, D., and Buddendick, M. (1984). Synthesis and processing of ribosomal RNA in the growing oocytes of *Calliphora erythrocephala*. *Insect Biochem.* **14**(5), 569–586.

Richards, G. (1976a). The *in vitro* induction of puffing in salivary glands of the mutant *l(2)gl* of *Drosophila melanogaster* by ecdysone. *Wilh. Roux's Arch.* **179,** 339–348.

Richards, G. P. (1976b). The control of prepupal puffing patterns *in vitro:* Implications for prepupal ecdysone titres in *Drosophila melanogaster*. *Dev. Biol.* **48,** 191–195.

Richards, G. (1976c). Sequential gene activation by ecdysone in polytene chromosomes of *Drosophila melanogaster*. IV. The mid prepupal puffs. *Dev. Biol.* **54,** 256–263.

Richards, G. (1976d). Sequential gene activation by ecdysone in polytene chromosomes of *Drosophila melanogaster*. V. The late prepupal puffs. *Dev. Biol.* **54,** 264–275.

Richards, G. (1976e). Keeping genes in order. *New Scientist,* **22,** 171–172.

Richards, G. (1978a). The relative biological activities of a- and b-ecdysone and their 3 dehydro derivatives in the chromosome puffing assay. *J. Insect Physiol.* **24,** 329–335.

Richards, G. (1978b). Sequential gene activation by ecdysone in polytene chromosomes of *Drosophila melanogaster*. VI. Inhibition by juvenile hormones. *Dev. Biol.* **66,** 32–42.

Richards, G. (1980a). The polytene chromosomes in the fat body nuclei of *Drosophila melanogaster*. *Chromosoma* **79,** 241–250.

Richards, G. (1980b). Ecdysteroids and puffing in *Drosophila melanogaster*. *In* "Progress in Ecdysone Research. Developments in Endocrinology" (J. A. Hoffmann, ed.), pp. 363–378. Elsevier, Amsterdam.

Richards, G. (1981a). The radioimmune assay of ecdysteroid titres in *Drosophila melanogaster*. *Mol. Cell Endocrinology* **21,** 181–197.

Richards, G. (1981b). Insect hormones in development. *Biol. Rev.* **56,** 501–549.

Richards, G. (1982). Sequential gene activation by ecdysteroids in polytene chromosomes of *Drosophila melanogaster*. VII. Tissue specific puffing. *Wilh. Roux's Arch.* **191,** 103–111.

Richards, G. (1985). Polytene chromosomes. *In* "Comprehensive Insect Physiology, Biochemistry and Pharmacology" (G. A. Kerkut and L. I. Gilbert, eds.), Vol. 2, pp. 255–300. Pergamon Press, Oxford/New York/Toronto/Sydney/Paris/Frankfurt.

Richards, G. (1986). Into the lair of the gene. *Archives of Insect Biochemistry and Physiology*, Suppl., 143–155.

Richards, G. (1987). Gene regulation during insect development. *Int. J. Invertebr. Reprod. Dev.* **12,** 115–144.

Richards, G. (1992). Switching partners? *Curr. Biol.* **2,** 657–659.

Richards, G. (1996). The ecdysone regulatory cascades in *Drosophila*. *In* "Advances in Developmental Biochemistry," Vol. 5, pp. 81–135.

Richards, G., and Ashburner, M. (1984). Insect hormones and the regulation of genetic activity. *In* "Biological regulation and development" (R. F. Goldberger and K. R. Yamamoto, eds.), Vol. 3b, pp. 213–253. Plenum, New York.

Richards, G., and Hoffmann, J.A. (1986). Introduction to the insect neuroendocrine system. *In* "Handbook of natural Pesticides. III. Insect Growth regulators" (E. D. Morgan and N. B. Mandava, eds.), Part A, pp. 1–14. CRC Press, Boca Raton, FL.

Richards, G., and Lepesant, J.-A. (1984). Ecdystroid regulation of the major transcripts of *Drosophila melanogaster* larval salivary glands and fat bodies. *In* "Biosynthesis, Metabolism and Mode of Action of Invertebrate Hormones" (J. Hoffmann and M. Porchet, eds.), pp. 273–284. Springer-Verlag, Berlin/Heidelberg.

Richards, G., Cassab, A., Bourouis, M., Jarry, B., and Dissous, C. (1983). The normal developmental regulation of a cloned Sgs-3 "glue" gene chromosomally integrated in *Drosophila melanogaster* by P element transformation. *EMBO J.* **2,** 2137–2142.

Riddiford, L. M. (1993). Hormones and *Drosophila* develoment. *In* "The development of *Drosophila melanogaster*" (M. Bate, and A. M. Arias, eds.), pp. 899–939. Cold Spring Harbor Laboratory Press, Cold Spring Harbor, NY.

Riddihough, G., and Pelham, H.R.B. (1986). Activation of the *Drosophila hsp27* promoter by heat shock and by ecdysone involves independent and remote regulatory sequences. *EMBO J.* **5,** 1653–1658.

Riddihough, G., and Pelham, H. R. B. (1987). An ecdysone response element in the *Drosophila hsp27* promoter. *EMBO J.* **6,** 3729–3734.

Ringborg, U., and Rydlander, L. (1971). Nucleolar-derived ribonucleic acid in chromosomes, nuclear sap, and cytoplasm of *Chironomus tentans* salivary gland cells. *J. Cell Biol.* **51,** 355–368.

Ringborg, U., Daneholt, B., Edstrom, J.-E., Egyhazi, E., and Lambert, B. (1970a). Electrophoretic characterization of nucleolar RNA from *Chironomus tentans* salivary gland cells. *J. Mol. Biol.* **51,** 327–340.

Ringborg, U., Daneholt, B., Edstrom, J.-E., Egyhazi, E., and Rydlander, L. (1970b). Evidence for transport of preribosomal RNA from the nucleolus to the chromosomes in *Chironomus tentans* salivary gland cells. *J. Mol. Biol.* **51,** 679–686.

Ris, H. (1957). Chromosome structure. *In* "The Chemical Basis of Heredity" (W. D. McElroy and B. Glass, eds.), pp. 23–69. The Johns Hopkins Press, Baltimore.

Ris, Y. (1960). Structure of the chromosomes. *In* "Khimicheskie osnovy nasledstvennosti," pp. 25–49. Inostrannaya literatura, Moscow. [in Russian]

Risau, W., Symmons, P., Saumweber, H., and Frasch, M. (1983). Nonpackaging and packaging proteins of *hn* RNA in *Drosophila melanogaster*. *Cell* **33,** 529–541.

Ritossa, F. (1962). A new puffing pattern induced by temperature shock and DNP in *Drosophila*. *Experientia* **18,** 571–573.

Ritossa, F. M. (1963). New puffs induced by temperature shock, DNP and salicilate in salivary chromosomes of *Drosophila melanogaster*. *Dros. Inform. Serv.* **37,** 122–123.

Ritossa, F. M. (1964a). Experimental activation of specific loci in polytene chromosomes of *Drosophila*. *Exp. Cell Res.* **35,** 601–607.

Ritossa, F.M. (1964b). Behaviour of RNA and DNA synthesis at the puff level in salivary gland chromosomes of *Drosophila*. *Exp. Cell Res.* **36,** 515–523.

Ritossa, F. M. (1968). Unstable redundancy of genes for ribosomal RNA. *Proc. Natl. Acad. Sci. USA* **60,** 509–516.
Ritossa, F. M. (1972). Procedure for magnification of lethal deletions of genes for ribosomal RNA. *Nature New Biol.* **240,** 109–111.
Ritossa, F. M. (1976). The *bobbed* locus. *In* "The Genetics and Biology of *Drosophila*" (M. Ashburner and E. Novitsky, eds.), Vol. 1b, pp. 801–846. Academic Press, London/New York/San Francisco.
Ritossa, F. M., and von Borstel, R. C. (1964). Chromosome puffs in *Drosophila* induced by ribonuclease. *Science* **145,** 513–514.
Ritossa, F. M., and Pulitzer, J. F. (1963). Aspects of structure of polytene chromosome puffing of *Drosophila busckii* derived from experients with antibiotics. *J. Cell Biol.* **19,** 60A.
Ritossa, F. M., and Scala, G. (1969). Equilibrum variations in the redundancy of rDNA in *Drosophila melanogaster*. *Genetics* **61**(Suppl.), 304–317.
Ritossa, F. M., and Spiegelman, S. (1965). Localization of DNA complementary to ribosomal RNA in the nucleolus organizer region of *Drosophila melanogaster*. *Proc. Natl. Acad. Sci. USA* **53,** 737–745.
Ritossa, F. M., von Borstel, R. C., and Swift, H. (1964). The action of ribonuclease on salivary gland cells of *Drosophila*. *Genetics* **50**(2), 279.
Ritossa, F. M., Pulitzer, J. F., Swift, H., and von Borstel, R. C. (1965). On the action of ribonuclease in salivary gland cells of *Drosophila*. *Chromosoma* **16,** 144–151.
Ritossa, F. M., Atwood, K. C., Lindsley, D. L., and Spiegelman, S. (1966a). On the chromosomal distribution of DNA complementary to ribosomal and soluble RNA. *Natl. Cancer Inst. Monogr.* **23,** 449–472.
Ritossa, F. M., Atwood, K. C., and Spiegelman, S. (1966b). On the redundancy of DNA complementary to amino acid transfer rNA and its absence from the nucleolar organizer region of *Drosophila melanogaster*. *Genetics* **54,** 663–676.
Ritossa, F. M., Atwood, K. C., and Spiegelman, S. (1966c). A molecular explanation of the *bobbed* mutants of *Drosophila* as partial deficiencies of ribosomal DNA. *Genetics* **54,** 819–834.
Ritossa, F. M., Scalenghe, F., Di Turi, N., and Contini, A. M. (1970). On the cell stage of X-Y recombination during rDNA magnification in *Drosophila*. *Cold Spring Harbor Symp. Quant. Biol.* **38,** 483–490.
Rizki, T. M. (1967). Ultrastructure of the secretory inclusions of the salivary gland cell in *Drosophila*. *J. Cell Biol.* **32,** 531–534.
Roap, A. K., Marijnen, J. G., van der Ploeg, M. (1984). Anti DNA''RNA sera specificity test and application in quantitative *in situ* hybridization. *Histochemistry* **81,** 517–520.
Roark, M., Raghavan, K. V., Todo, T., Mayerda, C. A., and Meyerowitz, E. M. (1990). Cooperative enhancement at the *Drosophila* Sgs-3 locus. *Dev. Biol.* **139,** 121–133.
Robert, M., and Kroeger, H. (1965). Localization zusatzlicher RNS-synthesi in Trypsin-behandelten Riesenchromosomen von *Chironomus thummi*. *Experientia* **21,** 326–327.
Roberts, B., Whitten, J. M., and Gilbert, L. I. (1976). Patterns of incorporation of tritiated thymidine by the dorsal polytene foot-pad nuclei of *Sarcophaga bullata* (Sarcophagidae: Diptera). *Chromosoma* **54,** 127–140.
Roberts, D. B., Brock, H. W., Rudden, N. C., and Evans-Roberts, S. (1985). A genetic and cytologenetic analysis of the region syrrounding the *LSP-1* b-gene in *Drosophila melanogaster*. *Genetics* **109,** 145–156.
Roberts, P. A. (1971). Chromosomes of functionally different malpighian tubes of *Drosophila gibberosa*. *Genetics* **68**(Suppl.), 54.
Roberts, P. A. (1972). Puff patterns in the larval fat body of *Drosophila gibberosa*. *Genetics* **71**(3), 52.
Roberts, P. A. (1983). Gene activity changes in the course of aging in *Drosophila gibberosa*. *Genetics* **104**(1), 60.
Roberts, P. A. (1988). Developmental changes in midgut chromosomes of *Drosophila gibberosa*. *Chromosoma* **97,** 254–260.

Roberts, P. A., and Jacobsen, J. (1991). An overview of gene activity in the fat body of *Drosophila gibberosa*. *Chromosoma* **101,** 115–122.
Roberts, P., and MacPhail, L. A. (1985). Structure and activity of salivary gland chromosomnes of *Drosophila gibberosa*. *Chromosoma* **92,** 55–68.
Robinson, J. S., Graham, T. R., and Emr, S. D. (1991). A putative zinc finger protein *Saccharomyses cerevisae* Ups18p, affects late Golgi functions required for vacuolar protein sorting and efficient alpha-factor prohormone maturation. *Mol. Cell. Biol.* **11,** 5813–5824.
Rodems, A. E., Hendrikson, P. A., and Clever, U. (1969). Proteolytic enzymes in the salivary gland of *Chironomus tentans*. *Experienta* **25,** 686–687.
Rodionov, A. V., Smirnov, A. F., and Smaragdov, M. Y. (1981). Effect of environmental factors on the chromosomes of eucaryotes. *In* "Issledovaniya po genetike," pp. 41–54. [in Russian]
Rodman, T. C. (1964). The larval characteristics and salivary gland chromosomes of a tumorigenic strain of *Drosophila melanogaster*. *J. Morphol.* **115,** 419–445.
Rodman, T. C. (1968a). Intranucleolar DNA of polytene chromosomes. *J. Cell Biol.* **39**(2), 113a.
Rodman, T. C. (1968b). Relationship of developmental stage to initiation of replication in polytene nuclei. *Chromosoma* **23,** 271–287.
Rodman, T. C. (1969). Morphology and replication of intranucleolar DNA in polytene nuclei. *J. Cell Biol.* **42,** 575–582.
Rodman, T. C., and Kopac, M. J. (1964). Alterations in morphology of polytene chromosomes. *Nature* **202,** 876–877.
Roehrdanz, R. L., Kitchens, J. M., and Licchesi, J. C. (1977). Lack of dosage compensation for an autosomal gene relocated to the X chromosome in *Drosophila melanogaster*. *Genetics* **85,** 489–496.
Rohm, P. B. (1947). A study of evolutionary chromosome changes in *Sciara* (Diptera). Chromosome C in the salivary gland cells of *Sciara ocellaris* and *Sciara reinoldsi*. *Am. Natur.* **81,** 5–29.
Roiha, H., and Glover, D. (1981). Duplicated rDNA sequences of variable lengths flanking the short type I insertions in the rDNA of *Drosophila melanogaster*. *Nucleic Acids Res.* **9,** 5521–5532.
Roiha, H., Miller, J. R. Woods, J. C., and Glover, D. M. (1981). Arrangements and rearrangements of sequences flanking the two types of rDNA insertion in *Drosophila melanogaster*. *Nature* **290,** 749–753.
Roiha, H., Read, C. A., Browne, M. J. and Glover, D. M. (1983). Widely differing degrees of secuence conservation of the two types of rDNA insertion within the melanogaster species subgroup of *Drosophila*. *EMBO J.* **2,** 721–726.
Romans, P., Hodgetts, R. B., and Nash, D. (1976). Maternally influenced embryonic lethality: Allelic specific genetic rescue at a female fertility locus in *Drosophila melanogaster*. *Can. J. Genet. Cytol.* **18,** 773–781.
Roote, J., Gubb, D., and Ashburner, M. (1996). Aberrations in the l(2)34Fc–l(2)35Bb region of chromosome arm 2L. *Dros. Inform. Serv.* **77,** 55–65.
Ross, E. B. (1939). The post-embryonic development of the salivary glands of *Drosophila melanogaster*. *J. Morphol.* **65,** 471–496.
Rothfels, K., Feraday, R., and Kaneps, A. (1982). A cytological description of sibling species of *Simulium venustum* and *S. verecundum* with standard maps for the subgenus *Simulium Davies* (Diptera). *Can. J. Zool.* **60,** 1110–1128.
Rougvie, A. E., and Lis, J. T. (1988). The RNA polymerase II molecule at the 5′ end of the uninduced *hsp70* gene of *Drosophila melanogaster* is transcriptionally engaged. *Cell* **54,** 795–804.
Rougvie, A. E., and Lis, J. T. (1990). Postinitiation transcriptional control in *Drosophila melanogaster*. *Mol. Cell. Biol.* **10,** 6041–6045.
Rowe, T., Couto, E., and Kroll, D. J. (1987). Camptothecin inhibits *hsp70* genes in *Drosophila*. *Natl. Cancer Inst. Monogr.* **4,** 49–53.
Roy, S., and Lakhotia, S. C. (1977). Synchrony of replication in sister salivary glands of *Drosophila kikkawai*. *Ind. J. Exp. Biol.* **15,** 794–796.
Roy, S., and Lakhotia, S. C. (1979). Replication in *Drosophila* chromosomes: part II—unusual

replicative behaviour of two puff sites in polytene nuclei of *Drosophila kikkawai*. *Ind. J. Exp. Biol.* **17,** 231–238.

Rubin, G. M., and Hogness, D. S. (1975). Effect of heat shock on the synthesis of low molecular weight RNAs in *Drosophila:* accumulation of a novel form of 5S RNA. *Cell* **6,** 207–213.

Rubinstein, L., and Clever, U. (1972). Chromosomal activity and cell function in polytenic cells. V. Developmental changes in RNA synthesis and turnover. *Dev. Biol.* **27,** 519–537.

Rudkin, G. T. (1955). The ultraviolet absorption of puffed and unpuffed homologous regions in the salivary gland chromosomes of *Drosophila melanogaster*. *Genetica* **40,** 593.

Rudkin, G. T. (1959). A comparison of the incorporation of tritiated thymidine and cytidine into giant chromosome puffs. *Genetics* **44,** 531–532.

Rudkin, G. T. (1964) The proteins of polytene chromosomes. "Nucleohistones" (J. Bonner and P. Ts'o, eds.), pp. 184–192. Holden Day, San Francisco/London/Amsterdam.

Rudkin, G. T. (1965). The ralative mutabilities of DNA in regions of the X chromosome of *Drosophila melanogaster*. *Genetics* **52,** 665–681.

Rudkin, G. T. (1972). Replication in polytene chromosomes. *In* "Results and Problems in Cell Differentiation" (W. Beermann, ed.), Vol. 4, pp. 59–85. Springer-Verlag, Berlin/Heidelberg/ New York.

Rudkin, G. T. (1973). Cyclic synthesis of DNA in polytene chromosomes of Diptera. *In* "Cell Cycle in Development and Differentiation" (M. Balls and F. Billet, eds.), pp. 279–292. Cambridge Univ. Press, London.

Rudkin, G. T., and Corlette, S. L. (1957). Disproportionate synthesis of DNA in a polytene chromosome region. *Proc. Natl. Acad. Sci. USA* **43,** 964–968.

Rudkin, G. T., and Duncan, G. D. (1984). The control of gene relocation and transcription. *Inst Cancer res. Sci. Rept.*, 34–35.

Rudkin, G. T., and Stollar, B. D. (1977a). High resolution detection of DNA-RNA hybrids *in situ* by indirect immunofluorescence. *Nature* **265,** 472–474.

Rudkin, G. T., and Stollar, B. D. (1977b). Naturally occurring DNA"RNA hybrids. I. Normal patterns in polytene chromosomes. *In* "Human Cytogenetics: ICN-UCLA Symposia on Molecular and Cellular Biology," Vol. 7, 257–269. Academis Press, New York.

Rudkin, G. T., and Woods, P. S. (1959). Incorporation of ^{3}H-cytidine and ^{3}H-thymidine into giant chromosomes of *Drosophila melanopgaster* during puff formation. *Proc. Natl. Acad. Sci. USA* **45,** 997–1003.

Russel, R. J., Healy, M. J., and Oakeshott, J. G. (1992). Molecular analysis of the *lethal(1)B214* region at the base of the X chromosome of *Drosophila melanogaster*. *Chromosoma* **101,** 456–466.

Russel, S., and Ashburner, M. (1996). Ecdysone-regulated chromosome puffing in *Drosophila melanogaster*. *Metamorphosis*, 109–143.

Russel, S. R. H., Heimbeck, G., Goddard, C. M., Carpenter, A. T. C., and Ashburner, M. (1996). The Drosophila *Eip78C* gene is not vital but has a role in regulating chromosome puffs. *Genetics* **144,** 159–170.

Rydlander, L. (1984). Isolation and characterization of the two giant secretory proteins in salivary glands of *Chironomus tentans*. *Biochemistry* **220,** 423–431.

Rydlander, L., and Edstrom, J.-E. (1980). Large size nascent protein as dominating component during protein synthesis in *Chironomus* salivary glands. *Chromosoma* **81,** 85–99.

Rydlander, L., Pigon, A., and Edstrom, J.-E. (1980). Sequences translated by Balbiani Ring 75S RN *in vitro* are present in giant secretory protein from *Chironomus tentans*. *Chromosoma* **81,** 101–113.

Rykowski, M. C., Parmelee, S. J., Agard, D. A., and Sedat, J. W. (1988). Precise determination of the molecular limits of a polytene chromosome band: regulatory sequences for the *Notch* gene arc in the interband. *Cell* **54,** 461–472.

Ryseck, R.-P., Walldorf, U., and Hovemann, B. (1985). Two major RNA products are transcribed from heat-shock locus 93D of *Drosophila melanogaster*. *Chromosoma* **93,** 17–20.

Ryseck, R.-P., Walldorf, U., Hoffmann, T., and Hovemann, B. (1987). Heat shock loci 93D of

Drosophila melanogaster and 48B of *Drosophila hydei* exibit a common structural and transcriptional pattern. *Nucleic Acids Res.* **15,** 3317–3333.
Sachs, R. I., and Clever, U. (1972). Unique and repetitive DNA sequences in the genome of *Chironomus tentans*. *Exp. Cell Res.* **74,** 587–591.
Saiga, H., Grond, C., Schmidt, E. R., and Edstrom, J.-E. (1987). Evolutionary conservation of 3' ends of members of a family of giant secretory proteins in *Chironomus*. *J. Mol. Evol.* **25,** 20–28.
Saiga, H., Botella, L., and Edstrom, J.-E. (1988). Subrepeats within the BR1b repeat unit in *Chironomus pallidivittatus* can be classified into different types depending on codon usage. *J. Mol. Evol.* **27,** 298–302.
Sakharova, M. N., Rapoport, I. A., and Beknazar'ants, M. M. (1971). Puffs induced by rodanide and a puff model for determining drug affected enzymes. *Doklady Akad. Nauk SSSR* **196**(5), 1217–1220. [in Russian]
Salpeter, M. M., Bachmann, L., and Salpeter, E. E. (1969). Resolution in electron microscope radioautography. *J. Cell Biol.* **41,** 1–20.
Salz, H., Davis, R. L., and Kiger, J. A., Jr. (1982). Genetic analysis of chromomere 3D4 in *Drosophila melanogaster:* The *dunce* and *sperm-amotile* genes. *Genetics* **100,** 587–596.
Salz, H. K., and Kiger, J. A., Jr. (1984). Genetic analysis of chromomere 3D4 in *Drosophila melanogaster*. II. Regulatory sites for the *dunce* gene. *Genetics* **108,** 377–392.
Sampedro, J., Galceran, J., and Izquierdo, M. (1989). Mutation mapping of the 2B5 ecdysone locus in *Drosophila melanogaster* reveals a long-distance controlling element. *Mol. Cell. Biol.* **9,** 3588–3591.
Sanchez-Herrero, E., Casanova, J., Kerridge, S., and Morata, G. (1985). Anatomy of the *bithorax complex* of *Drosophila*. *Cold Spring Harbor Symp. Quant. Biol.* **50,** 165–172.
Sanderson, C. L., McLachlan, D. R. C., and DeBoni, U. (1982). Altered steroid induced puffing by chromatin bound aluminium in a polytene chromosome of the blackfly *Simulium vittattum*. *Can. J. Genet. Cytol.* **24,** 27–36.
Sandstrom, D. J., Bayer, C. A., Fristrom, J. W. and Restifo, L. L. (1997). *Broad-Complex* transcription factors regulate thoracic muscle attachment in *Drosophila*. *Devel. Biol.* **181,** 168–185.
Sang, J. H. (1968). Lack of cortisone inhibition of chromosomal puffing in *Drosophila melanogaster*. *Experientia* **24,** 1064.
Santa-Cruz, M. C., and Diez, J. L. (1979). Galactose induced Balbiani-Ring-like structures in chromosomes I and II of *Chironomus thummi*. *Experientia* **35,** 48–50.
Santa-Cruz, M. C., Villanueva, A., and Diez, J. L. (1978). Effect of galactose treatment in the puffing pattern of *Chironomus thummi* Balbiani Rings. *Chromosoma* **69,** 93–100.
Santa-Cruz, M. C., Morcillo, G., Aller, P., and Diez, J. L. (1981). Heat shock induced puffing changes in *Chironomus thummi*. *Cell Differ.* **10,** 33–38.
Santelli, R. V., Machado-Santelli, G. M., and Lara, F. J. S. (1976). *In vitro* transcription by isolated nucleic of *Rhynchosciara americana* salivary glands. Characteristics of incorporation and inhibition by a-amanitin. *Chromosoma* **56,** 69–84.
Santos, J. F., dos, Lewgoy, F., and Valente, V. L. S. (1993). Heat shock puffs are induced by selenium in *Drosophila melanogaster*. *Dros. Inform. Serv.* **72,** 145–146.
Sass, H. (1980a). Features of *in vitro* puffing and RNA synthesis in polytene chromosomes of *Chironomus*. *Chromosoma* **78,** 33–78.
Sass, H. (1980b). Puffing und RNA-synthese in larvalen und imaginalen Polytanchromosomen aus verschiedenen Geweben von *Chironomus tentans*. *Biol. Zbl.* **99,** 399–428.
Sass, H. (1981). Effects of DMSO on the structure and function of polytene chromosomes of *Chironomus*. *Chromosoma* **83,** 619–643.
Sass, H. (1982). RNA polymerase B in polytene chromosomes: Immunofluorescent and autoradiographic analysis during stimulated and repressed RNA synthesis. *Cell* **28,** 269–278.
Sass, H. (1984). Gene identification in polytene chromosomes: some Balbiani Ring 2 gene sequences are located in an interband-like region of *Chironomus tentans*. *Chromosoma* **90,** 20–25.
Sass, H. (1990). P-transposable vectors expressing a constitutive and thermoinducible *hsp82-neo*

fusion gene for *Drosophila* germline transformation and tissue-culture transfection. *Gene* **89,** 179–186.

Sass, H. (1994). Interspecific transgenic analysis of basal versus heat-shock-induced expression of a *Drosophila pseudoobscura hsp82-neo* fusion gene in *D. melanogaster*. *Roux's Arch. Dev. Biol.* **204,** 101–111.

Sass, H. (1995). Transcription of heat shock gene loci versus non-heat shock loci in *Chironomus* polytene chromosomes: Evidence for heat-induced formation of novel putative ribonuclein particles hs RNPs in the major heat shock puffs. *Chromosoma* **103,** 528–538.

Sass, H., and Bautz, E. K. F. (1982a). Immunoelectron microscopic localization of RNA polymerase B on isolated polytene chromosomes of *Chironomus tentans*. *Chromosoma* **85,** 633–642.

Sass, H., and Bautz, E. K. F. (1982b). Interbands of polytene chromosomes: Binding sites and start points for RNA polymerase. *Chromosoma* **86,** 77–93.

Sass, H., and Meselson, M. (1991). Dosage compensation of the *pseudoobscura hsp82* gene and the *melanogaster ADH* gene at ectopic sites in *Drosophila melanogaster*. *Proc. Natl. Acad. Sci. USA* **88,** 6795–6799.

Sass, H., and Pederson, T. (1984). Transcription-dependent localization of U1 and U2 small nuclear ribonucleoproteins at major sites of gene activity in polytene chromosomes. *J. Mol. Biol.* **180,** 911–926.

Sauaia, H., Laicine, E. M., and Alves, M. A. R. (1971). Hydroxyurea-induced inhibition of DNA puff development in the salivary gland chromosomes of *Bradysia hygida*. *Chromosoma* **34,** 129–151.

Saumweber, H., Symmons, P., Kabisch, R., Will, H., and Bonhoeffer, F. (1980). Monoclonal antibodies against chromosomal proteins of *Drosophila melanogaster*. Establishment of antibody producing cell lines and partial characterization of corresponding antigens. *Chromosoma* **80,** 253–275.

Saumweber, H., Frasch, M., and Korge, G. (1990). Two puff-specific proteins bind within the 2,5 kb upstream region of the *Drosophila melanogaster Sgs-4* gene. *Chromosoma* **99,** 52–60.

Scalenghe, F., and Ritossa, F. (1976). Controllo dell'attivita genica in *Drosophila*. Il puff al locus ebony e la glutamina sintetasi I. *Atti della Acad. Naz. dei Lincei, Serie VIII,* **13,** 439–528.

Scalenghe, F., and Ritossa, F. (1977). The puff inducible in region 93D is responsible for the synthesis of the major "heat-shock" polypeptide in *Drosophila melanogaster*. *Chromosoma* **63,** 317–327.

Schafer, U., and Kunz, W. (1975). Two separated nucleolus organizers on the *Drosophila hydei* Y chromosome. *Mol. Gen. Genet.* **137,** 365–368.

Schalet, A. (1983). Vital loci located at the junction of polytene X chromosome sections 2B and 2C in *Drosophila melanogaster*. *Dros. Inform. Serv.* **59,** 107.

Schalet, A. P. (1986). The distribution of and complementation relationships between spontaneous X-linked recessive lethal mutations recovered from crossing long-term laboratory stocks of *Drosophila melanogaster*. *Mutat. Res.* **163,** 115–144.

Schalet, A., and Finnerty, V. (1968). The arrangement of genes in the proximal region of the X chromosome of *Drosophila melanogaster*. *Dros. Inform. Serv.* **43,** 128–129.

Schalet, A., and Lefevre, G., Jr. (1976). The proximal region of the X chromosome. *In* "The Genetics and Biology of *Drosophila*" (M. Ashburner, and E. Novitsky, eds.), Vol. 1b, pp. 848–902. Academic Press, London/New York/San Francisco.

Schaltmann, K., and Pongs, O. (1982). Identification and characterization of the ecdysterone receptor in *Drosophila melanogaster* by photoaffinity labelling. *Proc. Natl. Acad. Sci. USA* **79,** 6–10.

Scharrer, B., and Hadorn, E. (1938). The structure of the ring-gland (*corpus allatum*) in normal and lethal larvae of *Drosophila melanogaster*. *Proc. Natl. Acad. Sci. USA* **24,** 236–242.

Schedl, P., Artavanis-Tsakonas, S., Steward, R., and Gehring, W. J. (1978). Two hybrid plasmids with *Drosophila melanogaster* DNA sequences complementary to mRNA coding for the major heat shock protein. *Cell* **14,** 921–929.

Schin, K. S., and Clever, U. (1965). Lysosomal and free acid phosphatase in salivary glands of *Chisonomus tentans*. *Science* **150,** 1053–1055.

Schin, K. S., and Clever, U. (1968a). Ultrastructural and cytochemical studies of salivary gland regression in *Chisonomus tentans*. *Z. Zellforsch.* **86,** 262–279.

Schin, K. S., and Clever, U. (1968b). Ferritin uptake by salivary glands of *Chisonomus tentans* and its intercellular localization. *Exp. Cell Res.* **49,** 208–211.

Schin, K. S., and Laufer, H. (1972). Role of acid hydrolases in cell breakdown. *J. Cell Biol.* **55,** 230a.

Schin, K. S., and Laufer, H. (1973). Studies of programmed salivary gland regression during larval-pupal transformation in *Chironomus thummi*. I. Acid hydrolase activity. *Exp. Cell Res.* **82,** 335–340.

Schin, K. S., and Laufer, H. (1974). Uptake of homologous haemolymph protein by salivary glands of *Chironomus thummi*. *J. Insect Physiol.* **20,** 405–411.

Schlesinger, M. J. (1988). Function of heat shock proteins. *ISI Atlas Sci. Biochem.* **1,** 161–164.

Schlesinger, M. J., Ashburner, M., and Tissieres, A. (Eds.) (1982). Heat shock from bacteria to man. Cold Spring Harbor Laboratory Press, Cold Spring Harbor, NY.

Schmidt, E. C., and Goodman, R. M. (1976). Hormone-induced alterations in nonhistone protein methylation and phosphorylation in *Sciara coprophila*. *Differentiation* **6,** 177–186.

Schmidt, E. R., and Godwin, E. A. (1983). The nucleotide sequence of an unusual nontranscribed spacer and its ancestor in the rDNA in *Chironomus thummi*. *EMBO J.* **2,** 1177–1183.

Schmidt, E. R., and Godwin, E. A. (1987). Non transcribed spacer length heterogeneity and tandem-repetitive elements in the rDNA of Chironomids. *Mol. Genet. (Life Sci. Adv.)* **6,** 199–204.

Schmidt, E. R., Godwin, E. A., Keyl, H.-G., and Israilewski, N. (1982). Cloning and analysis of ribosomal DNA of *Chironomus thummi piger* and *Chironomus thummi thummi*. The nontranscribed spacer of *Ch. th. thummi* contains a highly repetitive DNA sequence. *Chromosoma* **87,** 389–407.

Schmidt, E. R., Keyl, H.-G., and Hankeln, T. (1988). *In situ* localization of two haemoglobin gene cluster in the chromosome of 13 species of *Chironomus*. *Chromosoma* **96,** 353–359.

Schneuwly, S., Kuroiwa, A., Baumgartner, P., and Gehring, W. J. (1986). Structural organization and sequence of the homeotic gene *Antennapedia* of *Drosophila melanogaster*. *EMBO J.* **5,** 733–739.

Scholnick, S. B., Morgan, B. A., and Hirsh, J. (1983). The cloned *dopa decarboxylase* gene is developmentally regulated when reintegrated into the *Drosophila genome*. *Cell* **34,** 37–45.

Schoon, H., and Reinsing, L. (1973). The effects of protein synthesis-inhibiting antibiotics on the puffing pattern of *Drosophila* salivary gland chromosome *in vitro*. *Cell Differ.* **2,** 97–106.

Schrons, H., Knust, E., and Campos-Ortega, J. A. (1992). The *Enhancer of split* complex and adjacent genes in the 96F region of *Drosophila melanogaster* are reguired for segregation of neural and epidermal progenitor cells. *Genetics* **132,** 481–503.

Schuldt, C., Kloetzel, P. M., and Bautz, E. K. F. (1989). Molecular organization of RNP complexes containing P11 antigen in heat-shocked and non-heat-shocked *Dosophila* cells. *Eur. J. Biochem.* **181,** 135–142.

Schulz, R. A., and Butler, B. A. (1989). Overlapping genes of *Drosophila melanogaster:* Organization of the *Z600-gonadal-Eip28/29* gene cluster. *Genes Dev.* **3,** 232–242.

Schulz, R. A., Cherbas, L., and Cherbas, P. (1986). Alternative splicing generates two distinct *Eip28/29* gene transcripts in *Drosophila* Kc cells. *Proc. Natl. Acad. Sci. USA* **83,** 9428–9432.

Schulz, R. A., Xie, X., Andres, A. J., and Galewsky, S. (1991). Endoderm-specific expression of the *Drosophila mex^1* gene. *Dev. Biol.* **143,** 206–211.

Schurin, M. F. (1959). Localized cytochemical and submicroscopical differentiation in *Drosophila virilis* salivary gland chromosomes. *Genetics* **44**(4), 534.

Schwartz, D. (1973). The application of the maize-derived gene competition model to the problem of dosage compensation in *Drosophila*. *Genetics* **75,** 639–641.

Schwartz, M. B., Imberski, R. B., and Kelly, T. J. (1984). Analysis of metamorphosis in *Drosophila melanogaster:* characterization of giant, an ecdysteroid-deficient mutant. *Dev. Biol.* **103,** 85–95.

Schwartz, Yu. B., Demakov, S. A., and Zhimulev, I. F. (1998). Cloning and analysis of DNA from interband regions 85D9/D10 and 86B4/B6 in polytene chromosomes of *Drosophila melanogaster*. *Genetica* **34** (N8), 998–1006. [in Russian]

Scott, M. P. (1987). Complex loci of *Drosophila*. *Ann. Rev. Biochem.* **56,** 195–227.

Scott, M. P., Weiner, A. J., Hazelrigg, T. I., Polisky, B. A., Pirrotta, V., Scalenghe, F., and Kaufmann, T. C. (1983). The molecular organization of the *Antennapedia* locus of *Drosophila*. *Cell* **35,** 763–776.

Scouras, Z. (1984). "Study of the polytene chromosomes of the cell of the salivary gands of *Drosophila auraria*." Ph.D. thesis, Aristotelian University of Saloniki. [in Greek].

Scouras, Z. G., and Kastritsis, C. D. (1984). Balbiani Rings and puffs of the polytene chromosomes of *Drosophila auraria*. *Chromosoma* **89,** 96–106.

Scouras, Z. G., and Mavragani-Tsipidou, P. (1992). Balbiani rings in *Drosophila:* the Balbiani ring 1 is a common characteristic in several *montium* species. *Cytobios* **69,** 97–100.

Scouras, Z. G., Karamanlidou, G. A., and Kastritsis, C. D. (1986). The influence of heat shock on the puffing pattern of *Drosophila auraria* polytene chromosomes. *Genetica* **69,** 213–218.

Scouras, Z. G., Milioni, D., Yiangou, M., Duchene, M., and Domdey, H. (1994). The beta-tubulin genes of *Drosophila auraria* are arranged in a cluster. *Curr. Genet.* **25,** 84–87.

Searles, L. L., and Voelker, R. A. (1986). Molecular characterization of the *Drosophila vermilion* locus and its supperssible alleles. *Proc. Natl. Acad. Sci. USA* **83,** 404–408.

Searles, L. L., Ruth, R. S., Pret, A.-M., Fridell, R. A., and Ali, A. J. (1990). Sructure and transcription of the *Drosophila melanogaster vermilion* gene and several mutant alleles. *Mol. Cell Biol.* **10,** 1423–1431.

Sebeleva, T. E. (1968). Quantitative determination of DNA during the formation of one of the puffs and one of the Balbiani rings in *Chironomus*. *Tsitologia* **10**(6), 765–769. [in Russian]

Sebeleva, T. E. (1972). A study of protein sinthesis in the salivary glands of *Chironomus*. *Tsitologia* **14**(1), 46–52. [in Russian]

Sebeleva, T. E., and Kiknadze, I. I. (1977). A cytochemical analysis of the mucopolyasccharide content of the secretion of the salivary glands of *Chironomus thummi* larvae. *Tsitologia* **19**(2), 147–153. [in Russian]

Sebeleva, T. E., Sherudilo, A. I., and Kiknadze, I. I. (1965). Quantitative estimation of DNA during puff formation in *Chironomus dorsalis*. *Genetika* **1**(2), 102–105. [in Russian]

Sebeleva, T. E., Kolesnikov, N. N. and Kiknadze, I. I. (1981). A comparative analysis of the secretion proteins of the salivary gland of *Chironomus thummi* larvae differing in the amounts of Balbiani rings. *Doklady Akad. Nauk SSSR* **256**(4), 975–978. [in Russian]

Sebeleva, T. E., Kolesnikov, N. N., Kopantsev, E. P., and Karakin, E. I. (1985). Secretory characteristics of the secretion proteins in salivary gland cells differing in the number of Balbiani rings. *In* "Organizatsia i ekspressia genov tkanespetsificheskoj funktsii u *Diptera*," pp. 109–115. Nauka, Novosibirsk. [in Russian]

Seecof, R. L., Kaplan, W. D., and Futch, D. G. (1969). Dosage compensation for enzyme activities in *Drosophila melanogaster*. *Proc. Natl. Acad. Sci. USA* **62,** 528–535.

Segal, D., and Gelbart, W. M. (1985). Shortvein, a new component of the decapentaplegic gene complex in *Drosophila melanogaster*. *Genetics* **109,** 119–143.

Segraves, W. A. (1991). Something old, some things new: The steroid receptor super family in *Drosophila*. *Cell* **67,** 225–228.

Segraves, W. A., and Hogness, D. S. (1984). Molecular and genetic analysis of the 75B ecdysone-inducible puff of *Drosophila melanogaster*. *Genetics* **107**(3), 96–97.

Segraves, W. A., and Hogness, D. S. (1990). The E75 ecdysone-inducible gene responsible for the 75B early puff in *Drosophila* encodes two new members of the steroid receptor superfamily. *Genes Dev.* **4,** 204–219.

Segraves, W. A., and Richards, G. (1990). Regulatory and developmental aspects of ecdysone-regulated gene expression. *Invertebr. Reprod. Dev.* **18,** 67–76.

Segraves, W. A., Louis, C., Schedl, P., and Jarry, B. P. (1983). Isolation of the *rudimentary* locus of *Drosophila melanogaster*. *Mol. Gen. Genet.* **189,** 34–40.

Segraves, W. A., Louis, C., Tsubota, S., Schedl, P., Rawls, J. M., and Jarry, B. (1984). The *rudimentary* locus of *Drosophila melanogaster*. *J. Mol. Biol.* **175,** 1–17.

Sehnal, F. (1985). Growth and life cycles. *In* "Comprehensive Insect Physiology, Biochemistry and Pharmacology" (G. A. Kerkut and L. I. Gilbert, eds.), Vol. 2, pp. 34–53. Pergamon Press, Oxford/New York/Toronto/Sydney/Paris/Frankfurt.

Sekelsky, J. J., McKim, K. S., Chin, G. M., and Hawley, R. C. (1995). The *Drosophila* meiotic recombination gene mei-9 encodes a homologue of the yeast excision repair protein Rad1. *Genetics* **141,** 619–627.

Semeshin, V. F. (1990). "Electron microscopic analysis of the chromomere organization of *Drosophila* polytene chromosomes." Doctor of Sciences thesis, Inst. Cytol. Genet., Novosibirsk. [in Russian]

Semeshin, V. F., and Szidonya, J. (1985). EM mapping of rearrangements in the 24–25 sections of *Drosophila melanogaster* 2L chromosomes. *Dros. Inform. Serv.* **61,** 148–154.

Semeshin, V. F., Zhimulev, I. F., and Belyaeva, E. S. (1979a). Electron microscope autoradiographic study on transcriptional activity of *Drosophila melanogaster* polytene chromosomes. *Chromosoma* **73,** 163–177.

Semeshin, V. F., Zhimulev, I. F., and Belyaeva, E. S. (1979b). Cytogenetic study on the 9E–10A region of the X chromosome of *Drosophila melanogaster*. I. Morphology of the region and mapping of deletions involving the 10A1–2 band. *Genetika* **15**(10), 1784–1792. [in Russian]

Semeshin, V. F., Baricheva, E. M., Belyaeva, E. S., and Zhimulev, I. F. (1982). Electron microscopical analysis of *Drosophila* polytene chromosomes. I. Mapping of the 87A and 87C heat shock puffs in development. *Chromosoma* **87,** 229–237.

Semeshin, V. F., Zhimulev, I. F., Shilov, A. G., Belyaeva, E. S., and Baricheva, E. M. (1983). RNP-bodies in puffs of *Drosophila melanogaster*. *Dros. Inform. Serv.* **59,** 108–109.

Semeshin, V. F., Baricheva, E. M., and Zhimulev, I. F. (1984). Morphology and genetic organization of the 67B region in 3L chromosome of *Drosophila melanogaster*. *Genetika* **20**(5), 794–799. [in Russian]

Semeshin, V. F., Baricheva, E. M., Belyaeva, E. S., and Zhimulev, I. F. (1985a). Electron microscopical analysis of *Drosophila* polytene chromosomes. II. Development of complex puffs. *Chromosoma* **91,** 210–233.

Semeshin, V. F., Baricheva, E. M., Belyaeva, E. S., and Zhimulev, I. F. (1985b). Electron microscopical analysis of *Drosophila* polytene chromosomes. III. Mapping of puffs developing from one band. *Chromosoma* **91,** 234–250.

Semeshin, V. F., Belyaeva, E. S., Zhimulev, I. F., Lis, J. T., Richards, G., and Bourouis, M. (1986a). Electron microscopical analysis of *Drosophila* polytene chromosomes. IV. Mapping of morphological structures appearing as a result of transformation of DNA sequences into chromosomes. *Chromosoma* **93,** 461–468.

Semeshin, V. F., Belyaeva, E. S., Zhimulev, I. F., Richards G., and Bourouis M. (1986b). Mapping of morphological structures arising during insertion of genetic material into the *Drosophila* chromosomes. *Genetika* **22**(9), 2220–2227. [in Russian]

Semeshin, V. F., Demakov, S. A., Perez Alonso, M., Belyaeva, E. S., Bonner, J. J., and Zhimulev, I. F. (1989a). Electron microscopical analysis of *Drosophila* polytene chromosomes. V. Characteristics of structures formed by transposed DNA segments of mobile elements. *Chromosoma* **97,** 396–412.

Semeshin, V. F., Demakov, S. A., and Zhimulev, I. F. (1989b). Characteristics of the structures of the polytene chromosomes of *Drosophila* formed by transposable DNA fragments. *Genetika* **25**(11), 1968–1978. [in Russian]

Semeshin, V. F., Demakov, S. A., Perez Alonso, M., and Zhimulev, I. F. (1990). Formation of interbands from DNA material of the P-element in *Drosophila* polytene chromosomes. *Genetika* **26**(3), 448–456. [in Russian]

Semeshin, V. F., Chernukhin, V. A., Shabel'nikov, I. V., Omel'yanchuk, L. V., Belyaeva, E. S., and Zhimulev, I. F. (1994). Cytogenetic analysis of interband insertions in polytene chromosomes of *Drosophila*. *Genetika* **30**(7), 927–933. [in Russian]

Semeshin V. F., Kritikou D., Zacharopoulou A., and Zhimulev I. F. (1995). Electron microscope investigation of Mediterranean fruit fly *Ceratitis capitata*. *Genome* **38,** 652–660.

Semeshin, V. F., Artero, R. D., Perez-Alonso, M., Galindo, I. M., and Paricio, N. (1997a). Electron-microscopic analysis of the 54AB region of *Drosophila melanogaster* 2L chromosome. *Tsitologia* **39**(1), 102. [in Russian]

Semeshin, V. F., Belyaeva, E. S., and Zhimulev, I. F. (1997b). Analysis of insertions with various level of expression of reporter gene in polytene chromosomes in *Drosophila*. *Tsitologia* **39**(1), 102–103. [in Russian]

Semionov, E. P. (1988). Multiplicity of nucleoli in larval polytene cells of *Drosophila melanogaster* Oregon R strain and its X"O derivative. *Dros. Inform. Serv.* **67,** 71.

Semionov, E. P., and Kirov, N. K. (1986). Increased number of nucleoli in the salivary gland cells of *Drosophila melanogaster* under conditions of rDNA dose compensation. *Chromosoma* **93,** 477–482.

Sengun, A. (1947). Uber intraindividuelle Variabilitat des 4ten Chromosoms bei *Chironomus*. *Rev. Fac. Sci. Univ. Istanbul Ser. B* **12,** 289–305.

Sengun, A. (1948a). Cytologische Untersuchungen bei einer *Culex*- und eiber *Aedes*-Art. *Rev. Fac. Sci. Univ. Istanbul Ser. B* **13,** 143–160.

Sengun, A. (1948b). Vergleichend-ontogenetische Untersuchungen uber die Riesenchromosomen verschiedener Gewebearten der Chironomiden. *I. Comm. Fac. Sci. Univ. Ankara* **1,** 187–248.

Sengun, A. (1949). Difference in sructure between the same giant chromosomes from the same larvae of *Drosophila repleta*. *Nature* **163,** 1002.

Sengun, A. (1951a). Vergleichend-ontogenetische Untersuchungen uber die Riesenchromosomen verschiedener Gewebearten der Chironomiden. *II. Rev. Fac. Sci. Univ. Istanbul Ser. B* **16,** 1–44.

Sengun, A. (1951b). Meiotische und somatische Chromosomen von *Chironomus*-Larven. *Rev. Fac. Sci. Univ. Istanbul Ser. B.* **16,** 15–27.

Sengun, A. (1952a). Uber die Zahl der Spiralwindungen, der Blockchen und sekundaren Querbander. *Cytologia* **17,** 1–5.

Sengun, A. (1952b). Uber die Wirkung der Temperatur auf die Riesenchromosomen der *Chironomus*-Larven. *Rev. Fac. Sci. Univ. Istanbul Ser. B.* **17,** 357–361.

Sengun, A. (1954). Variability of the banding patterns of giant chromosomes. *J. Hered.* **45,** 119–122.

Sengun, A. (1961). Incorporation of tritiated thymidine into the giant chromosomes of larvae *Chironomus*. *Pathol. Biol. Semaine Hop.* **9,** 753–755.

Sengun, A., and Kosswig, C. (1947). Weiteres uber den Bau der Riesenchromosomen in verschidenen Geweben von *Chironomus* larven. *Chromosoma* **3,** 195–207.

Serebrovsky, A. S., and Dubinin, N. P. (1930). X-ray experiments with *Drosophila*. *J. Hered.* **26,** 469–478.

Serfling, E. (1976). The transcripts of Balbiani Ring from *Chironomus thummi:* Giant RNA molecules with messenger characteristics. *Chromosoma* **57,** 271–283.

Serfling, E. (1978). Feinstrukturanalyse ribosomaler RNA-Gene nach Kloneierung Nucleolarer DNA in Bakterien. *Biol. Rdsch.* **16,** 222–231.

Serfling, E. (1979). Feinstruktur und Expression von Histon-Genen–eine Ubersicht. *Biol. Zbl.* **98,** 641–660.

Serfling, E., and Huth, A. (1978). Balbiani Ring RNA in the cytoplasm of *Chironomus thummi* salivary gland cells. *Chromosoma* **66,** 205–223.

Serfling, E., Panitz, R., and Wobus, U. (1969). Die experimentelle Beeinflussung des Puffmusters von Riesenchromosomen. I. Puffinduktion durch Oxytetracyclin bei *Chironomus thummi*. *Chromosoma* **28,** 107–119.

Serfling, E., Wobus, U., and Panitz, R. (1972). Effect of a-amanitin on chromosomal and nucleolar RNA-synthesis in *Chironomus thummi* polytene chromosomes. *FEBS Lett.* **20,** 148–152.

Serfling, E., Maximovsky, L. F., and Wobus, U. (1974). Synthesis and processing of ribosomal ribonucleic acid in salivary gland cells of *Chironomus thummi*. *Eur. J. Biochem.* **45,** 277–289.

Serfling, E., Meyer, L., Rudolph, A., and Steiner, K. (1983). Secretory proteins and Balbiani Ring gene activities in salivary glands of *Chisonomus thummi* larvae. *Chromosoma* **88,** 16–23.

Shaaya, E., and Karlson, P. (1965). Der Ecdysontiter wahrend der Insektenentwicklung. II. Die postembryonale Entwicklung der Schmeissfliege *Calliphora erythrocephala* Meig. *J. Insect Physiol.* **11,** 65–69.

Shaffer, C. D., and MacIntyre, R. J. (1990). The isolation of the *acid phospatase-1* gene of *Drosophila melanogaster* and a chromosomal breakpoint inducing its position effect variegation. *Mol. Gen. Genet.* **224,** 49–56.

Shakhbazov, V. G., and Taglina, O. V. (1990). Feature of the dynamics of heat shock puffs in highly inbred strains and heterozygous hybrids of *Drosophila melanogaster*. *Genetika* **26**(1), 43–48. [in Russian]

Shamshurin, A. A. (1972). Hormones of insects. *Priroda* **4,** 53–63. [in Russian]

Shannon, M. P., Kaufman, T. C., Shen, M. W., and Judd, B. H. (1972a). Lethality patterns of *zw* mutations in *D. melanogater*. *Dros. Inform. Serv.* **48,** 92.

Shannon, M. P., Kaufman, T. C., Shen, M. W., and Judd, B. H. (1972b). Lethality patterns and morphology of selected lethal and semi-lethal mutations in the *zeste-white* region of *Drosophila melanogaster*. *Genetics* **72,** 615–638.

Shapiro, N. (1937). The method of studying the process of mutation in a limited region of the chromosome. *Dros. Inform. Serv.* **7,** 94–95.

Sharp, P. A. (1994). Split genes and RNA splicing. *Cell* **77,** 805–815.

Sharp, Z., Gandhi, V., and Procunier, J. (1983). X chromosome nucleolus organizer mutants which alter major type I repeat multiplicity in *Drosophila melanogaster*. *Mol. Gen. Genet.* **190,** 438–443.

Shcherbakov, D. Yu., Karakin, E. I., and Kokoza, V. A. (1985). Use of peptide marking of the secretory proteins of the salivary glands of *Drosophila melanogaster* in analysis of the degree of their structural homology. *In* "Organizatsia i ekspressia genov tkanespetsificheskoy funktsii u *Diptera*," pp. 59–70. Nauka, Novosibirsk. [in Russian]

Shcherbakov, E. S. (1965). Spontaneous translocation of a part of the nucleolar organizer in a natural population of *Simulium nolleri* Fried (Diptera). *Vestvik LYU Biol.* **21**(4), 154–155. [in Russian]

Shcherbakov, E.S. (1966a). On nucleolar polymorphism in a natural population of the midge *Olagmia ornata*. *Tsitologia* **8**(4), 510–513. [in Russian]

Shcherbakov, E. S. (1966b). Structure of the polytene chromosomes of the salivary glands of *Simulium nolleri*. *Tsitologia* **8**(6), 703–713. [in Russian]

Shcherbakov, E. S. (1968). On the question of heterozygosity of puffs and structure of individual bands of giant chromosomes of midges (Similidae, Diptera). *Genetika* **4**(5), 60–69. [in Russian]

Shea, M. J., King, D. L., Conboy, M. J., Mariani, B. D., and Kafatos, F. C. (1990). Proteins that bind to *Drosophila* chorion cis-regulatory elements: a new C_2H_2 zinc finger protein and C_2C_2 steroid receptor-like component. *Genes Dev.* **4,** 1128–1140.

Sheen, F.-m., Lim, J. K., and Simmons, M. J. (1933). Genetic instability in *Drosophila melanogaster* mediated by *hobo* transposable elements. *Genetics* **133,** 315–334.

Shepherd, B., Garabedian, M. J., Hung, M.-C., and Wensink, P. C. (1985). Developmental control of *Drosophila yolk protein 1* gene by cis-acting DNA elements. *Cold Spring Harbor Symp. Quant. Biol.* **50,** 521–526.

Shermoen, A. W., and Beckendorf, S. K. (1982). A complex of interacting DNAase I-hypersensitive sites near the *Drosophila* glue protein gene, *Sgs4*. *Cell* **29,** 601–607.

Shermoen, A. W., Jongens, J., Barnett, S. W., Flynn, K., and Beckendorf, S. K. (1987). Developmental regulation by an enhancer from the *Sgs-4* gene of *Drosophila*. *EMBO J.* **6,** 207–214.

Shestopal S. A., Makunin I. V., Belyaeva E. S., Ashburner M., and Zhimulev I.F. (1997). Molecular characterization of the *deep orange (dor)* gene of *Drosophila melanogaster*. *Mol. Gen. Genet.* **253,** 642–648.

Shibata, H., and Yamasaki, T. (1995). Molecular evolution of the duplicated *Amy* locus in the *Drosophila melanogaster* species subgroup: Concerted evolution only in the coding region and an excess on nonsynonymous substitutions in speciacion. *Genetics* **141,** 223–236.

Shilova, I. E., Razmakhnin, E. P., Zakharenko, L. P., Mertvetsov, N. P., Ammosov, A. D., and Gasevsky, Yu. V. (1983). Study on the nucleolar organizer in stretched polytene chromosomes by DNA/RNA hybridization. *Tsitologiya* **25**(7), 754–758. [in Russian]

Shirk, P. D., Roberts, P. A., and Harn, C. H. (1988). Synthesis and secretion of salivary gland proteins in *Drosophila gibberosa* during larval and prepupal development. *Wilhelm Roux's Arch. Dev. Biol.* **197,** 66–74.

Shore, E. M., and Guild, G. M. (1986). Larval salivary gland secretion proteins in *Drosophila* structural analysis of the *Sgs-5* gene. *J. Mol. Biol.* **190,** 149–158.

Siegel, J. G. (1981). Genetic characterization of the region of the *Drosophila* genome known to include the histone structural gene sequences. *Genetics* **98,** 505–527.

Silva, F. J. (1984). ^{3}H-uridine labelling patterns and chromosomal polymorphism in *Drosophila subobscura:* J and U chromosomes. *Genetica* **63,** 147–152.

Silva, F. J., Botella, L. M., and Edstrom, J.-E. (1990). Functional analysis of the 3′-terminal part of the Balbiani Ring 2.2 gene by interspecies sequence comparison. *J. Mol. Evol.* **31,** 221–227.

da Silva, M. J. L., (1980). Estudo morphologico e citoquimico das glandulas salivares de *Bradysia spattitergum* (Hardy) (Diptera, Sciaridae). *Rev. Brasil. Biol.* **40,** 447–461.

Simakov, Yu. Y., Nikichin-Nikiforov, A. Z., and Kuznetsova, I. B. (1990). Biotesting of the toxicity of compounds in an water medium on polytene chromosomes of chironomids. *Eksperimental'naya vodnaya toksikologia* **14,** 246–250. [in Russian]

Simeone, A., La Volpe A., and Boncinelli, E. (1985). Nucleotide sequence of a complete ribosomal spacer of *D. melanogaster*. *Nucleic Acids Res.* **13,** 1089–1101.

Simoes, L. C. G. (1970). Studies on DNA synthesis during larval development of *Rhynchosciara* sp. *Rev. Brasil. Biol.* **30,** 191–199.

Simoes, L. C. G., and Cestari, A. N. (1969). Behavior of polytene chromosomes "*in vitro*." *Genetics* **61** (Suppl., 1(2)), 361–372.

Simoes, L. C. G., and Cestari, A. N. (1981). Morphological and physiological aspects of salivary gland chromosomes of *Rhynchosciara in vitro*. *Rev. Brasil. Genet.* **4,** 495–511.

Simoes, L. C. G., Amabis, J. M., and Boni, J. A., de. (1970). Studies on the development of a DNA puff in *Rhynchosciara* sp. *Cien. Cult.* **22,** 176–181.

Simoes, L. C. G., Amabis, J. M., and Cestari, A. N. (1975). Puffs in the heterochromatin in chromosomes of *Rhynchosciara*. *Cien. Cult.* **27,** 159–161.

Simoes, L. C. G., Cestari, A. N., and Uemura, G. (1981). Morphological and physiological aspects of the salivary gland chromosomes of *Rhynchosciara in vitro*. Autolysis. *Rev. Brasil. Genet.* **3,** 281–308.

Simon, J. A., and Lis, J. T. (1987). A germline transformation analysis reveals flexibility in the organization of heat shock consensus elements. *Nucleic Acids Res.* **15,** 2971–2988.

Simon, J. A., Sutton, C. A., and Lis, J. T. (1985a). Localization and expression of transformed DNA sequences within heat shock puffs of *Drosophila melanogaster*. *Chromosoma* **93,** 26–30.

Simon, J. A., Sutton, C. A., Lobell, R. B., Glaser, R. L., and Lis, J. T. (1985b). Determinants of heat shock-induced chromosome puffing. *Cell* **40,** 805–817.

Sin, F. Y. T. (1987). Relationship between "early" and "late" protein induced by 20-hydroxyecdysone in imaginal wing discs of *Drosophila melanogaster*. *Mol. Biol. Rep.* **12,** 7–13.

Sin, Y. T. (1975). Induction of puffs in *Drosophila* salivary gland cells by mitochondrial factor(s). *Nature* **258,** 159–160.

Sin, Y. T., and Leenders, H. J. (1975). Studies on the mitochondrial NAD-dependent isocitrate dehydrogenase of *Drosophila* larvae after induction of gene activity by anaerobiosis. *Insect Biochem.* **5,** 447–458.

Singh, A. K., and Lakhotia, S. C. (1983). Further study of inducibility of 93D puff by homogenate of heat shocked cells. *Ind. J. Exp. Biol.* **21,** 363–366.

Singh, A. K., and Lakhotia, S. C. (1984). Lack of effect of microtubule poisons on the 93D or 93D-like heat shock puffs in *Drosophila*. *Ind. J. Exp. Biol.* **22,** 569–576.

Singh, A. K., and Lakhotia, S. C. (1988). Effect of low-temperature rearing on heat shock protein synthesis and heat sensitivity in *Drosophila melanogaster*. *Dev. Genet.* **9,** 193–201.

Singh, O. P. (1988a). *Melanagromyza obtusa*—A suitable system for the study of polytene chromosomes. *Curr. Sci.* **57,** 48–49.

Singh, O. P. (1988b). Absence of 93D-like locus of *Drosophila melanogaster* in *Melanagromyza obtusa*. *Ind. J. Exp. Biol.* **26,** 558–559.

Singh, O. P. (1991). Chromomeric patterns and photo-maps of polytene chromosomes of *Melanagromyza obtusa* in three larval tissues. *Cytobios* **65,** 187–198.

Singh, O. P., and Gupta, J. P. (1985). Differential induction of chromosome puffs in two cell types of *Melanagromyza obtusa*. *Chromosoma* **91,** 359–362.

Sinha, P., and Lakhotia, S. C. (1980). Replication in *Drosophila* chromosomes: V. Polytene chromosome replication after *in vitro* culture of larval salivary glands. *Ind. J. Exp. Biol.* **18,** 1059–1165.

Sinha, P., and Lakhotia, S. C. (1983). Replication in *Drosophila* chromosomes. IX. Stimulation of initiation of polytene replication cycles *in vitro* by juvenile hormone. *Cell Differ.* **12,** 11–17.

Sinha, P., Arati Mishra, and Lakhotia, S. C. (1987). Chromosomal organization of *Drosophila* tumors. I. Polytene chromosome organization and DNA synthesis in ovarian pseudonurse cells in *otu* mutants of *Drosophila melanogaster*. *Chromosoma* **95,** 108–116.

Sinibaldi, R. M., and Cummings, M. R. (1981). Localization and characterizarion of rDNA in *Drosophila tumiditarsus*. *Chromosoma* **81,** 655–671.

Sirlin, J. L. (1960). Cell sites of RNA and protein synthesis in the salivary gland of *Smittia* (Chironomidae). *Exp. Cell Res.* **19,** 177–180.

Sirlin, J. L., and Jacob, J. (1962). Function, development and evolution of the nucleolus. *Nature* **195,** 114–117.

Sirlin, J. L., and Jacob, J. (1964). Sequential and reversible inhibition of synthesis of ribonucleic acid in the nucleolus and chromosomes: Effect of benzamide and substituted benzimidazoles on dipteran salivary glands. *Nature* **204,** 545–547.

Sirlin, J. L., and Schor, N. A. (1962). Macromolecular synthesis in isolated polytene nuclei. *Exp. Cell Res.* **27,** 165–167.

Sirlin, J. L., Jacob, J., and Kato, K.-I. (1962). The relation of messenger to nucleolar RNA. *Exp. Cell Res.* **27,** 355–359.

Sirlin, J. L., Tandler, C. J., and Jacob, J. (1963). The relationship between the nucleolus organizer and nucleolar RNA. *Exp. Cell Res.* **31,** 611–615.

Sirlin, J. L., Jacob, J., and Birnstiel, M. L. (1966). Synthesis of transfer RNA in the nucleolus of *Smittia*. *Natl. Cancer Inst. Monogr.* **23,** 255–270.

Sirotkin, K. (1982). Developmentally regulated transcription at the 67B heat shock cluster. *In* "Heat Shock from Bacteria to Man" (M. L. Schlesinger, M. Ashburner, A. and Tissieres, eds.), pp. 69–75. Cold Spring Harbor Laboratory Press, Cold Spring Harbor, NY.

Sirotkin, K., and Davidson, N. (1982). Developmentally regulated transcription from *Drosophila melanogaster* chromosomal site 67B. *Dev. Biol.* **89,** 196–210.

Skaer, R. J. (1977). Interband transcription in *Drosophila*. *J. Cell Sci.* **26,** 251–266.

Skaer, R. J. (1978). Transcription in the interbands of *Drosophila*. *Phil. Trans. R. Soc. London* **B283,** 411–413.

Skoglund, U., and Daneholt, B. (1986). Electron microscope tomography. *Trends Biochem. Sci.* **11,** 499–503.

Skoglund, U., Andersson, K., Bjorkroth, B., Lamb, M. M., and Daneholt, B. (1983). Visualization of the formation and transport of a specific hnRNA particle. *Cell* **34,** 847–855.

Skoglund, U., Andersson, K., Strandberg, B., and Daneholt, B. (1986). Three-dimensional structure of a specific pre-messenger RNP particle established by electron microscope tomography. *Nature* **319,** 560–564.

Slavicek, J. M., and Krider, H. M. (1987). The organization and composition of the ribosomal RNA gene non-transcribed spacer of *Drosophila busckii* is unique among the Drosophilids. *Genet. Res. Cambr.* **50,** 173–180.

Sliter, T. J., and Gilbert, L. I. (1992). Developmental arrest and ecdysteroid deficiency from mutations at the *dre4* locus of *Drosophila. Genetics* **130,** 555–568.

Sliter, T. J., Sedlak, B. J., Baker, F. C., and Schooley, D. A. (1987). Juvenile hormone in *Drosophila melanogaster*. Identification and titer determination during development. *Insect Biochem.* **17,** 161–165.

Sliter, T. J., Henrich, V. C., Tucker, R. L., and Gilbert, L. I. (1989). The genetics of the Dras 3-Roughened-ecdysoneless chromosomal region (62B3-4 to 62D3-4) in *Drosophila melanogaster:* Analysis of recessive lethal mutations. *Genetics* **123,** 327–336.

Slizynski, B. M. (1944). A revised map salivary gland chromosome 4 of *Drosophila melanogaster. J. Hered.* **35,** 322–325.

Slizynski, B. M. (1950). *Chironomus versus Drosophila. J. Genet.* **50,** 77–78.

Slizynski, B. M. (1963). Functional changes in the polytene chromosomes of *Drosophila melanogaster. In* "Genetics Today: Proc. XI Intern. Congr. Genet." (S. J. Geerts, ed.), Vol. 1, pp. 108. Pergamon Press, Oxford/London/Edinburgh/New York/Paris/Frankfurt.

Slizynski, B. M. (1964a). Addition to the characteristic of *ft. Dros. Inform. Serv.* **39,** 102.

Slizynski, B. M. (1964b). Functional changes in polytene chromosomes of *Drosophila melanogaster. Cytologia* **29,** 330–336.

Smith, H. S., Bona, M., Abbott, A. G., DiBartolomeis, S., Mayes, S. R., Edstrom, J.-E., Jackle, H., and Gerbi, S. A. (1985). Middle repetitive DNA may be mobile and comes under DNA amplification control in DNA puffs of *Sciara. J. Cell Biol.* **101**(5), 3.

Smith, P. D., Koenig, P. B., and Lucchesi, J. C. (1968). Inhibition of development in *Drosophila* by cortisone. *Nature* **217,** 1286.

Smith, W. G., and Witkus, E. R. (1970). Secretory development in larval *Drosophila* salivary glands. *J. Cell Biol.* **47,** 196a–197a.

Smith, W. J. (1969). Effects of ribonuclease on nuclear ultrastructure in larval *Drosophila* salivary glands. *J. Ultrastruct. Res.* **27,** 168–181.

Snyder, M., and Davidson, N. (1983). Two gene families in a small region of the *Drosophila* genome. *J. Mol. Biol.* **166,** 101–118.

Snyder, M., Hirsh, J., and Davidson N. (1981a). The cuticle genes of *Drosophila:* A developmental regulated gene cluster. *Cell* **25,** 165–177.

Snyder, M., Hunkapiller, M., Hood, L., and Davidson, N. (1981b). The larval cuticle genes of *Drosophila. J. Supramol. Struct.* **4,** 418.

Snyder, M., Hunkapiller, M., Yuen, D., Silvert, D., Fristrom, J., and Davidson, N. (1982). Cuticle protein genes of *Drosophila:* structure, organization and evolution of four clustered genes. *Cell* **29,** 1027–1040.

Sokoloff, S. (1977). Replication of polytene chromosomes in the dipteran *Phryne*. Disser. Grad. Doktors Naturwiss. Univers., Tubingen.

Sokoloff, S., and Zacharias, H. (1977). Functional significance of changes in the shape of the polytene X chromosome in *Phryne. Chromosoma* **63,** 359–384.

Solovyova, I. V. (1992). Molecular mapping of insertional mutations at the ecs locus. *Genetika* **28**(2), 63–71. [in Russian]

Solovyova, I. V. and Belyaeva, E. S. (1989). A cytogenetic analysis of the 2B1-2–2B9-10 region of the X chromosome of *Drosophila melanogaster*. VIII. A genetic analysis of mutations produced in the P-M system of hybrid dysgenesis. *Genetika* **25**(7), 1209–1217. [in Russian]

Somme-Martin, G., Colardeau, J., and Lafont, R. (1988). Metabolism and biosynthesis of ecdysteroids in the *Drosophila* development mutant *ecd*[1]. *Insect Biochem.* **18,** 735–742.

Sondermeijer, P. J. A., and Lubsen, N. H. (1978). Heat shock peptides in *Drosophila hydei* and their *in vitro* synthesis. *Eur. J. Biochem.* **88,** 331–339.

Sondermeijer, P. J. A., and Lubsen, N. H. (1979). The activity of two heat shock loci of *Drosophila hydei* in tissue culture cells and salivary gland cells as analyzed by *in situ* hybridization of complementary DNA. *Chromosoma* **72,** 281–291.

Sonnenblick, B. P. (1950). The early embrylogy of *Drosophila melanogaster*. *In* "Biology of *Drosophila*" (M. Demerec, ed.), pp. 62–167. Wiley, New York.

Sorger, P. K. (1991). Heat shock factor and the heat shock response. *Cell* **65,** 363–366.

Sorsa, M. (1969). Ultrastructure of puffs in the proximal part of chromosome 3R in *Drosophila melanogaster*. *Ann. Acad. Sci. Fenn. Ser. A* **150,** 1–21.

Sorsa, M., and Pfeifer, S. (1972). Puffing pattern of O-hour prepupal of *Drosophila melanogaster*. *Hereditas* **71,** 119–130.

Sorsa, M., and Pfeifer, S. (1973a). Response of puffing pattern to *in vivo* treatments with organomercurials in *Drosophila melanogaster*. *Hereditas* **74,** 89–102.

Sorsa, M., and Pfeifer, S. (1973b). Effects of cadmium on development time and prepupal puffing pattern of *Drosophila melanogaster*. *Hereditas* **75,** 273–277.

Sorsa, V. (1974). Organization of replicative units in salivary gland chromosome bands. *Hereditas* **78,** 298–302.

Sorsa, V. (1975). A hypothesis for the origin and evolution of chromomere DNA. *Hereditas* **81,** 77–84.

Sorsa, V. (1976). A cytological view of the functional organization of chromomeres. *Hereditas* **82,** 63–68.

Sorsa, V. (1979). Electron microscopic localization and ultrastructure of certain gene loci in salivary gland chromosomes of *Drosophila melanogaster*. *In* "Specific Eukaryotic Genes Alfred Benzon Symposium" (J. Engberg, H. Klenow, and V. Leick, eds.), pp. 55–71. Munksgaard, Copenhagen.

Sorsa, V. (1984). Electron microscopic mapping and ultrastructure of *Drosophila* polytene chromosomes. *In* "Insect Ultratructure" (R. C. King amd H. Akai, eds.), Vol. 2, pp. 75–107. Plenum, New York.

Sorsa, V. (1988). "Polytene Chromosomes in Genetic Research." Eoois Horwood Ltd., Chichester, West Sussex.

Sorsa, V., Green, M. M., and Beermann, W. (1973). Cytogenetic fine structure and chromosomal localization of the *white* gene in *Drosophila melanogaster*. *Nature New Biol.* **245,** 34–37.

Sorsa, V., Saura, A. O., and Heino, T. I. (1984). Electron microscopic map of divisions 61, 62 and 63 of the salivary gland 3L chromosome in *Drosophila melanogaster*. *Chromosoma* **90,** 177–184.

Southgate, R., Ayme, A., and Voellmy, R. (1983). Nucleotide sequence analysis of the *Drosophila* small heat shock gene cluster at locus 67B. *J. Mol. Biol.* **165,** 35–57.

Spadoro, J. P., Copertino, D. W., and Strausbaugh, L. D. (1986). Differential expression of histone sequences in *Drosophila* following heat shock. *Dev. Genet.* **7,** 133–148.

Spear, B. (1974). The genes for ribosomal RNA in diploid and polytene chromosomes of *Drosophila melanogaster*. *Chromosoma* **48,** 159–179.

Spear, B. B., and Gall, H. G. (1973). Independent control of ribosomal gene replication in polytene chromosomes of *Drosophila melanogaster*. *Proc. Natl. Acad. Sci. USA* **70,** 1359–1363.

Speiser, Ch. (1974). Eine Hypothese uber die functionelle Organization der Chromosomen hoherer Organismen. *Theoret. Appl. Genet.* **44,** 97–99.

Spencer, C., Gietz, R. D., and Hodgetts, R. B. (1986a). Analysis of the transcription unit adjacent to the 3′-end of the DOPA-decarboxylase gene in *Drosophila melanogaster*. *Dev. Biol.* **114,** 260–264.

Spencer, C., Gietz, R. D., and Hodgetts, R. B. (1986b). Overlapping transcription units in the DOPA-decarboxylase region of *Drosophila*. *Nature* **322,** 279–281.

Spierer, A. (1984). A molecular approach to chromosome organization. *Dev. Biol.* **4,** 333–339.

Spierer, A., and Spierer, P. (1984). Similar level of polyteny in bands and interbands of *Drosophila* giant chromosomes. *Nature* **307,** 176–178.
Spierer, P., Spierer, A., Bender, W., and Hogness, D. S. (1983). Molecular mapping of genetic and chromomeric units in *Drosophila melanogaster*. *J. Mol. Biol.* **168,** 35–50.
Spradling, A. (1980). The structure and expression of *Drosophila* chorion genes. *Carneg. Inst. Wash. Yrbk.* **79,** 73–78.
Spradling, A. (1981). The organization and amplification of two chromosomal domains containing *Drosophila* chorion genes. *Cell* **27,** 193–201.
Spradling, A. (1987). Gene amplification in Dipteran chromosomes. *In* "Result and Problems in Cell Differentiation," Vol. 14, 199–212.
Spradling, A., and Orr-Weaver, T. (1987). Regulation of RNA replication during *Drosophila* development. *Ann. Rev. Genet.* **21,** 373–403.
Spradling, A. C., and Rubin, G. M. (1981). Drosophila genome organization: Conserved and dymamic aspects. *Ann. Rev. Genet.* **15,** 219–264.
Spradling, A. C., and Rubin, G. M. (1983). The effect of chromosomal position on the expression of the *Drosophila* xanthine dehydrogenase gene. *Cell* **34,** 47–57.
Spradling, A. C., Penman, S., and Pardue, M. L. (1975). Analysis of *Drosophila* mRNA by *in situ* hybridization: Sequences transcribed in normal and heat shocked cultured cells. *Cell* **4,** 395–404.
Spradling, A. C., Pardue, M. L., and Penman, S. (1977). Messenger RNA in heat shocked *Drosophila* cells. *J. Mol. Biol.* **109,** 559–587.
Spradling, A. C., Karpen, G., Glaser, R., and Zhang, P. (1992). Evolutionary conservation of developmental mechanisms: DNA elimination in *Drosophila*. In "Evolutionary Conservation of Developmental Mechanisms," pp. 39–53, Wiley-Liss, New York.
Spruill, W. A., Hurwits, D. R., Lucchesi, J. C., and Steiner, A. L. (1978). Association of cyclic GMP with gene expression of polytene chromosomes of *Drosophila melanogaster*. *Proc. Natl. Acad. Sci. USA* **75,** 1480–1484.
Srivastava, J. P., and Bangia, K. K. (1985a). Specific activation of puff 93D in the polytene chromosome of salivary glands of *Drosophila melanogaster* by paracetamol. *J. Curr. Biosci.* **2,** 48–50.
Srivastava, J. P., and Bangia, K. K. (1985b). Effect of caffeine on puffing pattern in the polytene chromosomes of salivary glands of *Drosophila melanogaster*. *Curr. Sci.* **54,** 651–652.
Staiber, W. (1982). Induction of a special Balbiane Ring by position effect in the salivary gland chromosomes of *Acricotopus lucidus*. *Experientia* **38,** 1490–1491.
Staiber, W. (1986). Intermediate puffing patterns as evidence of spontaneous fusions of two different polytene cell types in the *Acricotopus* salivary gland. *Eur. J. Cell Biol.* **42,** 171–175.
Staiber, W., and Behnke, E. (1985). Developmental puffing activity in the salivary gland and Malpighian tubule chromosomes of *Acricotopus lucidus*. *Chromosoma* **93,** 1–16.
Stathakis, D. G., Pentz, E. S., Freeman, M. E., Kullman, J., Hankins, G. R., Pearlson, N. J., and Wright, T. R. F. (1995). The genetic and molecular organization of the Dopa decarboxylase gene cluster of *Drosophila melanogaster*. *Genetics* **141,** 629–655.
Staub, M. (1969). Veranderungen im Puffmuster und das Wachstum der Riesenchromosomen in Speicheldrusen von *Drosophila melanogaster* and spatlarvalen und embryonalen Spendern nach Kultur *in vivo*. *Chromosoma* **26,** 76–104.
Steel, C. G. H., and Davey, K. G. (1985). Integration in the insect endocrine system. *In* "Comprehensive Insect Physiology, Biochemistry and Pharmacology" (G. A. Kerkut and L. I. Gilbert, eds.), Vol. 8, pp. 1–35. Pergamon Press, Oxford/New York/Toronto/Sydney/Paris/Frankfurt.
Steele, M. M., Young, W. J., and Childs, B. (1969). Genetic regulation of glucose-6-phosphate dehydrogenase activity in *Drosophila melanogaster*. *Biochem. Genet.* **3,** 359–370.
Steffensen, D. M. (1963a). Evidence for the apparent absence of DNA in the interbands of *Drosophila* salivary chromosomes. *Genetics* **48,** 1289–1301.
Steffensen, D. M. (1963b). Localization of deoxyribonucleic acid exclusively in the bands of

Drosophila salivary chromosomes. Genetics today. *Proc. XI Intern. Congr. Genet. The Hague* **1,** 109.

Steffensen, D. M., and Wimber, D. E. (1971). Localization of tRNA genes in the salivary chromosomes of *Drosophila* by RNA: DNA hybridization. *Genetics* **69,** 163–187.

Steffensen, D. M., and Wimber, D. E. (1972). Hybridization of nucleic acids to chromosomes. *Results prob. Cell Differ.* **3,** 47–63.

Stein, H. (1976). RNA synthesis with giant chromosomes of isolated *Drosophila hydei* nuclei as a correlate to the experimentally altered chromosome morphology. *Exp. Cell Res.* **103,** 1–14.

Steinemann, M. (1980). Chromosomal replication in *Drosophila virilis*. I. Diploid karyotype of brain cells. *Chromosoma* **78,** 211–223.

Steinemann, M. (1981a). Chromosomal replication in *Drosophila virilis*. II. Organization of active origins in diploid brain cells. *Chromosoma* **82,** 267–288.

Steinemann, M. (1981b). Chromosomal replication in *Drosophila virilis*. III. Organization of active origins in the highly polytene salivary gland cells. *Chromosoma* **82,** 289–307.

Steinemann, M., and Steinemann, S. (1990). Evolutionary changes in the organization of the major LCP gene cluster during sex chromosomal differentiation in the sibling species *Drosophila persimilis, D. pseudoobscura* and *D. miranda*. *Chromosoma* **99,** 424–431.

Steinemann, M., and Steineman, S. (1992). Degenerating Y chromosome of *Drosophila miranda:* A trap for retrotransposons. *Proc. Natl. Acad. Sci. USA* **89,** 7591–7595.

Steinemann, M., Steinemann, S., and Pinsker, W. (1996a). Evolution of the larval cuticle proteins coded by the secondary sex chromosome pair: *X2* and *Neo-Y* of *Drosophila miranda*. I. Comparison at the DNA sequence level. *J. Mol. Evol.* **43,** 405–412.

Steinemann, M., Steinemann, S., and Pinsker, W. (1996b). Evolution of the larval cuticle proteins coded by the secondary sex chromosome pair: *X2* and *Neo-Y* of *Drosophila miranda*. II. Comparison at the amino acid sequence level. *J. Mol. Evol.* **43,** 413–417.

Steinemann, M., Steinemann, S., and Turner, B. M. (1996c). Evolution of dosage compensation. *Chrom. Res.* **4,** 185–190.

Steiner, E. K., Eissenberg, J. C., and Elgin, S. C. R. (1984). A cytological approach to the ordering of events in gene activation using the *Sgs-4* locus of *Drosophila melanogaster*. *J. Cell Biol.* **99,** 233–238.

Steitz, J. A. (1988). "Snurps." *Sci. Am.* **258,** 36–41.

Stepanova, N. G., Nikitin, S. M., Valeeva, F. S., Kartasheva, O. N., Zhuze, A. L., and Zelenin, A. V. (1985). Application of 7-amino-actinomycin D for the fluorescence microscopical analysis of DNA in cells and polytene chromosomes. *Histochem. J.* **17,** 131–142.

Stern, C. (1960). Dosage compensation-development of a concept and new facts. *Can. J. Genet. Cytol.* **2,** 105–118.

Stevens, B. J. (1964). The effect of actinomycin D on nucleolar and nuclear fine structure in the salivary gland cell of *Chironomus thummi*. *J. Ultrastruct. Res.* **11,** 329–353.

Stevens, B. J., and Swift, H. (1966). RNA transport from nucleus to cytoplasm in *Chironomus* salivary gland. *J. Cell Biol.* **31,** 55–77.

Stewart, B., and Merriam, J. R. (1974). Segmental aneuploidy and enzyme activity as a method for cytogenetic localization in *Drosophila melanogaster*. *Genetics* **76,** 301–309.

Stewart, B. R., and Merriam, J. R. (1980). Dosage compensation. *In* "The Genetics and Biology of *Drosophila*" (M. Ashburner, T. R. F. Wright, eds.), pp. 107–140. Academic Press, London/New York/Toronto/Sydney/San Francisco.

Stewrd, R., and Nusslein-Volhard, Ch. (1986). The genetics of the dorsal-bicaudal-D region of *Drosophila melanogaster*. *Genetics* **113,** 665–678.

Stich, H., von. (1956). Bau und Funktion der Nucleolen. *Experientia* **12,** 7–14.

Stocker, A. J., and Kastritsis, C. D. (1972). Developmental studies in *Drosophila*. III. The puffing patterns of the salivary gland chromosomes of *Drosophila pseudoobscura*. *Cromosoma* **37,** 139–176.

Stocker, A. J., and Kastritsis, C. D. (1973). Developmental studies in *Drosophila*. 7. The influence of ecdysterone on the salivary gland puffing patterns of *D. pseudoobscura* larvae and prepupae. *Differentiation* **1,** 225–239.

Stocker, A. J., and Pavan, C. (1974). The influence of ecdysterone of gene amplification, DNA synthesis, and puff formation in the salivary gland chromosomes of *Rhynchosciara hollaenderi*. *Chromosoma* **45,** 295–319.

Stocker, A. J., and Pavan, C. (1977). Developmental puffing patterns in salivary gland chromosomes of *Rhynchosciara hollaenderi*. *Chromosoma* **62,** 17–47.

Stocker, A. J., Pavan, C., and Charlton, C. (1972). Induction of DNA puffs in *Rhynchosciara* by ecdysterone. *Genetics* **71**(Suppl.), 563.

Stocker, A. J., Fresquez, C., and Pavan, C. (1973). Ecdysteron-induced gene amplification and general DNA synthesis in *Rhynchosciara*. *Genetics* **74**(Suppl.), 266.

Stocker, A. J., Fresquez, C., and Luntzios, G. (1978). Banding studies on the polytene chromosomes of *Rhynchosciara hollaenderi*. *Chromosoma* **68,** 337–356.

Stocker, A. J., Troyano-Pueyo, M., Pereira, S. D., and Lara, F. J. S. (1984). Ecdysteroid titers and changes in chromosomal activity in the salivary glands of *Rhinchosciara americana*. *Chromosoma* **90,** 26–38.

Stocker, A. J., Amabis, J. M., Gorab, E., Elke, C., and Lezzi, M. (1997). Antibodies against the D-domain of a *Chironomus* ecdysone receptor protein react with DNA puff sites in *Trichosia pubescens*. *Chromosoma* **106,** 456–464.

Stockert, J. C. (1975). Uranyl-EDTA-Hematoxylin: a new selective staining technique for nucleolar material. *Histochemistry* **43,** 313–322.

Stockert, J. C. (1977). Osmium tetroxide"phenylenediamine staining of nucleoli and Balbiani rings in *Chironomus* salivary glands. *Histochemistry* **53,** 43–56.

Stockert, J. C. (1985). Cytological effects of berberine sulfate on *Chironomus* salivary gland nuclei. *Chromosoma* **93,** 21–25.

Stockert, J. C. (1990). The normalized Balbiani Ring size as a quantitative parameter for the morphological analysis of transcription activity in polytene chromosomes. *Biol. Zbl.* **109,** 139–146.

Stockert, J. C., and Colman, O. D. (1975). Nucleolar patterns in *Chironomus* salivary glands. *Naturwissenschaften* **62,** 439.

Stockert, J. C., and Diez, J. L. (1979). The formation of ribonucleoprotein droplets in *Chironomus* salivary gland nuclei. *Chromosoma* **74,** 83–103.

Stokes, D. G., Tartof, K. D., and Perry, R. P. (1996). CHD1 is concentrated in interbands and puffed regions of *Drosophila* polytene chromosomes. *Proc. Natl. Acad. Sci. USA* **93,** 7137–7142.

Stollar, B. D. (1970). Double-helical polynucleotides: immuno chemical recognization of differing conformations. *Science* **169,** 609–611.

Stollar, B. D. (1975). The specificity and applications of antibodies to helical nucleic acids. *CRC Crit. Rev. Biochem.* **3,** 45–69.

Stone, B. L., and Thummel C. S. (1993). The *Drosophila* 78C early late puffs contains E78, an ecdysone-inducible gene that encodes a novel member of the nuclear hormone receptor superfamily. *Cell* **75,** 1–20.

Storti, R. V., Scott, M. P., Rich, A., and Pardue, M. L. (1980). Translational control of protein synthesis in response to heat shock in *Drosophila melanogaster* cells. *Cell* **22,** 825–834.

Strashnyuk, V. Yu., and Shakhbasov, V. G. (1989). Developmental changes of electrokynetic characteristics of cell nuclei in *Drosophila* salivary glands. *Mol. Gen. Biophys.* **14,** 58–62. [in Russian]

Strashnyuk, V. Yu., Taglina, O. V., and Shakhbazov, V. G. (1991). Ecdysone-dependent changes in the activity of developmental puffs in the salivary glands of *Drosophila* cultured *in vitro* with reference to heterotic effect and selection for adaptively significant traits. *Genetika* **27**(9), 1512–1518. [in Russian]

Strausbaugh, L. D., and Kiefer, B. I. (1979). Genetic modulation of RNA metabolism in *Drosophila*.

III. requirement for an rDNA-deficient X chromosome in $Y^{bbSuVar-5}$-mediated increases in RNA synthesis. *Genetics* **93,** 411–422.

Strobel, E., Pelling, C., and Arnheim, N. (1978). Incomplete dosage compensation in an evolving *Drosophila* sex chromosome. *Proc. Natl. Acad. Sci. USA* **75,** 931–935.

Struhl, K. (1996). Chromatin structure and RNA polymerase II connection: Implications for transcription. *Cell* **84,** 179–182.

Stuart, W. D., and Porter, D. L. (1978). An improved *in situ* hybridization method. *Exp. Cell Res.* **113,** 219–222.

Stuart, W. D., Bishop, J. G., Carson, H. L., and Frank, M. B. (1981). Location of the 18"28S ribosomal RNA genes in two Hawaiian *Drosophila* species by monoclonal immunological identification of RNA, DNA hybrids in situ. *Proc. Natl. Acad. Sci. USA* **78,** 3751–3754.

Sumegy, J., Wieslander, L., and Daneholt, B. (1982). A hierarchic arrangement of the repetitive sequences in the Balbiani Ring 2 gene of *Chironomus tentans*. *Cell* **30,** 579–587.

Surdej, P., Got, C., and Miassod, R. (1990a). Developmental expression pattern of a 800 kb DNA continuum cloned from the *Drosophila* X-chromosome 14B-15B region. *Biol. Cell* **68,** 105–118.

Surdej, P., Got, C., Rosset, R., and Miassod, R. (1990b). Supragenic loop organization: Mapping in *Drosophila* embryos, of scaffold-associated regions on a 800 kilobase DNA continuum cloned from the 14B-15B first chromosome region. *Nucleic Acids Res.* **18,** 3713–3722.

Surdej, P., Brandli, D., and Miassod, R. (1991). Scaffold-associated regions and repeated or cross-hybridizing sequences on an 800 kilobase DNA stretch of the *Drosophila* X chromosome. *Biol. Cell* **73,** 111–120.

Sutherland, J. D., Kozlova, T., Tsertzinis, G., and Kafatos, F. C. (1995). *Drosophila* hormone receptor 38: A second partner for *Drosophila* USP suggests an unexpected role for nuclear receptors of the nerve growth factor-induced protein B type. *Proc. Natl. Acad. Sci. USA* **92,** 7966–7970.

Swaroop, A., Sun, J.-W., Paco-Larson, M. L., and Garen, A. (1986). Molecular organization and expression of the genetic locus *Glued* in *Drosophila melanogaster*. *Mol. Cell. Biol.* **6,** 833–841.

Sweeny, T. L., Guptavaij, P., and Barr, A. R. (1987). Abnormal salivary gland puff associated with meiotic drive in mosquitoes (Diptera: Culicidae). *J. Med. Entomol.* **24,** 623–627.

Swida, U., Lucka, L., and Kress, H. (1990). Glue protein genes in *Drosophila virilis:* Their organization, developmental control of transcription and specific mRNA degradation. *Development* **108,** 269–280.

Swift, H. (1959). Studies on nuclear fine structure. *Brookhaven Sympos. Biol.* **12,** 134–152.

Swift, H. (1962). Nucleic acids and cell morphology in Dipteran salivary glands. *In* "Molecular Control of Cellular Activity" (J. M. Allen, ed.), pp. 73–125. McGraw-Hill, New York/Toronto/London.

Swift, H. (1963). Cytochemical studies on nuclear fine structure. *Exp. Cell Rec.* **9**(Suppl.), 54–67.

Swift, H. (1964). The histones of polytene chromosomes. *In* "The Nucleohistones" (J. Bommer, P. Ts'o, eds.), pp. 169–183. Holden Day, San Francisco/London/Amsterdam.

Swift, H. (1965). Nuclear morphology of the chromosome. *In* "The Chromosome *in Vitro*" (G. Gerganian, ed.), Vol. 1, pp. 26–79. Williams and Wilkins, Baltimore, Maryland.

Swift, H. (1973). Molecular cytogenetics: A symposium summary. *In* "Molecular Cytogenetics" (B. A. Hamkalo and J. Papaconstantinou, eds.), pp. 325–348. Plenum, New York/London.

Swift, H. (1974). The organization of genetic material in eukaryotes: Progress and prospects. *Cold Spring Harbor Symp. Quant. Biol.* **38,** 963–979.

Swift, H., Adams, B. J., and Larsen, K. (1964). Electron microscope cytochemistry of nucleic acids in *Drosophila* salivary glands and *Tetrahymena*. *J. R. Microsc. Soc.* **83,** 161–167.

Szidonya, J., and Reuter, G. (1988). Cytogenetic analysis of the *echinoid (ed)*, *dumpy (dp)* and *clot (cl)* region in *Drosophila melanogaster*. *Genet. Res. Cambr.* **51,** 197–208.

Tagliasacchi, A. M., Forino, L. M. C., Frediani, M., and Avanzi, S. (1983). Different structure of polytene chromosomes of *Phaseolus coccineus* suspensors during early embryogenesis. II. Chromosome pair VII. *Protoplasma* **115,** 95–103.

Talbot, W. S. (1993). Structure, expression and function of ecdysone receptor isoforms in *Drosophila*. Ph.D. thesis, Stanford University, Stanford, California.

Talbot, W. S., Swyryd, E. A., and Hogness, D. S. (1993). *Drosophila* tissues with different metamorphic responses to ecdysone express different ecdysone receptor isoforms. *Cell* **73,** 1323–1337.

Tanguay, R. M., and Vincent, M. (1979). Intracellular distribution of heat shock proteins in *Chironomus tentans*. *Can. J. Genet. Cytol.* **21,** 583.

Tanguay, R. M., and Vincent, M. (1981). Biosynthesis and characterization of heat shock proteins in *Chironomus tentans* salivary glands. *Can. J. Biochem.* **59,** 67–73.

Tanguay, R. M., and Vincent, M. (1982). Intracellular translocation of cellular and heat shock induced proteins upon heat shock in *Drosophila* Kc cells. *Can. J. Biochem.* **60,** 306–315.

Tanouye, M. A., Ferrus, A., and Fujita, S. C. (1981). Abnormal action potentials associated with the *Shaker* complex locus of *Drosophila*. *Proc. Natl. Acad. Sci. USA* **78,** 6548–6552.

Tanzer, E. (1922). Die Zellkerne einiger Dipterenlarven und ihre Entwiklung. *Z. wiss. Zool.* **119,** 114–153.

Tartof, K. D. (1971). Increasing the multiplicity of rRNA genes in *Drosophila melanogaster*. *Science* **171,** 294–297.

Tartof, K. D. (1973). Regulation of ribosomal RNA gene multiplicity in *Drosophila melanogaster*. *Genetics* **73,** 57–71.

Tartof, K. D. (1974a). Unequal mitotic sister chromatid exchange and disproportionate replication as mechanisms regulating ribosomal RNA gene redundancy. *Cold Spring Harbor Symp. Quant. Biol.* **38,** 491–500.

Tartof, K. D. (1974b). Unequal mitotic sister chromatid exchange as the mechanism of ribosomal RNA gene magnification. *Proc. Natl. Acad. Sci. USA* **71,** 1271–1276.

Tartof, K. D. (1975). Redundant genes. *Ann. Rev. Genet.* **9,** 355–385.

Tartof, K. D., and Dawid, I. B. (1976). Similarities and differences in the structure of the X and Y rRNA genes of *Drosophila*. *Nature* **263,** 27–30.

Tartof, K. D., and Perry, R. P. (1970). The 5S RNA genes of *Drosophila melanogaster*. *J. Mol. Biol.* **51,** 171–183.

Tautz, D., Tautz, C., Webb, D., and Dover, G. A. (1987). Evolutionary divergence of promoters and spacers in the rDNA family of four *Drosophila* species: Implications for molecular coevolution in multigene families. *J. Mol. Biol.* **195,** 525–542.

Tautz, D., Hancock, J. M., Webb, D. A., Tautz, C., and Dover, G. A. (1988). Complete sequences of the rRNA genes of *Drosophila melanogaster*. *Mol. Biol. Evol.* **5**(4), 366–376.

Terol, J., Perez-Alonso, M., and Frutos, R. de. (1991). *In situ* localization of the *Antennapedia* gene on the chromosomes of nine *Drosophila* species of the obscura group. *Hereditas* **114,** 131–139.

Terol, J., Perez-Alonso, M., and Frutos, R. de. (1995). Molecular characterization of the *zerknullt* region of the Antennapedia complex of *D. subobscura*. *Chromosoma* **103,** 613–624.

Terra, W. R., and de Bianchi, A. G. (1974). Chemical composition of the fly *Rhychosciara americana*. *Insect Biochem.* **4,** 173–183.

Terra, W. R., and de Bianchi, A. G. (1975). Rate of cocoon production by the larva of the fly *Rhychosciara americana* and storage sites of precusors. *J. Insect Physiol.* **21,** 1547–1550.

Terra, W. R., Ferreira, C., and de Bianchi, A. G. (1975). Distribution of nutrient reserves during spinnig in tissues of the larva of the fly, *Rhychosciara americana*. *J. Insect Physiol.* **21,** 1501–1509.

Tewari, R. R., and Agrawal, U. R. (1984). Asynchronous puffing in the foot pad chromosomes of *Parasarcophaga*. *Experientia* **40,** 750–751.

Thierry-Mieg, D. (1982). *Paralog,* a control mutant in *Drosophila melanogaster*. *Genetics* **100,** 209–237.

Thomas, C. A., Jr. (1971). The genetic organization of chromosomes. *Annu. Rev. Genetics* **5,** 237–256.

Thomas, C. A., Jr. (1973). As the conference opens. *In* "Molecular Cytogenetics" (B. A. Hamkalo and J. Papaconstantinou, eds.), pp. 1–7. Plenum, New York/London.

Thomas, C. A., Jr. (1974). The rolling helix: A model for the eukaryotic gene? *Cold Spring Harbor Symp. Quant. Biol.* **38,** 347–352.

Thomas, G. H., and Elgin, S. C. R. (1988). Protein/DNA architecture of the DNAse I hypersensitive region of the *Drosophila hsp26* promoter. *EMBO J.* **7,** 2191–2201.

Thomas, H. E., Stunnenberg, H. G., and Stewart, A. F. (1993). Heterodimerization of the *Drosophila* ecdysone receptor with retinoid X receptor and *ultraspiracle*. *Nature* **362,** 471–475.

Thomas, S. R., and Lengyel, J. A. (1986). Ecdysteroid-regulated heat-shock gene expression during *Drosophila melanogaster* development. *Dev. Biol.* **115,** 434–438.

Thomopoulos, G. N. (1988). Ultrastructure of the *Drosophila* larval salivary gland cells during the early developmental stages. I. Morphological studies. *J. Morphol.* **198,** 83–93.

Thomopoulos, G. N., and Kastritsis, C. D. (1979). A comparative ultrstructural study of "glue" production and secretion of the salivary glands in different species of the *Drosophila melanogaster* group. W. Roux's Arch. *Dev. Biol.* **187,** 329–354.

Thomopoulos, G. N., Neophytou, E. P., and Kastritsis, C. D. (1989). An ultrastructural and histochemical developmental study of *Drosophila auraria* salivary gland cells during the third-instar period. *Can. J. Zool.* **67,** 421–429.

Thomopoulos, G. N., Neophytou, E. P., and Limberi-Thomopoulos, S. (1991). Rickettsiae-like structures in the larval salivary gland cells of *Drosophila auraria*. *J. Morphol.* **207,** 17–21.

Thomopoulos, G. N., Neophytou, E. P., Alexiou, M., Vadolas, A., Limberi-Thomopoulos, S., and Derventzi, A. (1992). Structural and histochemical studies of Golgi complex differentiation in salivary gland cells during *Drosophila* development. *J. Cell. Sci.* **102,** 169–184.

Thomson, J. A. (1969). The interpretation of puff patterns in polytene chromosomes. *Curr. Modern Biol.* **2,** 333–338.

Thummel, C. S. (1989). The *Drosophila E74* promoter contains essential sequences downstream from the start site of transcription. *Genes Dev.* **3,** 782–792.

Thummel, C. S. (1990). Puffs and gene-regulation molecular insights into the *Drosophila* ecdysone regulatory hierarchy. *BioEssays* **12,** 561–568.

Thummel, C. S. (1992). Mechanisms of transcriptional timing in *Drosophila*. *Science* **255,** 39–40.

Thummel, C. (1995). From embryogenesis to metamorphosis: The regulation and function of *Drosophila* nuclear receptor superfamily members. *Cell* **83,** 871–877.

Thummel, C. S. (1996). Flies on steroids—*Drosophila* metamorphosis and the mechanisms of steroid hormone action. *Trends Genet.* **12,** 306–310.

Thummel, C. S., Burtis, K. C., Jones, C. W., and Hogness, D. S. (1984). Isolation and characterization of an ecdysone-inducible gene located at 74EF in the polytene chromosomes. *Genetics* **107,** 106.

Thummel, C. S., Burtis, K. C., and Hogness, D. S. (1990). Spatial and temporal patterns of *E74* transcription during *Drosophila* development. *Cell* **61,** 101–111.

Thuroff, E., Stoven, S., and Kress, H. (1992). *Drosophila* salivary glands exhibit a regional reprogramming of gene expression during the third larval instar. *Mech. Dev.* **37,** 81–93.

Tiepolo, L., and Laudani, U. (1972). DNA synthesis on polytenic chromosomes of *Anopheles atroparvus*. *Chromosoma* **36,** 305–312.

Tiepolo, L., Diaz, G., and Laudani, U. (1974). Differential DNA synthesis in homologous regions of hybrid polytenic chromosomes (*Anopheles atroparvus* × *A. labranchiae*). *Chromosoma* **45,** 81–89.

Tikhomirova, M. M., Vatti, K. V., Mamon, L. A., Barabanova, L. V., and Kutskova, Yu. A. (1994). Mechanisms underlying the resistance of genetic material of the animal cell to stress treatment. *Genetika* **30**(8), 1097–1104. [in Russian]

Tissieres, A., Mitchell, H. K., and Tracy, U. M. (1974). Protein synthesis in salivary glands of *Drosophila melanogaster*: Relation to chromosome puffs. *J. Mol. Biol.* **84,** 389–398.

Tiwari, P. K., and Lakhotia, S. C. (1984). Replication in *Drosophila* chromosomes. XIII. Comparison of late replicating sites in two polytene cell types in *D. hydei*. *Genetica* **65,** 227–234.

Tobler, J., Bowman, J. T., and Simmons, J. R. (1971). Gene modulation in *Drosophila:* Dosage compensation and relocated v^+ genes. *Biochem. Genet.* **5,** 111–117.

Todo, T., Roark, M., Raghavan, K. V., Mayeda, C., and Meyerowitz, E. (1990). Fine structure mutational analysis of a stage- and tissue-specific promoter element of the *Drosophila* glue gene *Sgs-3*. *Mol. Cell. Biol.* **10,** 5991–6002.

Tokumitsu, T. (1968). Some aspects on effects of *Drosophila tissue* extracts on the puffing pattern of incubated *Drosophila* salivary glands. *J. Fac. Sci. Hokkaido Univ. Ser. VI Zool.* **16,** 525–530.

Toledo, S. M. de, and Lara, F. J. S. (1978). Translation of messages transcribed from the "DNA puffs" *Rhynchosciara*. *Biochem. Biophys. Res. Commun.* **85,** 160–166.

Topol, J., Ruden, D. M., and Parker, C. S. (1985). Sequences required for *in vitro* transcriptional activation of a *Drosophila hsp70* gene. *Cell* **42,** 527–537.

Torok, I., and Karch, F. (1980). Nucleotide sequences of heat shock activated genes in *Drosophila melanogaster*. I. Sequences in the regions of the 5′ and 3′ ends of the *hsp70* gene in the hybrid plasmid 56H8. *Nucleic Acids Res.* **8,** 3105–3123.

Torok, I., Mason, P. J., Karch, F., Kiss, I., and Udvardy, A. (1982). Extensive regions of homology associated with heat-induced genes at loci 87A7 and 87C1 in *Drosophila melanogaster*. *In* "Heat Shock from Bacteria to Man" (M. L. Schlesinger, M. Ashburner, and A. Tissieres, eds.), pp. 19–25. Cold Spring Harbor Laboratory Press, Cold Spring Harbor, NY.

Torres, M., and Sanchez, L. (1989). The *scute (T4)* gene acts as a numerator element of the X:A signal that determines the state of activity of *Sex-lethal* in *Drosophila*. *EMBO J.* **8,** 3079–3086.

Toshev, L. B., and Semionov, E. P. (1987). The compensatory response locus deletion increases the number of nucleoli in *Drosophila melanogaster* polytene cells. *Chromosoma* **95,** 258–262.

Travers, A. A. (1994). Chromatin structure and dynamics. *BioEssays* **16,** 657–662.

Trepte, H.-H. (1976). Das Puffmuster der Borstenapparat—chromosomen von Sarcophaga barbata. *Chromosoma* **55,** 137–164.

Trepte, H.-H. (1977). Uber die Beziehungen zwischen dem Wachstum der Kerne und dem Wachstum und der Puffaktivitat der Polytanchromosomen wahrend der Entwicklung des Borstenapparates bei der Fleischfliege *Sarcophaga barbata*. *Biol. Zbl.* **96,** 551–570.

Trepte, H.-H. (1980). Autonomous puffing patterns in thoracic and abdominal polytene bristle cell chromosomes of the flesh fly *Sarcophaga barbata*. *Dev. Biol.* **75,** 471–480.

Trepte, H.-H. (1993). Ultrastructural analysis of Balbiani ring genes of *Chironomus pallidivittatus* in different states of Balbiani ring activity. *Chromosoma* **102,** 433–445.

Tryselius, Y., Samakovlis, C., Kimbrell, D. A., and Hultmark, D. (1992). *Cec* C, a cecropin gene expressed during metamorphosis in *Drosophila* pupae. *Eur. J. Biochem.* **204,** 395–399.

Tschudi, C., and Pirrotta, V. (1980). Sequence and heterogeneity in the 5S RNA gene cluster of *Drosophila melanogaster*. *Nucleic Acids Res.* **8,** 441–451.

Tsukiyama, T., Becker, P. B., and Wu, C. (1994). ATP-dependent nucleosome disruption at a heat shock promoter mediated by binding of GAGA transcription factor. *Nature* **367,** 525–532.

Tulchin, N., and Rhodin, J. A. G. (1970). Of microtubules and membranes: the development of the salivary gland imaginal disc. *J. Ultrastruct. Res.* **32,** 443–457.

Tulchin, N., Mateyko, G. M., and Kopac, M. J. (1967). *Drosophila* salivary glands *in vitro*. *J. Cell Biol.* **34**(3), 891–897.

Turberg, A., Spinder-Barth, M., Lutz, B., Lezzi, M., and Spindler, K.-D. (1988). Presence of an ecdysteroid-specific binding protein ("receptor") in epithelial tissue culture cells of *Chironomus tentans*. *J. Insect Physiol.* **34,** 797–803.

Turner, B. M., Birley, A. J., and Lavender, J. (1992). Histone H4 isoforms acetylated at specific lysine residues define individual chromosomes and chromatin domains in *Drosophilar* polytene nuclei. *Cell* **69,** 375–384.

Turner, S. H., and Laird, C. D. (1973). Diversity of RNA sequences in *Drosophila melanogaster*. *Biochem. Genet.* **10,** 263–275.

Uchida, S., Uenoyama, T., and Oishi, K. (1981). Studies on the sex-specific lethals of *Drosophila melanogaster*. III. A third chromosome male-specific lethal mutant. *Jpn. J. Genet.* **56,** 523–527.

Udvardy, A., and Schedl, P. (1993). The dynamics of chromatin condensation: Redistribution of topoisomerase II in the 87A7 heat shock locus during induction and recovery. *Mol. Cell Biol.* **13,** 7522–7530.

Udvardy, A., Sumegi, J., Toth, E. C., Gausz, J., Gyurkovics, H., Schedl, P., and Ish-Horowicz, D. (1982). Genomic organization and functional analysis of a deletion variant of the 87A7 heat shock locus of *Drosophila melanogaster*. *J. Mol. Biol.* **155,** 267–280.

Udvardy, A., Maine, E., and Schedl, P. (1985). The 87A7 chromomere. Identification of novel chromatin structures flanking the heat shock locus that may define the boundaries of higher order domains. *J. Mol. Biol.* **185,** 341–358.

Ukil, M., Chatterjee, K., and Mukherjee, A. S. (1984). Cytophotometric analysis of *in situ* binding of non-histone protein to the chromatin in *Drosophila melanogaster*. *Dros. Inform. Serv.* **60,** 201–202.

Ukil, M., Chatterjee, K., Dey, A., Ghosh, S., and Mukherjee, A. S. (1986). Affinity binding of non-histone chromatin proteins to the X chromosome of *Drosophila* by *in situ* chromatin reconstitution and its significance. *J. Cell Sci.* **86,** 35–45.

Umbetova, G. H. (1991). Immunofluorescent localization of transcriptionally active regions in the polytene chromosomes of *Drosophila melanogaster*. Ph.D. thesis, Novosibirsk. [in Russian]

Umbetova, G. H., Kokoza, E. B., and Zhimulev, I. F. (1988). Localisation of the DNA fragment from the *Drosophila melanogaster* X chromosome 10A1-2 band in polytene chromosomes of some ofher species. *Genet. (Life Sci. Adv.)* **7,** 43–46.

Umbetova, G. H., Vlassova, I. E., and Zhimulev, I. F. (1994). Immunofluorescent localization of DNA-RNA hybrids in polytene chromosomes of *Drosophila melanogaster* during development. *Dros. Inform. Serv.* **75,** 126–131.

Urness, L. D., and Thummel, C. S. (1990). Molecular interractions within the ecdysone regulatory hierarchy: DNA binding properties of the *Drosophila* early ecdysone-inducible E74A protein. *Cell* **63,** 47–61.

Valencia, J. I., and Plaut, W. (1970). Further studies on temporal autonomy of DNA replication in chromosome rearrangements. *J. Cell Biol* **47**(2), 216a.

Valente, V. L. S., and Cordeiro, A. R. (1991). Comparative study of puffing patterns of *Drosophila willistoni* and *Drosophila cubana* salivary gland chromosomes. *Rev. Brasil. Genet.* **14,** 299–313.

Valentin, M., Bollenbacher, W. E., Gilbert, L. I., and Kroeger, H. (1978). Alterations in ecdysone content during the post-embryonic development of *Chironomus thummi:* Correlations with chromsomal puffing. *Zeitschr. Naturforsch.* **33C,** 557–560.

Valeyeva, F. S. (1975). Effect of pilicarpine on puffing of the polytene chromosomes of *Chironomus thummi* salivary glands. *Tsitologia* **17**(9), 1032–1036. [in Russian]

Valeyeva, F. S. (1991). Binding of actinomycin D and effects of its different concentrations on transcription and morphology of polytene chromosomes. Ph.D. thesis, Novosibirsk. [in Russian]

Valeyeva, F. S. (1995). Cycloheximide-induced local decompactization of polytene chromosomes. *Doklady Russ. Akad. Nauk* **343,** 272–274. [in Russian]

Valeyeva, F. S., Kiknadze, I. I., Panova, T. H., and Perov, N. A. (1979). Effect of high doses of actinomycin D on structure, puffing and transcription of the polytene chromosomes of *Chironomus thummi* salivary glands. *Tsitologia* **21**(12), 1411–1418. [in Russian]

Valeyeva, F. S., Stepanova, N. G., Nikitin, S. M., Zhuze, A. L., Kiknadze, I. I. and Zelenin, A. V. (1984). Fluorescent microscopic and autoradiographic analysis of the differential reaction of different regions of the polytene chromosomes of *Chironomus* on long-term effect of 7-amino-actinomycin D and ^{3}H-actinomycin D. *Tsitologia* **26**(11), 1255–1261. [in Russian]

Valeyeva, F. S., Vlassova, I. E., and Zimmermann, V. G. (1997). An immunofluorescent study of a pseudopuff in the centromeric heterochromatin region of *Chironomus thummi* polytene chromosomes. *Genetica* **33**(N9), 1313–1315. [in Russian]

Varute, A. T., and Sawant, V. A. (1971). b-Glucuronidase during embryogenesis, growth and metamorphosis of the larvae of blowfly *Crysomyia rufiferacis*. *Comp. Biochem. Physiol. B* **38,** 211–223.

Vaslet, C. A., O'Connel, P., Izquierdo, M., and Rosbash, M. (1980). Isolation and mapping of a cloned ribosomal protein gene of *Drosophila melanogaster*. *Nature* **285,** 674–676.

Vaury, C., Chaboissier, M. C., Drake, M. E., Lajoine, O., Dastugue, B., and Pellison, A. (1994). The *Doc* transposable element in *Drosophila melanogaster* and *Drosophila simulans:* Genomic distribution and transcription. *Genetica* **93,** 117–124.

Vazquez, J, and Schedl, P. (1994). Sequences required for enhancer blocking activity of *scs* are located within two nuclease-hypersensitive regions. *EMBO J.* **13,** 5984–5993.

Vazquez, J., Pauli, D., and Tissieres, A. (1993a). Transcriptional regulation in *Drosophila* during heat shock: A nuclear run-on analysis. *Chromosoma* **102,** 233–248.

Vazquez, J., Farkas, G., Gaszner, M., Udvardy, A., Muller, M., Hagstrom, K., Guyrkovics, H., Sipos, L., Gausz, J., Galloni, M., Hogga, I., Karch, F., and Schedl, P. (1993b). Genetic and molecular analysis of chromatin domains. *Cold Spring Harbor Symp. Quant. Biol.* **58,** 45–53.

Vazquez-Nin, G., and Bernhard, W. (1971). Comparative ultrastructural study of perichromatin- and Balbiani Ring granules. *J. Ultrastruct. Res.* **36,** 842–860.

Vazquez-Nin, G. H., Echeverria, O. M., Fakan, S., Leser, G., and Martin, T. E. (1990). Immunoelectron microscope localization of snRNPs in the polytene nucleus of salivary glands of *Chironomus thummi*. *Chromosoma* **99,** 44–51.

Vazquez-Nin, G. H., Echeverria, O. M., Carbajal, M. E., Tanguay, R. M., Diez, J. L., and Fakan, S. (1992). Immunoelectron microscope localization of Mr 90000 heat shock protein and Mr 70000 heat shock cognate protein in the salivary glands of *Chironomus thummi*. *Chromosoma* **102,** 50–59.

Velazquez, J. M., and Lindquist, S. (1984). *hsp70:* nuclear concentration during environmental stress and cytoplasmic storage during recovery. *Cell* **36,** 655–662.

Velazquez, J. M., DiDomenico, B., and Lindquist, S. (1980). Intracellular localization of heat shock proteins in *Drosophila*. *Cell* **20,** 679–689.

Velazquez, J. M., Sonoda, S., Bugaisky, G., and Lindquist, S. (1983). Is the major *Drosophila* heat shock proteins present in cells that have not been heat shocked? *J. Cell Biol.* **96,** 286–290.

Velissariou, V., and Ashburner, M. (1980). The secretory proteins of the larval salivary gland of *Drosophila melanogaster*. Cytogenetic correlation of a protein and a puff. *Chromosoma* **77,** 13–27.

Velissariou, V., and Ashburner, M. (1981). Cytogenetic and genetic mapping of a salivary gland secretion protein in *Drosophila melanogaster*. *Chromosoma* **84,** 173–185.

Velkataraman, V., O'Mahony, P.J., Manzcak, M., and Jones, G. (1994). Regulation of juvenile hormone esterase gene transcription by juvenile hormone. *Dev. Genet.* **15,** 391–400.

Vidal, B. C. (1977). Variation in dry mass concentration and protein contents in DNA puffs. *Caryologia* **30,** 69–76.

Vidal, O. R., Spirito, S., and Riva, R. (1971). Heterogeneous ultratructure of the granules of *Drosophila* salivary glands. *Experientia* **27,** 178–179.

Viinikka, V., Hannah-Alava, A., and Arajarvi, P. (1971). A reinvestigation of the nucleolus-organizing regions in the salivary gland nuclei of *Drosophila melanogaster*. *Chromosoma* **36,** 34–45.

Vincent, A., Colot, H. V., and Rosbash, M. (1985). Sequence and structure of the *serendipity* locus of *Drosophila melanogaster*. *J. Mol. Biol.* **186,** 149–166.

Vincent, M., and Tanguay, R. M. (1979). Heat-shock induced proteins present in the cell nucleus of *Chironomus tentans* salivary glands. *Nature* **281,** 501–503.

Vincent, M., and Tanguay, R. M. (1982). Different intracellular distributions of heat-shock and arsenite-induced proteins in Drosophila Kc cells: Possible relation with the phosphorylation and translocation of a major cytoskeletal protein. *J. Mol. Biol.* **162,** 365–378.

Vincent, W. S., III, Goldstein, E. S., and Allen, S. A. (1990). Sequence and expression of two regulated transcription units during *Drosophila melanogaster* development: *Deb-A* and *Deb-B*. *Biochem. Biophys. Acta* **1049,** 59–68.

Visa, N., Gonzalez-Duarte, R., and Santa-Cruz, M. C. (1988). A cytological and molecular analysis of *Adh* gene expression in *Drosophila melanogaster* polytene chromosomes. *Chromosoma* **97,** 171–177.

Vlad, M., and Macgregor, H. C. (1975). Chromomere number and its genetic significance in lampbrush chromosomes. *Chromosoma* **50,** 327–347.

Vlassova, I. E., and Kiknadze, I. I. (1975). Effect of cycloheximide on the incorporation of ^{3}H-thymidine into salivary gland chromosomes of *Chironomus thummi* at early stages of development of larvae. *Tsitologia* **17**(5), 518–523. [in Russian]

Vlassova, I. E., and Zhimulev, I. F. (1988). Effect of transcription inhibitors on RNA synthesis in the polytene chromosomes of *Drosophila melanogaster*. *Tsitologiya* **30**(5), 568–572. [in Russian]

Vlassova, I. E., Belyaeva, E. S., and Zhimulev, I. F. (1983). Induction of giant heat-shock puffs in polytene chromosomes of *Drosophila melanogaster* by 20-OH ecdysone and ethanol. *Dros. Inform. Serv.* **59,** 134–136.

Vlassova, I. E., Umbetova, G. H., Zimmermann, V. H., Alonso, C., Belyaeva, E. S., Zhimulev, I. F. (1984). Immunofluorescent localization of DNA-RNA hybrids in *Drosophila melanogaster* polytene chromosomes. *Doklady AN SSSR* **274**(1), 189–192. [in Russian]

Vlassova, I. E., Umbetova, G. H., Zimmermann, V. H., Alonso, C., Belyaeva, E. S., and Zhimulev, I. F. (1985a). Immunofluorescence localization of DNA:RNA hybrids in *Drosophila melanogaster* polytene chromosomes. *Chromosoma* **91,** 251–258.

Vlassova, I. E., Umbetova, G. H., Belyaeva, E. S., and Zhimulev, I. F. (1985b). Identification of transcriptionally active regions in polytene chromosomes of *Drosophila melanogaster* by the method of immunofluorescent localization of DNA-RNA hybrids. *Genetika* **21**(3), 424–432. [in Russian]

Vlassova, I. E., Zhimulev, I. F., Belyaeva, E. S., and Semeshin, V. F. (1987). The effect of DRB and a-amanitin on the RNA synthesis in *Drosophila melanogaster* polytene chromosomes. *Dros. Inform. Serv.* **66,** 151–154.

Vlassova, I. E., Demin, S. Yu., Ilyinskaya, N. B., and Zhimulev, I. F. (1991). Immunofluorescent localization of DNA/RNA hybrids in polytene chromosomes with different degree of compaction in *Chironomus plumosus* larvae. *Tsitologia* **33**(3), 57–60. [in Russian]

Voelker, R. A., Langley, C. H., Leigh Brown, A. J., Ohnishi, S., Dickson, B., Montgomery, E., and Smith, S. C. (1980). Enzyme null alleles in natural populations of *Drosophila melanogaster:* Frequencies in a North Carolina population. *Proc. Natl. Acad. Sci. USA* **77,** 1091–1095.

Voelker, R. A., Wisely, G. B., Huang, S.-M., and Gyurkovics, H. (1985). Genetic and molecular variation in the RpII215 region of *Drosophila melanogaster*. *Mol. Gen. Genet.* **201,** 437–445.

Voelker, R. A., Huang, S.-M., Wisely, G. B., Sterling, J. F., Bainbridge, S. P., and Hiraizumi, K. (1989). Molecular and genetic organization of the suppressor of *sable* and *Minute(1)1B* region in *Drosophila melanogaster*. *Genetics* **122,** 625–642.

Voellmy, R., and Rungger, D. (1982). Heat-induced transcription of *Drosophila* heat-shock genes in *Xenopus* oocytes. *In* "Heat Shock from Bacteria to Man" (M. L. Schlesinger, M. Ashburner, and A. Tissieres, eds.), pp. 49–56. Cold Spring Harbor Laboratory Press, Cold Spring Harbor, NY.

Voellmy, R., Goldschmidt-Clermont, M., Southgate, R., Tissieres, A., Levis, R., and Gehring, W. (1981). A DNA segment isolated from chromosomal site 67B in *Drosophila melanogaster* contains 4 closely linked heat shock genes. *Cell* **23,** 261–270.

Vogt-Kohne, L. (1961). Quantitative cytochemische Untersuchungen an Nukleolen aus Speicheldrunsenkerner von *Chironomus thummi*. *Chromosoma* **12,** 382–397.

Vogt-Kohne, L., and Carlson, L. (1963). Cytochemische Untersuchungen an Balbiani-ringen des 4 Speicheldrusenchromosomen von *Chironomus tentans*. *Chromosoma* **14,** 186–194.

Vossen, J. G. H. M., Leenders, H. J., Derksen, J., and Jeucken, G. (1977). Chromosomal puff induction in salivary glands from *Drosophila hydei* by arcenite. *Exp. Cell Res.* **109,** 277–283.

Wadsworth, S. C. (1982). A family of related proteins is incoded by the major *Drosophila* heat shock gene family. *Mol. Cell. Biol.* **2,** 286–292.

Wadsworth, S. C., Craig, E. A., and McCarthy, B. J. (1980). Genes for three *Drosophila* heat-shock-induced proteins at a single locus. *Proc. Natl. Acad. Sci. USA* **77,** 2134–2137.

Wakimoto, B. T., and Kaufman, T. C. (1981). Analysis of larval segmentation in lethal genotypes associated with the *Antennapedia* gene complex in *Drosophila melanogaster*. *Dev. Biol.* **81,** 51–64.
Wakimoto, B. T., Lewis, R. A., and Kaufman, T. C. (1980). Genetic analysis of the *Antennapedia* gene complex: Mutant screen of proximal 3R, bands 84A-84B1. *Dros. Inform. Serv.* **55,** 140–141.
Walker, A. R., Howells, A. J., and Tearle, R. G. (1986). Cloning and characterization of the *vermilion* gene of *Drosophila melanogaster*. *Mol. Gen. Genet.* **202,** 102–107.
Walker, V. K., and Ashburner, M. (1981). The control of ecdysterone-regulated puffs in *Drosophila* salivary glands. *Cell* **26,** 269–279.
Walker, V. K., Watson, K. L., Holden, J. J. A., and Steel, C. G. H. (1987). Vitellogenesis and fertility in *Drosophila* females with low ecdysteroid titres, the $L(3)3^{DTS}$ mutation. *J. Insect Physiol.* **33,** 137–142.
Walldorf, U., Richter, S., Ryseck, R.-P., Steller, H., Edstrom, J.-E., Bautz, E. K. F., and Hovemann, B. (1984). Cloning of heat-shock locus 93D from *Drosophila melanogaster*. *EMBO J.* **3,** 2499–2504.
Wallrath, L. L., Lu, Q., Granok, H., and Elgin, S. C. R. (1994). Architectural variations of inducible eukaryotic promoters: Present and remodeling chromatin structures. *BioEssays* **16,** 165–170.
Walshe, B. M. (1947). Feeding mechanisms of *Chironomus* larvae. *Nature* **160,** 474.
Walter, L. (1973). Syntheseprozesse an den Riesenchromosomen von *Glyptotendipes*. *Chromosoma* **41,** 327–360.
Wang, M., Champion, L. E., Biessmann, H., and Mason, J. M. (1994). Mapping a mutator, *mu2*, which increases the frequency of terminal deletions in *Drosophila melanogaster*. *Mol. Gen. Genet.* **245,** 598–607.
Ward, C. L. (1949). Karyotype variation in *Drosophila*. *Univ. Texas Publ.* **4920,** 70–79.
Waring, G. L., and Pollack, J. C. (1987). Cloning and characterization of a dispersed, multicopy, X chromosome sequence in *Drosophila melanogaster*. *Proc. Natl. Acad. Sci. USA* **89,** 2843–2847.
Warmke, J. W., Kreuz, A. J., and Falkenthal, S. (1989). Colocalization of chromosome bands 99E1–3 of the *Drosophila melanogaster* myosin light chain-2 gene and a haplo-insufficient locus that affects flight behavior. *Genetics* **122,** 139–151.
Watanabe, T., and Kankel, D. R. (1990). Molecular cloning and analysis of *l(1)ogre*, a locus of *Drosophila melanogaster* with prominent effects on the postembryonic development of the central nervous system. *Genetics* **126,** 1033–1044.
Wattiaux, R. (1969). Biochemistry and function of lysosomes. *In* "Handbook of Molecular Cytology" (A. Lima-de-Faria, ed.), pp. 1160–1178. North-Holland, Amsterdam, London.
Weber, F., Mahr, R., Meyer, B., Eppenberger, H. M., and Lezzi, M. (1983). Cell-free translation of Balbiani Ring RNA (75S) of *Chironomus tentans* salivary glands into high molecular weight products. *Wilhelm Roux's Arch. Dev. Biol.* **192,** 200–203.
Weeks, J. R., Hardin, S. E., Shen, J., Lee, J. M., and Greenleaf, A. L. (1993). Locus-specific variation in phosphorylation state of RNA polymerase II *in vivo*: Correlations with gene activity and transcript processing. *Genet. Dev.* **7,** 2329–2344.
Wegmann, I. S., Quack, S., Spindler, K.-D., Dorsch-Hasler, K., Vogtli, M., and Lezzi, M. (1995). Immunological studies on the developmental and chromosomal distribution of ecdysteroid receptor protein in *Chironomus tentans*. *Arch. Insect Biochem. Physiol.* **30,** 95–114.
Wei, L.-H., Erlanger, B. E., Eastman, E. M., Miller, O. J., and Goodman, R. (1981). Inverse relationship between transcriptional activity and 5-methyl-cytosine content of DNA in polytene chromosomes of *Sciara coprophila*. *Exp. Cell Res.* **135,** 411–415.
Weideli, H., Schedl, P., Artavanis-Tsakonas, S., Steward, R., Yuan, R., and Gehring, W. J. (1978). Purification of a protein from unfertilized eggs of *Drosophila* with specific affinity for a defined DNA sequence and the cloning of this DNA sequence in bacterial plasmids. *Cold Spring Harbor Symp. Quant. Biol.* **42,** 693–700.
Weirich, G., and Karlson, P. (1969). Distribution of tritiated ecdysone in salivary glands and other

tissues of *Rhynchosciara* and *Chironomus* larvae. An autoradiographical study. *Wilhelm Roux' Arch.* **164,** 170–181.

Wellauer, P. K., and Dawid, I. B. (1977). The structural organization of ribosomal DNA in *Drosophila melanogaster*. *Cell* **10,** 193–212.

Wellauer, P. K., Dawid, I. B., and Tartof, K. D. (1978). X and Y chromosomal ribosomal DNA of *Drosophila:* Comparison of spacers and insertions. *Cell* **14,** 269–278.

Welshons, W. J. (1965). Analysis of a gene in *Drosophila*. *Science* **150,** 1122–1129.

Welshons, W., and Keppy, D. O. (1975). Intragenic deletions and salivary band relationships in *Drosophila*. *Genetics* **80,** 143–155.

Wemmer, T., and Klambt, C. (1995). A genetic analysis of the *Drosophila* closely linked interacting genes *bulge*, *argos* and *soba*. *Genetics* **140,** 629–641.

Wen, W.-N., and Hague, D. R. (1979). Localization of 5S gene loci on polytene chromosomes of *Glyptotendipes barbipes* (Staeger) and transcription of RNA from these sites. *Cytologia* **44,** 487–503.

Wen, W.-N., Leon, P. E., and Hague, D. R. (1974). Multiple gene sites for 5S and 18 + 28S RNA on chromosomes of *Glyptotendipes barbipes* (Staeger). *J. Cell Biol.* **62,** 132–144.

Westwood, J. T., Clos, J. and Wu, C. (1991). Stress-induced oligomerization and chromosomal relocation of heat-shock factor. *Nature* **353,** 822–827.

Wharton, K. A., Johansen, K. M., Xu, T., and Artavanis-Tsakonas, S. (1985). Nucleotide sequence from the neurogenic locus *Notch* implies a gene product that shows homology with proteins containing EGF-like repeat. *Cell* **43,** 567–581.

Wharton, L. T. (1943). Analysis of the metaphase and salivary gland chromosome morphology with the genus *Drosophila*. *Univ. Tex. Publ.* **4313,** 282–319.

White, K. P., Hurban, P., Watanabe, T., and Hogness, D. S. (1997). Coordination of *Drosophila* metamorphosis by two ecdysone-induced nuclear receptors. *Science* **276,** 114–117.

White, M. J. D. (1948). The cytology of the Cecidomyidae (Diptera). IV. The salivary gland chromosomes of several species. *J. Morphol.* **87,** 53–80.

White, M. J. D. (1954). "Animal Cytology and Evolution, Second Edition." *Cambridge Univ. Press,* London.

White, M. J. D. (1973). "Animal Cytology and Evolution. Third Edition." Cambridge Univ. Press, London/New York/Melbourne.

White, R. L., and Hogness, D. S. (1977). R loop mapping of the 18S and 28S sequences in the long and short repeating units of *Drosophila melanogaster* rDNA. Cell **10,** 177–192.

Whitmore, T. (1986). Localization of a "housekeeping" gene by *Chironomus thummi thummi*. *Cytobios* **46,** 193–200.

Whitten, J. M. (1965). Differential deoxyribonucleic acid replication in the giant foot-pad cells of *Sarcophaga bullata*. *Nature* **208,** 1019–1021.

Whitten, J. (1969). Coordinated development in the foot pad of the fly *Sarcophaga bullata* during metamorphosis: changing puffing patterns of the giant cell chromosomes. *Chromosoma* **26,** 215–244.

Whitten, J. M. (1975). Possible transfer of salivary secretion to mid-gut at the time of pupation in Diptera *Cyclorrhapha*. *Ann. Entomol. Soc. Am.* **69**(2), 273–274.

Widmer, R. M., Lucchini, R., Lezzi, M., Meyer, B., Sogo, J. M., Edstrom, J.-E., and Koller, Th. (1984). Chromatin structure of a hyperactive secretory protein gene (in Balbiani ring 2) of *Chironomus*. *EMBO J.* **3,** 1635–1641.

Widmer, R. M., Lezzi, M., and Koller, Th. (1987). Structural transition in inactive Balbiani ring chromatin of *Chironomus* during *micrococcus* nuclease digestion. *EMBO J.* **6,** 743–748.

Wiener, J., Spiro, D., and Loewenstein, W. R. (1964). Studies on an epithelial (gland) cell junction. II. Surface structure. *J. Cell Biol.* **22,** 587–598.

Wieslander, L. (1975). The presence of 5S RNA genes in two consecutive chromosome bands in *Chironomus tentans*. *Mol. Biol. Rep.* **2,** 189–194.

Wieslander, L. (1979). Number and structure of Balbiani ring 75S RNA transcription units in *Chironomus tentans*. *J. Mol. Biol.* **134,** 347–367.

Wieslander, L., and Daneholt, B. (1977). Demonstration of Balbiani Ring RNA sequences in polysomes. *J. Cell Biol.* **73,** 260–264.

Wieslander, L., and Lendahl, U. (1983). The Balbiani Ring 2 gene in *Chironomus tentans* is built from two types of tandemly arranged major repeat units with a common evolutionary origin. *EMBO J.* **2,** 1169–1175.

Wieslander, L., Lambert, B., and Egyhazy, E. (1975a). Localization of 5S RNA genes in *Chironomus tentans*. *Chromosoma* **51,** 49–56.

Wieslander, L., Lambert, B., and Wobus, U. (1975b). The location of repeated DNA sequences in the chromosomes of *Chironomus tentans*. *Chromosoma* **52,** 159–173.

Wieslander, L., Sumegi, J., and Daneholt, B. (1982). Evidence for a common ancestor sequence for the Balbiani Ring 1 and Balbiani Ring 2 genes in *Chironomus tentans*. *Proc. Natl. Acad. Sci. USA* **79,** 6956–6960.

Wieslander, L., Hoog, C., Hoog, J.-O., Jornvall, H., Lendahl, U., and Daneholt, B. (1984). Conserved and nonconserved structures in the secretory proteins encoded in the Balbiani Ring genes of *Chironomus tentans*. *J. Mol. Evol.* **20,** 304–312.

Wieslander, L., Hoog, C., Lendahl, U., and Daneholt, B. (1986). An example of satellite-like evolution of coding sequences. *Chem. Scr. B* **26,** 159–163.

Wigglesworth, V. B. (1955). The breakdown of the thoracic gland in the adult insect, *Rhodnius prolixus*. *J. Exp. Biol.* **32,** 485–491.

Willard, H. F., and Salz, H. K. (1997). Remodeling chromatin with RNA. *Nature* **386,** 288–229.

Williams, S. M., Furnier, G. R., Fuog, E., and Strobeck, C. (1987). Evolution of the ribosomal DNA spacers of *Drosophila melanogaster:* different patterns of variations on X and Y chromosomes. *Genetics* **116,** 225–232.

Williams, S. M., Robbins, L. G., Cluster, P. D,. Allard, R. W., and Strobeck, C. (1990). Superstructure of the *Drosophila* ribosomal gene family. *Proc. Natl. Acad. Sci. USA* **87,** 3156–3160.

Williamson, J. H., and Bentley, M. M. (1983). Dosage compensation in *Drosophila:* NADP-enzyme activities and cross-reacting material. *Genetics* **103,** 649–658.

Williamson, J. H., Procunier, J. D., and Church, R. B. (1973). Does the DNA in the Y chromosome magnify? *Nature New Biol.* **243,** 190–192.

Wilson, E. (1936). The cell and its role in development and heredity. *Mosc. Leningr. Biomedgiz.* [in Russian]

Wiltenburg, J., and Lubsen, N. H. (1976). A differential effect of the incubation temperature on the inhibition of RNA synthesis by DRB in cells of *Drosophila hydei*. *FEBS Lett.* **70,** 17–22.

Wimber, D. E., and Steffensen, D. M. (1970). Localization of 5S RNA genes on *Drosophila* chromosomes by RNA-DNA hybridization. *Science* **170,** 639–641.

Wimber, D. E., and Steffensen, D. M. (1973). Localization of gene function. *Ann. Rev. Genet.* **7,** 205–223.

Wimber, D. E., and Wimber, D. R. (1977). Sites of 5S ribosomal genes in *Drosophila*. I. The multiple clusters in the *virilis* group. *Genetics* **86,** 133–148.

Winter, C. E., de Bianchi, A. G., Terra, W. R., and Lara, F. J. S. (1977a). The giant DNA puffs of *Rhynchosciara americana* code for polypeptides of the salivary gland secretion. *J. Insect Physiol.* **23,** 1455–1459.

Winter, C. E., de Bianchi, A. G., Terra, W. R., and Lara, F. J. S. (1977b). Relationships between newly synthesized proteins and DNA puff patterns in the salivary glands of *Rhynchosciara americana*. *Chromosoma* **61,** 193–206.

Winter, C. E., de Bianchi, A. G., Terra, W. R., and Lara, F. J. S. (1980). Protein synthesis in the salivary glands of *Rhynchosciara americana*. *Dev. Biol.* **75,** 1–12.

Wobus, U., and Serfling, E. (1977). The replication frequency of DNA in Balbiani Ring 2 of *Chironomus thummi*. *Chromosoma* **64,** 279–286.

Wobus, U., Panitz, R., and Serfling, E. (1970). Tissue specific gene activities and proteins in the *Chironomus* salivary gland. *Mol. Gen. Genet.* **107,** 215–223.

Wobus, U., Panitz, R., and Serfling, E. (1971a). a-amanitin: its effect on RNA synthesis in polytene chromosomes. *Experientia* **27,** 1202–1203.

Wobus, U., Serfling, E., Baudisch, W., and Panitz, R. (1971b). Chromosomale Structurumbauten bei *Acricotopus lucidus* korreliert mit Anderungen im Proteinmuster. *Biol. Zbl.* **90,** 433–442.

Wobus, U., Serfling, E., and Panitz, R. (1971c). Salivary gland proteins of a *Chironomus thummi* strain with an additional Balbiani Ring. *Exp. Cell Res.* **65,** 240–245.

Wobus, U., Popp, S., Serfling, E., and Panitz, R. (1972). Protein synthesis in the *Chironomus thummi* salivary gland. *Mol. Gen. Genet.* **116,** 309–321.

Wobus, U., Baumlein, H., Panitz, R., Serfling, E., and Kafatos, F. C. (1980). Periodicities and tandem repeats in a Balbiani Ring gene. *Cell* **22,** 127–135.

Wobus, U., Baumlein, H., Bogachev, S. S., Borisevich, I. V., Panitz, R., and Kolesnikov, N. N. (1990). A new transposable element in *Chironomus thummi*. *Mol. Gen. Genet.* **222,** 311–316.

Wohlwill, A. D., and Bonner, J. J. (1991). Genetic analysis of chromosome region 63 of *Drosophila melanogaster*. *Genetics* **128,** 763–775.

Wolf, B. E., and Sokoloff, S. (1976). Changes in the form of the polytene X chromosome in *Phryne (Sylvicola) cincta* - causes and functional significance. *In* "Chromosomes Today," Vol. 5, pp. 91–108. Israel Universities Press, New York, and Wiley, Jerusalem.

Wolffe, A. P., and Pruss, D. (1996). Targeting chromatin disruption: Transcription regulators that acetylate histones. *Cell* **84,** 817–819.

Wolfner, M. (1980). Ecdysone-responsive genes of the salivary gland of *Drosophila melanogaster*. Ph.D. thesis, Stanford University.

Wolstenholme, D. R. (1965). The distribution of DNA and RNA in salivary gland chromosomes of *Chironomus tentans* as revealed by fluorescence microscopy. *Chromosoma* **17,** 219–229.

Wolstenholme, D. R. (1973). Replicating DNA molecules from eggs of *Drosophila melanogaster*. *Chromosoma* **43,** 1–18.

Woodard, C. T., Baehrecke, E. H., and Thummel, C. S. (1994). A molecular mechanism for the stage-specifity of the *Drosophila* prepupal genetic response to ecdysone. *Cell* **79,** 607–615.

Woodcock, D. M., and Sibatani, A. (1975). Differential variations in the DNA of *Drosophila melanogaster* during development. *Chromosoma* **50,** 147–173.

Woodruff, R. C., and Ashburner, M. (1979a). The genetics of a small autosomal region of the structural gene for alcohol dehydrogenase. I. Characterization of deficiencies and mapping of *ADH* and visible mutations. *Genetics* **92,** 117–132.

Woodruff, R. C., and Ashburner, M. (1979b). The genetics of a small autosomal region of the structural gene for alcohol dehydrogenase. II. Lethal mutations in the region. *Genetics* **92,** 133–148.

Woods, P. S., Gay, H., and Sengun, A. (1961). Organization of the salivary-gland chromosome as revealed by the pattern of incorporation of ^{3}H-thymidine. *Proc. Natl. Acad. Sci. USA* **47,** 1486–1493.

Wright, K. A. (1969). The anatomy of salivary glands of *Anopheles stephensi* Liston. *Can J. Zool.* **45,** 579–587.

Wright, L. G., Chen, T., Thummel, C. S., and Guild, G. M. (1996). Molecular characterization of the 71E late puff in *Drosophila melanogaster* reveals a family of novel genes. *J. Mol. Biol.* **255,** 387–400.

Wright, T. R. F. (1987). The genetic and molecular organization of the dense cluster of functionally related, vital genes in the DOPA decarboxylase region of the *Drosophila melanogaster* genome. *Results Prob. Cell Differ.* **14,** 95–120.

Wright, T. R. F., Beermann W. (1981). Gene to band relationships in dopa decarboxylase region of *Drosophila melanogaster*. *Genetics* **97,** 115–116.

Wright, T. R. F., Beermann, W., Marsh, J. L., Bishop, C. P., Steward, R., Black, B. C., Tomsett, A.

D., and Wright, E. Y. (1981). The genetics of dopa decarboxylase in *Drosophila melanogaster*. IV. The genetics and cytology of the 37B10–37D1 region. *Chromosoma* **83,** 45–58.

Wu, C. (1980). The 5' ends of *Drosophila* heat shock genes in chromatin are hypersensitive to DNase I. *Nature* **286,** 854–860.

Wu, C. (1982). Chromatin structure of Drosophila heat-shock genes. In "Heat Shock from Bacteria to Man" (M. L. Schlesinger, M. Ashburner, and A. Tissieres, eds.), pp. 91–97. Cold Spring Harbor Laboratory Press, Cold Spring Harbor, NY.

Wu, C. (1984). Two protein-binding sites in chromatin implicated in the activation of heat-shock genes. *Nature* **309,** 229–234.

Wu, C., Wilson, S., Walker, B., Dawid, I., Paisley, T., Zimarino, V., and Ueda, H. (1987). Purification and properties of *Drosophila* heat shock activator protein. *Science* **238,** 1247–1253.

Wu, N., DiBartolomeis, S. M., Liang, C., Smith, H.S., and Gerbi, S. A. (1990). Expression and developmental profile of DNA puffs in *Sciara coprophila*. *J. Cell Biol.* **111**(5), 123.

Wuhrmann, P., and Lezzi, M. (1980a). Chromatin in living cells decondenses with an increase in the nuclear ionic strength. *Experientia* **36,** 763.

Wuhrmann, P., and Lezzi, M. (1980b). K^+- and Na^+-activity measurements in larval salivary glands of *Chironomus tentans*. *Eur. J. Cell Biol.* **2,** 473.

Wuhrmann, P., Ineichen, H., Riesen-Willi, R., and Lezzi, M. (1979). Change in nuclear potassium electrochemical activity and puffing of potassium-sensitive salivary chromosome regions during *Chironomus* development. *Proc. Natl. Acad. Sci. USA* **76,** 806–808.

Wunsch, S., Schneider, S., Schwab, A., and Oberleithner, H. (1993). 20-OH-ecdysone swells nuclear volume by alkalinization in salivary glands of *Drosophila melanogaster*. *Cell Tissue Res.* **274,** 145–151.

Wurtz, T., Lonnroth, A., and Daneholt, B. (1990a). Higher order structure of Balbiani Ring premessenger RNP particles depends on certain RNase a sensitive sites. *J. Mol. Biol.* **215,** 93–101.

Wurtz, T., Lonnroth, A., Ovchinnikov, L., Skoglund, U., and Daneholt, B. (1990b). Isolation and initial characterization of a specific premessenger ribonucleoprotein particle. *Proc. Natl. Acad. Sci. USA* **87,** 831–835.

Xiao, H., and Lis, J. T. (1988). Germline transformation used to define key features of heat-shock response elements. *Science* **239,** 1139–1142.

Xiao, H., and Lis, J. T. (1989). Heat shock and developmental regulation of the *Drosophila melanogaster hsp83* gene. *Mol. Cell. Biol.* **9,** 1746–1753.

Xiao, H., and Lis, J. T. (1990). Closely related DNA sequences specify distinct patterns of developmental expression in *Drosophila melanogaster*. *Mol. Cell. Biol.* **10,** 3272–3276.

Xiao, H., Perisic, O., and Lis, J. T. (1991). Cooperative binding of *Drosophila* heat shock factor to arrays of a conserved 5bp unit. *Cell* **64,** 585–593.

Yagura, T., Yagura, M., and Maramatsu, M. (1979). *Drosophila melanogaster* as different ribosomal RNA sequences on X and Y chromosomes. *J. Mol. Biol.* **133,** 533–547.

Yamamoto, H. (1970). Heat-shock induced puffing changes in *Balbiani Rings*. *Chromosoma* **32,** 171–190.

Yamamoto, H., and Miklos, G. L. G. (1984). Molecular cloning of extra organs locus of *Drosophila melanogaster*. *Ann. Rep. Natl. Inst. Genet. Jpn.* **34,** 41–42.

Yanicostas, C., and Lepesant, J.-A. (1990). Transcriptional and translational cis-regulatory sequences of the spermatocyte- specific *Drosophila janus* B gene located in the 3' exonic region of the overlapping *janus* A gene. *Mol. Gen. Genet.* **224,** 450–458.

Yanicostas, C., Vincent, A., and Lepesant, J.-A. (1989). Transcriptional and posttranscriptional regulation contributes to the sex-related expression of two sequence-related genes at the *janus* locus of *Drosophila melanogaster*. *Mol. Cell. Biol.* **9,** 2526–2535.

Yankulova, E. D., Poluektova, E. V., and Korochkin, L. I. (1986). Relationship of puffing and esterase pattern in different organs of *Drosophila virilis* during development. *Ontogenez* **17**(6), 613–619. [in Russian]

Yannoni, C. Z., and Petri, W. H. (1984). Localization of a gene for a minor chorion protein in *Drosophila melanogaster:* A new chorion structural locus. *Dev. Biol.* **102,** 504–508.
Yao, T.-P., Segraves, W. A., Oro, A. E., McKeown, M., and Evans R. M. (1992). *Drosophila ultraspiracle* modulates ecdysone receptor function via heterodimer formation. *Cell* **71,** 63–72.
Yao, T.-P., Forman, B. M., Jiang, Z., Cherbas, L., Chen, J.-Don, McKeown, M., Cherbas, P., and Evans, R. M. (1993). Functional ecdysone receptor is the product of *EcR* and *Utlraspiracle* genes. *Nature* **366,** 476–479.
Yen, P. H., and Davidson, N. (1980). The gross anatomy of a tRNA gene cluster at region 42A of the *Drosophila melanogaster* chromosome. *Cell* **22,** 137–148.
Yen, P. H., Sodja, A., Cohen, M. Jr., Conred, S. E., Wu M., Davidson, N., and Ilgen, Ch. (1977). Sequence arrangement of tRNA genes on a fragment of *Drosophila melanogaster* DNA cloned in *E. coli. Cell* **11,** 763–777.
Yoshimatsu, H. (1962). Electron microscopy of the salivary gland cells of the *Chironomus* larva. *Ann. Zool. Jpn.* **35,** 89–94.
Yoshimatsu, H., and Uehara, M. (1968). On some fibrin-like properties of the secretion products of the salivary gland of the *Chironomus* **larva.** *Ann. Zool. Jpn.* **41,** 107–112.
Yost, H. J., and Lindquist, S. (1986). RNA splicing is interrupted by heat shock and is rescued by heat shock protein synthesis. *Cell* **45,** 185–193.
Young, M. W., and Judd, B. H. (1978). Nonessential sequences, genes and the polytene chromosome bands of *Drosophila melanogaster. Genetics* **88,** 723–742.
Yurov, Yu. B. (1982). Determination of the distance between the points of the initiation of DNA replication forks movement in the polytene chromosomes of *Drosophila. Tsitologia* **24**(3), 334–337. [in Russian]
Zachar, Z., Garza, D., Chou, T.-B., Goland, J., and Bingham, P. M. (1987). Molecular cloning and genetic analysis of the *suppressor-of-white-apricot* locus from *Drosophila melanogaster. Mol. Cell. Biol.* **7,** 2498–2505.
Zacharias, H. (1979). Under-replication of a polytene chromosome arm in the Chironomid *Prodiamesa olivacea. Chromosoma* **72,** 23–51.
Zacharopoulou, A. (1987). Cytogenetic analysis of mitotic and salivary gland chromosomes in the medfly *Ceratitis capitata. Genome* **29,** 67–71.
Zacharopoulou, A. (1990). Polytene chromosome maps in the medfly *Ceratitis capitata. Genome* **33,** 184–197.
Zacharopoulou, A., Bourtzis, K., and Kerremans, Ph. (1991). A comparison of polytene chromosomes in salivary glands and orbital bristle trichogen cells in *Ceratitis capitata. Genome* **34,** 215–219.
Zaha, A., Loencini, O., Hollenberg, C. P., and Lara, F. J. S. (1982). Cloning and characterization of the ribosomal RNA genes of *Rhynchosciara americana. Chromosoma* **87,** 103–116.
Zaha, A., Leoncini, L., Stocker, A. J., Hollenberg, C. P., and Lara, F. J. S. (1984). Hybridization of poly $(A)^+$ RNA salivary glands of *Rhynchosciara americana* to restriction DNA fragment and polytene chromosomes. *Brasil. J. Med. Biol. Res.* **17,** 257–264.
Zainiev, G. A., and Shilova, I. E. (1984). Estimation of the length of transcribed DNA of Balbiani rings - the giant puffs of the polytene chromosomes of *Chironomus. Tsitologiya* **26**(2), 196–202. [in Russian]
Zainiev, G. A., Baumlein, H., Wobus, U., Kolesnikov, N. N., Kiknadze, I. I., Zakharenko, L. P., Panova, T. M., and Blinov, A. G. (1985). Microcloning of DNA of the A1-2 region of chromosome IV of *Chironomus thummi* containing Balbiani ring BRa. *Tsitologiya* **27**(5), 528–534. [in Russian]
Zajonz, M., Ramesh, S. R., and Kalisch, W.-E. (1996a). Chromosomal linkage of 130 kD secretion protein fractions in *Drosophila sulfurigaster sulfurigaster* of the *Drosophila nasuta* subgroup. *Dros. Inform. Serv.* **77,** 47–73.
Zajonz, M., Ramesh, S. R., and Kalisch, W.-E. (1996b). Homologous fractions of larval salivary secretion in various species of *Drosophila. Dros. Inform. Serv.* **77,** 73–74.

Zajonz, M., Ramesh, S. R., and Kalisch, W.-E. (1996c). Sex-specific fractions in salivary plugs of *Drosophila hydei*. *Dros. Inform. Serv.* **77,** 74–76.

Zajonz, M., Ramesh, S. R., and Kalisch, W.-E. (1996d). Deglecosylation of larval secretion protein fractions in the *Drosophila nasuta* subgroup. *Dros. Inform. Serv.* **77,** 76–78.

Zakharenko, L. P., Onishchenko, A. M., Mikichur, N. I., and Maksimovsky, L. F. (1976). A microbiochemical study on nucleic acids in the structures of the Dipteran salivary gland cells. *Tsitologiya* **18**(8), 975–980. [in Russian]

Zakharenko, L. P., Maksimovsky, L. F., and Kiknadze, I. I. (1978). A comparative characterization of the content and synthesis of RNA in the cell structures of *Chironomus thummi* salivary glands during metamorphosis. *Ontogenez* **9**(3), 253–262. [in Russian]

Zakharenko, L. P., Kiknadze, I. I., Kobzev, V. F., and Kumarev, V. P. (1988). Study on DNA homology in Balbiani rings of two *Chironomus thummi* subspecies with the use of synthesis polydeoxynucleotides. *Tsitologia* **30**(1), 110–112. [in Russian]

Zegarelli-Schmidt, E. C., and Goodman, R. (1981). The Diptera as a model system in cell and molecular biology. *Int. Rev. Cytol.* **71,** 245–363.

Zelentsova, E. S., Vashakidze, R. P., Krayev, A. S., and Evgen'ev, M. B. (1986). Dispersed repeats in *Drosophila virilis:* Elements mobilized by interspecific hybridization. *Chromosoma* **93,** 469–476.

Zelhof, A. C., Yao, T., Evans, R. M., and McKeown, M. (1995). Identification and characterization of a *Drosophila* nuclear receptor with the ability to inhibit the ecdysone response. *Proc. Natl. Acad. Sci. USA* **92,** 10477–10481.

Zhao, K., Hart, C. M., and Laemmli, U. K. (1995). Visualization of chromosomal domains with boundary element-associated factor BEAF-32. *Cell* **81,** 879–889.

Zhimulev, I. F. (1973a). Description of a new type of secretion in the larval salivary gland of *Drosophila melanogaster*. *Dros. Inform. Serv.* **50,** 46.

Zhimulev, I. F. (1973b). Comparison of puffing in proximal and distal parts of the salivary gland in *Drosophila melanogaster*. *Dros. Inform. Serv.* **50,** 96–97.

Zhimulev, I. F. (1974a). Comparative study of the function of polytene chromosomes in laboratory stocks of *Drosophila melanogaster* and the *l(3)tl* (lethal tumorous larvae). I. Analysis of puffing patterns in autosomes of the laboratory stock Batumi-L. *Chromosoma* **46,** 59–76.

Zhimulev, I. F. (1974b). Cytological aspects of the transcriptional activity of the polytene chromosomes of *Drosophila melanogaster*. Ph.D. thesis, Novosibirsk. [in Russian]

Zhimulev, I. F. (1975). Hormonal regulation of the secretion of mucoproteids by the salivary gland cells of *Drosophila melanogaster* larvae during metamorphosis. *Ontogenez* **6**(3), 257–262. [in Russian]

Zhimulev, I. F. (1982). Chromomere organization of the polytene chromosomes. Doctor of Science thesis, Novosibirsk. [in Russian]

Zhimulev, I. F. (1992). "Polytene Chromosomes: Morphology and Structure," pp. 1–480. Nauka, Novosibirsk. [in Russian]

Zhimulev, I. F. (1993). "Heterochromatin and Position Effect Variegation," pp. 1–490. Nauka, Novosibirsk. [in Russian]

Zhimulev, I. F. (1994a). "Chromomeric Organization of Polytene chromosomes," pp. 1–564. Nauka, Novosibirsk. [in Russian]

Zhimulev, I. F. (1994b). Report on *Drosophila melanogaster* new mutants. *Dros. Inform. Serv.* **75,** 33.

Zhimulev, I. F. (1996). Morphology and structure of polytene chromosomes. *Adv. Genet.* **34,** 1–490.

Zhimulev, I. F. (1997). Polytene chromosomes, heterochromatin and position effect variegation. *Adv. Genet.* **37,** 1–555.

Zhimulev, I. F., and Belyaeva, E. S. (1974). Incorporation of ^{3}H-uridine into the chromosomes of the salivary glands of *Drosophila melanogaster*. *Genetika* **10**(9), 71–79. [in Russian]

Zhimulev, I. F., and Belyaeva, E. S. (1975a). Proposals to the problem of structural and functional organization of polytene chromosomes. *Theor. Appl. Genet.* **45,** 335–340.

Zhimulev, I. F., and Belyaeva, E. S. (1975b). ^{3}H-uridine labelling patterns in the *Drosophila melanogaster* salivary gland chromosomes X, 2R and 3L. *Chromosoma* **49,** 219–231.
Zhimulev, I. F., and Belyaeva, E. S. (1975c). On the structural and functional organization of polytene chromosomes. *Genetika* **11**(2), 175–182. [in Russian]
Zhimulev, I. F., and Belyaeva, E. S. (1976). Changes in the structure of the polytene chromosomes of *Drosophila* during long-term culture of larval salivary glands in the abdomen of adult flies. *Tsitologiya* **18**(1), 5–9. [in Russian]
Zhimulev, I. F., and Belyaeva, E. S. (1977). Variation in the banding pattern of the polytene chromosomes of *Drosophila melanogaster* larvae. *Genetika* **13**(8), 1398–1408. [in Russian]
Zhimulev, I. F., and Belyaeva, E. S. (1985a). On the information content of polytene chromosome chromomeres. *Biol. Zentralbl.* **104,** 633–640.
Zhimulev, I. F., and Belyaeva, E. S. (1985b). Informational content of polytene chromosome chromomeres. *In* "Molekulyarnye mekhanizmy geneticheskikh protsessov," pp. 34–43. Nauka, Moscow. [in Russian]
Zhimulev, I. F., and Belyaeva, E. S. (1991). Chromomeric organization of polytene chromosomes. *Genetica* **85,** 67–72.
Zhimulev, I. F., and Feldman, M. S. (1982). Genetic loci in the *Lyra*-deficiency region of the *D. melanogaster* 3L chromosome. *Dros. Inform. Serv.* **58,** 152.
Zhimulev, I. F., and Grafodatskaya, V. E. (1974). A simple method of induction of anaerobiosis puff in *Drosophila melanogaster*. *Dros. Inform. Serv.* **51,** 96.
Zhimulev, I. F., and Ilyina, O. V. (1980). Localization and some characteristics of *sbr* in *Drosophila melanogaster*. *Dros. Inform. Serv.* **55,** 146.
Zhimulev, I. F., and Kolesnikov, N. N. (1975a). Synthesis and secretion of mucoprotein glue in the salivary gland of *Drosophila melanogaster*. *Wilhelm Roux's Arch.* **178,** 15–28.
Zhimulev, I. F., and Kolesnikov, N. N. (1975b). Methods for collecting synchronously developing *Drosophila melanogaster* larvae. *Ontogenez* **6**(6), 635–639. [in Russian]
Zhimulev, I. F., and Kulichkov, V. A. (1977). The region of breaks in *Drosophila melanogaster* polytene chromosomes: Localization and peculiarities of replication. *Genetika* **13,** 85–94. [in Russian]
Zhimulev, I. F., and Lychev, V. A. (1970). A study on the variability in the size of puffs in the left arm of the third chromosome of *Drosophila melanogaster*. *Ontogenez* **1**(3), 318–324. [in Russian]
Zhimulev, I. F., and Lychev, V. A. (1972a) RNA synthesis in the left arm of the third chromosome of *Drosophila melanogaster* during the last five hours of larval development. *Ontogenez* **3**(3), 289–298. [in Russian]
Zhimulev, I. F., and Lychev, V. A. (1972b). Increase in the mass of RNA-containing bodies in the salivary gland cell nuclei of *Drosophila melanogaster* lethal larvae of *ltl*(lethal tumorous larvae) strain. *Genetika* **8**(6), 51–55. [in Russian]
Zhimulev, I. F., and Lychev, V. A. (1972c). Functioning of the salivary gland chromosomes of *Harmandia loewi* larvae (Diptera, Cecidomyiidae). I. Changes in the morphology of the giant cell chromosomes during development. *Ontogenez* **3**(2), 194–201. [in Russian]
Zhimulev, I. F., and Mal'ceva, N. I. (1997). Action of ecdysterone on salivary gland and nurse cell polytene chromosomes of *Drosophila melanogaster otu* mutant *in vivo*. *Dros. Inform. Serv.* **80,** 77–82.
Zhimulev, I. F., and Szidonya, J. (1991). In vivo and in vitro puffing of lethal *ft* mutations of *D. melanogaster*. *Dros. Inform. Serv.* **70,** 234.
Zhimulev, I. F., Belyaeva, E. S., and Lychev, V. A. (1974). Changes in banding pattern in the polytene chromosomes of the *l(3)tl* mutant of *Drosophila melanogaster*. *Genetika* **10**(10), 73–79. [in Russian]
Zhimulev, I. F., Belyaeva, E. S., and Lychev, V. A. (1976). Comparative study of the function of polytene choromosomes in laboratory stocks of *Drosophila melanogaster* and the *l(3)tl* mutant

(lethal tumorous larvae). II. Changes of banding pattern and transcriptional activity in the salivary chromosome of l(3)tl. *Chromosoma* **55,** 121–136.

Zhimulev, I. F., Belyaeva, E. S., Khudyakov, Yu. E., and Pokholkova, G. V. (1980a). Report on *Drosophila melanogaster* new mutants. *Dros. Inform. Serv.* **55,** 211.

Zhimulev, I. F., Bgatov, A. V., Fomina, O. V., Kramers, P. G. N., Eeken, J., and Pokholkova, G. V. (1980b). Genetic loci and the *ras-dsh* interval of the X chromosome of *Drosophila melanogaster. Doklady Akad. Nauk SSSR* **225**(3), 738–742. [in Russian]

Zhimulev, I. F., Belyaeva, E. S., and Aizenzon, M. G. (1980c). Cytogenetical analysis of the 2B1-2–2B9-10 region of the X chromosome of *Drosophila melanogaster*. III. Changes in puffing in the salivary gland chromosomes in homozygotes for the $l(1)pp\text{-}l^{t10}$ mutation. *Genetika* **16**(9), 1613–1631. [in Russian]

Zhimulev, I. F., Belyaeva, E. S., Pokholkova, G. V., Semeshin, V. F., Bgatov, A. V., and Baricheva, E. M. (1980d). Cytogenetic study of the 9E–10A region of the X chromosome of *Drosophila melanogaster*. II. Mutations affecting viability and certain morphological traits. *Genetika* **16**(8), 1404–1424. [in Russian]

Zhimulev, I. F., Belyaeva, E. S., Semeshin, V. F. (1981a). Informational content of polytene chromosome bands and puffs. *CRC Crit. Rev. Biochem.* **11,** 303–340.

Zhimulev, I. F., Belyaeva, E. S., Pokholkova, G. V., Kochneva, G. V., Fomina, O. V., Bgatov, A. V., Khudyakov, Yu. E., Patzevich, I. V., Semeshin, V. F., Baricheva, E. M., Aizenzon, M. G., Kramers, P., and Eeken, J. (1981b). Report on *Drosophila melanogaster* new mutants. *Dros. Inform. Serv.* **56,** 192–196.

Zhimulev, I. F., Pokholkova, G. V., Bgatov, A. V., Semeshin, V. F., and Belyaeva E. S. (1981c). Fine cytogenetical analysis of the band 10A1–2 and the adjoining regions in the *Drosophila melanogaster* X chromosome. II. Genetical analysis. *Chromosoma* **82,** 25–40.

Zhimulev, I. F., Semeshin, V. F., and Belyaeva, E. S. (1981d). Fine cytogenetical analysis of the band 10A1–2 and the adjoining regions in the *Drosophila melanogaster* X-chromosome. I. Cytology of the region and mapping of chromosome rearrangaments. *Chromosoma* **82,** 9–23.

Zhimulev, I. F., Izquierdo, M. L., Lewis, M., and Ashburner, M. (1981e). Patterns of protein synthesis in salivary glands of *Drosophila melanogaster* during larval and prepupal development. *Wilhelm Roux's Arch. Dev. Biol.* **190,** 351–357.

Zhimulev, I. F., Belyaeva, E. S., Semeshin, V. F., Pokholkova, G. V., and Grafodatskaya, V. E. (1981f). On the structural and functional organization of polytene chromosomes. *In* "Molecular bases of genetic processes: Proc. XIV Intern. Congr. Genet.," Vol. *III*, Book 2, pp. 271–283. Mir, Moscow. [in Russian]

Zhimulev, I. F., Belyaeva, E. S., Semeshin, V. F., Pokholkova, G. V., Grafodatskaya, V. E. (1981g). On the structural-functional organization of polytene chromosomes. *In* "Molekulyarnye osnovy geneticheskikh protsessov," pp. 403–413. Trudy 14-go Mezhdunarodnogo genet. kongr. Nauka, Moscow. [in Russian]

Zhimulev, I. F., Belyaeva, E. S., Pokholkova, G. V., Kochneva, G. V., Fomina, O. V., Bgatov, A. V., Khudyakov, Yu. E., Patzevich, I. V., Semeshin, V. F., Baricheva, E. M., Aizenzon, M. G., Kramers, P. G. N., and Eeken, J. C. J. (1982a). *Drosophila melanogaster* new mut. *Dros. Inform. Serv.* **58,** 210–214.

Zhimulev, I. F., Semeshin, V. F., Kulichkov, V. A., and Belyaeva, E. S. (1982b). Intercalary heterochromatin in *Drosophila*. I. Localization and general characteristics. *Chromosoma* **87,** 197–228.

Zhimulev, I. F., Vlassova, I. E., and Belyaeva, E. S. (1982c). Cytogenetic analysis of the 2B3-4–2B11 region of the X-chromosome of *Drosophila melanogaster*. III. Puffing disturbance in salivary gland chromosomes of homozygotes for mutation $l(1)pp1^{t10}$. *Chromosoma* **85,** 659–672.

Zhimulev, I. F., Semeshin, V. F., Kochneva, G. V., Fomina, O. V., Baricheva, E. M., and Belyaeva, E. S. (1982d). A cytogenetic study on the 9E–10A region of the X chromosome of *Drosophila melanogaster*. IV. Isolation and features of chromosomal rearrangements in *ras-dsh* interval. *Genetika* **18**(4), 596–612. [in Russian]

Zhimulev, I. F., Pokholkova, G. V., Bgatov, A. V., Semeshin, V. F., Umbetova, G. H., and Belyaeva, E. S. (1983). Genetic interpretation of polytene chromosomes banding pattern. *Mol. Biol. Rep.* **9,** 19–23.

Zhimulev, I. F., Pokholkova, G. V., Bgatov, A. V., Umbetova, G. H., and Solovjeva, I. V., Belyaeva E. S. (1987a). Genctic loci in the 9E–10B region in the *Drosophila melanogaster* X chromosome. *Dros. Inform. Serv.* **66,** 194–197.

Zhimulev, I. F., Pokholkova, G. V., Bgatov, A. V., Umbetova, G. H., Solovjeva, I. V., Khudyakov, Yu. E., and Belyaeva, E. S. (1987b). Fine cytogenetical analysis of the band 10A1-2 and the adjoining regions in the *Drosophila melanogaster* X-chromosome. V. Genetic characteristics of the loci in the 9E-10B region. *Biol. Zentralbl.* **106,** 699–720.

Zhimulev, I. F., Belyaeva, E. S., Mazina, O. M., and Balasov, M. L. (1995a). Structure and expression of the *BR-C* locus in *Drosophila melanogaster* (Diptera: Drosophilidae). *Eur. J. Entomol.* **92,** 263–270.

Zhimulev, I. F., Belyaeva, E. S., Semeshin, V. F., Pokholkova, G. V., Kokoza, E. B., Kozlova, T. Yu., Demakov, S. A., Mal'ceva, N. I., Demakova, O. V., Balasov, M. L., Koryakov, D. E., Makunin, I. V., and Belousova, N. V. (1995b). Molecular-cytogenetic organization of polytene chromosomes. *Izvestiya Akad. Nauk (seriya chimich.)* **9,** 1622–1638. [in Russian]

Zhimulev, I. F., Semeshin, V. F., Umbetova, G. H., Vlassova, I. E., and Belyaeva E. S. (1996). Transcription in the 100B3/4–5 interband of *Drosophila melanogaster* polytene chromosomes. *Dros. Inform. Serv.* **77,** 133–135.

Zhou, S., Yang, Y., Scott, M. J., Eisen, A., Koonin, E. V., Fouts, D. L., Wrightsman, R., Manning, J. E., and Lucchesi, J. C. (1995). *Male-specific lethal-2*, a dosage compensation gene of *Drosophila*, undergoes sex-specific regulation and encodes a protein with a RING finger and a metallothionein-like cysteine cluster. *EMBO J.* **14,** 2884–2895.

Ziegler, R., and Emmerich, H. (1973). Phosphorylation of chromosomal proteins in *Drosophila hydei*. *Dros. Inform. Serv.* **50,** 180–182.

Zimarino, V., and Wu, C. (1987). Induction of sequence-specific binding of *Drosophila* heat shock activator protein without protein synthesis. *Nature* **327,** 727–730.

Zimmerman, J. L., Fouts, D. L., and Manning, J. E. (1980). Evidence for a complex class of nonadenylated mRNA in *Drosophila*. *Genetics* **95,** 673–691.

Zimmerman, J. L., Petri, W., and Meselson, M. (1983). Accumulation of a specific subset of *Drosophila melanogaster* heat shock mRNAs in normal development without heat shock. *Cell* **32,** 1161–1170.

Zuchowski, C. I., and Harford, A. G. (1976a). Ribosomal genes not integrated into chromosomal DNA in *Drosophila melanogaster*. *Chromosoma*. **58,** 219–234.

Zuchowski, C. I., and Harford, A. G. (1976b). Unintegrated ribosomal genes in diploid and polytene tissues of *Drosophila melanogaster*. *Chromosoma* **58,** 235–246.

Index